"Painstakingly and exhaustively lays out the history of the human quest for space. . . . Provides much new information from the Russian side of the race, having been given unprecendented access to records there."

—*New York Newsday*

"Arguably the most comprehensive history of rocketry and space travel ever."

—*Booklist*

"Burrows brings to the effort a style that is by turns eloquent, witty, sardonic, and simple. He is never dull, not even when describing complicated technology or bureaucratic mucking about. *This New Ocean* is a dandy piece of writing. And it tells a dandy tale." —*St. Louis Post-Dispatch*

"An incisive and lucid account of humanity's leap into space . . . an illuminating study of Cold War machinations." —*The Philadelphia Inquirer*

"An encyclopedic history of space exploration by an insider and veteran reporter who has lost nothing in his enthusiasm and respect for what humankind has wrought. . . . Likely to be the bible for those tracking a unique period in Earth history—the 'first' space age." —*Kirkus Reviews*

"*This New Ocean* is a valued resource for the reader fascinated by the topic of spaceflight and eager to catch up as the second age gets under way."

—*Houston Chronicle*

"A very solid and weighty tome detailing our fascination with space and the great beyond." —*The San Diego Union-Tribune*

"Displays a refreshing wit and a somewhat ironic attitude about humanity's struggle to escape the confines of Planet Earth. This is no puff piece from NASA. . . . Frustration, failure, tragedy, and bitter competition are intertwined in this story with the glorious moments when something worked as planned. My guess is that readers will be amazed, as I was, by the fact that there is so much to learn about the story of space exploration, and by the sheer number and quality of people who have contributed so far."

—*Sunday Republican* (Springfield, MA)

"Distinguished by its comprehensiveness—the author's extensive research is obvious—and its evenhandedness . . . likely to stand as a worthy history of our 'first space age' well into our second." —*Savannah Morning News*

ALSO BY WILLIAM E. BURROWS

Critical Mass: The Dangerous Race for Superweapons in a Fragmenting World (with Robert Windrem)

Exploring Space: Voyages in the Solar System and Beyond

Mission to Deep Space: Voyager's Journey of Discovery

Deep Black: Space Espionage and National Security

On Reporting the News

Vigilante

Richthofen

THIS NEW OCEAN

THIS
NASA
TOM NEWSOM

NEW OCEAN

The Story of the First Space Age

William E. Burrows

THE MODERN LIBRARY
NEW YORK

1999 Modern Library Paperback Edition

TITLE PAGE ILLUSTRATION: In this hypothetical future scene, a camera-carrying astronaut in a Manned Maneuvering Unit is on a maintenance visit to the Hubble Space Telescope. (*To Extend Our Vision*, by Tom Newsom, NASA Art Program)

Originally published in hardcover by Random House, Inc., in 1998.

Grateful acknowledgment is made to the following for permission to reprint previously published material:

AMERICAN ASTRONAUTICAL SOCIETY: Excerpts from *History of Rockets and Astronautics.* This material was originally presented at the 17th History Symposium of the International Academy of Astronautics, Budapest, Hungary, 1983, and was originally published in the AAS publication *History of Rocketry and Astronautics,* American Astronautical Society History Series, vol. 12, John L. Sloop, editor, pp. 70–71, 1991. Copyright © 1991 by American Astronautical Society. This material is reprinted here by permission of the AAS.

Aviation Week & Space Technology: Excerpt from the editorial "Voyager the Intrepid" from the August 28, 1989, issue of *Aviation Week & Space Technology.* Reprinted by permission of *Aviation Week & Space Technology.*

FARRAR, STRAUS & GIROUX, INC. AND INTERNATIONAL CREATIVE MANAGEMENT: Excerpts from *The Right Stuff* by Tom Wolfe. Copyright © 1979 by Tom Wolfe. Rights in the British Commonwealth are controlled by International Creative Management. Reprinted by permission of Farrar, Straus & Giroux, Inc. and International Creative Management.

MRS. MILDRED K. LEHMAN: Excerpts from *This High Man: The Life of Robert H. Goddard* by Milton Lehman [New York: Farrar, Straus & Giroux, Inc., 1963 (hardcover) and New York: Da Capo Press, 1988 (paperback)]. Reprinted by permission of Mrs. Mildred K. Lehman.

THE MCGRAW-HILL COMPANIES: Excerpts from *The Papers of Robert H. Goddard,* edited by Esther Goddard and Edward Pendray (New York: McGraw-Hill, Inc., 1970). Copyright © 1970. Reprinted by permission of The McGraw-Hill Companies.

WILLIAM MORROW & COMPANY, INC.: Poem from pages 57–58 of *A Baker's Nickel* by William Cohen. Copyright © 1986 by William Cohen. Reprinted by permission of William Morrow & Company, Inc.

NATIONAL GEOGRAPHIC SOCIETY: Excerpt from "Columbia's Landing Closes a Circle" by Tom Wolfe from the October 1981 issue (vol. 160, no. 4, p. 474) of *National Geographic.* Reprinted by permission of National Geographic Society.

The New York Times: Excerpt from "The New Soviet Arms Build-up in Space" by Robert Jastrow (October 3, 1982). Copyright © 1982 by The New York Times Company. Reprinted by permission.

Library of Congress Cataloging-in-Publication Data
Burrows, William E.
This new ocean: the story of the first space age/William E. Burrows.
p. cm.
Includes bibliographical references and index.
ISBN 0-375-75485-7
1. Outer space—Exploration. 2. Astronautics. I. Title.
QB500.262.B87 1998
629.4—dc21 98-3252

Modern Library website address: www.modernlibrary.com
Printed in the United States of America
9 8
Book design by Victoria Wong

To Galileo, Magister of the glass eye.
Thanks for everything.

Foreword

Carl Sagan once rhapsodized about having lived at the perfect time. If he had been born fifty years earlier, the eloquent astronomer explained, he would have missed planetary exploration entirely because it would have been only "figments of the speculative imagination." Born fifty years later, he would have been just as unlucky because he would have missed the thrill of the beginning of travel to other worlds.

"In all the history of mankind there will be only one generation which will be the first to explore the solar system, one generation for which, in childhood the planets are distant and indistinct discs moving through the night, and for which in old age the planets are places, diverse new worlds in the course of exploration," Sagan said in a lecture in 1970. "There will be a time in our future history when the solar system will be explored and inhabited by men who will be looking outward toward the first trip to the stars. To them and to all who come after us, the present moment will be a pivotal instant in the history of mankind."

Sagan spoke for many, including myself, who believe that along with the inevitable afflictions of advancing age comes the infinitely greater reward of having been privileged to witness one of the truly great and lasting human endeavors: the beginning of the migration to space.

Sagan's words came back to me on the evening of May 15, 1997, as I sat in the darkened IMAX auditorium of the National Air and Space Museum in Washington and watched American astronauts and Russian cosmonauts, standing side by side under a spotlight, being heartily applauded—congratulated—by eight hundred people who knew they had bravely led humanity off the planet for the first time in what no one in the room doubted was the beginning of the same kind of endless odyssey that drives other creatures on

Earth to their own immense journeys. Everyone who was there that night knew that they were sharing a historic moment. The first space age was over. The second was under way.

There were Glenn and Titov, Shepard and Leonov, Stafford and Shatalov, Lucid, Popovich, Conrad, and others who, with many in the audience and countless others who were not there, had dueled in space and in the process accomplished the beginning. The occasion was a formal dinner that launched a permanent exhibition, called "Space Race," which traced four decades of competition and some cooperation between the two superpowers. That race did nothing less than open a new dimension of infinite size and endless wonder for all mankind. "We all turn out to be cosmonauts," said Glenn, the gracious master of ceremonies and a U.S. senator, "and Earth is our spaceship." It was like being in Barcelona when Columbus returned from his first expedition to the New World.

I went back to the museum the next morning, back to the spot where I had stood in black tie the night before, watching as Glenn showed his wife the silvery space suit he wore when he made his historic flight in *Friendship 7* in February 1962. Schoolchildren whose parents hadn't even been alive when Glenn became the first American to fly an orbit around the world (in the process scoring a propaganda victory on the order of Doolittle's brazen raid from the *Hornet* to Tokyo in 1942) took a cursory look at the old relic and raced on, as they always do. Gagarin's orange space suit was beside Glenn's. The helmets were there, too, and so were Wernher von Braun's and Sergei Korolyov's slide rules, both of them made by the same German company, which figured. There was Kennedy being quoted by *U.S. News & World Report* as saying that "We are behind" the Communists in the race to space. There were scale models of the huge Saturn 5 Moon rocket and the Soviet Union's ill-fated N-1 counterpart. And there, too, was a once-top-secret spy satellite picture shot from more than a hundred miles up that clearly showed a line of ants in Red Square waiting to enter the Lenin Mausoleum to pay their respects to the great Bolshevik. Diary entries by Vasily Mishin, the chief designer who succeeded Korolyov and who therefore lost the race to the Moon, were displayed, too.

For the kids who were swarming around me that morning, the flights of *Friendship 7* and *Apollo 11* might as well have been John Cabot's voyage in the *Matthew* or Admiral Byrd's trek to Antarctica. They were ancient history, dead as the Peloponnesian War and Trafalgar. But I had been around even before the cold war. So I remembered Sputnik and the shock and humiliation that swept the country as one American rocket after another turned into a greasy fireball down in Florida. I was seven days into my first newspaper job as a clerk on the Foreign Desk of *The New York Times* when the Cuban missile crisis erupted and the possibility of my premature demise hung in the air like the sword of Damocles. Seven years later, I was a reporter who covered part of the Apollo story and who, in any case, witnessed the momentous process that led to the historic landing. Like Sagan and most of the rest of cognizant humanity, I actually got to watch the first travelers from my planet set foot on another world in living black and white. And I remember it vividly to this day. It lives in my memory, not as a story that was told to me or as words on paper, but as a series of real events that I witnessed.

I was also privileged to be alive and paying attention when NASA was born, when the aerospace industry walked off a cliff after the Moon was abandoned, when the shuttle and the station were hammered into the space program's bedrock, when Reaganomics nearly killed solar system exploration, when Voyager 2 made it to Neptune against all odds, when Galileo reported that it had found ice water, possibly containing life, on Jupiter's moon Europa, when the Hubble Space Telescope sent back pictures of stars being born billions of years ago in gas clouds that were trillions of years high, and when little Sojourner romped around Mars.

I came to understand, at firsthand and up close, that however the politicians postured and however many people got killed in wars in Asia and Africa and while trying desperately to make it over the Berlin Wall, the cold war was an unmitigated blessing for both sides' rocketeers and the rest of the space fraternity because they fed off and were nourished by the competition. Rockets were almost always developed for the wrong reason during the cold war. But they were developed. And most important of all, four trips to Russia after the war and some reflection about what I had witnessed during those years convinced me that there were startling similarities, parallels, not only between the two programs, but between the individuals who propelled them. I learned that those who led the greatest odyssey in history were ultimately indivisible. They were simply earthlings. Seen from that perspective—from a vantage point that includes a visceral feeling for the compulsion to reach ever higher, a clear recollection of how it was finally realized, and some understanding of the powerful political currents that energized it—the journey to space becomes a collective human enterprise that I believe is leading to a planetary civilization and the spread of the human seed to other places.

That's what struck me at the National Air and Space Museum that night: eight hundred people had come together to celebrate a perfect time. This is the story of how that time happened.

Stamford, Connecticut
January 20, 1998

Acknowledgments

What follows is the result of a collaborative effort between me and a couple of hundred other people over the course of almost three decades. I am proud to say that many of them—scientists, engineers, and scribes—have become my friends over the long course of this story. They are American and Russian, military and civilian, and they are the best in the world at what they do. Knowing them, including those who are no longer here, has enhanced my life beyond measure.

There are some 175 taped interviews alone, going back to Margaret Mead in May 1969, each representing a generosity with time that was taken from crowded schedules. Some of the interviews went directly into this book; others provided a wider background from which it grew.

Jeffrey T. Richelson, undoubtedly the country's leading civilian expert in overhead reconnaissance and other technical intelligence collection and a very generous individual, voluntarily supplied a stream—no, a mountain—of documents relating to military intelligence operations for more than a decade. More recently, so has Dwayne A. Day, the bright, energetic military space analyst at the Space Policy Institute at George Washington University. Things just kept coming in the mail from the two of them, like presents from guiding angels.

Four others have been my wise men and muses for a decade or more, guiding me through the technical and political shoals whenever they were called. Bruce Murray, Caltech geologist, former Jet Propulsion Laboratory director, and as astute an observer of space culture as there is, shared his valuable insights many times and became an unwitting and valuable guru. So did Mert Davies at RAND, a true American hero for his role in the creation of the nation's space reconnaissance capability, who is as endlessly generous as he is congenial. At a time in his life when most people would have headed for retirement, Mert headed for Jupiter on the Galileo mission and is still there. "Robert," the Third Man, has led me through a series of tutorials on intelligence from the political, philosophical, and occasionally technical standpoints since the *Deep Black* days. They have been invaluable for my understanding of the culture of intelligence collecting, processing, and use. Now he is my respected friend. Mac Bundy was always there, too, sometimes filling in historical detail, other times pointing me in new directions, and still others just cheerfully massaging ideas and poking at the world over lunch at our common establishment. He is missed. These four trusty horsemen have my deepest, lasting gratitude.

Saunders B. Kramer, the ubiquitous presence in the U.S. space program and one of the true experts on the subject of this book, is in a class by himself. He is that rare individual whose broad technical knowledge is leavened with a soaring imagination, a wonderful sense of humor, daunting wisdom, and short patience for knaves and buffoons. His promptly reading a manuscript I only half jokingly said I couldn't pick up, and then returning a pile of corrections, additions, and helpful suggestions, greatly enhanced this work. Similarly, Dave Morrison, the director of space at NASA's Ames Research Center, yet again volunteered his services and took on the task of scrutinizing the two chapters that concentrate on solar system exploration, as he did with all of *Exploring Space.* A hint of what my astronomer and friend knows about the subject can be gleaned in the sources section. Gratitude does not express my lasting debt to these wise and pervasively knowledgeable gentlemen. And as usual, any mistakes that made it past them belong to me alone.

I also deeply appreciate the generosity of many others who shared their time, energy, and spirit. As innumerable science writers and JPL scientists and engineers know, the legendary Jurrie van der Woude was always there. He has long provided material I did not know I needed with the unfailing dry humor for which he is widely loved and respected. Like his friend Mert Davies, Jurrie's vocabulary does not contain the word "no." And like Mert, he is a valued friend. So is Brenda Forman, who loosed me on an innocent U.S. Space Foundation in 1992, and has given me the benefit of her extensive knowledge of spaceworld, dazzling intelligence, and guileless ethic. I will always be grateful to Pete Scoville, who is no longer here, for grounding me in space reconnaissance (if I can put it that way); to Nick Johnson, a former senior analyst of the Soviet space program and now NASA's orbital debris expert, for doing the same; and to Sergei Khrushchev for providing a movable feast in Bodø, Oslo, and Providence on what it was like on the other side of the curtain. Grigori S. Khozin did the same in Moscow. Gresha also put in many hours setting up interviews there, being my faithful native guide from Kaluga to Star Town, and helping me navigate the Leninsky Prospekt bus route and Metro. Meanwhile, his wife, Tanya, kept my body and soul together with the infinite inventiveness that is the hallmark of the Russian cook.

No Russian ever said "*nyet.*" I am especially grateful to Boris and Gilberta Kantemirov of the Memorial Museum of Cosmonautics in Moscow for kindly sharing their expertise, their photographs, and their delightful dispositions; Sergei Krichevsky, a poetic cosmonaut, for clearheaded appraisals of his country's space program in Kaluga and Star Town; Larisa A. Filina for the tour of Korolyov's home; Valentina Ponomareva and Yuri Biryukov for their insights into the politics of cosmonautics; and Sergei Kapitsa, Aleksandr Basilevsky, and Roald Sagdeev for their appraisals of Soviet space science as it fit into the larger scheme. Vladimir V. Lytkin, vice director of the Tsiolkovsky State Museum, took time out of his own busy schedule to make certain I had a thorough tour of the facility and a wonderful lunch afterward. I am also indebted to two of my graduate students, Julia Ferrara and Karina Chobanyan, and to Stanislav A. Gushchenko, a student at Rostov State University, for translating large tracts of material with care and intelligence.

Past and present editors at *Air & Space/Smithsonian,* including Tom Huntington, George Larson, and Linda Shiner, are owed a special word of appreciation. A significant proportion of what is in here is the rich residue of articles I have done for them in my capacity as a contributing editor. The collaboration with these professionals has been one of the most rewarding of my life.

Mark R. Taylor of the National Air and Space Museum's Film Archive earned my gratitude for screening *Challenge of Outer Space.* Martin Collins and Dave DeVorkin, archivists at the museum, are bright and knowledgeable individuals who have given me more over the years than they know and who I am also proud to call friends. Daniel Jensen and Sam Welch of the National Archive, who tracked down CORONA N-1/L-3 imagery, also deserve a word of appreciation, as do Colin Fries and Gwen G. Pitman in the NASA History Office for helping with all the U.S. photographs, and Chris Faranetta at Energia for doing the same with the N-1 and its stablemates. Bert Ulrich, the personable curator of the NASA Art Program, proudly shared hundreds of pictures and a generous amount of time, in the process turning up the splendid painting that graces this book's jacket and goes to the heart of the story inside. I am also deeply grateful to Bob Bunim, my hawkeyed old friend and station chief in California, who followed the space program there and kept reporting back, as usual. Art Dula provided documents from his collection of Soviet space material and shared his own valuable insights on the subject, so I must thank him, too.

Bob Loomis, my editor, is in a category by himself. This is our third project together, and I must say I am profoundly grateful that an individual of his caliber appreciates my work and has been so enthusiastic, supportive, and diligent over the years. I am ever mindful that serious work cannot be published unless an editor responds to it, believes in it, and wholeheartedly supports it. The synergy between us has therefore been a source of deep strength, pride, and contentment over the years. "My main value to you is my ignorance," says the Socrates of book publishing, straight-faced.

I must also thank Sarah Whitworth, who once again held the fort at NYU while I roamed in search of information and hid in my bunker in Connecticut trying to turn the notes, tapes, articles, papers, pictures, handouts, clippings, and the rest into something coherent.

I am deeply grateful for the loyalty, dedication, love, and sense of excellence Joelle, my wife, has given this, its predecessors, and me. She repeatedly abandoned her own work to pore over the manuscript, not only using her pencil the way a surgeon uses a scalpel but bringing fresh and conscientious ideas to the larger story. How very fortunate I am to have such total, uncompromising, and loving support.

Finally, I was and remain profoundly inspired and humbled by LJBMD, whose courage, energy, tenacity, and sense of purpose are beyond admiration. They also serve who draw the pericardium and mons veneris with colored pencils and eat their yogurt alone.

Contents

PART ONE
The New Ocean

NASA

The human space program has existed in the collective unconscious of humanity since the dawn of awareness.

—FRANK WHITE, *The Overview Effect*

1

Bird Envy

The bronze archer is called the *Saltine Warrior.* He stands on a pedestal of marble, muscled legs spread defiantly apart, at Syracuse University. Most of his weight is on the right leg; the other is extended far out. His arched torso is also heavily muscled, and his long hair hangs down, nearly touching his shoulders, while he gazes straight up with complete concentration. The Indian's left arm points to where he is looking. His hand grasps a bow pulled so far back that it is bent almost in a semicircle, almost to the breaking point. The fore and middle fingers of his right hand hold an unseen arrow to a taut, unseen string. This bowman is absolutely focused on a distant, unseen target.

He is not coiled for a quick shot at a formation of geese passing overhead, honking their melancholy cadence, on some long migration. This man's target is beyond birds. It is altitude. He is poised to fire his arrow as high as he can; as high as the force that holds him to Earth will let it go. The Saltine Warrior and every other archer knows in his soul that there is a primal urge to shoot at the sky, to defy the relentless force that imprisons him by riding the arrow to freedom. "I am with you," this archer and all of the others say to their creator. "I too can look down on Earth and touch other worlds."

So did Dr. Faustus. "I'll leap up to my God. Who pulls me down?" he cried in frustration and anguish. Konstantin Tsiolkovsky, the great Russian visionary of aerodynamics and rocketry, was an obsessed gravity hater who believed implicitly that only floating in space could unburden the mind and bring freedom to the soul. Tsiolkovsky rode the arrow. And when the people controlling a spacecraft that is approaching Mars announce that "*We* are thirty thousand kilometers from encounter," it is because they believe on some subliminal level that they're on the spacecraft; that they are riding the arrow too.

When the history of this century is written, the story of mankind's first breaking gravity's relentless hold and touching places beyond Earth will be one of its most exciting and important chapters, together with the development of the computer, the discovery of the genetic code and the engineering that soon began to follow, and only a handful of other remarkable advances. The story of the first space age is set in a cauldron whose soup is laced with altruists and scoundrels, visionaries and parasites, heroes and knaves. It is the story of political ideologies that fought relentlessly above the atmosphere, as they did on the planet beneath it, in a dangerous duel for a global supremacy that was more illusion than serious possibility. Using space to help humanity communicate, forecast the weather, and accomplish other beneficial things, and traveling there for the ultimate adventure and refuge, was the central core of the idea. It was the great philosophical driving force, the engine, that seemed to make a presence in space imperative.

Yet space could be reached only by rockets and the shame of it was that rockets were conceived, and reconceived time and again, expressly to wreak unparalleled destruction and kill large numbers of the very people whose enlightenment and salvation they promised. So there are dark shadows here. Every rocket that has ever carried instruments to record solar activity or gone to Mars to search for life was on a family tree whose true ancestors were designed and built to be people killers with intercontinental reach. Used as they were intended to be used—and fanatics on both sides who longed for that day are part of the story—they would have ended life here before they found it somewhere else. Yet the warriors and those who sold them the means to commit what was euphemistically called mass destruction with re-entry vehicles, as Frank White has indicated, were only the late philosophical aberrations of a much older and deeper drive. What makes them important in this story is the fact that they, and only they, were able to turn fantasy into reality because they could spend vast amounts of government money to do so. But they, too, were riding the arrow. And that arrow, which does indeed go back to the beginning of awareness, remains after the greatest danger this planet has ever faced subsided. That, too, is part of the story.

An Art Unknown

Daedalus, the master artist and craftsman of Greek myth, was the Leonardo da Vinci of the ancient world (he also became history's first genetic engineer by creating the Minotaur when he coupled Pasiphae and the Cretan bull). But Daedalus is best remembered for moving the possibility of flying from fantasy to accepted theory by setting up an actual flight regime. Having been imprisoned by King Minos in the same labyrinth that he had built to hold the dreaded Minotaur and then freed only to find he could not get off Crete because its waters were heavily patrolled, Daedalus "turned his mind to arts unknown," as the poet Ovid later explained in *Metamorphoses.* "Minos may control the land and sea," Daedalus decided, "but not the regions of the air. I will try that way."

Daedalus devised a mission plan and a technology to carry it out and in doing so anticipated reality. He came up with a bold plan to break out of the blockade by flying to Sicily, a distance of more than five hundred miles over the Aegean, the way a bird would. He therefore built birds' wings to fit himself and his son, Icarus. Carefully attaching small feathers together with wax, he fashioned them into a gently curved surface, then stitched ever-larger ones in place. Next, Daedalus became history's first test pilot by beating the wings until he could feel himself rising. He taught Icarus to do the same. Finally, father and son took off in formation and headed west. As all the world knows, Icarus was lost at sea because he abandoned the flight plan and, as scriptwriters would put it many centuries later (however erroneously), left his wingman. In an apparent attempt to reach Heaven, Icarus climbed so high that the Sun melted the wax in his wings, sending him plummeting to a watery death. All he left behind were some feathers floating on the sea. But the more conservative Daedalus, who stuck to the mission profile and did not push his vehicle beyond its design capability, reached his destination safely and hung up his wings.

The father and son's mythical journey became a timeless metaphor for many aspirations, afflictions, and qualities. By Homer's time, Daedalus had become synonymous with artistry; Socrates, himself a sculptor, traced his lineage back to the flying "artificer." That seems to have also made Daedalus history's first science fiction hero. More to the point, the flight out of the maze and into the unknown not only joined art and science, but combined both with courage. And it enlisted all three in the cause of freedom. With Daedalus, flight became a metaphor for spiritual as well as physical liberation. It showed what imagination and daring could accomplish. But it also represented a visceral longing to escape from dangerous or unhappy situations, a motivation that would become one of several leitmotifs running through the reach for space. For his part, poor Icarus only proved the folly of imprudence, hubris, and disobeying orders.

The later Greek writers, such as the second-century satirist Lucian of Samosata, thought of Daedalus as a master astrologer who taught the "science" of star-reading to Icarus. Astrology would slowly, grudgingly, give way to astronomy in the Middle Ages, and as it did so, pure fantasy would evolve into fiction based increasingly on hard, provable science about the world as it actually exists. Lucian believed that the Moon was another heavenly body, not a symbolic chariot that traveled across the night sky.

But this flew in the face of a giant's teachings. Aristotle, who taught the importance of the direct observation of nature, nevertheless stubbornly rejected any idea having to do with a "plurality of worlds" or the possibility that Earth moved. The great philosopher insisted that since all matter is contained in one world, there can be no others; that, in any case, other planets would be so heavy they would come together as one. It therefore followed that Earth was the center of the universe. As Willy Ley would note in his fine early chronicle of rocketry, Aristotle's profound mistake was wholeheartedly accepted by Christian teachers well over a thousand years later. "Not only was it simply forbidden to teach anything that contradicted or diverged from Aristotle's statements," Ley wrote, "it was also denied that there was anything that Aristotle had not known." God's creating Earth as the center of everything made perfect sense to the church.

But Lucian was not persuaded by Aristotle's unsupported argument. And he had his own agenda. In A.D. 160, he wrote the first known story about a voyage to the Moon. It was actually two stories, collectively called *Vera historia* (*True History*) and it was intended to poke fun at Phoenician and other storytellers who concocted gruesome accounts of the mysterious region west of Gibraltar that was known as the Pillars of Hercules. (The wily Phoenicians spread the terrifying tales to frighten other sailors and traders so they could ply the eastern Atlantic without competition.) Since Lucian was more interested in spinning an outrageous tale than in where it took place, he decided to get fiction decisively off the ground.

At one point Icaromenippus, the hero of one of the stories, gazes at the sky and sees "the starres scattered up and down the heavens, carelessly." Yearning for knowledge about the Moon and the Sun, Icaromenippus decides to fly to the Moon from Mount Olympus using one wing from an eagle and another from a vulture (a design improvement that avoided the wax which killed poor Icarus). "At first," he says from his far-off vantage point, "me thought I saw a very little kind of Earth, far less than the Moon; and thereupon stooping down, could not yet find where such mountains were, or such a Sea. . . . At the last, the glittering of the Ocean by the Sun beams shining upon it, made me conjecture it was the Earth I saw." This made Lucian the first known writer to send his imagination to the Moon and use it to view Earth. Not satisfied with being on the Moon, though, Icaromenippus extends his reach (just as his partnamesake did), continuing on to the stars and finally arriving in Heaven three days later. But since he is considered to be an intruder by the immortals who live there, Icaromenippus is taken back to Earth by Mercury and stripped of his wings because of his impudence.

The other tale concerns a ship carrying athletes beyond the Pillars of Hercules that is caught in a colossal whirlwind, lifting it from the sea and carrying it high in the sky for seven days and nights. On the eighth day, the passengers and crew discover that they are on a "shining Island": the Moon. In a narrative worthy of the *Odyssey,* the travelers meet warriors called Hippogypi who ride giant three-headed birds. They are astonished to learn that a war is looming between the Moon and the Sun involving 60 million men, 80,000 of whom are in the aerial cavalry. The fierce combatants are joined by spiders the size of islands, gigantic ants, 30,000 flea riders, a legion of garlic throwers, and lots more. But Lucian has no stomach for gore-strewn battlefields so he sanitizes them. Dead combatants do not litter the Moon, the Sun, or the space between; they conveniently dissolve into vapor and smoke.

"For Lucian, a trip to the Moon was simply an extension of the traditional traveller's tale in which the fictional description of strange and unfamiliar lands was nothing more than an opportunity for satire," air and space historian Thomas D. Crouch has explained. "It was the purest fantasy," and could as easily have been set on Earth as on the Moon. As such, he added, it led the way to *Gulliver's Travels* and *Alice in Wonderland,* but it was not science fiction. As much could have been said of *The Wizard of Oz,* in which Dorothy and Toto were swept up in a tornado that could have been spawned by Lucian himself.

Still, *True History* was a full-blown adventure that probably described for the first time a voyage to another world, a landing on it, encounters with alien creatures in harrowing and bizarre circumstances, and a return home. It was indeed fiction without science. Yet it suggested that there could be many worlds out there, some of them inhabited by aliens. And it took the reader to those worlds, in the process setting the stage for other tales that would adapt well to whatever science could offer. Lucian of Samosata not only set the stage for earthlings to encounter horrific creatures on other worlds, but for the creatures themselves to find and threaten Earth itself, as would happen when H. G. Wells's Martians started *The War of the Worlds* in 1898.

Fireworks

While Lucian was using only his pen to get earthlings off their planet, alchemists in Cathay were combining physical elements to do it for real, though they didn't know it. They were producing an elixir that supposedly prolonged life and turned ordinary materials into gold and silver. The second quality was helpful for the first, since it was believed that eating and drinking from gold utensils not only extended life but ensured immortality. A succession of emperors who wanted both supported the alchemists, probably making the Han Dynasty the first government to subsidize drug research.

The concoction's two key ingredients, sulfur and saltpeter, were also known to become explosive when combined with a carbon such as realgar or charcoal. In his book *Chen yuan miao yao lueh,* the noted alchemist Cheng Ssu-yüan warned against heating the sulfur, saltpeter, and carbon mixture and cited instances in which people were hurt and houses burned because the dangerous compound had been mishandled. Firecrackers were first used in China sometime between 200 B.C. and A.D. 200 (at precisely the same time that astronomers there were following eclipses, planetary movements, comets, and other doings in the night sky) and were quickly recognized as potential terror weapons. While their practical use in warfare would have been negligible, the psychological impact on the enemy of seeing and hearing explosions created at will must have been incredible, at least at first. At some point, however, it began to dawn on the alchemists or their leaders that the mixture's explosive force could be used to deliberately blow up things and propel objects. This was good news to a succession of dynasties that faced long-term, unrelenting attacks by marauding Mongol armies and that sought a high-tech weapon for the defense of the empire.

Over the course of the next thousand years, the elixir of life was gradually turned into the compound of death and destruction: black powder, or gunpowder, as it is more commonly called. This explains why the Chinese picked the name *huo yao* as a general term for the explosive, fire-producing mixture; it means "fire drug."

Rocketry, together with guns of great size and range, grew out of the spread of gunpowder. The Chinese developed a "fire arrow" and then a six-foot-long "flying fire lance" propelled by burning gunpowder. Wu-ching Tsung-yoa's *Complete Compendium of Military Classics,* published in 1054, said that a hundred-thousand-man army laying siege to the city of Tzu T'ung was routed in 994 by regular war machines and fire arrows. A treatise written in 1628 by Mao Yüan-i noted that the larger flying fire lance was propelled by two fire-spraying cylinders connected in tandem and operating in succession, with the first igniting the second. This made the flying fire lance a two-stage rocket.

In Korea, the powder was first made in 1377, though it is not known whether the invention was indigenous or spread from China. Within twelve years, Koreans were producing a weapon called *ju-hwa,* or "running fire," which could be shot out of a bamboo tube held by a cavalryman. Although the actual nature of running fire has been lost, it seems to have been a rocket-propelled arrow. "It is detrimental to the enemy," according to one contemporary account. "Its loud noise and shape instill fear and incite surrender. Once used at night, its exhaust flame lights the fields and shakes the enemy's spirits."

Less than a century later, running fire was replaced by the more advanced magical machine arrow, which came in three sizes and could be fired in salvos of fifteen from magical mechanical arrow launchers that held as many as a hundred of them. The launchers looked something like the multiple rocket launchers used in World War II. The arrows were propelled by paper or bamboo cylinders tightly packed with powder—the propellant—and sealed at the forward end. When lighted at the open end, the burning powder would turn to expanding hot gas, pushing the rocket forward as it escaped through the open end. The key was to control the rate at which the powder burned so that it propelled the rocket rather than exploded. Extra charcoal slowed the burn rate, while making the size of the hole just right allowed the gas to escape instead of turning the rocket into a bomb. (Rocket designers six centuries later would face the same problem calculating the diameter of their own nozzles.)

Rocketry took hold in Europe by the end of the thirteenth century. Several formulas for "flying fire" are in the *Book of Fires for Burning Enemies,* attributed to Marcus Graecus (Mark the Greek), who lived at the time. They have been found, sometimes duplicated word for word, in a number of fourteenth-century manuscripts. Here is one of the most popular:

> Take 1 libre [pound] of native sulphur, 2 libra of charcoal of linde or willow, 6 libra of slapetrosum [saltpeter or potassium nitrate] which are all three well mixed on a marble stone. Afterwards place the powder at pleasure in an envelope for flying [a rocket] or for making thunder [a firecracker]. Note that the envelope for flying should be long and slender and filled with powder well packed. But the envelope for making thunder should be short and thick and half-filled with the said powder and strongly tied at both ends with iron wire.

Rockets were probably first used in Europe for warfare in Silesia in 1241 during a battle between Henryk Pobozny, a Polish prince, and Tartars. Some of the earliest rockets were fired in Italy, where they got their name: *rochetta.* The word comes from the Latin for "spindle," a tapered cylinder that was used to twist yarn in hand spinning and that still twists it on a spinning wheel. Similarly, the Italian word *rocchetto* means "spool of thread," which is also a cylinder. Rocket-propelled arrows were fired by Bologna against Forli, a rival city-state, in 1281. By 1488, as a weapons inventory from Aragon shows, they had reached Spain. By the end of the fifteenth century, however, military rockets were sliding into a long decline. They were being eclipsed by conventional firearms and artillery, which were far more reliable and efficient.

Ships to Sail the Heavenly Breezes

Meanwhile, the urges to explore Earth, fly off it, and visit other worlds were coming together like tributaries that formed one of the deep streams flowing into the river that was the Enlightenment. And that particular stream ran into yet another: a scientific revolution of literally astronomical proportion. Courage, imagination, and intellect seemed to be surging everywhere.

In 1492, Christopher Columbus, a courageous sailor and undoubtedly the greatest navigator of his age, had sailed westward toward an unknown world and ran into Earth's other hemisphere.* He found a place that was not only inhabited, which was profoundly important in its own right, but which was alive with indigenous people and other life forms that Columbus found were culturally and physically different from those he and his men had left behind.

While Columbus probed the Atlantic, Leonardo da Vinci dreamed of doing the same in the atmosphere. In 1495, he came up with an idea for extending the cannon's range to three miles straight up by firing rockets from it. It is the earliest known marriage between a gun and a rocket and it formed a clear link between the Saltine Warrior's arrow and the colossal cannon that Jules Verne used more than three and a half centuries later to fire his heroes to the Moon. The irrepressible Florentine genius studied birds and bats in detail and then came up with a richly detailed treatise on how they fly, correctly noting that aerodynamics and hydrodynamics are intimately related: "A bird makes the same use of wings and tail in the air as a swimmer does of his arms and legs in the water." This led directly to three inventions that were supposed to give man birdlike capabilities: an ornithopter, which a pilot was supposed to fly by flapping its wings; a helicopter, which was intended to screw itself upward into the air; and a parachute, which looked like a tent. Since he understood that his invention was strictly experimental, Leonardo warned potential test pilots that the ornithopter "should be tried over a lake, and you should carry a long wineskin as a girdle so that in case you fall you will not be drowned." He had in effect designed a fifteenth-century hang glider to be flown by a man wearing a fifteenth-century Mae West.

Three years after Leonardo's death, the remnants of Ferdinand Magellan's crew completed the first circumnavigation of Earth, another of the greatest feats of exploration in recorded history. Before he died in 1521, the stalwart Portuguese seaman ran into enough new and exotic varieties of flora and fauna to bring to mind Lucian's *True History.*

Eight years after Magellan's *Victoria* became the first vessel to circle Earth, Nicolaus Copernicus became the first modern scientist to assert that the planet itself circled the Sun and so did the other planets. Aristarchus of Samos, a pre-Christian Greek astronomer, had actually said it first. In what was undoubtedly the most original scientific work of the time, Aristarchus claimed that Earth made one rotation on its axis every day and circled the Sun once a year.

* He did not make the voyage to prove that the planet is round. As early as the beginning of the fifth century B.C., and possibly earlier, Greek philosophers concluded that they lived on a globe after hearing reports from travelers who noticed that the height of stars in the north increased as they journeyed north, then sank as they rode or walked south. Furthermore, objects such as ships gradually lowered out of sight on the horizon as distance increased. "We sail out of the harbor," Virgil wrote in *The Aeneid,* his epic masterpiece, "and the countries and cities recede." No educated person in Columbus's time believed that Earth was flat, let alone that huge dragons devoured ships that went over the edge.

He also said that the other planets similarly moved around the Sun and that the Sun itself, as well as the stars, were actually motionless. Although there is no written record of Aristarchus's theory, it apparently became widely known because Cleanthes, the preeminent Stoic philosopher, said that he ought to have been indicted for impiety.

That and more: Aristarchus not only challenged God, but he, and then Copernicus, propounded a theory that seemed so stupid on the face of it that even a fool would not believe it. If Earth were spinning, there would be a sense of motion, but there was none. And simple physics would cause people and other things not attached to the ground to fly off into the air. Finally, if the planet and everything on it were spinning, why would an object dropped from, say, the tower in Pisa, fall to the tower's base when the tower itself was moving away from the object? If the fools were right, the object would hit the ground half a mile behind the moving tower. It was obvious that they made no sense.

Aristarchus had therefore been overshadowed by Claudius Ptolemy, the second-century mathematician, geographer, and astronomer, whose thirteen-volume *Almagest* stated flatly that Earth—a globe—was the center of the universe and that the Sun, the Moon, and the stars revolved around it in uniform, circular, orbits. That seemed to make sense. Ptolemy's teachings were accepted throughout much of Europe until Copernicus intruded.

Copernicus explained the heliocentric theory in his immortal *De revolutionibus orbium coelestium* (*Concerning the Revolutions of the Heavenly Bodies*). He calculated that Earth did more than simply rotate. In common with the other planets, it had "several motions." If the other planets were spheres that had gravity and that moved through the heavens, he asked, why would Earth, alone, *not* move? And there was this: "That it [Earth] is not the center of all revolutions is proved by the irregular motions of the planets, and their varying distances from Earth, which cannot be explained as concentric circles with the Earth at the center." This radical and philosophically dangerous treatise, which was published in 1543 as Copernicus lay on his deathbed, was dedicated to Pope Paul III. Before long a humanist named Giordano Bruno and another astronomer, Galileo Galilei, would learn to their dismay that the Holy See was anything but honored by such gestures.

As Copernicus thought about the space beyond Earth, the chief of the artillery arsenal at Sibiu, Transylvania (now Romania), thought about extending the rocket's range and of using it to get people to Copernicus's imagined places. Conrad Haas had to deal with big guns; that was his job. But he was obsessed by big rockets. So it occurred to him that as powder burned away inside a rocket's chamber, the diminishing amount of remaining powder had to carry the dead weight of the entire rocket. In other words, as the amount of propellant decreased, the load it had to carry increased, causing a sharp loss of velocity.

The way to solve the problem, Haas finally decided, was to shed the deadweight. He published a paper in 1529 that described the first true multiple-stage rocket. The idea was to connect two or three rocket cylinders, or, as they would later be called, stages. The bottom stage would be ignited and would push the whole contraption off the ground. Its powder would not only turn to hot gas as the stacked cylinders climbed, it would burn the first stage's paper wall in the process, lessening the rocket's overall weight. The last of the burning powder in the first stage would ignite the powder at the base of the second stage, and the process would then repeat itself. At the top of the stages there would be what Haas called a "payload" (a term that would stick). Being an artilleryman, he designed a military missile that carried a small powder keg in its nose. Being a space dreamer, he also designed a small cylindrical house, complete with a round roof, windows, and door as another kind of payload. It was the first known design for a space station.

Studying the firmament—the night vault that sparkles with thousands of flickering lights—is older than recorded history. By its nature it was a very precise business, even by the time of Ptolemy, and it has remained so. If the purpose of astronomy is to describe the relationship of every object in the heavens to every other object in order to make sense of it all, then the utmost care has to be taken in establishing their relative positions.

Astronomers have to track a "space race" that started with the Big Bang or earlier and which continues unabated. The planets in this solar system race around the Sun. The system itself is only a minuscule dot in a galaxy among uncounted billions of other galaxies that are also in constant motion. Some, like Andromeda, spin in great spirals like pirouetting starfish or pinwheels; others form enormous cosmic ellipses; still others, like the Magellanic Clouds, are irregular explosions of star systems that defy neat description. None of this, let alone the fact that the galaxies interact through gravitational pull, was known to sixteenth-century astronomers. But they did know that the planets and star formations moved across the sky along routes that could be predicted. And the predictions rested squarely on the excruciatingly patient notation of planet and star positions over very many nights. During the Middle Ages, Arab caliphs built observatories and studied the night sky continuously and systematically, not only out of intellectual curiosity, but also to set feasts and fasts according to the position of the Moon. They got so good at it that they corrected and extended Greek astronomical charts.

As the end of the sixteenth century approached an astronomer and nobleman named Tycho Brahe, working on the Danish island of Hven, made a series of historic observations of Mars and other planets. Lacking a telescope, which had not yet been invented, Brahe used a quadrant—a four-squared mural—to track and plot the altitudes at which planets and stars crossed the meridian. The measurements were unprecedentedly precise and showed that earlier observations failed to adequately predict planetary positions. The technical rigor of his observations created a vast, valuable library of data. Then his funding dried up. Deprived of a research allowance by King Christian IV, Brahe did what other scientists, artists, and philosophers did and continue to do in similar circumstances. He moved to a more nourishing environment. Under the benevolent patronage of the Holy Roman Emperor Rudolph II, Brahe began publishing his observations in Prague in 1599. But the Dane disagreed with Copernicus's heliocentric theory, believing instead that while the other planets circled the Sun, all of them, including the Sun, circled Earth. Brahe, too, thought he knew enough about physical law to understand that people would be thrown off a spinning planet.

One of Brahe's assistants was a young German astronomer who also dabbled very successfully in astrology. His name was Johannes Kepler and he had a reputation as an excellent seer who successfully predicted plague, famine, and Turkish invasions (all of which came regularly and fairly often).

Kepler started to pore over Brahe's observations of Mars in 1601, right after his master died, making detailed calculations that led to a profoundly important discovery: the planets orbit the Sun, not in the perfect circles so beloved by Aristotle and others, but in ellipses. This, the first and most important of three laws postulated by Kepler, would forever form the basis of mankind's understanding of the way the planets in this system move. He published it and a second law having to do with the speeds that planets travel in their orbits in 1609. The book, immodestly but accurately titled *A NEW ASTRONOMY, or A PHYSICS OF THE SKY Derived from the Investigations of the MOTIONS OF THE STAR MARS Founded on Observations of the NOBLE TYCHO BRAHE,* contained a dangerous idea: celestial bodies are physical objects whose motions are produced by natural causes.

On the morning of Saturday, February 19, 1600—the year the twenty-nine-year-old Kepler started to work for Brahe—Giordano Bruno, the eclectic Dominican friar and Italian Renaissance philosopher, was taken from the cell in Rome's Nona Tower he had occupied for seven years, stripped naked, gagged, tied to a stake, and paraded through the cramped streets at the head of a hooded group of chanting inquisitors known as the Company of Mercy and Pity. But they had neither. Bruno's tormentors told him that a last-minute recantation for his sins would save him from eternal damnation as a heretic. Bruno could not have expressed contrition even if he had wanted to, since his jaw was clamped shut, a spike pierced his tongue, and another spike stuck in his palate. There was no way the men in the hoods would allow the man they were about to murder to tell the crowd at the Campo de' Fion in front of the Theater of Pompeii, where the procession finally stopped, what he had told

Cardinal Robert Bellarmine, the Catholic Church's greatest intellectual: "I neither ought to recant, nor will I. I have nothing to recant, nor do I know what I should recant." Later, he sealed his fate by showing contempt for his accusers. "In pronouncing my sentence," said Bruno, who had taught in Paris, Oxford, and Wittenberg, "*your* fear is greater than mine in hearing it." With a torch in one hand and a crucifix in the other, one of Bruno's killers demanded repentance one last time. The condemned man disgustedly turned his face away. The fire was lit and one of history's most profound and original thinkers was burned alive.

The "monstrosities" that Bruno advocated in debates and in writing had to do with astronomy, the least practical of the sciences—least practical but most threatening to men whose power rested on religious dogma and superstition. Bruno described an infinite universe with stars that were suns like Earth's and where other earths circled those suns. Worse, he taught that if God could create life on Earth, He could create it on the other planets, too. Therefore, Earth was Heaven for the creatures on the other planets. And if that was true—if there were indeed Heaven on Earth—then the power to say who went to Heaven and who did not would be lost, and with it, the power of the church. That is why a dangerous rabble-rouser such as Giordano Bruno had to be murdered.

Bruno's execution—which was condemned by at least one Roman tabloid of the time—was still fresh in the collective memory nine years later when Galileo turned a new invention, the telescope, on the universe. He had come to the conclusion in the waning years of the sixteenth century that Copernicus was right. Then, during a trip to Venice, he heard that a Dutchman had invented a "miracle tube," or "optick stick," sometime before or in 1608.

Galileo quickly returned to the University of Padua, where he lectured in mathematics, and built his own "spyglasses," as he first called the optical devices (they were not referred to as "telescopes" until 1611). The first dedicated astronomical telescopes were wooden tubes covered with paper or leather with hand-ground convex lenses on one end and a concave eyepiece on the other. He calculated that his first telescope made objects nine times larger than could be seen with the eye alone. But it was only a test model. "I shortly afterwards constructed another telescope with more nicety, which magnified objects more than sixty times," he later reported. "At length, by sparing neither labor or expense, I succeeded in constructing for myself an instrument so superior that objects seen through it appear magnified nearly a thousand times, and more than thirty times nearer than if viewed by the natural powers of sight alone." These were the tools, primitive as they were, that would begin a revolution in how mankind saw itself relative to the rest of creation: not at its center, but as an infinitesimal component. The new tools would bridge the void between distant worlds. And they would eventually do even more than that. They would become time machines that could see far into the past. But first they would draw other worlds nearer to this one. And what could be drawn nearer could be closely studied. And what could be studied might be touched. It was only a matter of improving the bridge.

On January 7, 1610, one of the telescopes—"a very excellent instrument," in Galileo's estimation—was pointed at Jupiter when he discovered something he had not noticed before because of the limitations of the other telescopes. Three "little stars, small but very bright," seemed to flank the great planet and were at varying distances from each other. The formation was so peculiar that it piqued his curiosity. What is more, they lined up on Jupiter's ecliptic, an imaginary plane running through its equator, and appeared to be brighter than the other "stars" in the region. Galileo carefully drew a diagram of what he saw: two of the stars were east of Jupiter and one rested, seemingly suspended, to its west. He went to bed believing that the three objects were, indeed, stars. But the image of the strange formation would not go away.

Led by "some fatality," Galileo therefore scrutinized Jupiter with particular care the following night. What he discovered in the black velvet that seemed to surround it would soon do nothing less than revolutionize astronomy, change forever the way the inhabitants of this planet conceived the universe beyond it, and simultaneously land him in the pantheon of immortal scientists and make him the scourge of the Church of Rome. What Galileo noticed on the night of the eighth was that the formation had changed radically. Although the objects that the astounded astronomer saw as he squinted though his eyepiece remained on Jupiter's ecliptic, all three were now due west of the planet, closer together than they had been the night before, and separated by equal distances. The next night, they were seen to have moved again, two by now having returned to the east of Jupiter and one having disappeared altogether (he correctly assumed that it was "hidden" by the planet).

Galileo himself should describe what happened next. "I discovered that the interchange of position which I saw belonged not to Jupiter, but to the stars to which my attention had been drawn, and I thought therefore that they ought to be observed henceforth with more attention and precision." More carefully diagrammed observations convinced the jubilant astronomer that "there are three stars in the heavens moving about Jupiter, *as Venus and Mercury around the Sun;* which at length was established as clear as daylight by numerous other subsequent observations. These observations also established that there are not only three, but four, erratic sidereal bodies performing their revolutions round Jupiter" (italics added). Painstaking diagrams made on successive nights left no doubt that all four "stars" were circling the great planet.

Galileo had spotted four of Jupiter's largest moons. So, apparently, had a German astronomer known as Simon Marius who is often credited with first seeing them on December 29, 1609, ahead of his Italian rival. Marius is the one who named the circling retinue after Jupiter's mythical lovers: Io, Europa, Ganymede, and Callisto. But Galileo is generally credited with making the discovery first because he published first. (That much has not changed in almost four centuries.) For his part, Galileo wanted to name the moons the Medicean Planets in honor of his patron, Cosimo de' Medici, grand duke of Tuscany. That practice would end when only national treasuries, not princely coffers, could support such research.

The master astronomer also turned his telescope on his own planet's lone satellite. There he saw, up close, a world similar to parts of his own. In a book titled *Message from the Stars,* which was published in 1610 and which made him an instant celebrity, he described regions of the cratered moonscape that reminded him of Bohemia.* Galileo, who was a trained watercolorist, made wash drawings in ink of what he saw. And this, in his words, is what that was: "On the fourth or fifth day after a new moon, when the moon was seen with brilliant horns, the boundary which divides the dark part from the light . . . traces out an uneven, rough, and very wavy line. . . . There is a similar sight on Earth about sunrise, when we behold the valleys not yet flooded with light though the mountains surrounding them are already ablaze with glowing splendor on the side opposite the Sun."

Unaided eyes had watched the Moon pass through its phases for uncounted millennia, but the advance and retreat of light on its surface, let alone details on the surface itself, was imperceptible because of the distance. Galileo, however, brought the Moon so close that he (and the princes and compatriots who were soon privileged to peer in awe through his optical tube) could actually see light from the Sun creep across the ancient orb's ghostly mountains, valleys, and craters. "With his first astronomical telescope," the physicist Philip Morrison and his wife, Phylis, have said, "Galileo healed the ancient split between the heavens and the earth."

By observing that Jupiter's "stars" were actually moons engaged in an intricate but predictable dance around it, Galileo established the fact that Earth was not the only planet to have a moon and that other centers of motion were out there. More important, he quickly came to believe that what he saw—smaller bodies in orderly orbits around a larger one—was a model of the solar system itself and that Copernicus had therefore been right all along.

But the most significant observational leap, at least where mankind's eventual move to space was concerned, was onto the Moon's surface. By portray-

* The treatise is often translated as *The Starry Messenger.* The Latin title, *Sidereus nuncius,* can mean either, but *Message from the Stars* more closely reflects Galileo's meaning. (Galilei, *Dialogue on the Great World Systems,* p. xi.) (*Note:* All titles cited in short form in footnotes can be found listed in full in the list of sources at the end of the book.)

ing it as being like the Bohemian plains, complete with surrounding mountains, Galileo in effect connected Heaven and Earth. When his friend, the Florentine painter Lodovico Cigoli, was commissioned by Pope Paul V to provide a painting in the dome of a papal chapel, the artist decided to depict a popular representation of the Virgin Mary in *Assumption of the Virgin.* This time, though, she was not standing on a two-dimensional crescent Moon, as was often the case. She stood on the cratered, mountainous, *earthlike* Moon that Cigoli had seen through Galileo's telescope. The implication was obvious to everyone who now courted Galileo, whose fame was quickly spreading, as well as to many others. If the Moon was similar to Earth, and if a way could be found to get to it, people could live there. If they could live there, it was entirely possibly that other places existed where they could also live. In fact, if there were such worlds, it was conceivable that there might even be living creatures already on them, as poor Bruno had insisted, meaning that life was not unique to Earth.

This was not news the church wanted to hear. Dogma had it that Earth, which was created by God for humanity, was the center of creation and was a unique masterpiece. It was therefore the only place that could be inhabited by His creatures. Any theory to the contrary therefore literally challenged the Vatican's interpretation of the world. Furthermore, an unending series of scientific revelations about outer space, one after another, did "not sound very biblical," in the Morrisons' words. A religion that claims divine knowledge and omniscience cannot afford to have its followers surprised and delighted by developments it has neither generated nor understands. In addition, the Pontiff and his cardinals felt beleaguered by the Protestant Reformation, which had by then captured half of Europe. (They had hated Bruno so much they labeled him a despised "Lutheran," which was generic slang for heretic. Ironically, he was a Dominican, and therefore a member of the Vatican's own thought police. "Dominican" translates to "dogs of God.") As a consequence, they were in no mood for new challenges from another quarter.

Nor did the church stand alone. Sides were drawn from Rome to London. While each new discovery was related with awe from town squares to princely dinner tables, Galileo was vilified, even by Giovanni Magini, the professor of astronomy at Bologna, who swore that the new planets would be "extirpated from the sky." There was a sense among many learned individuals—philosophers, physicians, and poets among them—that the combined doctrine of Copernicus, Kepler, and Galileo, three revolutionaries, threatened chaos. Their theory of force and change challenged any semblance of permanence and regularity, of worldly order and stability. John Donne reacted by proclaiming that if the new astronomers were right, the world was plunging toward dissolution because they had "replaced the circle of perfection with the straight line of mortality." The Copernicans, as they were pejoratively called, were repeatedly attacked for being "men of no intellect."

Was it right, one historian of science would ask rhetorically, almost three and a half centuries later, "to subvert the vast and documented discourse of the Schools [including those of Athens], so well based on the familiar evidence of the senses, which allows us to account in an orderly manner for nature and life and the soul itself—and fits in so handsomely with revealed truth—to launch ourselves in a sea of paradoxes and unnatural conclusions, simply because a man has come forward with two lenses and a length of pipe?"

As the furor built, Galileo sent a copy of *Message from the Stars* to Kepler as soon as it was published. Kepler was by then solidly in the Copernican school and was quick to grasp the importance of his colleague's new tool, which turned his very soul rhapsodic. "What now, dear reader, shall we make out of our telescope?" he wrote. "Shall we make it a Mercury's magic wand to cross the liquid ether with, and, like Lucian, lead a colony to the uninhabited evening star? Or shall we make it Cupid's arrow, which entering by our eyes, has pierced our innermost mind, and fired us with a love of Venus?"

It was obvious to Kepler that what could be seen could be reached, so his thoughts instinctively turned to great feats of exploration already accomplished or under way: to the daring voyages of Columbus and Magellan, Balboa and da Gama, Cabot and Drake. Then he replied to Galileo in an open letter on April 19, 1610. *Conversation with the Messenger from the Stars,* as he called it, predicted the flight of *Apollo 11* to the Moon 359 years later in prose that was as hauntingly beautiful as it was prescient:

> Who would have believed that a huge ocean could be crossed more peacefully and safely than the narrow expanse of the Adriatic, the Baltic Sea or the English Channel? Provide ship or sails adapted to the heavenly breezes, and there will be some who will not fear even that void [of space]. . . . So, for those who will come shortly to attempt this journey, let us establish the astronomy: Galileo, you of Jupiter, I of the Moon.

Kepler now believed implicitly that the great obstacle was not inhabiting the Moon, but reaching it. While he is best known for the three laws of planetary motion, his other trade, astrology, gave him quite another dimension. He used his vivid imagination and fluid writing skill to craft stories about magic and mysticism.

One such work, *Somnium* (*The Dream*), was another important forerunner of science fiction and was heavily footnoted with astronomical data. In it, a young Icelander named Duracotus returns home after studying with the great Brahe and is lectured by a demon on the nature of the Moon and its relation to

Earth. The demon knows about the Moon because he is able to fly there during eclipses and can therefore describe its seas, mountains, long nights, weather, snakelike inhabitants, and what Earth looks like from its surface. *Somnium* was probably drafted in 1593 and revised in 1609, while Galileo was grinding glass, but it was not published until 1634, four years after its author's death. Since a footnote in that edition says explicitly that it was written to refute arguments against the movement of Earth, and therefore supported Copernicus, it amounted to another challenge to papal dogma.

Kepler's descendents, after all, would have known that in 1616 Galileo had been summoned by the Inquisition in Rome and forced to renounce Copernican theory, which was declared false and heretical, and then banned. (The ban was not lifted until 1835.) During the ensuing seventeen years, the astronomer was ensnared in vicious intrigues, including the planting of false evidence in his files, that would have done credit to both the Medicis and the Borgias. The Jesuits, who were becoming the general staff of the Church of Rome, showed the pope that the educational system they had carefully built around Aristotle's teachings, and which they hoped would spearhead a counterreformation, was in great danger because of Galileo. In fact, they warned Urban VIII—who was not totally unsympathetic to Galileo—that he might be more dangerous "than Luther and Calvin themselves."

In 1632, two years before Kepler died, the publication of Galileo's monumental *Dialogue on the Great World Systems* pushed the Vatican and Galileo's other enemies to their limit. The book, one of a handful of truly towering scientific masterpieces, opened with a frontal attack on the Aristotelian and Ptolemian model of the solar system and used all of the motion in the universe—the spinning planets, exploding stars, racing comets, and whirling moons—to challenge the stubbornly held belief that change was bad and that a lack of it was not only noble but was a sign of perfection. The following year, 1633, the astronomer's now-exasperated inquisitors again forced him, on his knees, to recant Copernican teachings on pain of being burned at the stake. (The church was aware that it had in effect gotten some very bad press because of Bruno, and it was therefore bluffing, but there was no way for Galileo to know that.)

Dialogue on the Great World Systems was immediately banned, and its author was placed under house arrest in Siena. He was eventually allowed to live in seclusion in a villa outside of Florence, where he died in 1642, still under house arrest. In the same sort of irony that caused Beethoven to go deaf, the father of modern astronomy became blind before he died. But blinder still were his enemies in the church and other reactionaries. *Dialogue on the Great World Systems,* considered to be a kind of scientific pornography, was condemned to the index of forbidden books until 1823. It took 359 years for the

Vatican to admit that Galileo had been right after all. On October 31, 1992 (ironically, a pagan holiday in much of the Christian world), Pope John Paul II told the Pontifical Academy of Sciences that Galileo's persecutors "unduly" mixed the doctrine of faith with scientific investigation.

Galileo left a rich legacy to the scientists and explorers who came after him. Perhaps the most important of these, as the Morrisons have said, was the telescope itself. "The telescope continues, it lives. . . . [T]he point is that out of new experience come new questions, new explanations, then an extended theory, with new concepts," they have written. "Fifty years after the telescope, twenty years after Galileo's death, there may have been a dozen active observatories in Europe. They had better instruments, able to see more detail, able to find fainter objects, able to measure more carefully, and able to understand much more of what they saw."

Copernicus, Brahe, Kepler, and Galileo changed mankind's concept of its place in the world. So would Isaac Newton. Galileo's purging and public embarrassment had the effect of moving the scientific revolution out of Italy's chilly environment and sending it north to France and England. Newton's *Philosophiae naturalis principia mathematica* (*Mathematical Principles of Natural Philosophy*), published in 1687, is generally considered to be the single greatest scientific work ever written. It contained two sets of laws that would profoundly influence the development of rocketry and space travel. The first was the law of universal gravitation. This explained the motion of all of the heavenly bodies found by the astronomers and how they affected each other in a vastly complex and infinitely subtle relationship that even extended to the caressing of Earth's oceans, which causes the tides.

The second set were the three laws of motion. The first said that objects remain at rest or in motion until some other force causes change. Applied to space, it would mean that a body moving through a vacuum would head in one direction and at a constant speed until its direction or speed or both were changed. The second law had it that a force acting on an object will make it accelerate in the same direction as the force and that the amount of acceleration will be proportional to the force and inversely proportional to the mass of the object. This would quantify the amount of force a rocket engine would have to exert to move an object. The third, and most famous of the three, said that for every action, there is an equal and opposite reaction. In other words, a rocket engine firing at one end of a projectile will, if powerful enough, move the projectile in the opposite direction (and it will keep going in that direction, according to the first law, until another force—say, gravity—changes it).

"These great achievements mark the closing of an epoch in the history of the thought of the world and the beginning of a new, for they entirely overthrew earlier views respecting the nature of the cosmos and established others which were entirely different," the eminent British astronomer and mathematician Forest Ray Moulton has written. "They permanently removed man from his proud position at the center of creation and placed him on a relatively insignificant body; but, as a compensation, they rescued him from a universe of chance and superstition and gave him one of unfailing and majestic orderliness."

And they did more than that. They bridged the ancient world and the modern, in the process showing humanity the way out of its confinement and pointing to a place that in comparison is awesome in its size, grandeur, process, and possibility. The body of work that accomplished that feat—astronomy and the laws of the planets—is the least practical for the public at large. Yet it is the most supremely enlightening of the sciences. Copernicus, Kepler, Galileo, and Newton stand with Bach, Mozart, and Wagner; Dante, Shakespeare, and Goethe; Michelangelo, Rembrandt, and Cézanne in their towering nobility. They did nothing short of start a scientific revolution that moved Earth away from the center of the universe and put it where it belongs, in the process launching a whole new way of understanding this world and all of those beyond it. This was fundamental to the move to space. And so too, of course, was the means of reaching it.

Riding Skyrockets

Princes of theory such as Brahe, Copernicus, Kepler, Galileo, Newton, and other ingenious astronomers and mathematicians always had less intellectual counterparts, the blacksmiths, jugglers, and acrobats of spaceflight's prehistory, who just wanted to get off the ground. They popped up (or at any rate tried to) in many places throughout much of recorded history, which, if nothing else, shows that the urge to fly is universal. If the beginning of astronomy is lost in the mist of time, so is the beginning of the daring business of test piloting. But its patron saints (who came to be called daredevils until technology brought a degree of predictability to their profession) certainly spent time considering where burning gunpowder and other devices could take them. The question would have been obvious enough: If a small amount of gunpowder could propel an arrow or a lance high into the air, could not a much larger amount propel a man?

Legend has it that a Chinese mandarin named Wan Hu took the first rocket ride, or at any rate tried to, in A.D. 1500. As he sat on a chair, the story goes, forty-seven coolies lit forty-seven black-powder rockets that were attached to it. Wan Hu traveled a short distance and then disappeared in an explosion and a cloud of smoke. If it really happened, Wan Hu had the triple distinction of being the first person to ride a rocket, the first to fly on a self-propelled, heavier-than-air device, and the first rocket pilot to get killed during a test flight.

Frank H. Winter, a historian at the National Air and Space Museum, has called the experiment highly improbable for several compelling reasons and has suggested that since Wan Hu is roughly translated as "crazy fox," the story is merely a fable warning against the kind of overexuberance that got Icarus in trouble. Whatever the truth, Wan Hu remains the first Chinese to have a crater on the Moon named after him.

Then there was a Turk named Larari Hasan Celebi, who, it is claimed, was shot into the sky by fifty-four pounds of gunpowder to celebrate the birth of Sultan Murad IV's daughter, Kaya Sultan, in 1623. Before his skyrocket blasted off on the banks of the Bosphorus, Celebi is said to have told the sultan: "Your Majesty, I leave you in this world while I am going to have a talk with the prophet Jesus." The rocket then carried Celebi high into the air, where he opened several "wings," and then glided to a safe landing in front of the royal palace. "Your Majesty," the young Turk reportedly told the sultan while crowds cheered, "Prophet Jesus sends his greetings to you." Celebi was rewarded with a pouch of gold, made a cavalry officer, and is said to have been killed in combat in the Crimea. At least two respected Turkish sources give this account. Norwegian and Turkish scientists took the story seriously enough to suggest in 1986 that the bottom of the Bosphorus be searched for the remnants of Celebi's rocket. There is no evidence that anyone actually did so, however.

Another somewhat credible account, titled "On Fire Balloons," appeared in a London newspaper in September 1784. It described a celebratory launching at a reception given by the king of Siam for a new French ambassador. "The inventor of this firework, sitting himself down on the end of one of these rockets, ordered it to be fired and was whisked up into the air higher than any four steeples in the world could reach were they set upon one another," the correspondent reported. "The rocket having spent its strength, and being ready to fall down, all luminous with the infinite number of stars that broke from it every moment, the engineer opened a sort of umbrella he had carried with him, when it was extended, was less than 30 feet in diameter. This umbrella was made of feathers. . . . The engineer, by his great umbrella, came to the ground, surrounded with stars, as gently as if he had wings." As late as 1978, according to Winter, very large, gunpowder-filled bamboo rockets were in fact used at a festival in northeast Thailand in an ancient attempt to persuade the rain god to help grow rice. The historian himself witnessed the building and firing of the rockets and reported that the average flight time was twenty seconds, easily enough to get one of them to an altitude of four steeples. "It is conceivable that one of the expert 17th century rocket-makers from the northeastern part of the realm was sent south to dazzle the *Ferengs* ('foreigners') with a special treat," he has written.

Then there was the strange, but fully documented, case of Frederick Rodman Law, or just Rodman Law, alias the "Human Fly" and the "Human Bullet." Law was the brother of Ruth Law, a famous pilot, and was a movie stuntman who specialized in parachuting out of airplanes and burning balloons and off bridges, skyscrapers, and other edifices, including the Statue of Liberty. On March 13, 1913, Law climbed into the top of a forty-four-foot-long skyrocket that was pointed straight up at the end of Williams Avenue in Jersey City, New Jersey. The thing had been packed with nine hundred pounds of gunpowder by its manufacturer, the International Fireworks Company (later the Unexcelled Fireworks Company of New York). Law, who sat in the nose of the rocket wearing goggles and what appeared to be a leather football helmet, announced that he intended to fly to 3,500 feet and then bail out. With a crowd of 150 people, including a movie cameraman and reporters from the *Herald Tribune* and *The New York Times* looking on, the "Human Bullet" told Samuel L. Serpico, the manager of International Fireworks, to "light the fuse when ready." Serpico did as he was told. "After a few seconds," according to one witness, "there was a terrific explosion with a shock that threw most of the crowd to the ground [as] the big projectile burst into a thousand pieces." Law was thrown thirty feet and landed unconscious, his hands and face scorched and his clothes torn, muddied, and charred. But he soon recovered from his injuries and continued to perform stunts, though never again in a rocket.

Rocket Building Takes Shape

Rocket theory in England goes back at least to the *Epistola,* a Latin treatise on gunpowder written by the monk Roger Bacon in the middle of the thirteenth century. The problem with Bacon's formulas, as well as those concocted elsewhere in Europe, the Middle East, and Asia between 1100 and 1800, was that the rockets they powered were literally hit or miss. The range and velocity of any model—all of them crafted by hand—could vary quite a bit depending on the powder formula, how the powder was packed, the weight of the casing, the size of the exhaust nozzle, and other factors.

The first major battles with rockets that involved Europeans occurred during a revolt against the British which began in 1781 in the Mysore region of southwest India and lasted through 1799. The Indians fired crude but effective rockets against British regulars during battles at Seringapatam in 1792 and 1799. "No hail could be thicker," a young English officer named Bayly lamented in his diary. "Every illumination of blue lights was accompanied by a shower of rockets, some of which entered the head of the column, passing through to the rear, causing death, wounds, and dreadful lacerations from the long bamboos of twenty or thirty feet, which are invariably attached to them."

The Royal Laboratory at Woolwich Arsenal was therefore ordered to design and develop a dependable war rocket that could be produced in large quantities as standard equipment for the artillery. This was done by William Congreve, a Cambridge-educated socialite who was an intimate of the Royal Family and whose father was commandant of the Royal Artillery and Woolwich's comptroller. Congreve had studied law and run a newspaper. As the eighteenth century turned into the nineteenth, and in the aftermath of the battles in India (and in anticipation of others with France), he responded by turning his keen intellect and imagination to inventing a better rocket.

After at least three years of experiments, Congreve published *A Concise Account of the Origin and Progress of the Rocket System,* in November 1807. Even then there were those who fretted about national security and the danger of leaks, and since Congreve was one of them, he happily "sanitized" his report. "In the following pages I have cautiously avoided any disclosure which might lead to a discovery of the interior structure and combination of the rocket, on which all powers depend, this rule I have observed for obvious reasons," the inventor wrote with evident pride.

Noting that the Indian rockets had had a range of less than a thousand yards, Congreve designed one that traveled twice as far. It was an iron cylinder stuffed with seven pounds of compressed powder, and it weighed thirty-two pounds. The breakthroughs were using metal "carcasses" instead of paperboard; refining the powder through granulation machines to give more predictable results; and using pile driver presses to compact the powder so it was a denser and therefore more even-burning charge. He also incorporated noses into his design—warheads, in today's jargon—that could carry a variety of munitions, including incendiary, shrapnel, explosive, or shot. Other models would follow in relatively quick succession.

Congreve realized early that rockets were particularly suited to naval warfare because, unlike cannons, they did not recoil and destabilize the ship. He therefore suggested that his 2,000-yard model be used as part of a plan, soon accepted, "for the annoyance of Boulogne" by the Royal Navy. Ten boats were fitted with incendiary rockets for an attack on the French port city on November 21, 1805, but a fierce storm prevented the attack. A second attempt, on October 8, 1806, was successful. "In about half an hour above 2,000 rockets were discharged," Congreve reported with evident relish. "The dismay and astonishment of the enemy were complete—not a shot was returned—and in less than ten minutes after the first discharge, the town was discovered to be on fire." The rockets were used with even greater success to shell Copenhagen in 1807 and then other European cities. And Congreve was at least indirectly responsible for the national anthem of the United States. On the night of September 13–14, 1814, his ubiquitous rockets were used to shell Baltimore's

Fort McHenry, causing the "red glare" that inspired Francis Scott Key to write "The Star-Spangled Banner."

By 1828 Congreve had invented rockets for use in whaling and lifesaving, and his military models had been adopted by armies around the world. But as a means of propulsion, black powder had gone as far as it was going to go, at least for the time being.

Science and Fiction Converge on the Moon

In August 1835, *The New York Sun* ran a series of articles reporting some truly extraordinary observations of the Moon. The newspaper claimed that Sir John Herschel, a distinguished British astronomer and the son of Sir William Herschel, who had discovered Uranus in 1781, had made an even more important discovery. While using a South African telescope that weighed seventy-seven tons and could magnify the surface of the Moon to a few hundred yards, reporter Richard Adams Locke explained, Herschel had actually spotted life there. Quoting *The Edinburgh Journal of Science,* Locke wrote in impressive detail that the astronomer had seen huge precious stones, beaches, a bestiary of strange animals that included miniature bison, horned bears, and tail-less beavers. Most amazing of all, Locke added, the gigantic telescope had revealed intelligent moonmen who "averaged four feet in height and were covered except on the face with short and glossy copper-colored hair and had wings composed of thin membrane."

Except for the fact that Herschel really was in South Africa at the time, everything else Locke reported—even the name of the journal—was a hoax. As intended, the *Sun*'s circulation shot up almost overnight until it hit twenty thousand copies, making it the most widely read newspaper on Earth by the end of the month. A pamphlet version of the series sold sixty thousand copies to people who were intrigued and delighted to learn that life abounded on their planet's cosmic companion. Even after Locke's articles were exposed for the lies they were, many people continued to believe in the existence of intelligent life on the Moon.

At a time when even Europe's middle and lower classes basked in the dimming glow of the Enlightenment and before the Industrial Revolution's machines raised the standard of living of the masses—while alienating or frightening many individuals—people delighted in thinking that they were not alone in the extended world. The nineteenth century would close with humanity projecting its own darkest side to conjure evil invaders from space who tried to inflict technological terror on Earth, but it opened with a generally hospitable attitude about possible neighbors elsewhere in the solar system. Rather than assume that intelligent life was scarce beyond the precinct of

Earth, or even that it was unique here, as Christian fundamentalists believed and continue to believe, liberal intellectuals took it for granted that human beings were sprinkled almost everywhere.

In *Celestial Scenery, or the Wonders of the Planetary System Displayed,* an introduction to the solar system published in 1838 that was written for an educated readership, a learned individual named Thomas Dick happily speculated about the possibility of life on other planets and even in the Sun. He estimated that there were more than four billion "Lunarians" on the Moon (while describing how Earth looked to them) and billions of other individuals who not only lived on the other planets, but on Saturn's rings and on the moons of Saturn, Jupiter, and Uranus. Dick calculated that Mercury was capable of supporting more people than Earth because, unlike Earth, it had no oceans and therefore had relatively more room. And after noting that most astronomers did not believe the Sun was populated, he quoted no less an authority than the elder Herschel himself as saying that "it is not altogether improbable that the Sun is peopled with rational beings" who lived in an "interior stratum" that protected them from the intense fire and light on the surface.* "We ought not to set limits on the wisdom and arrangements of the Creator by affirming that rational beings could not exist and find enjoyment on such a globe as the Sun," wrote Dick, adding that the dark appearance beneath the fire indicated that the temperature of the interior could be low enough to allow life to exist.

Described, too, were several tantalizing, telltale signs of life on Mars, all of them deduced through the blur of the Herschels' telescopes. There was a high degree of probability that dark spots on the Martian surface were oceans, Dick reported. John Herschel was invoked yet again as having explained that the large bodies of water "by a general law in optics, appear *greenish,* and form a contrast to the land."† In addition, Mars had clouds and an atmosphere; a variety of seasons similar to Earth's; snow and ice at its poles; and days of about the same length. In other words, Mars was the most similar to Earth of all the planets. And that led to an inescapable conclusion. "Were we warranted from such circumstances to form an opinion respecting the physical and moral state of the beings that inhabit it, we might be apt to conclude that they are in a condition not altogether very different from that of the inhabitants of our globe."

In 1877, Italian astronomer Giovanni Virginio Schiaparelli did indeed see what appeared to be dark areas connected by faint lines on Mars. He called the

* Sir David Brewster, who described "continents and oceans and green savannas" on Mars in the 1850s, stated emphatically that Herschel "never asserted and never did believe that the children of the Sun were human beings, but, on the contrary, 'creatures fitted to their condition as well as we on this globe are to ours' " (Feinberg and Shapiro, *Life Beyond Earth,* pp. 258 and 380).

† The inference that water was necessary for life as it existed on Earth was correct. But it also contradicted the rationale for life on Mercury.

lines *"canali."* The word's first meaning is "channels," or "grooves," but it also translates to "canals." Percival Lowell, the Harvard-educated scion of Boston Brahmins, decided that the lines drawn by Schiaparelli could represent nothing less than actual water-carrying canals. He taught himself astronomy, built an observatory near Flagstaff, Arizona, in 1894, and devoted the rest of his life to proving that the canals were built by Martians in a desperate effort to stave off a deadly planetwide drought by moving precious water from the planet's icy poles to its parched equatorial region.

Lowell doggedly wrote articles and delivered lectures in defense of his theory into the twentieth century. Not many astronomers agreed with him, but the public remained fascinated by the prospect of intelligent life on a neighboring planet. The first spacecraft to reach Mars in the 1960s would find no trace of life at all, let alone intelligent life, and the last vestige of the canal theory would finally be put to rest once and for all.

But Lowell deserves to be remembered for more than his obsession with Martians. His observations also produced a great deal of solid data on the Red Planet and innovations in the use of the telescope. More important, Lowell popularized Mars by turning it into Earth's alter ego: an exotic and accessible place where fiction writers such as H. G. Wells and Edgar Rice Burroughs could conjure fantastic creatures and provide a ready target for the rocket builders-to-be who would read them. Fact and fiction would come together in a long, synergistic relationship on the Martian plains. Lowell, whose own work was part fact, part fiction, would stimulate the search for life there; a search that continues to this day. It would also fire the imagination of Robert Hutchings Goddard, a youngster from Worcester, Massachusetts, who was transfixed by Lowell's public lectures and who wanted to reach the sky so badly that he virtually hung out in a tree in his backyard to do so. Goddard would become one of rocketry's true fathers and surely one of its most enigmatic characters. So the dream spread.

Jules Verne Goes to the Moon

True to the spirit of Joseph and Étienne Montgolfier, the French brothers who launched the first man-carrying balloon in 1783 after having used a sheep, a duck, and a rooster in flight tests, Jules Verne's own imagined trip to the Moon began in the gondola of a balloon. Having come from his home in Nantes to Paris to study law in 1848, the aspiring young writer moved into salon society and gradually began to think about the effects of science and technology in general, and on how they drove the Industrial Revolution, in particular.

By 1861, Verne had joined a club, the Cercle de la Presse Scientifique, and befriended Félix Tournachon. Originally a journalist and artist, Tournachon had moved into photography, where, under the name "Nadar," he became one of its true pioneers. Nadar photographed the Parisian sewers by electric light and later would virtually invent the photo-interview. But he is best remembered as the first person to take aerial photographs of Paris while suspended under a balloon.

Verne, who had been fascinated by balloons at least as early as 1851, thought that was just marvelous. He would find an enduring place in literature, not by developing complex personalities the way Dostoyevsky did, but by sending otherwise ordinary characters on truly epic journeys that were made plausible because they were held together by real science and technology. Today, Verne would be called an unabashed technofreak whose genius was in generating excitement from the facts themselves. And like Daedalus before him and Tsiolkovsky after, he used his imagination to escape from his own labyrinth: an unhappy marriage.

By 1863, Nadar, Verne, Guillaume de La Landelle, Ponton d'Amencourt, and others had started a Society for Aerial Locomotion in Nadar's studio. It was there that d'Amencourt demonstrated a working model of a steam helicopter while everyone debated the merits of lighter-than-air versus heavier-than-air flying machines. De La Landelle invented the word *aviation* for the title of a book that included a picture of a two-masted helicopter. But ballooning, not airscrewing, was the thing in the mid-1860s as compulsive adventurers like Verne and Nadar became enthusiastic aeronauts. *Five Weeks in a Balloon,* Verne's first novel, describes a journey of exploration over Africa in which the airship *Victoria* is carried by trade winds and uses a special device that heats hydrogen to stay aloft. It was patterned after *Le Géant,* a huge balloon that Nadar and the others built and in which Verne flew in 1863.

From the Earth to the Moon was published two years later and was destined to become a classic. Contrary to common belief, however, it was born in a crowded field. That same year, 1865, Achille Eyraud published his *Voyage à Vénus.* Although this small novel made no literary splash, it did introduce the "spaceship" and used real physics to explain in some detail the working of a *moteur à réaction* that operated on recycled water. Meanwhile, Alexandre Dumas *fils,* another of Verne's acquaintances, published *Voyage à la lune;* Henri de Parville turned out *Un Habitant de la planète Mars;* and the well-known French astronomer Camille Flammarion produced *Mondes imaginaires et mondes réels.* The last was a factual survey of popular astronomy that took up the issue of the existence of other worlds and the possibility that there were other habitable planets. In addition, an anonymous Englishman published *A History of a Voyage to the Moon* and an equally anonymous Frenchman cranked out *Voyage à la lune.*

But Verne's contribution had two qualities that the others lacked. His characters—clearheaded composites of scientists, engineers, and explorers—were modern, truthful, and volitional. "Instead of yielding to the traditional modesty of being 'insignificant sons of great ancestors,' they act with the full knowledge that their time has surpassed any preceding time," Willy Ley would write. "They know that they know more than their ancestors did." In order to exploit the truly big questions in science and exploration—peering inside Earth itself, racing around it in record time, sailing beneath its ocean surface, and flying off of it in search of other worlds—he had to pore over his beloved science and invest a great deal of time making the necessary calculations. In the case of *From the Earth to the Moon,* he even went to astronomers to have his calculations checked. No wonder the book's heroes have to come to grips with the three basic problems in all space launches: developing enough velocity to break gravity's hold, launching in the right direction, and doing so at the right time.

From the Earth to the Moon takes place in the United States immediately after the Civil War. Its principal characters belong to the Gun Club, a fictitious fraternity of artillery lovers in Baltimore, many of whom lost arms and legs on the battlefield to the very cannons they worship. "The Yankees, the first mechanics in the world, are engineers—just as the Italians are musicians and the Germans metaphysicians—by right of birth," Verne explained puckishly. "Nothing is more natural, therefore, than to perceive them applying their audacious ingenuity to the science of gunnery."

And not just to gunnery. The amputees and others decide to take the grandest shot of all; to fire a huge cannonball at the Moon just to prove that they can hit it. But then a young man named Ardan appears (no cosmopolitan Frenchman would have failed to notice that Ardan was Verne's good friend, the famous Nadar) who convinces the Gun Club to send him on the journey in a padded, furnished projectile shot out of a gigantic cannon called the Columbiad. Eventually there will be a crew of three. Here is what Verne says about the target:

> From the time of Thales of Miletus, in the fifth century B.C., down to that of Copernicus in the fifteenth and Tycho Brahe in the sixteenth century A.D., observations have from time to time carried on with more or less correctness, until in the present day the altitudes of the lunar mountains have been determined with exactitude. Galileo explained the phenomena of the lunar light produced during certain of her phases by the existence of mountains, to which he assigned a mean altitude of 27,000 feet. After him Hevelius, an astronomer of Dantzic, reduced the highest elevations to 15,000 feet; but the calculations of Riccioli brought them up again to 21,000 feet.*

* Johannes Hevelius (1611–87) was indeed a noted Polish astronomer who studied the Moon's surface and published the first detailed map of it in *Selenographia* in 1647. Giambattista Riccioli (1598–1671) proposed names for lunar features in his *Almagestum Novum* (1651). Most are still in use.

If Verne did not read *Mathematical Principles of Natural Philosophy,* he at least gave every indication of understanding how its three laws of motion related to his story. Here, for example, is what he says about the spacecraft:

> The travelers' sleep was rendered more peaceful by the projectile's excessive speed, for it seemed absolutely motionless. Not a motion betrayed its onward course through space. The rate of progress, however rapid it might be, cannot produce any sensible effect on the human frame when it takes place in vacuum, or when the mass of air circulates with the body which is carried with it. What inhabitant of the earth perceives its speed, which, however, is at the rate of 68,000 miles per hour? Motion under such conditions is "felt" no more than repose; and when a body is in repose it will remain so as long as no strange force displaces it; if moving, it will not stop unless an obstacle comes in its way. This indifference to motion or repose is called inertia.

And about the optimum time to fire the massive cannon and the projectile's trajectory:

> It is clear you should choose a time when the moon is at its perigee. But it should also be a time when the moon is crossing the zenith, because that would further shorten the trip by a distance equal to the Earth's radius, or 3,919 miles. The final trajectory then need be only 214,976 miles. . . . [Y]ou must aim the cannon at the zenith: thus the trajectory will be perpendicular to the plane of the horizon and the projectile will be able to escape from gravity much more rapidly. But to be able to aim at the moon at the zenith, your gunsite must be situated in a latitude no greater than the moon's declination: that is, the site must be between 0 and 28 degrees latitude, north or south.

The three heroes are in turn shot out of Columbiad (with water softening the sudden acceleration) and ride the projectile toward the Moon, showing off their prodigious science and mathematical expertise to each other all the way. Only a chance encounter with a second moon, whose gravitational pull changes the aluminum projectile's direction, saves them from crashing into its big brother. The encounter happens in *Around the Moon,* a sequel that came out in 1867 and that ended with the trio splashing down in the sea.

Partly because of poor translations, reviewers of the English-language editions of *From the Earth to the Moon, Twenty Thousand Leagues Under the Sea,* and other Verne science fiction tended to dismiss them as children's stories, at least until the space age. In fact, *From the Earth to the Moon* was not only brilliant scientific prophesy, but it was a biting social and political commentary and satire about an America that seemed to Verne to have more brawn than brain; on its military mentality (he was appalled by the carnage in the Civil War); and on jingoistic expansionism (including his own country's imperial ambitions).

Rocketry's three giants—Tsiolkovsky, Goddard, and Hermann Oberth—not only were profoundly inspired by *From the Earth to the Moon* but actually learned from it. Far from being a simplistic tale, *From the Earth to the Moon* and its sequel were serious attempts not only to use science to whet the public's appetite for adventure but to convince it that nature, science, and technology were now and forever inseparable. Science and technology were opening the New World and altering the Old. And Verne made himself their harbinger. Work had begun on the Suez Canal in 1859; Louis Pasteur announced his germ theory in 1861; James Clerk Maxwell's electromagnetic theory of light was published in 1862; Joseph Lister began antiseptic surgery in 1865; and in 1866, halfway between *From the Earth to the Moon* and *Around the Moon,* Alfred Nobel invented dynamite. Walter James Miller, a professor of English at New York University, has written that Verne's description of the pouring of the Columbiad—a triumph of metallurgy—amounted to his officiating at the marriage of nature and technology:

> The ground trembled as these cascades of molten metal, sending whorls of smoke toward the sky, volatilized the moisture in the core mold and sent it through the vent-holes in the stone revetment in the form of dense vapors. The artificial clouds spiraled toward the zenith, reaching a height of 3,000 feet. A savage, wandering on the other side of the horizon, would have thought some new crater was being formed in the heart of Florida, but there was neither eruption, nor tornado, nor tempest, nor clash of the elements, none of those terrible catastrophes nature is capable of producing. No! It was man alone who had created these reddish vapors, these gigantic flames worthy of a volcano, these loud tremors like the shock of an earthquake, these reverberations rivaling the sound of hurricanes. It was his hand that had flung—into an abyss he had created—a whole Niagara of molten metal.

"He seems most inspired when he can portray the works of man and the great natural wonders on the same scale," Miller observed. "And such imagery signals a definite turning point in man's relation with Mother Earth. Until now, even for Verne, Niagara has stood mainly as a symbol of the powerful forces of nature with which pioneers must cope. But here Verne makes Niagara serve as a metaphor not for the forces of nature but for the forces of man." And in doing so, he set the basis for man to create whatever forces he needed to go to space.

From a technical standpoint, the two Moon books heralded the space age; the imaginative leap that connected an ancient fantasy with a reality they themselves helped to shape. As might be expected, Verne got a few things wrong. The air-pressure buildup inside the cannon as the projectile races toward its mouth, combined with the immense explosive force behind the vehicle, would have caused it to disintegrate, for example. And even if it hadn't been crushed by being caught between compressed air and the explosion of 400,000 pounds of guncotton, the force of the blast itself would have turned the projectile's occupants into pulp.

Yet this does not diminish Verne's extraordinary vision. Tampa, the launch site, is less than 120 miles from Cape Canaveral and is on the correct latitude. The projectile's change in direction near the little Moon anticipated the discovery of the gravity-assisted course change technique in 1961 that would make exploration of the outer planets possible. The projectile's landing in water suggested the splashdowns that would be used in the Mercury, Gemini, and Apollo programs. His calculation that it would take the projectile four days to reach the Moon given an initial velocity of twelve thousand yards a second was similarly on the mark. Michael Ardan's seemingly outrageous prediction that "light or electricity" would one day be used to propel spacecraft was realized with the discovery that the solar wind—Kepler's heavenly breezes—could push interplanetary spacecraft and so could charged particles shot out of an ion engine. The sixteen-foot-diameter aperture telescope used by the Gun Club to find the projectile on the Moon became the Mount Palomar telescope, which went into operation in California in 1948.

But Verne's greatest and most self-fulfilling prophesy had to do with the use of rockets to break the projectile's descent to the lunar surface. Perhaps more than any of the other innovations that he used for the expedition to the Moon, the rocket engine itself inspired the fathers of the space age.* Tsiolkovsky would have this to say about it: "For a long time I thought of the rocket as everybody else did—just as a means of diversion and of petty everyday uses. I do not remember exactly what prompted me to make calculations of its motion. Probably the first seeds of the idea were sown by that great fantastic author Jules Verne: he directed my thoughts along certain channels, then came a desire, and after that, the work of the mind." Oberth would recall that during his formative years, he "always had in mind the rockets designed by Jules Verne." And Goddard was so impressed by *From the Earth to the Moon* that he noted it in his diary, along with Wells's *The War of the Worlds*. The latter, serialized in magazines in 1897 and brought together as a book the following year, described a terrifying invasion of Earth by Martians who eventually suc-

* He neglected to use a rocket as the main propulsion system apparently because a sufficiently powerful one was inconceivable in 1865. And the cannon served as a logical metaphor for what came to be called the military-industrial complex.

cumb to terrestrial diseases. It was the prototypical expression of the way earthlings imagine evil aliens and it set the basis for countless dime novels and grade-B movies.

From Buffalo to the Moon

As the nineteenth century drew to a close, interest in other worlds grew. This was partly because astronomers, including Schiaparelli, Lowell, and Asaph Hall, who discovered Mars' two small moons in 1877, were opening the sky to the public. But it was also because of those who chronicled the infinite possibilities of what could happen out there. To be sure, Verne and Wells were unquestionably the most influential popularizers of space. But a legion of lesser lights with decidedly uneven abilities came behind them. There were New World aristocrats such as John Jacob Astor, who published *A Journey in Other Worlds* in 1894 to establish a literary credential. And there were anonymous hacks throughout Europe and from Greenwich Village to the Tenderloin who pounded typewriters and scrawled on tablets, grinding out cheap thrills by the pound for drinks and rent money. First the writers predicted the rocketeers. Then they heralded them. Then they played counterpoint to them. Ultimately, they would adulate and then turn on them.

As early as 1869, Edward Everett Hale described the first Earth satellite in a serial in *The Atlantic Monthly* that would be published under the title *The Brick Moon.* Twelve years later, Hermann Ganswindt described an interplanetary spaceship and created artificial gravity by spinning it. Percy Greg's two-volume *Across the Zodiac,* which came out in 1880, had a colossal spaceship named *Astronaut* cruise to Mars with an anti-gravity device. Kurt Lasswitz's *Auf zwei Planeten* (*Two Planets*), published in 1897, sent good-natured, highly intelligent Martians to Earth (they were very different from Wells's nasties, who invaded the same year). Garrett P. Serviss, an astronomer, described for the first time massed armadas of interplanetary spacecraft in *Edison's Conquest of Mars.* The three-volume *Aventures extraordinaires d'un savant Russe* (*The Extraordinary Adventures of a Russian Scientist*) by G. Le Faure and H. de Graffigny, which appeared in 1889, amounts to a tour of the Moon, comets, asteroids, and the planets with a variety of spacecraft ranging from projectiles like Verne's to those using solar sails.

Meanwhile, dime novels and magazine stories proliferated, many of them illustrated with fantastical pictures of lunar and Martian cities and spectacular, often lush, spacescapes. Pulp fiction such as *Six Weeks in the Moon* and *The Rocket; or Adventures in the Air* were sold everywhere. In 1877, the play *A Trip to the Moon,* based on Verne's masterpiece, was staged in New York's Booth Theater. It was followed by *A Trip to Mars,* a musical performed by the

Lilliputians—a company of midgets—in New York's Niblo Theater in 1893. Midgets were taken to be freaks well into the twentieth century and therefore made especially attractive extraterrestrials.

Frederick Thompson certainly thought so. He used them in his own extravaganza called "A Trip to the Moon," which was a major attraction at the Pan American Exposition in Buffalo, New York, in 1901. Thompson was an architect by training and a P. T. Barnum by instinct. He staged his show in an eighty-foot-high building that had the grandeur of a railroad station. For fifty cents—twice the price of any other attraction on the midway—customers of Thompson's Aerial Navigation Company could ride a thirty-seat spaceship named *Luna* all the way to the Moon. The craft was a brilliantly lighted green and white cigar-shaped ship, "the size of a small lake steamer," according to one account, with a large cabin in the middle and huge red canvas wings.

At the sound of a gong and chains—the nautical motif that would partly define the space age—*Luna* started to gradually rock and then seemed to lift into the sky with its wings slowly flapping. The passengers, sitting on steamer chairs, saw first clouds floating by, then a model of Buffalo, complete with the exposition itself and hundreds of blinking lights. They could even see Niagara Falls. As they looked through the spaceship's railing, the city seemed to fall away, while the whole planet itself came into view. Then even Earth itself shrank into a dot. As *Luna* sailed on to the Moon, the voyagers saw a twinkling sky. Next, they sailed into a fierce electrical storm with lightning, thunder, and whistling wind, followed by a flight through a bluish atmosphere.

Finally the passengers spotted the Moon, with the face of the Man in the Moon clearly visible. Next, *Luna* slowed, banked to the right, and soon struck something solid. It was the lunar surface. After the captain announced that their destination had been reached, the passengers walked off the spaceship and into an amazing scene. They were greeted in a crater by "Selenites," sixty midgets with spiked backs, who escorted them past stalactites and "crystallized mineral wonders" and then down an illuminated avenue to the City of the Moon. As they walked through the city, they were offered samples of green cheese by some Selenites and saw strange bazaars, souvenir shops, and mooncraft demonstrations. The journey ended in the castle of the Man in the Moon. There, in a large throne room decorated with glass columns and bronze griffins, the king sat on a mother-of-pearl throne, giant bodyguards at his sides. He treated the earthlings to a view of his electric fountain, which showed all of the colors of the spectrum through splashing water, and to a dance performed by the maids of the Man in the Moon. Then the curtain fell and the show was over. The first electrical and mechanical space extravaganza, the genesis of amusement park space rides around the world, was over in twenty minutes.

Some 400,000 people used the Aerial Navigation Company to tour the Moon. President William McKinley, Secretary of War Elihu Root, Secretary of State John Hay, and apparently every member of the cabinet, plus several governors and most of the justices of the Supreme Court, traveled to Buffalo for a trip to the Moon. So did Thomas A. Edison. After returning to Earth, Root was asked if he thought airships could be used in warfare. "If they are all as successful as this one," he is reported to have answered, "they would work very well."

When the Pan American Exposition closed, Thompson moved his navigation company to George C. Tilyou's Steeplechase Park at Coney Island on the Brooklyn shore, where it became another hit. Indeed, the "trip" so captivated the public that Thompson staked out his own area of Coney Island. In 1903 "A Trip to the Moon" became the sensational main attraction at the new Luna Park. That same year, two brothers from Dayton finally got a man-carrying heavier-than-air machine into the air.

"When the time comes for some historian of the far-distant future to survey critically the technical achievements of the nineteenth and twentieth centuries and to weigh the comparative economic importance of those achievements," a chronicler of aviation wrote seven years after the Wright brothers' flight at Kitty Hawk,

> it may be that the invention of the aeroplane flying-machine will be deemed to have been of less material value to the world than the discovery of Bessemer and open-hearth steel, or the perfection of the telegraph, or the introduction of new and more scientific methods in the management of our great industrial works. To us, however, the conquest of the air, to use a hackneyed phrase, is a technical triumph so dramatic and so amazing that it overshadows in importance every feat that the inventor has accomplished. If we are apt to lose our sense of proportion, it is not only because it was but yesterday that we learned the secret of the bird, but also because we have dreamed of flying long before we succeeded in ploughing the water in a dug-out canoe. From Icarus to the Wright Brothers is a far cry.

It was a cry that was also heard by Tsiolkovsky, who published at his own expense a seminal work on reaching beyond the atmosphere the same year the Wright brothers flew into it.

2

Rocket Science

Konstantin Edvardovich Tsiolkovsky had good reason to hate gravity; it imprisoned him in a world of pain. He was born on September 17, 1857, in Izhevskoye, a village in the province of Ryazanakaya, southwest of Moscow. His father's income as a forester, and then as a clerk, was meager. Yet the family got by and young Konstantin's first years were happy ones. "I was passionately fond of reading and devoured everything I could get hold of," he would later recall. "I loved to indulge in daydreams. . . . In my imagination, I could jump higher than anybody else, climb poles like a cat and walk ropes. I dreamed that there was no such thing as gravity."

But the happiness turned to deep sorrow when at the age of thirteen he lost his mother. At the same time, worsening ear trouble left him almost totally deaf and dependent on a tin ear horn. "It estranged me from others and compelled me, out of boredom, to read, concentrate, and daydream. Because of my deafness, every minute of my life that I spent with other people was torture. I felt I was isolated, humiliated—an outcast. This caused me to withdraw deep within myself, to pursue great goals so as to deserve the approval of others and not be despised."

Turning inward, Tsiolkovsky took refuge in the quiet world of studying mathematics, physics, and astronomy. These led to his inventing balloons and designing lathes and windmills. By sixteen, he had outgrown Izhevskoye's intellectual resources, so enough money was scraped together to send him to Moscow. There, he rented a corner of a room from a laundress and virtually camped in the public library, where he taught himself differential and integral calculus and trigonometry. He also attended lectures on science and mathematics and used most of his allowance to buy chemicals for experiments. By his own account, he lived on ninety kopeks worth of brown bread a month. "My pants were covered with yellow stains and holes, thanks to acids," he said in his autobiography. "Boys in the streets used to cry, 'Have the mice eaten your pants?' " It was about then that the afflicted young man began thinking about escaping to space.

At nineteen, he left the city for a job teaching mathematics and physics in the town of Borovsk, fifty miles to the southwest. Soon, his home became a small library crammed with papers and books related to rocketry. In 1881 he met Dmitri Mendeleyev, the chemist who had become famous in 1869 for inventing the periodic table of elements. Mendeleyev took enough interest in Tsiolkovsky to teach him advanced chemistry. A year later, the prodigy took another teaching job, this one in Kaluga, fifty miles farther from Moscow. Even by the standards of imperial Russia, Kaluga was a backwater. Yet Tsiolkovsky, by then an impoverished family man, read as much science as he could and became fixated on finding a way of getting into the air and then beyond it.

There was more to the germination of space travel in Russia than engineering and physics, however. Tsiolkovsky and other intellectuals, including a fiction writer named Fyodor Mikhailovich Dostoyevsky, were deeply influenced by Nikolai Fyodorov, the chief cataloger in the Moscow library and a well-known Christian theorist and mystic. Fyodorov led a monkish existence and by many accounts spent most of his salary buying books for students such as Tsiolkovsky. He also provided Tsiolkovsky with space in the library, stacked that space with philosophy books, and tutored his poor young charge in his own philosophy.

Fyodorov believed that Earth is not humanity's natural home; rather, human beings are organisms who are more properly at home in the entire cosmos. He also taught that everything in the firmament, from gigantic suns in other galaxies to the tiniest pebbles on Earth, was alive in some state and had a degree of consciousness. But as creatures of the highest consciousness, he told his disciples, humans had the special task of bringing design and purpose to the chaos of life on Earth and throughout the cosmos. In other words, he was convinced that it was mankind's task to "regulate nature." Ironically, that notion would find a home in Communist philosophy, with disastrous environmental results. Fyodorov also believed that science would eventually find a way to bring back to life everyone who had ever lived. But in order to do that, he would explain, all of the dispersed atoms of the dead humans would have to be collected from around the universe and then be given a place to stay. The obvious problem, however, was to find living space for all of the untold billions of resurrected dead. The equally obvious answer, at least to Fyodorov, was to send them to other planets.

Tsiolkovsky's absorption with reaching beyond this world was therefore strictly utilitarian and was only part of a monist and panpsychistic philosophy. That is, rocketry's first genius believed that the whole universe was alive with intelligent beings and that their existence constituted a single, unifying principle that defined everything. He continued to write about all this in his later years in works, some fictionalized, such as *The Monism of the Universe, Synopsis of Cosmic Philosophy, The Planets Are Occupied by Living Beings, There Are Also Planets Around Other Suns,* and *The Wealth of the Universe.* "There is no substance which cannot take the form of a living being," Tsiolkovsky would write in *Synopsis of Cosmic Philosophy.* "The simplest being is the atom. Therefore the whole universe is alive and there is nothing in it but life. But the level of sensitivity is endlessly various, and depends upon the combinations of which the atom is part."

All but oblivious to tsarist oppression and the tides of politics, yet fixated by the chimera of universal life, Tsiolkovsky pored over the science of the time, including the fundamentals of solid propellant rocketry, and let his prodigious imagination soar. In the spring of 1883 he wrote his first work, *Free Space,* which was cast as a diary but which was actually a systematic study of what happens to a human being and a mechanical system in a gravity- and resistance-free environment. The hundred-and-forty-nine-page monograph provided the first serious description of weightlessness, which clearly enthralled its author because it suggested disembodiment:

> The observed body does not, in free space, exert any force on what lies beneath it, and vice versa. Therefore, if dwellings were needed in free space, they could not—however great their size—collapse because of their instability.
>
> Entire mountains and castles of arbitrary shape and size could keep their position in free space without any support, or any connection with a support.
>
> There is neither up nor down, there. For example, there is no such thing as down, because "down" is in the direction in which bodies move at accelerated speeds. . . . Just as the moon hovers above the earth without falling down to it, so a man there can hover over a chasm which would be frightening to earthlings. He is not, of course, suspended by ropes but hovers like a bird; or rather, like a counterpoised aerostat, since he has no wings.

The means of getting to space was also described:

> Consider a cask filled with a highly compressed gas. If we open one of its taps the gas will escape through it in a continuous flow, the elasticity of the gas pushing its particles into space will continuously push the cask itself. The result will be a continuous change in the motion of the cask. Given a sufficient number of taps (say, six), we would be able to regulate the outflow of the gas as we liked and the cask (or sphere) would describe any curved line in accordance with any law of velocities.

The rural schoolteacher had used Newtonian physics to conceive a space station and a throttleable rocket engine.

A stream of monographs and books that eventually would number more than five hundred, including science fiction, followed. Tsiolkovsky began teaching himself aeronautics in 1885. This resulted in *The Theory and Practice of the Aerostat,* a highly detailed work that provided the theoretical basis for dirigibles made of corrugated metal. He remained convinced throughout his life that metal dirigibles filled with hydrogen would wear better than rubberized fabric and would be more gas-tight. There was also a series of papers on airplanes. Rejecting Leonardo's wing-flapping ornithopter as impractical, Tsiolkovsky described a monoplane with a cantilever wing, gasoline engine, enclosed cockpit, and retractable landing gear in "The Aeroplane or Bird-like (Aviation) Flying Machine," an article published in 1894 in the magazine *Science and Life.*

But Fyodorov's seeds were deeply planted; it was space that would always most captivate Tsiolkovsky. *On the Moon,* a novella published in 1887, described Earth as seen from the Moon and recounted the effect on the human body of getting there, much as he had done in *Free Space.* Another work of fiction, *A Change in the Earth's Relative Gravity,* described the lifestyle of the inhabitants of Mercury, where a gravitational force half that of Earth's resulted in "none of those disorders and conflicts among nations from which our poor earth suffers" and "which makes one [individual] the slave of another." *A Dream of the Earth and Sky,* published in Moscow in 1895, described artificial earth satellites built and inhabited by "sky-dwellers" in two-hundred-mile-high orbits who were also free of Earth's meanness. The father of Russian rocketry believed implicitly that gravity imprisoned the spirit as well as the body—that it was a metaphor for captivity, the shackles of servitude, oppression, and all dark deeds—and that only a soul floating weightless in space could truly be free.*

But the very notion that ordinary mortals could find individual freedom by shedding the bounds of Earth—by abandoning society on their own volition—annoyed both the church and the tsar's censors. The censors therefore stubbornly held up publication of *A Dream of the Earth and Sky.* For Tsiolkovsky, who was one of Verne's foremost disciples, fact and fiction were inseparable because they sprang from the same intellectual well. There had to be a dream before it could be realized. Fiction was therefore the blueprint of fact.

Unlike Verne, however, Tsiolkovsky's fiction rested, not partly on fact, but squarely on it. All of the years of study and calculations convinced him that black powder was too inefficient to propel a spacecraft. Liquids were the an-

* It was and remains a vision shared by many otherwise disparate individuals and groups. The thirty-nine members of the Heaven's Gate cult who committed suicide in California in March 1997 believed that their bodies were merely "containers" that existed on a wretched planet headed for Armageddon and that a spaceship flying near the Hale-Bopp comet would take them to Heaven. (See "Videotapes Left by 39 Who Died Described Cult's Suicide Goal," *The New York Times,* March 28, 1997.)

swer. What was needed was a highly combustible liquid fuel such as hydrogen that could be made to burn even more efficiently by being mixed with an oxidizer such as liquid oxygen. Hydrogen had the added benefit of being the lightest of the elements. But of course it was oxygen that made burning possible. Even Tsiolkovsky's youngest pupils knew that a lighted candle would go out when covered with a glass as soon as the trapped oxygen was burned up. Since there was no air in space, a rocket would have to carry its own oxygen in order to operate. He concluded that hydrogen, or any volatile liquid, could be combined with liquified oxygen in a tight metal chamber and ignited. If the expanding gases that developed because of that continuing explosion were vented through an exhaust hole of the right size at high velocity, the rocket and anything it carried would be propelled in the opposite direction, just as Newton said.

Nor would the reaction be haphazard. A rocket's performance could be precisely determined by whatever fuel-oxidizer combination was used and by its quantity. The force produced by a rocket would come to be called "thrust" and would be measured in kilograms or pounds. Performance would be measured by the amount of thrust produced by one pound of propellant in one second. This would come to be called "specific impulse." All other things being equal, a rocket's velocity would depend on the specific impulse of its propellant.

It would also depend on a positive thrust-to-weight ratio and on the velocity of the burned gases leaving the rocket. The thrust-to-weight ratio is the relationship between the amount of thrust a rocket produced and the weight of whatever it was supposed to propel. Tsiolkovsky spent untold hours calculating how much thrust would be required to launch a given weight. He concluded that in order to break gravity's hold and get to an altitude suitable for orbiting Earth, a rocket would have to reach a velocity of close to 5 miles a second, or about 300 miles a minute, or nearly 18,000 miles an hour. And the best way to achieve that, he figured, would be with a multistage rocket. Two or more stages, each with a rocket and its propellant, would be attached and launched together. When the propellant in the first stage had been burned, the stage itself would be jettisoned, and the second stage would take over. The process could be repeated with a third and even a fourth stage. The vehicle would therefore in effect be shedding dead weight rather than hauling an empty, hitchhiking carcass as it strained for altitude, just as Conrad Haas had suspected.

All of this and more was meticulously described in "Exploring Space with Reactive Devices," a seminal work that appeared in the journal *The Scientific Review* in 1903, the year the airplane made its debut. Proving that wood, wire, and fabric could be pushed into the sky was profoundly liberating, even if the source of power was a loud, smelly, and fickle gasoline engine. Balloons were quiet and beautiful and soothing. But they were also fragile, hostage to the capricious wind, and maddeningly slow. Flying in a balloon was like riding under a wandering elephant; it could be stopped and even turned slightly, but basically it did what it wanted to do. Airplanes, on the other hand, went where they were pointed, could be landed quickly, and were very fast, at least compared to the elephants: pachyderms versus Pegasus. And in a subtle way, using an engine, or "airscrew," to pull a heavier-than-air machine off the ground quietly validated rocketry because it proved that heavy machines could leave Earth under their own power. How high they went depended only on their shape, mass, and power. No one seriously believed that balloons could take to space. But metal driven by the right kind of motor was another matter. Airplanes and rockets would have a long, intimate relationship that would culminate in their being crossbred.

"Exploring Space with Reactive Devices" was truly significant because it substituted hard, scrupulously devised data—actual numbers—for all of the wild speculation, abstraction, and flights of fancy. It crystallized Tsiolkovsky's theories into a single work that was packed with hard information: material grounded in science and described in the immutable language of mathematics. It amounted to a handbook that proved that reaching space was not only possible, but practical.

Yet the science was inseparably swattled in the old dream. "Mikhail Mikhailovich," Tsiolkovsky wrote to *The Scientific Review*'s editor in his query letter,

> I have worked out several aspects of the problem of ascending into space by means of a reactive device. . . . My mathematical conclusions, based on scientific data verified many times over, show that with such devices it is possible to ascent into the expanse of the heavens, and perhaps to found a settlement beyond the limits of the earth's atmosphere. . . . People will take advantage of this to resettle not only all over the face of the earth but all over the face of the universe.

That goal remained foremost in Tsiolkovsky's mind. All of the calculations and physical narrative had as their core goal the single-minded objective of providing the means for people to live in space, use it to help Earth, and set off on immense voyages of cosmic exploration and habitation. The prophet of the rocket considered almost every aspect of human activity and related all of them to a life in space that he believed would be utopian. He envisioned using energy captured from the Sun to power colonies of people living and working in orbit for the benefit of a planet whose environment and resources had been taxed to their limit.

Understandably, mysticism and bald speculation did not play well with Marxists whose fundamental doctrines, including dialectical materialism, rested on their own "science," a ubiquitous term that claimed to explain the world with ironclad rationality. No wonder Tsiolkovsky's mystical musings were carefully ignored during his lifetime and were kept buried in private and state archives until after communism itself dissolved into its own constituent fragments.

Yet the part about controlling all of nature—the entire universe—for the benefit of humanity found a receptive audience in a relentlessly expansionist and all-controlling creed such as Marxism, which worshipped machines for precisely that purpose. It was natural grist for the party's hack philosophers and other writers. Aleksandr Bogdanov's *Red Star,* published in 1908, described a Communist utopia on Mars.

The combination of Marxism and Marsism—of using science to master, manipulate, and endlessly expand mankind's environment for the everlasting benefit of a completely communized society—would produce a stream of futuristic literary pap about a world megacity that ran on selfless brotherhood. Mountains and oceans would be moved, nature shoved aside or bulldozed to oblivion, to suit humanity's needs. A member of the new world order would be able to walk or ride from Australia to Russia on a continuous highway whose pillars themselves housed thousands of dwellers. Others would be carried around socialism's global village by ubiquitous airbuses and skycars. Everyone would be happy and secure in the bright, brave, new world.

"As he thought of future generations, of a better fate for mankind, and of human happiness, Tsiolkovsky's imagination carried him out into the universe," Evgeny Riabchikov, a Soviet journalist, wrote. "The realm of the stars," he went on, "so amazingly well-organized, fascinated him—because of the boundlessness of space, the gigantic reserves of energy, and the astounding accuracy and harmony in the movement of the planets. As he dreamed of the future, Tsiolkovsky mentally peopled the heavens with colonies of human beings." Riabchikov's prose was cast in a shade of purple that was often used to describe Soviet heroes during the Communist years. But it was nevertheless essentially accurate.

Ever the restive theoretician, Tsiolkovsky published supplements to "Exploring Space with Reactive Devices" in 1911, 1912, and 1914 and put out an enlarged and updated edition in 1926. Everything published before the revolution in 1917 was done at his own expense. The monarchy showed no interest in helping him; the only grant he ever received—a paltry 470 rubles—was awarded in 1899 by the Academy of Science for experiments with a primitive wind tunnel. But the Bolsheviks, who were as afflicted by a national inferiority complex as most other Russians, saw things very differently. They grasped the fact that Tsiolkovsky's work could be used to change the world under their own aegis and that it could bolster both the prestige of the Communist Party and of the nation (provided the screwball stuff about rocks being alive that he inherited from Fyodorov could be suppressed). So they pulled the sixty-two-year-old schoolteacher out of near obscurity in 1919 and appointed him to the Socialist Academy (eventually replaced by the Soviet Academy of Sciences). Official recognition of his accomplishments was supplemented by a state pension, which allowed him to stop teaching and devote all of his time to developing his theories of space travel. The old man somehow fit into the scheme of using science not to enhance nature but to conquer and transform it; to break its will until it was nothing more than a slave to the requirements of socialism.

Stalin would become so taken by the vision of using science to make Earth itself subservient to his new society that he would order scientists into an All-Union Program for the Transformation of Nature. They were ordered to do nothing short of change the climate and the face of the planet by eliminating deserts and constructing gigantic hydroelectric dams, canal systems, mechanized farms, nuclear power plants, and other components of the socialist utopia he believed was inevitable. The assault that followed would cause environmental damage so massive and pervasive that its full extent has yet to be measured.

In 1920, turning once more to fiction, Tsiolkovsky published *Beyond the Planet Earth.* This epic adventure, crafted in the tradition of *From the Earth to the Moon,* followed the exploits of an international team of space adventurers as they journeyed to the Moon and beyond. Like his hero, Verne, nuance was not novelist Tsiolkovsky's strong suit. So Ivanov, a mild-mannered Russian, conceived of a detailed plan to use rockets to escape from Earth and convinced Newton, an Englishman; Galileo, an Italian; Franklin, an American; and others that the plan was feasible. It led to the exploration of the Moon and the establishment of gigantic, life-supporting colonies—"mansion-conservatories"—in Earth orbit that were inhabited by individuals who would be judged best in breed if they were pedigreed dogs: "people who were easy to live with, gentle, resourceful, hard-working, physically tough, not too old, and if possible without domestic ties . . . angels in human form."* Verne used science to make his tale more plausible; Tsiolkovsky used his tale to make science more plausible.

It is unlikely that any of the Communist Party drones who recognized and then elevated the Kaluga schoolteacher studied his calculations, let alone understood them. But at least some of the firebrands in Moscow, no doubt want-

* The spiritual quality of Tsiolkovsky's spacefarers almost certainly had its origin in Fyodorov's teaching.

ing to see how the news of their young revolution was playing abroad, did study the foreign press. The newspapers of 1919, the very year Tsiolkovsky was elevated to the academy, told of the Paris peace conference and the Treaty of Versailles, which would set the stage for Woodrow Wilson's League of Nations, unwittingly prepare Germany's political soil for the seeds of Nazism, and hasten the birth of the ballistic missile.

Of far less enduring interest, or so it seemed, was the story of a thirty-seven-year-old professor of physics at Clark University in a place called Worcester, Massachusetts, who had made himself a laughingstock by asserting in a respectable publication that it was possible to hit the Moon with a rocket. His name was Robert Hutchings Goddard.

The "Moon-Rocket Man"

Robert Goddard had not heard of Konstantin Tsiolkovsky by 1919. Nor had the Russian heard of him. Yet both shared the spiritual and intellectual bloodline that ran straight as an arrow from Galileo to Newton to Verne to Wells. Like Tsiolkovsky, Goddard had been a sickly adolescent who loved science and who was captivated by the possibility of reaching worlds beyond his own. His ailments, diagnosed as colds, pleurisy, and bronchitis, as well as undefined stomach and kidney problems, kept him out of school so much that he turned to self-education and escape as better remedies than medication. If he could not be in the world beyond his front door for much of the time, young Goddard would carefully construct another world within his mind; a world that was confined by his home but that was as boundless as the universe. He read the physical and chemical sciences voraciously, just as Tsiolkovsky had done, borrowing textbooks from the Worcester Public Library and poring through his own copies of Cassell's encyclopedic *Popular Educator* and *Technical Educator.* He climbed a cherry tree in his backyard to get as close to the sky as he could by day, and he tried to bridge the gap at night with a telescope.

And like Tsiolkovsky, young Goddard feasted on science fiction. He loved and was inspired by Verne. But he was electrified by Wells's *The War of the Worlds,* which was serialized by a Boston newspaper in 1898: "No one would have believed in the last years of the nineteenth century that this world was being watched keenly and closely by intelligences greater than man's and yet as mortal as his own," Robbie Goddard and thousands of other fascinated kids and their parents read, ". . . that as men busied themselves about their various concerns they were scrutinized and studied, perhaps almost as narrowly as a man with a microscope might scrutinize the transient creatures that swarm and multiply in a drop of water." Great stuff. Really keen. And what followed was even more amazing: a battle to the death between the terrified earthlings and the disgusting invaders.

The story so thrilled Goddard that in 1932, at the age of fifty, he wrote a fan letter to Wells:

> In 1898, I read your *War of the Worlds.* I was sixteen years old, and the new viewpoints of scientific applications, as well as the compelling realism . . . made a deep impression. The spell was complete about a year afterward, and I decided that what might conservatively be called "high altitude research," was the most fascinating problem in existence. . . . There can be no thought of finishing [work on rockets], for "aiming at the stars," both literally and figuratively, is a problem to occupy generations, so that no matter how much progress one makes, there is always the thrill of just beginning.

Goddard submitted the first of his own two great works to the Smithsonian in 1919. It was based on intensive theorizing he had done while a research instructor at Princeton University in 1912–13. The paper, called *A Method of Reaching Extreme Altitudes,* was issued as part of the Smithsonian Miscellaneous Collections early in January 1920. The sixty-nine-page document, modestly bound in brown paper, was a richly detailed grant proposal that described how to build a reusable two-stage solid-propellant rocket—down to exhaust nozzle diameters and intricate component data—and use it for atmospheric research well beyond the nineteen-mile-high limit of balloons. The idea was to send up instruments to measure the density, chemistry, and temperature of the upper atmosphere, "radioactive rays from matter in the sun," and the ultraviolet light coming from it. To do that, as well as to collect data for weather forecasting at lower altitudes, Goddard designed his rocket so that it would float back to the ground under a parachute. He also briefly mentioned liquid-fuel rocket propulsion.

That was fine as far as it went. But Goddard, like Tsiolkovsky, was looking beyond the atmosphere. Undoubtedly reflecting on Verne, he went on to calculate the velocity required to reach "infinite altitude." Given enough acceleration over enough distance, he went on, "the mass [of the rocket] finally remaining would certainly never return." And, still echoing the French novelist, he chose the darkened surface of the Moon as a likely target for a rocket carrying a small amount of flash powder. The powder would ignite on impact and be visible through telescopes, Goddard declared, proving that the trip had been accomplished. While the physicist carefully explained that a shot at the Moon had "important features" (which he neglected to specify) and was generally interesting, he also made it clear that such a mission had "no obvious scientific importance." On January 11, 1920, the Smithsonian itself issued a news release that outlined Goddard's theory and experiments in *A Method of Reaching Extreme Altitudes* and, almost as an aside, briefly mentioned the Moon shot as an "interesting speculation."

Whatever the substance of his paper—and there was a great deal of substance to it, complete with equations, tables, diagrams of rockets, and even the probability of a rocket being hit by a meteor—green-eyeshaded newspaper editors from Providence to Portland thought it was about as interesting as watching a wad of tobacco juice hit a spittoon. But the Moon angle was something else, so they played it for all it was worth. "Modern Jules Verne Invents Rocket to Reach Moon," a headline in *The Boston American* read, while *The Milwaukee Sentinel* trumpeted the "Claim Moon May Soon Be Reached" and *The San Francisco Examiner* told its readers that "Savant Invents Rocket Which Will Hit Moon."

The New York Times ran the story on the twelfth, providing a reasonably thorough report on the substance of Goddard's paper under a page-one headline that described him as a man who "Believes Rocket Can Reach Moon." Seeing that account, one enterprising organization among several wrote to the Smithsonian that very day to offer its services. "As undoubtedly you will desire some special starting point from which to start this rocket, and one at which the greatest number of people could have the opportunity to observe its departure, the Bronx Exposition, Inc. offers the use of Starlight Amusement Park for the purpose, and at the same time will be happy to provide all the facilities needed for the occasion."

That would have been hideous enough for Goddard. But the following day, the thirteenth, as reporters from the Associated Press and other news organizations bore down on Worcester, the shaken and embarrassed physicist saw to his sorrow that he had not escaped the *Times*' editorial page either. There, under "Topics of The Times," an anonymous writer not only said that the notion of hitting the lunar surface with a rocket was poppycock, but in the process smeared its now-infamous harbinger with oily contempt.

"That Professor Goddard, with his 'chair' in Clark College and the countenancing of the Smithsonian Institution, does not know the relation of action to reaction, and of the need to have something better than a vacuum against which to react—to say that would be absurd. Of course he only seems to lack the knowledge ladled out daily in high schools." What every high school student on the planet knew (perhaps even those whose craniums were decorated with dunce caps) and what poor Goddard clearly did not know, the *Times* man explained with venomous contempt, was that a rocket would not work in the vacuum of space for the simple and obvious reason that there would be nothing against which to push. A "romancer" such as Verne could get away with ignoring Newton's immutable laws, he added, but a "savant" like Goddard could do no such thing. The smugness, a common affliction among editorial writers, essayists, and others whose opinions overwhelm their knowledge, was misplaced. A rocket engine flying in a vacuum pushes, not against the vacuum, but against itself.

Goddard reacted to the ridicule by being "as uncommunicative as possible," as he himself put it, during the barrage of newspaper interviews that followed. Seeing that the tactic was not working, and now anxious to prevent embarrassment at the Smithsonian, he tried to parry the attack on the eighteenth by issuing a signed statement to the Associated Press insisting that rockets could take photographs above the atmosphere and collect valuable data at high altitude. "Too much attention has been concentrated on the proposed flash powder experiment," he complained angrily, "and too little on the exploration of the atmosphere." The real point was that the serious business of collecting information about very high altitudes would have to come before a Moon shot, Goddard continued, since "any rocket apparatus for great elevations must first be tested at various moderate altitudes." He also proposed raising $50,000 to $100,000 from the public to do high-altitude rocket experiments.

But the rebuttal was lost in a chorus of cackling that horrified the quiet, prim physicist. "Well, Robert," the chairman of Clark's Chemistry Department would tease in the hall between classes, "how is your Moon-going rocket?" Some reporters called him "the Moon-rocket man" and goaded him about his theory. And the public was not far behind. During the next few years, the idea of dispatching rocket ships to the Moon became a cultural staple, along with speakeasies, flappers, the Charleston and Black Bottom, barnstorming pilots, Charlie Chaplin, and the exploits of record-breaking daredevils in planes and balloons. Tin Pan Alley cranked out tunes such as "Oh, They're Going to Shoot a Rocket to the Moon, Love." A press agent at the Mary Pickford Studios in Hollywood sent Goddard this galling telegram: WOULD BE GRATEFUL FOR OPPORTUNITY TO SEND MESSAGE TO MOON FROM MARY PICKFORD ON YOUR TORPEDO ROCKET WHEN IT STARTS. And Captain Claude R. Collins, a veteran Army Flying Corps pilot and the president of the Aviator's Club of Pennsylvania, was one of more than a hundred men and women who volunteered for lunar excursions. The would-be passengers included three Russian physics students who had read that, if successful, the rocket carrying the flash powder would be followed by a second carrying people. They were convinced, however, that the stampede for seats on the second flight would preclude them and they therefore sought an edge. In a letter written to Goddard in French on May 26, 1929, they begged for "permission to leave on the first flight. We do not know if there is a cabin for the passengers; and the apparatus—can one steer it?"

Goddard's deep anger at what he took to be the trivialization of his work would lead him to exile and near seclusion in New Mexico's more hospitable physical and emotional environment. He would have been better off in Germany, where at least there were kindred spirits, but he had no way of knowing that, either.

Visions of Rockets in Planetary Space

There was, of course, a time when Hermann Julius Oberth had heard of neither Tsiolkovsky nor Goddard. But like them, he had heard of Verne, who was the wellspring of his obsession. Oberth was born in Hermannstadt, Transylvania, on June 25, 1894, the son of German-speaking parents in a German-speaking region of the Austro-Hungarian Empire. He was given both of Verne's Moon novels when he was eleven and became so captivated by the dream that played out on their pages that he practically learned them by heart. But even as a youngster, Oberth knew that any creature shot out of a cannon at escape velocity would, in his words, be "crushed without pity by the enormous acceleration." He therefore racked his brain to come up with a safer method, in turn considering and then rejecting a magnetic device in a long tunnel from which the air had been removed, an airplane with silk propellers, and a large wheel whose spinning produced high centrifugal force. Finally, reluctantly, he settled on the rocket, which he rightly considered too dangerous, unwieldy, and expensive.

Bowing to the wishes of his father, who was a physician, Oberth enrolled for medical studies at the University of Munich just as Verne, his guiding light, had suffered through the tedious minutia of law school at his own father's insistence. Yet the nineteen-year-old's fascination with rocketry would not end. In 1917, with Germany mired in its third year of trench warfare, Oberth suggested to the kaiser's War Department that a liquid-propelled, long-range bombardment missile be used to inflict damage far behind enemy lines. He was ignored. But Oberth was not put off. He remained so fascinated by the idea of taking to space that he finally stood up to his father, abandoned medicine, and went to Heidelberg University after the armistice to do graduate work in physics.

Oberth soon came to the same conclusion about high-energy liquid propellants as Tsiolkovsky. In May 1922, he wrote to Goddard asking for a copy of the Smithsonian paper after reading about it in the newspaper. Fearing that the American would think he was trying to steal his ideas, though, Oberth carefully explained that he himself had been working on liquid-propellant rockets for many years. And he added that he would be honored to send his own research results back across the Atlantic because "only by the common work of scholars of all nations can be solved this great problem." Goddard mailed his notorious treatise to Oberth with misgivings. He loathed the idea of working with or even helping anyone else, and particularly with a man whose adopted country, Germany, he considered to be warlike. The American rocket pioneer would always reveal his own insecurity by referring to the physicist from Heidelberg as "that German Oberth."

Meanwhile, Oberth set down his own theories in a doctoral dissertation that, reflecting his medical studies, also contained heavy doses of material on how spaceflight would affect the human body. His dissertation adviser, an astronomer named Max Wolf, had nothing but praise for the document, which he called scientifically correct and ingenious. But he rejected it anyway, explaining to his disgruntled student that it was mainly about physical-medical subjects, not pure physics, and he was therefore not competent to judge it. Why Oberth was not sent to a different dissertation adviser is not clear, though he would have done no better with a biologist, since he himself was inventing what would eventually be called space medicine. So instead of leaving Heidelberg with a coveted doctorate in physics, Hermann Oberth was awarded a consolation prize: an elaborate certificate testifying that while he had failed to make the grade, he had at least given it a pretty good shot.

But Oberth was as stubborn as he was daunted. Determined to get his work in print, he took it and the certificate to several publishers. Only one, Oldenbourg in Munich, agreed to publish it, and then on condition that the author himself pay a generous part of the printer's bill for what would now be called a vanity book (Oberth was easily as vain as the next physicist). But the gamble paid off. *Die Rakete zu den Planetenraumen* (*The Rocket into Planetary Space*) was published in 1923 with mixed reviews and moderately good sales. Within two years, though, it would spread the concept of rocketry throughout Europe and beyond and become a classic.

The book was only ninety-two pages long, but those pages were crammed with calculations on virtually every aspect of advanced rocketry, including its applications and the physiological stuff that had gotten him into trouble with Professor Wolf. The introduction stated flatly that the author would prove four assertions: the science and technology of the time were capable of producing rockets that could climb beyond the atmosphere; such vehicles could reach high enough velocities to break out of Earth's gravity altogether; rockets would be able to carry people ("probably without endangering their health"); and manufacturing rockets might be profitable within a few decades. Using two theoretical rockets—the Model B and the Model E—the author laid out the principles of rocket motion using calculations and physical concepts that might as well have been Arabic to the average German and then went on to describe their technical applications. Even Oberth would later admit that most of the book had to be "explained" to the general public by Max Valier, Otto Willi Gail, Karl August von Laffert, Willy Ley, and others who were both knowledgeable in the physical sciences and able to fathom the dense snarls of rocket mechanics for ordinary intellectuals.

But the equations amounted to a Rosetta Stone for the cognoscenti. They showed that rockets could indeed operate in a vacuum; that they could move faster than the velocity of their own spent gases; and that, given enough velocity, they could actually haul payloads into orbit around Earth and beyond. The challenge of developing enough velocity to do the job led Oberth, like Goddard and Tsiolkovsky, to consider a variety of propellants, including alcohol and hydrogen. And the Model E rocket, a 115-foot-high bullet that stood upright on four large fins, was the classic shape of the rockets to come.

The last part of *The Rocket into Planetary Space* described a spaceship, space travel, and even sketched a crude space station in plain enough German so that bright and attentive readers could grasp the ideas behind them. Finally, Oberth appended three pages to the end of his work, which said that "one can easily see that I have worked completely independently" and that his book and Goddard's "supplement one another." There is little doubt that Oberth was telling the truth, since Tsiolkovsky's work was all but inaccessible and Goddard's experiments with liquid propellants, which began as early as 1920, were locked away by their intensely secretive and suspicious author.

Goddard himself was dismayed when he read *The Rocket into Planetary Space,* partly because he believed that the German Oberth had indeed helped himself to generous portions of his own work, and probably also because of his rival's obvious boldness and brilliance. After all, while the forlorn Yankee was still living down his timid suggestion that a powder-driven rocket could reach the Moon, his Romanian counterpart was describing vehicles, missions, and spatial trajectories that could get humanity to the planets on liquid turned to fire.* The difference between how Goddard's and Oberth's ideas were received said much about their respective cultures' openness to what would eventually be called breakthrough technology.

Whatever their shared goal, Goddard felt anything but professional camaraderie toward Oberth. "This monograph develops a theory of rocket action which is practically the same theory as that presented in my article," he complained to the secretary of the Smithsonian on August 1, 1923. "As the matter stands, Oberth has presented a well-written and comprehensive paper, which sets forth the general method as his own (as he claims to have been working independently), and which supposedly demonstrates that what I have been working upon cannot be developed for the ultimate uses for which his device

* Oberth did not formally obtain German citizenship until World War II.

can be applied." Oberth "has himself overlooked a few important matters," Goddard snickered, adding petulantly that he was sending copies of all of his unpublished research to both Clark's trustees and to the Smithsonian to prove the quality and originality of his own work. "I am not surprised that Germany has awakened to the importance and the development possibilities of the work," Goddard concluded, "and I would not be surprised if it were only a matter of time before the research would become something in the nature of a race." He was closer to the mark on that score than even he could know, though the race would be heavily one-sided.

Four years after the end of the Great War, the weary and grieving citizens of Germany and much of the rest of Europe at last began to shed some of the angst that marked the course of the conflict and put it behind them in favor of a renewed interest in the arts and sciences. "In short," Willy Ley would note years afterward, "it was again possible to think." And what many of them were thinking as a result of *A Method of Reaching Extreme Altitudes* and *The Rocket into Planetary Space* was that scientists finally seemed ready to deliver what Verne and the other storytellers had promised.

It takes nothing from Tsiolkovsky and Goddard to say that Oberth's treatise, followed by a 423-page sequel called *Wege zur Raumschiffahrt* (*The Roads to Space Travel*), published in 1929, did more to spread the concept of going to space than any other scientific work to that time. Tsiolkovsky's articles and books were printed in relatively small numbers—his basic works were published in quantities of about four thousand, while Kaluga editions were printed in batches of about two thousand and distributed by the author—and were virtually unavailable outside Russia. Goddard, who was shy to begin with, turned into a near recluse after the Moon fiasco. He kept his distance from the space societies that were coalescing in England, Germany, France, and elsewhere in Europe, as well as in the United States and the Soviet Union, while conducting advanced experiments with ever-larger but finicky liquid-fired rockets as far as possible from the scrutiny of others.

Oberth, on the other hand, was a relentless popularizer who did everything he could to draw attention to rocketry in general and to his own work in particular. In 1929, for example, he agreed to become the technical adviser for *Frau im Mond* (*The Girl in the Moon*), a silent film made by Fritz Lang, the director of *Metropolis* and one of the world's most celebrated moviemakers. Oberth used his Model E spacecraft as the basis for the one in the film, which had an impressively detailed interior and a spectacular blastoff, complete with smoke and sparks coming out of what was supposed to be its engine. He also agreed to construct a real liquid-fueled rocket and launch it as a publicity stunt on the day the film premiered. But there was at least one insurmountable problem. "He had no idea how to go about it," his friend Willy Ley would recall years later. "He was—a point I wish to emphasize—the greatest authority on rocket propulsion at that time . . . but he was a theorist, not an engineer." Recognizing as much after the deal with Lang and the UFA Film Company had been made, Oberth finally hired Rudolf Nebel, a Bavarian who passed himself off as an engineer, and Alexander Borisovich Shershevsky, a Russian aviation student and a dedicated expatriate Bolshevik. But it was to no avail. Nearly four months of frustrating work that included a near–nervous breakdown and at least two explosions (one of which almost blinded Oberth in one eye) produced nothing beyond the successful static test of a small version of the motor that was supposed to have been used on the actual rocket. The film, unheralded by a real rocket launch and facing competition from new "talkies," premiered on October 15, 1929, and was moderately successful.*

But Oberth's scientific accomplishments and high visibility invigorated Germany and to a lesser extent the world beyond its borders in at least three ways. It rekindled national pride in the aftermath of the Great War and stimulated intellectualism and free thought in a nation that had long been dominated by conservative Junker governments. It challenged other scientists and engineers to think and write about going to space. And, most important, it fired the imagination of serious amateurs—"enthusiasts," as the British called them—who came together to form the rocket and space societies.

By 1924, the competition Goddard had predicted had turned into a rivalry on low boil. Germany was celebrating Oberth as the "true father of Rocketry," while Goddard remained in isolation and was generally ignored. Max Valier, an Austrian science writer, gadabout, adventurer, and author of the popular (and error-prone) *Der Vorstoss in den Weltenraum,* or *A Dash into Space,* dismissed Goddard as a mere tinkerer who was "afraid to handle" liquid fuels while championing Oberth with nationalistic fervor. But beyond that, the former Austrian pilot implored true Germans to prepare themselves to sacrifice for the sake of breaking free of the home planet in tones that evoked Wagner. "There is no doubt: the moment is here, the hour has come, in which we may dare to undertake the attack on the stars with real prospects of results. It is clear that the armor of the earth's gravity will not lightly be pierced, and it is to be expected that it will cost, to break through it, much sacrifice of time, money, and perhaps also human life."

Ironically, Valier himself would become the liquid rocket's first known fatality, though not in an attack on the stars. He was killed in May 1930 when

* Using rocket power for publicity also extended to the automobile business. Early in 1928, Fritz von Opel conceived of using rockets to propel one of his cars. On May 23, 1928, twenty-four solid-propellant rockets pushed his "Opel Rak II" to almost 125 miles an hour at the Avus Speedway near Berlin. The auto magnate had himself photographed on that occasion and delivered a speech on the radio promising even more advanced automobiles. The project soon fizzled, however. (Ley, *Rockets, Missiles, and Space Travel,* pp. 119–23.)

the liquid oxygen–gasoline engine in a car he had built exploded, firing a large steel shard into his aorta. An alliance with Fritz von Opel, the automobile magnate, had led to the construction of cars propelled by black powder that were used to generate publicity. Valier wanted to push the technology even farther. Similar stunts with black powder were soon being done with railway cars, gliders, bicycles, and even a rocket-propelled sled.

Liquid Fire at Auburn

Valier's snipe at Goddard was misplaced. He had no way of knowing that, far from being afraid to handle liquids, the only real experimenter in the trio was at that moment wringing reality out of all the theory. What both Tsiolkovsky and Oberth were unable to do—actually fly what they put on paper—Goddard was working hard to accomplish. He finally succeeded on the afternoon of March 16, 1926, at his Aunt Effie's farm at Auburn, just south of Worcester.

The rocket itself was nothing like what Tsiolkovsky or Oberth envisioned. Far from being streamlined, it consisted of a small motor connected to an oxygen tank and a gasoline tank by rigid tubes. The oxidizer and fuel fed upward through the tubes into the combustion chamber, where they were ignited, and exploded out of a small nozzle as hot gas. Goddard knew that covering his skeletal rocket with bullet-shaped metal would add so much weight that it would never fly. And it was not going to have to slice through much atmosphere anyway, so streamlining was unnecessary. The improbable device, essentially two cylinders arranged vertically and connected by tubing, weighed ten and a quarter pounds loaded with liquid.

The minimal rocket and the pyramid-shaped launch cradle on which it rested early that afternoon, with its erstwhile inventor posing solemnly beside it for his wife's new French "Sept" motion-picture camera ("Seven" was the number of seconds it would run before it had to be rewound), looked more like a light jungle gym in some snowy schoolyard than the ancestor of the mighty machines that would send men to the Moon and their robotic proxies across the solar system. And its performance matched its appearance. Goddard would note drily in his diary that evening that his rocket reached an altitude of 41 feet and landed 184 feet from the launch point, according to an assistant who measured the flight. Total flight time was two and a half seconds. Well, the Wrights hadn't done all that much better at Kitty Hawk.

Goddard allowed himself more emotion the following day. "The first flight with a rocket using liquid propellants was made yesterday at Aunt Effie's farm in Auburn," he wrote.

> Even though the release was pulled, the rocket did not rise at first, but the flame came out, and there was a steady roar. After a number of seconds it

> rose, slowly until it cleared the frame, and then at express-train speed, curving over to the left, and striking the ice and snow, still going at a rapid rate. It looked almost magical as it rose, without any appreciably greater noise or flame, as if it said, "I've been here long enough; I think I'll be going somewhere else, if you don't mind."

Goddard, undoubtedly aware that he was writing for posterity, cleaned up what he thought he heard his little rocket say. His wife, Esther, heard it this way: "I think I'll get the hell out of here!"

He would publish his second great Smithsonian monograph, *Liquid-Propellant Rocket Development,* in March 1936 based on that first, historic firing and on a stream of improvements that followed.

Samaras on the Wind

While Goddard's historic achievement at Aunt Effie's farm went unnoticed beyond his own small circle, *The Rocket into Planetary Space* set off an explosion of interest in Europe and the Soviet Union that soon turned into a near frenzy of creativity. It focused on a whole new place where the frustrated inventors, engineers, closet chemists and physicists, and other dreamers could use their often-prolific imaginations. The possibilities, after all, were as infinite as the medium.

Walter Hohmann, an architect and the city engineer of Essen-on-the-Ruhr, published a technically dense tome called *Die Erreichbarkeit der Himmelskörper* (*The Attainability of the Celestial Bodies*) in 1925. It was dense but important. Hohmann worked out the most efficient ways to change the orbit of a satellite circling Earth or send a spacecraft from Earth to another planet. The first involved firing the satellite's rocket to increase its speed, and therefore its altitude, and then again to change what was an elliptical orbit into a circular one before it headed toward open space.

The second required the spacecraft to leave Earth flying in the same direction as Earth was moving and arrive at the other planet in the same direction *it* was moving in by flying a very long ellipse. This is roughly the same as a quarterback leading his receiver while both are running forward: the ball and the receiver have to converge at the same point, and so must the spacecraft and the planet. The opportunity to do that would come to be called the "launch window" and would depend on a number of factors, the most important of which would have to do with the relative positions of Earth and the other planet (both moving on the same side of the Sun and in the right formation, for example). The window for launches to Venus—the perfect time to send a spacecraft there—would be calculated as once every nineteen months; for

Mars, once every twenty-six months. To this day, such maneuvers are called the Hohmann Transfer or Trajectory maneuver.

In France, aviator Robert Esnault-Pelterie was among the first to conclude that whatever the rocket's use in reaching other worlds, firing it on long ballistic trajectories could cause devastating damage on this one. He figured, in other words, that rockets would be the artillery of the future and would be cheaper than using armadas of manned bombers. And Esnault-Pelterie said as much to his government in a secret report and request for research money in May 1928.* After some preliminary work by Esnault-Pelterie and a general of artillery named J. J. Barre, the project was summarily dropped because, they were told, it was not considered that "a study of rockets is worthy of the activity of an officer." That bit of arrogance guaranteed France would not be the first to develop the long-range guided missiles that might have helped prevent its painful and humiliating defeat in 1940. Germany would have no such qualms.

Esnault-Pelterie—"REP" to everyone who knew him—published *L'Astronautique* in 1930 and a supplement, *L'Astronautique-Complement,* four years later. The books were not so much innovative as they were brilliant compilations of virtually everything that was known about spaceflight at the time, including interplanetary travel and the conditions in which such trips could be made. The word *"L'Astronautique"* in the titles was especially significant, since "astronautics"—literally, "navigating the stars"—was invented by the Belgian science fiction writer J. H. Rosny Sr. on December 26, 1927, in Esnault-Pelterie's presence. The occasion was a dinner whose purpose was the founding of a committee in the French Astronomical Society to promote spaceflight. At the end of the meal, Esnault-Pelterie and a banker friend decided to award a cash prize for outstanding work in rocketry and space travel and thought hard for the right word to describe such work. Rosny gave it to them. Oberth was the first winner.

Mit Raketenpost

Even little Austria was producing rocketeers, including at least one who would have understood Esnault-Pelterie perfectly, out of all proportion to its size. They were Eugen Sänger, Friedrich Schmiedl, and Hermann Potocnik.

* Esnault-Pelterie is generally credited with being first to consider the use of rocket-propelled long-range ballistic missile artillery. But in 1927, an Italian aviation pioneer and rocketeer, General G. Arturo Crocco, gave a secret lecture on the subject to his government's General Staff. Crocco's presentation on solid-fuel rockets with unlimited terminal velocities was so impressive that he was given research funding out of what would now be called the black budget. Crocco did impressive research on solid propellants and successfully launched miniature missiles beginning in 1928. (Crocco, "Early Italian Rocket and Propellant Research," in Durant and James, *First Steps Toward Space,* pp. 33–48.)

Sänger was a highly original aerodynamicist who wanted to crossbreed airplanes and rockets. He became convinced that airplanelike lifting bodies propelled by rockets would serve mankind better than vertically fired rockets. By the end of the 1920s he conceived of a rocket-propelled "aerospace plane" that burned oxygen and gasoline and that could reach speeds of 6,200 miles an hour and altitudes above forty miles. Sänger's rocket plane evolved into a so-called antipodal bomber that could theoretically have struck the United States from Europe by coasting through the stratosphere like a flat rock on a pond. The Nazis even picked New York, Washington, and Pittsburgh, the nation's steel producing center, as targets. The bomber was never built, but its descendants, the rocket-propelled X-15 high-altitude research aircraft and then the space shuttle, would.

Like many others, Schmiedl fantasized about building "space rockets," and also like the others, he could not afford it. So he lowered his sights and in July 1928 came up with solid-propellant, parachute-equipped rockets that could deliver mail. Schmiedl calculated that rockets could get mail from Europe to North America in forty minutes and to any place on the planet in an hour. So he built a small version, the first public postal rocket, and on February 2, 1931, shot 102 letters from Schoeckl Mountain to Radegund. That September, he used an improved model to carry 333 more letters (in what may have been the biggest understatement in the history of rocketry, some were marked "Special Delivery"). The idea, as usual, was to attract investors. In this case, Schmiedl was looking for a backer who would invest in a rocket that could carry mail across the English Channel. But there were no takers among Schmiedl's devoutly conservative countrymen.

"I continued building postal rockets until 1935," he wrote years later. "They were constructed as group [cluster] and step [multistage] as well, but they never exceeded the weight of 30 kg because I could not afford to build bigger ones. Though each was built differently, all of them were intended to advance the cause of space flight. Later, I destroyed nearly all of my research notes and photographs of rocket launches and proceedings, for fear they might be used by the military. I abandoned space research entirely after World War II."

Potocnik, writing under the pseudonym Hermann Noordung, published a book, *The Problem of Spaceflight,* in 1929 that described a space station—a *Wohnrad,* or "living wheel"—in unprecedented detail. Potocnik's station was to be a hundred feet across and consist of three parts: the wheel itself, which would house the crew; a "power house" that used huge mirrors to collect sunlight for heat, hot water, and electrical energy; and an observatory. All three were to be connected by air hoses and electrical cables. The author, a civil engineer, had his wheel spin around a hub so that centrifugal force created artificial gravity. Seeing that none of his contemporaries were taking him seriously, Potocnik wrote to Oberth and was in turn encouraged to do the book. He had the pleasure of seeing it published just before he died of tuberculosis at the age of thirty-six. "Noordung" thus gave the world the basic engineering study for a space station. The concept of the living wheel would long outlive its inventor.

Rockets to Defend the Revolution

There was no other towering genius in the Soviet Union to match Tsiolkovsky, whose own considerable ego compelled him to note in the 1926 edition of *Exploring Space with Reactive Devices* that Goddard's and Oberth's works appeared "many years" after his own. Instead, a critical mass of scientists and publicists (often, as with Oberth, they were one and the same) formed to do research, sound the call to space in lectures and print, and germinate their new cause with the creation of societies. This was done within a brutally repressive system that put far more emphasis on industrial production and collectivization than on innovation in the laboratory. Individuality, which ran counter to state dogma, was severely discouraged.

In the Soviet Union, as in Germany, abstraction soon gave way to gross reality. Rocketeers in both countries during the period between the two world wars could dream about space travel all they liked, but they were shackled to a common and unavoidable dilemma: their lifeblood, their financial sustenance, flowed from politicians and generals who had their own desiderata. No politician or general wanted to go to Saturn to count rings.

While Tsiolkovsky, the master rocketeer and space visionary, remained almost unknown to the world outside the USSR, he was venerated at home and unquestionably had a profound influence on his country's early infatuation with the rocket. Tsiolkovsky's mantle was passed to several first-rate scientists and designers, most of whom worked for state-operated organizations. The widespread belief in the United States following the launch of Sputnik that it was captured German scientists who provided the enemy with the means to orbit a satellite was as wrong as it was arrogant. Both the rocket launcher and the spacecraft were strictly homegrown and were the products of a national commitment that went back to the years immediately after the 1917 revolution. Captured Germans and some of their rockets certainly helped. But the feat would have been impossible without a relatively large number of brilliant Soviet citizens.

As early as 1894, a chemical engineer named Nikolai Ivanovich Tikhomirov, thinking along the lines that would later attract Esnault-Pelterie, was experimenting with powder rockets that he thought would have military applications. His appeals for official support were ignored until after the revolution and the civil war that followed. After an appeal to Lenin himself in May 1919, Tikhomirov received permission from the Revolutionary Military Council in March 1921 to start a laboratory in Moscow to develop "smokeless powder" war rockets. To the Bolsheviks' credit, it was the first time a government anywhere gave official approval for rocket research. The name of the place was a tribute to comrade Tikhomirov. It also reflected the breadth of the imagination of the socialist bureaucrat who came up with it: the Laboratory for the Development of Engineer Tikhomirov's Invention.

The lab was moved to Leningrad in 1925, enlarged in 1928, and finally given a new and weighty name by the now-entrenched party bureaucrats: the Gas Dynamics Laboratory of the Military-Scientific Research Council of the Revolutionary Military Council of the USSR. In shorthand, it was called the Gas Dynamics Laboratory, or simply GDL. But its full name was as accurate as it was long. GDL was a dedicated research arm of the Red Army.

While the number of people who worked at GDL in the late 1920s could be counted on two hands, its contribution to Soviet rocketry was profound. By 1931, GDL scientists were working on booster rockets for aircraft, solid-propellant missiles, and most important, the country's first liquid-propelled rocket engines. Valentin P. Glushko, one of the leading second-generation rocket men, did basic combustion and ignition research at GDL throughout the 1930s, getting ever more thrust out of his engines. The laboratory would lead the nation in rocket design for years. Tikhomirov died in 1930 at the age of seventy, and a crater on the far side of the Moon was named after him many years later.

In 1924, a Central Bureau for the Study of the Problems of Rocketry was formed, and so was an All-Union Society for the Study of Interplanetary Flight. The society was the first serious space club and understandably it made Tsiolkovsky a member (though this was strictly honorific, since he now rarely left Kaluga and preferred to keep to himself). Both semiofficial institutions put on two major space exhibitions in 1927, the second of which featured lectures, rocket designs, possible spaceship configurations, and material on Tsiolkovsky, Oberth, Goddard, Valier, and Esnault-Pelterie. The idea of traveling to space was by then so popular that groups to discuss it were forming around the whole country. Modernization was the thing. Domination of the air and the conquest of space would help to guarantee socialism's salvation and validate its doctrine. Left unmentioned, but as the old man himself knew so well, other worlds held a special allure for the denizens of one who were trapped in poverty, oppression, and obsessive regimentation.

Between 1929 and 1931, the Central Bureau and the All-Union Society were combined into a Group for the Study of Reaction Motion (GIRD) with two distinct and virtually independent branches in Moscow (popularly known as MosGIRD) and in Leningrad (LenGIRD). Soon, other branches sprouted at Kharkov, Baku, Tbilisi, Rostov-on-Don, Gorky, Kiev, Dniepropetrovsk, Bryansk, and elsewhere, all of them supervised by MosGIRD. Almost nothing is known about the offshoots. But the two big ones were important. Their members published original research on rocketry and space travel, translated foreign publications, and organized a number of conferences. While both GIRDs were ostensibly civilian, they were in fact affiliated with the larger Society for the Promotion of Defense and Aero-Chemical Development, a "voluntary" paramilitary organization that also had ties to the Red Army. To "volunteer" in the new socialist patois meant to do one's patriotic duty or suffer the consequences. Every aspect of aviation, and now rocketry, was woven into the motherland's defense fabric. Given what was taken to be fascism's never-ending threat—Winston Churchill had talked about strangling socialism in its cradle, and an allied army had actually tried to do so when it occupied Archangelsk from 1918 to 1920—any other course amounted to suicide.

LenGIRD was founded by two irrepressible dreamers: Nikolai Aleksevich Rynin and Yakov Isidorovich Perelman. Starting in 1928 and ending in 1932, Rynin, a professor of aerial transport, came out with an incredible nine-volume encyclopedia of space called *Interplanetary Communications* ("communication" being synonymous with "flight"). The massive work started with the early legends and ended with the state of the art. In preparing it, Rynin wrote to Goddard in January 1926 to ask for a copy of the American's famous monograph, which by then was well known in the Soviet Union. After reading it, he sent Goddard this review: "I have read very attentively your remarkable book 'A Method of Reaching Extreme Altitudes' edited *in* 1919 and I have found in it quite all the ideas which the German Professor H. Oberth published *in* 1924 as new about flying in 'infinite' altitude." The flattery must have warmed Goddard's soul.

Perelman did for Tsiolkovsky what Valier, Ley, and the others had done for Oberth: he interpreted the old man, often eloquently, for a large general audience. Perelman's own book on spaceflight, *Interplanetary Travels,* sold 150,000 copies in ten editions. And several dozen others came out in the 1930s as rocketry's popularity spread. During the same period, only three books on spaceflight were published in Great Britain and the United States.

MosGIRD contained two of the second generation's stellar rocket men: Fridrikh Arturovich Tsander and Sergei Pavlovich Korolyov. Tsander, in particular, was fixated on spaceflight. According to one account, he loved Mars so much that he referred to it the way he would a person and would try to inspire his colleagues by shouting "On to Mars!" when he entered MosGIRD's basement office at 19 Sadovo-Spasskaya Street. Asked why he had chosen the Red Planet as his primary target, he is reported to have answered, "Seemingly,

Mars has an atmosphere, and it is therefore possible that life exists there." He nevertheless named his son Mercury and his daughter Astra.

Tsander developed the OR-1 (*opytnaya raketa,* or "experimental rocket"), the Soviet Union's first liquid-propelled engine, and ran more than fifty OR-1 tests between 1929 and 1932. It produced eleven pounds of thrust. He also got the idea, eventually to be widely used in his own country and elsewhere, of running the liquid oxygen pipes around the combustion chamber to cool it. Groups led by Tsander, Glushko, and Mikhail K. Tikhonravov, another giant in Soviet rocketry, built and flight-tested a series of liquid-propelled rockets in the early 1930s. One of them, a bullet-shaped missile powered by a 150-pound-thrust OR-2 developed by Tsander and called the GIRD-X, reached an altitude of more than three miles on November 25, 1933.

But the honor of developing and launching the Soviet Union's first liquid-fuel rocket, named simply 09, went to Tikhonravov and MosGIRD's then-leader, the twenty-six-year-old Korolyov, who was an engineer from Ukraine. It was launched three months before GIRD-X, on August 17, 1933, and seven years after Goddard's own rocket went up.* The 09's flight lasted eighteen seconds, and it reached an altitude of 1,300 feet. Korolyov penned a report written within minutes of the flight. "She really flew," he said with characteristically quiet happiness. "She really flew. We didn't do all that work for nothing."

Within hours, the occasion was marked by this notice, which was posted at the launch site: "The first Soviet rocket operating on liquid propellant has been launched. The 17th of August is undoubtedly a portentous day in the life of the GIRD, and starting with this moment Soviet rockets must fly over the Union of Republics. . . . Soviet rockets must conquer space!"

Korolyov was a protégé of Tsander. An experienced pilot before coming to MosGIRD, he started working on rocket engines to propel gliders soon after he got there and continued to design rocket-propelled aircraft and winged rockets throughout the 1930s. One of them, called the Model 212, was a rail-launched flying bomb that first flew in 1939 and was the counterpart of Germany's V-1. Korolyov considered it to be a miniature version of an eventual man carrier.

To improve efficiency, the best minds in the Gas Dynamics Laboratory and the two GIRDs were brought together on orders from the Revolutionary Military Council and General Mikhail N. Tukhachevsky, the head of the Red Army's Ordnance Department, deputy people's commissar for Army and Navy affairs, and GDL supervisor. The new organization, which came into

* It was about seven feet long, weighed forty-two pounds including fuel, and was propelled by an engine that developed sixty-five pounds of thrust. (Baker, *The Rocket,* p. 79.)

being on September 21, 1933, was called the Reaction Propulsion Research Institute, or RNII. It fell under the jurisdiction of the People's Commissariat of Heavy Industry and was destined to figure very importantly in the nation's space program. RNII was ordered to work on theoretical and practical rocketry for "military technology and the national economy." Specifically, it was instructed to develop and test engines using solid, liquid, and gaseous propellants, experimental jet and rocket "systems" for artillery and aircraft armament, jet aircraft, and to study other organizations' gas dynamics projects. By the end of the decade, RNII would have a large cadre of "technical-engineering workers," "technical scientists," and engineers, most of them drawn from GIRD and from the Gas Dynamics Laboratory. Tikhonravov was made head of the department that developed liquid-propellent rockets. Some accounts put the number of scientists and engineers at 260; others at about a thousand. It would also have five experiment stations, each with its own workshop, and six special test stands for running jet and rocket engines. Under Tikhonravov, RNII produced a series of successful rockets, one of which reached 9,000 feet on August 13, 1937.

Whatever Tsiolkovsky and the brightest and most imaginative of his heirs wanted, their government told them to prepare to go to war, not to the planets. And their own ranks would be severely decimated by Stalin's brutal purges in 1937–38. Then-Marshal Tukhachevsky himself was executed in 1937 on groundless charges of "passing military intelligence to a foreign power [Germany]," sabotaging the Red Army, and trying to restore capitalism.* He was followed in 1938 by Ivan Kleimenov, head of RNII, Georgi Langemak, his deputy, and a large number of others. Mstislav V. Keldysh was one of a number of outstanding mathematicians and scientists who were summarily branded "reactionary" and "bourgeois." He would survive the purge to head RNII after the war, make important contributions to his country's space program, and eventually head the prestigious Academy of Sciences of the USSR.

Even the preeminent physicist Lev Landau was arrested by Lavrenty P. Beria's dreaded secret police, the NKVD, only to be miraculously rescued by Pytor L. Kapitsa, one of the outstanding Soviet physicists of the time, who happened to be on favorable terms with Stalin. Kapitsa was an expert on how gases such as oxygen, nitrogen, hydrogen, and helium behave at low temper-

* In Marxist-Leninist doctrine as practiced in the Soviet Union at the time, every theoretical or practical "error" was interpreted as "arming the enemy" and therefore amounted to sabotage. That crime was punishable by long imprisonment and very often by execution. Zbigniew K. Brzezinski has made the point that purges tend to be a permanent part of totalitarian regimes, since they prevent the formation of "too rigid lines of power demarcation within the system," meaning that they frighten, intimidate, or altogether eliminate any entrenched challenge to the top leadership in a kind of periodic housecleaning. In addition, they demonstrate the strength of the regime by showing that it can survive even by transforming its most ardent supporters into venal vermin. (Brzezinski, *The Permanent Purge,* p. 30.)

atures. He had worked with Ernest Rutherford at Cambridge until 1934, when Stalin ordered him home. Compressing oxygen and hydrogen were crucial to firing large rockets (and heavy warheads) into space, just as Tsiolkovsky had calculated, and Kapitsa played a major role in the research to accomplish it.

Beria's own involvement in high technology, first with the rocket program and then with the development of nuclear weapons, was similar to the role played by Heinrich Himmler and his SS in Nazi Germany. Both were vicious thugs who routinely committed atrocious crimes, including torture and assassination, and who were neither knowledgeable of nor interested in science. Yet each was clever enough to understand that whoever controlled the "geniuses" who controlled the weapons of mass destruction wielded formidable power. "They wanted to run these brains. They had an understanding that these brains would produce rockets and all this high technology," Pytor's son, Sergei Kapitsa, would say many years later. "It was the key to power in the future; it was a power game and they were part of it." The younger Kapitsa was himself an aeronautical engineer turned physicist.

Beria relished his involvement with the nuclear and rocket programs, but his participation was ordered by his master. Stalin, a dedicated practitioner of what would later be called "creative tension," thought that his scientists would bend to their tasks with particular fervor and be less inclined to spill secrets if they worked in constant terror of a murderous psychopath like Beria.

Based in part on damning testimony by Glushko, who by then had become Korolyov's bitter rival, Korolyov himself was picked up by the NKVD in 1937, interrogated about his dangerous "pyrotechnics and fireworks," branded a traitor for collaborating with an anti-Soviet organization in Germany—technically, "subversion in a new field of technology"—and shipped in a sealed boxcar to Magadan on the Pacific coast. There, he was put to excruciating work in the infamous Kolyma gold mines.*

But he was saved by Andrei N. Tupolev, a leading aircraft designer who was himself arrested in 1938 and sentenced to five years in prison. Korolyov had been one of Tupolev's students at the Moscow Higher Technical Institute and had impressed his lecturer. Now, the lecturer intervened to pull his talented and energetic former pupil out of Hell. Unlike Korolyov, Tupolev and his staff were sent to the NKVD's Central Design Bureau 29, a special prison, or *sharashka,* on Beria's orders and were put to work designing military aircraft for a war that seemed more likely every day. Tupolev's bombers, starting with a rugged twin-engine model called the TU-2, would become mainstays of the Soviet Air

* Although the exact number of people who were arrested between 1929 and 1953 will probably never be known, one knowledgeable Russian, General Dmitri Volkogonov, estimated the figure to be 21.5 million, a third of whom were shot to death. ("Cleaning Up Stalin's Act," *The New York Times,* Op-Ed page, May 8, 1996.)

Force during World War II and throughout the cold war. Korolyov was pulled from the jaws of almost certain death to design airplanes in a relatively posh jail—a "golden cage," in Sergei Kapitsa's words—first near Moscow and, after the Nazi invasion, at Omsk. He was then transferred to yet another prison at Kazan, where he was ordered to help develop small rocket packages that would push airplanes off the ground faster. His boss, it turned out, was his old nemesis, Valentin Glushko. Aleksandr Solzhenitsyn's *The First Circle* was about life in such a *sharashka* outside Moscow. The novel took its title from the First Circle of Dante's *Inferno,* where wise and philosophical pagans were kept separate from sinners, who were tortured. One prisoner, or *zek* in Russian slang, explained the reason for *sharashkas* this way: "You need only remember the newspaper piece that said: 'It has been proved that a high yield of wool from sheep depends on the animals' care and feeding.' "

The prison years profoundly affected Korolyov, both physically and mentally. He developed a heart condition at Kolyma, was so malnourished that he lost all of his teeth, and somehow his jaw was broken. Never especially ebullient to begin with, the shortish, stocky man with glistening brown eyes came out of the gulag a committed cynic with a gloomy sense of the future. "We will all vanish without a trace" became his favorite expression. "Sometimes," he told friends years later, "I wake up at night and lie there thinking: at any moment the order might be given and the very same guards could burst in and shout: 'Come on, scum, get your things together and on your way!' "

Throughout the war and afterward, the wary Korolyov understood that the dangerous and manipulative men in the Kremlin and their apparatchiks on every level cared about rockets only to the extent that they could be used for purposes that were practical and that would directly benefit the defense of the motherland: to kill people. Even pioneering work on meteorological sounding rockets in 1934 and 1935 had been done by the Society for the Promotion of Defense and Aero-Chemical Development, which put the project solidly in the military's fold.

Korolyov therefore learned, as Goddard and a number of German rocketeers had learned, to hide the Moon and the planets in the attic of his imagination. "I only wish that in your future works . . . you would pay more attention not to interplanetary problems, but to the rocket engine itself, to a stratospheric rocket, etc., since all this is now closer to us, more understandable, and more necessary," he wrote to Perelman as early as April 1935. "A great deal of nonsense has been written on interplanetary themes, and it is now hurting us badly. Just a few days ago, they said directly to me in one journal: 'We are avoiding publishing material on rocketry, because they are all lunar fantasies, etc.' Now I have even more difficulty in convincing people that this is not so, *that rockets are a defense and a science*" (italics added).

The Societies

Korolyov and his colleagues may have had to avoid mentioning their fantasies, but ordinary citizens elsewhere delighted in the great space dream, and carried rocketry during the late 1920s and into the 1930s through a network of societies. They showed, at least for a while, that a new technology could be forged by dedicated amateurs. Like their counterparts in astronomy, archaeology, and paleontology, they made their own notable discoveries. But there was one important difference. Unlike buying a backyard telescope, which was a onetime investment, rocketry's cost curve went up in direct proportion to the rockets' size and complexity. Keeping expenses down by limiting the performance of the rockets was out of the question because it would have defeated the groups' purpose. Yet advancing the technology would become so expensive so quickly that only national treasuries could afford to do so. The first and most famous of the societies to run into that conundrum was Verein für Raumschiffahrt, the Society for Space Travel, or simply the VfR.

Of all the amateur groups, the VfR, founded in a saloon in Breslau on July 5, 1927, was unquestionably the most important. The idea for the VfR came from an engineer named Johannes Winkler who was then working as a church administrator. Its ten or so founding members, who included Willy Ley and Max Valier, set two goals for the organization: popularize space travel and do serious experiments in rocketry. But those were amorphous targets. Asked what the VfR's "program" was, Ley would respond this way:

> We did have a program of sorts, but while we knew precisely what we were *not* going to do, we could not formulate clearly what we were going to do. On the negative side we were certain that we would not touch solid fuels in any form. We also were not going to stick a rocket motor for liquid fuels on a car, railroad car, or glider. We were, in short, not going to do anything but build rockets. But how those rockets would look and what they might be used for was something we could not tell because we did not know. I wish I had known in those days the answer Michael Faraday is said to have given to visitors to his laboratory who asked him about the use of his experiments with electricity. Once, when a lady asked the question, Faraday is said to have replied, "Madam, what use is a newborn baby?"

The VfR was anything but a club in a casual sense. It was a formal organization that soon found a permanent home, called the Raketenflugplatz, or Rocketdrome, an abandoned military garrison on three hundred acres on the northern edge of Berlin. It published its own newsletter, *Die Rakete.* The organization claimed five hundred members within its first year, and nine hundred a year later, including Oberth, Walter Hohmann, Esnault-Pelterie, and the energetic and infinitely resourceful Rudolf Nebel. They were men who considered the development of rockets for space travel to be not merely a hobby but a serious, often full-time, undertaking. The Raketenflugplatz's sturdy buildings were therefore used for both work and living. And there was plenty of open space for testing. Nebel and his pals spent a great deal of time scrounging for free tools, metal rods, sheet aluminum, a drill press, welding equipment, two lathes, benches, paint, pipes, screws, and other hardware. Some of them lived at the Raketenflugplatz rent-free, using welfare checks to pay for their food or getting fifteen-pfennig meals at the Siemens welfare kitchen. Yet everyone apparently owned a dark suit, which, bending to the European formality of the time, was worn to formal meetings. Winkler, who was president of the organization, decided to lend extra dignity to a meeting in the spring of 1931 by wearing a tuxedo.

Dmitri Marianoff, Albert Einstein's stepson-in-law, said after a visit to the rocket facility that "the impression you took away with you was the frenzied devotion of Nebel's men to their work. Most of them were [like] officers living under military discipline. Later, I learned that he and his staff lived like hermits. Not one of these men was married, none of them smoked or drank. They belonged exclusively to a world dominated by one single wholehearted idea." That same year, 1931, the VfR added a bright young engineer to its rolls. He was an eighteen-year-old aristocrat named Wernher von Braun.

Testing of one kind or another, all of it frustrating, went on regularly. Nebel, who was an endlessly manipulative zealot, another member named Klaus Riedel, and a third man built a small liquid-fueled Mirak, or "Minimum Rocket," and tested it at Nebel's grandparents' farm in Saxony. It exploded. Overheated combustion chambers, uneven propellant flow and mixing problems, leaks, premature or late ignition, and other mechanical afflictions were always turning rockets into duds or bombs. Getting the fuel to flow into the combustion chamber, for example, was one of the most nagging problems for rocketeers everywhere. A rocket engine would not fire effectively if fuel dripped into its combustion chamber like rainwater plopping into a barrel. The fuel had to be forced into the chamber under pressure. That required pumps, which were notoriously inefficient, or a compressed gas such as carbon dioxide or nitrogen, to push the fuel out of its tank, through the tube, and into the chamber. The liquid rocket's intricacy increased both performance and the possibility of disaster, and that would never change.

Then the break came. On May 10, 1931, an improved Mirak, running on liquid oxygen and gasoline that was pressurized by the same kind of carbon dioxide cartridge used in seltzer water bottles, broke free from its test stand and reached sixty feet. "You know the secret baby we discussed?" Riedel told Ley by telephone the following day as if they were conspirators in some drama. "Well, I took it out yesterday to make a test run. I didn't expect anything; you know, I used those heavy valves and the fat struts along the fuel lines. And the damn beast flew! Went up like an elevator, very slowly, to eighteen meters. Then it fell down and broke a leg."

The first intended flight came three days later with the same rocket, now rechristened Repulsor 1. Willy Ley has to describe it himself:

> In spite of all the makeshift the "flying test stand" took off with a wild roar. It hit the roof of the building and raced up slantwise at an angle of about seventy degrees. After two seconds or so it began to loop the loop, rose some more, spilled all the water out of the cooling jacket, and came down in a power dive. While it was diving the wall of the combustion chamber—being no longer cooled—gave way in one place, and with two jets twirling it the thing went completely crazy. It did not crash because the fuel happened to run out just as it pulled out of a power dive near the ground; actually it almost made a landing. Examination showed that it was intact save for the motor. We could not examine it at once, since we were dizzy from watching and jumping out of the way and had to sit down in the grass for a moment. In retrospect it seems as if all this took place very close to the ground, but my notes say that we estimated altitude as about 200 feet immediately afterward."

By the spring of 1932, the rockets had flown three miles horizontally and nearly a mile straight up. More important, there had been 270 static tests and 87 actual flights, giving the rocketeers a wealth of experience.

In spite of donations by a few sympathetic businessmen and members such as Valier, Winkler, and Ley, who frequently dug deep into their own pockets, the VfR was always in tough financial shape. So the members of the club became inventive. They enrolled a Who's Who of rocketry in addition to their own notables: Esnault-Pelterie, Perelman, Rynin, Potocnik, Sänger, and others. This not only cross-pollinated the organization with creative talent, but it attracted a legion of dues-paying lesser lights who wanted to mingle with the stars of the new rocket world.

And it worked. Members came from Czechoslovakia, Denmark, France, Poland, the Soviet Union, Spain, and South America. There were incentives for enrolling new members that included autographed pictures of Valier and autographed copies of a book, *Die Fahrt ins Weltall* (*The Trip into the Universe*), written by Ley. In 1930, with the Depression biting ever deeper into Germans' pockets, the first in a series of "rocket shows" were staged for admission fees.

But the Depression ground relentlessly on. *Die Rakete* folded for lack of funds, and with no way of keeping the far-flung membership informed of the organization's activities, the organization itself began to fold. Nebel's desperate, crackpot attempt to pull in funds by building a man-carrying rocket whose passenger was supposed to bail out, called Project Magdeburg, failed. Then the municipality of Reinickendorf, where the Raketenflugplatz was located, presented the VfR with an enormous water bill that it could not pay. Ironically for a group whose membership boasted many engineers, the charges were for years of leaking faucets. The lease was then canceled, and the old drill field was turned over to a new paramilitary group; a kind of air force reserve unit whose young men wore, in Ley's words, "beautiful powder blue uniforms." The uniforms signaled the end of the VfR. Hitler had come to power. Nebel, who had already flirted with the military for fund-raising purposes, was wearing a swastika. And young von Braun had gone to work for the Army.

To Leave "A Lousy Planet"

Even before the demise of the VfR, two other rocket groups had been born: the American Interplanetary Society, founded in New York in April 1930 (it became the American Rocket Society in 1934), and the British Interplanetary Society, founded in Liverpool in September 1933 (it moved to London three years later).

The dozen founders of the American organization were all either science fiction writers (several of whom worked for Hugo Gernsback, a magazine mogul who was himself a founding member) or s-f buffs, as they preferred to call themselves. They included David Lasser, the managing editor of *Science Wonder Stories* and the AIS's first president; Fletcher Pratt, a science fiction writer who also chronicled the Civil War; and Charles W. Van Devander, a newsman who was the first editor of the AIS *Bulletin,* a future press secretary to New York Governor W. Averell Harriman, and a well-known radio commentator. Another founding member, G. Edward Pendray, was a New York *Herald Tribune* reporter and another totally dedicated dreamer. Pendray even steamed to Germany to establish a liaison with the VfR in April 1931. When he returned to the States the following month, he wrote a detailed report on developments at the Raketenflugplatz, complete with descriptions of the Miraks and hand-drawn diagrams, for the society. Pendray would one day help edit Goddard's diaries (Goddard himself accepted an invitation to join the AIS but always kept his distance).

Lasser, the son of Russian immigrants, was another space dreamer and a character in his own right. A high school dropout who managed to graduate from MIT with a degree in engineering, he wrote *The Conquest of Space,* which appeared in 1931 and which Arthur C. Clarke would call "the first book in the English language to explain that space travel wasn't just fiction."

Clarke, himself a preeminent space visionary, credited Lasser's book as being "one of the turning points in my life, and I suspect, not only of mine." In common with Tsiolkovsky and others, Lasser had a utopian streak that turned to socialism and which led him to become a labor organizer during the Depression. His union activities earned him the respect of both Franklin D. Roosevelt and W. Averell Harriman, but also the wrath of the far right, which would smear him as a Communist after World War II. His *Private Monopoly: The Enemy at Home,* which appeared in 1945, fueled the fire by arguing that nearly all modern misery was caused by greedy economic policies.

Though far smaller than the German group, at least at the outset, the members of the American Interplanetary Society shared the VfR's lofty goals: the popularization of rocketry and the research to get it off the ground. Like their British, French, German, Russian, and other counterparts, the Americans were irrepressible inventors and dreamers who brought a pseudoreligious zeal to their vision of the future.

They were far ahead of their government on that score. The vision included embracing not only Mars the planet, for example, but Mars the god of war. "It is my contention that an agent ideal to the use of the scientific militarist, for both the air raid and the long distance bombardment is now in the process of development; that its eventual perfection is but a matter of time; and its use in warfare is certain to occur," the somewhat verbose David Lasser told AIS members at a meeting in the American Museum of Natural History on October 22, 1931. "I refer to the rocket. The perfection of the rocket," he added with a flourish worthy of Wells, "in my opinion will give to future warfare the horror unknown in previous conflicts and will make possible destruction of nations, in a cool, passionless and scientific fashion."

The AIS's inaugural *Bulletin,* a four-page typewritten newsletter run off on plain paper, came out in June 1930 and carried news of Valier's death, the summary of a paper on interplanetary travel that had been read by Pratt at one of the biweekly meetings, a prediction by a Princeton astronomer that a "space-navigating cruiser" could reach the Moon by 2050, and a report noting that a Czech inventor named Ludvik Ocenašek was experimenting with twenty-inch-long rockets as a first "serious attempt to shoot a rocket to the Moon." Later issues carried all kinds of tidbits, some prescient, some dead wrong, but all generating the kind of excited speculation that kept the membership stimulated. The September 1930 issue, for example, reported that Guglielmo Marconi had told the Italian Society for the Advancement of Science that radio signals could in theory "echo" off distant bodies. He was predicting radio astronomy. It also carried news of a Soviet professor named V. V. Stratonoff who had assured an audience in Prague that the overpopulation of Earth would lead to the colonization of another planet, probably Venus.

A debate on whether there was life on the Moon was going on as late as 1931. That September's *Bulletin* reported that the eminent Harvard astronomer William H. Pickering believed that swarms of insects on the Moon could have accounted for apparent changes in the shape of the crater Erastesthenes. "There is no reason to suppose that the lunar insects resemble any terrestrial animals," the authoritative academician was quoted as having written. "Since their environment is so different, we may say further, that it is extremely improbable that they do resemble any terrestrial form of life, but since it is not worthwhile to coin a new name for them, and since they must be of the same order of size as our insects, we will simply call them by that name, whether they have two legs, four, six or fifty." And the *Bulletin* carried an article in January 1932 suggesting the possibility of "luxuriant vegetation" on Mars along what may or may not have been Lowell's canals.

And like its opposite numbers, the American Interplanetary Society tried hard to grow by attracting the leading lights of the period, even as most of them chased wider audiences for the sake of their own careers. (Esnault-Pelterie drew two thousand listeners for a lecture on interplanetary flight given at the museum on January 14, 1931.) The AIS evolved into the American Rocket Society and then into the Washington-based American Institute of Aeronautics and Astronautics, a prestigious engineering organization. Yet most of those who forked over three dollars for annual dues in its Depression-ridden formative years were quiet believers whose imaginary ride to space carried them away from reality's drudgery or, as in the case of Tsiolkovsky, reduced pain.

Bernard Smith, to take one example, joined the AIS in 1932. An impoverished handyman in the best of times, he was now constantly out of work. The meetings at the museum and the rocket firings provided entertainment that lifted him, even temporarily, out of his misery. "It was a lousy planet," he would recall, and "the rocket ship was the only way to get off it." Here is his zany description of a launch on Staten Island:

> The firing procedure was as follows: After the gasoline and liquid oxygen tanks were charged and closed, a stick was slid into a horizontal slot cut into the cock handles. The outer end of the stick was fastened to a lanyard that stretched to the firing pit. I had designed the handle to hold the stick until the lanyard had turned the cocks a quarter turn, whereupon the stick would fall away. With everything in readiness a gasoline-soaked rag placed under the nozzle was ignited and everyone ran to the pit to watch me pull the lanyard. But the angle of pull was not quite right and the stick fell away before the cock was turned. My professional reputation being at stake I knew what I had to do. I ran back to the launcher, replaced the stick, reignited the rag, payed out the lanyard in the right direction and pulled it, thereby becoming on 14 May 1933, a day before my 23rd birthday, the first lad in America to publicly launch a liquid fuel rocket.

Judging by the criteria exercised decades later at Cape Canaveral, wherein any rocket that rose at least six inches off the pad was pronounced a success, this one was a resounding achievement. It rose a couple of hundred feet before it exploded. A man in a rowboat offshore saw it fall and retrieved it for us in triumph. Good thing the fins were of balsa; otherwise it would be at the bottom of lower New York Bay to this very day.

During the post mortem it was pointed out that I could have gone up instead of the rocket.

Smith went on to design air-launched missiles for the U.S. Navy, perhaps with the knowledge that he had something in common with Konstantin Tsiolkovsky that was even more profound than a love of rocketry. For both of them and for countless others the possibility of going to space offered hope and the promise of salvation. Thinking about that distant time, being part of a great, predestined effort that must one day succeed in finally breaking humanity's chains, was a shared experience and a profound calling. It was Kepler's calling. It was Galileo's calling. And it was Bernie Smith's calling, too.

"All the Universe or Nothing"

The possibility of space travel was an established belief by 1935. But that wasn't only because of the societies. Rather, a synergy developed between fact and fantasy. Where the general public was concerned, the fact had far less to do with the societies—where the converted preached to one another—than with the spread of aviation and ordinary people's participation. The hard part of traveling to space, as far as most laymen were concerned, was getting off the ground in the first place. By the mid-1930s flying was broadly accepted around the world because it had evolved from a deadly sport practiced by a privileged few to a form of popular entertainment and serious transportation.

Starting after the war and going well into the 1930s, fliers were everywhere, breaking distance, speed, and altitude records over land and sea. John Alcock and Arthur Whitten-Brown, Richard E. Byrd and Floyd Bennett, Charles A. Lindbergh, Amelia Earhart, Wiley H. Post, John Macready and Oakley Kelly, Sir Alan Cobham, Dieudonné Costes, Albert F. Hegenberger and Lester Maitland, and scores of others set distance records that captivated millions of readers and listeners. So did the barnstormers and mail pilots.

Nor did the air belong to the daredevils alone. Throughout the 1930s, commercial airlines sprang up around much of the world, also leaping over ever-longer routes. In 1936, nine passengers paid $1,438 (about $20,000 in today's dollars) for seats on the first Pan American Airways China Clipper flight from San Francisco to the Far East. Ford Tri-motors, succeeded by Donald Douglas's great DC-3, were carrying as many as twenty-one passengers on routes throughout the country and their Fokker counterparts were doing the same in Europe. No one seriously believed that airplanes could fly to space, any more than they thought balloons could. But that was not the point. The heralds of manned flight—the storytellers—had been right when they foretold of machines that would carry people into the sky. They had also told of machines that would carry people to outer space. And sure enough, along came Tsiolkovsky, Goddard, Oberth, Esnault-Pelterie, Tsander, and the others, proving, however tentatively, that the heralds were right again.

So the synergy built. Science fiction directly addressing other worlds and how to get to them exploded during the 1920s and '30s. Hugo Gernsback, who is often called the "father of science fiction," launched a publication called *Modern Electrics* as early as 1908 that carried some s-f, including his own. In a story called "Ralph 124C 41 +," the hero (Ralph) tracks a Martian spaceship making off with his girlfriend with what would eventually turn out to be radar. Gernsback's description of the device was so detailed that years later the United States Patent Office refused to consider Sir Watson Watt's application for the actual invention. In April 1926, Gernsback introduced the first all sci-fi magazine, *Amazing Stories,* which became an instant success by featuring such popular writers as Wells and Burroughs. He also founded *Science Wonder Stories* and *Air Wonder Stories,* two more science fiction magazines, followed by *Science Wonder Quarterly.* As the 1930s continued, the field became crowded, not only with more magazines such as *Astounding Stories of Super-Science* and *Astounding Science Fiction,* but with pulp science fiction and hardcover page-turners such as S. Fowler Wrights's *Deluge,* which was made into a motion picture.

Things to Come, a film that was based on Wells's novel *The Shape of Things to Come,* was released in 1936 and was probably the most notable science fiction film of the prewar period because it seriously tried to articulate the driving force behind space exploration. The screenplay, which was written by Wells himself, began by showing how the scourge of a long war led to the rule of a fur-clad warlord-oaf called "The Boss" (played by Ralph Richardson) who presided over a civilization that was primitive and savage. It then jumped ahead to 2036, a better time when Earth had evolved into a self-assured techno-utopian global city vaguely reminiscent of the Marxist ideal. But Theotocopulos, a sculptor (Cedric Hardwicke), tried to get the citizens of "Everytown" to destroy a giant gun that was poised to shoot a young couple to the Moon. This pitted art, which was portrayed as sentimentally irrational, contemptuously subjective, and pitifully reactionary, against science, which was the opposite: rational, orderly, and progressive. (Verne's lost novel, *Paris in the Twentieth Century,* which was discovered only in 1989, also predicted that the arts would be despised and that even his friend Victor Hugo would be forgotten in a world that worshiped technology.)* This amounted to the latest installment in the old backlash against the Age of Reason and the Industrial Revolution. Both triggered Romanticism's return to nature and the veneration of the artist, not the scientist/mechanic, as nature's supreme creator (it is still playing out, however feebly, in postmodernism).

Forty years later, George Lucas, the audio-visual generation's Jules Verne, would wisely choose the middle ground in his sensationally successful *Star Wars* trilogy by looking backward as well as forward; by bringing history and familiar, living creatures like Chewbacca, an affable ape, together with ultra-high technology. Luke Skywalker was schooled in the ways of Jedi knighthood by a wise Arthurian character whose light saber suggested Excalibur, while the tired and scuffed X-Fighters looked war-surplus, and the racist barman's refusing to serve R2-D2 and C3PO with a "Hey, we don't serve their kind here; you're Droids" went back to how Indians had been treated in clichéd Hollywood Westerns. Indeed, the strictly mental "Force" used by the good and evil protagonists suggested that people wanted to control technology, not the other way around.

At any rate, the young woman in *Things to Come* was the daughter of a leader named John Cabel (Raymond Massey) and her love interest was the son of a character named Passworthy (Edward Chapman). In the end, Theotocopulos was foiled, and the gun fired the star-crossed couple to space.

Their fathers watched them speed away, Cabel with exhilaration, Passworthy with the contemptible sorrow and foreboding that were the sure mark of a wimp:

> "There—there they go! That faint gleam of light," Cabel says excitedly.
> "I feel that what we've done is monstrous," replies Passworthy.
> "What they've done is magnificent."
> "Will they come back?"
> "Yes, yes. And go again and again, until a landing is made and the Moon is conquered. This is only a beginning."
> "If they don't come back—my son and your daughter—what of that, Cabel?"
> "Then, presently, others will go." [On the heavenly breezes.]
> "Oh, God, is there ever to be any age of happiness?" Passworthy moans. "Is there never to be any rest?"

* Some late-twentieth-century anarchists think the opposite. They are technophobic in that they believe that unbridled technology—machine worship—enslaves mankind because it is antithetical to freedom and, in a fundamental sense, inhuman. The economist and social critic Thorstein Veblen made the point in *The Theory of the Leisure Class* (1899) that intellectuals tend to hate science and despise engineering because they are practical disciplines.

"Rest enough for the individual man: too much, and too soon, and we call it death. But for Man, no rest and no ending. He must go on, conquest beyond conquest. First this little planet with its winds and ways, and then all the laws of mind and matter that restrain him. Then the planets around him, and at last out across immensity to the stars. And when he has conquered all the deeps of space and all the mysteries of time, still he will be beginning."

"But . . . we're such little creatures," Passworthy whines. "Poor humanity's so fragile, so weak. Little, little, animals."

"Little animals. If we're no more than animals we must snatch each little scrap of happiness and live and suffer and pass, mattering no more than all the other animals do or have done. It is this—or that: all the universe or nothing," Cabel says, sternly, pointing to the cosmos. "Which shall it be, Passworthy? Which shall it be?"

It had been only ten years since Goddard's little rocket had sputtered off the farm in Auburn and half that time since the first Mirak had worked. Now the film was taking on the same grand endeavor addressed by Kepler, Verne, and the others. It was a lofty one, and he had a suitably lofty philosophy. The problem was that the first steps on the way to the universe would be taken not by John Cabel but by The Boss.

3

Gravity's Archers

VIENNA, Jan. 30—The arrival of letters in New York which had left Vienna but a half hour earlier and the destruction of the entire population of great cities from an equal distance and with equal rapidity, are two seeming miracles which will be practicable possibilities in the near future, according to Professor Hermann Oberth, Hungarian pioneer in space-rocket flight experiments, who lectured tonight at the Vienna Meteorological Institute of Urania.

Professor Oberth related the results of twenty years of experimenting with the problems of rocket flight, which were first ridiculed, then exploited by a motion picture company, finally are accepted today as important scientific research work. . . .

Practical developments in rocket employment about to be realized, according to Professor Oberth, are . . . analysis of the upper strata of the atmosphere . . . photographing war-time enemy positions by cameras carried by rockets, which the enemy would be unable to shoot down but which would return, boomerang-like, behind the firer's lines . . . map-making of inaccessible and untraversable territory . . . the dispatch by a single rocket from Vienna of thirty kilograms of mail, which by shooting up beyond the earth's air envelope and consequently dropping attached to a parachute would reach New York less than half an hour later . . . as the propellant of airplanes traveling at great heights . . . for bombing from the other side of the earth an enemy country with a murderous rain of rockets carrying poison-gas containers capable of exterminating whole populations in a few minutes.

The murderous rockets would arrive before the mail (poor Schmiedl, who was practicing his own kind of special delivery in the area at the time, would no doubt have agreed). By the time the above dispatch appeared in *The New York Times* on January 31, 1931, Germany was in chaos. Widespread unemployment gripped the country, thugs from the left and right were spilling blood in the streets, and the embattled Weimar government was growing weaker by the day.

William L. Shirer, a correspondent who was there, described the situation in *The Rise and Fall of the Third Reich:*

> The hard-pressed people were demanding a way out of their sorry predicament. The millions of unemployed wanted jobs. The shopkeepers wanted help. Some four million youths who had come of voting age since the last election wanted some prospect of a future that would at least give them a living. To all the millions of discontented Hitler in a whirlwind campaign offered what seemed to them, in their misery, some measure of hope. He would make Germany strong again, refuse to pay reparations, repudiate the Treaty of Versailles, stamp out corruption, bring the money barons to heel (especially if they were Jews) and see to it that every German had a job and bread. To hopeless, hungry men seeking not only relief but new faith and new gods, the appeal was not without effect.

The National Socialist (Nazi) Party surprised even itself when it captured one hundred and seven seats in the Reichstag in the national election in September 1930, turning it from the ninth and smallest party in parliament to the second largest. Adolf Hitler crystallized his vision of an invigorated fatherland by promising to turn it into a successor to the Holy Roman and Hohenzollern empires. It would last a thousand years, he swore, and would be called the Third Reich. But national salvation and renewal depended upon rearmament. There could never be strength without the means of defending the nation. The major obstacle to this was the Treaty of Versailles, which not only forced Germany to make the reparations, but severely curtailed the size of its Army and Navy, forbade an Air Force, and prohibited the manufacture of warships, military aircraft, artillery, and other major weapons. It also made a specific point of dissolving the renowned General Staff.

But the essence of the General Staff and its virulently fundamentalist Prussian doctrine, like some precious sperm, was carefully protected from extinction. (Soldiering was to Prussia what opera was to Italy. Karl von Clausewitz, the great military strategist, had been Prussian, as had Manfred von Richthofen, the legendary Red Baron.)

Long before Hitler came on the scene, General Hans von Seeckt, who succeeded Field Marshal Paul von Hindenburg as chief of the General Staff in 1919, vowed to preserve and nourish the General Staff and subvert the treaty, which would not even take effect until the following January. "The chains of Versailles which curtailed Germany's freedom of action must not be allowed to bind her spirit," he would write. Accordingly, Seeckt posed two targets for the remnants of the Army: achieve the highest possible level of technical proficiency within the treaty's restrictions, and quietly evade and overcome those restrictions in preparation for "the Day" when the Reichswehr would awaken like Wotan, their beloved Wagner's virtuous and powerful god, to save Valhalla from rape and plunder by a world of sinister and greedy Alberichs. "The form changes," he pledged to the Prussian officer corps, which had been born in the Thirty Years' War and made arrogant by beating Napoleon a century and a half later, "but the spirit remains as of old."

Technical proficiency translated to training and developing and producing war-fighting machines. This was far more challenging than disguising generals in Tyrolean hats and lederhosen and hiding their monocles and swagger sticks because it would have to be done with utmost secrecy. The treaty prohibited the manufacture of dreadnoughts and U-boats, pursuit planes and bombers, and all manner of armor and artillery. The Luftwaffe, to take only one example, would get around this by purchasing "civilian" planes that could be easily converted to fighters and bombers and having its pilots learn to fly in sport glider clubs and in Russian aircraft. But rockets were no problem. They were not so much as mentioned in the treaty for the simple reason that none of its framers had thought seriously about them one way or another, let alone as weapons.

That is why Professor-Colonel Karl Emil Becker, the coauthor of a respected textbook on ballistics and a practiced artilleryman in an army that traditionally loved huge guns, took an interest in developments at the Raketenflugplatz. Becker believed that big guns had reached their limit with the Paris Gun, a howitzer that in 1918 had been used by Germany to lob shells into the French capital from eighty miles away. Rocket-propelled missiles, on the other hand, had a number of advantages. They would eliminate barrels and their massive supporting equipment while greatly increasing range and destructive capacity. They could not be attacked on the way to the target as airplanes could. And their surprise deployment and use in large numbers would hurt and possibly even destroy enemy morale. Or so Becker thought.

Accordingly, the colonel and a highborn subordinate in the Army Ordnance Department's Ballistics and Munitions Branch, Captain D'Aubigny Ritter von Horstig, decided to learn all about rockets. They were soon joined by other artillery officers. One of them, Captain Walter Dornberger, was a dedicated rocket enthusiast who devoured Oberth. At about the time the holder of the certificate of competence from Heidelberg delivered his lecture to the Austrian weathermen, in fact, the Army was quietly experimenting with its own solid-propellant rockets at the artillery proving ground at Kummersdorf-West, a secluded spot protected by pine trees at the edge of the Brandenburg Forest, south of Berlin. (The facility's official name was the suitably ambiguous Kummersdorf-West Experimental Station.)

But the soldiers also hedged their bet by visiting the VfR. Early in 1932, Becker, Horstig, and Dornberger quietly paid several visits to the Raketenflugplatz, always dressed as civilians so they wouldn't draw attention. There were things about the VfR that the soldiers deeply disliked. One of them was Rudolf Nebel, who never seemed to pay for anything and whom they regarded as a sleazy, self-promoting, opportunistic jackal. Another was the fact that the civilians were preoccupied with promoting and publicizing space travel. The soldiers considered any thought of traveling to the Moon or Mars absurd and irrelevant to the fatherland's pressing requirements. And they loathed publicity. Becker and his aides were determined to keep rocket work top secret for at least three reasons. First, and most obviously, they did not want to be discovered as even appearing to circumvent the treaty. Second, they were convinced that the element of surprise would help to confound an enemy under rocket attack. Finally, being the good soldiers they were, they didn't want to plant ideas about the virtues of long-range rocket bombardment in the heads of potential enemies. They therefore found the VfR's rocket extravaganzas to be anathema.

"The value of the sixth decimal place in the calculation of a trajectory to Venus interested us as little as the problem of heating and air regeneration in the pressurized cabin of a Mars ship," Dornberger would say many years later. "We wanted to advance the practice of rocket building with scientific thoroughness. We wanted thrust-time curves of the performance of rocket motors. We intended to establish the fundamentals, create the necessary tools, and study basic conditions." No wonder that, in von Braun's words, the visitors deplored "what they called our 'circus-type' approach."

On the other hand, the VfR's growing expertise could not be ignored. Becker therefore overcame his misgivings about Nebel and on April 23, 1932, invited him to demonstrate a Repulsor at Kummersdorf-West. Nebel assured Becker that he would be able to launch one of the rockets without difficulty. But the officer set strict rules. Nebel would be paid 1,367 reichsmarks if he got the thing to pop its parachute and fire a red flare at the top of its trajectory. If that did not happen, he would receive nothing.

Nebel, von Braun, and Riedel drove to the proving ground in the early-morning darkness on June 22 and got their rocket off the ground. It even climbed to 200 feet. But then it swung into an almost horizontal trajectory and crashed less than a mile away, its chute unopened and its flare unlit. Nebel received not a pfennig.

The demonstration figured importantly for Nebel because it convinced Becker and the others that in addition to being a scoundrel he was also an unreliable liar. And it convinced them that, ultimately, the Army was going to have to develop its own rockets and missiles. Still, it would need civilians to do that. Von Braun, who had caught both Becker's and Dornberger's eye as a bright, highly knowledgeable, and irrepressibly enthusiastic engineer, was summoned to meet with Becker soon after the Repulsor fiasco and was sent back to the Raketenflugplatz with offers of jobs for himself and several of the others. This touched off at least one heated debate. His nationalistic fervor notwithstanding, Nebel argued against the Army's offer because, he insisted, it meant that most of the best minds in the VfR would leave. (And leave him behind.) He vehemently explained that since the society was the sole repository of knowledge about liquid-propellant rockets in the country, the Army would sooner or later have to support it with funding on his own terms. Riedel also rejected the offer. He wanted to remain true to the VfR's credo and, in addition, start his own rocket and spaceflight corporation. But a number of members of the society did abandon civilian rocketry to create weapons for paychecks that came regularly. And the first was Wernher von Braun.

"A" for Effort

Wernher von Braun was an aristocrat, a *Freiherr,* or baron, though he rarely used the title. He was born on March 23, 1912, the second of three sons sired by Baron Magnus von Braun, a well-off Junker gentleman farmer. His mother was a baroness in her own right and an amateur astronomer. In 1932, the year Wernher went to work at Kummersdorf-West as a twenty-year-old engineer, his father was appointed minister of agriculture in the ultraconservative government of Franz von Papen. The "Cabinet of Barons," as it was called, had close ties to both the Army and the bankers, industrialists, and landed gentry that were Germany's ballast. Young von Braun was therefore part of a sophisticated elite that was quietly but proudly nationalistic yet had little sympathy for a Nazi street rabble that he himself considered "vulgar." What drove him was the spirit behind Cabel's soliloquy at the end of *Things to Come.* He believed implicitly that the time was coming for humanity to finally leave Earth, that it was the most important undertaking in all of history, and that his role in the most majestic of all conceivable adventures was to help lead the way.

Yet like Sergei Korolyov, who would one day be his relentless adversary in the most intense technological race in history, von Braun was under no illusion about what his new masters wanted. The noble vision, whose sight he never lost, would therefore be subordinated to a lower requirement so that he could get on with his research. Serious rocketry was evolving from a dangerous hobby to a state enterprise that no individual or group could afford. And the state wanted the capacity to blow up its enemies. Period. Scruples would therefore bend to an expedience whose evil consequences could be sensed in 1932 but not clearly foretold. One learned to see but not observe. Or one simply looked the other way. The Faustian deal between the rocketeers and the butchers who were their benefactors would forever haunt von Braun and some of his colleagues. And elaborate ways to rationalize the arrangement would be found in years to come.

"There has been a lot of talk that the Raketenflugplatz finally 'sold out to the Nazis,' " von Braun would one day write in an unpublished article.

> In 1932, however, when the die was cast, the Nazis were not yet in power, and to all of us Hitler was just another mountebank on the political stage. Our feelings toward the Army resembled those of the early aviation pioneers, who, in most countries, tried to milk the military purse for their own ends and who felt little moral scruples as to the possible future use of their brainchild. The issue in these discussions was merely how the golden cow could be milked most successfully.

That, as events would show, was somewhat glib.

Work on rocketry was taken so seriously by the Army in 1934 that the impoverished VfR was encouraged to shut down altogether while work at Kummersdorf-West was speeded up. If the government had valued the civilian operation, it could easily have kept it going. But it was the VfR's penchant for publicity that finally did it in, since the more serious the Army became about rocket research, the more the imperative of secrecy took hold. On April 6, Paul Josef Goebbels's Ministry of Propaganda issued a decree prohibiting all public discussion of rocketry that had to do with either the military or with technical details. Officially, there *was* no rocket program.

Meanwhile, Dornberger, von Braun, and a handful of others labored on their small rockets. Starting with VfR designs in October 1932, von Braun painstakingly built a series of water-cooled, alcohol–liquid oxygen motors that were endlessly modified according to whatever accident or shortcoming developed. They in turn led to the A-1 (for *Aggregat-1,* or Assembly-1), the drawings for which were ready by the summer of 1933.

Like *Protarchaeopteryx,* a theropod that is thought to be an ancestor of the first true bird, the A-1 spawned a series of rockets that led directly to the first ballistic missile and then to space launchers. And like that unfulfilled reptilian chicken, now mostly wrapped in the shroud of time, the A-1 itself could not quite get off the ground. Yet it marked the true beginning of modern rocketry.

It was clear by 1933 that the more powerful engines would have to go at the bottom of the launch vehicle rather than have their high velocity flame destroy the propellant tanks that fed them. So engines were built into the bottom ends of cylinders that held alcohol. The upper end supported a liquid-oxygen tank with separate containers that held the nitrogen that would force the propellants under pressure into the engine's combustion chamber. It was a measure of Dornberger's and von Braun's ceaseless work that in less than a year they developed an engine that could generate almost 650 pounds of thrust. Burn time was calculated at sixteen seconds. A gyroscope was mounted in the vehicle's nose for stability. An electric motor outside the A-1 would get the gyroscope spinning at a rate of nine thousand rotations a minute before launch. Then, like a child's top, it would hold its position during flight and therefore keep the A-1 on course. The whole vehicle was a little more than four feet high and weighed 330 pounds, giving it the required thrust-to-weight ratio of two to one. There was a series of bench tests, some of them promising, during the winter of 1933–34. But field tests were another matter. Three A-1s blew up before they got off the ground.

But the A-1's immediate successor, an A-2 named Max after one of the two precocious characters in the popular *Katzenjammer Kids* comic strip, streaked 5,000 feet above the North Sea island of Borkum on December 19, 1934. It carried its gyroscope between its fuel and oxidizer tanks, which moved the center of gravity back from the nose and therefore increased stability. Moritz, the other "kid," went up the following morning on a second, thrilling, letter-perfect flight.

The Army's rocketeers had secretly eclipsed the civilians and proven that long-range, liquid-propelled artillery rockets were feasible. Von Braun, whose emerging brilliance was in large part responsible for the success, had in the meantime transferred from engineering studies at the Technical University of Berlin to a doctoral program in applied physics at the University of Berlin. His work at Kummersdorf-West related directly to his dissertation, which he successfully defended before the A-2s flew. The document was so secret that even its title was classified. The deliberately vague reference to "combustion experiments" on his diploma meant everything. And it meant nothing.

The Rocket Team Evolves

Hitler in effect publicly ripped up the Versailles Treaty in March 1935. As he expected, the move was not challenged by the nations that were supposed to enforce its provisions. That done, he brought his country's growing economic strength to bear on its one glaring weakness: the military. Conscription was begun, warship construction was started, and the Luftwaffe was unveiled. This, of course, worked to the advantage of the rocket men. Repudiating the treaty was like smashing shackles. The suddenly exhilarated military establishment and its scientific and engineering support system could now come out of hiding and get to the real business of developing weapons that befitted what they believed was their country's inherent greatness.

There were several important developments within the rocket fraternity between 1935 and 1937. In January 1935 the Army and the still-nascent Luftwaffe agreed to develop rockets jointly. Following a visit to Kummersdorf-West that month, Wolfram von Richthofen, a cousin of Manfred and himself an ace in the Great War, became convinced that rocket propulsion would vastly increase aircraft performance. He saw the day when bombers would attack his country from altitudes above 33,000 feet and he believed that rocket-propelled interceptors would be the best way to shoot them down. That vision would lead to the first flight of a rocket-powered aircraft, the Heinkel HE-176, on June 30, 1939, and to the production of its successor, the Messerschmitt ME-163 Komet. Starting in 1944, the ME-163B, which looked like a cross between a barrel and a bat, would be used to attack waves of Allied bombers over the Third Reich, though without much effect.*

It was also obvious that artillery rockets were going to have to grow substantially if they were to be useful weapons. The technological basis for developing bigger missiles was in place. But by mid-1935, it was also clear that the rocketeers had outgrown Kummersdorf-West. There was the problem of security. Not only did testing rockets create a thunderous racket, but sending them flying above the treetops of a Berlin suburb was not conducive to secrecy. And there was an obvious safety problem: a runaway missile coming down in a populated area and going off like a bomb would be a nightmare. Finally, and most obviously, missiles that were big enough to fly long distances had to be tested over long distances. That could not be done over land as densely populated as mainland Germany. All of this made a compelling case for a design, development, and test facility sited in a remote, coastal area. That meant the Baltic.

Von Braun spent Christmas 1935 at his parents' farm in Silesia and related the news from Berlin. He also told them that his service and the Air Force were looking for a joint facility along the coast from which missiles could be fired on trajectories that went hundreds of miles. "Why don't you take a look

* The rocket plane would have dazzling performance, being able to climb to thirty thousand feet in just over three minutes. The 279 ME-163Bs that were built served on both fronts. Exploding engines and landing mishaps made them so dangerous to fly that more airmen were killed flying them than were killed in front of their guns. (Baker, *The Rocket,* pp. 260–61.)

By 1945, with Allied bomber formations infesting the skies over Germany day and night, matters would became so desperate that a disposable rocket interceptor was built. Called the Natter (Viper), it was supposed to be shot straight up from a mobile launcher, fire solid-propellant rockets in its nose into the underside of the attacking bombers like pellets from a shotgun, and then turn away so the pilot could bail out. The Luftwaffe knew that an attack by a Viper would be slipshod, but hoped that sending up salvoes of the cheap, mass-produced interceptors would decimate the attackers by sheer weight of numbers. They never saw combat. A Viper was tested on March 1, 1945, by a lieutenant named Sieber, who was knocked unconscious after launch when the plane's canopy flew off. He was killed moments later in a fiery crash. (Sparks, *Winged Rocketry,* pp. 50–62.)

at Peenemünde?" Baroness von Braun asked her son. "Your grandfather used to go duck hunting up there." The place she had in mind was a wilderness area of marshes, dunes, and forests at the northern end of the island of Usedom, due east of Rostock and a little more than a hundred miles north of the capital. That suited Wernher and his colleagues just fine.

By mid-January of the new year the sprawling area at the end of Usedom had been divided (at least on paper) into an Army facility called Peenemünde-East and an Air Force installation, Peenemünde-West. The Air Ministry was glad to put up 5 million reichsmarks for clearing and construction. Becker, now a general, thought he saw an attempt by the Luftwaffe to get early control of the whole place and angrily pledged 6 million marks on behalf of the Army. But he also quietly complained to Dornberger that their intensely conservative service was inherently more careful—too careful—about spending money than the cocky and aggressive airmen. "If you want more money," he warned, "you have to prove that your rocket is of military value."

So that March, Dornberger invited the commander in chief of the Army, Major General Werner von Fritsch, to visit Kummersdorf-West, which was still the proving ground, for a carefully orchestrated briefing on the rocket's "unexcelled potential" as an artillery weapon. The monocled officer was then escorted to the test stands, where 660-, 2,200-, and 3,300-pound-thrust motors were run up in an ear-splitting finale. When the thunder ended and the smoke cleared, Fritsch is reported to have asked only one important question: "How much do you want?"

Now a synergism was taking hold as part of a fundamental role reversal. The rocket builders had spent years unsuccessfully begging for money from a poor, distracted, and uninterested government. Suddenly, the government was finally committing to rocket research "big time" (as von Braun put it). A heavy investment would be made: more than $70 million on Peenemünde-East alone by August 1939, when the "Army Research Center" was declared fully operational with a staff of nearly two thousand scientists and technicians. But that meant it was put-up-or-shut-up time. The dream facility had to produce a very long range artillery missile that could be turned out in large quantities. The consequences of failure were almost unimaginable.

The A-2 being an unqualified success, Dornberger, von Braun, an engineer named Walter Riedel (no relation to Klaus), and the others began thinking about its successors. The first, the A-3, would have the 3,300-pound-thrust engine and an advanced three-axis gyroscope guidance and control system. This represented a giant leap beyond its predecessor. The A-3's long, narrow fins provided "arrow stability," making certain that its nose kept swinging back into the airflow—in the right direction—however turbulent the air was. The A-3 was tested three times in early December 1937, each of them a partial failure because the parachutes that were supposed to float them back down after their engines burned out released prematurely. The A-4, whose colossally powerful 50,000-pound-thrust engine was then already in its design phase, would be a very different matter. Very different indeed.

Goddard in the High Lonesome

The Third Reich's largess was not matched in the United States, where the rocket builders were forever scrounging for private contributions. One day in the spring of 1930, for example, a handsome twenty-eight-year-old gentleman called on Daniel Guggenheim at his Hempstead, Long Island, mansion. The multimillionaire was known for his philanthropy, and the ruddy-faced young man had therefore come to ask for a substantial contribution. The two men were acquaintances who respected each other, and the elder of the two—Guggenheim was seventy-three—knew that the younger had not come to request the money for himself.

"You believe these rockets have a real future?" Guggenheim wanted to know.

"Probably. Of course one is never certain."

"But you think so. And this professor of yours, he seems capable?"

"As far as I can tell, he knows more about rockets than anybody else in this country."

"How much money does he need?"

"For a four-year project, he would need twenty-five thousand dollars a year."

"Do you think it's worth my investing a hundred thousand?"

"Well, of course it's taking a chance. But if we're ever going beyond airplanes and propellers, we'll probably have to go to rockets," the young man answered, staring into the flames in the living room's large hearth. "Yes," he said at last, "I think it's worth it."

The professor was Robert H. Goddard. His champion was Colonel Charles A. Lindbergh, the hero who three years earlier had flown the *Spirit of St. Louis* alone across the Atlantic to Paris. The Lone Eagle had met Goddard in his office at Clark in November 1929 and been impressed by his work. Understandably, Goddard had in turn complimented Lindbergh for his vision. Then he spelled out a crucial fact of life for the well-connected young hero: research cost money. Having had to pass his own hat among St. Louis businessmen for the Ryan monoplane that he flew to immortality, Lindbergh understood the problem and sympathized with Goddard. He had been further impressed when they met again at a conference at the Carnegie Institution in Washington on December 10, 1929. Carnegie itself had sent $5,000 to the Smithsonian for

Goddard's use. The celebrated aviator, already convinced that airplanes had about reached their speed limit, had become increasingly captivated by Goddard's description of his work and its challenges.* The two men, each a shy but resolute visionary, developed a deep mutual respect.

Goddard consulted a meteorologist at Clark about the best place in the country to fire rockets year-round. Then he and Esther struck out for southeastern New Mexico's warm, dry climate on July 15, 1930, with Guggenheim's first installment tucked in his pocket. Four loyal assistants and their families would follow, filling an entire freight car with personal possessions, Goddard's own launch tower and other equipment, and hardware that had been liberated from Clark's physics shop. By the end of the summer, he had rented a pueblo-style ranch house and eight acres of land from Miss Effie Olds, a spinster whose family made automobiles someplace in Michigan. Mescalero Ranch, as the place was called, was at the end of a secluded road three miles from Roswell. That would insure privacy. At a crossroads village they had passed on the way to Roswell, in fact, the Goddards saw a sign hanging on rusted hooks at an abandoned gas station: "High Lonesome." The name would prove to be a metaphor for Robert Goddard's remaining years.

While Goddard was in no sense greedy, neither did he fit the caricature of the detached scientist who suffered poverty in grim resignation as the price of doing what he loved. To the contrary, he thought about money a great deal and hammered out contracts like a lawyer. He would take out forty-eight patents during his lifetime and be issued thirty-five others after his death. The first was issued as early as July 1914. Goddard was as forthright about being paid for his work as he was secretive about its substance. "The Bureau of Ordnance is to pay all development costs, such as drafting, materials and labor, manufacturing of test samples, and testing," he wrote to an admiral in 1920 about a shipboard gun he designed to fire depth charges at submarines. "During the period of development work under this contract . . . the Department [of the Navy] will pay me the sum of one hundred dollars monthly. During such times as I may be called upon by the Bureau to perform service exclusively for the Bureau . . . a further sum of fifteen dollars per diem will be paid, together with reasonable travel expenses exclusive of living expenses while traveling, and while engaged in such work."

* That year's prestigious Schneider Cup was won by a Supermarine S.6 floatplane that reached 328.63 miles an hour. It was designed by R. J. Mitchell, who would later design the Spitfire. In any case, Lindbergh did not seem to have anticipated jets even though a great deal of theoretical work had already been done on the subject in Hungary, Germany, Italy, and elsewhere. Goddard himself applied for a patent on a "thermal air jet-propulsion system" in February 1931. (Gohlke [*sic*], "Thermal-Air Jet-Propulsion," *Astronautics,* May 1942, p. 12.)

By 1935, Goddard had gradually turned into a recluse whose experiments were done in virtual isolation. This worried his friend Lindbergh, who warned that if he didn't put at least some time into self-promotion, his funding would dry up and there would be other unpleasant consequences. The colonel suggested that he publish something about his research on liquid rockets.

The result was *Liquid-Propellant Rocket Development,* the second and last of Goddard's great papers. The monograph, which was published by the Smithsonian in March 1936, was a concise description—a virtual handbook—of his advanced experiments. It was also a clear claim to an achievement made a decade earlier for which he stubbornly believed he had been given too little recognition. "The first flight of a liquid oxygen–gasoline rocket was obtained on March 16, 1926, in Auburn, Massachusetts, and was reported to the Smithsonian Institution May 5, 1926. This rocket is shown in the frame from which it was fired. . . . Pressure was produced initially by an outside pressure tank." The narrative and accompanying equations would be read with great interest at Kummersdorf-West and elsewhere.

By 1935, Goddard was testing large rockets that looked like rockets were supposed to look. His A Series was the first to demonstrate self-stabilization using a gyroscope. (He disdained naming rockets, at least at that time, because he thought it was overly dramatic and led to the sort of ridicule that had severely embarrassed him sixteen years earlier.) On July 12 one of the sleek machines achieved "beautiful stabilization," he wrote to Harry F. Guggenheim, his new benefactor. "The rocket corrected itself, moving each side of the vertical, during the entire period of propulsion." It also reached an altitude of about a mile and a quarter. Reaching higher altitudes would require incremental advances—higher-thrust engines and more powerful pumps, for example—not breakthroughs.

The Suicide Club

Meanwhile, Goddard's publications were being pored over in Pasadena, California, among other places. In February 1936, with the concept of real rocketry spreading, a graduate student at the Guggenheim Aeronautical Laboratory of the California Institute of Technology, or GALCIT, and some friends decided to design their own high-altitude liquid or solid sounding rocket. Frank J. Malina and the others carefully reviewed Tsiolkovsky's, Goddard's, Oberth's, Esnault-Pelterie's, and particularly Sänger's work on high-exhaust velocities. And as they read, they became obsessed by the possibility of building a rocket that would go really high. So Malina went to one of his professors and proposed that he, Malina, be allowed to write his dissertation on propelling sounding rockets to extreme altitudes. Clark B. Millikan's response was to advise

Malina to go out and get a good job in the aircraft industry, which was bustling, and not waste his time on harebrained ideas. GALCIT was renowned for the quality of its aeronautical research, but it had no interest whatsoever in rockets; the very word was taken to be synonymous with pulp fiction, war with Mars, and other frivolous entertainment.

War with Mars

Mars itself was an enduring talisman for the rocket crowd; a totem that focused the public imagination on an exotic but as yet unreachable place and therefore provided a little respectability for space travel's evangelists. Lowell's stubbornly held belief that there were canals on Mars dug by a dying civilization continued to be politely ignored or refuted by other astronomers. Yet the publicity he had generated at the turn of the century had an inertia that worked like an amorphous mucilage holding the concept of space travel together and validating it. Lowell's ghostly gospel was still out there, waiting to hatch in the collective psyche like alien larvae. Mars had delicious secrets, probably frightening ones, that were an elixir for the imagination.

And so it happened that on the evening of Sunday, October 30, 1938, a brilliant and imaginative twenty-two-year-old actor and director named Orson Welles fired the psyches of millions of radio listeners and gained instant fame in a way that even he had not been able to anticipate. Between eight and nine that evening—prime time—he staged an adaptation of *The War of the Worlds* on the Columbia Broadcasting System's *Mercury Theatre on the Air.* Listeners were told at the beginning of the program that they were about to hear a rendering of H. G. Wells's fiction classic. But countless thousands had tuned their dials to NBC's *Chase & Sanborn Hour* hoping to hear the popular Edgar Bergen and his two delightful dummies: the tuxedo-clad, wise-cracking Charlie McCarthy and a dull-witted bumpkin named Mortimer Snerd. Instead they got a guest singer who was not very good.

Meanwhile, on CBS, the actors stood around the big boxy microphones and read their roles. A weather report was followed by dance music that purportedly came from Ramon Raquello and his orchestra playing in the Meridian Room of New York's Hotel Park Plaza. Suddenly, one of the songs was interrupted by a bulletin—common enough in those days—saying that a professor at an Illinois observatory had reported seeing explosions of incandescent gas going off at regular intervals on Mars. Anything that went off at regular intervals hinted of a planned cadence, or sequence, which suggested intelligence and could mark the start of an invasion. Raquello's music was soon interrupted again, this time by a Professor Pierson (played by Welles) who described Mars and, then again, by news that a flaming object had just crashed on a farm at Grovers Mill, New Jersey.

At that point, the remote from the Meridian Room was seemingly abandoned. The action moved to rural New Jersey, where an apparently horrified reporter, his voice cracking, stood in a darkened field and gave a hair-raising description of what had come to Earth. It was about then that Bergen's frustrated fans switched to CBS. This is what they heard:

> Ladies and gentlemen, this is the most terrifying thing I have ever witnessed—wait a minute! Someone's crawling out of the hollow top—someone or . . . *something.* I can see peering out of that black hole two luminous discs. Maybe eyes, might be a face. . . . Good heavens, something's wriggling out of the shadow like a gray snake. Now it's another one. . . . They look like tentacles. . . . I can see the thing's body now—it's large as a bear. It glistens like wet leather. But that face, it's . . . it's . . . ladies and gentlemen, it's indescribable; I can hardly force myself to keep looking at it, it's so awful. The eyes are black and gleam like a serpent; the mouth is . . . kind of V-shaped, with saliva dripping from its rimless lips that seem to quiver and pulse. . . . Wait! Something's happened! (Hissing sound followed by a humming that increases in intensity.) A humped shape is rising out of the pit. I can make out a small beam of light against a mirror. What's that? There's a jet of flame springing from that mirror and it leaps right at the advancing men. It strikes them head on. Good Lord, they're turning into flame.

The "Martians" (who had had to make a 35-million-mile journey at near–light speed to hold the story together) then neatly incinerated all but 127 of 7,000 armed soldiers. The reporter himself went off the air so suddenly that there could be no doubt he had been either zapped by a death ray, immolated by a fire mirror, or instantly asphyxiated by poison gas. "Flash" bulletins followed in rapid succession, describing the vanguard of the extraterrestrial invasion force coming down in an armada of heavy pods. Central New Jersey was quickly occupied. A voice sounding like President Franklin D. Roosevelt's issued emergency instructions, while an announcer reported that communication lines were down from Pennsylvania to the Atlantic, railroad service was interrupted, and highways were clogged with families trying to escape the Martian terror. CBS said three times that what was coming over the airwaves was only a "play," but many of the 3 million to 6 million people in North America who were sitting by their wooden radios either did not hear it or were too panic-stricken to understand.

"In Newark, in a single block at Heddon Terrace and Hawthorne Avenue, more than twenty families rushed out of their houses with wet handkerchiefs and towels over their faces to flee from what they believed was to be a gas raid," *The New York Times* would report the next morning. Hysterical New Yorkers ran to police stations for evacuation instructions. Rhode Islanders demanded that their state's electricity be turned off so as not to attract the invaders. In Pittsburgh, a man barely prevented his wife from taking poison. People ran to church to pray for their souls.

Welles, who combined formidable acting talent with an instinctive genius for self-promotion, staged a press conference the next day at which he made believe he was sorry for having caused the most frightening night in the history of radio. "Of course, we are deeply shocked and deeply regretful about the result of last night's broadcast," the thoroughly disingenuous wide-eyed wonder, his brow furrowed, told a crowd of reporters. "It rather came as a great surprise to us."

The Martian invasion sent Welles to Hollywood, where he would soon create *Citizen Kane,* the controversial classic film about William Randolph Hearst, and provoked a short investigation by the Federal Communications Commission. It also no doubt further proved to the aeronautical engineers at GALCIT and elsewhere that the rocket freaks were crackpots. Yet ultimately *The War of the Worlds* in its several reincarnations, including a film, helped to drive serious science because it made Mars alluring to a public that would ultimately be called upon to fund the expeditions that went there.

The Suicide Club Finds a Godfather

Meanwhile, Malina, undeterred, appealed to Theodor von Karman, the urbane Hungarian expatriate and world-famous aerodynamicist, for help in building his high-altitude rocket. Von Karman had been lured to Caltech in 1926 by its president, the Nobel Prize–winning physicist Robert A. Millikan. Robert also happened to be the father of Clark, who had advised young Malina to build airplanes. Von Karman instantly responded to the enthusiasm of Malina and his friends John W. Parsons and Edward Forman and promised support. Parsons dabbled in black magic and the occult, was a self-taught chemist with an encyclopedic knowledge of explosives, and was a driven rocketeer. Forman was an expert mechanic and a dedicated rocket builder, too. Not only were they not Caltech students, however, but neither of them had ever even gone to college. Yet von Karman was moved as he listened to them eagerly explain how they had corresponded with their German counterparts (most likely at the Raketenflugplatz) and brag about the number of pockmarks caused by rocket explosions in their backyards. Before long, the three of them were joined by two other honest-to-goodness Caltech students: Apollo Milton Olin Smith and Hsue-shen Tsien, a Chinese-born research engineer and a brilliant and tenacious mathematician.

After a few successful engine tests in a dry riverbed in the nearby Arroyo Seco, the members of the Rocket Research Project, as they now called it, asked for and received laboratory space on the quiet Pasadena campus. Their inaugural experiment, however, ended no differently than everyone else's. The engine misfired, causing a loud explosion and filling the building with a putrid odor from methyl alcohol and nitrogen tetroxide. A follow-up experiment produced two more powerful explosions, jarring mechanical equipment and the nerves of the more sedate aeronautical engineers. Disdainful to begin with, the regular residents now named the newcomers the "Suicide Club" and banished them back to the Arroyo Seco.

Whatever Clark Millikan's initial reaction to the Rocket Research Project, his father seems to have liked the idea and, with von Karman, supported it. It happened that the elder Millikan was on Guggenheim's board of advisers and therefore evaluated Goddard's work for the foundation. So early in August 1936, within six months of the start of the Caltech rocket project, Robert Millikan wrote to Goddard saying that Malina would be vacationing with his parents in Texas and would appreciate the opportunity of stopping in Roswell for a meeting. "He is an able and scientifically minded chap, of good judgment," the president of Caltech said, adding that Malina "would not abuse any confidence that you may repose in him."

The previous year, Goddard had tried to help the Pasadenians solve minor challenges. Parsons had written to him on September 19, 1935—the day Tsiolkovsky died in Kaluga at the age of seventy-eight—to ask for help with a solid-propellant problem and was answered a week and a half later.

But Goddard drew a sharp distinction between sharing what he considered to be basic data and giving away technical secrets. Millikan's assurance therefore meant little or nothing to him. Malina arrived in Roswell on Sunday, August 30, and was taken to dinner by the Goddards, who, he would remember, treated him cordially. But the rocket research might as well have been classified for all Goddard wanted to talk about it. In a manner of speaking, in fact, it *was* classified.

The young Caltech student was given a tour of Goddard's machine shop the next morning, was driven to the launch site, and had a general discussion with his host that left out all work done after the publication of *Liquid-Propellant Rocket Development,* which he had already read. He was also shown the sixteen-year-old *Times* column. It was clear to Frank Malina that Robert Goddard despised the news media. As he left, he was offered a job after graduation, but it was also clear that, in his words, the forty-seven-year-old physicist "felt that rockets were his private preserve, so that any others working on them took on the aspect of intruders."

The day after Malina left, Goddard wrote to Millikan to explain what had happened: "I wish I could have been of greater assistance to him, but it happens that the subject of his work, namely the development of an oxygen-gasoline rocket motor, has been one of the chief problems of my own research work, and I naturally cannot turn over the results of many years of investigation, still incomplete, for use as a student's thesis."

Goddard had a point. Yet his unwillingness to share the fruits of his advanced research with Malina, or with anybody else for that matter, deeply angered von Karman because it ran counter to the scientific gospel that says sharing research purifies and strengthens the process of discovery. Airing research results, and therefore allowing others to repeat the experiments for themselves—to "replicate" them, in the jargon of science—is supposed to separate what is wrong from what is right the way nuggets are separated from sand in panning for gold. Withholding research results therefore subverts the process.

Malina's distinguished mentor and protector would therefore make his own feelings known about the secretive scientist in an autobiography that appeared twenty-two years after Goddard's death. "Goddard, it was said, was the first man to fire a liquid rocket," von Karman would write with undisguised contempt. "However interesting this is historically, I have never equated the characteristic of being first with being the greatest. . . . Naturally we at Cal Tech wanted as much information as we could get from Goddard for our mutual benefit. But Goddard believed in secrecy. . . . The trouble with secrecy is that one can easily go in the wrong direction and never know it." The recipient of his adopted country's first National Science Medal, presented by John F. Kennedy in 1963, went on to explain how Goddard's obsessive secrecy caused him to waste three or four years developing a sophisticated, but unnecessary, gyroscope for a sounding rocket.

He then recounted a meeting with Guggenheim, Lindbergh, Clark Millikan (who had by then changed his mind about rockets), Goddard, and himself in 1938 at which the rocket's use as a weapon was discussed. Goddard, he wrote, refused to share complete information about one of his own rockets. "There is no direct line from Goddard to present-day rocketry," Theodor von Karman wrote caustically. "He is on a branch that died."

There was much to what he said. Goddard's achievements were substantial. But they were accomplished in as sure a vacuum as the one he wanted his rockets to penetrate. The real core of German and Russian research was indigenous and state-sponsored. That first firing of a liquid-fuel rocket in 1926 remained largely unknown to outsiders for years; *Liquid-Propellant Rocket Development* did not contain the essential specifics that would have helped either the Germans or the Russians, though it was certainly useful to them for checking their own experiments. His many patents were unavailable to them.

But it was a different story at home. Goddard's place in history is as enigmatic as he was. It is defined by subtleties that are not clearly evolutionary but that are nonetheless valid. Irrespective of his neurotic secrecy, what did get out—*A Method of Reaching Extreme Altitudes* and *Liquid-Propellant Rocket*

Development, for example—showed the dedicated few that their tribe had a future in space. He was the inspiration and guiding light for a generation of American rocket builders in clubs and societies from coast to coast. The studiously read *Bulletin* of the American Interplanetary Society and its successor organization carried stimulating articles about Goddard's work and occasionally even a letter from him. Perhaps more important, the few outsiders with whom he did share his work—Harry Guggenheim, Charles Lindbergh, and longtime Smithsonian Secretary Charles G. Abbot, to name a few—were so captivated by it that they in turn influenced the nation's space program out of all proportion to their number.

Nor was Frank Malina unaffected by Goddard. Goddard's achievements, as well as Sänger's, motivated him enough to want to go to Roswell in the first place. Who can say what subliminal effect talking with the great man and being on his premises had on the impressionable young student? Within two years, Malina was studying the thermodynamic characteristics of rocket engines with his pals and testing the results in the Arroyo Seco almost within sight of the Rose Bowl. Six years after that, they founded Caltech's Jet Propulsion Laboratory, which turned out rockets for war and peace and then evolved into the world's leader in solar system exploration.

Weapons from Roswell

Meanwhile, Goddard himself tried to help the Army Air Corps and the Navy wage war against the foreigners who were turning into his political as well as his professional rivals. (In common with them, though, he continued to lobby for lucrative government contracts.) He tried unsuccessfully to interest the Army in long- and short-range missiles and in a rocket-propelled aircraft. "Except for the possible use for planes, there appeared to be no particular interest in rockets or rocket propulsion," he wrote after a fruitless sales meeting at the War Department on May 28, 1940. "The Army Ordnance representative stated that he considered that rockets would not be useful for Army Ordnance, and that if any development work was to be carried out, it should be in the improvement of trench mortars."

However eminent Goddard was, his eccentricities and secrecy were beginning to haunt him, just as Lindbergh had warned. Neither Goddard nor anyone else could contain the inventive process, so rocketry was beginning to blossom at Caltech and elsewhere. He was particularly piqued when he learned that the Caltech group had gotten an Air Corps contract to develop Jet Assisted Take Off. Its name notwithstanding, JATO was an externally mounted solid-propellant rocket system that shoved planes off land or water faster than they could get off on their own. It was the same basic system that Korolyov had developed in the gulag.

"I feel that every effort should be made to put my work, turned to national defense lines, on as powerful a footing as possible, but feel that there are forces in existence which may make it difficult," he complained to Abbot following the meeting of May 28. "The Cal. Tech situation was clarified when I learned that the Army Air Corps had, a year or so ago, given a list of desirable researches to the National Research Council to be turned over by them to the best agencies for obtaining results. One problem was given to the Massachusetts Institute of Technology, and jet propulsion was given to the California Institute of Technology. Apparently," Goddard added testily, "my experience and years of work, including early sponsorship by the Smithsonian Institution, were given no consideration by the National Research Council, or at least by those who had charge of this arrangement."

Goddard had worked on the "rocket application to aircraft" as far back as the 1920s. But he was afraid his discoveries would be stolen and that he would be denied the fame and fortune he believed were rightfully his. He had taken the trouble to brief Henry du Pont and three of his company's scientists on his research in the presence of Lindbergh in November 1929, for example, but was met with a silence that fed his paranoia. "I realized soon that the object of this questioning was not so much to determine what could be done on airplanes as to find every last detail of the rocket I have developed during the last nine years," he wrote to Abbot,

> and after I saw this I evaded further questions as to these constructional details as much as possible. Even at that, I said more than I wish I had. . . . The affair gave me a decidedly unpleasant impression. Nothing was said about the desirability or undesirability of supporting any of my work. Several of those present, including the Colonel, talking the thing over out of my hearing, and the conference ended. In the course of the conversation, one of the assistants let drop the remark that a knowledge of what I had done would "save them a lot of grief" in the development work.

Goddard finally won a joint contract with the Navy's Bureau of Aeronautics and the Army to develop liquid-propelled JATO for patrol planes. That might have calmed him about Caltech. But then another Army contract went to von Karman and his young charges. On August 6, 1941, Homer A. Boushey, an Air Corps lieutenant and one of von Karman's students, successfully tested JATO for the first time in a single-engine Ercoupe light plane. Besides being a fine flier, the thirty-one-year-old Boushey was an aspiring inventor, a serious student of aeronautics, and a dedicated reader of Verne, Wells, and the space fiction magazines of Hugo Gernsback and others. He had also read *A Method of Reaching Extreme Altitudes* and idolized Goddard. Boushey's response to an offhand invitation to visit Mescalero Ranch in the summer of 1940 was therefore to fly there in a pursuit plane. The eager young aviator somehow penetrated Goddard's reserve on that first visit and was invited back a few weeks later for what turned out to be an astonishing experience.

Goddard led Boushey to the machine shop, pulled a cloth shroud off "Nell," an experimental rocket that lay on two sawhorses, and then described how the rocket worked. (He had evidently changed his mind about naming rockets.) He shared details about Nell's pumps, turbines, gas generators, and gyroscopes with the awestruck flier, who would note appreciatively that there were "exquisite hand-tooled parts." After dinner the Goddards and their guest went out to the veranda to take the night air. It was a warm, starry evening and there was a lightning storm on the horizon; a perfect setting in which to talk about the vision that Goddard shared with von Braun and Korolyov and which he kept as carefully hidden as they did.

"I remember how hesitant I was about mentioning interplanetary flight," Boushey would recall. "The doctor exchanged glances with Esther and then went on to speak not just of travel to the planets, but of interstellar flight as well. He spoke as if it were merely a matter of work and experiments. . . . He wasn't sure how long it would take for people to understand and support his work, but he felt they would do so in time." Speaking quietly, Goddard drifted into a reverie about humanity's need to explore the cosmos, adding that in the distant future the Sun and this system's planets could grow too cold to support life.

Sounding eerily like Tsiolkovsky, Goddard told the lieutenant that the time would come when Earth people would conquer their "inertia" and become universal explorers, spreading their culture to other solar systems. And the means of doing so, he added, would not necessarily require chemical rockets. The enthralled aviator would later describe that night's conversation in detail. "Future space craft, he said, might be equipped with a very light, mirror-like sail. It would be opened up in the vacuum of space like a parasol, and there would be nothing to damage it. It would lock on to the sun's rays, and the energy it received might operate a small boiler which would emit a jet of steam, perhaps, and the ship would sail on and on." Goddard was seeing the same ship Kepler, Tsiolkovsky, Tikhonravov, von Braun, Korolyov, and the others saw. He shared their religion.

Homer Boushey would go on to become a lifelong Goddard missionary and a brigadier general in the Air Force. He would also become an early, vocal proponent of an American base on the Moon that would function in an artificial environment and be equipped with missiles that could strike any nation that attacked the United States. Goddard died of cancer on August 10, 1945, and was buried at the family plot in Worcester on the fourteenth. That was the day World War II ended and, therefore, the day the cold war began.

Vengeance Weapon Number 2

In the meantime, Germany had bankrolled its own rocket program, and the results were showing. Reflecting on the creation of the A-4, Walter Dornberger would tell American interrogators after World War II that the challenges "were similar to those which would have faced Wilbur Wright if one had demanded of him in 1902 that he build in three years a completely automatic [B-17] Flying Fortress," the American heavy bomber that carried the air war to Germany. Dornberger, by then a general, was engaging in a bit of hyperbole. Yet it is hard to overstate the A-4's place in history, or that of the supremely innovative research and development team that produced it. They and their A-4s were destined to become the river of technology that fed the two rival ballistic missile and space programs that grew out of the war their own country was destined to lose.

The world's first liquid-propelled ballistic missile and the true ancestor of all space launch vehicles was born out of the marriage of heavy funding and creative brilliance. The project was given a top priority by the Army High Command, which meant that the Ballistics and Munitions Branch had a serious financial commitment from its own service. The Army missile was also nurtured by a partnership with the Luftwaffe. Hitler remained indifferent to it until his bombers were shot down in large numbers by the Royal Air Force during the Battle of Britain in the summer and autumn of 1940. He then began to consider other ways of punishing the British, particularly after Britain's Bomber Command and the American Eighth Air Force started pummeling the Third Reich day and night in 1943. But the missile had strong early support from Fritsch and then from Field Marshal Walther von Brauchitsch, who replaced him on Hitler's orders in February 1938.* Dornberger insisted on having all A-4 work done "under one roof," which stimulated creativity among designers who could freely communicate with each other. It also made the operation more efficient, since engineers could leave their drawing boards and go right into laboratories or workshops to see firsthand how their drawings were being bent into metal (as the saying goes).†

* Hitler made his only appearance at a rocket facility on a wet, chilly day in March 1939, when he visited Kummersdorf. The Führer got the full treatment, including a run-up of the 660-pound-thrust engine, which made him grimace even though his ears were packed with cotton. Later, after finishing a vegetarian dinner and mineral water in the officers' dining room while Dornberger kept pitching, he bade his hosts good-bye with an ambiguous remark: "*Es was doch gewaltig!*" ("Well, it was grand!") Hermann Göring, on the other hand, became ecstatic when he was given the same treatment at Peenemünde five years later. "This is colossal," the rotund reichsmarshal proclaimed, bear hugging Dornberger. "We must fire one at the first post-war Nuremberg Party rally!" (Ordway and Sharpe, *The Rocket Team*, p. 27.)

† The identical technique was used by Lockheed's Advanced Development Company, popularly known as the Skunk Works, after the war to create the U-2 and SR-71 reconnaissance aircraft and the F-117 Nighthawk stealth fighter-bomber.

The Raketenflugplatz's irregulars were supplanted by a legion of scientists, engineers, and draftsmen from universities and technical institutes, and direct contact with those institutions was begun. The men who designed Peenemünde's rockets would no longer fly by the seats of their pants and neither would their weapons. There were specialists to design everything from pumps to radio receivers. By the summer of 1939, for example, Peenemünde could boast (though it would not have dared) a world-class aerodynamics institute with a staff of sixty; within four years, the number would grow to two hundred.

Politics was thornier. Von Braun was right when he explained that aviation pioneers courted the military for funding. They included the Wright brothers, Glenn Curtiss, and a slew of subsequent American and European aircraft designers and manufacturers who depended on war departments and ministries to finance their projects. So did other rocketeers, including Goddard, Tsander, and Korolyov. Certainly Goddard invested plenty of time in demonstrating his black-powder rockets, some suitable for use as trench mortars and what would now be called bazookas, for the Army at the Aberdeen Proving Ground in Maryland as early as October and November 1918.

What he left out, however, was his own and his colleagues' ambiguous relationship with the Nazis themselves. In 1937—the year the twenty-five-year-old prodigy was chosen to play a key role at the world's first full-blown rocket facility at Peenemünde—he was officially ordered to join the Nazi Party and readily did so. Far from being alone, he was only one of many on the rocket team and elsewhere who entered the Führer's political fold. Some, such as Rudolf Hermann, Anton Beier, Ernst Steinhoff, and Arthur Rudolph, were ideologically dedicated Nazis. "We didn't want to build weapons; we wanted to go to space," Arthur Rudolph would say years later. "Building weapons was a stepping stone. What else was there to do but join the War Department? Elsewhere there was no money." If Rudolph was only a reluctant participant in the development and production of rocket weapons, however, he was able to overcome his misgivings with remarkable fervor. He joined the Nazi Party in mid-1931—two years before Hitler even came to power and at a time when the "party" was full of violent thugs—supposedly because he was anti-Communist. And he would spend part of the war as the head of production at a concentration camp called Dora, where an estimated thirty thousand slave laborers perished—many in ritualistic public hangings—for such crimes as making spoons while building ballistic missiles for the Third Reich.

But the evidence indicates that most of the German rocketeers, like their Soviet counterparts who joined the Communist Party, were ideologically apathetic. Their party cards were tickets to work and to social acceptance in an otherwise severely restrictive society. The cards put them safely on the inside in a place and at a time when being on the outside could be abidingly dangerous. It was easier to look at their drawing boards than inside their souls. Nor was this unique to the rocket establishment. Herbert von Karajan, the celebrated leader of the Berlin Philharmonic, joined the Nazi Party and frequently played the martial music associated with it. Ferdinand "Ferry" Porsche, the automobile designer and industrialist, joined in 1937 because Hitler enthusiastically supported the idea of a mass-produced "people's car," eventually called the Volkswagen, and it was therefore good for business. Porsche had no apparent qualms about using Hungarian Jewish slave laborers and Russian prisoners on his production line, a book detailing the history of the company reported in 1996, even though he himself seemed indifferent to Nazism. "He walked through the crimes like a sleepwalker," according to the author.

Dornberger himself never joined the party, though he claimed to believe in its goals, insisting that his sole aim was to provide Hitler with enough guided missiles to win the war. Leo Zanssen, the commandant of what was to become the Army facility at Peenemünde, represented a singular minority at the other extreme. He was an ardent Catholic who turned against the Nazis after serving on the eastern front and remained openly hostile to them throughout most of the war. The fact that he and Dornberger kept their positions until 1945 indicates that one did not have to be a Nazi to survive at Peenemünde. It is a tribute to Zanssen that he was despised by Schutzstaffel (SS) General Hans Kammler, a particularly vicious killer who built Auschwitz and who would dominate the Army rocket program by the end of the war.

Von Braun also joined the SS in 1940. The organization was formed in 1928 from a gang of fanatical pro-Nazi teenagers and grew to become the nation's ubiquitous political enforcer, the operator of the concentration camps, and a close ally of the Gestapo. Von Braun later claimed that he enlisted in the murderous group on orders from its chief, Heinrich Himmler, and with Dornberger's blessing. "He [Dornberger] informed me that the SS had for a long time been trying to get their 'fingers in the pie' of the rocket work. I asked him what to do. He replied on the spot that if I wanted to continue our mutual work, I had no alternative but to join." Himmler did in fact want to discredit Dornberger and take control of the rocket program, though that was not until 1943. Von Braun went into the SS as a lieutenant and would advance to the rank of major. There is nothing to say that the promotion was not strictly honorific, particularly since he was working on his rockets with total dedication and would have had little time and less inclination for politics. "Ferry" Porsche, among other notables, also "accepted" Himmler's offer to become an honorary officer in the murderous organization because, as he later said, "there was no way I could refuse."

It is hard for an outsider, and particularly one looking back decades later, to fathom why von Braun did not believe that his position as technical director of the ballistic missile program made the need to join the party and the SS unnecessary. Yet there were two facts he and Dornberger would have known that put them both in a potentially perilous position. Whatever Himmler and his henchmen thought about their ability, there was no evidence that he believed they were irreplaceable, since he had no technical grasp of rocketry. Worse, the SS was stocked with sadistic killers whose collective reputation rested, as did Beria's NKVD, on terror that could be irrational. Zanssen would have been purged and probably worse had Dornberger not intervened. And von Braun himself was threatened with death for his interest in space and spent two weeks in jail in March 1944 after being arrested by the Gestapo on fabricated charges that he associated with Communists and was working too slowly on the ballistic missile program. Himmler was pressuring him to turn the program over to the SS, but he refused to do so. Dornberger got von Braun out of Himmler's clutches, too. It is hard to believe that the missile program's guiding genius—and a Junker aristocrat at that—would have been harmed or killed for keeping his distance from the Nazis. But, as was the case in the Soviet Union, reason was overwhelmed by ferocious, often irrational, force. Von Braun had every reason to believe that his position insured his safety. Yet state-sanctioned murder was in the air, so the price of miscalculation could have been terrible.

At any rate, Peenemünde's nucleus consisted of first-rate designers and administrators. Von Braun was by all accounts a charismatic leader whose charm, tact, and manners almost, but not quite, concealed a will of steel. He was also a gifted draftsman, an excellent engineer, and a brilliant conceptualist. Walter Thiel, an outstanding chemical engineer, headed propulsion development, which for the most part meant producing the A-4's daunting engine. He was the prototypical German university professor: pale, intense, authoritarian, arrogant, and apt to become depressed under stress. Thiel was too absorbed in his work to think much about politics and, unlike many of his colleagues, also seems to have successfully dodged the Nazi Party. Krafft Ehricke, a former panzer captain who had been wounded at Dunkirk and who fought in the titanic Battle of Moscow, was one of Thiel's brightest propulsion engineers. Rudolf Hermann, an energetic and talented aerodynamicist, ran the wind tunnel in which the missile's flight characteristics were tested on models shaped like bullets with fins. Hermann Steuding of the Technical University of Darmstadt made fundamental discoveries in guidance theory. It was no good sending a missile hundreds of miles, after all, if it couldn't hit its target. Their ghosts would ride every rocket that followed the A-4 into the space age.

The only size restriction placed on the missiles was that they had to be able to get through railroad tunnels strapped to the flatcars that would take them to their launch sites. The A-4 stood forty-six feet high and weighed fourteen tons loaded. Its main section consisted of two propellant tanks set in tandem above the engine. The upper tank held 9,180 pounds of fuel, a mixture of ethyl alcohol and water; the lower contained 12,170 pounds of liquid oxygen. Fuel and oxygen were forced into Thiel's engine by two pumps that were driven by steam which was itself generated by mixing hydrogen peroxide and sodium permanganate. The two chemicals had a hypergolic relationship: like von Karman and Goddard, Korolyov and Glushko, they ignited on contact. The pumps fed the fuel and oxidizer into a large round combustion chamber through eighteen pots, or "roses," at a rate of thirty-three gallons a second. The squirting propellant turned to a misty soup that was lighted by an external igniter shoved into the chamber like an enema syringe. The engine itself consisted essentially of the combustion chamber and an exhaust nozzle to which it was welded. The nozzle provides a handy example of the sort of painstaking work that went into the creation of the A-4. By then, accepted engineering had it that the cone-shaped nozzle was most efficient when its edge flared ten to twelve degrees, making it long and narrow like a pencil point. But repeated calculations and bench tests showed Thiel that a thirty-degree opening—a wider, shorter cone—reduced friction between the exhaust gases and the wall and therefore limited turbulence at the end of the nozzle. This in turn substantially increased the engine's exhaust velocity, which increased its thrust. The more efficient nozzle, the eighteen-pot injection system, and the bulbous (as opposed to long and narrow) shape of the combustion chamber itself were revolutionary.

So were scores of other innovations, led by the guidance system, which fundamentally pushed the frontier of rocketry. The A-4's large fins would stabilize it only when they could work in the air through which it sped like a flat hand extended out of a moving car. The fins would therefore become increasingly useless as the missile arced out of the thick air on a flight that would take it to an altitude of fifty miles. So four small carbon steering rudders were positioned to stick into the exhaust gas stream that blew out of the engine. Controlled by two gyroscopes in the A-4's forward section (which were themselves hard-won technological triumphs), the rudders would correct the three axes on which the missile moved: roll, pitch, and yaw (the last two being up and down and sideways).

The forward section, a truncated cone right behind the missile's warhead, was the skull that held the weapon's brains. The gyroscopes were in there. So were two radios. One kept the missile on course by following a signal from the ground. The other would stop the propellant pumps at a point in the A-4's flight when cutting off the engine would start it into a long arcing trajectory that ended at its target. There were also batteries to run the gyroscopes and radios, ground power-control sockets, a device that armed the warhead two minutes after liftoff, and three compressed-air bottles that were used to pressurize the fuel tank before liftoff. All of this was designed to toss a 2,200-

pound explosive up to 200 miles, where it would drop on its target without warning at a speed close to 2,500 miles an hour.

By August 1941, with the British Isles still impregnable and the Luftwaffe working on a winged, pilotless flying bomb—a much cheaper competitor of the A-4 that was eventually called the V-1—Hitler began to take the A-4 seriously. By then the men of Peenemünde were even considering two- and three-stage rockets, called the A-10 and A-11, which presaged true intercontinental ballistic missiles that could be launched from Western Europe and hit cities along the U.S. East Coast. But that technology was well beyond even their grasp, and as the war progressively worsened, there were many more important things to do than perfect missiles that could reach a quarter of the way around the world. The A-10 was seriously studied and then shelved in 1942. The A-11 was strictly a dream machine: a three-stage spaceship. "With slightly improved mass ratio and better propellants," von Braun would say years later, "this combination might easily have projected the pilot of an A-9 [a manned rocket] into a permanent satellite orbit around the Earth." The A-11 never made it beyond lunchtime banter at Peenemünde. But it would be resurrected on the pages of an American magazine called *Collier's,* on television, and elsewhere at another time and in another place.

Peenemünde-West was an exciting place for the Luftwaffe's engineers. Besides the V-1 and ME-163, the facility designed and often even tested a variety of rocket-propelled weapons, the most promising of which was Wasserfall (Waterfall, or Cascade), a large winged rocket interceptor designed to knock down enemy bombers. Used in quantity and in conjunction with the ME-262, the world's first operational jet aircraft, Wasserfall could have blunted the Allied bombing campaign. But the missile had steering and guidance problems that were never overcome and Hitler decided to waste the ME-262, a thoroughbred of a fighter, as an ineffective bomber.

"A Fiery Sword Going into the Sky"

After mishaps on June 13 and August 16, 1942, an A-4 wearing a "Frau im Mond" logo was successfully tested on October 3. It blasted off and arced high over the Baltic late in the afternoon of a beautiful day. Dornberger and the others watched the glowing dot disappear behind a white plume in the clear blue sky. Then he and Zanssen hugged each other and wept. Krafft Ehricke and some of his colleagues screamed. "It looked like a fiery sword going into the sky," he would later recall. "There came this enormous roar and the whole sky seemed to vibrate; this kind of unearthly roaring was something human ears had never heard. It is very hard to describe what you feel when you stand on the threshold of a whole new era; of a whole new age. . . . It's like those people must have felt—Columbus or Magellan—that for the first time saw entire new worlds and knew the world would never be the same after this. . . . We knew the space age had begun."

Frau im Mond's engine fired for fifty-eight seconds before it was turned off. Four minutes later, having touched the edge of space and easily surpassed every world altitude and velocity record, the missile splashed into the sea 120 miles from where it had lifted off. Ehricke would remember Dornberger, too, as saying, "This is the first day of the space age." At a celebration at the officers' club that night, he proclaimed that "the spaceship is born," and most likely added that it would also prove to be a weapon capable of altering the course of the war for the fatherland. He was right on the first point. But if he mentioned the possibility that it would ensure victory, he could not have been more wrong.

Early in July 1943, Albert Speer, the architect who had become minister of Armaments and War Production, invited Dornberger and von Braun to brief Hitler on the A-4. The Reich was by then under almost constant attack from the air. After the meeting, which included color moving pictures of one of the rockets being launched, Hitler became ecstatic, according to Speer. "The A-4 is a measure that can decide the war," he informed his minister. "And what encouragement to the home front when we attack the English with it! This is the decisive weapon of the war, and what is more it can be produced with relatively small resources. Speer, you must push the A-4 as hard as you can! Whatever labor and materials they need must be supplied instantly." It was yet another of the Führer's blunders.

The Ballad of the A-4

In late August 1944, 150 copies of a slim booklet stamped "Top Secret" were distributed to the troops who would begin firing A-4s within a couple of weeks. The *A-4 Fibel* (*A-4 Primer*) was a user-friendly technical manual written in simple form and containing jaunty verse, cartoons of bosomy girls in bathing suits or negligees, and drawings of German villages under snow at Christmas. It began like this:

LISTEN EVERYBODY!

Here, dear Reader, is
the new A-4 primer,
This dry material is presented
in an easy manner
So that it will become part
of your flesh and blood.
However, always remember one thing

ALL THIS MATERIAL ABOUT THE A-4
IS TOP SECRET. REMEMBER THAT!

. . . on this planet where you live
In an age of guided missiles
A sky ship in the universe—
A long dream of mankind—
May someday fascinate our century.
But today you must master a weapon still
Unknown because it is classified top secret.
It is called, for short, the A-4 Device. . . .
Whoever talks about it commits treason
and damages himself and the State.
First of all, remember, do not enter
into any debate.
Should an outsider, an informer or
a wise guy question you
Tell him with a stupid expression on your face
I don't know anything.

YOU ARE A MEMBER OF THE LONG-RANGE
ROCKET SQUAD

You will help launch the A-4. You will work with a projectile that flies higher and farther than any known projectile. The A-4 has a detonation effect unequaled until now by any missile or bomb. . . .

FASTER THAN SOUND

Only five minutes will elapse between launching and impact. In those five minutes, however, everything must operate perfectly. Every single element of the A-4 has to be tested carefully and set prior to launching to insure that the missile hits its target. Little things may cause misfiring. . . .

REMEMBER:

Every miss will help the enemy, damage us through the loss of valuable materials, and endanger the lives of you and your comrades.

MORAL:

The A-4 will hold it against you if you don't study this manual carefully. If you do, the enemy will be troubled by each of your well-placed shots.

"Troubled" was not the word for it. The first A-4s—now generally known as Vergeltungswaffe 2 (Vengeance Weapon 2, or V-2), the name given to them by the Propaganda Ministry—to be fired in anger were pointed at London and

Paris and launched on September 7 and 8, 1944. The weapon's new name was apt. Bombs by the thousands were falling on Germany, and Paris had been liberated by the Allies only two weeks earlier. So nineteen V-2s were fired at the French capital from mobile launchers for sheer spite. They caused very little damage.

But it was a far different story in England and Belgium. A total of 1,403 V-2s would be launched against various targets in London and the south of England, causing both damage and casualties. Antwerp, and particularly its docks, would suffer a rain of 1,610 of the murderous rockets, while 86 struck Liège. An estimated 12,685 people—more than two thirds of them civilians—were killed by V-1s and V-2s, with the V-2s alone accounting for some 5,400.

Relative to the aerial bombardment of London, though, the death and destruction caused by the V-2s were not great, and neither was their effect on resilient British morale. Yet they amounted to a dark omen for the future, a fact that was apparent to any scientist, soldier, or journalist who was paying attention.

"The significance of this demonstration of German skill and ingenuity lies in the fact that it makes complete nonsense out of strategic frontiers, mountain and river barriers," the Columbia Broadcasting System's Edward R. Murrow reported from London on November 12, 1944, while the missiles were falling silently out of the sky and exploding in eruptions of brick and timber, mortar and glass. "And, in the opinion of many able scientists, it means that within a few years present methods of aerial bombardment will be as obsolete as the Gatling gun. It serves to make more appalling the prospect or the possibility of another war." Murrow also accurately predicted the reaction of the victors to the weapon of the future when the guns finally fell silent: "German science has again demonstrated a malignant ingenuity which is not likely to be forgotten when it comes time to establish controls over German scientific and industrial research."

An Antechamber of Hell

A photograph showing heavy construction at Peenemünde taken by an RAF Spitfire pilot on a reconnaissance run over Usedom in mid-May 1943 led to a series of high-altitude flyovers the following month. One batch of pictures, taken on June 23 by a Mosquito making repeated runs, revealed two V-2s lying on trailers at Peenemünde-East. Photographs of the Luftwaffe's Peenemünde-West facility taken by the speedy twin-engine aircraft on the same sortie turned up four small, tailless aircraft: the rocket-propelled ME-163s.

The pictures, as well as loose talk in Germany and POW interrogations, led to Operation Hydra, a three-hundred-plane raid against Peenemünde by

Bomber Command on the night of August 17. The battle was fierce. Fifty of the attackers were shot out of the sky by anti-aircraft fire and interceptors. But more than a thousand tons of high explosives blasted the sprawling rocket facility, killing an estimated 735 people. Among them were more than 100 Russian prisoners of war and Polish slave laborers who were trapped in barbed-wire enclosures that were incinerated by incendiaries and blown apart by thousand-pounders. Those who tried to climb through the barbed wire were machine-gunned or attacked by SS guard dogs. Walter Thiel, his wife, and their four children died when a direct hit turned their house into a flaming mausoleum. Their deaths, as well as those of other civilians, were no accident. "The first bombs were to be aimed, not on the test stands or even the laboratories," according to one account, "but on the living quarters of the scientists: these men were irreplaceable, the other targets were secondary."

Within two days, Himmler suggested to his Führer that the V-2s be built by slave labor in an underground concentration camp. Such a facility would be absolutely secret and impervious to air attack, the chief of the SS explained. What he did not say was that the use of slave labor, which the SS controlled in its system of concentration camps, meant that his own organization would finally get its fingers into Dornberger's pie. In the process, more people would die producing V-2s than were killed by them.

Hitler's own reaction to the attack was to order work started on a so-called high-pressure pump battery on the coast of France. This most sinister of German weapons, also known as the V-3 or the London Gun, was to be a set of four 116-foot-long guns that were to be built near Calais and that would fire nine-foot-long explosive "darts" at London at a rate of up to six hundred an hour. The monster weapons were never built.

Within ten days of the attack on Peenemünde, inmates from Buchenwald were trucked into a lonely valley near Nordhausen in the Harz Mountains to begin expanding a series of tunnels that had been dug before the war for the storage of military chemicals. Technically, the facility was named Mittelwerk (Central Works), an innocuous term that derived from the company that was set up on September 24 by Speer's ministry to operate it; less formally, it was known as Dora. Both Dora, which was located at Niedersachswerfen, and another camp at Nordhausen, two and a half miles away, were destined to become infamous. A French Resistance leader who survived Dora called it one of the Schutzstaffel's "antechambers of hell."

Prisoners—mostly Russian, Polish, and French at first, and, beginning in the summer of 1944, Jews as well—lived and worked under conditions that actually made other camps such as Auschwitz and Buchenwald seem relatively benign. Those who prepared the large tunnels and then built the V-1s and V-2s were crowded into cold, wet chambers where dysentery, tuberculosis, and pneumonia were rampant. Hygiene was virtually nonexistent; water was inadequate, and in place of toilets, oil drums were cut in half and covered with boards. The resulting stench, mixed with that of urine-soaked clothing and other filth, was incredible. An official total of 2,882 prisoners perished at Dora between October 1943 and March 1944 alone, many of them starved to death or arbitrarily executed. Half of the estimated sixty thousand prisoners who entered Dora did not leave it alive. They suffered excruciating deaths to build 5,789 V-2s and a few thousand V-1s.

As happened elsewhere, the tempo of the slaughter became even more furious as *Götterdämmerung* approached. "In January 1945, new guards and officers arrived from the extermination camp of Auschwitz, which had just been evacuated," Resistance leader Jean Michel later wrote. "These born criminals were the cause of the deterioration of already appalling conditions. Dora became a hell even more atrocious than before. The mass hangings began. Up to fifty-seven deportees a day were hung. An electric crane, in the tunnel, lifted twelve prisoners at a time, hands behind their backs, a piece of wood in their mouths, hung by a length of wire attached at the back of their necks to prevent them crying out. All prisoners had to watch these mass hangings."

Not all of the hangings were arbitrary. Some of the prisoners risked their lives by trying to sabotage the V-2s. They would pull out or urinate on wires, for example, hoping that even so small an act would be enough to prevent the missile from reaching its target. Others were hanged for such offenses as making a spoon during work. The secretary to Dora's commander would testify at U.S. Army war crimes trials that sabotage reports had been written on the doomed prisoners before they were turned over to the SS for execution. Hannelore Bannasch would swear under oath that it was Arthur Rudolph who had performed that service.

If Dornberger and von Braun did not approve of the ongoing tragedy, they certainly knew that they were making a dreadful deal for the sake of their rockets. They and Rudolph, who was more deeply implicated because he supervised production, remained intimately involved with Dora and showed no apparent qualms about using slave labor except as it affected the quality of work.

"During my last visit to the Mittelwerk," von Braun wrote to Dora's director on August 15, 1944, "you proposed to me that we use the good technical education of detainees available to you and Buchenwald to tackle . . . additional development jobs." "Detainee" was his euphemism for "slave laborer." He added that he had gone to Buchenwald "to seek out more qualified detainees" and had "arranged their transfer to Mittelwerk."

Robert C. Baldridge, who was in the U.S. 9th Infantry Division, was among the first GIs to liberate Dora and later wrote about the experience in a book called *Victory Road.* "Five thousand of them were dead, lying around in stacks, unburied," the appalled soldier would recall. "The living were mostly skin and bones, maybe seventy or eighty pounds. . . . Some were in their bunks, filthy and smelly in their striped jackets, wasted away with sicknesses like diarrhea, awaiting certain death with vacant stares. . . . They were blank, dull, expressionless, and their eyes were glazed over. The stench of the place was almost unbearable, and on that quiet spring day the buzzing sound of flies humming around the bodies strewn in piles was nauseating."

Nor did it take a high-ranking intelligence officer to fit von Braun and his colleagues squarely into the picture. Even Baldridge, an artilleryman, knew that the baron had made many inspection trips to the caves of agony. "I am not saying that he condoned the conditions at Nordhausen—I do not know whether he did or not—but I am saying he must have known in some way about the perverted and deadly situation there and did nothing about it," he wrote.

Jean Michel would heartily agree. "I do not hold it against the scientists that they did not choose to be martyrs when they discovered the truth about the camps," he wrote of Dora and the men of Peenemünde after the war. "No. Mine is a more modest objective. I make my stand solely against the monstrous distortion of history which, in silencing certain facts and glorifying others, has given birth to false, foul, and suspect myths." He went on to rail against a book that contained an "enthusiastic chapter" on the V-2; *I Aim for the Stars,* a film that glorified von Braun without mentioning Dora;* and a reference in one of von Braun's own books to his possessing a "profound and benign sense of humanity."

"Sense of humanity! . . . Journeys to the moon ought not to force the sufferings of man on earth into oblivion," Michel protested. "Scientists," he quoted another author as saying, "are the mercenaries of modern warfare." And he quoted Dornberger himself: "I had made rockets my life's work. Now we had to prove that their time was come, and to this duty all personal considerations had to be subordinated. They were of no importance."

Both prisoner and overseer would agree on only one thing: the anguish was for nothing. "If, in the end, the V-2 was to permit man to go to the moon and to instigate a nuclear balance of terror, this futuristic programme did not grant Nazi Germany the possibility of winning the war. . . . On the contrary. The obsession with rockets contributed decisively in the precipitation of the irreversible fall of the Third Reich," Michel continued, because it sapped resources that could have been used to produce tanks, airplanes, submarines, and other far more effective weapons.

* It was a potboiler starring the suave Curt Jurgens as von Braun. The film's title prompted one of von Braun's detractors to quip, "and sometimes I hit London." (Gray, *Angle of Attack,* p. 37.)

Speer would make the same point in his own memoir. "Our most expensive project was also our most foolish one," he wrote. "Those rockets, which were our pride and for a time my favorite armaments project, proved to be nothing but a mistaken investment. On top of that, they were one of the reasons we lost the defensive war in the air." This was probably an allusion to the fact that the V-2's estimated total cost—two billion wartime dollars—would have been spent more effectively on perfecting the Wasserfall anti-aircraft rocket and producing it in large numbers, along with the interceptor version of the ME-262.

While the V-2's contribution to the war was as negligible as it was ruinously expensive, it was the product of extraordinarily gifted and disciplined minds and of a government that was willing to commit large amounts of resources to its creation. The combination produced an array of what would now be called amazingly inventive high-technology weapons. The ME-163 was the first and only operational rocket fighter; the ME-262 was the first operational jet aircraft; the Arado AR-234 was the world's first usable jet bomber-reconnaissance plane; the Henschel HS-293, a 1,700-pound radio-controlled flying bomb that was actually fired with some success against enemy merchant vessels, was the first operational cruise missile. These, as well as the V-1 and V-2, represented an unparalleled burst of technological creativity.* And so did the experimental launching of missiles from submerged submarines.

National Rocketry

The war was the crucible in which heavy rocketry was born. Germany's reliance on vengeance weapons, which was grounded in the belief that salvoes of ballistic missiles could be decisive against a well-armed opponent, was strategically flawed and terribly expensive. Yet there was a deeper reality. The "circus" had become deadly serious. The comical tubular contraptions that were fed liquid oxygen from coffee pots and lighted with gasoline-soaked rags had grown into towering bullet-shaped monsters that could carry payloads weighing as much as a Volkswagen with considerable dependability. They did not win the war. But they repeatedly demonstrated that, as Murrow predicted, a distant enemy could now be obliterated at the touch of a button.

* While the Peenemünde rocketeers were extraordinarily creative, they in no sense held a monopoly in that area. Army Ordnance preferred in-house development as opposed to contracting with industry from the beginning, but many firms nonetheless produced technologically advanced weapons on their own. Rheinbote (Messenger from the Rhine), for example, was the first three-stage rocket to go into service. Developed by Rheinmetall-Borsig, it was more than thirty-seven feet long and had a range of 137 miles. The problem was that the pencil-shaped missile carried only forty-four pounds of explosives. Twenty of them were fired from the Netherlands into Antwerp in November 1944, but hardly anyone paid attention. (Ley, *Rockets, Missiles, and Space Travel,* pp. 232–33.)

And there was another reality. Even before the war it had become apparent that the guided missile's size and complexity required a degree of management and funding so large that only national treasuries could afford to produce the weapons. The societies of amateurs therefore gave way to professionals who worked for the state or for industries that contracted with it. And behind the new rocket professionals there were cadres of scientists, mathematicians, specialist engineers, draftsmen, technicians, accountants, and secretaries. The rockets required an infrastructure that could be supported only by a powerful command economy.

In addition, the rocket would forever serve two masters at the same time, or rather a single master with two dispositions: one for war and one for peace. The fundamental principles of rocketry and most of its applications are invariable, and so are its support systems. This would mean that civilian and military rockets, the methods of flying them, and most of their missions would evolve in superficially separate programs whose distinctions were far more apparent than real. Refining the rocket that delivered mail, in other words, would simultaneously refine the weapon that could wipe out everyone who read mail. That much was clear before the war ended.

And something else was clear. The United States had come out of World War I as a triumphant upstart and nascent superpower: a vast, churning, capitalist citadel that straddled a frontier and adorned itself with art, advertising, frenzied investment around the world, and an unbridled belief in its inherent goodness and greatness. It then fought World War II as the wary partner of a radically socialist regime, the lesser of two singular evils, which it had considered an enemy at least since 1921.

Nor had the Allied occupation of Archangel and Churchill's remark about strangling communism in its cradle done anything to calm the famously xenophobic (and often invaded) Russians. There had never been illusions about where they fitted into Western strategy: they were cannon fodder. If Marxist doctrine was right, the heavily armed imperialists would attack them after the Germans were out of the way.

So now, with Germany nearly crushed and the Nazis all but vanquished, the two most powerful victors each saw a new enemy—really an old enemy that had only been deferred—looming as the smoke of battle began to fade.

4

Missiles for America

By mid-January 1945, the Red Army had swept through the Baltic and was moving toward Peenemünde while American and British forces were advancing in the west, squeezing what was left of German defenses in a relentlessly closing vise between the Oder and the Rhine. Seeing that the war was all but over and believing that they sat on a treasure more valuable than all the looted gold and art in the Third Reich, von Braun and his most trusted colleagues met in secret at a hotel near Peenemünde and decided to move to an area likely to be overrun by the U.S. Army. One of his engineers would later explain the reasoning this way: "We despise the French; we are mortally afraid of the Soviets; we do not believe the British can afford us, so that leaves the Americans."

The SS finally had its prize, though it was anything but one that would bring the killers power and prestige. Hans Kammler, now in charge, ordered the rocket team to evacuate Peenemünde and head south to the missile production facility at Nordhausen. The first train left on February 17, carrying 525 passengers, including wives and children. It was four days after Dresden had been turned into an inferno by a massive firestorm that was started and fanned by Allied bombs. A last V-2 was tested at Peenemünde on the nineteenth. Then many of the facility's larger machines and workshops were systematically blown up to make them useless to the Russians who were advancing on the region. Enough would survive to make the place valuable to its conquerors anyway.

On March 19, Hitler ordered all of the nation's research facilities and their records destroyed. Where Peenemünde's rocketeers were concerned, however, getting rid of the only card they had with which to barter their futures would have been unthinkably stupid. The Führer's command was therefore ignored. Instead, Dieter Huzel, von Braun's assistant, was instructed to find a secure hiding place for the thirteen years' worth of reports and drawings that constituted a priceless record of their work.

"These documents were of inestimable value," Huzel would explain after the war. "Whoever inherited them would be able to start in rocketry at that point at which we had left off, with the benefit not only of our accomplishments, but of our mistakes as well—the real ingredient of experience. They represented years of intensive effort in a brand-new technology, one which, all of us were still convinced, would play a profound role in the future course of human events."

That was precisely the point. Von Braun knew that those records and the experience the members of his team carried in their heads were a treasure trove of data on the world's operational ballistic missile technology and the starter set for going to space. Keeping his team together as a usable commodity and protecting its records and brains would therefore very likely insure survival in a dangerous and uncertain time. Neither von Braun nor any of the others had illusions about what was in store for Germans who had nothing with which to bargain. Some of the team joked bitterly that while the new Germany would have no need for rocket scientists, all of their missiles' war surplus aluminum could be turned into pots and pans and sold to housewives. But Wernher *Freiherr* von Braun had no intention of ending up as a tinker or a door-to-door salesman. He had plans, even as the last-ditch defense of the fatherland raged, to turn the rocket team into a phoenix that would fly first to America and then to space.

So Huzel, an engineer named Bernard Tessmann, and a small contingent of annoyed soldiers collected, crated, and carefully marked fourteen tons of documents from Nordhausen and another rocket center nearby and drove them in three trucks to Dornten, fifty miles to the northwest. Throughout the night of April 4 and well into the next morning, Huzel, Tessmann, and the cursing, sweating soldiers carried the crates into a vaulted room three hundred or four hundred yards inside an abandoned mine shaft. Dynamite charges were then set off, turning the chamber into a sealed tomb.

Meanwhile, as the U.S. 3rd Armored Division and other units approached Nordhausen, Kammler ordered as many as five hundred of the rocket team's senior members farther south to Oberammergau, a town in the Bavarian Alps. The location could not have been more appropriate; it was the site of the famous passion play that is put on every ten years in memory of the great plague of 1633. They traveled without their families and were protected (and watched) by armed guards on Kammler's private train, which contained a dozen crowded sleeping cars and a well-stocked dining car. Its rueful passengers promptly named it Vergeltungs-Express: the Vengeance Express.

Kammler had been the overseer of several concentration camps and was well aware of what could happen to him after being captured. So he decided to hold the rocketeers hostage in case he had to make a deal with the Americans. Meanwhile von Braun, clearly the most valuable of the hostages, had broken his left arm and shoulder in an automobile accident and headed for Oberammergau by car.

Once there, the watchful Kammler suggested that the likelihood of enemy air attack made it imperative that his valuable flock be moved out of the barbed-wire compound into which it had been herded by his SS guards and dispersed in hamlets and villages around the ancient town. For his part, von Braun was more worried about the trigger-happy guards than about the boyishly exuberant U.S. Army P-47 fighter pilots who constantly crisscrossed the area looking for targets to shoot up. Kammler, meanwhile, seems to have concluded that whatever the rocket men were worth, he was worth nothing, and would therefore be locked up or worse if caught. He therefore slunk off and disappeared into the countryside.

American tanks and infantrymen entered Nordhausen on April 11 and, like Robert Baldridge, were sickened by what they saw. Just northwest of Nordhausen, outside of what appeared to be a mine entrance at Niedersachswerfen, enfeebled skeletons in filthy striped pajamas babbled and gestured excitedly to their saviors about "something fantastic underneath the mountain . . . important." They were talking about Dora, whose cavernous tunnels contained parts for more than one hundred complete V-2s. Correspondent Ernie Pyle's celebrated dogfaces had dug up an extraordinary bone.*

Special Mission V-2

No one in the Allied camp seriously believed that V-2s were going to win the war or ever could win it. But that did not alter the fact that the Nazis had birthed a major new weapon with a potential that was becoming obvious to the few who were paying attention. All of the major armies were by then using batteries of solid rockets, such as the Soviet Union's Katyusha, to extend the range and, more important, the intensity of their artillery barrages.

Still, if liquids were not exactly in the popular culture during the spring of

* Mittelwerk's presence had first become known early in the previous September, when a German prisoner of war who had been an electrician at Peenemünde confirmed to British interrogators that there was a "central Works" and gave its location and other characteristics. It was photographed two weeks later and again in October by the RAF. Analysis of the photographs showed extensive production and storage facilities and railroad flatcars whose tarpaulin-covered freight threw shadows that suggested V-2s. A plan to bomb the place with napalm, no doubt incinerating the prisoners the way Japanese soldiers were then being burned to death on islands in the Pacific, was abandoned in favor of using ground forces. It is interesting to ponder whether preserving the rockets took precedence over preserving the people. (Irving, *Mare's Nest,* p. 312.)

1945—despite ample news coverage by Murrow and others who drew vivid pictures of the "death from the stratosphere" that the vengeance weapons had wreaked on England—they certainly had developed a small, dedicated, and secretive subculture. As early as 1936, for example, the Soviet Union's GIRD had bench-tested an engine called the 12-K that produced an impressive 660 pounds of thrust and had a specific impulse of 210 seconds. Research on liquids had progressed steadily until the German invasion in 1941, when it was stopped because of the need to design and build the tactically more useful solids.

It was a far different story in the United States, however, where secure borders insured that research steadily continued. In Pasadena, GALCIT's strict orders from the Army Air Forces to concentrate only on Jet Assisted Take Off for airplanes had been countermanded in the summer of 1943, when von Karman was asked to study and comment on three British intelligence reports on reaction propulsion devices for planes and "projectiles." Much of the material was from German POWs and was inaccurate or exaggerated, Frank Malina would later report. "The fact that our conclusions bore little resemblance to actual German missile and aircraft developments . . . is irrelevant to their impact on the 1943 military scene in the U.S.A.," he wrote.

The most immediate result of that impact had been a request by the Army Air Forces liaison officer at Caltech for a study on the possibility of using rocket engines to propel ballistic missiles. The study, called "Memorandum on the Possibilities of Long-Range Rocket Projectiles," was completed by Malina and Hsue-shen on November 20, 1943. It concluded that missiles could not go more than a hundred miles with existing engines, but that more powerful ones would be able to send rockets over greater distances and with heavier payloads. It was as optimistic as it was solidly researched (von Karman had looked over his young charges' calculations before they went out), and made the case for weapons of enormous, perhaps unlimited, range.

The report was an important catalyst for serious Army interest in long-range rocketry. It was also the first document to carry the small study group's new name: the Jet Propulsion Laboratory. None of JPL's founding fraternity could have known at the time how important a role it was going to play in the space age. What they did know, though, was that "jet" was infinitely more respectable than "rocket," which was still often equated with firecrackers.

The liaison officer forwarded Malina's and Hsue-shen's report to then-Captain Robert B. Staver, the Army Ordnance officer on campus, who passed it on to Colonel Gervais Trichel at the Pentagon. Trichel headed the Ordnance Department's Rocket Development Branch, which was founded that year as a direct consequence of the V-2s and which was trying to determine whether liquid-propelled ballistic missiles were really feasible.

The report led Trichel to award GALCIT/JPL a contract to develop a "long-range rocket missile and launching equipment" for the Army. The program was soon given yet another acronym: ORDCIT, for Ordnance/California Institute of Technology. If there was any doubt in Trichel's mind that the age of the ballistic missile was about to dawn, the first V-2 attacks on London put it to rest. So in December 1944, with the alarming possibility of liquid-propelled German missiles turned to reality, he decided to double his chance to develop an American ballistic missile by awarding a second contract to the General Electric Company to do its own study. That project was given the code name "Hermes," after the son of Zeus and Maia in Greek mythology. Not only was Hermes the messenger of the gods, but he conducted souls to Hell. It is unlikely that the U.S. Army knew that Hermes's most typical monument, the *herma,* or herm, was usually a stone pillar carved in the shape of the head on the top of a penis.

Trichel also decided that Project Hermes could best be helped by rounding up the brains behind the vengeance weapons and picking them. So in February 1945, with Allied armies rolling ever deeper into Germany, he sent Staver to Europe to find as many of the leading German rocket scientists as he could, either before the shooting stopped or as soon afterward as possible. With the help of British intelligence, the newly promoted major compiled a long blacklist. The name at the top was, of course, Wernher von Braun.

Staver's first stop was London, where he quickly learned that not everyone agreed on the value of the V-2 or of its creators. His immediate superior, Colonel Horace B. Quinn, expressed his own thoughts on the matter with characteristic succinctness. "I don't care if the Russians get all those krauts," Quinn snarled. "I say good riddance." If Staver needed proof that Quinn was missing a larger point, he got it shortly afterward, when he was knocked out of bed by one of several V-2s that exploded nearby. It or another slammed into Hyde Park, near his hotel, killing sixty-two people in a grisly eruption of earth and shrapnel.

But head-hunting was not all that Trichel had in mind. He was also after the real article: V-2s, and lots of them. In December 1944—the month after he drew up the contract with General Electric and three months after the first V-2 struck England—Trichel decided it was vital that Project Hermes (General Electric) get as many of the weapons as the U.S. Army could lay its hands on. Four months later, he asked Colonel Holger N. Toftoy, the Paris-based chief of Army Ordnance Technical Intelligence, to find one hundred V-2s in usable condition so they could be shipped to the White Sands Proving Ground in New Mexico. There, they would be test-launched by the Army and General Electric.

It is an indication of the seriousness with which the U.S. Army took ballistic missiles that by the end of 1944, while the Battle of the Bulge raged in the

Ardennes, it contracted with GALCIT/JPL and General Electric for separate missile development work, picked a permanent site where the prototypes would be tested, and resolved to capture and bring to America both the inventors of the vengeance weapons and at least a hundred of the missiles themselves.

So within days of the capture of Mittelwerk, Toftoy, carrying out Trichel's orders, put together an ad hoc unit called Special Mission V-2 whose job it was to grab both men and missiles and get them out of a disintegrating Third Reich. But there were three obstacles. Since the V-2s were meant to be launched soon after they were made rather than to be stored in large numbers, the soldiers who liberated the huge tunnel factory found mainly components, not finished missiles. They would therefore have to be sent home in pieces and put together at White Sands. That did not figure to be an insurmountable problem, especially since the rest of the war booty—Quinn's "krauts"—was supposed to be sent to the United States with all the hardware.

Another obstacle dismayed Toftoy because it *was* potentially insurmountable. Nordhausen and Niedersachswerfen were due to be turned over to the Red Army on or about June 1 to become part of the eastern occupation zone. (The Soviets would not arrive until July, but he had no way of knowing that.) Toftoy therefore sped to the underground production site with a company of ordnance trucks brought in from Cherbourg, 770 miles to the west, and went right to the tunnel complex to grab the parts and get them out before their rightful owners showed up.

That was when the third obstacle became apparent: Toftoy had no V-2 parts list, and there was no longer anyone at Mittelwerk who knew enough about the finished missiles to tell him what to steal. So he took the precaution of ordering his men to take a hundred of everything. With a company of heavily armed infantrymen guarding the huge tunnel, the missile parts were loaded onto the trucks and driven to the town of Nordhausen, where they were transferred to German freight cars. The first train, with forty cars full of V-2 components, chugged out on May 22, 1945, bound for the port of Antwerp. Another shipment of equal size would leave every day for the next nine days.

While all that was going on, Staver showed up and immediately began looking for the men on his blacklist. He turned up a few, including Walter Riedel, who had remained behind and settled in the area. He also uncovered a complete Wasserfall guidance system in a barn and other equipment elsewhere. But Staver's greatest accomplishment was tricking a local official into revealing where the Peenemünde archive was buried. He did that by pretending von Braun and others had already provided the information. The official, wanting to ingratiate himself to his country's conquerors anyway, told the American officer about the bonanza that lay behind the dirt and rocks in the mine shaft at Dornten. Staver and a work detail rushed to the site and frantically dug through the debris. The precious cargo they found was quickly loaded onto trucks on May 27 and rushed safely west.

Colonel Korolyov Goes to Germany

Peenemünde was overrun by the Second White Russian Army on May 5. On the twenty-sixth, while the precious V-2 parts were being spirited out of Mittelwerk, a group of Soviet officers arrived there for a preliminary inspection. It had been firmly established that the facility was in the Soviet occupation zone and that it and its contents were therefore to have been left intact for their new owner. But that was unthinkable. The Russians were politely received. But their attention was diverted, and they missed the evacuation that was taking place illegally, and therefore very quickly, right under their noses. Mittelwerk was formally occupied by the Red Army on July 5, 1945. The relatively few rocket shells, engines, propellant tanks, fins, and other stuff that remained on the underground assembly line and in the network of adjacent tunnels was a paltry haul compared to what had been evacuated. Yet even the remnants astonished the Russians. Their potential certainly did not escape Sergei Korolyov, who had been released from Kazan a year earlier and made deputy chief designer of the People's Commissariat of Aviation Industry. As fate would have it, his boss, the chief designer, was none other than Valentin Glushko, the rocket-engine genius who had been instrumental in sending him to prison in the first place.

Korolyov was commissioned a colonel in the Red Army that summer and, with several of his country's best-regarded rocket specialists, went to Germany in early September to tour the captured facilities in the Harz Mountains and north to what was left of Peenemünde. A young and gifted engineer named Boris Y. Chertok led the contingent at Nordhausen and would spend the next two years directing the joint Soviet-German Raabe Institute in Bleicherode, where he would carefully turn Germany missile technology into Soviet missile technology.

Korolyov had in the meantime left Glushko and now worked for the newly created State All-Union Design Bureau of Special Machine Building, an organization whose ambiguous title camouflaged the fact that its assignment was to produce nuclear-tipped missiles that could, in the words of one of its members, "meet the challenge of the American atomic bomb." He and many of his comrades who visited Germany while the embers of war were barely out would form the critical mass of their country's space program. They included Glushko, Vasily P. Mishin, Vladimir P. Barmin, Viktor I. Kuznetsov, Leonid A. Voskresensky, Nikolai A. Pilyugin, Arvid Pallo, Georgi A. Tyulin, Viktor V. Kazansky, and Mikhail K. Tikhonravov. Barmin, Chertok, Glushko, and Tikhonravov in particular would emerge from the postwar chaos as stalwarts of their nation's rocket program, some to challenge Korolyov, others to contribute to his luster.

In fact, they were not the first Russians to study the V-2. Responding to a request for intelligence on the weapon from Churchill himself in July 1944, Stalin had allowed British agents to inspect a test site in Poland while fighting was still going on in the area. He also sent Chertok and Tikhonravov, a top engine specialist named Aleksei Mikhailovich Isayev, Viktor Bolkhovitinov, and a few other Russians to do their own snooping. Chertok would recall many years later walking into a factory and seeing the lower part of Isayev's body and legs sticking out of the bottom of one of the bulb-shaped V-2 engines while his head was inside. "What is this?" he asked Bolkhovitinov, who was also watching Isayev. "This is what cannot be," the astonished Bolkhovitinov answered. "One and a half tons was the limit of our dreams," Chertok would say. "Yet here we quickly calculated, based on the nozzle dimensions, that the engine thrust was at least twenty tons."

The British themselves were left with what amounted to table scraps. These included just eight captured V-2s that were ready to fire from mobile batteries in the northeastern Netherlands and in the Hachenberg region of Germany, east of Bonn. Having rounded up personnel who knew how to launch the things, it was decided to demonstrate them—to see whether they could be launched and learn how they performed—near the northern port of Cuxhaven in a project code-named "Backfire." The launchings took place during the first two weeks of October 1945 for the benefit of American, British, French, and Soviet officers and the news media. Glushko, also a colonel, was one of the Russians who accepted an invitation to witness the event. Korolyov, who by then was working with captured V-2s of his own and some of their midlevel designers and technicians, was apparently not invited. But he went anyway in his own colonel's uniform and become incensed when he was kept on the other side of the fence. There seem to have been more colonels at Cuxhaven that day than at a Confederate Air Force convention. Theodor von Karman, disguised in a crumpled colonel's uniform and looking about as military as Albert Einstein, was also there, and so was Caltech's William H. Pickering, who would one day follow von Karman as head of JPL.

The Soviets may have gotten short shrift where the rockets were concerned, but they apparently had good intelligence that reached beyond that short visit to Poland. Lieutenant M. S. Hochmuth, who assisted Staver, got into a fascinating conversation with yet another colonel named Yuri A. Pobedonostsev, who also happened to be a rocket specialist. Hochmuth would remember the encounter years later:

> He knew my name and that I had been there [at Mittelwerk]. He told me the stuff [taken from Mittelwerk] was going to White Sands (this was supposed to be a secret). We began to discuss engineering. I asked him how things were at Nordhausen, and he said he was having a hell of a time because we had cleaned the place out. He was a very technical guy and said if they were able to see White Sands, we could see Peenemünde. Fat chance.

Overcast and Paperclip

On the morning of May 2—two days after Hitler blew his brains out in a Berlin bunker and five before the Third Reich collapsed—the eminently aristocratic Magnus von Braun of Greifswald, Pomerania, and Peenemünde surrendered to Private First Class Fred P. Schneikert of Sheboygan, Wisconsin, and the 324th Infantry Regiment of the 44th Infantry Division.

Magnus, impeccably attired under a leather coat, pedaled a borrowed bicycle down a lonely Bavarian road in front of the resort hotel in which his older brother, Walter Dornberger, and most of Germany's rocket elite were holed up, playing cards. Seeing a platoon of edgy GIs in front of him, Magnus stopped. Schneikert, in a culvert, suspected an ambush and quickly dropped into a crouch. Then, ever so cautiously, he approached the dapper young nobleman while his buddies covered him. Now they were face to face. Schneikert's own M-1, its safety off, pointed right at Magnus. "*Komm vorwärts mit die Hände hoch!*" the PFC ordered. Magnus slowly approached Schneikert, as he was told, his hands in the air. Alternately speaking English and German, Magnus tried to explain that Wernher von Braun and others who had invented the famous V-2 rocket wanted to be captured by the Americans. To emphasize the importance of the catch, and undoubtedly to impress Schneikert, Magnus told him that he and his brother should be taken to "see Ike" as soon as possible. "You are a nut," Private Schneikert said to Baron von Braun.

Both von Brauns and the others were interrogated in the beautiful Bavarian ski resort of Garmisch-Partenkirchen. Scores of Americans and Britons, only one or two of whom knew enough to ask informed questions, grilled them without any coordination whatever. "They didn't know what to ask," Dornberger would complain. "It was like they were talking Chinese to us." The Germans, knowing their own worth, were as brazen as their inquisitors were ignorant and naive:

> "Did you ever try to kill Hitler?"
>
> "Did you ever try to kill Roosevelt?"
>
> "Suppose you were working for us and we had another war with Germany."

"Not likely in my lifetime."
"Suppose it happened, though."
"Then I am still a German."
"You'll never make it to America talking like that."

To the contrary, the issue of patriotism was beside the point. The Germans were a prize of immense value in the currency of military power and their feelings for the fatherland were therefore all but irrelevant where their captors were concerned. The British tentatively offered von Braun and some of the others refuge, but having them in the United Kingdom was not taken to be a national imperative. In fact, where the big rockets von Braun and the others wanted to build were concerned, the Peenemünde engineer had been quite right: the British could not afford them. The Crown did get Walter Riedel and a few others who were not wanted by von Braun. Those whom the master rocketeer prized would accompany him to the land of opportunity.

The Russians, with their historic grounding in rocketry and paranoia, were a different story. Not only did they broadcast offers of excellent positions and high salaries to the former Peenemündans from a radio station in Leipzig, they even tried to bribe the kitchen staff at heavily guarded Garmisch to smuggle a job offer to von Braun and, failing that, try to kidnap him. Relatively few of the first string took the bait. Helmut Grottrup, a guidance and control specialist who had worked for Steinhoff, was the most prominent exception. And he did so, not for political reasons, but because he didn't want to be separated from his family and, in any case, didn't like the deal von Braun was cutting with the Americans. Grottrup was rewarded with his own rocket institute near Mittelwerk and was eventually moved to Moscow. The Russians bragged to the press that "for every German rocket expert in American hands we got four or five." They were talking mainly about six thousand or so production engineers, technicians, and section managers who were second- and third-rate compared to Peenemünde's elite; they could and would put V-2s together and test them. But they were lackluster, not dynamically creative, and could not hope to form the kind of innovative nucleus that was headed for America.

Von Braun, Dornberger, and others at Garmisch who wanted to go to the States played a subtle game for high stakes. During the interrogation sessions they carefully provided just enough information to convince their captors that they were valuable, but not so much that they would be released either to go to the feared and contemptible Russians or to disband altogether and pick their separate ways through the shambles of their country, perhaps forced to make or sell aluminum pots after all. Herbert Axster, Dornberger's former chief of staff, would later admit that there was a plot to withhold information in order to get to America. "I realized more and more that they wanted something from us," he said of his inquisitors. "And of course that has to be paid for. We had to sell ourselves as expensively as possible."

"Our primary concern was 'stability' and continuity," von Braun would say later. "We were interested in continuing our work, *not* just being squeezed like a lemon and then discarded." But it was crucial that the work be done in America.

Early in May, von Braun was asked to write a lengthy report on what he saw as the future of rocketry. The request came from none other than one of von Karman's prized protégés, Dr. Hsue-shen Tsien, who had been given a commission specifically so he could talk with the Germans in the military zone. The document provided a perfect opportunity for von Braun to whet his captors' appetite.

The *Survey of Development of Liquid Rockets in Germany and Their Future Prospects* carefully made the point that the A-4—"known to the public as V-2," as he put it to distance himself from the Führer's deadly vengeance—was really a stratospheric rocket "conditioned by this war" and that it represented rocketry's infancy. In other words, the master rocketeer and his companions had tried to produce a vehicle for scientific uses, but forces over which they had no control had turned it into a deadly weapon. It was no time to be subtle. The point was that the V-2 carried the basics of weapons that could be decisive in wars and also for the exploration of space, and the rocket's own genetic makeup was and could only be known to its creators.

"We are convinced that a complete mastery of the art of rockets will change conditions in the world in much the same way as did the mastery of aeronautics and that this change will apply both to the civilian and military aspects of their use," von Braun predicted, adding that large amounts of funding would be required and setbacks had to be expected. Specifically, he wrote, long-range, high-speed commercial and military aircraft were on the distant horizon, and so were "multi-stage piloted rockets," and a space station. "When the art of rockets is developed further, it will be possible to go to other planets, first of all to the moon." The document was certainly self-serving. But it also reflected von Braun's deeply held beliefs and was loaded with concrete plans, not wild speculation, all of them grounded in elegant engineering.

Von Braun's prodigious imagination, his vision, was always tempered by a pragmatic understanding that the rocket could serve many masters and that the more of them it served the more valuable it (and its creators) would be. Included in the recovered cache of Peenemünde documents, for example, were plans for the construction of solar-powered space stations orbiting 5,000 miles high that would use mirrors two miles in diameter to focus the Sun's rays on Earth as an alternate energy source.* This was a refinement of Oberth's and Noordung's work. But solar power beamed to Earth, like fire and guns, could be used for peace or war: to boil water for electrical power or as a directed energy weapon against enemies. Similarly, von Braun readily agreed to use the V-2 to launch an atmospheric science package called Tonne that was designed by Erich Regener, one of Germany's best-known experimental physicists. Regener wanted to get his barographs, spectrographs, thermographs, and other instruments far higher than balloons could take them. Peenemünde's technical director had no problem with this, but he stipulated that Regener include instruments to measure the atmosphere's temperature, pressure, and density at various altitudes. Such data would be useful for meteorology. But they would also help V-2s hit their targets. The Tonne never flew on a V-2, but an American version would.

Dornberger, von Braun, and the others need not have worried about their value. The products of their work—the most advanced military aircraft and rockets in the world—guaranteed their tickets to America. A sifting process was going on at Garmisch by the time von Braun wrote his report. More than three hundred of his subordinates had already been released, and others would soon be let go to return to their families and the uncertainties of life in their broken, divided, and occupied country. But a far different fate was in store for the most accomplished of the others.

On July 20, 1945, an operation classified as secret and code-named "Overcast" was put into effect for the temporary exploitation of the cream of the rocket crop. The idea was to bring some 350 Germans, about a hundred of whom specialized in missiles, to the United States on six-month or possibly year-long contracts without their families. They would be used in the war with Japan, at least according to the formal plan, and would then be repatriated. The nice thing about ballistic missiles, though, was that they could be used in a war with anyone.

Yet even so short a stint for so righteous a cause worried some officials who considered von Braun and his colleagues to be permanently tainted because of their close association with the Nazis. Undersecretary of War Robert Patterson acknowledged, for example, that everything possible had to be done to defeat Japan, including using "all information that can be obtained from Germany or any other source." But the Germans "are enemies," he warned, and "bringing them to this country raises delicate questions, including the possible strong resentment of the American public, who might misunderstand the purpose of bringing them here and the treatment accorded them."

* Saunders Kramer pointed out that placing the mirrors at that altitude would cause them to orbit Earth every few hours, creating a serious aiming problem as far as receiving stations on the ground were concerned because they would always be moving relative to the stations. The only place to put them where they would remain at the same spot relative to Earth would be at geosynchronous range: 22,300 miles over the Equator. This was well beyond the Germans' capabilities, of course, but not beyond their prodigious imaginations.

But more expedient voices prevailed. Some, horrified by the sudden appearance of a new threat, the dark Red menace, thought that ignoring any weapon that could counter it was ridiculous at best and insane at worst. Nine years later, a commission headed by former President Herbert Hoover would reflect the nation's fear of communism and try to justify all-out, no-holds-barred competition with a determined and sinister adversary. In a secret report to the White House in 1954, the commission said: "It is now clear that we are facing an implacable enemy whose avowed objective is world domination. . . . There are no rules in such a game. Hitherto accepted norms of human conduct do not apply. . . . If the United States is to survive, long-standing American concepts of fair play must be reconsidered." The report did not refer specifically to von Braun. But it might as well have. It took the position that the ultimate act of immorality was to perish without a struggle and that any means necessary to win against a ruthless enemy was fair. This certainly included using former enemies such as Arthur Rudolph and others who had knowingly participated in horrendous atrocities.

Dr. von Braun, like Dr. Faustus and Dr. Frankenstein, was one of the most enigmatic of souls: the genius of vision who is corrupted by his own hubris. Whatever the war had cost civilians and combatants on both sides, including the unspeakable misery at Dora, it had been a blessing for the rocket team. (The always hard-up Goddard, who tried hard but with little success to sell his own wares from self-imposed exile, was not so lucky.) Death and unimaginable suffering were the down payments for the space age and the currency was expedience. The moral dilemma faced by the United States in hiring von Braun and the others—though it was not much of a dilemma and it did not last very long—would lead to a debate that continued throughout the cold war and still lingers. One of the first to draw blood would be Tom Lehrer, the Harvard mathematics lecturer turned cabaret balladeer, who in the 1960s would make von Braun the subject of one of his earliest lampoons:

> Don't say that he's hypocritical,
> Say rather that he's apolitical,
> "Once the rockets are up, who cares where they come down?
> That's not my department," says Wernher von Braun.
>
> Some have harsh words for this man of renown,
> But some think our attitude should be one of gratitude,
> Like the widows and cripples in old London town
> Who owe their large pensions to Wernher von Braun.
>
> You too may be a big hero,
> Once you've learned to count backwards to zero,
> "In German oder English I know how to count down,
> Und I'm learning Chinese," says Wernher von Braun.

Alden Whitman, *The New York Times*' renowned obituary writer, was another who scorned von Braun. Whitman published thirty-seven of his own favorite obits in a volume that came out in 1971. He noted in a foreword that, where selecting scoundrels for obituaries was concerned, "the more majestic the scale the better." Lumped in with Lizzie Borden, Al Capone, and "Legs" Diamond were "former Nazis who have cleansed themselves by building weaponry for the United States." But the Germans' advocates in the defense establishment, in the spirit of the Hoover Commission, would have maintained, with justification, that the responsibility for protecting the United States rested on their shoulders, not on those who wrote musical ditties and obituaries in newspapers. Any means necessary to enhance the national security of the United States (not to mention the attendant security of the respective armed services) was therefore acceptable.

Nor was the expedience restricted to the hiring of retreaded Nazis. It applied to the Japanese as well. Even before Dora opened, the only known use of bacteriological warfare in World War II had taken place in occupied China, where the Japanese Army tested cholera, dysentery, typhoid, plague, anthrax, and other agents on Americans, Britons, Chinese, and other "human guinea pigs" before the chemicals were used against Chinese and Russian forces by a special group known as Unit 731. Infected prisoners, tied naked to beds, were dissected without anesthetic in order to see how the various diseases worked while plague bombs were dropped on Chinese cities. The Imperial Army even planned to send the plague to the United States on balloons floating across the Pacific. (They never got around to that, but they did send explosives, some of which started forest fires in Washington State and elsewhere.) In return for handing over all of the records of the atrocities so the U.S. Army could benefit from them, however, no Japanese involved in the work was prosecuted for war crimes and the matter was quietly dropped.

It is conceivable that some in the War Department were considering the use of V-2s or their successors against Japanese-held islands or against the home islands themselves. But the atomic bombings of Hiroshima and Nagasaki and Japan's almost immediate surrender made moot any argument that the Germans could be used to fight their former ally. Yet there was a more traditional reason for bringing the Germans to the United States, even before the Soviet Union emerged as a full-blown menace: it jump-started the Army's ballistic missile program at a time when rivalry with the Navy over the weapons was just taking hold. If the ballistic missile was the long-range artillery weapon of the future, and if a powerful new enemy replaced the Axis, the imperative of moving ahead with the development of the weapons was clear.

On September 29, 1945, seven German rocket specialists were therefore spirited into the United States without the customary stop at Fort Strong, an island in Boston Harbor, to have their entry papers examined and recorded. They in fact had no entry permits and did not go through the regular immigration process. Instead, they were classified as "wards of the Army," which was wholly responsible for them. That meant they had something even better than entry permits; they had the quiet acquiescence of the nation's political leadership.

Six of the seven were taken to the Aberdeen Proving Ground in Maryland, where they were put to work translating, cataloging, and evaluating the massive number of documents from Dornten. The seventh, von Braun, was sent to Fort Bliss, Texas, as the advance man for the rest of the rocket team. Operation Overcast was renamed "Operation Paperclip" on March 16, 1946. Before the German brain drain was finally ended, at least 1,600 scientists, engineers, research specialists, and managers would be quietly fed into America's warfighting machine. Only a minority, a little more than a hundred, were von Braun's missile men. At least thirty jet and rocket aircraft experts went to Wright Field, south of Dayton, to work for the Army Air Forces. Submarine and torpedo specialists were sent to the Navy.

The records of many of the "prisoners of peace" (as von Braun called his contingent) were politically laundered. Evidence linking them to war crimes and fervent Nazism was quietly removed from their files or altered. This had the effect of transferring all responsibility for the anguish that attended the birth of the Third Reich's ballistic missiles away from the Arthur Rudolphs to high-profile professional killers such as Himmler and underlings such as Hans Kammler.

Walter Dornberger was taken to England in 1945 to stand trial for the missile attacks on British civilians. He was released from prison two years later, however, for lack of a case. If attacking civilians in England was a war crime, so was attacking civilians in Germany, which the RAF and the Eighth Air Force did with devastating consequences. Dornberger nevertheless remained a stubborn advocate of the fatherland's cause. He had "extreme views on German domination," in the words of one British interrogator, "and wishes for a third world war." He was finally put to work at Wright Field and then headed for the Bell Aircraft Corporation in Buffalo, New York, where Krafft Ehricke also found employment. One of Rudolph's interrogators wrote this about him: "100% NAZI, dangerous type, security threat . . . !! *Suggest internment.*" The memorandum was dropped into Rudolph's file and all but forgotten until a time came when it was convenient to find it.

The hiring of Rudolph and the others by the United States and the Soviet Union showed clearly that ethical considerations were irrelevant compared to the insatiable requirements of the national war apparatus. Hermann Göring, head of the Luftwaffe, plunderer of European art treasures, and Hitler's intimate and onetime chosen successor, was sentenced to be executed by the International Military Tribunal at Nuremberg for crimes against humanity. Qualitatively, Göring's crimes were no worse than Arthur Rudolph's. But the bloated reichsmarshal, a World War I hero turned depraved stooge and toady, was assigned a prominent role in the victors' morality play because he was both highly visible and a useless relic. A brilliant and less visible Göring—say, the designer of the ME-262 jet fighter or some other technological wonder that was thought to be militarily valuable—would in all probability not have been fated to swallow poison two hours before he was scheduled to hang.

Science and technology in the service of war-fighting capability emerged as the most important military lesson of World War II. And any means of acquiring them was eminently acceptable. However important ethical considerations may have been in theory, they carried relatively little weight where the survival of the state was concerned. The soldiers and civilians who embraced that doctrine would argue with considerable justification that provoking a nuclear war through weakness was the most unethical course of all.

Rockets for America

Rockets, as well as more traditional artillery, would be proven to work at proving grounds. A facility to test air-to-air and small sounding rockets was set up on Wallops Island, off the Virginia coast, with the first launch taking place on June 27, 1945. But rocketry's most important proving ground was the one set up at White Sands, New Mexico. It was roughly 120 miles from Goddard's Mescalero Ranch, about 45 miles north of El Paso, and practically a stone's throw from Alamogordo, where the first atomic bomb was proven to work on July 16, 1945. The White Sands facility, technically an annex of the Army's Aberdeen Proving Ground in Maryland, was established seven days before the Trinity explosion at Alamogordo in order to test GALCIT/JPL's missiles, the Hermes V-2s, and whatever rockets followed. Its location, on a sea of sand flanked by the San Andreas Mountains on the west and the Sacramento Mountains on the east, guaranteed the kind of seclusion that had sent the German rocketeers from Kummersdorf-West to Peenemünde. But unlike Usedom, the southwest had dry air that practically eliminated corrosion, and its predictably clear skies guaranteed some of the best flying and rocket-testing weather in the world.

The V-2 components that Toftoy had spirited away from Mittelwerk—more than 360 metric tons of them—were loaded onto Liberty Ships at Antwerp and started for the United States by the end of May 1945. They were taken off the vessels in New Orleans, put onto railroad cars, and finally rented flatbed trucks, and hauled to White Sands. There, the disassembled missiles waited for the arrival of the Germans who knew how to put them together. The first group, handpicked by von Braun, arrived in October. By then, the nation that was to be their new home was already working hard on its own rockets.

The Army was pouring $3 million into the Jet Propulsion Laboratory, specifically for the ORDCIT project, which was supposed to develop a line of tactical ballistic missiles. One of them, Corporal, was forty feet long and closely matched the V-2. Meanwhile, the Massachusetts Institute of Technology was mastering a difficult but indispensable technique for getting missiles to go where they were supposed to go: inertial guidance. Whereas V-2s rode radio beams from the ground for guidance to their targets, MIT's Instrumentation Laboratory was using its experience designing Sperry gyroscopes for military aircraft to perfect a so-called six-degree-of-freedom system that was entirely inertial. That is, it was self-contained. The guidance system would be able to tell a missile or a spacecraft precisely where to go, in any direction, with no outside help. Inertial guidance would free rockets forever from their radio leashes.

Bell Aircraft had already designed and built a full-scale mockup of the first of a projected series of X (for Experimental) rocket planes that would carry pilots to the edge of space at record-breaking speeds. The first, shaped like a .50-caliber bullet with stubby wings, was called the X-1. On October 14, 1947, an orange X-1 named *Glamorous Glennis* would become the first aircraft to fly faster than the speed of sound with an all-but-unknown fighter ace named Chuck Yeager at the controls. Glennis was Mrs. Yeager.

The Machismo Factor

The Navy saw the Army's aggressive maneuvering with the Germans and their missiles, the construction of Wallops Island and White Sands, and the funding of ORDCIT and Hermes as clear attempts to control the nation's embryonic missile program. The chief of naval operations responded to the threat by ordering that a Guided Missile Section be formed within the Navy Department. Its mandate was straightforward: to develop guided missiles for use in war.

Less straightforward was a decision taken on October 3, 1945, to establish a Committee for Evaluating the Possibility of Space Rocketry within the Navy's research-driven Bureau of Aeronautics. This amounted to a giant step beyond mere missilery. One of the committee's main goals was to study the feasibility of using Earth-orbiting satellites to relay naval communication signals. Only twenty-six days later the committee concluded that, given sufficient propulsion, it would be possible to send a single-stage "satellite rocket" or missile into orbit around Earth. This quickly led to an Earth Satellite Vehicle program. JPL was awarded a contract to study the problem and concluded that a liquid hydrogen–liquid oxygen engine could indeed get such a satellite into a 150-mile-high orbit.

It soon became apparent to the eager sailors, however, that appropriations were going to be hard to come by. Costs for the engineering and preliminary satellite design work alone was estimated in November 1945 to be between $5 million and $8 million, which was real money in those days. But budget cutbacks because of the demobilization made that kind of funding virtually impossible, especially for a pie-in-the-sky project that had no clear bearing on ocean warfare as it was traditionally conducted.

So the Navy suggested to the Army Air Forces on March 7, 1946, that a joint satellite committee be formed in which resources would be pooled. The Army Air Forces' initial response was positive. But then the shadow of rivalry spread across its soul like some dark cloud. Navy involvement in missiles, it was felt, could negate all of the V-2 and Hermes work in order to help a service that had no business taking to the air or space anyway. Major General Curtis E. LeMay, recently appointed deputy chief of the Air Staff for Research and Development, finally turned the Navy down, undoubtedly with the concurrence of the Army Air Forces chief of staff, General Carl A. Spaatz, who had commanded the U.S. strategic bombing of Europe. If Earth-circling satellites were to be constructed, the fliers told the sailors, the Army Air Forces would be the one to do it.

It was a fateful decision. LeMay was a no-holds-barred competitor: a command pilot who had thought up the strategy of firebombing Tokyo and who had fought against the United States Navy even before he took on Japan. In August 1937 he and other airmen had located, photographed, and "bombed" the battleship *Utah* to prove that the Army Air Corps, not the Navy, should be responsible for long-range reconnaissance. That dispute was, in turn, part of a larger one that pitted land-based strategic bombing against aircraft carriers.

Nor did battles over turf end there. A cat's cradle of competition existed in the predawn of the space age. The American military establishment had concluded in the waning stages of World War II that the new enemy out there, the Union of Soviet Socialist Republics, was large and immensely powerful. That was reason enough to want to pack rockets the way cowboys packed six-shooters.

The Army's motivation to grab von Braun, his colleagues, and the V-2s lay in what was essentially the single-minded belief that guided missiles, no matter what their range, were really long-range artillery. Artillery's objective was to protect friendly soldiers by obliterating their opponents. Missiles were therefore an obvious Army weapon, at least where the Army was concerned. The Navy disagreed. Its role was to control the high seas. Powerful rockets and whatever satellites might be shot into "outer space" (as it was called in 1946) would therefore figure importantly for waging war at sea because the satellites could spot, track, and target enemy naval vessels so the missiles could destroy them. Aerial reconnaissance had proved itself at the Battle of Midway, reaching far out to obliterate the enemy before it got close enough to do the same. Missiles were the obvious successors to aircraft to do that. The airmen—aggressive junior partners in the national military establishment—insisted that missiles were really pilotless bombers whether they had wings or not and that they therefore properly belonged to their own branch of service.

The U.S. Air Force, which was formally created by the National Defense Act of 1947 (along with the National Security Council and the Joint Chiefs of Staff), would quickly evolve as the nation's first line of defense, its chief deterrent. This happened partly by default. Matching the Red Army man for man, particularly across expanses of land that had swallowed Napoleon and Hitler, was unthinkable. Similarly, there was no way the Navy could realistically penetrate and fight its Soviet counterpart in the waters off Eurasia. The Air Force, on the other hand, could strike quickly over great distances. Speed and penetration were the thing. The only way to defeat an opponent that had an overwhelming ground game was to use a long, explosive pass, scoring in the end zone while the lines pushed, blocked, grunted, and hacked at each other.

The air was the new military frontier, Spaatz said in a lecture to Congress. "The next war will be preponderantly an air war," he predicted, adding that "attacks can now come across the Arctic regions, as well as across the oceans, and strike deep . . . into the heart of the country." General James H. Doolittle, the popular hero who had led B-25s from the carrier *Hornet* in the first air attack on Japan in April 1942, made the point dramatically by using a polar map projection that showed North America in a virtual embrace of the Russian bear's paws. Less obviously, the Air Force's role was also enhanced because the hungry airmen conferred upon themselves a cachet that smacked both of high-speed derring-do and an intimacy with high technology. The fact of the matter was that it took more skill to fly an airplane than it did to carry a carbine or drive a jeep. Finally, they aggressively forged ties with industry at a time when defense contracts were drying up.

There was intraservice rivalry too, part of it having to do with budgets, and part with machismo. LeMay's own service, for example, was dominated by "silk-scarf" or "blue-sky" bomber men like himself, and others who came from the fighter wings. They took pride in the fact that the war had first been carried to the enemy's heartland by air-combat crews, not soldiers slogging through mud. There was an implicit, though never articulated, belief that real men risked their lives flying combat missions or testing high-performance aircraft; they didn't push launch buttons from behind thick concrete walls or hide in underground missile silos like chipmunks, gophers, and other rodents.

Everyone in the Air Force, from the lowliest airman to the chief of staff, shared one unspoken secret: the service was and would forever be divided into two fundamental castes that had nothing to do with rank: those who wore fliers' silver wings on their chests and those who did not. Similarly, the pride of the Navy sported either the aviator's wings of gold or the submariner's dolphins. Naval aviators were further distinguished by their brown shoes; water sailors wore black, and everyone knew the difference.

The machismo factor that came with flying into, and then beyond, the atmosphere would sometimes be flagrant and sometimes supremely subtle. But it would always be there, just as it had been when Daedalus and Icarus had taken to the sky, and it would be pervasive. Like the desire to escape from a problem-plagued planet, the classic compulsion to control the "high ground," and the competitive instinct itself, flirting with death, as a means of both expression and self-definition, would be one of the enduring leitmotifs of the space age. It would provide the real high.

Project RAND

In March 1946, when LeMay turned down the satellite committee idea, the Navy was demonstrating an aggressiveness in missile and space research that alarmed the airmen. It had contracted for a variety of feasibility studies with JPL, the Glenn L. Martin Company, North American Aviation, and the Douglas Aircraft Company. To throw in his lot with the Navy, LeMay concluded, would very likely guarantee the sailors a long lead in high-altitude rocketry and effectively shut out the Army Air Forces. That was unthinkable, even to a silk-scarf guy like LeMay, who knew that he would have to close ranks with his service's missile people in the face of apparent poaching by another service.

What the airmen needed was their own satellite research study. More to the point, they felt that it was time to formally claim what they saw as their legitimate turf. They therefore let it be known early in 1946 that they wanted "primary responsibility for any military satellite vehicle, considering such activity to be essentially an extension of strategic air power." This was most likely the first time that LeMay and his fellow officers claimed space as a continuation of their traditional operational environment.

So LeMay contracted with Douglas at El Segundo, California, for a feasibility study on launching objects into Earth orbit. A team of scientists and engineers was assigned to the Research and Development project, or Project RAND, as it was called, on a "crash" basis. The group's first report was completed on May 2 and was on LeMay's desk in the Pentagon on the twelfth.

Preliminary Design of an Experimental World-Circling Spaceship, which had a hand-drawn missile blasting off from Earth on its cover, ran to 326 pages. A short introduction made two somewhat grandiose predictions: that satellites would eventually become "one of the most potent scientific tools of the Twentieth Century" and that America's sending a satellite into orbit "would inflame the imagination of mankind, and would probably produce repercussions in the world comparable to the explosion of the atomic bomb." They were right about the repercussions. But they had the wrong country.

The RAND study was remarkable, especially given the speed with which it was completed. Drawing on von Braun's Garmisch paper of the previous May, a distillation of the V-2 data, and their own relatively extensive knowledge of both aerodynamics and astrodynamics, the Douglas team calculated that getting a five-hundred-pound payload to orbit would require the rocket carrying it to be moving at 24,500 feet a second once orbital altitude had been reached. There was no way a single-stage launcher like the V-2 could do that, even with liquid hydrogen. The problem was damnable but it was not new. The higher the velocity, the longer the required burn. The longer the burn, the more propellant would be needed. More propellant meant larger tanks to hold it. Larger tanks meant less payload. Keeping the payload the same would require building a larger rocket to accommodate the larger tanks. The larger the rocket, the heavier it would be. The heavier it would be, the more thrust would be required. More thrust meant a bigger engine and more propellant. . . . It was brain food for the engineers. They loved it.

The answer had been provided by Conrad Haas in 1529 and certified correct by Tsiolkovsky 374 years later: go to multiple stages. The RAND team found that using alcohol and oxygen would require a four-stage rocket; liquid hydrogen and oxygen would require only two stages. The engineers were so thorough that they even pondered—in thirty-two impressively detailed pages—the possibility of their spaceship colliding with a meteoroid. They calculated, to take only one example, that the satellite's skin would have to be just over two inches thick to resist being penetrated by a meteoroid less than half the size of a dime that was traveling at 220,000 feet a second at an altitude of a hundred miles.

Louis N. Ridenour, a radar expert and a member of the University of Pennsylvania's Nuclear Physics and Electronics Department, listed a spate of jobs that Earth satellites could do. Having noted that radar and anti-aircraft weapons were improving, the professor predicted that air offensives would eventually be carried out by "high-speed pilotless missiles" that could be guided to their targets by satellites. He went on to say that satellites could also be used as orbiting missiles and that they could perform attack assessment, help predict the weather, and relay communication.

Furthermore, Ridenour maintained that an orbiting spacecraft would be "virtually undetectable" by the radar of the time and that it therefore "offers an observation aircraft [*sic*] which cannot be brought down by an enemy who has not mastered similar technologies." Earth observation—space-based military reconnaissance—would assume an increasingly important, and contentious, role as the cold war intensified. A February 1947 RAND follow-on study would provide the first detailed analysis of reconnaissance satellites and missions. There would be a flurry of others.

The following year, the RAND Corporation was spun off from Douglas as an independent nonprofit think tank that contracted with the Air Force as the representative of all three armed services' satellite requirements. This effectively helped to ensure Air Force dominance in space. So did a ruling in 1949 by the Department of Defense's Research and Development Board that space was to be the Air Force's exclusive domain.

One of the most interesting, yet largely overlooked, parts of that first RAND study was the way in which Ridenour ended his chapter. He concluded by showing that he, too, was no less star-struck than all of the other space enthusiasts. "The most fascinating aspect of successfully launching a satellite would be the pulse quickening stimulation it would give to considerations of interplanetary travel," he wrote. "Whose imagination is not fired by the possibility of voyaging out beyond the limits of our earth, traveling to the Moon, to Venus and Mars? Such thoughts when put on paper now seem like idle fancy. But, a man-made satellite, circling our globe beyond the limit of the atmosphere is the first step. The other necessary steps would surely follow in rapid succession. Who would be so bold as to say that this might not come within our time?"

White Sands

The Navy, meanwhile, steered its own course. On January 16, 1946, the month before the Bureau of Aeronautics rocketry committee approached LeMay, Ernest H. Krause, head of the Naval Research Laboratory's Communications Security Section and the architect of its guided missile program, invited a number of military officers, astronomers, and physicists to meet at the NRL. The lab, together with the Bureau of Aeronautics, was spearheading the Navy's move to space. It also had an active Rocket Sonde Research Section that had come into existence before the end of 1945 with an assignment to "investigate the physical phenomena in, and the properties of, the upper atmosphere."

Krause had exciting news. The Army, he said, had agreed to allow qualified scientists to stow experiments as ballast in the empty warhead sections of the V-2s at White Sands. The idea had been put forward by Toftoy, who by then had been promoted to chief of the Army Ordnance Rocket Branch.

Krause, who had been trained in nuclear physics, had a unique angle on the marriage of rockets and research. "Now, this is a good way to do some experimentation," he would remember thinking years later. "We're going to get away from this business of having a complicated, costly set of apparatus in a physics laboratory in a basement in some university, and because it is complicated and costly, it lasts for fifty years and generation after generation grinds out theses on that same equipment because it's expensive and new equipment is more expensive. We've got a set-up here which by its very definition is going to get destroyed each time. How good can you have it?"

James Van Allen, who was at the meeting, worked in the Applied Physics Laboratory at the Johns Hopkins University and was eager to study cosmic rays, as well as the structure and dynamics of the upper atmosphere. He also wanted to take pictures of Earth from very high altitude. Unlike von Braun, Goddard, Ridenour, and all of the others who became rhapsodic at the thought of people sailing around the solar system, however, Van Allen was a frustrated utilitarian who was interested in rockets solely because they could get his instruments higher than anything else could. Even Jules Verne left him unmoved. If Van Allen had a soul mate, it was Regener, the German who was fascinated by the properties of the space that immediately encloses Earth and who thought of rockets as dumb slaves whose only real purpose was to carry measuring devices to where they were needed. For both of them and a number of other physicists, using precious rockets to propel humans around Earth or to the Moon was a stupid and wasteful distraction. As the space age accelerated, they and those who wanted to send people to space would gravitate into two contentious camps, with resources an ever-present issue.

The idea of forming a V-2 Upper Atmosphere Research Panel to facilitate and coordinate the experiments on the limited number of missiles at White Sands was mentioned that day and formalized at a meeting at Princeton University on February 27, 1946. Representatives came from the Army Air Forces, the Signal Corps, the Naval Research Laboratory, Van Allen's lab, Harvard, Princeton, Michigan, the National Bureau of Standards, and General Electric. Fred L. Whipple, the Harvard astronomer who would figure prominently in the early development of space-based astronomy, was one of the nine founders. Their number would quickly grow to fifty-eight of the most respected scientists in the country as word spread of the amazing new opportunity to finally climb out of the atmosphere and take a look around. Army Ordnance was to be responsible for assembling, firing, and tracking the missiles and would contract work to General Electric.

Elaborate techniques, some similar to those used by the Germans, were devised to track the V-2s and monitor their vital signs. The monitoring, which would soon become relatively sophisticated, was done by using radios to transmit back to the ground what the instruments on the missile registered. The connection between space and Earth would be called a downlink (naturally, an uplink went from Earth to the spacecraft or missile). The messages from the instruments themselves—the readings—would be referred to as telemetry. It would be done by relaying raw coded data to a ground station, where it would be made intelligible through a process called data reduction. Such radio signals would either come back in so-called real time—as the readings were made or the data were collected—or be stored on audiotape until there was a more favorable time for transmission. Every missile launch, every manned and unmanned satellite that would go into Earth orbit, and every space probe that streaked across the solar system and beyond in the years to come would report back through telemetry. And intercepting and deciphering the opposition's missile and satellite telemetry would develop into a black art by both sides' intelligence communities.

As far as the scientists were concerned, the joint arrangement for the experiments was wonderful because the participants had wide latitude to explore whatever questions piqued their curiosity. "Basically, it was a very permissive policy as far as the details of how we went about it," Van Allen explained many years later. Yet his relationship with Toftoy paralleled Regener's with von Braun. The scientists accepted the fact that knowledge gained by their experiments would serve more than one purpose; their data would automatically go to the military, certainly for the design of ballistic missiles. "They definitely had that in the back of their minds," Van Allen added. "For example, the structure of the atmosphere, the density and temperature of the atmosphere as a function of altitude, those are things of obvious importance to guided missiles. The measurement of radiation . . . above the atmosphere was of obvious interest for radiation effects on components, on human flight."

Holger Toftoy was disarmingly honest about allowing civilian experiment packages to hitchhike on military rockets. "The primary aim of the Ordnance rocket program is to produce free rockets and controlled missiles that meet military requirements, but the field is so new we are placing emphasis on fundamental and basic research," he explained. "Our program has been planned not only to provide the necessary basic knowledge from all fields of science and apply this knowledge to missile design, but also to conserve personnel and funds and to prevent unwarranted duplication."

Toftoy wanted to do more than simply use scientists to perfect guided missiles, though. He wanted to establish a permanent relationship between the defense establishment and civilian scientists. That is precisely what had happened in Germany and was happening at the time in the Soviet Union. The civilians and the soldiers, bonded by patriotism, mutual dependence, complementary technologies, and the national treasury, were being forged into a single bedrock infrastructure.

The expatriate Germans, who were quartered at Fort Bliss, would help assemble their old weapons and prepare them for launch. And while they did that, they would continue to have their brains picked. On July 1, 1946, von Braun was grilled on details of the design of the V-2 and on the hardware that affected its accuracy. He was also questioned about the general design and the combustion chamber of the highly advanced, but unrealized, A-9 and A-10. His interrogators wanted to know how the rockets were to have been connected and how their upper stage would fire, for example. The rockets were held together and released by explosive nuts and bolts, von Braun answered, but there was no reason why a "mouse trap" system would not work as well. And the firing mechanism was simplicity itself. The igniter for the A-9 would have been mounted in the A-10's nose, he continued, and would have automatically blown a flame into the upper stage's propellant-soaked combustion chamber when the A-10's acceleration dropped to a certain level.

The interchange was not esoteric. The man who asked most of the questions was James E. Lipp, the head of Project RAND's Missiles Division and the chief of its Satellite Study Section. Lipp was beginning a follow-on to the May 2 study that was supposed to contain enough details so that specifications and contracts could be drawn up for the building of actual Earth satellites. Transcripts of all of the interviews were classified confidential and circulated not just at RAND but within the military and several companies.

Sixty-seven V-2s flew from White Sands between April 16, 1946, and September 19, 1952, when the program ended. David H. DeVorkin, a historian at the National Air and Space Museum in Washington, captured the spirit of the period when he wrote that "preparing for a V-2 launch was comparable to fighting a war. The enemy was time, procurement red tape, aging V-2s, balky equipment, and an infrastructure unable to cope, unsure of its mission. . . . For the scientists, a win required not only good diagnostics, but a data stream that revealed something new about the high atmosphere."

And the research was done with equipment and in an environment that were designed for a far different purpose. "Somewhat similar to the experiences of 19th-century naturalists who depended on, but often were compromised by, the direction of naval commanders whose missions were different from theirs, scientists preparing for flights at White Sands had to organize their tasks around a vehicle not built for research," DeVorkin added, "yet many were captured by the fascination of it."

Most of the V-2 launches were successful. Yet firing big rockets remained a daunting, frustrating, and a potentially deadly business because they were basically very complicated bombs. That much would never change. The launch record was littered with failures, as fragments from it clearly show: "cutoff at 19 seconds . . . exploded at 28.5 seconds . . . steering trouble at 2 seconds . . . propulsion trouble at 36 seconds . . . premature valve closure . . . propulsion failure . . . steering vane failure . . . tail explosion at 10.7 seconds . . . separation explosives detonated at liftoff . . . tail explosion at 8.0 seconds."

On May 29, 1947, one of the V-2s decided to take a hair-raising detour, narrowly missing El Paso, crossing the Rio Grande, and crashing into a hillside outside of Ciudad Juárez, which was thronged for a fiesta. It barely missed the building in which the city's construction companies stored their explosives. No one was injured. "We had already been called the only German task force that managed to invade United States territory and penetrate as far westward as El Paso," Ehricke quipped years later. "Now, we were known as the only German team that also managed to attack Mexico from their base in the United States."

Following a tradition that went back to the Raketenflugplatz, and which would continue indefinitely, those who launched missiles at White Sands and elsewhere did so from behind barriers that would protect them from violence when the creatures of their creation turned on them (Goddard had a protective barrier at Roswell). The blockhouse at Launch Complex 33, from which the V-2s and several other rockets were fired, therefore had nine-foot-thick concrete walls and a pyramid-shaped roof of reinforced concrete with a maximum thickness of more than twenty-six feet. This would in theory ensure the survival of the people inside even if a V-2 slammed into the structure at 1,500 miles an hour. It was never tested.

The missiles returned piles of data on the electrification of the ionosphere, radiation at the edge of space, and other phenomena. Number 12 broke 100 miles on October 10, 1946, for example, while sending back unprecedented ultraviolet information about the Sun that even made page one of *The Washington Post* and was praised as "an event of far-reaching astrophysical consequences" by *Sky & Telescope.*

It was not generally appreciated then, nor would it be later, but most of the leading scientists who did experiments in space had to design and build their own instruments because of the innovative nature of their research. This could of course be intensely complicated. But it could also be amazingly simple when imagination was brought to bear. Ralph Havens of the Naval Research Laboratory knocked the tip off an automobile headlight, for example, and used it as a gauge to measure atmospheric pressure on Number 6, which went up on June 28, 1946. One of the V-2s carried a mouse and four others carried monkeys (all of them named Albert) because some researchers were anticipating sending people to space. The Soviets would do the same with dogs.

The last V-2 launched from White Sands in the final hours of the summer of 1952 was part of a growing stable of rockets designed for war and peace. In place of a warhead, *Army Ordnance Magazine* explained cutely, a research rocket would carry a "peace head." One of the most important of such rockets was the so-called Bumper-WAC, which proved the feasibility of multiple staging by combining the V-2 with a far smaller JPL-developed WAC Corporal missile.* Six Bumper-WACs were launched at White Sands, one of them soaring to a record 244 miles in February 1949.

Aerobee, a sleek twenty-five-foot-long Navy sounding rocket, was designed and produced by von Karman's own firm, Aerojet Engineering Corporation. It would evolve through a series of modifications into one of the most versatile—and beautiful—research rockets in the world. The first fully functional version shot into the sky on November 24, 1947; the last went up on January 17, 1985. One thousand thirty-seven of the sleek rockets poked at the edge of space from launch sites around the globe.

A New Perspective on Earth

Only five months after the German migration to the United States was turned from Project Overcast to Paperclip, the Glenn L. Martin Company won a contract from the Naval Research Laboratory to design and build ten (later fourteen) high-altitude research rockets called Viking, no two of which were to be exactly alike. The first Viking was launched on May 3, 1949, and reached an altitude of fifty miles. Number 8 had the dubious distinction of reaching the lowest altitude, four miles, but it had an inspiring reason: it was apparently so anxious to fly that it broke loose from its restraints and took off during a static test.

Viking, touted as an "all-American" innovation—an unmistakable dig at the V-2, its Teutonic stablemate—was the embodiment of the two-master rule. Technically, it was designed for the ubiquitous scientific research. But two years before the first launch, the Navy boasted that "the major portion of the information obtained from this program will contribute directly to the design and control of long-range guided missiles." That was true enough, but the data would not be cheap to come by; a Viking cost $450,000, while an Aerobee cost only $25,000.

The most dramatic return from the high-altitude tests was undoubtedly their photographs. V-2s, Vikings, Aerobees, and other sounders began stealing peeks at bigger chunks of the planet as they climbed higher. To be sure, the pictures that floated down on parachutes provided a bounty of scientific data, most of it about land features and the weather. But they did a great deal more than that. They continued the process of extending humanity's perspective of its home and the world beyond. The rockets that carried cameras—electronic eyes—to the rarified regions over White Sands, Wallops Island, and elsewhere

* WAC was designed as a test vehicle for the larger Corporal ballistic missile, and also as a sounding rocket in its own right that could be carried by a Corporal. Technically, WAC stood for "without attitude control." But since its male designers considered it to be Corporal's "little sister," WAC was also said to stand for "Women's Auxiliary Corps," a World War II distaff branch of the Army. (Koppes, *JPL and the American Space Program*, p. 23.)

represented a new stage in the long process of groping for breadth of view that had begun with balloons and airplanes and that would continue around Earth, at the Moon, across the entire solar system, and beyond: to the edges of the universe itself. The pictures would gradually reveal that Earth itself was a spaceship and that the larger world of which it was a part amounted to the ultimate conundrum: it was both finite and boundless. And the images, which were the easiest to understand of all the scientific data, were routinely sent to the news media. The pictures from the threshold of space were used, as they always would be, to bring the people who paid for them along for the ride.

Viking 11 broke the photographic altitude record on May 7, 1954, when it carried a fifteen-pound K-25 aircraft camera to 158 miles. Two basic kinds of pictures were taken by that rocket as it climbed straight up and then started to tumble back down. Most showed the terrain below in splendid detail. One, taken at the top of the trajectory, clearly showed El Paso, Las Cruces, the Rio Grande, and three railroad lines, all spread beneath cotton-puff clouds whose shadows dappled the desert floor.

Others, taken obliquely, captured terrain and clouds extending all the way to a gently curved horizon more than a thousand miles away. Two of the pictures, fitted together as a composite, made a portrait of sixty degrees of horizon southeast of White Sands. It showed a 1,036-mile-long crescent at the end of 600,000 square miles of parched land and, beyond, a mottled blanket of languid white vapor covering the Gulf of Mexico. Above the horizon, out beyond Earth's marvelously crisp edge, there was the stark blackness of deep space. (A camera-carrying Navy V-2 had broken 100 miles on March 7, 1947, but the pictures had been nowhere near as clear as Viking's.)

This particular white arrow returned an unexcelled view of Earth. The value of the thirty-nine frames that floated down that day, together with all of the others in files that had begun to swell, was wasted neither on the meteorologists nor on the intelligence collectors. Pictures from space could help predict the weather. They could also help predict an enemy's intentions.

Military Missiles

On September 3, 1949, the specially fitted external air filter on an Air Force reconnaissance plane flying over the North Pacific picked up radioactive particles that had been carried eastward from Central Asia. An intensive analysis of the particulate matter by specialists, including J. Robert Oppenheimer, the Manhattan Project's chief scientist, led to the conclusion that the Soviet Union had tested its first atomic bomb sometime between August 26 and 29. The Reds had long since been regarded as implacable enemies of the United States. But now, suddenly, they were alarmingly dangerous enemies.

The Communist bomb caused a fundamental change in U.S. strategic doctrine. For the first time in the nation's history it was vulnerable to crippling, if not devastating, attack from both within and without. A National Intelligence Estimate completed in September 1951 raised the specter of clandestine atomic, biological, and chemical attacks against the United States with weapons that were smuggled into the country or fired from off shore. The top secret report, put together by the civilian and military intelligence services, considered the possibility that such weapons could be dropped by disguised TU-4 bombers,* carried into seaports by merchant ships, smuggled in by diplomats or as commercial shipments, and even carried ashore at secluded places. "In its struggle with the non-Soviet world," the report said, "the USSR will have no scruples about employing any weapon or tactic which promises success in terms of overall Soviet objectives. Clandestine attack with atomic, chemical, and biological weapons offers a high potential of effectiveness against a limited number of targets, particularly if employed concurrently with, or just prior to, the initiation of general hostilities." In other words, the Russians would time the explosions and other havoc to coincide with a massive bomber and missile attack. Vasily Mishin, who succeeded Korolyov as the chief designer of his country's space program, would admit many years later that far from being a concoction of paranoid American intelligence analysts, such a plan was indeed considered as a way of offsetting the overwhelming advantage the United States had in warheads, bombers, and ballistic missiles.

But it was long-range atomic bombers, even though there were relatively few of them, that caused the most alarm. Even one could obliterate Washington. The feeling of vulnerability to a nuclear "sneak" attack—the first time in its history that the United States actually faced the possibility of sudden, widespread destruction—led to an unprecedented preoccupation with defense in the 1950s. It led directly to a civil defense program and fallout shelters, a vast, automated computer-driven radar fence along both coasts and across Canada, the building of a variety of fighter-interceptor jets specifically designed to attack Soviet bombers out at sea and over Canada, and ground-to-air missiles that were supposed to knock out any attackers that made it past the fighter screen. The anti-aircraft missiles would lead directly to ballistic missile defense, including President Ronald Reagan's Strategic Defense Initiative more than three decades later.

The leading anti-aircraft missile of the period was a mobile, white, pencil-

* The most obvious disguise would have been as Air Force B-29 Superfortresses, of which the Tupolev Design Bureau had made virtually exact copies.

shaped rocket named after the Greek goddess of victory and bearing sharklike fins. Nike was developed in successive versions—Nike-Ajax, Nike-Hercules, and Nike-Zeus—for the Army by Western Electric and became the first American anti-aircraft missile to knock down a bomber when one of the sleek weapons destroyed an unmanned B-17 drone in November 1951. The Air Force had its own anti-aircraft missile: a more ambitious, winged, bullet-shaped weapon named Bomarc that was propelled by a ramjet and carried a nuclear warhead. Seven hundred of them were built by the Boeing Aircraft Company and were deployed in limited numbers. The Navy protected its carriers and other vessels with still other missiles.

At the very outset and for decades to come, rocket design was driven by war-fighting requirements just as surely as it had been in the Third Reich. Scientists and engineers worked furiously to increase their weapons' reach and marry them to nuclear warheads. An arsenal of offensive missiles, some ground-launched with wings to increase their range, some developed as so-called standoff weapons that could be fired from bombers well out of range of anti-aircraft fire, were developed with varying success. They had names like Hound Dog, Mace, Matador, Navaho, and Skybolt. JPL developed a series of tactical missiles for the Army that were named, successively, Private, Corporal, and Sergeant. Most of them had, or would have, counterparts in the Soviet Union.

Von Braun and most of the other V-2 veterans were pulled out of Fort Bliss in 1950, after their usefulness in Hermes had ended, and resettled at the Army's Ordnance Guided Missile Center at Redstone Arsenal near Huntsville, Alabama. There they designed an improved V-2 called Redstone, which would become their adopted country's first medium-range ballistic missile and the first rocket to carry astronauts. More important, it was the first in a family of increasingly large lifters that would include an intermediate range missile named Jupiter, the rocket that would get the United States into space, and Saturn, which would get it to the Moon.

Consolidated Vultee (soon to be Convair and then General Dynamics) received a contract from the Army Air Forces in April 1946 to study a long-range ballistic missile. The project, known as MX-774, was withdrawn fifteen months later. Undaunted, Convair used its unspent funds to build and launch three test models that were somewhat smaller than the V-2. MX-774 would lay dormant for several years before emerging as Atlas, the nation's first intercontinental ballistic missile. Douglas Aircraft was in the meantime working on a 1,725-mile intermediate-range ballistic missile it named Thor after the Norse god of thunder. The aerospace community's penchant for naming its products after gods and goddesses, which would culminate in the Apollo spacecraft that went to the Moon, was a clear reflection of what it thought of itself and its contribution to society. No weapon system or space program would be named after Metis, the god of wisdom.

For all of the energy generated by the civilian and military rocket clique, though, the whole subject of rockets and missiles was only a sideshow where the Pentagonians were concerned. The big military battles during the Truman presidency were over bread and butter weapons: big Navy ships versus long-range Air Force bombers—supercarriers and submarines versus Stratojets and Stratofortresses—not over intercontinental ballistic missiles, space launchers, and Earth satellites. Those would come shortly.

The Enemy Imperative

The prospects for a coherent space policy in the early 1950s were bleak. If anything was going to move the United States out of its torpor, some of the rocket diehards believed, it would have to be foreign competition. In May 1952 the American Rocket Society reviewed suggestions it had received from its members on how to get the nation seriously committed to going to space. At least two members maintained that, in their opinion, the chances of accomplishing that were so pathetically poor that only a major achievement in that direction by the Soviet Union would get the United States moving. Richard Porter, who had run General Electric's part of Hermes, echoed the ARS members. After making the point that the first nation to get satellites into orbit around Earth would gain an obvious military advantage, the frustrated executive added that "the greatest utility with respect to civilization might be if the Russians were to build it instead of ourselves." In the same vein, and at about the same time, a Bell rocket engineer named Kurt Stehling pointed out that building a highway across Alaska was said to be impossible until the Japanese landed a small force in the Aleutians in World War II. Then, he added, the highway was completed in less than a year and a half so it could be used to repel a full-scale invasion.

Destination Moon, a 1951 film based on a novel by the politically conservative science fiction writer Robert A. Heinlein, was even more explicit. Faced with a government that is unwilling to seize the initiative because it is technically at peace with a relentless and diabolical foe, a small group of corporate visionaries decides to build a Moon rocket on their own. But first, they have to talk their fellow chief executive officers into backing the expensive project. "We are not the only ones who know the Moon can be reached," a retired general named Thayer warns a roomful of the business leaders during a lengthy pitch. "We're not the only ones who are planning to go there. The race is on and we'd better win it because there is absolutely no way to stop an attack from outer space. The first country that can use the Moon for the launching of missiles will control the Earth. That, gentlemen, is the most important military fact of this century." The rocket was hurriedly built and did reach the Moon. But in the tradition of *Frau im Mond,* it was too heavy to make it back to Earth unless one of the crew was left behind. The unlucky victim was a guy named Joe who had doubted all along that the moonship would make it, so it probably served him right. Whatever such a mission would have contributed to America's exploring the solar system, it would not have helped it militarily, since General Thayer was wrong. The time it would take missiles to strike Earth from the Moon would make them both vulnerable to interception and too late to help in all-out war.

The man in the White House, like his opponent in the Kremlin, gave no evidence whatever of thinking about outer space, let alone wanting to send machines and people there. Harry Truman's response to a preliminary plan to send an American satellite into orbit for the International Geophysical Year, or IGY, of 1957–58 was typically blunt. He called it "hooey."

Pop Culture Takes to Space

If space policy languished during Truman's presidency and into the beginning of Eisenhower's, it was not von Braun's fault. As the technical director of Germany's ballistic missile program, his job had been to design ballistic missiles and keep the Moon and Mars to himself. Now he had opposite priorities. The technical director of the U.S. Army Ordnance Guided Missile Development Group lived in a democratic country whose citizens, he believed, cared more about space than their government did. That being so, there was everything to gain by shelving the old rule against drumming up publicity and instead courting a public that was receptive and that voted.

The increasingly vocal members of the American Rocket Society were working on a satellite proposal that would go to the National Science Foundation, which was in the early stages of considering a satellite for the IGY. Meanwhile, Professor S. Fred Singer of the University of Maryland was inventing MOUSE, the Minimum Orbital Unmanned Satellite of Earth, which he would present at the Fourth International Astronautical Congress in Zurich in 1953.

An Annual Symposium on Space Travel, conceived and coordinated by the prolific Willy Ley as an answer to similar meetings being held in Europe, was launched at New York's Hayden Planetarium on Columbus Day 1951 (for reasons that were unabashedly symbolic as well as logistical).* That first meeting in the planetarium, which drew a full house of invited-only guests, was

* Columbus was called "that outstanding conqueror of space and earth" by one speaker. Ley referred to the symbolism in *Rockets, Missiles, and Space Travel.*

one of the watershed events in the American space program because it brought together under one astrodome an otherwise amorphous group of amateurs and professionals who shared the same goal. Luminaries such as Harvard's Fred Whipple gave papers on the scientific underpinnings of space travel, space medicine, and the embryonic status of space law. Robert P. Haviland, a General Electric engineer and a veteran of Hermes, was to deliver a paper about using satellites to relay television and other communication signals. But it was withdrawn at the last minute by his own company because it was allegedly too "speculative." Militarily valuable was more like it. Taking it all in at the planetarium was an editor at *Collier's* magazine named Cornelius Ryan.

Meanwhile, saucer-shaped unidentified flying objects—the ubiquitous UFOs—seemed to be hovering almost everywhere. The UFO phenomenon started in June 1947, when an alien spacecraft and its occupants were reported to have crash-landed at Corona, New Mexico, seventy-five miles northwest of Roswell. A rancher named W. W. Brazel reported having found a bright wreckage of rubber strips, tinfoil, heavy paper, and sticks in a field. His account made a local newspaper early the following month. It soon turned into the "Roswell Incident." Within a couple of years, stories circulated that aliens from the craft had been seen in the desert and that some of the creatures' burned bodies had actually been taken to the Roswell Army Air Field hospital. The month the UFO supposedly crashed at Corona, an Idaho businessman named Kenneth Arnold claimed to have seen a formation of alien spacecraft flying near Mount Rainier "like a saucer would if you skipped it across water." Six weeks before that first Hayden symposium, eighteen-year-old Earl Hart Jr. photographed what was purported to be a formation of flying saucers on a night flight over Lubbock, Texas. The picture, which appeared in *Life,* came to be known as the "Lubbock Lights."

Thousands of UFO sightings would be reported after the Roswell episode. A growing legion of cultists would not only describe vivid images of extraterrestrials and their spacecraft, but would claim to have been abducted by the aliens, in some cases hundreds of times, for experimental purposes, to be studied, or to be sexually abused. Throughout, the cultists insisted that the U.S. government was engaged in a massive cover-up because it didn't want people to panic, and that the remains of the spacecraft that crashed near Roswell, its occupants, and several other crashed saucers were hidden at a secluded and heavily protected place called Area 51, at Groom Lake, Nevada.

What Brazel saw was in fact the remains of a crashed Air Force balloon. The Air Force was experimenting with high-altitude balloons in the area in two programs: one in which airmen and realistic-looking, life-size dummies parachuted out of them at very high altitude, and the other to send camera-carrying balloons called U-1s over the Soviet Union to collect photographic intelligence. And Area 51, which in fact was and continues to be used to test top-secret high-performance reconnaissance planes, unmanned aircraft, and other advanced systems, made a convenient locale for the alien hardware.

There was no cover-up. "UFOlogy" was created as a modern myth shortly after the start of the atomic age, partly because frightened people wanted to believe that nuclear weapons can be countered by a higher intelligence, partly because many Americans did not trust government and were therefore prone to believe in conspiracies, and partly because they didn't want to think they were alone in the universe. A Gallup Poll taken fifty years after the Roswell Incident showed that 42 percent of college graduates still believed that flying saucers had visited Earth in some form. The UFO phenomenon, like the Earth-centered universe and Lowell's Martian canals, had absolutely no basis in fact and rested on superstition and irrationality. But it was great for space because it made it alluring for ordinary people. Given the number of individuals who claimed to have been abducted and either sexually abused or used for experiments, it is remarkable that none of them returned with so much as a towel as a souvenir.

At the same time, films such as *The Thing* and *The Day the Earth Stood Still* used alien invaders and emissaries, respectively, to send audiences around the country into bloodcurdling fear or pseudoreligious ecstacy. Michael Rennie, playing the alien Klaatu in *The Day the Earth Stood Still,* came to Earth (in a saucerlike spaceship) to warn its violence-prone inhabitants to curb their dangerous aggressiveness or risk annihilation by the galactic police. The film, too, had a cathartic effect on the first audiences in history to live with the prospect of nuclear annihilation. Unlike *Satan's Satellites, Battle of the Worlds, The War of the Worlds,* and most other films about encounters with aliens, Klaatu portrayed a kind, wise visitor. And God was the very model of cool logic: a robot. *When Worlds Collide,* which came out in 1951 (the year Klaatu and Gort, his faithful robot, landed on Earth and that doubter was left stranded on the Moon), described an impending collision between Earth and a runaway star. An arklike spacecraft with a severely limited number of earthlings and animals left just in time to begin a new civilization on another planet. Tsiolkovsky, the gravity hater, would have loved the theme of escape and transcendence. (Edward Teller, the so-called father of the hydrogen bomb, would probably have despised the fleeing fraidy cats. He and his colleagues at the Lawrence Livermore National Laboratory would hatch a plan in the mid-1990s to blast with a nuke any asteroid that threatens Earth.) The doomsday angle continued into 1998, when two film thrillers, *Deep Impact* and *Armageddon,* described horrific encounters between the home planet and "potentially hazardous objects," as astronomers euphemistically call asteroids, comets, and other dangerous objects that fly around the neighborhood at high speed.

At any rate, into 1951's fertile environment stepped Wernher von Braun, the budding savant of space, and another confirmed gravity hater. He appeared at the Hayden Planetarium's second symposium in October 1952 armed with a twelve-foot-high cutaway model of what was supposed to be a colossal three-stage rocket, the top of which was a winged space shuttle intended to return its crew to Earth.* The irrepressible visionary, whose paper addressed "The Early Steps in the Realization of a Space Station," had by then worked out his own multistage move to "conquer" space in impressive detail.

At one point von Braun debated Milton Rosen, who directed the Viking program, on the feasibility of his plan. Von Braun claimed that a station in space "will be the most fantastic laboratory every devised," as well as a "springboard to man's further ventures into outer space." Rosen rebutted that such speculation was fine if it was labeled as such and warned that funding "fantastic projects" could harm the country's defense effort. It was an Army-Navy game, and, like its football equivalent, both sides benefited from the publicity.

By then, a disparate band of rocket prophets, with the tireless von Braun at their head, was working the media to promote the cause. It was to be one of the first, if not *the* first, multimedia campaigns. "Connie" Ryan had gotten the idea of running a space extravaganza on the pages of *Collier's* during the first symposium. It would play out in eight spectacular issues, collectively called "Man Will Conquer Space *Soon*," starting in March 1952 and ending in April 1954. Ryan himself would go on to document and popularize the Normandy invasion in his best-selling book *The Longest Day* (which was turned into the longest movie).

The first *Collier's* piece featured lavish pictures of the shuttle and a roulette wheel–shaped space station, both as seen from space with Earth below. Many of the illustrations were done by Chesley Bonestell, whose training as an architect provided some technical knowledge to complement his enthusiasm for space. Bonestell's special effects, painstakingly researched for technical accuracy, were used for the landscape scenes in *Destination Moon* and *The War of the Worlds*. His diagrams of spacecraft and people floating in Earth orbit,

* The configuration was not exclusively von Braun's. Arthur C. Clarke, the respected space writer, described a "multistep rocket with a winged final stage, which is used to take material up to the orbit round the Earth, and then returns by atmospheric braking" in 1951. He called it a "ferry" or "tanker" ship and added that similar spacecraft could be used to land on planets with atmospheres, like Mars and Venus. He described his shuttle in *The Exploration of Space*, which had been published the year before.

along with haunting views of other worlds in the solar system—a half-lit Mars seen from Phobos, a ghostly Saturn viewed from Titan—provided a feast for space enthusiasts.

Likening the move to space to the Manhattan Project, the *Collier's* series sounded a clarion call to establish an American presence around Earth before the opposition occupied it. An editorial in the first installment did what *Destination Moon* would not quite do: it singled out America's opponent by name. The series, the editorial claimed, would constitute "an urgent warning that the U.S. must immediately embark on a long-range development program to secure for the West 'space superiority.' If we do not, somebody else will. That somebody else very probably would be the Soviet Union." Soviet scientists, like their American counterparts, concluded that a space station was feasible, the article continued, adding that the first nation to build such a station would control Earth. What followed bears repeating because it amounted to one of the earliest and most influential examples of boilerplate cold war space rhetoric:

> A ruthless foe established on a space station could actually subjugate the peoples of the world. Sweeping around the earth in a fixed orbit, like a second moon, this man-made island in the heavens could be used as a platform from which to launch guided missiles. Armed with atomic warheads, radar-controlled projectiles could be aimed at any target on the earth's surface with devastating accuracy.

An American space station, on the other hand, promised to "guarantee the peace of the world" and effectively eliminate iron curtains because of its occupants' ability to scrutinize every inch of Communist territory. As would be the case from then on, the likely reaction of the Soviet Union and its allies to having a manned and armable enemy outpost coasting over them every two hours was studiously avoided. At an estimated cost of $4 billion, the editorial continued, such a vehicle would be cheap. Von Braun would not have written those words, but his hand was surely there when Ryan or another editor did.

What followed in that issue and the others to come were richly detailed descriptions, complete with cutaway drawings, of a 250-foot-diameter space station whose slow spinning would create artificial gravity 1,075 miles above Earth. There were also the three-stage shuttle-booster combination; Whipple's description of an orbiting telescope and observatory; space taxis and satellites; astronaut clothing, including what would come to be called a manned maneuvering unit for work outside of the spacecraft; emergency capsules for quick ejection from a crippled shuttle; and comprehensive plans for expeditions to the Moon and Mars. The detail was extraordinary. The lunar explorers would live and work in the main body of a cargo rocket that had been turned into lunar Quonset huts, while a pair of seventy-five-foot-by-thirty-six-foot half cylinders, set in a deep chasm, protected their occupants from meteorites and cosmic rays.

It is hard to overstate the importance of the *Collier's* series. With von Braun as its architect and credible specialists filling in details based on real science and engineering rather than fanciful speculation, the articles constituted a blueprint for the U.S. space program. The order of the events laid out in the magazine would change when President John F. Kennedy decided to send men to the Moon, but the essential sequence would remain: use a shuttle to build a station, then use the station to push off for the Moon and Mars. Each step would be interlocked with, and therefore dependent upon, the others.

What is more, the technology was deftly written and larded not only with a new jingoism, but with the ever-seductive allure of possibly discovering life beyond Earth. "How can we say with absolute certainty that there isn't a *different* form of life existing on Mars—a kind of life we know nothing about?" Whipple teased. "There is only one way to find out for sure what is on Mars—and that's to go there." Finally, this was a scheme that was deposited, not in some inaccessible government office or in an ivy-twined academic mausoleum, but in subscribers' mailboxes and on newsstands across the country. And at fifteen cents an issue, the whole plan could be had for a dollar and twenty cents.

The essence of "Man Will Conquer Space *Soon*" was expanded and published in a series of four books, starting with *Across the Space Frontier* in 1952, followed by *Conquest of the Moon, The Mars Project,* and *The Exploration of Mars,* which appeared in 1956.

Meanwhile, television's graphic possibilities and growing audience were not wasted on von Braun either. On Wednesday, March 9, 1955, the fatherland's former chief rocket designer took his place with Goofy, Tinker Bell, Jiminy Cricket, the Seven Dwarfs, and Davy Crockett on *Disneyland,* the weekly network series that had begun the previous October. "Tomorrowland," one of the program's four permanent segments, was promoted as "science factual" and used animation, drawn from the *Collier's* series, that was reminiscent of *Fantasia* in its thunderous tonality, vivid sound effects, and dramatic score. A shirtsleeved von Braun, correctly billed as "one of the foremost exponents of space travel," gamely used his three-stage shuttle rocket and other props to sell space travel to Mouseketeers and their tail-fin-addicted parents.

"If we were to start today on an organized and well-supported space program," von Braun said, looking the camera straight in the eye, "I believe a practical passenger rocket can be built and tested within ten years." He was referring to his shuttle, which would have had a crew of ten. But it was not to be, at least not in a decade. Ten years from that night two so-called superpowers would be locked in a fierce competition not to send civilians around Earth but to land soldiers on the Moon.

Meanwhile, remembering the most important of the lessons learned at the Raketenflugplatz, von Braun assiduously played to the Musketeers as well as to the Mouseketeers. On October 20—seven months after he made his debut for Disney—he delivered a lecture on the "Challenge of Outer Space" in a smoke-filled room at the Armed Forces Staff College. The forty-minute briefing, followed by questions, outlined the United States' participation in the upcoming International Geophysical Year and gave a remarkably detailed description of what space rocketry entailed, including elliptical versus circular orbits, multistage launch vehicles, the use of centrifuges to simulate g forces, the nature and purpose of space suits, the Redstone's XR-1 upper stage, which was designed to return crews to Earth and which was therefore another precursor to the shuttle, and the possibility of propelling rockets with nuclear explosions.*

Gone was the folksy fellow with rolled-up sleeves and Disneyesque props. He was replaced by a grim-faced individual in a dark suit who puffed on cigarettes from behind a desk. This von Braun carefully explained that a satellite in polar orbit would pass over every place on Earth every twenty-four hours, a perfect route for robotic espionage. He noted that maps of Eurasia were five hundred yards off and added that the error could be reduced to twenty yards by using spacecraft instead of airplanes (a point later confirmed by RAND's Merton E. Davies). The implication was wasted on no one in the room: taking photographs of Earth from space not only would create an intelligence bonanza but would vastly improve targeting accuracy. To further make the point, von Braun told the Army, Navy, and Air Force officers that satellites, which he claimed were impervious to attack from the ground, could also guide ballistic missiles to their targets with unerring accuracy. He was so attuned to his audi-

* This would eventually be intensively researched at Princeton by physicists Theodore Taylor, Freeman Dyson, and others, as well as at General Electric by Dandridge Cole, in a project called Orion. The scientists who conceived of using nuclear explosions to propel spacecraft wanted to use them for solar system exploration. But like so many other scientists, they were funded by the military—in this case the Air Force—and were soon told by a captain that nukes could be used to keep spaceborne battleships on patrol as part of a Deep Space Bombardment Force that would preserve Pax Americana well into the twenty-first century by cruising majestically in orbits beyond the Moon. The captain believed that the force would stand between the despots in the Kremlin and the Free World. Following a briefing to other Air Force officers who did not seem to share their young colleague's imperial ambitions, funding for Orion was withdrawn and the project died. (Dyson, *Weapons and Hope,* pp. 65–66.)

ence that at one point he mistakenly referred to a civilian space "ship" as a "missile" and then quickly corrected himself. And, von Braun warned the officers, as he had warned the readers of *Collier's,* the Russians were "well aware of the capability of rockets" and he therefore hoped that the United States would conquer space while it still was able to do so. The space race had begun.

5

The Other World Series

The cold war would become the great engine, the supreme catalyst, that sent rockets and their cargoes far above Earth and worlds away. If Tsiolkovsky, Oberth, Goddard, and the others were the fathers of rocketry, the competition between capitalism and communism was its midwife.

The Second World War had been straightforward enough: a traditional contest between nation-states over territory. To be sure, the Axis shared a radically conservative rightist doctrine and fought ferociously at home and abroad to spread it. But the ideology was used to strengthen the state itself; it was never conceived as a political philosophy that would be carried by missionaries. Since both Germany's and Japan's fascism had in part been based on racial superiority, there could have been no Nigerian Nazis for Hitler, no Ethiopian Fascists for Mussolini, and certainly no Americans for Tojo.

Communism was fundamentally different because it called for the nation to strengthen a crusading ideology that in theory was transnational. Marxist doctrine had it that the nation-state existed solely to protect, nurture, and promulgate the ideal of international communism, and once that ideal had been reached, the state would wither and die. It was the workers and peasants of the world against their bourgeois, capitalist exploiters, and borders were therefore as meaningless politically as they were meteorologically. That, at least, was the true faith as delivered by Marx and evangelized by Lenin.

Josef Stalin quietly shelved the theory in the 1930s, however, and so did his country's lackeys in Eastern Europe and the Far East after World War II. Turning Marxism on its head, Stalin expediently ordained that the Union of Soviet Socialist Republics would remain the impregnable bulwark of socialism, protecting and nourishing it in the face of a hostile and dangerous outside world. Defining the nation as socialism's indispensable incubator provided a new justification for old-time nationalism. Yet where many of communism's agents and sympathizers throughout the world were concerned, the theory remained valid.

The tenets of Marxist ideology had been understood by the intelligentsia since the publication of *The Communist Manifesto* in 1848 and by most of the rest of the world since the Russian Revolution in 1917. Communism was taken very seriously in the United States precisely because it could come from within as well as from without. Since its apparent strength rested not on a bayonet but on an ideal, it could in theory quietly infiltrate a system. "Deutschland über Alles" was a *national* anthem; all that it stood for could be contained and removed from the body politic in a conventional war the way a diseased gall bladder or kidney stone could be removed from a patient. "The International" was taken to be more insidious; it was the theme song of cancer, a disease that surreptitiously rode the bloodstream of the world, attacking and devouring every healthy organism in its path and growing bigger and more dangerous as it did so.

By August 1953, when the Kremlin announced it had exploded its own hydrogen bomb, there was no doubt in the United States that it was locked in a multidimensional war with an enemy that was potentially far more dangerous than those it had recently vanquished. The cancer had already spread to Greece in 1947, Berlin in 1948, China in 1949, and Korea in 1950. And insurrections were sprouting in the jungles of Malaya, Indochina, and the Philippines.

Meanwhile, the specter of a homegrown Red menace—the international Communist conspiracy, as Senator Joseph R. McCarthy put it in his relentlessly accusatory monotone—gripped much of the United States in something approaching national paranoia. They were the hoary days of Julius and Ethel Rosenberg, Alger Hiss, Judy Coplon, and David Greenglass; the Hollywood Blacklist and its hounded victims; the purging of the Department of State; Herbert Philbrick's spy-counterspy revelations in *I Led Three Lives;* Walter Winchell's unrelenting crusade against Stalin and his minions everywhere; films such as *I Married a Communist* (which was worse than contracting *Reefer Madness*); and *The FBI in Peace and War,* a dramatic weekly radio tribute to J. Edgar Hoover's dedicated sleuths that a certain generation will forever associate with its theme song, "The Love for Three Oranges." (The program's creators showed a certain amount of courage in using that particular music, since it comes from an opera by Sergei Prokofiev, a Russian. Or maybe they knew that McCarthy, Roy M. Cohn, and other zealous Red-baiters were not sophisticated enough to catch on.)

Democracy seemed to mock itself. Subversives camouflaged as "progressives" appeared to lurk everywhere. A legion of dedicated spies and fifth columnists was allegedly training behind the Iron Curtain to slip into the United States and steal it from its own people.

Jack Webb, an actor who played a tough Los Angeles police sergeant named Joe Friday on a weekly television series called *Dragnet,* made a documentary for the Department of Defense, *Freedom and You,* that showed what it called "a typical American town . . . shrouded in secrecy and protected by utmost security, deep behind the Iron Curtain." The place was stocked with Communist agents-in-training who looked and sounded exactly like the boy and girl next door. "Frightening, isn't it?" Webb asked in a tone he might have used to grill witnesses to a murder. "From all appearances, this community could be in Iowa, California, Tennessee. But appearances are deceptive. . . . You might call this a college town, Communist style," he added as the camera showed bobby-socked teenagers hanging out at a soda fountain. "As part of a long-range plan to destroy our free way of life, these young Communists are studying the economic, political, and religious institutions that are the very heartbeat of America. They're studying *you*" for a sudden, decisive takeover of the United States. Whew!

There was some basis in fact to the penetration theory, but it had to do with espionage, not real subversion. Documents discovered in the archive of the Communist International, or Comintern, after the end of the cold war showed that a conspiracy had indeed been under way. The papers revealed that, far from being homegrown and populist, the Communist Party of the United States had been financed by Soviet intelligence almost from its founding in 1919 and had a core of rigorously disciplined members who placed highest allegiance to their Soviet masters.

American Communists, proudly subversive and either unwilling to believe in Stalin's brutality or convinced that it was necessary, organized a dedicated, though tiny, conspiratorial network in 1935. Rudy Baker, a graduate of Moscow's International Lenin School and the leader of the American party's clandestine branch, set up a "Brother-Son Network" in which he, the "Son," secretly conspired with "Brother" Georgi Dimitrov, head of the Comintern, in an intensive—and partly successful—effort to steal U.S. atomic bomb secrets. The records show that some U.S. government agencies, including the Department of State, were penetrated by secret informants and "agents of influence" or were simply compromised by idealistic sympathizers. There was less danger at home than most Americans believed—and more than they knew.

The View from the Kremlin

At the same time, first Stalin and then the xenophobes who succeeded him nursed their own fears, many of them well founded. All of them knew, as every schoolchild in the Soviet empire knew, that their vast country was Eurasia's heartland and had therefore been a coveted strategic prize for more than a thousand years. In a real sense, Russia had been defined by almost constant warfare and the catechism of conflict was as well known as its religious counterpart had been before the revolution. There had been wars against the Vikings, wars against German and Swedish crusaders, against Bulgarians, Hungarians, Lithuanians, and Poles. There had been catastrophic invasions by Tartars and Mongols and a whole series of wars with Turks and Persians. Even during the height of the cold war it was Turkey and Muslims in general, not the United States, that most Russians took to be their traditional enemies; St. George slaying a dragon is Moscow's logo, and that mythical reptile is Islam. Moscow itself was almost burned to the ground during Napoleon's short occupation in a war that was immortalized by both Tolstoy and Tchaikovsky. And it was the Turks again, this time with the British and French, who attacked the Crimea four decades later. Russia was so worried about being attacked that its railroad tracks were made wider than others in Europe and Asia just to confound invading armies.

The surprise attack by Japan against the United States was less infamous (to use Franklin D. Roosevelt's angry adjective) than the German invasion of the socialist republics that took place the same year. Hitler had planned Barbarossa, as the assault was called, under cover of a nonaggression treaty that was signed in 1939. So the U.S.S.R. not only had been the victim of a surprise attack but had been betrayed as well. And the Great Patriotic War that followed—an inferno that lapped as far east as Stalingrad and the northwest suburbs of Moscow itself—caused plunder, destruction, and death beyond comprehension. The memorial flame to the unknown soldier who perished defending Moscow that burned at the Kremlin wall also eulogized a dozen "hero cities"—Smolensk, Murmansk, Stalingrad, Odessa, Sevastopol, Leningrad, Kiev, Minsk, and others—that paid the price for the catastrophic betrayal. There was no city or town that did not have a memorial honoring its fallen civilians and soldiers; no community that did not have its share of bent and mutilated survivors, rows of plastic medals and battle ribbons hanging from their ill-fitting jackets and dresses. Next to nothing would be made of the Soviet sacrifice in the United States, where cold war propaganda insulated most of the people against the staggering savagery that had been endured by their old allies: Slavic subhumans, by the definition of the Aryans who slaughtered them. Americans measured the liberation of Europe and the defeat of Japan almost exclusively by the blood they alone had shed. Their loss, some four hundred thousand killed, was terrible enough. Yet it was paltry compared to the well over a million who perished holding off the Nazis at Stalingrad and Leningrad alone. The total number of Soviets killed in the war will never be known, but it is usually estimated at 27 million.

Now the champion of the West, the only nation to have used nuclear weapons and the owner of enough of them to turn the Soviet Union into radioactive rubble, had an overwhelming monopoly of strategic bombers and was in the early stages of developing the Atlas ICBM. Soviet Air Force Lieutenant Colonel Grigori A. Tokayev reportedly claimed that his former leaders thought that Texas (probably meaning White Sands) had turned into a vast Peenemünde.

The profoundly cynical men who searched the horizon from behind the Kremlin's brooding red ramparts saw that they were surrounded by a well-armed, technologically advanced enemy that had come out of the Great Patriotic War economically invigorated and heavily armed. Their own land, in contrast, had been ravaged and bled to its limit. Stalin, his successors, and their closest advisers therefore believed implicitly that the new threat from the West had to be checked or the United States and its partners in the North Atlantic Treaty Organization would finish what the Nazis had started. But even they could have had no way of knowing how close to the truth they were.

What Stalin and the others knew about U.S. airpower alone after the war would have caused deep apprehension. While it was never articulated in so many words, at least publicly, the United States Air Force did not think of its core mission as having to do with dogfighting, interception, testing, and other relatively glamorous work. The Air Staff understood that the service's reason for existence—its core responsibility—was long-range strategic bombardment: sending high-explosive "iron bombs" or nukes over long distances to blast the enemy until it was so dysfunctional that it lost the will to fight. That was the goal: to obliterate the enemy both on the battlefield and behind it.

On September 19, 1945—only seventeen days after the end of the war with Japan—three Army Air Forces generals, including Curtis LeMay, were ordered to fly three stripped-down B-29s on a Great Circle route all the way from Japan to Washington (poor weather in the capital and low fuel convinced them to land in Chicago instead). The only significance the *Chicago Tribune* saw in the nearly 7,000-mile flight was that it heralded the time, not far off, when commercial airliners would be able to fly from Chicago to Tokyo in twenty-four hours. But LeMay knew better. "I would guess that this flight was dreamed up to demonstrate and dramatize . . . the long-range capability of the [B-]29 to the American people and to the world at large," he said. Indeed. It was no coincidence that on August 30—four days before the Japanese surrendered—General Leslie R. Groves, who headed the Manhattan Project, had been handed a "Strategic Chart of Certain Russian and Manchurian Urban Areas" that amounted to a target list crammed with whatever information could be scraped together about cities and industrial facilities in the Soviet Union. It estimated that six atomic bombs would be needed to destroy

Moscow, for example, and six more to demolish Leningrad. Saunders B. Kramer of the Lockheed Missiles and Space Company, which manufactured the Polaris ballistic missiles fired from submarines, recalled years later that he saw a list of 175 Soviet cities, all with accurate coordinates, that were targeted for obliteration by the ever-prowling subs.

The concept of truly long-range bombardment took hold in that pivotal year, 1954, when reliance on what were called "forward bases" in Europe, North Africa, and Asia began to diminish. To prove that the brand-new Boeing B-47 Stratojets had worldwide reach, three of them flew 6,700 miles nonstop from California to Japan by refueling in the air along the way. In August, other B-47s made the 10,000-mile flight from Georgia to French Morocco and back, refueling in flight four times. They were followed by still others that took off from the same Georgia base, flew a "simulated bombing mission," and landed in North Africa. Far from being secret, the marathon missions were staged to demonstrate to friend and foe alike that no place on Earth was safe from retaliation if the United States or its allies were attacked. The 308th Bomb Wing was even awarded a trophy for its long-distance operations, which were widely heralded as being part of the strategy of "deterrence." Similarly, annual reconnaissance, navigation, and bombing competitions between SAC units were held not only to sharpen the flying and give the crews something to strive for short of the real thing, but to project the Strategic Air Command's image of invincibility.

In 1955, SAC had 1,309 dedicated bombers, in addition to 379 others assigned to reconnaissance that could carry bombs if they had to.* Most of these were the sleek, swept-wing B-47s that could carry nuclear weapons at up to six hundred miles an hour. Eighteen of the others were the first giant B-52 Stratofortresses to become operational. The B-52s were not only faster than their predecessors, but they could wipe out targets five thousand miles away and make it back home without refueling. It was no secret that SAC's bomber crews practiced attacking simulated Soviet targets under radar and at very low altitude almost around the clock and that some bombers were in the air all the time. Not only was it not secret, in fact, but the very concept of deterrence required the Russians to know that the aircrews practiced attacking them repeatedly.

The strategy grew out of a belief shared by many U.S. Air Force generals, LeMay and Thomas S. Power foremost among them, that only a devastating first strike against the U.S.S.R. could prevent an eventual balance of nuclear forces—an atomic standoff—that would benefit Moscow's overwhelming su-

* The number of bombers increased each year from 1946, when the Air Force had 148 B-29s. A total of 446 B-36s, 2,048 B-47s, and 744 B-52s were delivered to the Air Force. They far outnumbered and outflew their Badger and Bear counterparts. (Polmar and Laur, *Strategic Air Command,* pp. 229, 236, 241.)

periority in land forces. As commanding general of the Strategic Air Command from 1948 to 1957, the longest tenure of anyone who has held that position, LeMay personally picked its heavily publicized motto: "Peace Is Our Profession." He might have added that "War Is Our Wherewithal," in that he assembled the greatest armada of destruction in history. Likening his bomber force to the Pax Romana and Pax Britannica, LeMay unabashedly called it the Pax Atomica and reportedly told one of his combat crews, "There are only two things in this world: SAC bases and SAC targets." On paper, the beetle-browed, jowl-jawed, cigar-chomping general did not make defense policy. But having him and his ever-ready bombers at hand was like having an angry Doberman pinscher on a leash.

A doctrine of "preemption" was quietly adopted by the air generals in 1954 in which Soviet forces would be beaten to the punch if intelligence indicated that they were getting ready to attack the United States. LeMay coolly calculated that 80 percent of the nation's nuclear weapons would have to be dropped on the Soviet Union in a surprise attack that lasted only hours if the United States was to win a decisive victory.

That year, LeMay's bomber crews were cocked to deliver as many as 750 atomic bombs in a simultaneous "Sunday punch" surprise attack that would strike from all sides of the Soviet empire and leave an estimated 60 million dead and another 17 million injured. A Navy officer who attended a SAC briefing on the plan on March 15 came away appalled. "The final impression," he noted, "was that virtually all of Russia would be nothing but a smoking, radiating ruin at the end of two hours."

LeMay would remain a deeply enigmatic character long after the end of the cold war. In *Dark Sun: The Making of the Hydrogen Bomb,* published in 1995, Richard Rhodes would report that LeMay and Power, who succeeded him as commander in chief of SAC, foresaw circumstances that could have forced them to take the responsibility for starting World War III. Challenged on that score by an emissary from President Eisenhower, LeMay was quoted as having answered with characteristic bluntness: "If I see that the Russians are amassing their planes for an attack, I'm going to knock the shit out of them before they take off the ground." Told that the plan violated national policy, the SAC commander is said to have shot back, "I don't care. It's my policy. That's what I'm going to do."

But the general's willingness to start a nuclear conflagration would also be vigorously disputed. At least two Air Force historians would take sharp issue with Rhodes. One, Major Mark J. Conversino, would write in a review of *Dark Sun* that LeMay had been unduly demonized as a two-dimensional character when the fact of the matter was that he was a very complex individual who had become the victim of his own one-liners. Whereas Rhodes would assert that it was only LeMay's oath that had prevented him from attacking the

Soviet Union unilaterally during the Cuban missile crisis in October 1962, Conversino would use precisely the same point to show that the general had indeed felt bound by that oath and respected it. In a conversation with the author in December 1995, McGeorge Bundy, who was John F. Kennedy's national security adviser, came out on Conversino's side. Asked whether he thought LeMay would have started a nuclear war, Bundy answered emphatically that "he was too good a soldier to do that."

The fact that SAC had the wherewithal to clobber the Soviet Union during LeMay's tenure, but didn't, makes the case for the general. Yet two salient points can be made about the situation. LeMay promoted himself as being so ferocious that he was taken to be exactly what he indicated he was by friend and foe alike, causing everything from unease to outright alarm on both sides, and particularly in a jittery Kremlin. More to the point, the line between deterrence and conflagration—accidental or not—narrows in proportion to the destructiveness of the weapons. With each side swelling its nuclear arsenal to keep the other from attacking, the likelihood of inadvertent war increases.

Nevertheless, when President Dwight D. Eisenhower proposed an "open skies" policy to Nikita Khrushchev in 1955, in which each side's reconnaissance aircraft would monitor the other side to prevent sudden, nasty, and potentially catastrophic surprises, the Soviet leader rejected it out of hand. Khrushchev was convinced (wrongly) that the plan was a masquerade to collect intelligence pictures for use in locating Soviet targets. His son Sergei would say decades later that his father had been convinced that the United States was "really looking for targets for a war against the U.S.S.R. When they understand that we are defenseless against an aerial attack, it will push the Americans to begin the war earlier . . . [and] if in this fear of each other the Americans realized that the Soviet Union would become stronger and stronger, but was weak now, this [intelligence] might push them into a preventive war." Khrushchev, then, took LeMay at his word. Each side's fear of a surprise attack mirrored the other's.

The answer, in Khrushchev's estimation, would not be to try to match LeMay and the others plane for plane. Rather, it would be to leapfrog the U.S. bomber force by mass-producing missiles that could reach the United States with punishing effect; missiles that would reach America before the bombers reached the Soviet Union and that would therefore protect the nation by forcing a standoff.

The Fear of Surprise Attack

The men who ran the government of the United States in the 1950s and the informed political, industrial, and academic establishment from which they came had two primal fears, both of which drove national defense: "appeasement" of the Communists and a miscalculated Soviet surprise attack that could come as a result of it. Eisenhower's generation and the one after it had lived through Neville Chamberlain's appeasement of Hitler and seen the consequences: unspeakable atrocities, nations overrun and brutalized, and then, inevitably, a long and bloody war of retribution. If that mistake was not to be repeated, it was believed, communism would have to be contained. Every transgression would have to be answered in its earliest stages and with unwavering resolve. McCarthy's purging of the Department of State and dogged attack on the Army, the sensational hearings in which largely unsubstantiated recriminations were aired on the new medium of television, and the question that hung over successive administrations like the sword of Damocles—"Who lost China?"—weighed heavily even after the senator from Wisconsin faded from the scene and then died in 1957. Regional alliances were put in place to check Communist military and political breakouts. Where deterrence did not work, as in Korea, Cuba, and Vietnam, the doctrine called for counterforce.

The Eisenhower administration gave no real evidence of worrying that the country was being threatened from within. But the Soviet military machine was another matter. Intelligence coming out of the Soviet Union during most of Eisenhower's first term was piecemeal at best. But the gaping holes fed fear.

"Our knowledge of what was going on inside the U.S.S.R. was desperately weak," Dr. George B. Kistiakowsky would later recall. "Kisty," as he was known to a generation of Harvard chemistry students, had worked on the atomic bomb and gone on to become Ike's science adviser. "Much information had originally come from German engineers who had worked in the U.S.S.R.," said Kistiakowsky, referring mainly to the thousands of engineers and technicians who had helped develop the V-2. "But they were never trusted very much, and as the Soviets got better at nuclear weapons and guided missiles, the Germans were separated and finally allowed to emigrate to the West," he explained. There were also repeated attempts to parachute spies into Siberia and elsewhere or land them by submarine. "But that was a total failure," Kistiakowsky added. "They were usually intercepted and liquidated." And the defectors, who came over by the thousands, were a decidedly mixed bag.

One of the best in the bag, Colonel Tokayev, told his interrogators the Politburo had ordered that priority be given to strengthening the Long-Range Air Force. By 1953, U.S. intelligence estimated that the Kremlin's strategic air arm had a thousand obsolescent TU-4 bombers and was receiving as many as twenty-five more a month.

Worse, that same year the prototype of an indigenous jet bomber called the Mya-4 took to the air. The plane was code-named "Bison" by NATO and was estimated to be able to carry 10,000 pounds of bombs 7,000 miles at up to 560 miles an hour. Others, including Andrei Tupolev's TU-16 Badger and a huge turboprop called the TU-95 Bear, were also in production in 1953. The Bear could haul 25,000 pounds of nuclear weapons 7,800 nautical miles. That put the lumbering mastodon within easy range of the continental United States.

Washington was desperate for hard intelligence about what was happening in the veiled and heavily protected Soviet military establishment, and particularly in its aviation, missile, and nuclear programs. Richard M. Bissell Jr., the mild-mannered Yale economics professor who would soon head the CIA's U-2 program, shared a conviction with others in the national defense establishment that the growing stockpile of Soviet nuclear bombs and large numbers of jets to carry them might tempt the men in the Kremlin to launch a preemptive first strike. "This was at a time when the Pearl Harbor surprise attack was still very much on everyone's mind," he would say later. The combination of obsessive Soviet secrecy and steadily growing military capability, capped by nuclear weapons and long-range bombers and missiles, created an unprecedented need for reliable, constantly updated, intelligence. What Bissell was getting at, put simply, was that for the first time in its history, the United States faced the possibility of being obliterated in a matter of hours. That is what drove a growing intelligence collection process to frantic extremes.

The Central Intelligence Agency, which was created by the National Security Act of 1947, used two basic kinds of intelligence collection methods. One, traditional spying, concentrated on getting information out of disaffected or greedy people on the opposition's side. It was, and remains, the netherworld of international relations: the land of the spook and the counterspook. The other was generically called technical intelligence, meaning that it used machines: many kinds of cameras and listening devices on land and sea and in the air to do the same thing.

As early as 1947, crews in bombers and patrol planes specially equipped for top secret electronic and photoreconnaissance missions flew in a shadow war. They prowled along the edges of Eastern Europe and the Soviet Union, intercepting communication traffic, locating and measuring radar signals, and photographing harbors, air bases, and anything else that could possibly help determine what the intelligence specialists would soon come to call the enemy's "capabilities." Air Force and Navy reconnaissance planes routinely dashed right at Soviet radar in an effort to activate them so their signals could be recorded for use by attacking bombers in the event of war. Others darted deep into the U.S.S.R. ("denied territory," as the Pentagon drily called it) in desperate attempts to penetrate the curtain and find out what, exactly, the opposition had in the way of weapons to defend itself and threaten the United States. The intrusions were as dangerous as they were illegal. Many U.S. reconnaissance crews returned from the sorties with hair-raising stories of encounters with cannon-carrying MiGs. Some crews never returned.

And by the middle of the 1950s, the fliers were supplemented by radar dishes and antennas designed to intercept communication that had begun to surround the Communist bloc. The Soviet Union itself was targeted by antennas, some of them huge, that sprouted from England to West Germany to Turkey to Iran to Alaska that tracked missile tests and listened to radio signals of every description around the clock.

By then, U.S. intelligence had already accumulated a great deal of information on the nascent Soviet ballistic missile program. Yet with all of the spies, informants, peripheral flights, penetrations, radar coverage, and communication intercepts, the knowledge gap remained maddeningly large.

The Technological Capabilities Panel

With no clear idea of what was happening behind the Iron Curtain, and with the specter of menacing goblins seeming to fill the void, in the summer of 1954 Eisenhower authorized the creation of a blue-ribbon committee whose job it was to find ways of meeting the Soviet threat. James R. Killian Jr., the president of MIT, chairman of the President's Science Advisory Committee, and the man chosen by Ike to head the group in late July, said later that Eisenhower attached a high priority "to reducing the probability of military surprise," adding that the fear of a Soviet surprise attack "haunted Eisenhower throughout his presidency." And were such an onslaught to come, Killian added as if echoing Spaatz while all but dismissing the clandestine attack possibility, it would come from the sky.

The Technological Capabilities Panel consisted of a steering committee, three project teams, a communications working group, and a military advisory committee. The forty or so individuals who participated in the study represented some of the leading minds in the nation's industrial, military, and scientific elite. The project teams addressed offensive, defensive, and intelligence capabilities. Their combined report, which depended heavily on the National Intelligence Estimates produced by the nation's foreign intelligence organizations, was placed on Eisenhower's desk on Valentine's Day 1955. It was marked "Top Secret."

Within the tiny fraternity of those who were cleared to know about it, the Killian group's report was officially called "The Report to the President by the Technological Capabilities Panel of the Science Advisory Committee." Less officially, it was called "Meeting the Threat of Surprise Attack" and, even less officially, the "Surprise Attack Study" or simply the TCP report. Whatever its name, it was one of the few documents that would have a truly profound influence on the military posture of the United States throughout the cold war and long afterward.

While acknowledging that the United States had an "air-atomic power" advantage over the Soviet Union, the report added that the nation was nevertheless vulnerable to a surprise attack. It went on to address a litany of shortcomings: no large "multimegaton" (hydrogen) bomb capability; no reliable early warning system; no really effective technical intelligence collection system. While the three major categories of problems were addressed—offense, defense, and intelligence—it was recognized that in most respects, they were intimately interrelated. The nation's offensive capability—its bombers and naval vessels (there were no long-range missiles to speak of yet—were directly dependent on the defensive system for protection and on intelligence. The last, boiled down, was supposed to provide timely warning of an impending attack and pinpoint enemy targets for a counterattack. During the period ending in 1960, the report said, the United States would be severely damaged by a surprise nuclear attack, but "would emerge a battered victor." After that, with both sides possessing large numbers of ICBMs tipped with multimegaton warheads, "an attack by either side would result in mutual destruction."

The offensive group recommended that the highest priority be given to the development by the Air Force of intercontinental ballistic missiles. It also called for the construction of atomic submarines that could cruise under water for long periods and fire ballistic missiles. The bomber force had to be dispersed and specially protected. Some armed bombers, the Killian panel added, had to be in the air at all times to guarantee a retaliatory strike even if all of the ones on the ground were blown up. The group responsible for planning the continental defense of North America came up with its own master plan. It suggested a virtual wall of radar around much of the country, beefing up fighter defenses to intercept invading bombers, studying anti–ballistic missile (ABM) systems, monitoring and controlling coastal waters to prevent an attack from that direction, and improving overseas communications. ABM work started in 1957 and, a treaty against it notwithstanding, would never really end.

It was fitting that of all the highly classified parts of the TCP report, those dealing with intelligence collection were the most closely guarded. The intelligence group was headed by Edwin H. "Din" Land, the inventor of the Polaroid Land camera, and included James G. Baker, a Harvard astronomer and optics expert. "If intelligence can uncover a new military threat, we may take steps to meet it," the report said. "If intelligence can reveal an opponent's specific weakness, we may prepare to exploit it. With good intelligence, we can avoid wasting our resources by arming for the wrong danger at the wrong time."*

Land and the others recommended that technology for intelligence collection be pushed to its limit. Initially, this meant development of the U-2 spy plane, which Clarence L. "Kelly" Johnson of Lockheed's Advanced Development Projects division—the renowned Skunk Works—was building for the CIA at that very moment. Its remarkable camera and lens would be designed by both Land and Baker to spot objects the size of a basketball from over thirteen miles in the sky. But however capable the high-flying reconnaissance plane would be, and it would prove to be very capable indeed, there was no doubt that the opposition would eventually find a way to shoot it down. Its triple Mach-busting successor, the SR-71, would be far less vulnerable. Yet any airplane was at risk to an opponent determined to stop it.

Earth satellites were different. Most obviously, their altitude would provide a far wider view than could be had by air-breathing planes. That view would include the entire planet if the satellites circled Earth over its poles so all of Earth rotated under them. It would expose the entire Communist bloc (and everyplace else) to their cameras. Second, the extreme altitude would vastly increase the safety margin, as the 1946 RAND report had noted. Finally, flying through the space above another country was not taken to be illegal, whereas penetrating airspace absolutely violated international law.

Indeed, as events would soon show, the Eisenhower administration was thinking hard about legitimatizing space reconnaissance and coming to some interesting conclusions. The TCP's intelligence panel put "freedom of space" high on its agenda. A covering letter that accompanied a copy of the report that went to the Department of State mentioned a "re-examination of the principles of freedom of space, particularly in connection with the possibility of launching an artificial satellite into an orbit about the earth, *in anticipation of use of larger satellites for intelligence purposes*" (italics added).

Besides recommending that a high-altitude reconnaissance plane be built, the panel knew that satellites to do that work were on the horizon, and it therefore also recommended that they be used to collect intelligence, too. Land and his colleagues knew, for example, about a RAND proposal for a nuclear-

* President Lyndon B. Johnson would echo this when he made the first official, though oblique, reference to clandestine space surveillance. "I wouldn't want to be quoted on this," he told a small group of government officials and educators in Nashville in March 1967, "but we've spent thirty-five or forty billion dollars on the space program. And if nothing else had come out of it but the knowledge we've gained from space photography, it would be worth ten times what the whole program has cost. Because tonight we know how many missiles the enemy has and, it turned out, our guesses were way off. We were building things we didn't need to build. We were harboring fears we didn't need to harbor."

powered reconnaissance satellite that carried a television camera described in a report called Project Feed Back.

Although most of the guts of "Meeting the Threat of Surprise Attack" was presented at a formal, expanded session of the National Security Council on February 14, 1955, Eisenhower ordered that the essentials of the intelligence segment be given to him privately by Killian and Land because he feared that the particularly sensitive satellite reconnaissance recommendations would be leaked.

Nor was unauthorized disclosure of technical collection plans Ike's only concern in the intelligence area. He was emphatic that what was collected—the "take," or "product," as it would soon be called—from air and space be done by the CIA, not the Air Force, since he believed that his former comrades in arms were hopelessly addicted to inflating the enemy threat in order to do the same to their budgets. The Air Force's "extraordinary intelligence assessment that the Russians had this great bomber fleet," which was untrue, created problems for Eisenhower that he could not forget, Killian would say later in reference to the infamous "bomber gap" that never really existed. "You must find a way to do it [the U-2 air reconnaissance program] that does not give primary responsibility to the Air Force," Ike emphatically instructed Killian and Land. Control of the U-2 program and the reconnaissance satellites that followed it would therefore be weighted in favor of the civilians in the CIA.* That would not prevent titanic battles from developing over control of both the collection system and what it brought back.

Missiles for Moscow

Sergei Korolyov would claim years later that Stalin showed keen interest in rocketry, and particularly its military applications, during a briefing in 1946:

> At first he listened in silence, practically without taking the pipe out of his mouth. As he got more interested he began to interrupt now and then with terse questions. You felt he had a grasp of what rockets were about. He wanted to know about speed, distance, altitude of flight and payload. He was particularly concerned about accuracy of delivery. . . . Obviously, it had become clear to Stalin and his military advisers that the first experiments in building jet aircraft and rocket systems would have far reaching consequences.

* He also had two other reasons for assigning the U-2 program to the CIA. He thought that the program would somehow be less provocative if it was run by a civilian agency, though the fact that the civilians were professional spies made that reasoning somewhat dubious. Second, he was afraid that the Air Force would somehow antagonize the Soviets, in the process raising tensions. That, too, is dubious.

Stalin probably reacted to the briefing as Korolyov described it. Tokayev bore out the designer, and so did Sergei Kapitsa, who maintained that Stalin was particularly interested in intercontinental weapons as a means of fending off the West. Hiroshima and Nagasaki were atom-bombed not only to bring Japan to its knees, it has sometimes been said, but also to demonstrate American military superiority to Stalin, who was even then moving aggressively in Eastern Europe. What the attacks demonstrated to "Uncle Joe" was that there really was a weapon that could cause unparalleled destruction and the United States was capable of using it. And the Soviet dictator also knew, as his generals certainly did, that one of the main reasons the Red Army had been so bloodied by the Nazis during the first year and a half of the war had been its overwhelming emphasis on armor and infantry at the expense of artillery. The truck-mounted Katyushas, solid-propellant rockets that could be fired in salvoes of forty-eight at a time, had gone into use later and helped turn the tide against the Germans.

Stalin had therefore became determined as the war was ending to obtain his own atomic bombs and develop the means of firing them great distances. And when the dictator became determined, things started to happen. "Don't make his eyebrows move," was the way those who served closest to him put it.

The beginning of postwar Soviet rocketry began, as it did in the United States, with the remnants of the German rocket team. During the spring of 1946, Helmut Grottrup and the other technicians and engineers who had been "liberated" by the Soviets worked in East Germany to re-create V-2 manufacturing, test, and support equipment. They set up an assembly line that would be shipped back to the Soviet Union so fifteen V-2s could be built and tested. They also carried out combustion tests with its engines, trained a German and a Soviet launch and test crew, and generally familiarized themselves with the weapon. Meanwhile, Brigada Osobogo Naznachenia, or the Brigade for Special Purposes, was set up at the East German town of Berka so that the most talented Soviet military officer-engineers could learn everything about the missiles. BON, as it was called, would develop into the nucleus of the Soviet Union's missile and space establishment. When Grottrup's and his top aides' work in East Germany was finished, they and their families were shipped off to Moscow. They left in the middle of the night of October 21, 1946.

Some of the Germans were put to work by the Soviet Union's first and preeminent missile research facility, the Scientific Research Institute and Experimental Factory 88 for Guided Missile Development (or simply NII 88) in Kaliningrad, about fifteen miles northeast of Moscow off the main Moscow–Yaroslavl highway. Grottrup and several of his colleagues were brought to Kaliningrad to set up a V-2 assembly line so that more of the missiles could be tested and studied by the Russians themselves. Meanwhile, Korolyov was appointed chief designer of a division of NII 88 that was responsible for producing long-range missiles. He started with sixty engineers and fifty-five technicians. The number grew to three hundred by the end of 1947.

Other Germans were shipped to a specially prepared work site on the secluded island of Gorodomliya, two hundred miles northwest of Moscow on Lake Seliger. Keeping the Peenemünde alumni on the island was not much of a problem because there were peasants and others on the lake shore and in the surrounding countryside who despised all Germans and who would gladly have massacred them. Gorodomliya was to be an incubator for advanced rocket designs. Unlike the United States, the Soviet Union did not invite the Germans to integrate with its own scientists and engineers under formal long-term contract. Instead, they were paid to fulfill their assignment in relative isolation. When everything worthwhile had been pulled out of them—when they had been squeezed like so many lemons, as von Braun would have put it—they were turned loose to head west, where they were interrogated as Kistiakowsky said.

Stalin's Germans submitted several proposals for both ballistic and antiaircraft missiles, all of them outgrowths of the V-2 or its Wasserfall derivative. These ranged from one designated as the R-10, basically an extended range V-2, to the two-stage R-15—a combination of an improved V-2 and a ramjet-powered cruise missile—that was designed to Armaments Minister Dmitri Ustinov's specifications: it had to carry a 6,600-pound warhead 1,600 nautical miles.

The missile that came closest to fruition was the R-14, the seventy-seven-ton monster that was designed to use a new type of rocket engine to lob a 7,600-pound warhead more than 1,800 miles. Work on the R-14 began on Gorodomliya Island in the spring of 1949 and ended by that December. A thirty-seven-page CIA report written in August 1953 was loaded with details about the weapon. It contained information as arcane as the weights of the motor mount, stabilizing rings, and turbines. It described the R-14's fuel consumption in kilograms per second, its engine's combustion period, the weight of the warhead, the launch weight of the missile, and its maximum target range. At a time when ablative materials to protect warheads from the intense heat of reentry were virtually unknown, a thick layer of wood was to be used to insulate the explosives. All of this and more came from its designers, who were closely questioned after they made the pilgrimage west. That particular report even mentioned a certain Colonel "Korolov."

The U.S.S.R.'s first missile test installation, known as the State Central Test Range, also went into operation in 1946. It was sited near a railhead outside of the small town of Kapustin Yar, roughly 650 miles southeast of Moscow, on a bend in the Volga River. The town's name means "Cabbage Patch." The first Soviet V-2 was launched from there in October 1947.

Before the Soviet rocket men selected Kapustin Yar, they looked at maps of their vast country, and came up with a unique way to find places to test their V-2s. They designed a special train in the form of a self-contained test facility that could go to any place where there were rails. Then they had it custom-built in East Germany. It provided the ultimate in flexibility because it made every railhead in the Soviet Union a potential test site. It had roughly eighty cars, some to carry the V-2s themselves and others to carry spare parts, fuel, a repair shop, and a laboratory. Still other cars were for living quarters, a restaurant, a movie theater, and even steam baths.

Sergei Korolyov and his colleagues were now embarked on a crash effort to develop indigenous ballistic missiles. Throughout the war and afterward, Glushko, Isayev, and other engine designers in the GDL's Experimental Design Bureau had worked hard to get more thrust out of their rockets' combustion chambers. This was done by improving nozzle and injector design, developing alloys that were more heat-resistant (which reduced burn-through), making the coolant and pressurization systems more efficient, and improving both the design of the chamber walls and the way they were soldered together.

During the summer of 1946, before the Germans had even arrived, the Russians had tested an engine chamber that produced nearly a ton and a half of thrust. Three years later, they were up to seven tons. Progress was being made fast enough to warrant the creation of a Ministry for Medium Machine Building in 1953. Whatever its ambiguous name, the organization's sole responsibility was to direct the nation's missile program.

At about the time the ministry was founded, Korolyov was invited to brief the Politburo on the Soviet Union's rocket program. He did so in his laboratory and at a launch pad. In his memoirs, Khrushchev described with disarming honesty what happened when he and others in the political elite ran into high technology head-on: "I don't want to exaggerate, but I'd say we gawked at what he showed us as if we were sheep seeing a new gate for the first time. When he showed us one of his rockets, we thought it looked like nothing but a cigar-shaped tube, and we didn't believe it would fly. . . . We were like peasants in a market place. We walked around the rocket, touching it, tapping it to see if it was sturdy enough—we did everything but lick it to see how it tasted."

Continuing refinements in engine design in the Experimental Design Bureau of Glushko's Gas Dynamics Laboratory between 1954 and 1957 produced a pair of remarkable engines called the RD-107 and RD-108. Used together, they would make history. Each contained four combustion chambers that burned a kerosene-type fuel and liquid oxygen forced into them by a single turbopump. The RD-107 had two small steering engines that were the equivalent of the small steering vanes on the V-2. Its near twin, the RD-108, had four steering engines. Both developed 102 tons of thrust. That would not be enough to get a big ICBM off the ground, let alone heave one of Andrei Sakharov's nuclear warheads over the Arctic on a 4,000-mile flight to the United States, where it would turn into a mushroom cloud.

But grouping engines was another matter. If five of the four-chamber engines were run together on one missile, the total of 510 tons—more than a million pounds—of thrust they produced would certainly do the job. Korolyov himself would be credited with coming up with that idea (however circuitously). But it actually came from Mikhail Tikhonravov, one of the true geniuses of the Soviet space program and an early champion of clustered stages. Tikhonravov had gone on to the Scientific Research Institute of the Ministry of Defense after his German inspection tour was finished and had then joined OKB-1, Korolyov's own design bureau. He had delivered a paper describing a multistage rocket that had a range of 3,000 miles at the Artillery Academy as early as July 1948. Evidence supports the conclusion that Tikhonravov was actually a better designer than the better-known chief designer. But the head of a design bureau was privileged to claim whatever came out of it as his own. And that is what invariably happened in the fiercely competitive world of military aviation and rocketry, both in the Soviet Union and in the United States.

Korolyov quickly adopted his engineer's suggestion. He used a single RD-108 as a central core engine, or sustainer, at the bottom of a missile. Then he grouped four RD-107s around the RD-108 to form a kind of expendable skirt. All five engines were supposed to be ignited together. The effect figured to be spectacular, with flame spewing out of twenty exhaust nozzles at the same time while the dozen steering rockets produced their own exhaust plumes. When the RD-107s' propellant was used up, the four rockets would drop away, leaving the RD-108, an upper stage, and the payload to continue the flight. That, at any rate, was the way things were supposed to happen. The new missile was called the R-7, or affectionately, Semyorka—Number Seven—by the Russians. It was named the SS-6 Sapwood by NATO and the A-1 by some Western observers, causing confusion for years.

In order to test the new rocket, construction of a major launch facility near the hamlet of Tyuratam, due east of the Aral Sea on the desolate steppes of Kazakhstan, was begun in 1955 and had concrete for its first launch pad poured by the end of March 1957. In keeping with the Kremlin's fetish for giving weapons development and other sensitive facilities the names of places many miles away to fool spies, saboteurs, and NATO's targeteers, the Tyuratam launch complex was officially called the Baikonur Cosmodrome after a small mining town two hundred miles to the northeast. It still is. But whatever they called it, U-2s following railroad tracks discovered the complex as it was being bulldozed in late 1956. Then, using captured German military maps of the area, Dino A. Brugioni, one of the CIA's ace photointerpreters, named the facility after the railroad station at the end of the nearest rail spur. Tyuratam would become the Soviet Union's major ballistic missile test facility and its only cosmonaut launching center. But as far as the outside world was concerned, it did not exist.*

The first R-7 exploded on the pad on May 15, 1957, followed by two more complete failures on June 9 and July 12. Like their German predecessors and American competitors, the Russians were learning the intricacies of big-time rocket development the hard way. The fourth missile, which was launched on August 21, actually got off the ground and managed to make it all the way to Kamchatka in the Pacific before its dummy warhead disintegrated into a hail of fragments six miles short of the impact area. The problem was actually with the missile, not the warhead. But the fact that the engines finally worked as they were supposed to, and so did the guidance system, indicated that a lighter payload could be shot, say, to space.

While his scientists and engineers were trying to increase engine thrust, refine guidance and communication systems, make pumps and the other vital innards reliable, and lower the weight of the warheads, Khrushchev himself was denigrating the manned bomber as so much antique junk and celebrating the virtues of his still balky intercontinental missile. He tried to show that an insurmountable gap existed between the huge Soviet missile stockpile and its anemic American counterpart. (His task was made a little easier when the first two Atlases, tested in June and September 1957, met the same miserable fate as their German and Russian predecessors.) During a state visit to Great Britain on April 23, 1956, for example, the first secretary boasted that his country was on the verge of having "a guided missile with a hydrogen warhead that can fall anywhere in the world." Six months later he threatened Britain, France, and Israel with nuclear retaliation for attacking Egypt after Gamal Abdel Nasser, Moscow's client, nationalized the Suez Canal. Anastas Mikoyan, a Foreign Ministry official, borrowed a page from SAC's playbook by warning that Soviet planes and rockets could deliver an atomic punch anywhere on the planet. But it was sheer bravado. The bomber force was pathetic, and that first R-7, which was doomed, was only then being readied for the train trip to Tyuratam.

Three years later, Marshal R. Y. Malinovsky would publicly thank Khrushchev for providing the armed forces with ICBMs; Khrushchev himself told the Twenty-first Party Congress, which was attended by dutiful TASS re-

* The secrecy was all the more pointless because the Soviets tracked U-2 flights and were well aware that they photographed the huge complex. The facility was a principal target of Francis Gary Powers's ill-fated mission on May Day 1960. (Brugioni, "The Tyuratam Enigma," p. 108.)

porters, that the weapons were coming off the assembly line "like sausages." Pachyderms would have been a better metaphor. For all its brute force, and despite its being ready just ahead of Atlas, the Semyorka was as unsuited to being a ballistic missile as was its American rival. Counting the time it took to pump in propellant and check out its system—the prelaunch sequence—it took seventeen and a half hours to prepare an R-7 for flight. That was good enough for launching a surprise attack, but it was no way to respond to one. Atlas was no better. Furthermore, the R-7 was too fat to be put into a protective hole in the ground (eventually to be called a silo). It was therefore as unprotected from a nuclear explosion's blast wave as a circus elephant balanced on a stool. Finally, it would have been terrifically expensive as a mass-produced ICBM, since its engines ran on kerosene and liquid oxygen. Kerosene was cheap enough, but producing and feeding liquid oxygen to thousands of R-107 and R-108 engines would have been a nightmare. The R-7 was an unmitigated failure as an ICBM. Yet it would prove to be an unexcelled satellite launch vehicle and the progenitor of a long-lived series of successors.

International Geophysical Rockets

By the time the fourth and most successful R-7 flight took place in August 1957, a confluence of events was taking place in the enemy camp that would figure decisively in creating what would amount to an "open space" policy to substitute for the skies that never opened for Eisenhower.

On May 20, 1955—thirteen weeks after "Meeting the Threat of Surprise Attack" reached the president with its call for legally opening space so U.S. reconnaissance could use it—the White House's National Security Council issued NSC 5520, "U.S. Scientific Satellite Program." The document endorsed what it referred to as the Killian panel's recommendation that "*intelligence applications warrant* an immediate program leading to a very small satellite in orbit around the earth, and that reexamination should be made of the principles or practices of international law with regard to 'Freedom of Space' from the standpoint of recent advances in weapon technology" (italics added). Eisenhower's National Security Council was ordering that a small satellite, flying under cover of noble science, be orbited to set a precedent that would apply to the spycraft that followed it.

Having first noted that Killian's men had called for the orbiting of "a very small satellite," NSC 5520 went on to add that

> From a military standpoint, the Joint Chiefs of Staff have stated their belief that intelligence applications strongly warrant the construction of a large sur-

> veillance satellite. While a small scientific satellite cannot carry surveillance equipment and therefore will have no direct intelligence potential, it does represent a technological step toward the achievement of the large surveillance satellite, and will be helpful to this end. . . . Furthermore, a small scientific satellite will provide a test of the principle of "Freedom of Space."

That freedom would, in turn, rest squarely on the ancient maritime principle of freedom of the seas. It was universally recognized in international law that once having left their own nation's territorial waters, ships were entitled to travel where they wished without interference. Denying that right constituted an act of war. The idea was to get that concept moved to space so the reconnaissance satellites could operate unchallenged, either politically or militarily.

"The national interest of the United States was believed by officials at the time to depend on using a superficially egalitarian programme of international scientific cooperation for the disproportionate benefit, much of it military, of their own country," one historian has written about the motives behind the IGY satellite resolution. "That arrangement would not succeed, it was argued, if the security benefits to the United States became too obvious."

And that, of course, suited Eisenhower perfectly. Ike was obsessed with the idea of keeping space under civilian control, at least publicly, of not being snared in legal controversies with the Russians over air and space rights, and with the specter of an intelligence fiasco that would embarrass the United States. Taken together, and within the broader context of the perceived threat from the Soviet Union and the need for accurate intelligence, the IGY provided a respectable, legitimate cover for the start of the U.S. space espionage program.

The following month NSC 5522, a top-secret progress report, was more specific:

> *Recommendation 9-b: "Freedom of Space.* The present possibility of launching a small artificial satellite into an orbit about the earth presents an early opportunity to establish a precedent for distinguishing between 'national air' and 'international space', a distinction which could be to our advantage at some future date when we might employ larger satellites for intelligence purposes." (See also Recommendation C-8.)

Recommendation C-8, which followed eighteen pages later, had this to say about intelligence-collecting satellites:

> Intelligence applications warrant an immediate program leading to very small artificial satellites in orbits around the earth. Construction of large surveillance satellites must wait upon adequate solutions to some extraordinary

> technical problems in the information gathering and reporting system and its power supply, and should wait upon development of the intercontinental ballistic missile rocket propulsion system. The ultimate objective of research and development on the large satellite should be continuous surveillance that is both extensive and selective and that can give fine-scale detail sufficient for the identification of objects (airplanes, trains, buildings) on the ground.

Even before all of the technical problems had been worked out of the spy satellites—and they were horrendous—it was decided that their smaller scientific cousins were going to work with them for the national security of the United States. This meant that whatever Eisenhower said in public about the separation of civilian and military space activities, and there is every indication that he wanted to be true to his word, they were really part of a single program. Ike approved NSC 5520 five days after he received it.

The excuse for launching the small scientific satellite was in the works and so, too, was the satellite itself. The excuse was originally called the Third International Polar Year, and it was conceived in James Van Allen's home in Silver Spring, Maryland, on the evening of April 5, 1950. The First and Second Polar Years, which had taken place in 1882–1883 and 1932–1933, had been unprecedented attempts to study Earth's polar regions through international cooperation. A number of factors had made a third effort seem worthwhile. For one thing, December 31, 1950, was the closing date for analysis and publication of data collected during the Second Polar Year, meaning that those findings were old and the scientific well from which they had been drawn was technically dry. For another, ionospheric physicists such as Van Allen who were interested in studying that region knew the Army was running out of V-2s and that, at least in his opinion, support for high altitude scientific work was "waning rather rapidly." Another major international scientific effort would guarantee funding for replacements such as Aerobee and Viking. Finally, improvements in scientific instruments—many of them, such as radar and sonar, developed during the war—meant that an unprecedented array of new and accurate measuring tools were available.

So Van Allen invited several of his colleagues to dinner to discuss the situation. If there was a guest of honor that evening, it was Sydney Chapman, the Sedleian Professor of Natural Philosophy at Oxford University and probably the greatest living geophysicist. Chapman was venerated as the creator of an ingenious flexible model of Earth's atmosphere and the space surrounding it and was now looking for ways to coordinate geophysical research worldwide. Lloyd V. Berkner, another of the men who sat around the Van Allens' table, was a world-renowned geophysicist. He was a veteran of the Navy Bureau of Aeronautics's Engineering Division, a colleague of Robert Haviland, the General Electric satellite communication expert, and in his own right a very early proponent of using satellites as communication relay platforms and scientific data collectors.* Berkner had been a radio technician on Admiral Byrd's first Antarctic expedition in 1929 and had been active in the Second Polar Year.

The three other geophysicists at Van Allen's table were J. Wallace Joyce, Ernest H. Vestine, and an associate professor of physics at the University of Maryland named S. Fred Singer whose impatience at not being able to get instruments to space would drive him to design his own rocket-satellite combination within three years. It was MOUSE, or Minimum Orbital Unmanned Satellite, Earth. It would send a small spinning spacecraft into a two-hundred-mile-high orbit, where it would stay for twelve days, sending down the first hard physical data about conditions there and within Earth itself, including information about the planet's magnetic field. The idea received a lot of press and was the subject of papers Singer gave at the Hayden Planetarium, in Zurich, and elsewhere, where other impatient physicists greeted it enthusiastically. MOUSE soon died of neglect and starvation as other rocket-satellite projects were developed by such powerful organizations as the U.S. Army and Navy. But it did show unquestionably that the time was finally ripening for a marriage between science and rocketry.

Everyone who had dinner with Van Allen that evening shared a frustration that was nicely expressed by *New York Times* science writer Walter Sullivan: "From the ground, our view into space is hardly more enlightening than the view of the heavens obtained by a lobster on the ocean floor." The rocket was supposed to carry the crustacean to the surface, where the satellite would finally give it a view of the world.

The frustration of not being able to see that world, at least with their instruments, prompted Berkner to suggest a Third International Polar Year to take place in 1957–1958: twenty-five years, not fifty, after the last. With Chapman's and the others' weight behind it, the idea took hold within the international scientific community during the next three years. By 1954, the Polar Year had been rechristened the International Geophysical Year so it would reflect a much more ambitious strategy that included studying Earth's oceans, its atmosphere, and, above all, the Antarctic and outer space. The intense international effort would be so sweeping, in fact, that it would require eighteen months: from July 1, 1957 to December 31, 1958.

With the U.S. government quietly desperate to launch a small science satellite and American physicists and their foreign colleagues just as anxious to do

* Arthur C. Clarke, then of the British Interplanetary Society, is credited with first proposing satellites as communication relays, some in geosynchronous orbit, in letters to the magazine *Wireless World* in February and October 1945. (Emme, "Presidents and Space," pp. 8, 31.)

so, it remained only to design, build, and launch the thing. But that task, which figured to bring both prestige and influence to its creator, would lead to one of the most confused and nastiest episodes in the history of the space program.

Vanguard

A series of meetings in the summer of 1954 produced an agreement between Army Ordnance and the Office of Naval Research for a jointly developed Minimum Launch Vehicle for the IGY. Army and Navy negotiators, including Toftoy and von Braun, decided to give the Army responsibility for developing a modified Redstone booster, while the Navy would produce a small satellite, as well as handle logistical support and data acquisition and analysis. The Air Force's assignment was only to provide satellite tracking facilities, which was fine with the airmen, since they were under strict orders from the National Security Council to do what they most wanted to do: develop an ICBM with a 5,500-mile range and a megaton warhead that would explode like a million tons of TNT.

The Army's launcher could have been designed by Rube Goldberg, a popular cartoonist of the time who constructed complicated contraptions to do simple things. The first stage would be an elongated, "uprated" Redstone. The second stage would consist of a brace of twenty-four small Loki solid-propellant anti-aircraft rockets. The third would be a bundle of six more Lokis that, in turn, would boost yet another Loki carrying the Navy's five-pound satellite. Copies of the proposal, which was called Project Orbiter, went to JPL and the Naval Research Laboratory. JPL suggested that a miniature version of one of its own tactical rockets, Sergeant, replace Loki. The idea was accepted by Donald Quarles, the assistant secretary of defense for research and development, a key individual who was well aware of why it was important to launch a scientific satellite and who also had a firm grasp of both the U-2 program and reconnaissance satellite work that was taking shape in California.

Then the Naval Research Laboratory weighed in with a stunner. The NRL's engineers came to the same conclusion about Loki as the Pasadenians.* But they did not let it go at that. Instead they nominated their own entry, an advanced version of their hugely successful Viking, to be their country's IGY entry, not Orbiter. Vanguard, as they called it, would use the tried-and-true Viking first stage, an improved Aerobee second stage, and an unnamed third stage to toss a forty-pound satellite into orbit. The renegades at NRL lobbied

* Except for a corner of a parking lot, JPL is actually in La Cañada–Flintridge, a community bordering on Pasadena. Perhaps because Caltech is in Pasadena, or because it is an easier name to remember, or it fits more easily into a dateline, Pasadena was and remains widely accepted as JPL's hometown.

hard for Vanguard and soon won over both the influential American Rocket Society, which was pushing hard for U.S. participation in the IGY, and the Upper Atmosphere Research Panel of the National Academy of Science's IGY Committee. The Vanguard proposal was given to Eisenhower in late March 1955.

With the United States not even officially committed to launching a satellite for the International Geophysical Year, the situation became politically snarled as both the Army and the Navy jockeyed furiously for position. To make matters even more complicated, a special committee representing all three armed services tried to smooth the situation by recommending that both Orbiter and Vanguard be developed and that the Air Force field its own entry, an Atlas topped with an advanced Aerobee. The Office of the Secretary of Defense killed that last suggestion out of hand, not only because the Air Force's priority was countering Khrushchev's ballistic sausages, but also because there were no funds for two scientific satellites, let alone three.

Quarles sent the competing satellite proposals to Homer Joe Stewart's Committee on Special Capabilities, one of the Pentagon's own elite, in-house think tanks. It came out for Vanguard. The Office of Naval Research reacted by abandoning the deal with the Army, since it could not possibly stay in a partnership against an organization in its own service.

What happened next was quintessential Washington. The Army reacted by getting the vapors. Major General Leslie E. Simon, its assistant chief of ordnance for research and development, shot a memo to Quarles complaining that Homer Joe's committee had made gross errors in both fact and reasoning. Quarles reacted by telling Homer Joe to take another look at Orbiter. NRL reacted to the news that it was not the winner after all, but that the competition had been reopened, by insisting on equal time to make a second proposal. (The sailors were afraid that the Army would make radical changes in its own design after having studied theirs.) Homer Joe's committee reacted to the Army's new Orbiter proposal by stubbornly voting 6–2 in favor of Vanguard. It cited a number of technical reasons for doing so.

But at the time and for years afterward, another reason was persistently mentioned that had nothing to do with science and engineering. It had to do with a lingering, visceral hostility toward the Peenemünde alumni. There were people in the American rocket fraternity who thought that the Germans were gifted bad boys who had fought for the wrong side and who had managed to kill a lot of civilians from Nordhausen to London in the process, whether they actually launched the things or not. Whatever was painted on their skins, Redstone and Jupiter had Nazi bloodlines, and that was a fact. In a political as well as physical sense, on the other hand, Vanguard was made in America. At any rate, the Navy reacted with jubilation. But not for long.

The situation began to sour soon after work on Vanguard started. The tone of the project was set by the Department of Defense, which gave it a low priority compared to the massive bomber and missile buildup that was getting under way, including development of Polaris missile–firing submarines. It became clear to the NRL that Vanguard's procurement program was a distant second to the weapons programs. Furthermore, most of the experienced engineers at the Glenn L. Martin Company, which was supposed to produce Vanguard on the strength of its having produced Viking, were transferred from Baltimore to Denver in the autumn of 1955 to work on the company's new (and really lucrative) Titan ICBM program. Anyone in the NRL who saw that and still did not get the picture had only to visit the Vanguard office at Martin. It was secreted in a loft balcony in the aircraft production facility, where its designers roasted in summer, endured biting cold in winter, and had to dodge the droppings of sparrows that flitted around the rafters year round.

Worse, there were disputes with the Pentagon's Vanguard program office over the component testing schedule, the replacement of a Reaction Motors engine with one built by General Electric, and suspicions by early 1957 that Vanguard would shake apart at liftoff. In what would come to be a recurring pattern in the space program, all of the miscalculations and technical problems, many of which could not have been foreseen, were turning Vanguard from a relatively economical program into a budget buster. The original estimates did not show that the Air Force would charge nearly $8 million for the use of its facilities at Canaveral and along the Atlantic test range, nor that the Navy's tracking system, which extended from Maryland to Chile, would soak up another $14.5 million. Generals and admirals who took the Department of Defense's mission to be the defense of the country, starting with the use of tight funding to buy missiles, bombers, carriers, and subs, were disgusted to see what started out as a $30 million program topping $111 million. And for what? Just to get a little prestige by putting up an instrumented cannonball. Or so they thought.

Vanguard's birth pains were not reflected in its publicity which, as a song popular at the time had it, accentuated the positive and eliminated the negative. In 1956, for example, a book about the scientific benefits that were going to come from sending Vanguard to space reflected the use of carefully sanitized information put out by both the Navy and Martin, not reality. Reporting that it "was not felt necessary or desirable to set up a large, complex organization to co-ordinate the project," for example, made the NRL look heroic. In reality, the indifference or hostility of higher-ups was so pervasive, and escalating costs so horrific, that there was no choice but to keep the size of the organization small. That and other puffery were sins of omission. But the book's glaring error came from taking prophecy for reportage. "The Earth satellites developed under Project *Vanguard,*" the authors stated with no trace of doubt in the first sentence of their introduction, "are to be the first space vehicles." Sergei Korolyov did not see it quite that way.

Don Quarles Plays "Let's Make a Deal"

The military and civilian elements of the Technological Capabilities Panel and the NSC recommendations came together in the person of Quarles, who ran nearly all defense research projects and who was therefore intimately acquainted with the report by Din Land and his colleagues, the development of the U-2, and the then-embryonic reconnaissance satellite program. On the day the TCP report was released, February 14, 1955, the National Academy of Science's IGY committee recommended to Alan Waterman, the director of the National Science Foundation, that the United States contribute a scientific satellite to IGY research. Quarles then began making his moves. First he asked Waterman to pass the satellite suggestion on to the National Security Council in the White House. Four days later, he sent a letter to Robert Murphy, deputy undersecretary of state and a member of the National Security Council, that carried the proposal forward. Murphy, dutifully noting that such a satellite would "undoubtedly add to the prestige of the United States . . . and have considerable propaganda value," signed off within the month. Next, Quarles steered Waterman to Allen W. Dulles, the director of the Central Intelligence Agency. The spymaster from Princeton readily agreed to support the project and, probably on the recommendation of Bissell, would eventually contribute $2.5 million when it soared ten times over budget. Percival Brundage, the director of the budget and the nation's preeminent Olympian, also agreed to cooperate.

By May 20, 1955, then, America's "civilian" satellite program, soon to be named Vanguard, had the support of the Departments of State and Defense, the CIA, and the Bureau of the Budget. On that day, the National Security Council whittled the matter to a fine point when it approved NSC 5520.

It is important to note, though, that Waterman and other scientists, including Berkner, were emphatic in insisting that getting to space first was less important than doing it right: that more prestige would come from a real scientific breakthrough than from simply tossing any old piece of hardware into orbit for the sake of beating the opposition.

Amazingly, even the National Security Council's Planning Board echoed this in November 1956. While the Russians might launch first, and while the United States should make every effort to orbit a satellite "as soon as practicable," the board noted, "a more effective and complete scientific program" would at least partially offset the advantage of the opposition's being first.

Conversely, if the United States went to orbit first, "but the USSR put on a stronger scientific program, the United States could lose its initial advantage."

Space Reconnaissance Takes Shape

At the same time, the last component of the strategy worked out by Killian's intelligence panel and endorsed by the National Security Council was taking shape in California.

The Air Force had come out of World War II with an attitude about strategic reconnaissance that was not adaptable to the cold war. Doctrine established during the war had it that reconnaissance was to be used to identify and pinpoint targets for bombardment, locate and destroy "anti-air" threats, and do bomb damage assessment. The concept of routinely watching and listening to whole nations in peacetime, which was to be the hallmark of reconnaissance during the cold war, was not exactly embraced by the air generals. Furthermore, even the reconnaissance they conducted was done almost grudgingly, since their core mission required buying bombers and fighters to protect the bombers, not planes that took pictures. By 1951, a handful of men at Wright-Patterson Air Force Base, including Colonel Richard S. Leghorn and Amrom Katz, had begun to see a wider use for strategic reconnaissance. They were still a distinct minority, however.

But RAND, the Air Force's think tank for space, was doing what it was paid to do: look farther ahead than its benefactor. What it saw in 1951 went beyond somehow clicking photographs in space and getting them to Earth. The oldest rule in the intelligence game—and it was as old as history—was that information about the enemy had to be "timely" or it was useless. There was no point in finding out something, in other words, if it was too late to use. Along with accuracy, intelligence agencies the world over therefore always put the highest premium on speed. RAND's engineers had therefore begun considering "the miracle of television" (as the commercial networks put it) to get intelligence down faster than could be done by snapping photographs the way parents did of their children in the park on Sunday.

That year, 1951, RAND had produced a two-volume study, one devoted to weather reconnaissance, the other to a dazzlingly innovative technique called "reconnaissance by television." It signed a contract with the Radio Corporation of America the following year to study the possibility of adapting television cameras, radiation recording devices, and other equipment to space surveillance. The threat of surprise attack, coupled with early Air Force hostility to "photoreconnaissance" from space, moved Jim Lipp, Mert Davies, and other pioneers in the newest of the black arts at RAND to study a variety of imaginative techniques, including the feasibility of relaying imagery from spacecraft to Earth by the television method. The joint RAND-RCA study resulted in a second report, code-named "Feed Back," which in March 1954 recommended that the Air Force develop a so-called electro-optical satellite system that used a high-resolution vidicon camera to read pictures taken by the spacecraft and radio them right down. While the idea was wonderful, the imagery's resolution—the size of objects that could be distinguished—was calculated to be a disappointing 144 feet. That was fine for spotting hurricanes but not for locating and studying airplanes, submarines, and missile sites. RAND nevertheless encouraged the Air Force to get contractors to compete to develop higher-resolution imaging systems.

In March 1956, RAND also proposed taking standard photographs from satellites and then dropping the film capsules out of orbit and recovering them as they floated down. The idea, which originally came from a physicist named Richard C. Raymond, was contained in a top-secret twenty-page report entitled "Photographic Reconnaissance Satellites." No sooner was it submitted than it was abruptly withdrawn, probably because RAND's Air Force benefactors had suddenly become so enthralled at the prospect of near-real-time television capability that they dismissed the capsule idea as obsolescent before it even got off the ground. The CIA would take a very different attitude than the Air Force about dropping "buckets" from space.

RAND's studies on the various space reconnaissance techniques, together with work by Leghorn, Katz, another pioneer named George W. Goddard (no relation to Robert), and others went directly into the first actual space surveillance program. It was sponsored by the Air Force and called WS-117L, or the Advanced Reconnaissance System. WS-117L consisted of three separate space surveillance systems. By 1957, Lockheed's newly formed Missiles and Space Company, then located at Palo Alto, California, and eventually moved to nearby Sunnyvale, was working on all three: vidicon, film recovery, and infrared observation for ballistic missile early warning. Each satellite would go to orbit on an upper stage that would carry its own restartable rocket engine and steering thrusters. The upper stage was called Agena, and it was a true space truck. The three observation systems, collectively called Pied Piper by Lockheed, set the basis for every space-based photoreconnaissance and surveillance system that followed. Together with TRW, located south at Redondo Beach, Lockheed would remain a principal U.S. spy satellite manufacturer throughout the cold war.

The Song of the Scarlet Angel

If the proverbial man in the street in the United States thought about the Union of Soviet Socialist Republics at all in 1957, it was as a far-off place where a cynical and dangerous elite manipulated and repressed a race of stoical and unimaginative folk who alternately loafed on the socialist dole or slowly toiled to the mournful strains of "Marche Slav" and "The Volga Boatman." With a massive Army and what they believed to be a vast armada of warships and planes facing them, it was somehow comforting for Americans to believe that the enemy's nuclear weapons program was shoplifted from their own country in the dead of night and that there was a swastika inside every missile. Claiming to have invented the airplane was bad enough. But taking credit for baseball, the steam engine, radio, television, tractors, and even decimal coinage made the USSR a country of self-deluded oafs. The Russians were considered scientifically and technologically hopeless (when they were considered at all).

Jules Verne could have set them straight. He had Russia contribute the "enormous sum" of 368,733 rubles to build the Columbiad. "No one need be surprised at this who bears in mind the scientific taste of the Russians," Verne explained, "and the impetus which they have given to astronomical studies—thanks to their numerous observatories."

The literature alone was staggering. There was a crescendo of writing about space travel in the Soviet Union after the war as the lingering utopian ideal, the techno-icon, the great scarlet angel, promised to carry Russians out of their dreary, soul-deadening conformity on wings and fins of steel. The babushkas who patiently waited in line for three hours or more for sausage and potatoes may not have known of Konstantin Tsiolkovsky, but in a manner of speaking, he was there with them. And he was with their husbands, sons, and daughters who trudged off to machine shops and factories all across the land to bolster the fabled forces of production and in the process protect the motherland from the newly emerging threat.

There were popular books about Russian space pioneers and about jets, rockets, and life in space. And there was a blizzard of technical material in the form of both books and journal articles. Yakov Perelman had left a rich legacy of popular books about rocketry in the 1930s, including *To the Stars by Rocket, Interplanetary Travels, Flight to the Moon,* and *Tsiolkovsky: His Life, Inventions, and Scientific Works.* Popular writing of the time ranged from the twenty-four-page pamphlet *The Problem of Interplanetary Travel in the Work of Native Scientists* to A. A. Kosmodemyansky's *Famous Man of Science: Konstantin Eduardovich Tsiolkovsky,* a monumental 1,936-page paean to the master.

F. J. Krieger, a RAND analyst who did extensive (but not, he claimed, exhaustive) research on the literature of Soviet rocketry, reported that by 1957 important monographs, articles, and books on the subject numbered in the hundreds. A relatively small number of them were translations of Oberth,

Sänger, Goddard, Esnault-Pelterie, Hohmann, and other foreigners. But the vast majority were distinctly homemade and ranged from specialized studies of weightlessness to a highly detailed description of a "tankette"—a caterpillar-treaded lunar roving vehicle—to flight procedures for cosmic exploration. And there were an increasing number of articles in newspapers and popular-science magazines about space, many of which were interviews with well-known scientists and engineers.

But the "indicators" (as the CIA called them) that the Soviets fully intended to go to space as soon as possible were not confined to literature. In November 1953, A. N. Nesmeyanov, president of the Academy of Sciences of the U.S.S.R., told the World Peace Council in Vienna that "science has reached a state when it is feasible to send a stratoplane to the Moon, to create an artificial satellite of the Earth."

On April 16, 1955, only six months after the satellite resolution was adopted by the IGY at a conference in Rome, the Moscow evening newspaper *Vechernyaya Moskva* reported Soviet plans to launch such a spacecraft. It told of the creation of an Interdepartmental Commission on Interplanetary Communication. If the commission's reporting to the supremely prestigious Academy of Sciences wasn't enough of an indication that it was serious about space, its membership roster alone would have put aside all doubts. The group's twenty-seven members included Viktor A. Ambartsumyan, an internationally renowned astronomer; Pyotr L. Kapitsa, the physicist whose pedigree included a nine-year stint at Cambridge, where he had worked closely with the great Ernest Rutherford; and Nikolai Bogolyubov, a brilliant mathematician who specialized in atomic energy. One of the commission's "immediate tasks," the article continued, would be the launching of a scientific Earth satellite to study the effects of weightlessness, ultraviolet and X rays from the Sun and stars, and observation of ice floes and clouds. The reference to weightlessness was a clear indication that there were plans to send a living creature to space despite the fact that biological research was irrelevant to IGY activities. The circumspect Soviets were all but ignored.

The White House's announcement on July 29, 1955, that the United States intended to launch a satellite during the International Geophysical Year, on the other hand, drew worldwide acclaim. It was taken for granted that America's unquestioned leadership in science and technology, supported by the planet's most robust economic and managerial base, would open the new frontier as it had the old one. Missing, but apparently not missed, was a steering group equivalent to the interplanetary commission.

Meanwhile, the Russians pressed on (without Khrushchev's approval). On January 30, 1956, the Academy of Sciences quietly made a firm commitment to launch a satellite later named Sputnik—literally "Companion" (of Earth)—

within two years as part of the IGY program.* Throughout the year a series of specialized conferences were held to decide what the spacecraft was to accomplish—measuring cosmic radiation and Earth's magnetic field, for example—and what the vehicle itself would be. By November, according to Tikhonravov, he and his colleagues had begun conceptualizing a "sputnik" that would carry a man and were considering how to design one that could land on the Moon.

But the first designs were less ambitious. The primary Sputnik, the nation's supreme contribution to the IGY, would be a scientific masterpiece: an antenna-sprouting cone that weighed almost a ton and a half and that was loaded with scientific instruments. It would carry a magnetometer, photomultipliers, a mass spectrometer, ion traps, a photon and cosmic-ray recorder, and other instruments to measure Earth's magnetic field and the particles and energy fields through which it was to fly.

Another IGY spacecraft would carry a passenger—a dog—and instruments to measure both the canine's vital signs and take radiation, micrometeoroid, and other readings. This "biosat" was designed to provide data on the dog's ability to adapt to space in anticipation of manned flights, which were then in the planning stage. It would weigh 1,118 pounds and, like its heavier stablemate, be roughly conical.

A third satellite, also conceived by the all-but-anonymous Tikhonravov in 1956, was the runt of the trio. It would be classicly simple: a 184-pound sphere that carried a pair of radio transmitters and four spring-triggered antennas designed to pop out when the satellite separated from its launch vehicle's pointed nose, or shroud. Unlike the primary spacecraft, which would be powered by sophisticated solar panels that converted the Sun's rays into energy the way leaves do on plants and trees, the aluminum ball would carry relatively primitive silver-zinc batteries powerful enough to guarantee fourteen days of continuous transmission. Radio amateurs around the world would therefore be able to hear it on two wavelengths as it streaked overhead. The radios, batteries, equipment for recording the pressure and temperature inside the spacecraft, a thermal control system to maintain an even temperature whether it was in sunshine or night, and related hardware and wiring would be contained in two hermetically sealed hemispheres. To aid tracking the spacecraft on ground radar, its carrier rocket—the launcher's upper stage—would be fitted with angle reflectors. The spent rocket would fly in formation with its payload, as would always be the case unless they were deliberately fired out of orbit.

* As originally used, the term meant "companion," or "fellow traveler," with Earth. But the word soon took on the generic meaning for any spacecraft. Whatever its proper name, Russians refer to any Earth satellite as a "sputnik."

The Astropound

According to the Soviet plan, dogs—the canaries of the space age—would be shot into orbit before humans. Alexei Pokrovsky, a member of the Soviet Committee for the IGY, announced at a news conference on June 18, 1957, that three dogs had already been rocketed straight up to an altitude of sixty-five miles.* Perhaps in anticipation of foreign reporters' notorious skepticism, Pokrovsky was careful to note that the canines had been filmed in flight. And for the benefit of Englishmen with their well-known, if occasionally eccentric, humanitarian concerns, he added that they were in excellent health. "I would like the British correspondents to inform the British Society of Happy Dogs about this," he said, "because the Society has protested to the Soviet Union against such experiments."

Tuning In

Indicators kept coming as the desirable turned into the possible for the Russians. That same month, Academician Ivan P. Bardin, the Soviet Union's leading metallurgist, sent a letter to the Special Committee for the International Geophysical Year in Brussels that spelled out his country's space plans in some detail. The U.S.S.R. Rocket and Earth Satellite Program for the IGY said explicitly that 125 meteorological rockets would be launched from three different "zones"—Arctic, central U.S.S.R., and the Antarctic—as well as an unspecified number of Earth satellites, he wrote.† All would study the structure of the atmosphere, cosmic rays, the ionosphere, micrometeors and meteorites, the physical and chemical properties of the upper astrosphere and more.

Within a month, *Radio,* a Russian amateur radio magazine, carried two articles that gave a fairly comprehensive description of the sputnik's intended orbit, how its appearance could be predicted, and how its twenty- and forty-

* In fact, nine were initially used, and the selection process was more complicated than might be imagined. All had to be mongrels because it was decided that they were more rugged than purebreds and therefore better able to withstand cold and deprivation. They also had to be female because the special spacesuits and sanitation apparatus were better suited to them than to males; be all or mostly white because that color showed up better on film and television; and weigh no more than sixteen pounds because of the rockets' lifting limitation. A succession of dogs with names such as Albina (Whitey), Kozyavka (Gnat), Malyshka (Little One), and Tsyganka (Gypsy) were sent up with instruments monitoring their pulses and respiration rates. Some parachuted back to Earth in their capsules, surviving both ascent and descent and emerging "completely normal," as Pokrovsky put it. Others did not. (Rhea, *Russians in Space,* pp. 140–41.)

† Typically, exact launch locations were not given for reasons of military security, which would irritate and frustrate Western scientists.

megacycle transmissions could be picked up. Diagrams were provided to vividly describe Earth, its equatorial plane, the inclination in degrees of the satellite's orbit to the Equator, and other "elements" of the mission. The July and August issues told readers how to build the right shortwave radio receiver to pick up Sputnik's signals and a direction-finding attachment for locating it.

Late that summer, after the first successful R-7 test, Khrushchev approved the space shot. However awed he had been four years earlier by the chief designer's briefing, Khrushchev had tended to see the space program (as opposed to the missile program) as a waste of precious resources. But a number of developments changed his mind. For one thing, he faced serious political opposition from a party elite that feared his growing power. Only adroit maneuvering, which included using the Air Force to fly loyalists into Moscow from around the country, prevented his ouster. For another, the military threat posed by the United States and its allies was real, and no amount of bluffing about missiles rolling off assembly lines could take the place of a credible deterrent. Khrushchev knew that the U-2s methodically following his country's railroad tracks were not turning up operational "sausages" because they were not there. But a rocket that could carry nearly 3,000 pounds to space could also lob a warhead onto America or any place else, so sending one of the machines into the distant sky would prove once and for all that no place on Earth was safe from Soviet missiles. Khrushchev therefore gave Korolyov the go-ahead to proceed with the nation's contribution to the IGY, though geophysics was entirely beside the point.

The Red Moon

A final, unmistakable signal that something was in the works came westward on August 27. That day, a TASS report in *Pravda* proclaimed that "successful tests of an intercontinental ballistic rocket and also explosions of nuclear and thermonuclear weapons have been carried out in conformity with the plan of scientific research work in the U.S.S.R." Actually, there were two ICBM tests, both of them with R-7s that made it to Kamchatka without serious trouble. Both were tracked by U.S. radar. The specter that haunted Eisenhower—the catalyst for the Technological Capabilities Panel—was finally coming true.

The conference to coordinate the IGY's final satellite and rocket plans started at the august National Academy of Sciences in Washington on September 30. The plenary meetings, attended by representatives of Australia, Britain, Canada, France, Japan, the Soviet Union, and the United States, were held in the building's domed lecture hall, where technical papers were delivered. The Soviet delegation was led by Anatoli A. Blagonravov, a member of his own country's Academy of Sciences and, not uncoincidentally, a lieutenant general of artillery.

In the library that adjoined the lecture hall, there was an impressive exhibit on Vanguard. The launch vehicle, as well as the tiny satellite of the same name that it was supposed to carry to space and onto the pages of history books, reflected their country's technoculture. They were clean-lined, wholesome, eager-looking, and stridently self-confident, like a turquoise-and-white '57 Chevy. The rocket was slim and graceful: an elongated, elegant white bullet that was supposed to carry what was then a twenty-one-pound spacecraft into orbit.

The exhibit at the National Academy reflected the optimism of the program's team. Besides the models, there was a mechanized miniature of the satellite that happily circled Earth, and even a stubby telescope of the sort volunteers in something called Operation Moonwatch were supposed to use to track it under the supervision of the Smithsonian Astrophysical Observatory at Harvard. That Vanguard would be the first object made on Earth to reach space was so beyond question that no one gave it much thought. At least none of the Americans did.

Vanguard's competitor, the smallest and lightest of the Sputniks designed by Tikhonravov, was at that moment bolted inside the shroud of an R-7 at Tyuratam, which was still under construction. The polished ball was Korolyov's last choice as a candidate for mankind's debut in space. But his first choice, the ton-and-a-half cone studded with antennas, carried so many science instruments and related equipment—was so complicated—that it had turned into a formidable engineering challenge. Some of the machines that constituted its complex innards did not work properly. Others simply resisted all attempts at communication. Finally, in exasperation, the big sputnik's managers pulled it off stage like a wheezing prima donna and sent in a less sophisticated but healthier understudy. The spacecraft that was supposed to fly as Sputnik 1 would go the following year as Sputnik 3.*

If there was one discordant note that ran through the weeklong meeting in Washington, it was over the Russians' insistence on keeping details of their satellite to themselves. The Americans gently chided them on their secrecy and they just as gently countered, saying that they considered it unseemly to "boast" about experiments until they were complete. "We will not cackle until we have

* The substitution was a carefully guarded secret during the cold war. In 1973, for example, Tikhonravov gave a paper at an international astronautical meeting at Baku. After reporting that a number of sputniks had been built, he added that "the first of them, the simplest in construction, was launched on 4 October 1957." This gave the impression that the first sputnik in space had gone up in its assigned order. (Tikhonravov, "The Creation of the First Artificial Earth Satellite: Some Historical Details.")

laid our egg," as one of the tight-lipped Russians put it. The disagreement still hung in the air on October 4, the last evening of the conference. But most Americans neither knew nor cared about what was happening inside the marble halls of the citadel of science on Constitution Avenue in their nation's capital that fateful day, let alone on a launch pad on the steppes of Central Asia.

American civilization had its collective heart and mind far away from smelly rockets on that autumn afternoon. The New York Yankees were tied with the Milwaukee Braves one game apiece in the World Series after Lew Burdette, a thirty-year-old West Virginian, had pitched the Braves to a 4–2 victory in front of 65,000 disgruntled fans in Yankee Stadium the day before. Arkansas Governor "Awful" Orval Faubus, a sworn segregationist, was telling people that he reminded himself of Robert E. Lee. The House Subcommittee on Un-American Activities ended four days of hearings on alleged subversive activity in Buffalo, New York, after three government witnesses accused an optometrist named Milton Rogovin of being the region's chief Communist. The espionage trial of Colonel Rudolf Abel, a real Soviet agent who plied his trade in Brooklyn, was postponed ten days so the KGB officer's lawyer would have more time to prepare his defense. The freighter *Santa Mercedes* had the distinction of becoming the two hundred thousandth deep water ship to make it through the Panama Canal. Francis J. McCarthy of San Francisco got a patent for an "item finder" that would allow shoppers in supermarkets to find what they wanted by pressing a button that lighted its location on a diagram of the store. The season premier of *Leave It to Beaver,* a sitcom about a precocious but lovable kid, aired on the Columbia Broadcasting System. The *Night Show,* on ABC, was running *Dracula.*

That day, a Studebaker Silver Hawk, "factory fresh" and complete with fins and "even" directional signals, was going for $1,995. And General Electric was promoting an eleven-cubic-foot-capacity rectangular refrigerator with a zero-degree freezer that held seventy pounds of food. It also boasted automatic defrosting, magnetic safety doors with removable and adjustable shelves, fruit and vegetable compartments, a frozen concentrated juice-can dispenser, and an automatic butter conditioner. The thing could be hung on a wall, set on a counter surrounded by cabinets, or even be used as a room divider. The idea, General Electric's advertising executives proudly explained, was to make food "easier to see and reach!" It was unquestionably the most advanced refrigerator on Earth.

On the steppes of Kazakhstan, the R-7's twenty engines came to life with a torrent of white fire at four minutes to midnight local time. Seconds later, it rose from its launch pad over a widening cascade of thunder, flame, and rolling clouds of thick smoke. Precisely two minutes after launch, with the rocket turning to a white speck that was vanishing in the darkness, the four boosters separated and came tumbling down through the night. The sustainer's RD-108 burned for another 150 seconds as it propelled its payload toward the northeast on an inclination of sixty-five degrees to the Equator. Moments later, the sustainer's shroud separated, shot forward, and then split in half, exposing its cargo to the frigid blackness. Then the satellite itself sprang forward, popping out its four spring-loaded antennas on cue. The first objects made by men to circle Earth—the shroud, the sustainer, and the sphere—were soon speeding through space at more than 17,000 miles an hour. The little armada was starting a journey that would take it over every place on Earth between Iceland and Cape Horn. It was, coincidentally, the hundredth-anniversary year of Tsiolkovsky's birth.

A Storm from Space

It was still midafternoon in Washington when Sputnik slid into orbit. Appropriately, the beeping spacecraft passed over the United States twice before anyone even realized it was there. The news was broken by Radio Moscow almost three hours after launch, followed eighteen minutes later by a dull TASS dispatch that was essentially repeated in *Pravda* the next morning:

"In the course of the last years in the Soviet Union, scientific research and experimental construction work on the creation of artificial satellites on the Earth has been going on," wrote some martinet whose wire service had no domestic competition and who was therefore unschooled in the art of the attention-grabbing hard news lead. "As the result of a large, dedicated effort by scientific-research institutes and construction bureaus, the world's first artificial satellite of the Earth has been created. On October 4, 1957, in the U.S.S.R., the first successful satellite launch has been achieved. According to preliminary data, the rocket launcher carried the satellite to the necessary orbital speed of about 8,000 meters per second." Then came the usual hackneyed bilge: "Artificial Earth satellites will pave the way to interplanetary travel and, apparently, our contemporaries will witness how the freed and conscientious labor of the people of the new socialist society makes the most daring dreams of mankind a reality."

Working as they did within a closed society that placed heavy emphasis on keeping its workers and peasants as complacent and tranquilized as possible unless ordered to do otherwise, the Russian journalists played the story that first day exactly as they would have played one on record tractor production or the opening of a hydroelectric power station someplace in Siberia. Without direction from the first secretary or anybody else in the Kremlin (none of whom really cared in the slightest about sending anything to space) professional inertia carried the day. But only that day.

Reuters, the BBC, and *New York Times* reporters in Moscow, on the other hand, knew a big breaking story when they saw one and scrambled to get it out. Hearing the news on Radio Moscow, the *Times'* bureau immediately cabled it to the Foreign Desk in New York, which relayed it to the Washington bureau, which passed it on to Walter Sullivan, the newspaper's senior science writer. Ironically, Sullivan got the word by telephone during an evening reception in the Soviet Embassy for IGY delegates to the Rockets and Satellites Conference. "Radio Moscow has just announced that the Russians have placed a satellite in orbit nine hundred kilometers above the Earth," an excited editor told him. Sullivan hurried from the downstairs telephone back up to the ballroom and related the news to Berkner and JPL's Pickering, both of whom instantly understood its significance. Lloyd Berkner then clapped his hands loudly to get everyone's attention. "I wish to make an announcement," he said after the room quieted. "I am informed by *The New York Times* that a satellite is in orbit at an elevation of nine hundred kilometers. I wish to congratulate our Soviet colleagues on their achievement." Sullivan had the pleasure of jotting down Berkner's response to the news he had delivered and then writing the story, which described the Russian delegation as beaming. And so they did, offering prodigious amounts of vodka and caviar in return, but not to the reporters, all of whom were by then looking for telephones. The elated Blagonravov, now clearly delighted that the secrecy was over and that his nation had staged an historic triumph, basked in the good wishes of his foreign counterparts and intimated that he had known preparations for the launch were complete when he left Moscow.

While the conference was supposed to have ended on October 5 with the routine endorsement of resolutions, it turned into an occasion at which the Russians accepted formal congratulations. It also provided the ebullient Blagonravov, chalk in hand, with the opportunity to brief Berkner and the others on the formerly secret operation. A photograph of him doing so would run in the *Times* two days later over one of the paper's "Man in the News" features headlined "Scientist and Soldier." Meanwhile, as Sullivan put it, "the model of the American Vanguard buzzed bravely, but a bit forlornly, around the globe elsewhere in the citadel of science on Constitution Avenue."

When the last of the smoke from the R-7 that carried Sputnik into history drifted away from its launchpad and faded into the Kazakh night, it left a rumor behind that itself drifted all the way back to Moscow and would linger there for decades. According to the story, a routine intelligence report had come into Moscow months earlier confirming that the American IGY Committee was planning on hosting the conference in Washington in late September and early October that would address launching a satellite. But the typist who transcribed the message for Korolyov dropped the word "conference."

The report that was handed to the chief designer therefore said that the United States was planning to launch a satellite in late September or early October. Korolyov, panicked, decided that he would have to launch first so as not to allow the United States to steal the limelight. "They [the Russians] wanted to make a big sort of hullabaloo about it," Sergei Kapitsa would recall many years later. "But there was no time. It was a question of days that the Americans could beat us. So up it went without any fanfare. As Korolyov was ready to fire . . . he did."

6

To Race Across the Sky

The New York Times hit the street on the morning of October 5, 1957, with an eight-column-wide banner headline shouting:

> SOVIET FIRES EARTH SATELLITE INTO SPACE;
> IT IS CIRCLING THE GLOBE AT 18,000 M.P.H.;
> SPHERE TRACKED IN 4 CROSSINGS OVER U.S.

Beneath the head were three Sputnik-related stories and a picture. The lead article came out of Moscow and reported the start of the space age. The others, including one by Walter Sullivan, ran under Washington datelines and supplied additional information. And there was a detailed diagram tracing one of the spacecraft's orbits. Sputnik was called both an "earth satellite" and a "Soviet-made moon." The placement of the stories on page one under a three-bank head showed the editors understood that Sputnik's flight was a major historical event. Curiously, however, the story yielded first place in the international section of the index to rioting in Warsaw and instead was relegated to news about health and science.

There was nothing tentative about reaction to the event, however, and it came swiftly. Joseph Kaplan, chairman of the U.S. IGY satellite program, called the feat "fantastic," the more so because Sputnik weighed eight times as much as any spacecraft then being planned by the United States. J. William Fulbright, the widely respected senator from Arkansas whose scholarships were sending some of the country's best minds to study abroad, reacted with studied calm. He gently dismissed Sputnik as a relatively inconsequential "trick." "It does not feed their people. . . . It does not convert anyone to communism. So far as real prestige is concerned, it is nothing." Representative James G. Fulton of Pennsylvania adamantly refused to concede the Russians anything. Instead, he turned their own often-used ploy on them by claiming that they had benefited from pioneering American rocketry. "There is a feeling in many quarters that it was like the canary that jumps on the eagle's back," he said. "When the eagle flies high as it can and then just as it reaches the top the canary jumps thirty feet higher and it has the record." But even in those first hours, it was obvious that Sputnik was no hitchhiking canary.

Columbia University sociologist C. Wright Mills, the acerbic left-wing author of *The Power Elite,* dismissed the Soviet spacecraft out of hand and warned against sliding into a blind, irrational race to space with the Russians. "Who wants to go to the Moon anyway?" he asked. Then, characteristically, he answered his own question by insisting that "the whole space gambol" was "a lot of malarkey."

William A. Holaday, a Pentagon guided missile expert, correctly observed that Sputnik's flight did not mean the Soviet Union was superior to the United States in missile and rocket development. And, he added, Vanguard was an open, as opposed to secret, program and was not being conducted on a "crash" basis. That was putting it mildly. The implication was that the opposition had poured enormous resources into the project to embarrass the United States, which was not true.

Donald Quarles, by then deputy secretary of defense, told the Science Advisory Committee, under whose auspices the Killian panel had written one of the most important and far-ranging foreign policy documents of the postwar period, that the Russians had in fact done the United States a great favor. They had ended the old legality question once and for all, unintentionally establishing the concept of freedom of international space. That included the freedom to send spy satellites there. Sputnik was doing what Vanguard had been supposed to do.

No Trespassing in Celestia

Quarles and the Department of State's lawyers may have taken the Soviet Union into consideration on the freedom of space issue, but they somehow overlooked its impact on James T. Magnan, a Chicagoan who had claimed nine years earlier that all of space was a nation called Celestia and that he was its absentee spacelord. He therefore immediately charged the Russians with trespassing. "I refuse to issue any license to Russia for use of outer space. Neither Russia, the United States, nor Great Britain has any claim to space except through my nation, Celestia," the angry industrial designer told United Press International.

The Sputnik Cocktail: Vodka and Sour Grapes

Rear Admiral Rawson Bennett, the chief of naval operations and therefore the Vanguard program's ultimate superior officer, expressed his own unambiguous feelings about the red moon. It was "a hunk of iron almost anybody could launch," he growled before drawing on his military expertise to state that the spacecraft's given weight seemed to be "erroneous." Clarence B. Randall, one of Eisenhower's special advisers, understandably belittled the Soviet accomplishment, calling Sputnik "a silly bauble."

But von Braun knew better. As it happened, he and Major General John B. Medaris, the commander of the Army Ballistic Missile Agency at the Redstone Arsenal in Huntsville, were entertaining Neil H. McElroy, the secretary of defense–designate, and Army Secretary Wilbur M. Brucker that very night. The U.S. Army's chief rocket designer had good reason to know that spacecraft were anything but easy to launch. And he especially did not believe they would be easy to launch by the Navy, a point he went out of his way to make to McElroy as soon as he heard the electrifying news about Sputnik.

"Vanguard will never make it," an exasperated von Braun told the new secretary of defense, undoubtedly thinking about the development problems the Navy's thoroughbred was having relative to his own proven Redstones and Jupiters. "We have the hardware on the shelf," he implored. "For God's sake, turn us loose and let us do something. We can put up a satellite in sixty days, Mr. McElroy! Just give us a green light and sixty days." He kept repeating "sixty days" until Medaris interrupted: "No, Wernher, *ninety* days." McElroy left for Washington without committing himself.

The father of the V-2 was not engaging in idle promises. The Army Ballistic Missile Agency had remained stubbornly committed to launching a satellite despite the president's preference for Vanguard. And the fact that the Navy project was experiencing problems only made Huntsville more frustrated and determined. On September 20, 1956, the Army launched a Redstone from Canaveral that carried two scaled-down Sergeant upper stages, both of which were set spinning before launch in a stabilization technique that worked perfectly. The Redstone-Sergeant combination tossed an eighty-four-pound payload 3,300 miles downrange over the Atlantic. A properly angled trajectory, von Braun would later point out, would have carried a satellite to orbit. It was on that basis that he would later complain bitterly, and with justification, that the Army could have beaten the Russians to space by a year.

What von Braun did not know, of course, was that there were people in the White House, the Department of State, and even the Pentagon who were not unhappy about the Soviet Union beating the United States to space. And Don Quarles, of course, was one of them.

The immediate reaction on most of Capitol Hill was that the Soviet spacecraft circling Earth was somehow an unfair surprise and, given the obvious power of the rocket that got it up there, a dangerous one. It was yet another manifestation of the Red Menace.

Senator Henry L. "Scoop" Jackson, a military hard-liner from Boeing's home state of Washington and the Democratic chairman of the Military Applications Subcommittee of the Joint Committee on Atomic Energy, called Sputnik nothing less than "a devastating blow to the prestige of the United States as the leader of the scientific and technical world." Stuart Symington, another cold warrior on the Senate Armed Services Committee, said that he considered the satellite "one more proof of growing Communist superiority in the all-important missile field. If this now known superiority develops into supremacy," he warned, "the position of the free world will be critical."

A shrill cacophony spread across the land like a prairie fire. And it didn't take long for it to lick at Dwight Eisenhower. The Democratic Advisory Council, which included Harry Truman and Adlai Stevenson, the party's foremost intellectual and its defeated presidential candidate, accused the administration of "unilateral disarmament. . . . The all-out effort of the Soviets to establish themselves as master of [the] space around us," the staunch liberal declared, "must be met by all-out efforts of our own."

Lyndon Baines Johnson, a onetime schoolteacher who rose to become the majority leader of the Senate, used a little history to construct a dire, if sophistic, historical picture with typically majestic flair. "The Roman Empire controlled the world because it could build roads," he drawled. "Later—when moved to the sea—the British Empire was dominant because it had ships. In the air age, we were powerful because we had airplanes. Now the Communists have established a foothold in outer space. It is not very reassuring to be told that next year we will put a better satellite into the air. Perhaps it will also have chrome trim and automatic windshield wipers."

The reference to satellites flying through "air" reflected the fact that the ambitious Texas Democrat had no apparent interest in space before Sputnik. But the Russians had handed him a superb cause; one which he instinctively recognized he could ride to glory. "The urgent race we are now in," he would tell a Democratic conference three months later,"—or which we must enter—is not the race to perfect long-range ballistic missiles, important as that is. There is something more important than any ultimate weapon. That is the ultimate position—the position of total control over Earth that lies somewhere out in space."

Neither Eisenhower nor Khrushchev shared that view, though, because their responsibilities transgressed political rhetoric. Both felt that whatever propagandistic, and ultimately strategic, value could be derived from the occupation of space, it was "delivery systems"—bombers and ballistic missiles—that were the true sine qua non of national defense. In a thoroughly defensive speech delivered in Oklahoma City after the launching of Sputnik and before he became ill, Ike reemphasized that it was the "retaliatory nuclear power" of the Air Force and the Navy that were the nation's first line of defense, not flinging small machines into Earth orbit. (He carefully avoided any reference to reconnaissance satellites.)

Echoing the Killian panel's recommendations, he called for strengthening the Strategic Air Command and dispersing its bombers; improving the radar early warning system; producing long-range ballistic missiles for the Air Force and the Navy; and developing a defense against missile attack.* The last was the seed that would germinate and grow into a succession of missile defense systems that sprouted during the Johnson and Nixon administrations and, as the Strategic Defense Initiative, or "Star Wars," took on a life of its own during Ronald Reagan's presidency. Characteristically, Eisenhower couched his plan for enhanced national defense within a framework of monetary restraint, calling for a "redoubled determination to save every possible dime." In the long run, he prophesied, "a balanced budget is one indispensable aid in keeping our economy, and therefore our total security, strong and sound." Ike was and would remain genuinely puzzled by what quickly turned into a panic.

In a sober assessment of the relative strengths of the U.S. and Soviet military systems, *The New York Times'* military analyst, Hanson W. Baldwin, declared that the Russians were "probably" ahead in ballistic missile development but that earlier estimates that the United States was hopelessly behind were exaggerated. But Baldwin also blamed Eisenhower's obsession with a balanced budget for the Soviet triumph. "Economy was the watchword, and it was economy—dictated by the President, by the National Security Council, and by the Bureau of the Budget—plus lack of high-level appreciation of the vital importance of the battle for men's minds that slowed the American [satellite] project," he wrote in a slender but heavily reported book called *The Great Arms Race.* Baldwin also made the point, which was fully shared by the National Security Council, the Air Force, and others, that the real significance of the Sputnik launch was not the satellite but its launcher. The rocket that had shot the satellite into orbit could also fling a hydrogen bomb all the way to Washington, D.C. That made the leader of the free world vulnerable to annihilation for the first time.

Killian, who was asked by Eisenhower to be his science adviser twenty days after Sputnik went up, calmly insisted that there was less to the feat than met the eye and that attacks on Ike were just shabby politics. As he had reason to know better than anyone else, there was actually more to it than met the eye:

* He also knew, but could not say publicly, that plans were under way to arm the nation's growing fleet of fighter-interceptors with nuclear-tipped air-to-air missiles that would be used to stop Soviet bombers hundreds of miles away from U.S. borders. The logical route for those bombers to take would have been over Canada, and everyone in the national command authority agreed that nuking them over Canadian territory was infinitely preferable to, say, losing Chicago or Detroit.

from the standpoint of international law, Sputnik was an unmitigated blessing. "The drama of space," he would note in his memoir, "stirred visions on the part of more than one politician that they might ride rockets to higher political ground." That certainly applied to LBJ.

Johnson's sarcastic reference to an American chromium-trimmed spacecraft was among the first stirrings of what would quickly grow into national self-flagellation: Americans seemed to have become so hedonistic, so addicted to frivolity, that their will had turned to mush. The corollary image was of a Soviet Union that was lean, tough, disciplined, and aggressive. "The time has clearly come to be less concerned with the depth of the pile on the new broadloom rug or the height of the tail fin on the new car," warned Senator Styles Bridges. The senior Republican on the Senate Armed Services Committee in effect called upon his bloated, complacent, and self-indulgent countrymen to push away their banana splits, hot fudge sundaes, malteds, cherry-lime rickeys, barbecued steaks, and hot dogs, drain their pools and martini glasses, abandon bikinis, poodle skirts, blue suede shoes, ukuleles, power boats, country clubs, T-Birds and Corvettes, canasta and mah-jongg, golf, juke boxes, psychiatrists, Sandra Dee, Elvis "the Pelvis," and Liberace (and everything they stood for), "and be more prepared to shed blood, sweat and tears if this country and the free world are to survive." Even Eisenhower told a reporter that he was beginning to wonder whether Americans had become too fond of the good things in life to make necessary sacrifices.

Bernard M. Baruch, the self-made millionaire and self-styled sage, wholeheartedly agreed. He, too, could not resist tossing Detroit on the sacrificial alter; of taking a shot at the country's most revered icon, its talisman of tomorrow: the gaudy, grinning, chrome-plated, tail-finned, wraparound, Dynaflowing embodiment of its moral and spiritual flatulence. "While we devote our industrial and technological power to producing new model automobiles and more gadgets, the Soviet Union is conquering space," the Will Rogers of Wall Street complained. The remedy, he added, was simple: "hard work and sacrifices and putting first things first."

Edward R. Murrow, by then the widely respected chief correspondent of CBS, mentioned that theme in a different context. "It is to be hoped that the explosion which flung the Russian satellite into outer space also shattered a myth," he said three days after Sputnik went up. "That was the belief that scientific achievement is not possible under a despotic form of government. . . . We failed to recognize that a totalitarian state can establish its priorities, define its objectives, allocate its money, deny its people automobiles, television sets and all kinds of comforting gadgets in order to achieve a national goal. The Russians have done this with the intercontinental ballistic missile, and now with the Earth satellite," he added before mentioning that the United

States should have learned "from the Nazis that an unfree science can be productive. . . . Now the Russians have carried the rocket and jet propulsion to their logical conclusion. They have made both the A-bomb and the H-bomb. They have proved that they can turn out as perfect instruments of destruction as our own, and in some cases do it quicker. They are ahead of us in ballistic missiles. They are flying their satellite. The White House, in the person of Mr. [Press Secretary James] Hagerty, may not be surprised or impressed, but the rest of the world certainly is."

Michigan Governor G. Mennen Williams was gentle on Eisenhower. Yet in common with many American of all stripes, he worried about what he mistakenly took to be Ike's lassitude. He even penned light verse that made the point by using the president's well-known love of golf:

> Oh Little Sputnik, flying high
> With made-in-Moscow beep,
> You tell the world it's a Commie sky
> And Uncle Sam's asleep.
>
> You say on fairways and on rough
> The Kremlin knows it all,
> We hope our golfer knows enough
> To get us on the ball.

Scientists, now basking in a new importance, had their own thoughts and were delighted to share them with everyone. One astronomer gloomily maintained that the game was over; that there would be no catching up. "No matter what we do now, the Russians will beat us to the Moon," declared John Rinehart of Harvard's Smithsonian Astrophysical Observatory. He added that he would not be surprised if the Russians reached the Moon "within a week."

Edward Teller, a dogged anti-Communist, invoked the now well-worn metaphor of surprise attack, telling a television audience that his adopted country had just lost "a battle more important and greater than Pearl Harbor." The hydrogen bomb's foremost advocate had a one-word answer to the question of what would be found on the Moon: "Russians."

America's torpor, in the opinion of some, was only matched by its naivete. Russians or their agents were depicted as plundering the secrets of a benevolent but naive giant. *U.S. News & World Report* claimed that Russian spies had stolen rocket data from the United States after World War II just the way the Rosenbergs and other traitors had pilfered the secret of the atomic bomb. The editors of *Look* resurrected the old "loose lips sink ships" idea to explain the calamity. A two-part series that would run early in 1958 blamed careless talk by V-2 people who "let the missile secrets get away."

Beep-Beepski

Whatever anyone else thought, however, a cartoonist working for the MIT humor magazine *Voodoo* refused to be awed by the Soviet achievement and instead advanced his own theory: that Sputnik was not only made by man but carried one. Donald J. Hatfield, a sophomore, drew a cutaway of the satellite, inside of which was a dopey, bearded Cossack curled in a fetal position, saying, "Beep . . . beep . . . beep." As was the case elsewhere in the U.S. scientific establishment, the student-artist found nourishment, if not solace, in sour grapes.

Welch's Nightmare

For a candy manufacturer turned political guru named Robert Welch, the revealed wisdom about Sputnik—the inside track—had it that the little spacecraft and its follow-ons were launched for no less a purpose than to destroy the United States. Welch abandoned his business in 1957 to devote all of his time to sounding the alarm against what he saw as a relentless and insidious Communist plot to enslave the American people and the rest of the world. Seeing himself as a combination Paul Revere and exorcist, Welch would start an organization called the John Birch Society in Indianapolis in December 1958 that was supposed to turn over the rocks beneath which the deadly parasites hid so they could be spotted and destroyed. His knowledge, he explained, came from "many years of intensive study of the methods, the progress, and the menace of the Communist conspiracy."

In Welch's view, the Communists shot machines into space with the sole intention of so alarming the United States that it overreacted to the threat, fatally weakening itself in the process. The chain of events, as he believed they were plotted by the diabolical geniuses in the Kremlin, went like this:

The U.S. government would respond to the Sputniks by running up a colossal weapons and foreign aid bill. That would send taxes through the ceiling and lead to an increasingly unbalanced budget, wild inflation, and then rigid government control of prices, wages, and other means of fighting the inflation. "Socialistic" control over the economy and every aspect of Americans' lives would come next, accompanied by a huge increase in the size of the federal bureaucracy, more centralization of power in Washington, and the virtual disappearance of states. That in turn would set the stage for the complete federalization of the educational system, a constant hammering into the American conscience of the horror of modern warfare and the necessity of peace. That would inevitably result in "the consequent willingness of the American people to allow the steps of appeasement by our government which amount to a piecemeal surrender of the rest of the free world and of the United States itself to the Kremlin-ruled tyranny."

All of this and more, including an assertion that every atomic bomb exploded in the Soviet Union had been made in the United States and smuggled out in parts by secret agents, was contained in *The Blue Book of the John Birch Society,* a combination bible-handbook that was written by Welch to warn about the Communist Menace. It is interesting to speculate how Welch would have responded to Eisenhower's near indifference to Sputnik. He might have taken it to mean that the president was even smarter than the Communists. But that would have effectively killed the very devil that provided Welch with his life's calling. More likely, he would have come to the grim conclusion that Washington was already so penetrated by Communists that the national spine was rotted beyond repair.

Khrushchev's Triumph

Whatever C. Wright Mills and others who were hostile or at least indifferent to the race Sputnik seemed to be starting may have thought, it was not malarkey to Nikita Khrushchev. Not by a long shot. It took the first secretary about a day to catch up to the Western press; to the adulation being heaped on his country and, by inference, on the all-but-anonymous Sergei Korolyov and colleagues such as Mikhail Tikhonravov. When first told that an R-7 had successfully carried Sputnik to space, Khrushchev is said to have reacted apathetically, putting the achievement down as "just another Korolyov rocket launch." But by the morning of October 6, with *Pravda*'s front page crowing that the "World's First Artificial Satellite of Earth [was] Created in the Soviet Nation," congratulations pouring in from around the world, and poems already being written to commemorate the historic event, Khrushchev awakened. And having awakened, he knew instinctively that the chief designer's triumph amounted to an undreamed of public relations bonanza. It would allow him to brandish the national sword, which suddenly seemed to be tempered by the R-7's flame and was glistening in its light.

Khrushchev would insist in the weeks and months to come that his country's triumph was proof not only that it could send nuclear warheads anywhere it pleased, but more important, that the political system that had spawned Sputnik was inherently superior to that of its adversary. It was proof the future belonged to the socialist republics and to the ideological principles on which they rested.

"When we announced the successful testing of an intercontinental rocket, some U.S. statesmen did not believe us," a smirking Khrushchev told James Reston of *The New York Times* a week after Sputnik went into orbit. "The Soviet

Union, you see, was saying it had something it did not really have. Now that we have successfully launched an earth satellite, only the technically ignorant people can doubt this. . . . We can launch satellites because we have a carrier for them, namely the ballistic rocket." The temptation to credit Marxism-Leninism with the triumph was irresistible. So Khrushchev did not resist. The "serial production" of ICBMs rolling off the assembly line "like sausages," he told a party congress, meant that "socialism has triumphed not only fully, but irreversibly." It was one of his patented bluffs. There were only four R-7 launchpads in the whole country at the time, three of them at Plesetsk, the third and last space center north of the capital, and one at Tyuratam.

A report issued three years later by the United States Information Agency on world reaction to Sputnik would seem to bear out the first secretary (who also became premier in 1958). It claimed that most Western nations believed that the two superpowers were at that point engaged in a race to space and that the Soviet Union would still be ahead a decade from then. "Within this rivalry," the study concluded, "space achievements are viewed as particularly significant because of the strong tendency for the popular mind to view space achievements as an index of the scientific and technological aspects of the rival systems, and to link space capabilities with military, especially missile, capabilities."

While that may have been true, at least superficially, it was not reflected in the United States. Half of Americans interviewed in a Gallup Poll between October 11 and 14 said they considered Sputnik to be a "serious blow" to their country's prestige, but 46 percent said it was no such thing. More than half claimed they were surprised that the Soviet Union had beaten the United States to space, while 44 percent reported that they had not been surprised, and 61 percent maintained that Earth satellites would "more likely be used for good purposes than bad."

Like fans at a cosmic marathon, politicians around the planet were beginning to lay odds on the outcome of the great race and quietly (or not so quietly) cheer their favorite team. There is no evidence, however, that they were raising serious questions about the race's real nature. The very term was inappropriate, since a competitive contest necessarily involves an end point determined by time or distance. But there could be no conclusion to this contest so long as either competitor kept launching spacecraft. And obviously, the "winner" of such an amorphous duel would be as undeterminable as the "prize" was undefinable. This almost certainly was known to all of the important participants—scientists, scholars, soldiers, and politicians alike—and was acknowledged by none of them because it was irrelevant.

What *was* relevant was that both powers were suddenly competing in a tremendously expensive contest that pitted elite groups of designers, technicians, managers, bureaucrats, politicians, scientists, and others against each other. And they dueled on a new and dangerous frontier, "outer" space, which was both unknown and inaccessible to the masses of people, including the dreamers, who would have to pay to stay in the race. Nobody who was taxed to finance the sending of colossally expensive machines and a handful of government employees to the "final" frontier would be able to stake a claim there (except for James Mangan and a few others, of course). There was nothing tangible in it for the vast majority of ordinary citizens. That being the case, having a race was the perfect thing to do precisely because it was ambiguous. If it was impossible to win, it was also impossible to lose, so neither elite would ever have to face the consequences of total failure as long as the cold war lasted.

Indeed, as both sides girded for combat in the heavens, Amitai Etzioni articulated a problem so fundamental to their respective space programs that it would haunt them forever. The eminent sociologist in effect took issue with the USIA study and concluded exactly the opposite: that the race to space was in reality a duel between opposing political and military establishments and would not benefit ordinary citizens on either side in any appreciable way. Even more to the point, he maintained that most people on planet Earth neither knew nor cared about who was in the space race, let alone who was winning it, and that, in any case, world "public opinion" counted for virtually nothing where the competing elites' political agendas were concerned.

Years later, Etzioni would cite a survey conducted by Gabriel Almond, a leading political scientist, showing that only 9 percent of Britons believed that Sputnik enhanced the position of the Soviet Union relative to the United States, while its position remained unchanged in Italy and actually declined in France and West Germany. At the same time, he added, polls in Mexico City and Rio de Janeiro in November 1957 showed that 33 percent of Mexicans and 49 percent of Argentineans had never even heard of Sputnik. "And that," Etzioni noted, pointedly, "did not include the countryside. Moreover, in the United States, the *Milwaukee Sentinel* had a revealing headline: 'Today, We Make History.' That was October 5, 1957; but the story relates to the Milwaukee Braves winning whatever they won. *Sputnik* was on page 3."

Then, whittling his case to a finer point, Etzioni questioned the significance of world opinion in the first place. "The world society is not an American democracy in which the average citizen follows the news through open media and has an opportunity to act on his opinion in the political realm. In most parts of the world . . . the horizon of people extends only as far as their village. The national capital does not exist. World events have no significance. People live in the world of their village; they are preoccupied by their next meal, not world affairs."

But the village to which Etzioni referred in what was then called "underdeveloped countries" had counterparts in the developed ones. There were infinite ways to define the "next meal." Even relatively well-off, media-drenched Americans had far more pressing needs than sending expensive machines into the boundless void of outer space. "Three squares," health care, a decent job, good education, safety, and other mundane—from the Latin *mundus,* or "world[ly]"—concerns, certainly including sports, were far more important than anything that sent machines or people off the planet. Next to the many challenges of daily existence faced even by educated, middle-class Americans, the entire space enterprise was all but irrelevant. And it would remain so throughout the century because, the competitiveness of their leaders notwithstanding, it did not seem to touch them in fundamental ways. In fact it did. Space reconnaissance and missile early-warning satellites, to take two examples, helped prevent a major war between the two superpowers. But the first were highly classified and the second were largely unknown beyond the military and the industry that built them.

Two groups, which to some extent overlapped, were to be directly affected by the great race. The larger one consisted of the thousands of civilians and soldiers whose livelihoods depended upon space. The other was the relatively small but irrepressible core of dreamers who would have turned out to see Columbus sail into the unknown.

The zealots, the dreamers on both sides, who formed the elites' critical mass, would continue to look beyond the space race as a ritualized political sport and see spaceflight as the inevitable fulfillment of a manifest destiny. They would continue to resolutely hold the vision. Sending a mechanical moon to orbit Earth, for example, created a particularly sweet moment for Mikhail Klavdyevich Tikhonravov, whose fertile mind spawned both the satellite and the rocket cluster that carried it onto the pages of history. So the venerable engineer, one of the truly unsung heroes in Korolyov's retinue, allowed himself some hyperbole when he looked at a sky he had helped to change forever; when he saw the speeding white dot sail past the stars. "This date," he said years later of October 4, 1957, "has become one of the most glorious in the history of humanity."

Laika's Historic Ride

The Russian engineer was undoubtedly as ecstatic a month later when Sputnik 2 went up carrying a slew of scientific instruments and a passenger. The event coincided with the fortieth anniversary of the Bolshevik Revolution and was announced by this simple story from TASS:

> In accordance with the IGY program for the scientific investigation of the upper atmosphere, and the study of physical processes and life conditions in cosmic space, a second artificial earth satellite was launched on 3 November.

A book on Soviet space science published within weeks of the satellite's flight described its instrumentation in considerable detail. It is not necessary to understand the physics of the solar radiation sensors—only one of the packages—to appreciate that the second Sputnik was sophisticated enough to make its predecessor look like an aluminum cannonball:

> In this [solar radiation] instrumentation, the radiation pickups were three special photoelectron-multipliers, spaced 120-degrees apart. Each photomultiplier is successively covered by several filters of thin metal and organic films and also of special optical materials, allowing the separation of various bands in the X-ray region of the solar spectrum and the hydrogen line in the far ultraviolet region. The electric signals emitted by the photomultiplier pointed at the sun were amplified by radio circuits and transmitted to the earth by means of a telemetering system.
>
> As a result of the fact that the satellite was continuously changing its orientation relative to the sun, and also spent part of the time on the part of its orbit that was not illuminated by the sun, in order to save electric power, the instrumentation circuits were turned on only when the sun fell in the field of view of one of the three light pickups.

But the second Sputnik carried more than sophisticated instruments. It carried the first living creature to go to space from Earth: a mixed-breed terrier named Laika (Barker), who made the historic journey in a specially designed, pressurized compartment, complete with automatic food and water dispensers, a sack to hold her waste, and electrodes that measured her pulse, breathing, and blood pressure. The data were radioed back home as a kind of canine electrocardiogram. Telemetry from the spacecraft told those who monitored Laika that her heart pounded during the high-g liftoff but that she survived the pressure and then seemed to adjust to life in orbit. A week later, she was killed by an injection of poison since, the Soviets claimed, her spacecraft could not be recovered. In fact, recovering the spacecraft with Laika alive would not have been an insurmountable problem had the engineers been under less pressure.

Sputnik 2 was important for a number of reasons. It put to rest once and for all a lingering feeling among some Americans that the earlier flight had been a hoax. It also eclipsed its predecessor in three fundamental respects. First, its weight—1,120 pounds—showed beyond doubt that the U.S.S.R. had a true intercontinental ballistic missile capability, however nascent. Second, its array of measuring instruments and their sophistication made it an unprecedented physical sciences platform and the true forerunner of Soviet space science missions, some of them first-rate. Finally, sending up a complicated mammal as a passenger not only proved that such a creature could survive a flight in space, but it clearly indicated that the Russians were actively planning to send people there. And if humans could orbit Earth, as seemed to be the case, they could reach other heavenly bodies. Korolyov's dog thereby joined Joseph and Étienne Montgolfier's sheep, rooster, and duck—the first creatures to fly in a balloon and therefore not under their own power—in helping to build Galileo's bridge to the Moon.

Eisenhower's Lament

The Soviet space triumphs caught Eisenhower in the classic conundrum of the national security process in a democratic society. Killian, who was also de facto head of the President's Science Advisory Committee, asserted in his memoir that the United States was in fact not trailing the Soviet Union in any way but superficially. The problem, he explained, was that the people of the United States were kept in the dark about their nation's true strength in ballistic missile technology. The "American panic" following Sputnik, he wrote, resulted from the fact that the public was "woefully ignorant of how much qualitatively advanced and forehanded rocket technology had been. . . . While the Soviets launched the first satellites, the rocketry we had under development in its qualitative aspect was potentially ahead of the Soviets. This ignorance—the result of excessive secrecy—undoubtedly contributed to the American people's frantic reactions to Soviet pronouncements and achievements."

That was true as far as it went. On the other hand, a slew of journalists, politicians, and generals were charging that it was precisely the country's excessive openness that had allowed atomic and missile secrets to be pilfered; that had led to the Sputnik debacle in the first place. Where was the line to be drawn? A military security system of some sort went back to Valley Forge and was at that moment evolving into one so stringent that even "Top Secret" was not the highest classification category.* Given the particularly insidious

* The highest level was, and remains, information that is compartmented. That is, it is accessible only to those with a so-called "need to know" specific material. "Sensitive Compartmented Information," or SCI, applies to all data collected by technical intelligence systems and the data themselves. Those who analyze aerial or satellite imagery, for example, are kept ignorant about details relating to the camera's operation or the working of the satellite. This has the effect of minimizing the damage from a leak, but it also inhibits sharing useful information. An SCI clearance requires more thorough background checks than does "Top Secret" and is more stringently enforced.

nature of the Communist threat, the military was understandably loath to reveal anything that it felt would be potentially helpful to the enemy. Meaningful details concerning the nation's missile development were certainly in that category.

So was intelligence, as Eisenhower well knew. The U-2 flights over the Soviet Union, then in their second year, were proving almost beyond doubt that the United States in no way lagged behind in either bombers or missiles, for example. Yet the flights had to be kept from the public for a number of reasons, some of which would also apply to their spaceborne successors. For one thing, although the Soviets tracked the aircraft and tried to shoot them down, making the missions public would have angered and embarrassed the Kremlin, which regularly protested that they were both intrusive and illegal (which was true). In addition, revealing the existence of the U-2s would have embarrassed the nations from which they staged: Norway, West Germany, Great Britain, Turkey, Pakistan, and Japan. Revealing that information, and necessarily the photointelligence lode the planes collected, would have simultaneously shown the opposition how good the cameras and other sensors were and ended the operation. That would have been politically expedient in the heat of the moment. But it would have been disastrous in the long run because a vital intelligence source would have been choked off. So the old general kept his secrets, even to the point of making his administration vulnerable to Democratic charges of bomber and missile "gaps" that would help John F. Kennedy win the presidency in 1960.

With it all, Ike remained deeply puzzled by the angst over Sputnik throughout the storm and didn't understand why Americans were "so psychologically vulnerable." He equated his fellow citizens with spoiled, petulant children who whine when they are denied instant gratification or lose a game. Like Johnson and Bridges, he fretted over what he took to be their apparent softness. He opposed a hasty, fear-driven response to the Soviet challenge and instead wanted the people of the United States to develop enough stamina and grit to compete "for years, even decades" if they had to.

Now a frontal attack on the nation's educational system began to develop. The battle over so-called progressive education, which had been simmering since the early 1950s, erupted as traditional educators reviled what they considered to be too much emphasis on "life adjustment" in the form of student clubs, yearbooks, newspapers, marching bands, and other irrelevant diversions at the expense of bread-and-butter basics: math, science, history, and foreign languages.

The traditionalists looked at their country's teenagers and woefully saw the Archie, Veronica, and Jughead cartoon characters come to life: distracted, egotistic know-nothings who whiled away their days preoccupied with proms, jalopies, saddle shoes, clothes, hairdos, meeting at the malt shop after school, outsmarting dopey parents and square teachers, listening to Presley, stealing boy- and girlfriends, pledging for fraternities and sororities, and looking great at the beach. Period.

The enemy's kids and university students, by contrast, seemed to labor obediently under piles of homework and bring unstinting seriousness to their work. Not only were the Reds militarily muscular and infinitely devious, but it turned out that they were apparently superbly educated as well, particularly in science and engineering. Senator Jackson announced that the Soviet Union was training more scientists than any Western nation and was accelerating the pace. An aide to Lyndon Johnson warned that Russia had 350,000 high school science and math teachers compared to a measly 140,000 in the United States. Admiral Hyman Rickover, the dour father of the atomic submarine, seized the most popular metaphor of the time and adopted it for his own message. He said that he hoped Sputnik would do in "matters of the intellect what Pearl Harbor had done in matters industrial and military."

The education gap was brought home by Health, Education and Welfare Secretary Marion Folsom, who weighed in with a report in early November—roughly coinciding with the second Sputnik launch—claiming that all Russian students took five years of physics, five of mathematics beyond arithmetic, and four of chemistry. In comparison, only one in four of their lackluster American counterparts even took a physics course, and only one in three took chemistry. "This [Soviet] system of education gives little freedom of choice to the individual," scolded *U.S. News & World Report.* "But it produced the scientists and engineers who built and launched *Sputniks I and II.*"

The implication seemed clear. Far from being an aberration, some kind of nightmare that would vanish by morning and never return, the Soviet triumph was real and was sticking. It seemed for all the world to be the result of a huge, dedicated, carefully thought out system of education that amounted to a deep well from which science and engineering talent could be drawn in almost limitless quantity. Whatever the obvious virtues of democracy, despots seemed to have an exasperating advantage because they could force the system to accomplish whatever goals they and their minions set.

A group of American chemists visiting the Soviet Union at the time was awed by what its members took to be almost bottomless reserves of character, energy, tenacity, and resources of both land and spirit: the components of an indomitable opponent. "One of the most important reasons Russia is a formidable adversary is the character of the Russian people themselves," they reported. "They are [a] vigorous, dynamic, intelligent, self-confident, well disciplined people, who would clearly give a good account of themselves in any conflict."

"In addition to the present military threat," one of the impressed scientists noted, "Russia gives promise of presenting a strong economic threat. . . . I was told that Russia plans to export large quantities of manufactured goods of all types within the next five to 10 years, including automobiles, cameras, watches, machine tools, electrical, radio and television equipment, chemicals, and textiles."

Whatever the chemist had been told, the Soviet Union would excel in space and military technology, not in consumer goods, which would remain notoriously shoddy. Nor could things have been otherwise because, in reality, the proletariat dictated nothing. Rather, the requirements of national security dictated an obsessive reliance on military hardware and technoglitz, both of which were fed into a voracious propaganda machine. That chemist would have learned more by looking at his host's shoes than by listening to his hollow predictions. In a manner of speaking, Soviet eyes fixed on the sky—on bravery, excellent science and technology, and the eternal tomorrow—would miss the profound shortcomings of life on the ground. That, of course, was precisely the idea.

All but lost in the loud litany of America's malaise and self-deprecation were dissenting voices, like that of a Russian-born engineer at MIT, who claimed that in reality the Soviet system was hopelessly elitist and narrowly focused on training mere technicians, not real scientists.

In any case, the furor over the supposedly sorry state of American education resulted in the National Defense Education Act, which Eisenhower signed on September 2, 1958. It authorized spending slightly less than $1 billion over a four-year period to get the nation's educational institutions back on track. One provision set aside money for low-interest loans to needy students who wanted to become science, math, or foreign-language teachers. A second provision set aside federal matching funds for public and private schools to buy equipment for the teaching of those three disciplines. And a third provision reserved nearly $60 million for 5,500 graduate fellowships related to national defense for the study of science, engineering, and foreign languages.

The Flight of Kaputnik

At a news conference on October 9, five days after Sputnik went up, Eisenhower tried to calm his countrymen by emphasizing that American missiles were in "top priority" production—the nation could take care of itself militarily—and that the United States was not in a race to space with the Soviet Union. The president almost simultaneously congratulated the Russians and then disparaged their achievement by calling their satellite "one small ball in the air." As a way of emphasizing that there was no race, he added that there would be no change in Vanguard because of Sputnik. The United States would stick to its "well-designed and properly scheduled" program in order to accomplish what it was supposed to as its contribution to the International Geophysical Year. The schedule, he added solemnly, called for launching a Vanguard test vehicle in December. Shrugging off Sputnik and maintaining that the nation was adequately protected calmed few either outside the administration or within it. One critic called Ike's reaction to the Soviet challenge "a curiously uncertain, even fumbling, apologia."

The president's apparent nonchalance also belied a sense of urgency inside the White House, where the race with the Russians and the need to get something into orbit fast were now finally acknowledged. John P. Hagen, who headed the Vanguard program, was therefore told that the Navy wanted a six-inch satellite on the rocket and that it would have to be launched in December, as the president had promised. The Navy itself was in turn ordered by the White House to tell the world that the flight would be a full-fledged attempt to orbit a satellite.

When Navy Secretary Thomas Gates expressed concern over what he saw as a dangerously hasty launch schedule, Neil McElroy spelled it out for him: "The Soviet's success with their satellite has changed the situation," he said, adding that the Vanguard launch had to take place "with deliberate speed." Gates knew that the December launch would be the first of a complete Vanguard and that only the first of the rocket's three stages had been successfully tested. So he had reason to worry.

Vanguard's public debut at Cape Canaveral was what reporters and editors call "a good story." The Russians were ahead 2–0, and the United States—the old favorite now afflicted with self-doubt, humiliation, and mild hysteria—was trying desperately to shake its malaise and even the score. If the thing did what it was supposed to do, Uncle Sam would be back in the game with a supremely patriotic flourish. If it did not, the national mortification would only intensify, probably turning into an orgy of anguish and self-recrimination from coast to coast. Any way an assignment editor looked at the launch of TV-3—Vanguard Test Vehicle 3—it figured to be a good story (unless the launch was scrubbed, as was to happen for two days because of bad weather, but even that would not go on forever).*

Cape Canaveral was therefore packed with reporters, photographers, and television cameramen on the morning of December 6 as Vanguard, gleaming in the dazzling sunshine like an elegant, black-tipped bullet, stood beside its

* TV-1, consisting only of the Vanguard third stage, was launched at Canaveral by Viking 14 on May 1 to test the separation system. TV-2, a whole Vanguard, successfully tested the first stage on October 23. It was not supposed to reach orbital altitude, and it did not.

slender gantry, poised to try to rescue the nation's honor. The satellite, a chrome sphere now shrunken to three and a quarter pounds, nestled inside TV-3's shroud.

By 11 A.M., the area had been cleared of extraneous traffic, including fishing boats, while thousands of people carrying binoculars, telescopes, and cameras crowded nearby beaches. J. Paul Walsh, Vanguard's youthful deputy director, sat in the blockhouse talking to Hagen in Washington. Another line was open to the White House. Outside, the public-address system was counting down for everyone at the Cape and for the millions who were watching on television. It was America's first nationally televised countdown: the space program's inaugural public performance, its debut.

"T minus fifteen and counting," resonated across the Banana River, over sparkling sand, and under a blue sky streaked with a few clouds. Ships and planes had already been cleared from the rocket's down-range flight path.

"Minitrack system clear," the metallic voice announced. That meant the electronic ears that were positioned to track Vanguard's ascent to orbit for the glory of the United States were paying attention.

"T minus three. All equipment on internal power." A flood of water began to pour into the flame pit under the first stage's engine.

"Satellite clear to launch." The flow of helium, which had been building pressure in the fuel-injection system, was cut off. TV-3 was now on its own as the final seconds ticked off.

"Three."

"Two."

"One."

"Zero."

"First ignition."

The GE engine blasted a torrent of flame and smoke into the pit. Vanguard, the focus of thousands of lenses and millions of eyes, began its tenuous climb to the sky. It took two seconds to reach its maximum altitude: four feet. Then there was an explosion accompanied by a sight that was already painfully familiar to everyone who worked with rockets: a fat, boiling, angry cloud of oily fire; an inferno that seemed to be held together by a spiderweb of dirty black veins and globs of rocket blood: its propellant. The beautiful cylinder, its nose cone shaken loose from the upper stage like a wobbly drunk losing his footing, slowly disappeared into the orange-and-black cloud.

"Explosion!"

There were gasps of horror on the beach and sickening disappointment inside the blockhouse and the White House.

Writer Tom Wolfe struck an irreverent, but pointed, note when he described the scene in *The Right Stuff.* "The first stage, bloated with fuel, explodes, and the rest of the rocket sinks into the sand beside the launch platform. It appears to sink very slowly, like a fat old man collapsing into a Barcalounger. The sight is absolutely ludicrous, if one is in the mood for a practical joke. Oh, Khrushchev had fun with that, all right! This picture—the big buildup, the dramatic countdown, followed by the exploding cigar—was unforgettable. It became *the* image of the American space program. The press broke into a hideous cackle of national self-loathing, with the headline KAPUTNIK! being the most inspired rendition of the mood."

The launcher was consumed by its own fire, but the grapefruit-size satellite somehow managed to roll clear, sending its melancholy message from terra firma. "Why doesn't somebody go out there and kill it?" asked *New York Journal American* columnist Dorothy Kilgallen in bitter exasperation.

The reporters were citizens too and, like many of their fellow countrymen and women, they felt vaguely betrayed. So they used their typewriters to punish the people who had done this to them; who had, however unintentionally, made them participants in a national humiliation. They described Vanguard's ignominious demise with a degree of angry cynicism that was fervid. "Disaster" was the word most used the next morning to report the calamity. But Kaputnik, Stayputnik, and Flopnik were irresistible. *The San Francisco News* ran a headline calling the explosion a "COLD WAR PEARL HARBOR." Khrushchev was reported to have suggested that Vanguard should have been named "Rearguard." One news account reported that the Soviet delegation to the United Nations asked whether the United States was interested in receiving aid from developing countries. Perhaps the unkindest cut of all came from *Time* magazine, which compared Sputnik's performance to Vanguard's and then wrathfully put Nikita Sergeyevich Khrushchev on its cover as its Man of the Year. One columnist wrote that many in Washington were referring to the White House as "the tomb of the well-known soldier."

Army to the Rescue

But Eisenhower was not as entombed as he appeared to be. On November 8—sixteen days after the successful but unheralded suborbital test flight by TV-2 and five after Sputnik 2 went into orbit—the Army was finally given permission to go to space. It had not been waiting idly for the order. The troubles plaguing Vanguard, which were no secret in the small rocket fraternity, moved the men at the Redstone Arsenal to ready their alternative to the Navy launcher in the likely event that it would not be able to deliver.

Long before the autumn of 1957, the Army Ballistic Missile Agency had refined the Redstone into an intermediate-range ballistic missile called Jupiter C and successfully tested three of the missiles out of a block of twelve. After completion of the last test, early in the year, John Bruce Medaris had shrewdly ordered that the evaluations be halted and the remaining nine missiles and related hardware be mothballed. His plan, as he described it, was to hold them "for other more spectacular purposes."

But nothing else at the ABMA was on hold. In the month between the second Sputnik's flight and Vanguard's destruction, Huntsville's Development Operations Division hurriedly came up with its own ambitious plan to leapfrog the Soviets. Claiming that the need for an integrated missile and space program was "accentuated by the recent Soviet satellite accomplishments and the resulting psychological intimidation of the West," the classified proposal went on to assert that space travel was nearly at hand and, more to the point, that the principles of warfare on Earth had to be extended to space. Not surprisingly, Huntsville's strategists maintained that the key to improved American "orbital and moon flight missions" lay in the rapid development of powerful boosters, and that those boosters be based on none other than the Redstone, Jupiter, and Juno, eventually to grow into a rocket that was as powerful as it was alliterative: the million-pound-thrust Super Jupiter. The mighty lifter, or its equivalent, would purportedly be used to accomplish three distinct kinds of missions: carrying people and cargo to orbit and back; controlling space for reconnaissance, scientific research, and interception of enemy satellites; and sending men to the Moon. The soldiers who wrote the proposal were no less imaginative than their competitors in the Air Force. The satellite interceptor would be flown by a two-man crew, for example, and the "personnel-carrying orbital payload" could be a platoon of combat infantrymen who would be shot to any spot on Earth on short notice and then parachute down inside an upper stage.

Meanwhile, the space shot was now a reality. Having gotten a green light on the launcher, von Braun and his colleagues assumed that they would get to develop the satellite as well. But they were wrong. Pickering, the chief of the Jet Propulsion Laboratory and a tough New Zealander who was educated at Caltech, believed that given the lab's involvement with Project Orbiter from the start, it had every right to continue participating in the push to reach space. He therefore convinced Medaris to award the satellite contract to JPL. The general undoubtedly believed that JPL, which by then had solid experience in both electronics and rocketry, could handle the assignment better than his own agency.

Whatever Medaris's motive, though, his decision would prove to be a momentous one. By allowing JPL to design and build what would turn out to be America's first spacecraft, he unknowingly played into the hands of a man who had already quietly decided that JPL needed to abandon work on tactical missiles and aim higher. Pickering wanted to build machines that would ex-

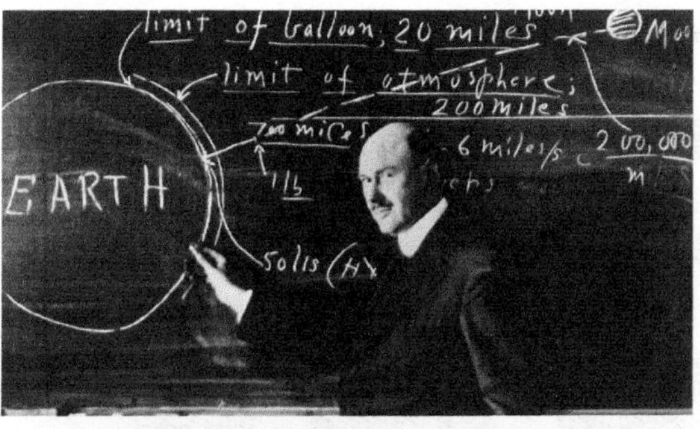

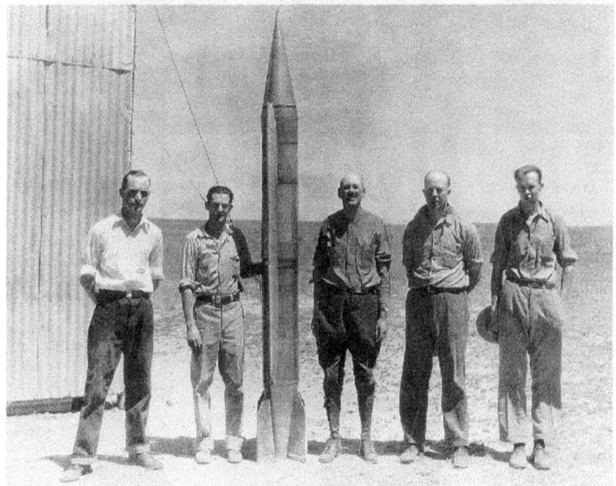

A stern-faced Robert H. Goddard posed in front of his blackboard at Clark University in 1924. He launched the first liquid-fueled rocket, a gawky contraption, two years later. By 1931, Goddard (center) and his assistants had moved to the High Lonesome near Roswell, New Mexico, and were reaching two thousand feet with sleek versions like the one at left. When he died in 1946, the reclusive and financially insecure "father" of American rocketry held more than two hundred patents. (NASA)

Goddard standing next to the rocket that made history on March 16, 1926, when it became the first to fly with liquid fuel. The rocket itself is the dark machine at the top of the frame. (NASA)

Collier's ran an eight-part series between 1952 and 1954 that was essentially a blueprint by Wernher von Braun, Willy Ley, and others for the U.S. manned space program. The winged aircraft shedding its second stage was von Braun's shuttle on its way to dock with a station. Cornelius Ryan, who attended a symposium on the subject at New York's Hayden Planetarium, masterminded the series. After reading the articles, Ward Kimball persuaded Walt Disney to put a "Tomorrowland" segment starring von Braun on his weekly network television series. Mouseketeers and rocketeers therefore had a synergistic relationship. (William E. Burrows collection)

The United States scored an early and important triumph when the Navy's Viking 11 broke the V-2's altitude record by reaching 158 miles on May 24, 1954. More than a record was at stake, however. This composite picture, looking from White Sands southeast across Mexico, and others showed enough detail to prove that space-based cameras had enormous potential for civilian and military observation. (Naval Research Laboratory)

With Sputniks 1 and 2 already in space, the Navy's attempt to get there literally went up in smoke on December 6, 1957, when Vanguard's first stage shut down two seconds after ignition in front of a large television audience. An embarrassed and beleaguered President Eisenhower insisted, correctly, that the United States was not trailing the Soviet Union in space technology. Few believed him. (NASA)

An ecstatic William Pickering, James Van Allen, and Wernher von Braun held a full-sized replica of Explorer 1 over their heads on the night of January 31, 1958, after it was confirmed that the Army-JPL launch into orbit was successful. Van Allen's cosmic-ray and micrometeorite detector discovered the radiation belt around Earth that bears his name. (NASA)

Some airplanes played key roles in leading the way to space. The wingless M2-F3 "Lifting Body" (above) was one of a number of similar designs that were used to test the concept of reentering the atmosphere from orbit like an aircraft, rather than having astronauts drop out of the sky under parachutes as if they were leaving the scene of an accident. The X-15 (below) was fundamentally important in research on handling, heating, and other characteristics at extremely high speeds and altitudes. USAF and NASA markings reflect the symbiotic relationship between the military and civilian space programs. (NASA)

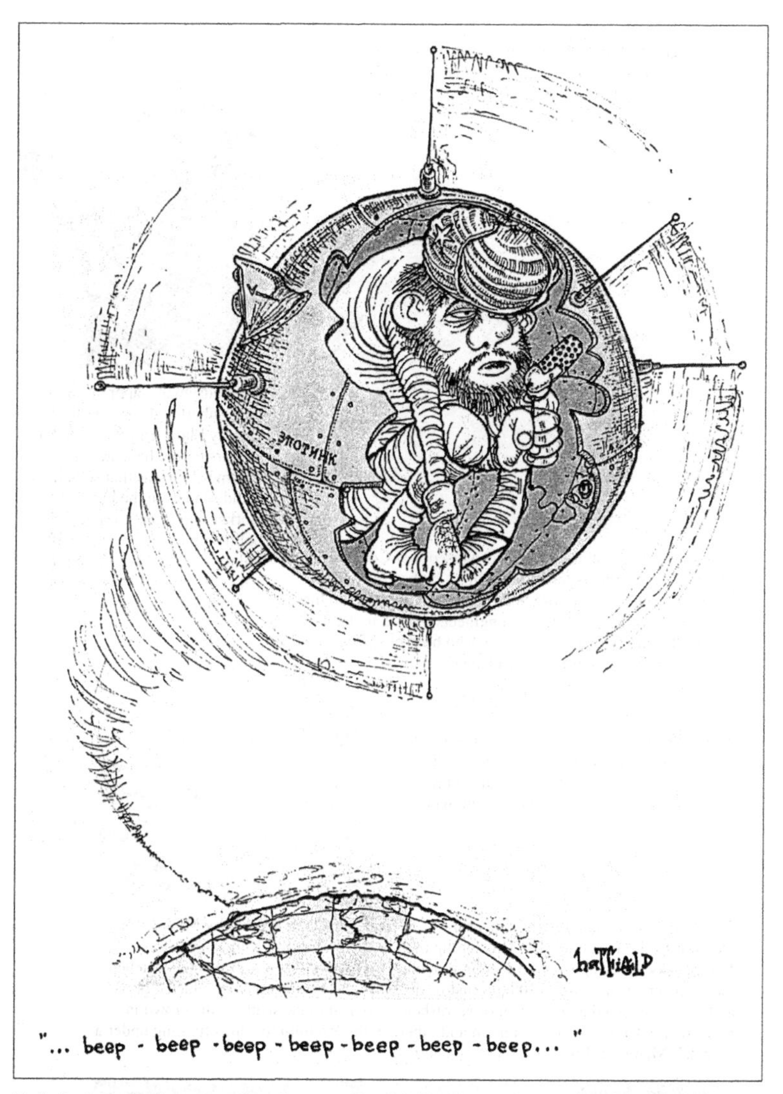
"... beep - beep - beep - beep - beep - beep - beep... "

Early Soviet space feats were often disparaged by disbelieving Americans. The "Sputnik cocktail"—one part vodka and two parts sour grapes—reflected the mood. Donald J. Hatfield, an MIT sophomore, drew this cartoon of a dopey-looking cossack radioing ". . . beep-beep-beep-beep . . ." from inside a spacecraft that looked as if it had been cobbled together in Gagarin's tractor factory. It appeared in November 1957 in *Voo Doo,* the campus humor magazine. (MIT *Voo Doo* magazine)

Yuri Gagarin, the first man in space, was deified as a symbol of Soviet technological prowess and the inevitable triumph of the proletariat and socialism. This is a hero's album.

A foundryman's apprentice at the Lyubertay Agricultural Machinery Plant, Yuri Gagarin made tractors before he made history. The message was that socialism could send even the humblest worker or peasant to space. This political ploy was dropped after Valentina Tereshkova's flight in 1963. (Memorial Museum of Cosmonautics)

This classic frieze at the Memorial Museum of Cosmonautics in Moscow shows the intrepid hero, bathed in sunshine and with hand over his pure heart, climbing the stairs to the stars. He is backed by scientist-engineers—none in the likeness of the anonymous Korolyov—who hold blueprints and Sputniks 1 and 3. (William E. Burrows)

The foremost "Little Eagle" under the adoring gaze of his benefactor and father figure, Sergei P. Korolyov, the renowned chief designer. (Memorial Museum of Cosmonautics)

Two days after his epic flight, Gagarin was honored on Lenin's Tomb by Minister of Defense Rodion Malinovsky. This telling picture (below) shows Khrushchev, to Malinovsky's left, apparently deep in thought with his eyes closed. Brezhnev watched the minister suspiciously. Mikhail Suslov, the party theoretician who conducted purges and was considered a bitter rival of Khrushchev, looked impassively at the crowd in Red Square. Churchill likened them and others in the Kremlin to "dogs fighting under a carpet." (Memorial Museum of Cosmonautics)

Here, as if poised for flight, he towers on a pillar over Leninsky Prospect in Moscow as socialist realism's tribute to the flight of *Vostok 1*. (William E. Burrows)

He was unusually melancholy when this picture, typical of innumerable others, was taken. His role as the nation's preeminent hero had no doubt become burdensome. The scar over his left eye was officially attributed to a fall taken while playing with his daughter, but that was far from true. (Memorial Museum of Cosmonautics)

Castro gave Gagarin this inscribed photo as a souvenir of his visit to Cuba and a symbol of the fraternal relations between their socialist states. The cosmonaut looked genuinely pleased. (Memorial Museum of Cosmonautics)

Valentina Gagarina mourned her husband at the Kremlin Wall, where he was interred, after his mysterious death at the controls of a MiG-15 trainer in 1968. His medals were displayed in front of his portrait. (Memorial Museum of Cosmonautics)

Valentina Tereshkova, or "Seagull," epitomized Khrushchev's insistence on firsts to be used for propaganda. She became the first woman in space when she flew in *Vostok 6* in June 1963. What went unreported was the fact that she became sick to the point of incoherence. She and fellow cosmonaut Andrian Nikolayev became the first spacefarers to wed; the first to have a child, Yelena; and the first to divorce. (Memorial Museum of Cosmonautics)

To underscore the stakes in the space race, the leaders of the U.S. Army's rocket team struck this ultraserious pose at Huntsville in 1956. Hermann Oberth sat in front and looked reproachfully at the camera. Behind and to his left, von Braun perched on the table. Seated across from von Braun was Ernest Stuhlinger. Behind him stood Major General Holgar N. Toftoy, who was largely responsible for getting the Germans and their V-2s out from under the noses of the Russians in a very dramatic operation. Eberhard Rees, another prominent Peenemünde alumnus, stood behind von Braun. (NASA)

The first astronauts were portrayed as a cross between latter-day Arthurian knights and, as one of them put it, Boy Scouts. Although their bravery was real enough, the Boy Scout image was grossly distorted, at least where several of them were concerned. The Mercury 7 astronauts repeatedly posed for pictures such as this one. They were (back row, left to right) Shepard, Grissom, and Cooper; (front row, left to right) Schirra, Slayton, Glenn, and Carpenter. (NASA)

Glenn really did live up to the title of "Mr. Clean Marine." His all-American-boy image got him elected to the U.S. Senate from Ohio. To complete a story-book life, he requested that he be allowed to return to space on a shuttle, and received permission from NASA to fly in October 1998 at the age of seventy-seven. He was fitted for his space suit the previous February. (NASA)

NASA, like its Soviet counterpart, commissioned artistic interpretations of its operations as a way of providing the public with a dimension beyond photography. Paul Calle portrayed *Gemini 4* astronauts James A. McDivitt and Edward H. White, who flew in June 1965. Their body language shows a relentless, nothing-can-stop-us attitude. (NASA)

The highs and lows of the Mercury Program are expressed in these pictures. President Kennedy presented Alan Shepard with NASA's Distinguished Service Award after the naval aviator became the first American to reach space when he sailed down range in *Freedom 2.* Hours later, JFK decided to send astronauts to the Moon. (NASA)

The next flight was an embarrassment. Gus Grissom is seen being hauled out of the Atlantic seconds after his *Liberty Bell 7* Mercury capsule slipped under the waves because its hatch mysteriously blew off. Ironically, his insistence on hatches that would not do that may have led to his and two other astronauts' deaths in a flash fire in Apollo Command Module 204 on January 27, 1967. (NASA)

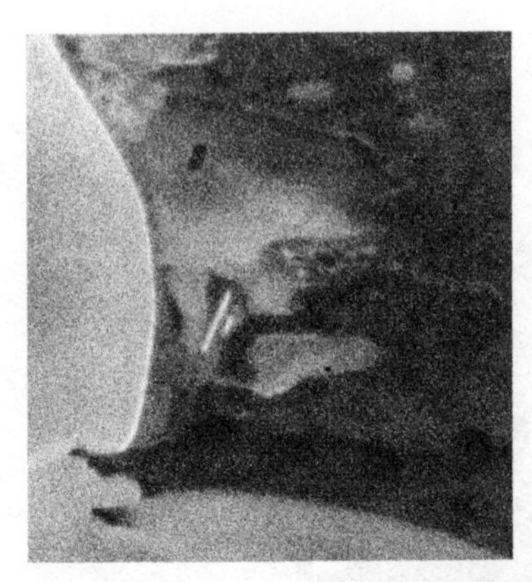

Collecting intelligence was the most pressing reason for the United States to take to space. This blurred picture shows the first target of the first space reconnaissance mission in history. The two white slashes are the apron and runway of a military air base near Mys Schmidta in the far northern U.S.S.R. It was taken on August 18, 1960, in a top-secret program code-named Corona. (CIA)

This excellent photo of the Severodvinsk Shipyard and Naval Base 402 on the White Sea was taken on February 10, 1969. The long shadows indicate that it was shot early in the day. They helped photo-interpreters estimate the size of the many structures. The large building with the grated roof in the center of the picture, for example, is a submarine construction hall roughly 1,230 feet long. What appears to be a maze in the upper third of the picture is apartment houses. (CIA)

The Pentagon was one of the subjects of the first KH-4B mission in late September 1967. Targets of known size in the United States were shot to calibrate the satellites' cameras and give interpreters a scale with which to calculate the size of foreign targets. Lines in the parking lot of the U.S. Embassy in Moscow were used for the same purpose. (CIA)

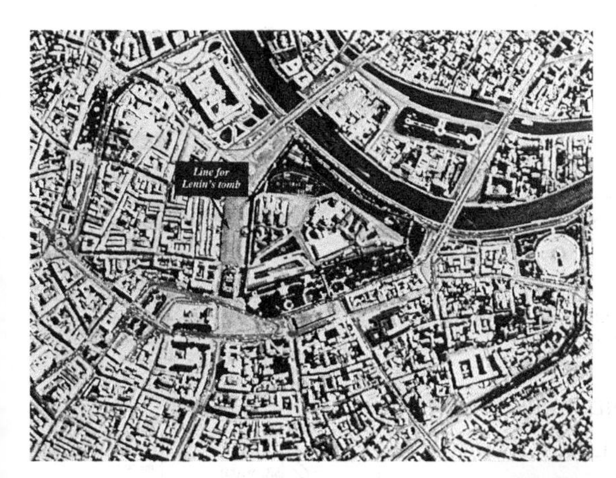

This image of Moscow, taken on May 28, 1970, has six-foot resolution. The rough triangle in the center is the Kremlin. Interpreters could pick out cars, trucks, a boat on the river, and even a line of people waiting to enter Lenin's tomb. (CIA)

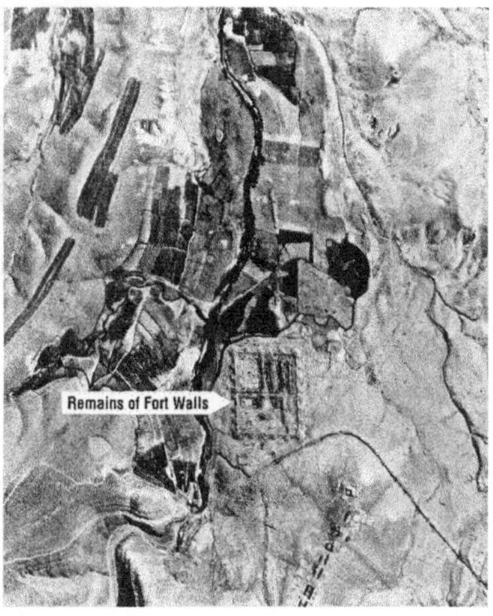

Space reconnaissance produced many thousands of pictures of ice floes, mountain ranges, deserts, bodies of water (like the shrinking Aral Sea), and other places that were and remain useful to scientists. This picture, taken over Lejjun, Jordan, shows the remains of the walls of a fort built by the Roman legion. (CIA)

This and two other photographs of the Soviet Union's first full-size nuclear-powered aircraft carrier, the *Admiral Kuznetsov,* under construction at a Black Sea shipyard in July 1984, were leaked to a British defense magazine. Seen here, looking straight down, are the front and rear sections of the vessel being built side by side under huge gantry cranes. Analysts concluded that the ship had two elevators, three catapults, phased-array radar, anti-aircraft missiles, and could carry seventy-five MiG and Sukhoi fighters. Ironically, the ship was put up for sale after the cold war and could have been bought for what it cost to take the pictures. (William E. Burrows collection)

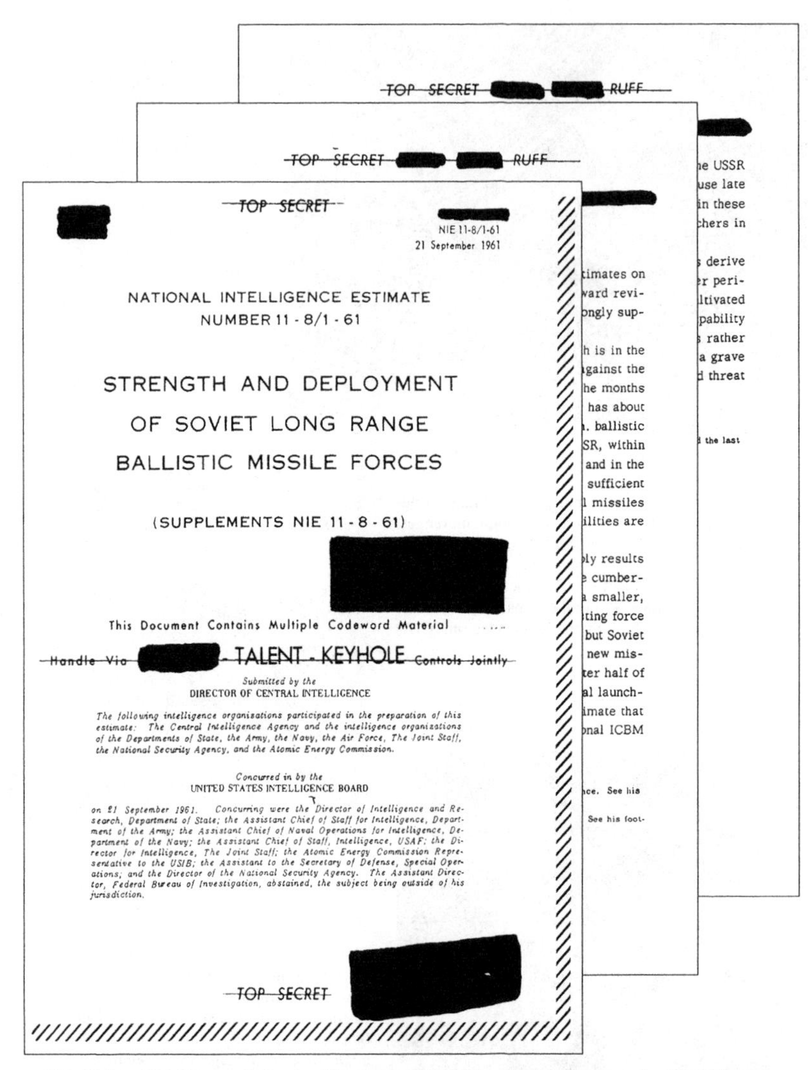

TOP SECRET

NIE 11-8/1-61
21 September 1961

NATIONAL INTELLIGENCE ESTIMATE
NUMBER 11-8/1-61

STRENGTH AND DEPLOYMENT OF SOVIET LONG RANGE BALLISTIC MISSILE FORCES

(SUPPLEMENTS NIE 11-8-61)

This Document Contains Multiple Codeword Material

Handle Via [redacted] - TALENT - KEYHOLE Controls Jointly

Submitted by the
DIRECTOR OF CENTRAL INTELLIGENCE

The following intelligence organizations participated in the preparation of this estimate: The Central Intelligence Agency and the intelligence organizations of the Departments of State, the Army, the Navy, the Air Force, The Joint Staff, the National Security Agency, and the Atomic Energy Commission.

Concurred in by the
UNITED STATES INTELLIGENCE BOARD

on 21 September 1961. Concurring were the Director of Intelligence and Research, Department of State; the Assistant Chief of Staff for Intelligence, Department of the Army; the Assistant Chief of Naval Operations for Intelligence, Department of the Navy; the Assistant Chief of Staff, Intelligence, USAF; the Director for Intelligence, The Joint Staff; the Atomic Energy Commission Representative to the USIB; the Assistant to the Secretary of Defense, Special Operations; and the Director of the National Security Agency. The Assistant Director, Federal Bureau of Investigation, abstained, the subject being outside of his jurisdiction.

TOP SECRET

This National Intelligence Estimate of September 21, 1961, written thirteen months after U.S. spy satellites began scouring the U.S.S.R. for long-range missiles and other strategic weapons, concluded that only "10–25 launchers from which missiles can be fired against the US" had been spotted and that the low number indicated "a smaller, second generation system" was being developed. Although "Soviet propaganda has assiduously cultivated an image of great ICBM strength," the NIE said, the real threat at the time came from bombers and submarine-launched missiles. The Air Force disagreed and, following procedure, put the number of launchers at fifty in an appendix. This document effectively ended the "missile gap" that haunted Eisenhower and helped deny Nixon the presidency. TALENT-KEYHOLE was the clearance level needed to see the satellites' highly classified imagery. (CIA)

The basic twenty-engine lifter was fundamental to the Soviet program from Sputnik to Soyuz and propelled the long-range ballistic missiles for which U.S. spy satellites searched. The four outer skirts, each containing four rockets, were jettisoned first, followed by the central sustainer, which pushed the payload toward orbit. (Memorial Museum of Cosmonautics)

Valentin P. Glushko, the great engine designer, ran his own design bureau and was a bitter rival of Korolyov. He was twice made a Hero of Socialist Labor, as signified by the medals. (Memorial Museum of Cosmonautics)

This Soyuz launcher-spacecraft was moved horizontally to the launch pad at Tyuratam and then erected for firing. The cavernous flame pit under the concrete pad directs smoke and scalding steam away from the rocket. (Memorial Museum of Cosmonautics)

Vladimir N. Chelomei, another of Korolyov's rivals, ran the bureau that produced the Proton heavy lifter, which is still in service. The wily rocket designer hired Khrushchev's son, himself a talented engineer, to gain political leverage. (Memorial Museum of Cosmonautics)

The sons of the desert pose in survival gear they made themselves in Nevada in 1963. From left are: Neil Armstrong, Frank Borman, Pete Conrad, James Lovell, James McDivitt, Elliot See, Thomas Stafford, Edward White, John Young, and "Deke" Slayton. No astronaut ever came down in the desert. But two of their rivals, Aleksei Leonov and Pavel Balyayev, spent a frigid night menaced by wolves near Perm after their *Voskhod 2* overshot the landing area by two thousand miles. (NASA)

Glenn had his lungs checked. (NASA)

Shepard had his temperature taken a few hours before he became the first American to reach space. (NASA)

Pete Conrad, who splashed down after the *Gemini 5, Gemini 11, Apollo 12,* and *Skylab 2* flights, had good reason to practice floating in his space suit. (NASA)

plore the solar system. As a university laboratory, he would explain years later, "It was quite clear to us that our future lay on the space program," not in weapons work, which offered little intellectual challenge. The Earth-hugging spacecraft, soon to be named Explorer, would be a first step in that direction.

Unlike Sputnik 1 and Vanguard, Explorer 1 would be elongated and contain its own tiny rocket. It would in fact be a miniature version of JPL's Sergeant missile. The satellite would be eighty inches long and six wide and would be divided into two sections: one containing a solid-propellant rocket engine, and another holding three measuring instruments and two radio transmitters.

The instruments and transmitters weighed eighteen and a half of the front section's thirty pounds. Two of the instruments were relatively mundane. One would record temperatures as Explorer 1 passed through night and day on its twelve daily orbits. Another would use a microphone to register micrometeorites slamming into the spacecraft. The third experiment, which belonged to Van Allen, was technically called a cosmic-ray and micrometeorite package. It was actually a super Geiger counter that was supposed to measure charged particle activity: the cosmic rays that the professor of physics from Iowa had spent years trying to catch at ever-higher altitudes with balloons, "rockoons" (rockets launched from balloons), V-2s, and Vikings. Now, if all went well, the intrepid scientist would in effect finally get to orbit Earth for a long look, not just a tantalizing peek, at his quarry.

In deference to Eisenhower's sensibilities about military rockets participating in the IGY, some canny individual with at least a little knowledge of Greek mythology decided that the hoarded Jupiter C missiles ought to function as part of a rocket combination named Juno, after Jupiter's wife and sister. And besides, it added a nice feminine touch to a brute of a lifter.

Juno 1's first stage was a Jupiter C. It in turn would support fourteen of the miniature Sergeants. All of the small rockets would be arranged in a round tub, or drum, with eleven forming an outer ring—the second stage—and three clustered inside the ring to act as the third stage. Jupiter C's engine would burn for 156 seconds, getting it up to sixty miles. It would then drop away from its upper stages. At a signal from the ground, the outer ring would fire for six seconds and then coast for two seconds. Then the three inner Sergeants, with Explorer on top of them, would ignite, yanking themselves and their payload away from the ring of dead rockets, which would follow the Jupiter C down to their watery end. The third stage would fire for six more seconds and then also fall back to Earth. Two seconds after that, the fourth and final stage—Explorer 1—would blast itself into orbit. It was essentially an improvement on the old Orbiter-Loki arrangement.

Before liftoff, that fourth stage—the fifteenth Sergeant—would be perched on top of the tub like a pencil standing on a coffee can. The tub was supposed to start spinning at a rate of about sixty rotations a minute just before liftoff and continue spinning during Juno's climb. It would act like a gyroscope to stabilize the three upper stages. The rapid spinning would also offset a thrust imbalance that would be caused if one of the clustered Sergeants either did not ignite or died after ignition. Otherwise, even one of the ring of rockets that did not contribute its share of thrust could send the whole stage tumbling catastrophically off course.

Having seen the Navy stung by the horrendous publicity after the Vanguard fiasco, Medaris issued orders that the Explorer program was to function under tight security. Explorer was therefore referred to in highly classified communications between Huntsville and JPL as "Missile 29," the idea being to make it appear that Juno's flight was to be just another advanced Redstone test. The upper stages would be shrouded in canvas while being hurried to the launchpad in the predawn darkness and they would remain hidden after they and the big rocket were in the vertical position. "I desire it well understood," Medaris warned, "that the individual who violates these instructions will be handled severely."* The precaution was justified, at least in Medaris's mind, because Juno would be carrying far more than Explorer 1. It would be carrying the Army's hopes for what Killian called "a central role in space."

Explorer Goes to Space

Under the tight security Medaris had ordered, a bulbous Air Force C-124 Globemaster transport plane landed at Canaveral on December 20, 1957, and disgorged the Juno. By January 17, the rocket was standing on its pad and being primed to go. Van Allen and his colleagues had in the meantime stuffed their cosmic ray package into Explorer, while others at JPL prepared the rest of the spacecraft. It was then delivered to Florida and bolted on top of Juno's third stage. By January 29, less than four months after Sputnik, everything was ready but the weather; the cape was being buffeted by some of the highest winds to have hit the area in anyone's memory.

On the thirty-first, the wind died down. Eisenhower spent that afternoon playing bridge in a cottage next to the Augusta National Golf Course, but Hagerty kept him in close touch with developments at Canaveral. That night

* Inadvertent explosions at Cape Canaveral, all of them witnessed by the news media, prompted *The New York Times* to take the unusual step of urging precisely the sort of secrecy that Medaris demanded. "Would it not be wiser—when malfunction and other obstacles are so likely and so frequent—to let the devoted men who are preparing Vanguard to do their work in quiet? When Vanguard is fired is time enough for the world to know what has happened," stated an editorial on March 14, 1958 (*Assault on the Unknown,* p. 92). The *Times'* stand flew in the face of frequent charges that the Fourth Estate cared more about running sensational stories, including about catastrophes, than it did about the nation's welfare.

Juno, bathed in blue floodlight, was poised to fly. Igniters were brought from storage and inserted in all fifteen Sergeants by one man. That would keep casualties down if something went wrong. The rocket's tanks were then topped off and, as liquid-oxygen vapor formed a pale lavender cloud in the artificial light, the tub started to spin. At 10:48 P.M., Juno 1 rose from her pad on top of blinding white fire, punctured a thin layer of overcast, and arced out over the Atlantic. Tight security or not, word had gotten out that a major launch was about to happen, so there were thousands on the beach that night to cheer as Juno and its payload headed for space.

A number of important government officials were there, too, and so were newsmen. The reporters had agreed to hold their stories—embargo them—until after the launch. They also agreed not to report that six days earlier the Navy had made one last, desperate attempt to get a Vanguard into space ahead of the Army, only to have the first-stage engine malfunction fourteen seconds before ignition.

It took more than an hour to establish that Explorer 1 had actually made it to orbit, so Ike reacted cautiously. "Let's not make too great a hullabaloo on this," he told one aide. But then, just after midnight, Brigadier General Andrew Goodpaster, his personal aide, told him that there was confirmation that the United States was in space. "That's wonderful," said the president. "I sure feel a lot better now."

Once Explorer reached orbit, controlling it passed from the Army, which had gotten it there, to JPL, which would run its experiments. When word reached Huntsville that Juno was on her way, the town went out of control. It exploded in celebration, with police cars and fire engines using their sirens and people parading in the streets. Some carried placards that said: "Our Missiles Never Miss" and "Move Over, Sputnik, Space Is Ours."

Appropriately, von Braun, Pickering, Van Allen, Secretary of the Army Brucker, General Lyman Lemnitzer, and some others were in the Pentagon's War Room when Explorer 1 reached space. They, too, were slow to get confirmation of the triumph because the satellite was temporarily out of radio range. There was, in Van Allen's words, an "exasperating lack of information. The clock ticked away, and we all drank coffee to allay our collective anxiety. After some ninety minutes, all conversation ceased, and an air of dazed disappointment settled over the room." Then, nearly two hours after launch, word came that Explorer 1 was actually in orbit. The place "exploded with exultation, and everyone was pounding each other on the back with mutual congratulations," the delighted physicist would recall.

Within minutes of the confirmation that Explorer 1 was in orbit, Van Allen, von Braun, and Pickering were whisked to the National Academy of Sciences in an Army car and taken in through a back door. After making a preliminary report to IGY staff members, the three jubilant men were led into the building's Great Hall, where excited reporters, photographers, and others waited. Then, beaming in triumph, they staged a photo opportunity: they held a full-scale model of Explorer 1 over their heads while flashbulbs popped. Pickering noted that his satellite's apogee, more than a thousand miles high, put it much farther out than either of the Sputniks. Medaris, who was also on hand, added that Explorer 1's record altitude showed that it was probably better than the Sputniks. That was nonsense, of course, and he would have known it. Yet the higher apogee did reflect a lifting capability that could send warheads the other way. Medaris certainly knew that much. And he knew that Nikita Khrushchev knew it, too.

The first secretary also would have known, however viscerally, that the course he set in moving his country to space was as seriously flawed as it was seemingly necessary. It was not only the Sputniks themselves that set the tone for a duel that would soon extend beyond Mars, but also the taunting braggadocio that accompanied them. If Khrushchev found it necessary to temporarily stave off his opponent by making the point that his rockets could reach Washington—a threat that hardly needed to be made in the first place—he could have done so in private and without publicly embarrassing the United States. Instead, he succumbed to the temptation to use his spacecraft as both evangelic tools and instruments of intimidation: to spread the gospel of socialist superiority while ridiculing his adversary and threatening it with annihilation.

Certainly the United States, which was well aware of the military potential of space in 1957, was headed there whether Khrushchev performed his bombastic rituals or not. Yet a show of statesmanship that autumn, perhaps an offer of cooperation in the peaceful uses of space while quietly acknowledging its military potential, might have avoided America's overreaction and the powerful backlash that followed. In Eisenhower, after all, Khrushchev had a man who wanted conciliation and the avoidance of colossally expensive competition.

Ruble for ruble, the Soviet Union's space program undoubtedly would generate more favorable propaganda than any other activity. The rockets would become the centerpiece of socialist technology and its most enduring legacy. But they were also props in what would become staggeringly expensive theater. Maintaining a program that was competitive with America's—that would eventually launch 101 spacecraft in a single year, or five times the rate of all other nations combined—would be so expensive that it would help undermine the very society it was supposed to reinforce.* Eisenhower, a fiscal conservative and an avowedly peace-seeking individual, could have provided Khrushchev with a

* The year was 1982. It would never even reach a hundred again. The total for the United States, Japan, and China was twenty. (Johnson, *The Soviet Year in Space: 1982,* p. 1.)

way out of that predicament. Yet the Soviet leader understood that his own country was dangerously weak relative to the United States and its North Atlantic allies, so he had to try to buy time by staging tricks until the balance of power fundamentally shifted. In doing that he thwarted Eisenhower, not only setting off an unprecedented economic and military competition, but preparing the way for a dramatic and decisive personal response by President Kennedy less than four years later.

Competition between the two space powers would stimulate feats of exhilarating accomplishment across a wide spectrum, from advances in high-speed computation to weather forecasting to worldwide communication to solar system exploration to manned spaceflight. But it would be the United States that most benefited from the competition, not the Soviet Union.

For their part, the news media were only too glad to report that the great race had finally been joined. "New Moon Made in U.S.," said a *Life* headline over a story saying that Explorer made up "to some extent, for the U.S. humiliation due to the Russian Sputniks." *Life*'s sister publication, *Time,* decided somewhat dramatically that "the 119 days between Sputnik 1 and Explorer were as important to the U.S. as any similar span in its history." But its disgruntled editors still managed to call the flight of Explorer 1 a "Promethean gift." The reaction among America's allies tended to be more enthusiastic. "U.S. honor has been saved, its dignity and prestige recovered," one newspaper in honor-bound Tokyo noted politely, while another in Buenos Aires rejoiced, "At Last, an Uncle Samnik."

The Discovery of the Van Allen Radiation Belt

Science would be one of the great benefactors of the space age, and the payoff began immediately, when Van Allen's cosmic-ray detector detected a blizzard of cosmic rays. These are protons, high-energy electrons, the nuclei of helium atoms, and other elements that radiate from the Sun and from its exploding relatives, the supernova, and stream through space at close to light speed. It was well understood by Van Allen and other astrophysicists that the intensity of cosmic radiation, which is lethal to life on Earth, increases with altitude since the planet's protective atmosphere thins as it gets higher. The sounding rockets had shown that much. Van Allen therefore assumed that the intensity of the particles slamming into Explorer 1's little gas-filled Geiger counter would increase progressively each time the spacecraft swung out toward its apogee. But that didn't seem to happen. Each time Explorer 1's transmitters "dumped" a batch of data to the ground stations over which it flew, there were indications that no hits were being registered at all. That was perplexing.

But the appearance was deceptive. It would turn out that the apparatus was so saturated by cosmic-ray hits that it became overwhelmed; mechanically speaking, it became paralyzed. The data from Explorer 1 and from one of its successors, Explorer 3, showed that there was so much radiation beyond four hundred miles that it could not be accounted for by simple cosmic-ray bombardment alone. Careful analysis of the data proved that there were two roughly doughnut-shaped belts, or a single gigantic one, of trapped radiation circling Earth. And measuring them in turn provided the first detailed picture of the shape of the planet's magnetic field; they defined the barrier that protects it from the deadly barrage of celestial radiation. There had been speculation about the existence of such a field for years. But direct sampling by means of a rocket turned a hypothesis into a certifiable fact. Van Allen's discovery would start a long chain of investigations of how solar radiation affects Earth's atmosphere.

And what a fancy Geiger counter could find around Earth it could find around other planets and their moons. Although it was not obvious at the time, scientists would come to understand that if such a shield could protect Earth, it could perform the same service on other worlds. The devices would become standard issue on the planetary explorers that were shortly to begin prowling around the Moon and far more distant places. James Van Allen's relentless climb for a long look far above the atmosphere was finally rewarded beyond even his imagination. The giant cosmic doughnuts, which extend out to about forty thousand miles, were named for their discoverer: the Van Allen Radiation Belt. He would be awarded the Crafoord Prize, comparable to the Nobel Prize, by the Royal Swedish Academy of Sciences in 1989.

Following the Leaders

Two days after Sputnik 1 went up, an astronomer at Philadelphia's Fels Planetarium estimated that it could have stayed in orbit for "thousands to a million years" because at an average altitude of more than 500 miles, there was virtually no atmosphere to slow it down. It was a wild guess and a bad one. In fact, the perigee of Sputnik 1, or its lowest orbital altitude, was 142 miles. There is residual atmosphere all the way up to about 500 miles. Plowing into even that very thin atmosphere at high speed during perigee created the drag that slowed the spacecraft and caused it to fall back to Earth ninety-two days after it was launched. Its American counterpart lasted 4,441 days, a little more than twelve years, because of its higher perigee.

Sputniks 1 and 2 and Explorer 1 were followed by others as the pace of the competition increased. The first successful Vanguard launch came on March 17, 1958, and the last on September 18, 1959. Three of the Navy spacecraft reached orbit out of eleven tries. It was hardly a distinguished launch record. Yet many of the NRL's goals were met. Vanguard's upper stage combination became the prototype for several successors. It was the first spacecraft to use solar cells to produce electrical power by displacing electrons with photons from the Sun. That is, the cells, which on most other spacecraft would be set in rows on large, flat, leaflike panels, converted energy absorbed from the Sun into usable power roughly the way chloroplasts on leaves use it to produce oxygen. And Vanguard's Minitrack system was the first extensive satellite radio tracking network.

Explorer 2 never made it to orbit because of a fourth-stage ignition failure. But its immediate successor went up on March 26, 1958, and sent back the cosmic-ray data that nailed down Van Allen's findings. The Explorer program, which ended with the launch of Number 55 on November 20, 1975, would become one of the longest and most successful undertakings of its kind.

Sputnik 3, its science instruments finally communicating with each other and a new cosmic radiation detector similar to Van Allen's on board, went into orbit on May 15, 1958. As the biggest and heaviest of the satellites sailed around Earth, Khrushchev could not resist taking another poke at the Americans, this time in an apparent reference to Vanguard. The United States would need "very many satellites the size of oranges," he told an Arab delegation in the Kremlin, "in order to catch up with the Soviet Union."

But Khrushchev neglected to mention that the score now stood even at three apiece. And there was something else he left out. The fact that the U.S. government made no objection to Soviet satellites flying over its territory had already led to the broad acceptance of such flights. That, in turn, would legitimatize the American reconnaissance satellites that would soon start the historic process of penetrating the veil of secrecy that shrouded his country.

Inventing NASA

As early as the summer of 1957 the members of the U.S. IGY satellite committee mulled over the notion of a "National Space Establishment" that would conduct unmanned space science and applications and the manned exploration of "outer space," all for the purpose of its "eventual habitation." The role of such an organization, they wrote in a formal proposal, would be to "unify and to greatly expand the national effort in outer space research, *specifically excluding areas of immediate military urgency*" (italics added). The document was formally called "A National Mission to Explore Outer Space" and was completed on November 21 in the aftermath of the two Sputniks. It was modified and renamed "National Space Establishment" on December 27, and sent to Killian, Caltech president Lee DuBridge, and Eisenhower's three-man Science

Advisory Committee, which included George B. Kistiakowsky, the Harvard chemist who would succeed Killian as the president's science adviser.

The call to start a national space agency was enthusiastically endorsed by the American Rocket Society. Only two weeks before the first of the two IGY proposals was released, Robert Truax, the president of the society, had written to I. I. Rabi, a Nobel laureate and a member of the President's Science Advisory Committee, to urge the creation of a civilian space agency that would spend $50 million a year. The rocket society, Truax said, was opposed to "stunts" and "stopgap measures" designed to match Sputnik and instead wanted to embark on a serious, long-term program to explore space. Truax also sent the influential physicist a copy of a report of the society's Space Flight Technical Committee, headed by Krafft Ehricke, who had by then moved to Convair in Texas. The report laid out a twenty-five-year program that included orbiting Earth, sending robotic spacecraft to the planets, and sending humans to the Moon. The substance of the rocket society's plan centered on the creation of a dedicated civilian space agency. It would be endorsed by the president's science advisers in an official memorandum written on December 30, 1957. That memorandum amounted to a declaration of war between Killian and Medaris.

On February 4, 1958, barely four days after Explorer 1's triumph and with the nation still exhilarated over it, Eisenhower announced the appointment of a blue ribbon panel within the President's Science Advisory Committee that was to outline a coherent space program and invent an organization to manage it. The group consisted of General James Doolittle, hero of the bomber raid on Tokyo in 1942; Polaroid's Edwin "Din" Land; Herbert York, an eminent physicist and head of the Pentagon's Advanced Research Projects Agency; and Edward Purcell, a Harvard physicist and Nobel laureate. Purcell chaired the group.

It was implicit, at least to the panel members, Killian, and the president, that the nation's preeminent space organization be civilian. It was anything but implicit where the Army and the Air Force were concerned, however. With the Communist threat growing, and with the battleground clearly extending ever higher, senior generals in both services tended to believe that turning the space program over to a bunch of civilians was outrageous, if not treasonous. Medaris and von Braun (the latter still wearing hero's garlands for Juno's triumph) campaigned "with fierce religious zeal" for their service, according to Killian. Medaris "vehemently proclaimed that military satellites should have greater priority than ballistic missiles, that the space program rightfully belonged to the Department of Defense, and that it would be a terrible mistake to give the responsibility for the U.S. space program to an independent civilian space agency," he would recall.

As a Washington insider and former university administrator, Killian was no stranger to tough politics. But Medaris's exceptionally forceful maneuvering so irritated him that he took an uncharacteristic parting shot at the general in his memoir. "I have noted with interest the post-army career of the dedicated, outspoken, belligerent, and pious general," Killian would write. "He retired, first to become president of the Lionel Corporation, builder of toy trains, and later to become an Episcopal priest, Father Bruce." The ordinarily taciturn Killian ended the riposte by scornfully quoting a comment attributed to Medaris in *People* magazine: "No human being, without the guidance of the Lord, could have been right as much as I was."

The Air Force, Killian went on, fought just as hard as the Army for control of the new agency. He noted that the air generals' argument that space was a continuum of the atmosphere was logical, but that it was negated by their "fantasies" about space wars and dropping bombs from satellites (Lyndon Johnson's bricks from the highway overpass).

Killian was not exaggerating. The move to arm the heavens had already begun. On January 28, while the Purcell panel was deliberating, Homer A. Boushey, now an Air Force brigadier general, delivered a speech to the Aero Club of Washington that sounded a messianic call to seize the high frontier. After describing the establishment of an American lunar outpost that would include an observatory, the man who as a young lieutenant had been captivated on Robert Goddard's darkened porch so long ago got down to the gritty basics.

"The Moon provides a retaliation [*sic*] base of unequaled advantage," Boushey said. "If we had a base on the Moon, the Soviets must launch an overwhelming nuclear attack toward the Moon two to two-and-one-half days prior to attacking the continental U.S.—and such launchings could not escape detection—or Russia could attack the continental U.S. first, only and inevitably to receive, from the Moon—some forty-eight hours later—sure and massive destruction." That, laid bare, was one general's ultimate use for the Moon: as a missile platform from which any place on Earth could be attacked and annihilated. As was often the case, however, the plan would have been as impossible to realize as it was fun to think about. Not only would 20 million pounds of building material, plus workers and the missiles, have had to be launched from Earth to construct the missile complex at a cost of trillions of dollars, but there would have been ample time to track the attacking missiles and pick them off during their trip to Earth.

Boushey's plan was not unique, however. R. D. Holbrook of the RAND Corporation and some of his colleagues wrote three different reports in 1958 on the feasibility of setting up a lunar outpost, one of them based on five student papers that came out of the Creative Engineering course at MIT. Meanwhile, engineers at the Lockheed Corporation and elsewhere were designing such a facility. And the Army fantasized about its own installation on the Moon, Project Horizon, which would appear as a four-volume report in June 1959. It was a seductive idea that would keep resurfacing throughout the cold war.

The President's Science Advisory Committee report, "Introduction to Outer Space," pleased Eisenhower, who used it to start a news conference. The document was a definitive policy statement that identified four factors that gave "importance, urgency, and inevitability to the advancement of space technology": the compelling urge of man to explore; national defense; prestige; and new opportunities offered by space technology for scientific observation and experiment. While acknowledging that there were roles for the military in space—reconnaissance, meteorology, and communication, for example—the report took a dim view of using space as a battlefield in the foreseeable future and went out of its way to knock down the notion of orbiting bomb-carrying satellites and building missiles bases on the Moon.

Congress was already headed for space. With Lyndon Johnson spurring it on, the Senate created a Special Committee on Science and Astronautics on February 6, 1958—two days after the appointment of the PSAC panel—and Speaker Sam Rayburn and John W. McCormack did the same thing in the House of Representatives a month later. Johnson and McCormack were made chairmen of their respective committees.

The clear intention was to draft legislation creating a space establishment. First, however, three core issues had to be settled: what the agency was supposed to accomplish; whether it would be civilian or military; and whether it would be grafted onto an existing organization or started from scratch.

Maneuvering for advantage that spring, as Killian noted, was intense. Homer E. Newell, a rocket scientist from the Naval Research Lab and a veteran of the Vanguard program who later played an important role in the space agency, would describe a "deluge" of proposals that poured into the committees. While the armed services worked their own agendas, the congressmen sifted through piles of documents (most notably "Introduction to Outer Space") and heard from lobbying groups like the American Rocket Society, the venerable members of the old V-2 Panel (now called the Rocket and Satellite Research Panel, since the V-2s were long gone), the National Academy of Sciences (which set up its own Space Science Board), and several leading academicians.

In the end, Congress voted in favor of a civilian agency that would pursue scientific and political goals, while leaving specific defense requirements to the Pentagon. There was a feeling, reflecting "General" Eisenhower's, that a peaceful program, open for the world to see despite the sort of embarrassment that plagued Vanguard, would be more impressive around the world than a military one, which would be seen as blatantly imperialistic.

Yet the arrangement was necessarily ambiguous. However neatly the administration and Congress might have wanted to categorize activities in space, missions would necessarily overlap. Keeping them separated was therefore all but ignored. Heavy boosters—the huge military rockets that would have to be used by the civilians—would turn out to be only one source of contention. There would be squabbling for years over their design, production, and allocation as the soldiers and civilians argued for their own institutions' fundamental interests.

Space observation, to take another example, would be called "remote sensing" by the civilians and "strategic reconnaissance" by the military. But both systems would perform the same basic function: watching Earth from orbit. The civilians' mandate to collect weather, geological, nautical, resource, and other data for open use would confound the military's penchant for cloaking its "spy" satellites, as well as others, in the absolute secrecy of so-called black programs. Geodesy—literally, sizing up Earth—would set the stage for one of the earliest battles, for example, as the military tried to keep the data secret because it related to ballistic missile targeting while civilian scientists wanted to release it to the public for civil engineering, transportation, and other purposes.

Sorting out and delineating the respective roles was all but impossible because they sprang from the same well. Historian Walter A. McDougall described the situation succinctly when he wrote that there was a growing realization "that separation of military and civilian activities was increasingly artificial in an age of scientific warfare and total Cold War. . . . The space program was a paramilitary operation in the Cold War, no matter who ran it."

Finally, there was the matter of creating a new agency or adapting an existing one to the new task. One body of opinion, which included the always conservative Bureau of the Budget, the Pentagon, and the White House, favored expanding the National Advisory Committee for Aeronautics to accommodate space activity.

Ironically, NACA's origin was similar to that of the embryonic space agency; it was conceived and born in a crisis atmosphere. In 1915, with Europe embroiled in a war in the air as well as on land and sea, aviation enthusiasts in the United States were appalled to see that their own country, which had given the world the airplane, was lagging behind other nations in its development. They took this to be both embarrassing and militarily dangerous. And they also worried, with justification, because "birdmen" and "aviatrixes" were getting killed in a large number of accidents. Calbraith P. Rodgers, for example, crashed nineteen times while making the first flight from the Atlantic to the Pacific in 1911 and died in a crash just four months later. There were fatalities in Europe as well, of course. But the Europeans attacked the problem by involving government in directed, applied research. It was an example that the American Aeronautical Society, the spiritual forerunner of the American Rocket Society, successfully lobbied to emulate in the United States.

Legislation creating NACA quietly slipped through Congress attached to a naval appropriation bill in March 1915. From then until the dawn of the space age, it conducted solid, low-profile aerodynamic research for industry and the military. It used wind tunnels to design hundreds of specialized airfoils, or wing cross sections, and other parts of aircraft. Yet by 1958, there was a feeling on the part of many scientists, as well as aircraft manufacturers and military officers themselves, that NACA had withered into a timid bureaucracy that was as afraid of making enemies as it was interested in pursuing research. This resulted, in Newell's estimation, in an agency that was too narrowly focused and too conservative to function in the seemingly boundless world of space. He cited NACA's missing out on a number of important aeronautical advances, notably the invention of the jet engine, as examples of its inherent exhaustion.

Its proponents' arguments notwithstanding, Congress finally decided to abandon NACA altogether because of both its conservatism and its ties to the military. Given the frantic nature of the competition that existed with the Soviet Union at the time, the agency delegated to meet that competition had to be bold and inherently dynamic. NACA did not look as though it could rise to that challenge.

The administration's space bill was submitted to Congress on April 2, 1958. Debate in both houses' committees resulted in a number of changes that were acceptable to Eisenhower, who signed the legislation—P. L. 58-568, the National Aeronautics and Space Act of 1958—into law on July 29. The agency that it mandated to replace NACA, called the National Aeronautics and Space Administration because it wanted to make a seamless bond between the atmosphere and the environs above it, would open its doors on October 1. But even before then, it would find itself entangled in its own thicket of political thorns.

7

War and Peace in the Third Dimension

There were greasy fireballs, dud upper stages, and large flaming cylinders that confounded their makers by veering off course, tumbling wildly out of control, then plummeting into the Atlantic and disappearing under great geysers like leviathans shot out of the sky by an invisible hunter.

Some rockets took wrong turns. Others refused to ignite. Still others separated prematurely, had a stage cut off too soon, were betrayed by their own amnesic brains, suffered strokes when their fuel or oxidizer blew through their containers' walls, or had their pumps seize into gummy paralysis. Those that were ornery enough to swerve off their flight path while still intact were destroyed by explosive charges to keep them away from the people who launched them and from the residents of and tourists in Cocoa Beach, Titusville, Melbourne, and other nearby communities who were becoming the new space culture's first groupies. The errant rockets, blown to pieces by somebody called a range safety officer, would disintegrate before they hit the water. Every potential success carried the seed of its own malfunction and demise. So it was, late in 1957 and throughout 1958, as the launch teams strained to push their technology to its limit east of the Indian River and due south of Mosquito Lagoon.

The score for 1958 was five successes—three Explorers, a Vanguard, and a whole Atlas that broadcast Eisenhower's Christmas message to the world—and thirteen failures or partial failures, four of which involved desperate, pathetic shots at the Moon by Air Force and Army probes called Pioneers. The name would be temporarily blackened but would survive to see a better day.

Meanwhile, the civilian and military rocket fraternity and those who chronicled what it did looked past Canaveral's charred launchpads and twisted carcasses, past the bubbling swells that were the Atlantic Missile Range's grave markers, and saw the future. The miserable launch situation notwithstanding, the fact that the space age had arrived was broadly accepted, and so was the competition that drove it.

Nuking the Moon

There was so much angst at JPL because of the Sputniks and what they seemed to represent that at one point the lab resurrected the idea that had heaped ridicule on poor Goddard in 1919: hitting the lunar surface with something that would go off just to prove that Americans could do it. This time, however, the instrument of public relations would not be flash powder. William Pickering and some of his scientists mulled over the possibility of setting off an eleven-kiloton atomic bomb on an Agena upper stage. It was thought that this first ISBM, or Interspatial Ballistic Missile, would "shower the Earth with samples of surface dust" and produce "beneficial psychological results" (one of which would have been to demonstrate to Khrushchev and the rest of the world that American missiles had fifty times the range of their Russian counterparts). The idea was floated to Lee DuBridge, Caltech's president, and Pickering said that it never came to more than "coffee table talk." That was not quite the case, however. The idea made its way north to Lockheed's Missiles and Space Division, which produced the Agena. There, Saunders B. Kramer was given the job of calculating whether the bomb would have time to explode before it was crushed on impact. The young scientist, who thought the idea was "nutty," found that it would indeed explode.

The idea of atom-bombing the Moon had also flared briefly in Santa Monica even earlier. It had been considered in 1956 by W. W. Kellogg at RAND, who had mentioned it in an unclassified report on observing the lunar surface.

No one took the idea seriously except some Air Force intelligence officers who felt obliged to warn that it would be just like the show-off Communists to try such a stunt. "The possibility of immense propaganda advantages make one event extremely attractive—that of exploding a Soviet atomic bomb on the moon," the service's secret *Air Intelligence Digest* reported in November 1959. "A 20 KT [kiloton] nuclear warhead exploded on or near the surface of the moon would be easily discernible to observers on earth. . . . From the penthouses of metropolitan areas to the nomadic shepherds of Afghanistan, the evidence would be irrefutable. This feat could demonstrate the Soviet claim to superiority and be supported by the personal experience of every observer." The article went on to claim that the political advantages would be "immense" before noting that there was not so much as a shred of evidence to indicate that the Kremlin was entertaining such an idea. The wily Khrushchev would certainly have understood that whatever power was demonstrated by being the first nation to spread nuclear weapons to another world, there would also be ignominious consequences.

Scouting the Opposition

Harebrained scenarios aside, the Soviet missile and space programs became the targets of the most massive (and expensive) intelligence operation in history. The reasons were compelling, and ranged from needing to know about Soviet military reconnaissance capability on Earth, to determining whether the enemy could rain nuclear bombs on the West from orbit, to figuring out whether a Soviet space feat was true or false, to calculating how (in the trade's sanitized technojargon) the Kremlin's "platforms" and other "assets" could be "negated" by ASATs (anti-satellite weapons) if war broke out.

Writing in the Fall 1961 issue of *Studies in Intelligence,* a classified CIA quarterly professional journal, veteran Soviet missile and space program analysts Albert D. "Bud" Wheelon and Sidney N. Graybeal drew a classic analogy between their work and that of a football coach:

> A college football coach, spurred by a vigilant body of alumni to maintain a winning team, is expected to devote a great deal of energy to what in a more deadly competition would be called intelligence activity. He must scout the opposition before game time and plan his own defense and offense in the light of what he learns. During a game he must diagnose plays as they occur in order to adjust his team's tactics and give it flexible direction in action. After the game he should be prepared with an appropriate analysis of what happened, both in order that his team may benefit from seeing its experience in clear focus and in order to placate or moderate the Monday-morning quarterbacks. Although both alumni and coach recognize that football has little to do with the true purpose of a college, the coach is under relentless pressure to win games because his team, in some intangible sense, stands for the entire college.
>
> It is much the same in the space race, a game which is similarly characterized by lively competition on the playing field and intense partisan interest among the spectators. In a way which is neither rational nor desirable, our stature as a nation, our culture, our way of life and government are tending to be gauged by our skill in playing this game. Because we should expect to lose as well as win matches in the series, our government must be provided by its intelligence services with reliable foreknowledge of the possibilities for Soviet space attempts and forecasts of probable attempts, with concurrent evaluations of all attempts as they are made, and with detailed reconstructions thereafter.

The notion that American culture would be judged by the world according to its space program, let alone that it would matter, would prove to be dubious. But that is the way it looked from the bench in the stadium as the cold war solidified and Soviet space feats multiplied. The supposed cultural danger from space was a shibboleth, and so was the military threat from that direction during all of Eisenhower's presidency, as Killian and the other science advisors repeatedly tried to explain to the public. The systems were so unreliable that a shot at Washington could easily have landed in Buenos Aires. ICBMs were another matter, however. A strategic intelligence system that was embryonic when the Korean War began in 1950 had therefore hurriedly been started by the time it ended three years later. And it was growing massively when Sputnik 1 went up.

By the end of 1957, detailed reports were being written about the expansion and improvement of Kapustin Yar and about three hundred or more short- and medium-range missiles that had already been tested there. A report on the top-secret NII 88 missile design bureau, issued in March 1960, credited Korolyov with being "the most talented Soviet engineer-designer" at the facility and noted that he had participated in V-2 firings as early as 1947 and had been a chief designer since 1951. Even in 1947, then, what passed for a U.S. intelligence apparatus was paying close attention to Korolyov and his colleagues. NII 88 itself was described as the principal center for guided missile development. "There is no more important known target of this type in the USSR," the report added. Presumably, the "target" reference had to do with intelligence collection, not possible obliteration.

The technical collection system used machines instead of agents—tracking radars, signals interception antennas, ships, planes, and eventually satellites—and a cadre of specialists in the military intelligence services, the CIA, the National Security Agency, and in private contracting firms and think tanks to make sense out of what the machines collected. In that capacity, the system depended mostly on cloaks, not daggers; that is, on "techint's" machines far more than "humint's" spies and traitors.* Machines were far better than spies (or "operatives" in their own sanitized jargon) for at least three reasons: they could collect infinitely more intelligence; they did not lie, distort, or betray; and, Francis Gary Powers and some other airmen aside, they could not be caught.

Although it was not known to outsiders, including those in government, there were and remain four distinct cultures (or tribes, as one insider called

* Technically, slang and program code names are all capitalized: TECHINT (technical intelligence), HUMINT (human intelligence), PHOTINT (photographic intelligence), SIGINT (signals intelligence), CORONA, KEYHOLE, GAMBIT, HEXAGON, RHYOLITE, CLASSIC WIZARD, and so forth. But historian Edward Jablonski has written, correctly, that such usage can make a book look like a coded message rather than a literary work.

them) at the CIA. There were tweedy scientists and engineers such as Wheelon, Graybeal, and Herbert R. "Pete" Scoville Jr., many with doctorates in physics from places such as Stanford and MIT, who were in the Directorate of Science and Technology and who designed very clever (and staggeringly expensive) machines that looked and listened, and who analyzed what the machines collected. There were the Ivy Leaguers with B.A.'s in literature from Yale, Brown, Princeton, and the rest of the "Ancient Eight" whose patron saint was the OSS's William J. "Wild Bill" Donovan and who drank straight-up martinis in Georgetown, ran the "dirty tricks" war, and plotted coups and assassinations from Guatemala to Iran.* They were in the ambiguously named Directorate for Plans (later the Directorate of Operations). The scientists and engineers tended to dismiss them as overgrown kids who on some level were acting out cloak-and-dagger movie roles, while the spooks often scorned any technology more complicated than the tiny Minox cameras that were standard issue to agents in those days. A third directorate, Intelligence, attracted Ph.D.'s in disciplines like economics and political science who analyzed what the other two collected and who were convinced that they belonged to a kind of superuniversity. Finally, there were the people who ran the agency as a whole, and who thought of all the others the way a rueful headmaster thinks of chronically untidy and undisciplined schoolboys.

Richard M. Bissell Jr., a Yale man, directed both the highly successful overhead reconnaissance operations and the ill-fated Bay of Pigs invasion and therefore had a foot in both the technical collection and so-called clandestine camps. So did Allen Welsh Dulles, a Princetonian and an OSS alumnus, who presided over both cultures as director of central intelligence from 1953 to 1961. Dulles never liked the U-2 program and had to be ordered to take it on by his president. Allen's brother, John Foster, ran the Department of State over in Foggy Bottom.

"Kidnapping" a Soviet Spacecraft

One of the truly notable spy capers occurred in the early 1960s, when American agents "borrowed" a Lunik, thoroughly photographed its interior, and returned it the following morning before it was missed. That brazen, and dangerous, operation happened when the Russians sent their Moon probe and other hardware to a trade fair in Mexico. Having learned to its astonishment that the Lunik was real, not a model, U.S. intelligence decided to see to it that the truck carrying the Moon explorer was the last to leave the fairgrounds on

* The Office of Strategic Services, or OSS, was headed by Donovan in World War II and became the CIA in 1947.

the way to the freight train that was to take it to its next destination. While the vehicle's driver was treated to a night's diversion in a local hotel, a substitute driver drove the crated spacecraft to a salvage yard. The crate was then pried open with painstaking care.

With lookouts from the local CIA station patrolling the neighborhood in cars with two-way radios, four agents working for something called the Joint Factory Markings Center got into the spacecraft in their stocking feet (so as not to leave telltale scuff marks) and photographed the payload section, a main antenna, engine mounting brackets, fuel and oxidizer tanks, electrical connections, and everything else of interest, including the serial numbers of the spacecraft's components and where they had been manufactured. The Factory Markings Center provided both civilian and military intelligence with detailed information about the serial numbers, other markings, and locations of the subcontractors that turned out the equipment. The information not only helped analysts calculate how many aircraft or spacecraft were being manufactured, but it was used to locate the factories for photoreconnaissance satellites and target them for destruction if war came.

The only unnerving event of the night happened just as two members of the team were prying open the crate's tongue-and-groove planks. Suddenly, street lamps came on, flooding the yard in light. "We had a few anxious moments until we learned that this was not an ambush but the normal lamp-lighting scheduled for this hour," one of the agents would recall. Indeed, the operation went so smoothly that the first roll of exposed film was quickly sent out to make certain they were getting the pictures they wanted. With the job finished by dawn, the Lunik was sent on its way, its owners apparently none the wiser.

Most of the pictures taken of earthbound Soviet missiles and spacecraft came from military attachés who snapped them at the annual Revolution Day parade in Red Square each November, and at trade shows and air shows such as the ones held in Paris and in Farnborough, England, home of the Royal Aircraft Establishment. The displays were valuable targets because they tended to feature the real article, not fakes, since the Russians were caught in a dilemma of their own. They could hardly project technological prowess by showing mere models, which would have invited ridicule and biting references to Potemkin spacecraft on phantom space missions. The notoriously thin-skinned Slavs therefore had to sacrifice a little technology for the sake of pride and politics.

By the same token, had Lunik's "kidnappers" (as they called themselves) really been ambushed, the spacecraft's owners undoubtedly would have tipped off the press and landed a political haymaker on its nefarious abductors. ("The imperialist criminals don't have to steal our technology," one could almost hear a gloating Nikita Khrushchev quip in one of his patented taunts. "All they have to do is ask us for it.")

Spies in the Skies

But Khrushchev could not embarrass the far more effective machines that began enclosing his vast and secretive country inside an even more vast electronic shell in the 1950s and that continued to grow into the 1980s. There were dish and later phased array radars and signals intercept antennas used singly or in combination in Norway, England, West Germany, Turkey, Iran, Pakistan, Saipan, Alaska, Hawaii, Antigua, Ascension Island, and Greenland, and at several locations in the United States. The early radars tracked the missiles and spacecraft; later, more advanced models would photograph them as well.

The operation in Norway was set up to monitor submarine-launched ballistic missile tests and weather and reconnaissance satellites launched up north at Plesetsk. The antennas in West Germany collected missile and space data without the knowledge of the host country. Those in Iran were at two sites code-named "Tacksman I" and "Tacksman II." One, perched on top of a 6,800-foot-high mountain at Mashhad, seven hundred miles due south of Tyuratam, provided the kind of coverage that warmed analysts' souls. It could track a missile as it blasted off from the launchpad, arced hundreds of miles over Siberia (shedding stages and transmitting telemetry as it went), and disappeared over the eastern horizon as it streaked toward Kamchatka, where its warhead, or a bunch of them, slammed down on make-believe U.S. missile silos or cities. The ouster of Prime Minister Mohammed Mossadeq, a popular Iranian nationalist, by the CIA in 1953 and the installation of the far more amenable shah was done in part because their country provided the best radar view of Tyuratam and was a dandy place from which to monitor its missile telemetry and other communication signals. Both of the Iranian sites would be lost when the mullahs staged their coup against the Pahlavis in 1979. The operation would then be moved far downrange, to western China.

The hardware on the ground paled compared to what was headed for space, at least qualitatively. By the time Sputnik 1 started the space age, a flock of dedicated military satellites was being designed. Some, such as Transit, were to provide navigational help for naval vessels. Others, called Notus, Courier, and Advent, were designed to break out of the electronic straitjacket of ground communication by relaying messages around Earth from the high vantage point of space. They were also supposed to set the basis for civilian communication satellites that would shortly shrink the world in a true communication revolution, though most were never launched. And still others, which at the time were far and away considered to be the most important of all, were invented for reconnaissance at a time when the United States seemed at serious risk of a devastating surprise nuclear attack.

The embryo of U.S. space reconnaissance was called WS-117L (for Weapons System). This was based on the RAND studies and was sponsored by the Air Force's Air Research and Development Command. Both organizations concluded during the frightening days in 1953 that work on a comprehensive space-based Earth observation system be started even though it was not yet possible to get it to space. The Lockheed Missiles and Space Division won a contract to develop WS-117L in June 1956. The idea was to create three different kinds of observation satellites, each using the company's versatile Agena upper stage to get to and stay in orbit. Agena would be the satellite that carried the particular observation equipment. It, in turn, was to be lifted toward orbit by specially adapted Thor intermediate-range ballistic missiles and Atlas ICBMs.

In addition to WS-117L, another program called for hitting the Moon. As early as February 1958, the Air Force was planning to use a Thor and a Vanguard upper stage to send a small probe crashing onto the lunar surface. "In addition to the scientific data that can be obtained from such a flight," Air Force Secretary James H. Douglas would say in a memorandum to Thomas S. Gates Jr., who became secretary of defense the following year, "the United States could make a major international psychological gain by beating the Russians to the Moon." Pioneer 1, launched in that direction by the Air Force on October 11, was supposed to orbit, but it did not make it, and neither did three successors.

One of the Earth observers, Midas (Missile Defense Alarm System), eventually would become the first ballistic missile early-warning satellite. It used infrared sensors and a telescopic lens to spot the heat plumes that signal a missile engine test or an attack. Midas itself, which worked in a 2,000-mile-high circular orbit, was a failure because its heat sensors kept picking up warm sunlight on cloud tops that sounded false alarms. But it would mature into a long-lived class of early-warning satellites called DSP, for Defense Support Program, that ringed Earth from geostationary orbit, 22,300 miles above the Equator. There, twelve-foot-long infrared telescopes that rotated six times a minute would stare at Eurasia continuously, looking for rocket tests, space shots, and the simultaneous launching of hundreds of ballistic missiles that would mean World War III was beginning. Far from being secret, DSP's existence was carefully publicized because it was important for the Kremlin to know that the United States would see an attack coming and retaliate. The warning from DSP and the mass missile launch against the Warsaw Pact that was sure to follow would become the linchpins of the doctrine of mutual assured destruction (MAD) that took hold in the 1960s.

The Soviet Union developed its own satellite warning system, though its spacecraft did not operate in geostationary orbit. Instead, nine at a time flew in highly eccentric orbits, coming in as low as 300 miles and then looping way out to 25,000 miles. Full, continuous coverage of North America meant that the Soviet spacecraft had to operate in threes and on planes 120 degrees apart and at an inclination to the Equator of 63 degrees. They were less reliable than their DSP counterparts, but they were good enough. Since both superpowers knew that the other would see an attack coming and would have time to counterattack before Armageddon, the ballistic missile early-warning satellites ensured a nuclear standoff. They were therefore called "confidence builders" in the distorted imagery of the cold war's political hall of mirrors.

Samos, a second class of observation satellite, seemed like Buck Rogers come true, at least on the drawing board. Like RAND's electro-optical spy satellite, it was designed to take photographs of Earth, develop the pictures on board, scan them electronically, and then radio them to Earth soon after an event occurred. Samos was therefore the dream machine for the general who wanted his fingers on the pulse of the planet. Bruce H. Billings, head of special projects in the Department of Defense's Research and Engineering Office, would tell presidential science adviser George B. Kistiakowsky that "there would be ten television channels to the ground and a library of information so complete that a general sitting in his easy chair in the Pentagon, just by pressing a button, will be able to see on a screen the complete display of current military activities in televised form anywhere in the world."

Billings's vivid imagination, however, was far ahead of the technology of the time. In reality, the general would have seen a single scene from someplace that was so blurry it was essentially useless. Opticians and engineers could not get the transmitted pictures clear enough for spy-quality imagery. And the problem was compounded by the fact that the system was too slow for the few minutes that the satellite passed over the target.

The third of the WS-117L satellite observation programs called for a recoverable capsule. That is, it used Richard Raymond's idea of rocketing exposed film out of orbit in canisters and parachuting them to Earth. Since it was nowhere near as fast as Samos promised to be, it was not given as high a priority as the near-real-time system. That, however, changed because of sharp Department of Defense budget cuts in Samos in 1956 and continuing development problems. Given the problems with radioing useful pictures from space, two RAND engineers in 1956 went back to the idea of either rocketing exposed film or both the film and the camera back down. After studying a whole range of problems, including firing the "package" out of orbit, protecting the film from searing heat during re-entry, and getting a parachute to work at high speed, they decided that the technique would work because it was simple and straightforward.

The matter might have stalled there except for Sputnik. On October 24, 1957—twenty days after the satellite had been shot into orbit by the immensely powerful R-7 ballistic missile—Eisenhower's top consultants on foreign intelligence activities submitted a semiannual report noting that plans were in the works for both a replacement for the U-2 spy plane and for the various WS-117L systems. The U-2's replacement would be the high-flying, lightning-fast A-12, originally code-named "Oxcart" and destined for the CIA. Its successor would be flown by the Air Force and called the SR-71. But there was no chance that either program could be operational within three years, the report concluded (the SR-71 would take eight years to get from concept to operational status). It therefore strongly suggested that an "interim" space-based photoreconnaissance system be developed sooner. RAND had come to the same conclusion. It sent a secret twenty-page report, "An Earlier Reconnaissance Satellite System," to Air Force headquarters in Washington on November 12. The document suggested that a 300-pound satellite be designed to go up on a Thor IRBM and that either the camera and film, or only its film, be returned to Earth.

Eisenhower formally agreed to go ahead with the bucket-dumper in February 1958, right after Explorer 1 reached space. The new program, which was intended to be only an interim one until all of WS-117L was in place, was named "Project Corona" at a meeting on March 10. Bissell whimsically picked that name because a related report was being typed on a Smith-Corona typewriter at the time. Its managers—the Air Force and the CIA—would refer to it as Corona. Those who used its "product"—who interpreted or otherwise used the pictures—would know the system by the Keyhole, or KH, designation given to the camera that took the photographs. The first two types of satellites in the Corona Program, for example, would be called the KH-1 and KH-2 after their Fairchild cameras. Eventually all of Corona's intelligence collectors, including the KH-1s, KH-2s, and KH-3s, would be retroactively called KH-4.

As worked out by Eisenhower's powerful science advisers, Corona was to be jointly run by the CIA and the Air Force. The agency would control funds for the covert operation, acquire the cameras and the satellite recovery vehicle—the capsules that were to carry the film back down—and handle the program's security. The Air Force would be responsible for getting the spacecraft built, launching them, and collecting the film capsules. Bissell, who was directing the fantastically successful U-2 program as Dulles's ambiguously named special assistant for planning and development, was put in charge of Corona as well. The program's deputy director was Air Force Brigadier General Osmund J. Ritland, who was vice commander of General Bernard Schriever's Air Force Ballistic Missile Division. Bissell and Ritland worked exceptionally well together. "The program was started in a marvelously informal manner," Bissell would say years later. "Ritland and I worked out the division of labor between the two organizations as we went along. Decisions were made jointly. There were so few people involved and their relations were so close that decisions could be and were made quickly and cleanly." The cozy relationship between their organizations was not to last.

Intelligence collection, in the words of General George J. Keegan Jr., one of the Air Force's intelligence chiefs, was "an immense source of power and control. It's at the center of the maelstrom." He was right. Reconnaissance was and remains immensely valuable because as the key means of watching and listening to the enemy, and as presented to the president and the National Security Council, it necessarily shapes the nation's perception of the threat.

What is actually seen, or what is interpreted as having been seen, is combined with the fruits of human spying and ordinary research to become National Intelligence Estimates. And NIEs, in turn, influence defense appropriations. It would therefore be tempting for a military service such as the Navy to place special emphasis on searching for enemy submarines and to interpret its findings as proving that they were being built in large quantities and were veritable masterpieces of design and construction. The intended effect would be to convince Congress that the Navy needed more money for both extra submarines and anti-submarine warfare equipment, including pricey sub-hunting patrol planes. Knowing this tendency all too well, and firmly believing that civilians had to control jointly used national programs, Eisenhower would act decisively in late 1960 to prevent the military from running space reconnaissance (except the tracking of foreign ships) or from having an exclusive role in interpreting the intelligence.

Like any spy operation worthy of the name, Corona had a cover name, Discoverer, and a cover mission: scientific research. Aside from obvious technical reasons for wanting to keep the space reconnaissance program under wraps, foremost among them being that secrecy was expected to turn up more intelligence than openness, there were important political ones as well. No one wanted to give the Kremlin, which was quietly but angrily protesting U-2 and other "provocative" flights over its territory, an excuse to publicly slap the United States with charges of being the first to plan space espionage at even higher altitudes. That kind of thing, it was felt, would weaken the elaborately constructed "freedom of space" argument in international law. And that, in turn, could cause the whole Corona project to be scrapped. Furthermore, trumpeting the fact that American spy satellites were operating over the Soviet Union would needlessly embarrass the Russians and probably force them to come up with a way of attacking the spacecraft.

A press release was therefore issued on February 28, 1958, claiming that Discoverer's mission was to collect "environmental" data in space and to carry biomedical specimens, including live animals, which were to be recovered and studied. A halfhearted attempt to make the story look real even included one highly publicized launch, that of Discoverer 3, which carried four "trained" black mice with tiny radios strapped to their backs. But the ruse was jinxed from the start. Just before the launch, while the Thor and its Agena-Discoverer spacecraft were still on the pad, telemetry from the capsule indicated that there was no mouse activity. What happened next is best described by the Central Intelligence Agency itself: "It was thought at first that the little fellows were merely asleep, so a technician was sent up in a cherry-picker to arouse them. He banged on the side of the vehicle and tried catcalls, but to no avail. When the capsule was opened, the mice were found to be dead. The cages had been sprayed with krylon to cover rough edges; the mice had found it tastier than their formula; and that was that."

Following the untimely deaths of the rodents, a "backup crew" was launched on June 3, 1959, but the Agena fired downward instead of horizontally for injection into orbit, sending its occupants to a watery grave. Discoverer 3 was the only spacecraft in the series to carry living creatures.

The Corona Program's managers were doubly irked because one of von Braun's Jupiters had carried two monkeys, Able and Baker, on a well-publicized 300-mile flight only six days earlier and both had survived. The Army did not come away unscathed, however. The very day the CIA's mice perished in the service of their country, Able died during minor surgery to remove an electrode that had been implanted under his skin, causing the ever-vigilant British Society Against Cruel Sports to lodge a formal protest with the U.S. ambassador in London.

During that period—1957 to 1959—the Air Force waged a public relations campaign designed to link itself to observation satellites. Besides leaking material to trade journals, for example, the Air Force released a nine-page pamphlet that described Discoverer in some detail. While the handout gave away no secrets, calling the operation an "open-end" research program, it did tease alert readers by identifying Midas and calling Samos a "photographic satellite." The releases and the pamphlet reflected an early Air Force ambivalence about space reconnaissance. Mostly, spying from space was taken to be a wasteful extravagance by a service that desperately wanted long-range bombers and ballistic missiles. It was not that the air staff did not want up-to-date strategic intelligence from space, particularly on targets that would justify all the bombers and missiles it wanted to buy; it just did not want to pay for it.* At the same time, with NASA taking over manned and other space op-

* One long-standing irritant for the Air Force over the years would be the fact that its reconnaissance budget was "fenced." That meant it could not be raided in lean years to pay for planes, missiles, or other hardware because it was an integral part of a larger, national program. This, in turn, created a certain amount of friction between reconnaissance officers and their comrades in arms in the fighter and bomber units.

erations, there was a clear feeling on the part of some generals that their service should keep as much of a hold on space as possible.

The Devil in the Details

Like Samos, Discoverer was designed to be stabilized on all three of its axes. There would be no pictures if the satellite vibrated, rocked, rolled, or tumbled. One of the hundreds of fundamental engineering challenges was therefore to fit it with a horizon sensor and tiny steering thrusters that would squirt gas to keep it in the right position. It would also fly a polar orbit so that an entire slice of the planet passed under its camera on each ninety-minute round-the-world flight. The camera itself, which would be improved over the years, was made by the Itek Corporation in Lexington, Massachusetts, and was a work of art in its own right. It had a twenty-four-inch focal length and used very high quality German glass that was taken through U.S. Customs without paying duty by faking contracts. That kept the CIA out of it.

Since the Corona flights were going to be near polar, they had to be launched from Vandenberg Air Force Base near Santa Barbara, California. The site was ideal for two reasons. First, jutting out into the ocean as it did, it provided a straight shot due south over the Pacific, which not only reduced the number of people who would witness the firings but ensured that a satellite that did not make it to orbit would fall into the sea, not into the center of some town. The thought of one of the spy satellites crashing into a community and not only causing death and destruction, but drawing curious reporters in the process, was nightmarish to the spy satellites' obsessively secretive managers. Second, with one notable exception, the desolate location offered excellent security. The one persistent security problem was to be caused by the Southern Pacific railroad, whose heavily traveled tracks passed through Vandenberg's launch complex. Corona launches would therefore have to be planned with an eye to the Southern Pacific timetable. For years, mission planners would be vexed by having to schedule the launches of one of their country's most secret and valuable intelligence collectors in the early afternoon and often with only minutes to spare between trainloads of tourists and commuters.

The idea was to launch the spacecraft almost due south so they would fly over or near the poles. Precise calculations told planners when the particular spot in the Soviet Union, China, or elsewhere would pass beneath the orbiting satellite, allowing its timer to be set so that it turned on the camera at precisely the right moment. Its mission accomplished, the Agena and its camera system would begin a series of maneuvers that were remarkable for their sophistication, particularly at the time.

As the Agena arced more than a hundred miles over the North Pole heading south toward Hawaii, it would turn 180 degrees, until its nose faced backward and it pointed down 60 degrees toward Earth. Then it would fire its satellite recovery vehicle: the re-entry capsule that was its nose, and that contained a spool of exposed film. The capsule's own tiny rockets would then fire to spin-stabilize it and a retro-rocket, now pointed toward Earth, would fire to slow its descent. All of this was particularly tricky because the capsule, like other spacecraft returning to Earth, had to hit the atmosphere at exactly the right angle: too high or low and it would stay in space, incinerate, or land in the wrong place. After it re-entered the atmosphere, the retro-rocket thrust cone, the ablative heat shield covering the nose, and a parachute cover would all separate.* A drogue chute would open, followed by a much larger orange-and-white-striped main chute, which lowered the recovery capsule through the air.

Meanwhile, six Air Force C-119s and one C-130 transport (eventually, all C-130s) would have arrived at a 200-by-60-mile rectangle called the "ballpark" a couple of hundred miles south of Honolulu. Three other transports would patrol an outer area called, appropriately, the "outfield." With the eighty-four-pound capsule sending a continuous signal as it floated down, the aircraft would close in on it. Then one of them would snare the parachute with a flying trapeze that looked like a giant version of the one at the circus. The capsule would be winched into the aircraft and flown to Hawaii. If the plane missed, according to the plan, the capsule would float on the sea and send out a signal until a salt plug dissolved, but long enough for it to be retrieved by naval vessels that had also been sent to the area ahead of time.

Corona had a troubled beginning, both technologically and politically. A baker's dozen launches, or attempted launches, were anything but promising, though they echoed what was happening east at Canaveral and, far beyond it, at Kapustin Yar and Tyuratam. An hour before the first attempted launch on January 21, 1959, tests were started on the Agena's hydraulic system. Suddenly, the explosive bolts that held it to its Thor booster, and which were supposed to detonate high in the sky to separate them, ignited. Sensing that separation had occurred, tiny rockets went off that were supposed to push the Agena fast enough so its propellant was pulled downward into its engine. The resulting explosion left the Agena and Thor smoldering on the pad. Like the Soviet censors who were infamous for airbrushing out-of-favor people from official photographs, Corona's dismayed managers named the twisted, smoking mess Discoverer O and simply refused to count it among the failures that followed.

* The heat shield, exactly like those on ballistic missile re-entry vehicles—warheads—was designed to peel off in layers as the intense heat that built up because of atmospheric friction increased in a process called ablation. The technique was made possible only by research on re-entering warheads. This is one example among countless others that shows how research in one area benefited another. And it was research that in part had to do with the IGY.

Throughout 1959 and well into 1960, the CIA, Lockheed Missiles and Space Company (no longer just a division), General Electric (which made the capsule), Fairchild, Itek, and others labored to perfect and integrate Corona's complex systems. Meanwhile, the demise of Discoverer O was followed by a dozen more frustrating, and budget-busting, mishaps that caused the CIA and the Air Force to dig ever deeper into their black—ultrasecret—pocket. The string of disasters stretched from that January to August 1960. Three of the Agenas never made it into orbit, and two others went into very erratic orbits; one capsule was dropped too soon; two cameras worked briefly and then failed, and a third failed altogether; one Agena's retro-rocket malfunctioned; and one of the spacecraft ran so low a temperature that it ruined its health by freezing components, including its batteries. Of five capsules that were actually shot out of the Agenas, two failed to transmit the signals that helped the planes find them, one overshot the recovery area by several hundred miles and was lost, and two overshot it by a somewhat wider margin: one was thought to have disappeared near the South Pole and the other was believed to have crashed somewhere in Scandinavia and was never found.

"It was a most heartbreaking business," Bissell would later recall. "If an airplane goes on a test flight and something malfunctions, and it gets back, the pilot can tell you about the malfunction and you can look it over and find out. But in the case of a recce satellite, you fire the damned thing off and you've got some telemetry, and you never get it back. There is no pilot, of course, and you've got no hardware. You never see it again. So you have to infer from telemetry what went wrong. Then you make a fix and if it fails again you know you've inferred wrong. In the case of Corona, it went on and on." That it did go on, that the Air Force and the CIA pressed grimly on, fixing first the booster, then the satellite and camera systems, and finally the re-entry capsule, was a measure of how desperately the United States needed the intelligence. All they had as of the spring of 1960 was film shot on the occasional, and increasingly risky, U-2 flights. And even that dried up on May 1, when Powers was blown out of the sky by an SA-2 anti-aircraft missile.

The break came on August 11, 1960, when the capsule from Discoverer 13 fell into the Pacific 330 miles northwest of the planned recovery area and was hoisted out of the ocean by a Navy helicopter.* Since the capsule hit the water instead of a trapeze, it was only a partial success. But that was trivial compared to the fact that the extraordinarily complex requirement of getting the Thor, the Agena, and the capsule to work together had finally been accomplished. And the wonder of it was that it was accomplished within thirty months of the first

* The backup technique of lifting the small capsules out of the water by helicopter would be perfected in time for use in the manned Mercury, Gemini, and Apollo programs.

American satellite having taken to space. The re-entry problem was fixed, and Bissell and his Air Force colleagues were absolutely ebullient.

The feat also provided propaganda-starved Americans with a first of their own, however minor: Discoverer 13 was the first man-made object to be recovered from space, and Ike was duly photographed inspecting the Stars and Stripes, which were said to have been the capsule's only passenger. The Soviet Union had tried to do the same thing with Sputnik 4's recovery capsule three months earlier but had failed because, like Discoverer 5, it had gone up instead of down. The intensity of the competition between the two spacefaring nations was accelerating as surely as some of their rockets.

The very next Corona flight was letter-perfect. Discoverer 14 went up on August 18 and became the first man-made object to be recovered in midair the following afternoon when an Air Force C-119 Flying Boxcar called *Pelican 9* snatched it on the third pass as it drifted 8,500 feet over the Pacific. Unaware of the true nature of what they had snared, the elated crewmen flew the capsule back to Hickam Air Force Base where the pilot was awarded the Distinguished Flying Cross and his crew received Air Medals on the spot. Their cargo, complete with a twenty-pound roll of processed negatives, was rushed to Washington. The first space reconnaissance picture showed two fuzzy white scars: the runway at Mys-Schmidta air base in the Soviet Far East and an adjacent apron. That picture, with a resolution on the order of twenty-five feet, revealed almost nothing. But it promised everything. The pictures that followed proved beyond doubt that the technique worked splendidly.

At 8:15 on the morning of the twenty-fourth, just before a National Security Council meeting, Killian, Land, National Security Adviser Gordon Gray, and George Kistiakowsky took a reel of the developed Discoverer 14 film into the Oval Office and, with Allen Dulles looking on, unrolled it across the carpet toward Eisenhower, who was standing beside his desk. The president was stunned by what he saw: "good" to "very good" photographs that covered 1.5 million square miles of Soviet and Eastern European territory and that turned up sixty-four airfields, twenty-six new surface-to-air missile sites, and a third major rocket-launch facility at Plesetsk. Discoverer 14's "take" was nothing short of phenomenal. It had collected more intelligence in seven orbits at an altitude of 115 miles than had been collected in four years of U-2 operations at fifteen. "For the analysts and estimators," Wheelon would later remark, "it was as if an enormous floodlight had been turned on in a darkened warehouse."

Eisenhower declared on the spot that no strategic reconnaissance pictures should be released to the public. It was a particularly statesmanlike decision, since doing so would have all but ended Democratic charges that the Soviet Union led the United States in ballistic missiles.

The science charade notwithstanding, the Discoverers were mostly taken to be what they were. Pied Piper had been described as "an earth reconnaissance satellite program by *Aviation Week* as early as mid-June 1958. The trade magazine matter-of-factly published a number of articles describing how the "advanced reconnaissance systems" were to be flown and made frequent references to the fact that WS-117L was a military reconnaissance system. It and all other "Weapons System" designations for satellites were therefore ordered dropped on October 20, 1958, in order to lessen the aggressive nature of the operation. WS-117L became Sentry and finally reverted to just plain Samos. No one was deceived. Following the first fully successful Discoverer operation in August 1960, for example, *Aviation Week* informed its readers, some of whom worked in the Soviet Embassy, that the event occurred "as a close examination of the satellite reconnaissance concept and its technical feasibility was in progress." *The New York Times* followed, reporting that "the technological feat marks an important step toward the development of reconnaissance satellites that will be able to spy from space."

Nor were the Soviets fooled. An article in the Soviet journal *International Affairs* correctly named all three American observation spacecraft—Midas, Samos, and Discoverer—and declared (as predicted) that "espionage satellites" were illegal. It went on to say that the Soviet Union had "everything necessary to paralyze United States military espionage both in the air and in outer space." Harry Truman would have called that "hooey," too, and he would have been right.

On August 18, 1960, the day Discoverer 14 began its epic flight, a Committee on Overhead Reconnaissance that functioned under the auspices of the CIA and that was chaired by the agency's James Q. Reber, issued a top-secret "List of Highest Priority Targets" in the Soviet Union. The thirty-two highest-priority targets for U.S. spy satellites' cameras reflected the continuing preoccupation with surprise attack:

> As in previous lists, the priority interest centers on: (a) The ICBM, IRBM, sub-launched missiles; (b) The heavy bomber; and (c) Nuclear energy. However, the principal emphasis is the ICBM and the questions of its deployment. At the moment, this objective transcends all others. In the main, it is expressed in this target list in terms of the search of sections of rail lines which are judged to be, among the total of USSR rails, the most likely related in some way to ICBM deployment and which are short enough in length to be considered as a terminal objective within operational capabilities.

The CIA's analysts correctly figured that R-7s and their successors would be moved by rail and that ballistic missile launch sites would therefore be at or near the end of rail spurs, so that following train tracks (which U-2 pilots had done for four years) would turn up the heavy weapons. Included in the ten-page memorandum were the huge Sary Shagan nuclear weapons test site in

Kazakhstan and the surface-to-air missile sites that would threaten SAC's bombers. Plenty of SAMs turned up, but they were not protecting many ICBMs. Not yet, anyway.

Within a couple of years, the entire U.S.S.R., its Eastern European proxies, Communist China, and most of the rest of the world would pass under the eyes of the orbiting spies. The Russians, who would begin retrieving their own space imagery from Cosmos 4 and 7 in April and July 1962, knew that they were under continuous surveillance. What they did not know was that the Discoverers and the spacecraft that succeeded them with ever-better cameras were calibrating those cameras—checking them out for photographic detail—not only on targets in the United States whose precise dimensions were known but in the parking lot behind the U.S. Embassy in Moscow, where parking spots were carefully marked off with paint lines of varying sizes and widths. It had the same effect as a television test pattern. The satellites also practiced photographing an automobile that was parked in front of the embassy on Novinsky Boulevard. Comparing the size of the car with the sizes of actual intelligence targets helped the photointerpreters calculate the exact dimensions of Soviet aircraft, armor, missiles, naval vessels, antennas, and even power cables that carried electricity to uranium enrichment facilities.

At the same time, U.S. intelligence feared that its space "assets" would be attacked. A Special National Intelligence Estimate issued on August 9, 1960, warned that the Soviets would theoretically be able to destroy American reconnaissance satellites, or at least "neutralize" communication with them, by 1963. In the meantime, the NIE predicted that the Kremlin would try to mobilize world support to put pressure on the United States to stop its snooping, but would not bring the issue to a political climax until its anti-satellite weapon was ready.

And besides all the other reasons for keeping the program top secret—largely to keep the Soviet Union from knowing what was being uncovered, the American public from knowing what it cost to do so, and the respective bureaucracies from knowing what the others were doing—the report repeated the old fear that publicizing space reconnaissance would only hurt the Russians' prestige and make them even more determined to end the threat to their secrecy: "[I]f the US Government refrained from officially avowing and attempting to justify a reconnaissance program, and perhaps explained the launching of new satellites on other ground such as scientific research, we believe that the chances are better than even that the Soviets would not press the issue until they were able either to destroy a vehicle, or to establish its mission by authoritative US acknowledgment or other convincing proof." In other words, fooling the Soviets was at that point totally out of the question, but embarrassing and goading them would be counterproductive.

Still, intelligence planners believed that no matter what the United States did by way of officially disguising Corona, the day would come when the Soviet Union "will probably seek to destroy US reconnaissance satellite vehicles." Wrong. Both sides' reconnaissance satellites would soon become vital to national security for intelligence collection, targeting, and arms control. The United States and the U.S.S.R. would also come to rely on them as "confidence builders" that reduced nasty surprises and verified agreed limits on strategic weapons. An attack on a spacecraft would be an act of war as surely as would an attack on a naval vessel. Both would invite immediate reprisal.

The Reconnaissance War

Soviet air bases, other intelligence targets, and the lines in the parking lot behind the embassy were not the only things that needed to be resolved. So did the war between the Air Force and the CIA over control of the program, which heated up in 1961. The irony was that whereas neither side had wanted reconnaissance to begin with—SAC's General Curtis E. LeMay, who never lost sight of the fact that his service's core mission was destroying the enemy, dismissed aerial reconnaissance units as "boutique" operations as early as 1954—they both began to covet it after Corona proved itself.

Richard Bissell left government service after the Bay of Pigs fiasco in April 1961, since he had run that failed operation, as well as the U-2 program and Corona. He was replaced by Herbert "Pete" Scoville Jr. and then by Wheelon, but neither of them enjoyed the clout of Bissell, who had forged Corona with Ritland. So there was a vacuum on the CIA's side of the alliance. Meanwhile, Robert S. McNamara, the new secretary of defense, and others in the Pentagon decided to consolidate all military space programs under the Air Force. The airmen knew that Soviet ICBMs were beginning to proliferate, as were their own missiles. Understandably, they wanted to be able to control the system that told them how many missiles the Soviets had and pinpointed them and other targets. They emphatically did not want the intelligence collection and evaluation process to be controlled by a bunch of civilians at CIA. McNamara told John A. McCone, who succeeded Dulles as director of central intelligence when John Kennedy succeeded Eisenhower, that the CIA should simply define requirements, perhaps do some advanced research, and examine the film that Air Force satellites collected.

But like Eisenhower, the CIA was convinced that a national program like strategic reconnaissance needed to be controlled by civilians, not a military establishment that would be tempted to serve its own, parochial interests. Wheelon ultimately persuaded McCone that the old relationship with the Air

Force was over. "There is no point in screwing another light bulb into a socket that is shorted out," he told his boss. Instead, he said, the nation needed the benefit of competition. McCone agreed, and so did Kennedy's science adviser, Jerome Wiesner, and Din Land. The vacuum at the CIA began to fill.

A National Reconnaissance Office

On June 10, 1960, Eisenhower had told Gates, his secretary of defense, to form a group to assess Samos, which was a poorly managed headache. Gates named Undersecretary of the Air Force Joseph V. Charyk, Director of Defense Research and Engineering Herbert F. York, and Kistiakowsky to take a hard look at the Air Force's now-cherished program, with Kistiakowsky in charge and Killian and others helping. They finished their evaluation in August and, on the eighteenth, met in Cambridge with Land, MIT's George W. Rathjens, Carl F. G. Overhage, who directed MIT's Lincoln Laboratory, and some others to consolidate the Samos findings and shape recommendations for an improved and streamlined management structure for Samos.

Kistiakowsky and the others concluded that space reconnaissance was absolutely vital to U.S. national security. They also agreed that it had to be conducted in total secrecy. The Air Force, they decided, was simultaneously mismanaging Samos and publicizing it and Corona in order to make the Air Force and the satellite programs seem inseparable. But in the process, the airmen were dropping tantalizing hints about the Samos system's intelligence function, which was exactly what Eisenhower did not want. They therefore decided that all responsibility for Samos be taken away from the generals and given to a new Department of Defense office that reported to the secretary of the Air Force, who was a civilian. He would in turn report to the secretary of defense. That, of course, had suited Eisenhower perfectly.

The tortoiselike momentum of the federal bureaucracy had made it a standing national joke for years. But this time, with the need to collect intelligence and assess it quickly, accurately, and fairly being of paramount importance, matters moved swiftly. Four days after the meeting in Cambridge, Kistiakowsky was back in Washington, where he briefed Bissell. The next day he did the same with Gates, who fully endorsed the plan. At that National Security Council meeting on August 24, after Ike had seen the first spy satellite imagery, he approved the Kistiakowsky group's recommendation and so, officially, did Gates.

On August 31, 1960, Secretary of the Air Force Dudley C. Sharpe issued a detailed memorandum creating an Office of Missile and Satellite Systems to oversee the Samos program. As undersecretary of the Air Force, Joe Charyk became its first director.

With the new office came a complete security clampdown and a conscious disinformation program that was intended to undo some of the damage from all the Air Force press releases about reconnaissance. On November 17, for example, United Press International sent out a story quoting an Air Force general in California as claiming that Discoverer 17 had successfully returned from space with artificially grown human cells and vegetation as part of the human spaceflight experiment program. It was deliberate disinformation. The spacecraft was really a reconnaissance satellite whose acetate film broke after less than two feet of its leader had gone through the camera. It was successfully plucked out of the air on November 14, but there was no useful imagery, let alone cell samples or chlorophyll.

A few weeks after the historic National Security Council meeting, Eisenhower ordered that the photointerpretation divisions of all three armed services and the CIA's Photographic Intelligence Division (which was created in 1953, later renamed the Photographic Intelligence Center, and hidden behind what appeared to be a Ford automobile dealership at Fifth and K Streets in downtown Washington) be combined into a single, civilian-controlled National Photographic Interpretation Center. NPIC (pronounced "EN-pic") moved into Building 213, a large windowless structure in the old Navy Yard in Anacostia, in 1963, where it received, processed, analyzed, and distributed imagery from space to its "customers." Pictures of Soviet submarines, for example, were routinely routed to ONI, the Office of Naval Intelligence. NPIC's first director was Arthur Lundahl, who came from the CIA, and who would report to the director of central intelligence.

A year after it was formed, the Office of Missile and Satellite Systems was renamed the National Reconnaissance Office, with Charyk still in charge. "National" was the operative word. Kistiakowsky would note emphatically in his memoir that the NRO was intended to be "of a national character, including OSD [Office of the Secretary of Defense] and the CIA, and not the Air Force alone." Its job would be to buy and operate the nation's spy satellites. And that operation quickly turned from dark gray to deep black. For thirty-two years, the NRO would be so secret that it officially did not exist; even its logo—an Earth-orbiting satellite shaped, ironically, like Sputnik 1—would be classified. Its heart for more than a quarter of a century would be on the fourth floor of the Pentagon's C Ring, behind a door that supposedly led to the ambiguously named Secretary of the Air Force, Office of Space Systems. The creation of the NRO permanently ended any pretense of direct Air Force control of space reconnaissance.

Charyk even established four separate programs to help delineate areas of responsibility and thereby keep the peace. The El Segundo, California, field office, which would coordinate spy satellite design with the two major contractors, TRW and Lockheed Missiles and Space, would operate within the Air Force Space Division's Special Projects Office. It was known within the NRO as Program A. Program B was the CIA's and was run by Leslie Dirks and Carl Duckett, whom Wheelon hired away from the Army. Program C represented the Navy and was therefore run by an admiral. And Program D was a short-lived aerial reconnaissance operation run by the Air Force. But keeping the Air Force and the CIA away from each other's throats would take more than the likes of Charyk or even the authority of the president of the United States.

Civilian Versus Military Intelligence

Whatever Eisenhower and his successors ordered, the duel over control of space reconnaissance and the interpretation of its product would go on for years between the Air Force and the CIA, with each developing deep, institutionalized antagonism for the other. Like gladiators in an arena, the soldiers and civilians hammered at each other relentlessly as they fought for control of strategic reconnaissance. Wheelon's decision in 1963 to expand the CIA's role in overhead reconnaissance by starting a Foreign Missile and Space Analysis Center, for example, was taken by the Air Force to be blatant poaching and sent General Schriever, head of the Air Force Systems Command, into a rage. For him, as well as for many other military officers such as George Keegan, the idea amounted to insidious poaching by civilians in an area that, properly speaking, rightly belonged to the military. There was a reflexive fear of having their own expert analyses of the threat being challenged, their turf overrun.

"The directive establishing FMSAC is sufficiently broad to give FMSAC the charter to clearly duplicate activities currently being carried out within the DOD and, more specifically, within the USAF," Schriever complained to Air Force Chief of Staff Curtis LeMay that December. "The establishment of this activity within CIA is most certainly the first step in competing with and possibly attempting to usurp the Services' capabilities. . . . We can no more rely upon CIA for critical technical intelligence than we can rely upon CIA for target intelligence. . . . I believe immediate action should be taken to slow down or block CIA action to duplicate DOD missile and space intelligence."

Bud Wheelon was taken by Brockway McMillan, who succeeded Charyk as the director of the NRO, to be an out-and-out menace. McMillan believed implicitly that the Air Force ought to run space reconnaissance so it could be streamlined. He considered the NRO to be primarily an Air Force activity and thought that the CIA, and notably Wheelon, were irrational and obstructionist. Certainly the CIA's deputy director for science and technology's emphasizing in-house developmental engineering, which was taken by the Air Force to be its exclusive preserve, infuriated McMillan. The feeling was mutual. Long after he was out of the CIA and heading the Arms Control Association, Pete Scoville, an otherwise avuncular individual, would scowl at the mere mention of McMillan's name.

Aversion to what the generals saw as unnecessary and dangerous civilian meddling in matters that were judged to be strictly military business would persist throughout the cold war. "Are they or are they not building submarines secretly?" Keegan asked rhetorically in 1981. "The judgment I want above all others is that of the foremost submariner in the United States Navy. I want *that* Navy captain evaluating *that* photography and making the primary judgment," he would say with unconcealed anger. "I *don't* want some GS-16 in CIA who's never been to war, never been to sea, never been on a submarine, and who knows nothing of the [military] operational arts. I don't want that guy to have a monopoly judgment going to the president that says the Soviets are not building submarines."

Nor was the CIA the only specter that seemed to threaten the Air Force. Pitching the idea for a dedicated Air Force weather satellite program, RAND's puckish Amrom Katz said in a 1959 memorandum that "if we claim this is a weather or meteorological satellite, various political and jurisdictional hackles at NASA and DOD, and US Weather Bureau levels will rise to the occasion. This we really don't need. We feel that sleeping hackles should be left lying." His solution, intended to keep civilian rivals at bay, was to refer to the program as "Cloud Reconnaissance."

Through the Keyhole

The spacecraft themselves—the crown jewels of U.S. intelligence—also disappeared in 1961 under the new security shroud. Aerial and space reconnaissance were combined into "overhead" reconnaissance, which covered both photographic and signals intelligence operations with satellites and aircraft. All photointelligence was part of the Talent-Keyhole system and was therefore given the KH designators.

During the first year of the Kennedy administration, even the words "Samos" and "Discoverer" vanished. To further complicate matters (which was the idea), Keyhole spacecraft were given "Byeman" code names such as Corona, Argon, Lanyard, Gambit, Hexagon, Kennan, and Crystal, plus individual four-digit designators that told the initiated what kind of satellite it was and its production number. The designator 5504, for example, would tell those cleared to know that the spacecraft was the fourth KH-11 (others knew it only as Kennan).

The last of 121 Corona program satellites, a KH-4B, went up in May 1972. Together, they returned more than 800,000 pictures, each of them exposing an area roughly 10 by 120 miles and changing forever what the United States knew and how it knew it. Two related satellites, an Army mapper called Argon (or KH-5) and an experimental, higher resolution intelligence collector named Lanyard (the KH-6), were launched between 1962 and 1964 with mediocre results and were abandoned.

Access to the imagery sent down by these spacecraft required the Talent-Keyhole, or T-KH, clearance, which was in turn part of the sensitive compartmented information system. By August 22, 1960, while the Discoverer 14 pictures were being analyzed in Washington, a memorandum of agreement among the three armed services and the CIA's Photographic Interpretation Center on procedures for handling the exposed film was already marked "Handle Via Talent-Keyhole Channels Only." Eisenhower, who carried the World War II "need to know" compartmented security system in his head, applied it to the National Reconnaissance Program. The day after the meeting in the Oval Office, when he pulled the security shroud over space reconnaissance, he issued a top secret memorandum saying that access to spy satellite data "is to be on a 'must know' basis related to major national security needs." The nondisclosure agreement that those with SCI clearances who worked for the government and contractors like Lockheed had to sign stipulated that the secrets revealed to them were then and forever the property of the U.S. government. Violation of the agreement, it said, could bring dismissal and criminal prosecution.

Analyzing the "Take"

The analysts themselves were the missile- and space-age descendants of the World War II photoanalysts and signals intelligence officers who had squinted at Peenemünde's V weapons and unscrambled codes with the famous captured Enigma encryption machine. The secretive fraternity of technosleuths took root and developed into grown-up Quiz Kids with clearances; men and women who invested their professional lives in knowing virtually all there was to know about phantom missiles and spacecraft they could never hope to see in person.

The harried analyst who works under "unbearable" pressure and who flirts with a nervous breakdown has been a staple of movie melodramas for years. But the opposite was usually the case. "I found it one of the most exciting chapters of my life," Wheelon would recall many years later. The Stanford- and MIT-educated physicist had to analyze what the opposition (the Soviets, not the U.S. Air Force) was doing in the realm of photoreconnaissance. "I found that doing this telemetry analysis and technical intelligence analysis was the closest thing to physics. It was a period of extraordinary excitement as we began to unravel this puzzle. They didn't have to pay us."

Nicholas L. Johnson, the Hercule Poirot of Soviet space-program analysts, readily agreed. He spent a great deal of time trying to divine the capabilities of the Soviet anti-satellite, or ASAT, program that would have been used to attack U.S. spacecraft in orbits under 600 miles in the event of war. (He also tried to find Soviet satellites' vulnerabilities so they, too, could be "taken out" if necessary.) "It's a remarkable kind of activity; it's a lot of fun," Johnson, too, would recall nostalgically. "I had some of my best fun in those years."

So did Dino A. Brugioni, the highly skilled photointerpreter who, with Art Lundahl, explained otherwise incomprehensible U-2 imagery to President Kennedy and his Executive Committee during the Cuban missile crisis in October 1962. Like soldiers who have scrawled graffiti on their enemies' walls since the time of the Roman Legion, those who were photographed from space also tried to tweak or insult their opponents. Brugioni saw imagery in which the inhabitants of Tyuratam used their heavy boots to stamp dirty words into the snow that they knew the reconnaissance satellites photographing their facility would pick up. The Chinese did the same thing with rocks at Shuang Tseng Tsu, their own missile launch facility, Brugioni said. And at the height of the missile crisis, he added, the tension in the Oval Office was momentarily broken when the Executive Committee was treated to a U-2 photograph that looked straight down on someone sitting on an outdoor latrine.

Their techniques evolved into a kind of black art that mixed science, engineering, and an intuition that developed only after untold hours peering through spectroscopes, and later at computer screens, or listening to the ribbons of taped telemetry and trying to separate the kernels of information from the signals upon which they rode. They lived with the raw intelligence so closely that after a while they knew the opposition's missiles as well as their own children's report cards, maybe better. They carried in their heads encyclopedic knowledge of particular missiles' size, range, propellant and oxidizer, vibration and acceleration rate, the intensity of their heat plumes, the working of their turbopumps, the stage separation sequence, an engine's specific impulse, and a great deal more.

Some Quiz Kids would get so good at massaging the telemetry that came down from the opposition's spacecraft that they could tell whether the occupant was a man, a woman, or a dog by minute differences in their heartbeats. Others looked down from their hundred-mile-high perches at the quarter-mile-long vehicle assembly building, launchpads, roads, railroad tracks, living quarters, machine shops, and other structures at Tyuratam so often that it was almost as if they lived there; as though it were their neighborhood. Others did the same thing at the Northern Fleet headquarters base at Severomorsk on the Barents Sea, where they could keep track of the comings and goings of submarines and destroyers and watch torpedoes and missiles being stored or loaded onto ships. Still others did it at air bases, where they counted bombers and watched the bunkers where the nukes were hidden.

Accurate analysis of a spacecraft's mission—eventually on the first orbit—soon became possible because of the development of a huge computer catalogue whose electronic brain bulged with data about every satellite's inclination, altitude, nodal period (the time it took to orbit Earth once), weight, shape, size, apogee and perigee, eccentricity (the shape of the orbit), time and date of launch, and more. And the data grew quickly. There was an average of ten thousand satellite observations a day worldwide by the mid-sixties and five times that number three decades later.

The satellite catalogue was kept by NORAD, the North American Aerospace Defense Command, deep in the bowels of Cheyenne Mountain outside Colorado Springs. It would become so comprehensive by 1967 that it yielded this (to take only one example from hundreds):

> The Soviet photoreconnaissance satellite program has continued at a high rate (about two launches per month) over the past two years. The program has enjoyed one of the highest priorities in the entire Soviet space effort, accounting for almost half of all Soviet space launchings during this period. . . . The program involved two basic types of reconnaissance vehicles. One payload weighs about 10,400 pounds and performs low resolution photographic missions (i.e., a ground resolution on the order of 10 to 30 feet under average conditions) and probably collects Elint (electronic intelligence) as well. The second type of payload weighs about 12,000 pounds; it conducts a higher resolution photographic mission, and we believe it achieves ground resolutions on the order of 5 to 10 feet. . . . The Soviets recover the photography acquired by both systems by deorbiting the entire spacecraft into the Kazakhstan recovery area after missions of about eight days.

The Soviets' NORAD counterpart did exactly the same thing, as this excerpt from a 1972 issue of the magazine *Aviatsiya i Kosmonavtika* (*Aviation and Cosmonautics*) indicates:

> The Recon Satellite Program long ago took a firm place in the Pentagon's global espionage system. . . . The USA's reconnaissance satellites are employed in the collection of photo, radio, electronic, television and radar intelligence, as well as the detection of nuclear detonations and ballistic missile launches, both from ground-based launch sites and submarines. These satellites are controlled by a Command Center in Sunnyvale, California, which guides the work of the entire complex, including the ground stations in the U.S. and overseas. . . . In order to obtain high-interest information the Pentagon employs both broad-scale [area surveillance] and pinpointed [close-look] photographic satellites launched into near-polar or-

> bits with perigees from 120 to 200 kilometers. . . . This low perigee, in combination with telescopic lenses, make it possible to obtain photographs with a resolution of 1.8 to 5.5 meters.

The reference to the "Command Center in Sunnyvale" was on the mark. The windowless Satellite Control Facility, commonly known as the Big Blue Cube because of its size and color, was and remains across the street from Lockheed Missiles and Space and for many years was used to control all of the reconnaissance satellites.

Those who interpreted the imagery on both sides, who spent most of their professional lives in effect peering down at enemy territory from fifteen to one hundred miles in the sky, reinforced their technical knowledge with an encyclopedic bag of arcane tricks. These consisted of what the interpreters called "signatures"; that is, features or patterns of behavior that gave away information. Brugioni could reel them off by the score: long, bowling alley–like lanes were usually tank firing ranges that betrayed the presence of an armored unit; the headquarters building of a military installation was the first to be cleared of snow (along with paths to the latrines); the occupied buildings were the ones with melted snow on their roofs; Sundays and holidays were the best days to inventory military equipment since it was usually parked or stowed; roads with wide turns were made that way to accommodate trucks carrying large missiles; railroad tracks at space centers generally led to launchpads; erectors, gantries, and flame pits were sized for the rockets they launched. Estimating what was inside crates by their size, shape, and construction—a specialty called "cratology" by the interpreters—paid off handsomely when Soviet freighters delivered MiGs to Iraq, MiGs and missiles to Cuba, and other weapons elsewhere.

Satellite pictures taken of the Lop Nor area of western China in December 1961 showed a unique circular road being cleared that was four thousand yards in diameter, followed by the construction of barracks, support facilities, an airfield, and finally a tower in the center of the round road that was connected to vans by communication and instrumentation cables. It was clear to the interpreters that the Chinese were preparing for their first nuclear test. And it was equally clear that Secretary of State Dean Rusk wanted to steal China's thunder by announcing on September 29, 1964, that the test was going to take place seventeen days before it did.

As the collection process grew during the 1960s, so did the infrastructure that used and supported it. Taped telemetry intercepts were originally flown to a processing center in the United States, copied, and distributed to the CIA and other users. Communication and electronic intelligence about missiles and space, which involved radio and microwave communication and the workings of radar, respectively, were channeled to the National Security Agency at Fort George G. Meade in Maryland. The NSA's Defense Space and Missile Analysis Center routinely asked for specific data on Soviet telemetry, analyzed it, and relayed it to facilities such as the Army Ballistic Missile Agency at Huntsville, where work on the nation's embryonic ballistic missile defense (BMD) program was going on. Imagery, of course, went to the National Photographic Interpretation Center and then to users.

But however tidy the collection and analytical process looked on paper, however neat the flowcharts, those who were responsible for predicting the course of the Soviet missile and space programs shared three afflictions with their colleagues in other areas of intelligence: the amount of data, much of it redundant or irrelevant, was becoming overwhelming; operational capabilities and intentions had to be divined from studying only technology, which strained interpretation; and the whole process was more often than not mired in politics.

The war between the Air Force and the CIA, for example, continued to grow to legendary proportions as the 1960s progressed. One early battle among innumerable others had to do with deciding whether the SS-8, a two-stage, eighty-five-foot-long ballistic missile code-named "Sasin," was powerful enough to carry the fabled and fearsome hundred-megaton warhead, a device so powerful that it would have poisoned cropland for many thousands of square miles, whatever it did to its target city. The Air Force, trying to justify its own Titan ICBM program, contended that the SS-8 was indeed a super–city killer, while the CIA insisted that it was quite a bit smaller. The CIA's analysis eventually prevailed, but only after a near–"fist fight" took place, one participant recalled many years later.

The mania to classify nearly everything, which continued from World War II, accelerated. It was supposedly done to keep valuable information out of the hands of the enemy. But the enemy's identity was always ambiguous. The so-called black programs were not only protected from the Soviet Union and other nations, but from Congress and competing agencies, since they were often horrendously expensive, duplicative, and potentially vulnerable to being either improved or sabotaged by a rival service. Invoking ultrasecret, highly compartmented "need-to-know" status on their juiciest operations—which many thoughtful participants condemned as impeding a healthy, synergistic exchange of information—was done to confound not only the Communists but each other as well.

The competition was endemic within the separate national space structures themselves and was waged on many levels, often vehemently. The rivalry between the Army and the Navy to launch America's first spacecraft was only one manifestation of a wider process that pitted military officers against civilians, advocates of manned flight against those who favored unmanned missions, and individual centers, commands, departments, and divisions within the civilian and military establishments against one another.

The Air Force in Space

There were (and remain) two fundamental aspects of the Air Force's collective psyche that played key roles in its approach to space in particular and to national security in general in the space age's formative years. And perhaps there was a third.

As members of the nation's junior service, air officers had a long tradition of either serving to support another service or being in conflict with one over missions and appropriations. As the members of the Army Air Service in World War I and the Army Air Corps and Air Forces in World War II, the fliers had often been relegated to fighting as appendages of ground forces. Only with the advent of strategic bombing and the bombers to carry it out during the last stages of World War II had Army aircrews come into their own as self-defined combatants. It was a role fought for by a long line of air generals, all of them in the service's pantheon of heroes: Billy Mitchell, "Hap" Arnold, Ira Eaker, Carl Spaatz, Hoyt Vandenberg, Thomas Power, Curtis LeMay, and others. Nor did the contention end when the U.S. Air Force became a separate service. It was soon challenged by the Army, which considered strategic missiles to be extended-range artillery. It was challenged by the Navy, which wanted to wage a nuclear air war with its supercarriers far from the United States. And it was challenged, too, by the CIA, which wanted to control technical intelligence systems that collected information from space. The wars have been described as "bloody" by some of their veterans. Whatever they were, they had the effect of hardening senior Air Force officers; of making them both defensive and aggressive.

At the same time, the devastation inflicted on cities in Germany and Japan during World War II—the atomic bombing of Hiroshima and Nagasaki, to be sure, but also the firebombing of Dresden and Tokyo, and the obliteration of other targets—proved beyond doubt that the U.S. Air Force was the most destructive long range war-fighting machine in history. The Army could not do that, and the Navy, however many nuclear weapons it had on carriers and in submarines, would be primarily responsible for controlling the seas. It was the Air Force that had the wherewithal to pulverize enemy countries quickly and at great distances. Its generals knew that, they were proud of it, and they wanted to carry out their mission without being undermined.

And, too, there was an unspoken cachet to flying that bonded those who did it. Unlike most soldiers and sailors, who were glued to the planet, pilots defied gravity, soared among the clouds, and fought in three dimensions. After World War II, they flew into combat alone or as part of very small groups, and they therefore tended to be individualistic, self-reliant, and aggressive. They considered themselves to be an elite.

The group's highest priority from the day the war ended, through Sputnik and beyond, was to ensure that it could fulfill its core mission. That meant creating and maintaining tactical and strategic strike forces with enough fighters, bombers, and missiles to wage any war against any foe over any distance. Slowly at first, and then more quickly, the airmen began to see the control of space as an extension of their just domain.

In a collection of "War Reports" published in 1947, Arnold called for the establishment of a ballistic missile program and the capability of launching the weapons from "unexpected directions," including from "true spaceships, capable of operating outside the earth's atmosphere." On January 15, 1948, Vandenberg became the first senior Air Force officer to go on record as extending his service's reach to space. "The USAF," he contended, "as the service dealing primarily with air weapons—especially strategic—has logical responsibility for the satellite." The bravado worked; the next day the Navy withdrew its own claim for control of satellite development, though that would be temporary.

A decade later, right after the Sputnik debacle, Air Force Chief of Staff Thomas D. White left no doubt that the Air Force would rise to the space challenge. "For all practical purposes air and space merge, form a continuous and indivisible field of operations," he wrote in *Air Force* magazine in March 1958.* The Air Force pounded that theme home continuously throughout the year. On February 3, 1959, White would use the term "aerospace" for the first time at a congressional hearing. It would be spelled out as official doctrine in an Air Force memorandum ten months later: "The aerospace is an operationally indivisible medium consisting of the total expanse beyond the earth's surface. The forces of the Air Force comprise a family of operating systems—air systems, ballistic missiles, and space vehicle systems. These are the fundamental aerospace forces of the nation." The Air Force's conception of its theater of operations, one of its historians would soon explain, was as simple as it was grandiose: it was "a continuum from the surface of the earth through . . . the limits of the solar system—or the universe." With the high ground having been staked out from the Army and the Navy, the Air Force would now be able to concentrate on its other opponent: the Soviets.

When Air Force planners thought of "space vehicle systems," they let their imaginations soar. Fifteen missions were soon conceptualized by Air Force

* This flew in the face of the Eisenhower administration's "Freedom of Space" concept, which depended on a clear distinction being made between national airspace and higher, orbital precincts, which were a separate entity.

planners as being their service's exclusive domain. These included operating a fleet of unmanned reconnaissance, communication, navigation, and weather satellites; an anti-satellite (ASAT) system that could kill enemy satellites; a manned maintenance and supply system "for outer space vehicles"; manned defensive spacecraft; manned space bombers; manned reconnaissance satellites; unmanned bombardment satellites; and lunar bases (plural). All of these and others were listed in a memorandum written by General Boushey within a year after he described the lunar missile base to members of the Aero Club of Washington. Most of them had no basis in reality, lent themselves to ridicule by politicians and others who caricatured the airmen as lunatics lost in space, and were embarrassing to the vast majority of Air Force officers.

Visions of Armed Spacemen

The most persistent of the Air Force's priorities for space, the most seductive of its dreams, was to get its pilots there: to send spaceplanes beyond airplanes. The idea came directly from the rocket-propelled antipodal bomber that Sänger designed in World War II to skip through the atmosphere, bomb U.S. targets, and then glide home. Flying to space in rocket-propelled aircraft captured the imaginations of civilians and the military alike in the early and mid-1950s precisely because it was a logical extension of aviation.

The rationale for developing the X Series aircraft, at least as it played out in public, was ostensibly to understand high-speed flight. But the celebrated duels between Chuck Yeager and A. Scott Crossfield, who pushed a Navy Douglas D-558-2 Skyrocket to Mach 2 in November 1953 in a blatant attempt by the Navy to surpass Yeager's record, partly obscured two important factors. The competition was only part of a broader, bitter Air Force–Navy rivalry that eventually included not only land-based versus sea-based ballistic missiles, but bombers versus aircraft carriers. The appropriations war was being fought across a wide front and public relations, no less than lobbying, was a key weapon.

More important, speed for its own sake was less relevant to combat flying than was altitude. Fighter pilots in combat in Korea and afterward depended more on "cornering"—maneuverability—and acceleration than on absolute speed, which caused high gs and required wide turns. Speed's main use was to gain altitude quickly. And high altitude conferred the ability to pounce on victims, as every combat pilot from Baron von Richthofen onward knew: be an eagle or be the eagle's prey. Knowing this, the Navy asked Douglas Aircraft in 1954 to study the feasibility of designing a piloted aircraft-type vehicle that could climb to a million feet: 190 miles. The engineers at Douglas reported that 133 miles was the best they could do. And besides, von Braun's three-stage rocket-shuttle notwithstanding, there was a respected body of opinion that said that rocket planes taking off from runways or launched in midair was the way to get to space.

X for "Xtraordinary"

The X-15 was such an aircraft; it got a flying start by being dropped from a B-52 bomber. The black, stubby-winged rocket plane was the distantly removed offspring of Walter Dornberger, who proposed a hypersonic research aircraft to the National Advisory Committee for Aeronautics—the venerable NACA—in January 1952, when he was with Bell Aircraft. The former chief of the V-2 program envisioned a plane with variably swept wings that could reach an altitude of up to seventy-five miles. But it was North American Aviation that barely won a joint NACA–Air Force–Navy production contract for three of the rocket planes in December 1955.

Harrison Storms, North American's chief designer, and others in the company and in NACA who created the extraordinary thoroughbred knew quite a bit about how to send men to space. But they knew a great deal less about how to get them back. One of the X-15's primary jobs was to find out. There was particular concern about the blistering effect of atmospheric friction. So rather than cool or insulate the X-15's skin to reduce the heat, North American and NACA wanted it to get at least as hot as 1,200 degrees Fahrenheit precisely so they could learn what happened to a machine that got that hot.

"Why was the X-15 built?" one of its pilots would ask, rhetorically. "The glib answer was 'to explore the hypersonic flight regime.' The real answer was to find out if we were smart enough to design an airplane that could fly and survive at hypersonic speeds. . . . Could we design an airplane that would be controllable outside the atmosphere and one that could successfully reenter that atmosphere at high speed and steep entry angles?" asked Milton O. Thompson, an X-15 test pilot. "More important, could a pilot survive and function adequately in this high-energy environment?"

The original design goal for the X-15 called for a minimum top speed of Mach 6.6, or 4,430 miles an hour, and an altitude of 250,000 feet: 47 miles. That would take it out of the ostensible atmosphere and into the fringe of space. Regular flying surfaces—ailerons, rudders, and elevators—would not work up there because there was no air for them to grab. So the X-15, which shot up and back down on a ballistic trajectory that lasted between ten and twelve minutes, was given a ballistic control system. This consisted of eight small jets in its nose and four on its wings. They were used to get the right reentry angle for the plunge back into the atmosphere at speeds so high some of the plane's skin heated to a cherry-red 1,300 degrees. Hypersonic aircraft, including the space shuttle, would have to re-enter the atmosphere at exactly the right angle or not survive the attempt, just as the Discoverer capsules had to do.

Typically, the first three months of X-15 operations were plagued with problems. The mishaps, including an engine that exploded during a ground test, a fuselage that split during a landing, and the death of a pilot in the program's only fatal crash, continued for years. Yet they came between triumphs that soon became legendary.

X-15s flew 199 times between June 8, 1959, when Crossfield, wearing a silver space suit, dropped from the B-52 in a shallow glide, to October 24, 1968, when William H. Dana flew one at a relatively tame 3,682 miles an hour up to 47 miles. One set an unofficial world altitude record of 354,200 feet—67 miles—in August 1963; another claimed a world speed record of Mach 6.7, or 4,520 miles an hour, four years later. The records still stand. Eight of its pilots won astronaut's wings for flying higher than 47 miles. Crossfield never flew high enough to win that honor. Neither did another X-15 pilot named Neil Armstrong, but he would get his astronaut's wings later.

Dyna-Soaring

Whatever its considerable research value, the Air Force had seen the X-15 as the precursor to its cherished goal of sending fighter, bomber, and reconnaissance pilots to space in what were then called "winged" or "manned satellites." A swept-wing research plane was not the only thing Dornberger proposed to the Air Force in 1952. With the antipodal bomber still in mind, he also envisaged a bomber-missile, or Bomi, which was eventually renamed the rocket bomber, or Robo; this would be boosted forty miles high. It was then to release a nuclear bomb that would decelerate, re-enter the atmosphere, and hopefully descend on its target. Meanwhile, Robo would swoop into a 10,000-mile-an-hour glide, make a 180-degree turn, start its own engines, and return to its original altitude for the trip home. When tests indicated that the thing would burn up trying to make such a turn, and that its own engine would be prohibitively heavy, the plan was scaled down until it looked something like the old round-the-world antipodal bomber.

As early as May 1955, the Air Force issued a general operational requirement that called for making a version of the X-15 large enough to perform the rocket bomber mission, a second variation to perform very high altitude manned reconnaissance in a system called Brass Bell, and a boost-glide research craft called Hywards.

But Air Force planners had a fundamental leap of imagination during the next two years that eclipsed even a souped-up X-15. They decided that they wanted nothing less than a dedicated, multirole space weapon system: a true spaceplane that would extend the X-planes to space. It would be a small, one-man, delta-winged glider that would be lifted into the sky not by a B-52 but by its own thundering rocket booster. The order to come up with a conceptual design for such a weapon, which was called Weapon System 464L, went out of Air Force headquarters on April 30, 1957.

Serious planning for the spaceplane accelerated after the launch of the first two Sputniks. On November 25, 1957, with the country beginning to writhe in self-doubt and apprehension, the Air Research and Development Command at Wright-Patterson Air Force Base was directed to give WS-464L, or Dyna-Soar (for "Dynamic Soaring") high priority. The following June, the Boeing Aircraft Company was chosen to develop the space weapon in conjunction with the Martin Company, which was to adapt its Titan ICBM for use as the booster. By then, Dyna-Soar was seen not only as a bombing, reconnaissance, and research platform, but as the precursor of spacecraft that would keep blue-suiters in orbit.

"Weapon systems that evolve from the DYNA SOAR development could operate as aerodynamic, boost-glide vehicles, as short-term satellites or satelloids, or as satellites in relatively stable orbits," the Air Force's deputy chief of staff told the director of the Air Research and Development Command in a letter in June 1958. "Further, they could be manned or unmanned, recoverable or unrecoverable. Combinations of any of these vehicles could be included in the final DYNA SOAR weapon system."

The Air Force was thinking of Dyna-Soar as the progenitor of a total space fighting system, one of whose missions could be ASAT: to attack enemy satellites. A "Boost-Glide Weapon Systems Application Studies" report made in June 1958 not only called for studies to determine Dyna-Soar's potential as an offensive, defensive, and reconnaissance spacecraft, but as an orbital "inspector" and interceptor. In other words, Dyna-Soar would in theory have been shot into orbit, where it would pull up to a Soviet satellite so its pilot could look it over and take detailed photographs for analysis on the ground. If the spacecraft under scrutiny seemed threatening—if it looked as if it was going to do something to prevent its picture from being taken—the Dyna-Soar pilot might shoot at it or otherwise immobilize it. Another report said that re-entry techniques learned in the Dyna-Soar program could be helpful in the design of "guided glide weapons," or orbiting bombs. Typically, no mention was made about the likely Soviet reaction to having their spacecraft looked over, photographed, and perhaps impaired by American astronauts acting like highway patrolmen or galactic vigilantes.

Meanwhile, the endlessly imaginative aeronautical engineers reacted reflexively to the Sputniks. Harrison Storms proposed to the Air Research and Development Command that an X-15 be fitted with rocket boosters, beefed up, and flown to orbit. If the Russians wanted to send a dog to space, he muttered, the Air Force should send a real rocket ship there with a man in it and then land at Edwards with a triumphant flourish. Rebuffed, Storms took the idea to LeMay. The patron saint of mass destruction reportedly had only one question: "Where's the bomb bay?"

That April, some of the general's hypercompetitive comrades in arms hatched a four-stage plan that was supposed to end by landing military astronauts on the Moon by 1965. The proposal, an "Air Force Manned Military Space System Development Plan," was put together by a hurriedly formed Man-in-Space Task Force in the Air Research and Development Command. It called for catapulting instruments, primates, and then men into space on top of ballistic missiles in an operation called Man-in-Space-Soonest, or MISS. This was to be followed by phase two, Man-in-Space-Sophisticated, in which a heavier capsule would orbit Earth for up to fourteen days. The next phase would involve lunar reconnaissance with soft-landing spacecraft carrying television cameras that were supposed to scout suitable landing sites for men. Finally, apes and then men would orbit the Moon, land on it, and then blast off for a safe (and tremendously impressive) return to Earth. The task force estimated that the whole mission would cost $1.5 billion and be completed by 1965. The idea ran absolutely counter to Eisenhower's temperate plan for space and, in any case, would be swallowed and chewed up by the civilian space agency.

Meanwhile, advocates of the airplane-spacecraft accelerated work on it during 1958 in troubled waters. NASA was born in October of that year with a mandate to run manned space operations. At about the same time, the Advanced Research Projects Agency was assigned all Air Force space projects, including Dyna-Soar. The program was then scaled back to a suborbital research program in 1960 by Herbert York, the Pentagon's director of defense research and engineering, who did not think that a space weapons system should be funded. (He was the same Herbert York who, with Kistiakowsky and Charyk, was simultaneously deciding that the Air Force should not control space reconnaissance, either.)

Dyna-Soar was a less tempting target in its subordinated research role. But that December a Titan 1 booster identical to the kind being considered to launch Dyna-Soar blew up at Vandenberg. And it did not just blow up. The hundred-ton ballistic missile exploded with such force that it destroyed an entire launch complex, providing a spectacular display of pyrotechnics that would have amazed and entertained any Southern Pacific commuters who happened to be passing by at the time.

Congress, on the other hand, liked Dyna-Soar. In a final report on the 1962 defense appropriation bill, the House Appropriations Committee signed off on "an operational, manned, military space vehicle over which the pilot has the greatest possible control" and declared that Dyna-Soar was "the quickest and best means of attaining this objective." It voted to pump $185.8 million into the project for fiscal year 1962, or $85.8 million more than President Kennedy himself requested. "The Dyna-Soar program is not competitive with the proposed space flight to the moon program announced by the President," the committee pointed out, adding that "for the remainder of this decade the space area close to the earth's surface will be of greater interest to military planners than will the area around the moon." That statement was made the month after Kennedy's famous Moon speech and two months after the first human being actually made it to space. He was Major Yuri Gagarin. And he was not American.

But Dyna-Soar, like its namesakes, was destined for extinction. Secretary of Defense McNamara did not like Dyna-Soar. He thought it was too expensive, particularly in light of the tremendous effort that was then under way to get Americans on the Moon. And in practice, he believed, it amounted to little more than a program to test re-entry techniques at a time when the broader question of whether Americans belonged in Earth orbit in the first place needed to be answered. On February 23, 1962, he ordered that the program be reoriented away from an operational system and toward the kind of experimental operation that characterized the X-planes. In fact, he even ordered that the name Dyna-Soar be replaced with an X number, so on June 26, 1962, it became the X-20.

By December 1963, its final design had long since been completed, a full-scale model had been built, and six pilots had been named to fly it, two of them veterans of the X-15 program. In addition, the Titan 3 had been picked as the space glider's booster. But the spaceplane was not to be. It was not just finding a billion dollars that troubled McNamara; it was finding all that money for a program that had no clear ultimate goal. Its end came at a hastily called news conference at the Pentagon on December 10, when the secretary of defense basically called Dyna-Soar a billion-dollar turkey that would have to be axed. And so it was.

It is important to note that Robo/Dyna-Soar/X-20 was in no sense a mainstream program in a service that remained wedded to conventional air and rocket bombardment. Rather, it represented the radical thinking of a handful of exuberantly imaginative airmen and contractors who believed implicitely that space was a vacuum in the politicomilitary, as well as physical, sense and that whoever controlled it would control Earth. That reasoning was not completely thought out, and, in fact, the actual role of warriors in space—what, exactly, they were supposed to do once they got there—would plague both superpowers for the remainder of the cold war. But sending soldiers to space seemed compelling in that highly competitive age. What is more, the Air Force was a service that prided itself on rewarding brave, adventuresome, and skilled warriors. Whatever the service's core bombing mission, the fact remained that cruising a hundred miles above Earth in delta-winged rockets would be a lot more exhilerating than flying dreary ten-hour practice missions in B-52s. The Air Force attracted dreamers who wanted to soar. That, at least for some of them, was the logic (and fun) of Dyna-Soar.

A Laboratory in Orbit

But the day McNamara pronounced Dyna-Soar dead, he announced that he was signing off on a consolation prize: another manned space program. The other high flyer, which had its origin two years earlier in the Air Force Space Plan of September 1961, was a space station that could perform military operations, and especially manned reconnaissance. With the cancellation of Dyna-Soar, McNamara signed off on a so-called Manned Orbiting Laboratory, or MOL, that was designed to be a research facility the size of a small house trailer that was supposed to orbit at 350 miles and do photoreconnaissance, meteorology, and other militarily useful things. As planned by the Air Force and Douglas Aircraft, the prime contractor, the station was supposed to be a forty-one-foot-long cylinder in which a crew of two would work in shirtsleeves for a month at a time. They would then return to Earth in a Gemini B, a military version of the two-man NASA spacecraft, and be replaced by another crew.

MOL's primary ostensible purpose, like Dyna-Soar's, was to keep airmen in space and see what they could accomplish there. Unlike Dyna-Soar, though, it would not need combat to prove itself useful and had no offensive capability whatsoever. (LeMay therefore rejected it out of hand.) MOL had a number of missions that did not require war-fighting to justify, however: physiological research over long periods in orbit, in-orbit structural repair outside the spacecraft, the development of space suits for long-duration missions, docking and undocking practice, and other things that came under the general rubric of establishing a manned presence in space. The Department of Defense repeatedly described MOL's goal: "to learn more about what man is able to do in space and how that ability can be used for military purposes." This was justified by Lieutenant General James Ferguson at a congressional hearing: "Man has certain qualitative capabilities which machines cannot duplicate. He is unique in his ability to make on-the-spot judgments. He can discriminate and select from alternatives which have not been anticipated. He is adaptable to rapidly changing situations. Thus, by including man in military space systems, we significantly increase the flexibility of the systems, as well as increase the probability of mission success." It would become a familiar argument over the next decades and would be invoked by NASA itself, particularly for manned lunar and planetary exploration. And it was true. Yet it also carefully ignored the fact that manned spaceflight was far more costly than using machines alone because of training, life support systems, redundancy, and other factors. Furthermore, MOL would duplicate much of what NASA's station, which was then being studied, was to do.

But there was far more to the concept of the Manned Orbiting Laboratory than simply establishing an amorphous human presence in space. The MOL itself was not planned to fly over the Soviet Union. Yet the advantages of its successors doing so were carefully considered. During the summer of 1963, the Department of Defense and the Air Force signed an agreement that explicitly listed five primary military missions for the station: general reconnaissance; reconnaissance of given spots on request; poststrike reconnaissance; continuous surveillance of an area; and ocean reconnaissance. The last, it was no doubt hoped, would appeal to the Navy. But as events would later show, the opposite was the case: the Navy wanted its own space surveillance capability. (The Navy had in fact proposed its own version of the MOL for a "Manned Earth Reconnaissance" mission using a cylindrical spacecraft as early as 1958, but it had gone nowhere.) Besides the primary missions, there were four secondary ones: inspection of unknown space vehicles; command and control of ground, sea, air, and space forces; operational support of ground, sea, air, and space forces; and bombardment. "It should be noted that the potentialities of bombardment from space can be studied *without* carrying weapons into orbit," *Air Force* magazine explained, adding, "Such studies are necessary if the advantages and disadvantages of such weapons are to be understood precisely and proper contingency defenses prepared."

President Johnson approved MOL on August 25, 1965, with the understanding that unmanned test flights were to begin in 1967 and manned test flights were supposed to start at the end of the following year. But as NASA's massive Apollo manned Moon program increased momentum, with its budget cresting in 1966, MOL's own fortune began a slow but inevitable decline. In addition, just about everybody accepted the fact that, however horrendously expensive unmanned space reconnaissance was, it was still cheaper than sending men to do it. By April 1969, budget cuts by the new Nixon administration pushed the first manned flight to mid-1972. It was finally killed later in 1969 after consuming $1.3 billion.

For projects that never got off the ground, Dyna-Soar and MOL could claim an important place in the history of the U.S. space program. Dyna-Soar was the true link between the X-15, which pioneered research in hypersonic flight and the mastery of high-speed landing techniques through the atmosphere, and the space shuttle, which would one day make such flights almost routine. And the MOL indirectly showed the way to Skylab, a highly successful Apollo follow-on program. While there was no direct connection between the two, MOL's ghost was a lingering reminder of what might have been, and therefore of what was possible.

Yet throughout, the Air Force was and would remain haunted by a problem that went to the core of its man-in-space doctrine: justifying the use of military, as opposed to civilian, astronauts. Inventing and embracing the term "aerospace" to legitimatize the extension of its realm into the heavens was both clever and understandable. Yet there was also a conundrum where men "in the loop" were concerned that would haunt the Air Force throughout the space age. Air and space might be physically contiguous. But it didn't follow that the requirement for military aircrews in the atmosphere was equally applicable to space. Military aviation had evolved through two world wars in a strictly hands-on environment; military airplanes had to be flown by pilots, and the fighters that opposed them likewise required humans at the controls in most cases. The gnawing problem where space was concerned had to do with the fact that this was not the case in orbit and beyond because of the development of smart and rugged machines, including computers, that could function there better than people. Indeed, James Van Allen and many other scientists believed that humans in general were not only superfluous in space but were horrendously expensive because of the training and support systems they required.

Laid bare—though this was not clearly articulated in the military literature of the time—the justification for using military officers in space had to do with extending the roles they played in the atmosphere. That is, they were in theory supposed to fly space fighters, space bombers, space reconnaissance craft, or combinations of them, as exact counterparts to the fighters, bombers, and reconnaissance planes they flew down below. But there was no need for such activities except as fabulously expensive propaganda. Manned U.S. space bombers would not be attacked by manned Soviet space fighters; they would be attacked by much cheaper remote-controlled ASATs. More to the point, what could a manned space bomber do that an unmanned bomber could not, and at far less cost? What was not actively discussed, at least publicly, during the tumultuous years that began in 1957 and extended through much of the cold war was this question, which went to the core of the military man-in-space issue: What, precisely, could human combatants do in space that machine-proxies could not do?

The Air Force tried hard to send its pilots to space, but it could never fully justify such missions as militarily significant, let alone necessary. Instead, it repeatedly fell back on patriotic abstractions. In January 1958, for instance, the Air Force ordered the Air Research and Development Command to "expedite" research on the manned space program because it was "vital to the prestige of the nation that such a feat be accomplished at the earliest technically practicable date—if at all possible before the Russians." But it was pure sophism. The Russians could never have afforded such a program, and there is no evidence that they would have pursued it even if they could.

An Agency for Space

The National Aeronautics and Space Administration's first administrator, T. Keith Glennan, was sworn in and handed his rolled-up commission by Eisenhower on August 19, 1958. Hugh L. Dryden, who headed NACA, was made the space agency's deputy administrator at the same ceremony. Glennan was a taciturn conservative who had served as one of five commissioners on the Atomic Energy Commission under Truman and who had temporarily left the presidency of the Case Institute of Technology in Cleveland for both that job and then to head NASA. It was hoped that Dryden's presence would help guarantee that NACA's subsumption by the space agency would be smooth.

Glennan had two essential mandates: to start a federal space agency from scratch and to use it to shape a space policy that would acquit itself honorably without indulging in what Eisenhower called reckless spending. Like his president, Glennan was a low-key optimist where space was concerned. He did not want to think of his country's space program as competing with the Soviet Union in some kind of race, and he believed implicitly that U.S. space technology would prevail in the long run.

The problem was that most of the scattered space establishment—certainly the military—did indeed think the country was in a race with the Soviet Union, the Soviet Union certainly thought so, and so did the news media and a significant number of their readers, listeners, and viewers. Whatever their true fears and patriotic motives, the scientists, engineers, and technocrats who were involved in space projects—many of them generals and admirals—reveled in racing with the Russians because it made them relevant and therefore both necessary and important. They also knew that they and their institutions, which had been suffering with anemic budgets since war's end, would thrive and grow by competing, not only with the Soviet Union but with each other. And by and large the media loved the idea of a race, too; sports of any kind was an immensely popular commodity because it was inherently dramatic and usually unpredictable.

Glennan, the low-key, congenial, midwestern conservative, was therefore destined to pull together and preside over an empire that heaved with restive, egotistical, ambitious, and jealous individuals—a few of them, like von Braun, prodigiously talented, and many others who were simply unexcelled in science and engineering—and get them to function together for a common purpose. Nothing, not even the famously vicious politics of academe, had prepared him for that task.

There would be no problem with the National Advisory Committee for Aeronautics thanks, in part, to Dryden and more fundamentally to the fact that it formed the true core of the new agency. On October 1, 1958, 177 NACA headquarters employees gathered in the courtyard of Dolley Madison House, across Lafayette Park from the White House, to hear Glennan proclaim the end of their forty-three-year-old institution and its metamorphosis into NASA. NACA's assets—nearly eight thousand people, three laboratories (the Langley and Ames Aeronautical Laboratories and the Lewis Flight Propulsion lab in Virginia, California, and Ohio, respectively), the High Speed Flight Station at Edwards Air Force Base, the Wallops Island sounding rocket facility, and a $100 million annual budget—were transferred intact to the new space agency. NACA's laboratories were given the more impressive title of research centers. Like NACA, NASA would occupy Dolley Madison House, but only until larger, more fitting, quarters could be built.*

That same day the White House issued an executive order that transferred the Vanguard project, its 150 personnel, and what was left of its budget from the Naval Research Laboratory to NASA. In addition, Army and Air Force lunar probe projects were turned over to Glennan, who immediately gave them back to the military services until he could get his agency up and running. All of that went relatively smoothly.

But neither the NACA facilities nor Vanguard and the embryonic lunar programs figured importantly as major components of the new space program. The really important elements would be the instruments for getting payloads into space in the first place: big rockets. And the most obvious place to look for big rockets was at the Army Ballistic Missile Agency at Huntsville. Whatever problems the Air Force was having with its own space program, its Ballistic Missile Division and the Strategic Air Command at least had unquestioned responsibility for the nation's ICBMs. They were therefore more firmly grounded than was the Army's missile program. More to the point, von Braun and the others were outstanding rocketeers, and Glennan was a smart enough administrator to go for the best talent he could find. He therefore decided to appropriate von Braun and the rest of the ABMA, and also the Army's successful spacecraft and tactical missile facility in Pasadena: Caltech's Jet Propulsion Laboratory.

* The building is at 1520 H Street, N.W. It was built in 1830 and was occupied by Mrs. Madison from 1837 to 1849. It was run by the exclusive Cosmos Club around the turn of the century and provided accommodations for the Wright brothers, among a cadre of other notables. (Van Nimmen, Bruno, and Rosholt, *NASA Historical Data Book*, Vol. 1, p. 250.)

Dryden, whose years in the capital had left him a veteran of the famous turf wars, warned his new boss that the Army would resist the takeover by every means possible. So Glennan took the precaution of getting Killian's and Eisenhower's approval before venturing into the Pentagon in late October. (A reconnoitering expedition to Huntsville the month before had convinced him that Killian was right about Medaris. Glennan would later recall that the general had treated him in a "cavalier" manner and that he was a "martinet, addicted to 'spit and polish,' never without a swagger stick, and determined to beat the Air Force.")

The Army, still flushed with pride because of the Explorer launchings and in beak-to-beak competition with both the Air Force and the Navy in rocketry and satellite technology, was in no mood to be raided by civilians who were seen as usurping military responsibility in the first place. Everybody in Huntsville knew that if NASA got the ABMA, the Army could kiss a role in space good-bye forever, and that seemed to be precisely the case. This was the Army's reward for bringing the German rocket team and its prize missiles to the United States and providing the scientists and engineers with a hospitable and creative work environment so they could launch America's space and ballistic missile programs. This was its reward for showcasing von Braun on television, in magazines and books, and on Capitol Hill to make space popular. This was its reward for bankrolling so many of the sounding rocket experiments and for allowing Van Allen's and other scientists' instruments to hitchhike beyond the atmosphere on the V-2s. Finally, and most ironically, this was its reward for coming off the bench to get the team back into the game after the opposition was allowed to score twice while the Navy dropped the ball—Vanguard—on television, yet.

Even his brush with Medaris did not prepare Glennan for his run-in with Wilbur Brucker, the secretary of the Army. After telling Brucker that NASA wanted both von Braun's team and JPL, Glennan was subjected to a tongue-lashing while two generals looked on "like wooden horses" without saying a word. The irate (in Glennan's words) secretary of the Army told the new NASA administrator that his service wanted to be helpful, but that he would not countenance the "breaking up of the von Braun team." Glennan told his diary that he had not realized until that moment how much of a "pet" of the Army von Braun and his operation really were. "He was its one avenue to fame in the space business," Glennan noted. So he countered by assuring Brucker that he understood that the Army needed the capability to work on its Pershing and Sergeant missiles, but that NASA needed von Braun and his group to work on the big lifters.

That hit a nerve. When the Air Force had finally wrested long-range ballistic missilery away from the Army and Navy in 1956, the Army decided to leapfrog the other services by leading the way in space exploration. It planned to do that by using its prize asset, von Braun and the other expatriate Germans, to develop a colossally powerful first stage. The ABMA's rocket team therefore began design studies for the advanced booster in April 1957: the big lifter that was first called Super Jupiter and eventually renamed Saturn. On August 15, 1958, the Advanced Research Projects Agency cut an order instructing the Army rocketeers to "initiate a development program to provide a large space vehicle booster of approximately 1,500,000-lb. thrust based on a cluster of available rocket engines." The Jupiter engine was certainly available, and putting eight of them together would produce the amount of thrust ARPA wanted. The order was historically important because it set the stage for von Braun and the others at Huntsville to fulfill their long-standing dream of producing a rocket that could propel men to the Moon and beyond.

And now, with Medaris, Brucker, assorted German expatriates, and everybody else, each thinking that his service had locked up that niche, along came Glennan out of nowhere to tell them that von Braun and his giant rocket were to be taken away, leaving the Army with tactical missiles whose stunted reach made them the tiddledywinks chips of ballistic rocketry. So Wilbur Brucker lost his self-control and, in a tirade, told Glennan emphatically that the rocket team was going to stay put and in place. By his own account, NASA's first administrator left the Pentagon with his tail between his legs. But he also knew there was no need to lose his own temper since he was going to win whatever Brucker said or did. Trumping the Army was straightforward and relatively easy because the transfer order came from the president.

JPL: "Spoiled Brats."

Caltech's Jet Propulsion Laboratory was another matter entirely. Pickering had made it clear that the lab wanted to get out of the missile business and into lunar and planetary exploration, and that it would welcome being courted by the new space agency. But since the Army was loath to see it go, too, Quarles argued forcefully that the Sergeant program in particular was too important to national security to be lost. Brucker and Medaris, desperate to buy time, had to use classic strategy: sacrifice a platoon to save a battalion. They wanted the ballistic missile agency far more than JPL, so a compromise was worked out in which JPL was turned over to NASA while it continued to work on Sergeant, and the ABMA stayed in the Army's fold for at least another year. Eisenhower approved JPL's transfer to NASA on December 3, 1958, to take effect on New Year's Day. Under the terms of the arrangement, the lab would remain the property of Caltech but would work for NASA under contract. That meant it would be in both camps, and in neither: the only outpost in the empire without civil service status. That arrangement was just one of the elements that would make the marriage a troubled one for years.

A deeper cause of conflict was the widely differing management styles and spirit of the two organizations and the temperaments of their leaders. Being an integral part of the prestigious California Institute of Technology (which was called "the campus" at JPL) and having spawned the nation's preeminent rocket research facility, invented its first homegrown ballistic missiles, and led the United States to space with Explorer 1, JPL tended to feel very good about itself. Some in Washington would say excessively good. Institutional pride, coupled with an informal lifestyle in keeping with southern California's benign climate and easygoing ways, made JPL free-spirited by the standards back east (as Californians called the place on the other side of the Continental Divide, the same way easterners referred to the Old World).

Jack N. James, an early JPL rocketeer, recalled years later how sweet life was with the Army in the 1950s, before NASA arrived. He reminisced like a henpecked spouse about the happy days before the vows were exchanged, salting his story with the kind of vignettes that are told and retold at college reunions about particularly wild fraternity parties. They were days, James remembered with dreamy nostalgia, when the engineers packed their own crates and rattled through the night on 850-mile pilgrimages to White Sands, where they fired their Corporals and Sergeants. "The Army seemed to love us," he said, recalling how one Corporal had rolled ninety degrees backward and almost hit Las Cruces. But there was no thought of punishment for the freakish accident: "no boards of inquiry, no investigations, no cost disallowances," he said.

James, a onetime roustabout, savored the memory of the time when somebody's office was completely boarded up—"entombed"—as a practical joke. And there was the time an engineer settled a long-overdue expense account by inventing a machine that showered the accounting department with a hailstorm of nickels. And there were the times when the guys on the Army's steering committee arrived at White Sands from Washington wearing loud Hawaiian-style shirts, just like Arthur Godfrey's, and got into food fights in the dining hall. And there was the time when. . . . They were great days, Jack James remembered wistfully; days when the people at JPL worked hard and played hard and were intensely proud of doing both.

Glennan came from a different world. He was born into North Dakota's long and bitter winters, grew up on Wisconsin's vast flatlands, graduated cum laude from Yale, directed the Navy's Underwater Sound Laboratory during the war, and eventually landed the presidency of the Case Institute. Now he was responsible for creating a cohesive national space agency, some of whose parts had yet to be conceived, let alone fitted into the whole. The job was like an orchestra leader's: he was under enormous pressure to assemble the musicians and then get them to play together under his baton while the audience—the president of the United States, Congress, the news media, and the nation—watched anxiously because most of them thought the Russians were ahead. But when Glennan looked past the Rockies at far-off Pasadena, he saw that Pickering and the others openly scorned what they considered to be a bunch of bloodless, lackluster, inexperienced civil servants back east. He saw to his dismay that JPL intended to stick to its own score, *solo voce.*

"[W]e won half a loaf and a lot of headaches," an exasperated Glennan would say about taking on the Pasadena operation. "JPL was not a well-managed laboratory—I doubt that it ever has been. Possessed of a bright and aggressive staff of young men, it lacked any sense of management techniques necessary for handling programs such as ours."

That may have been true from Glennan's perspective, but it did JPL a disservice. There were no food fights or nickel storms in Pasadena. The tactical missiles and the Explorers and the spacecraft that would follow had not been born in a fraternity house; they came from preeminent engineers and managers who deserved more credit than Glennan was willing to give. No one who met Bill Pickering would take him for someone who would tie a man's shoelaces together while he slept. The management styles were less different than their cultural contexts indicated. Yet whatever differences there were underscored the more fundamental, and inevitable, clash between unbridled creativity and the need for absolute institutional order. No one denied that both were vital to the space program, as they were vital elsewhere. The challenge was getting them to effect each other synergistically. Certainly no multifaceted program the size and scope of NASA could survive, let alone succeed, in the face of a chronically untidy organization. Glennan, Dryden, and the other organization men (as a popular book of the time called such corporate creatures) needed to consolidate NASA and point it in the right direction.

But what NASA was invented to do—send men to orbit Earth and touch the Moon before the Soviets did it (however officially unarticulated that was in 1958), launch daring and unprecedented voyages of exploration around the solar system, and collect information about the home planet and the universe around it with observatories in space and other advanced science satellites—rested squarely on the imaginations of the best scientists and engineers in the country and elsewhere. These people were to the space program what their engines were to the boosters; they provided the fire, the thrust, that made the beasts move. As every scientist and engineer in the world knew, it was implicit in the creative process that innovators needed special freedom to think beyond the bounds of the rules set by the technocrats. The outstanding aeronautical engineers of the period—Lockheed's legendary Clarence "Kelly" Johnson, Ben Rich, and North American's Harrison Storms foremost among them—

were dispositionally incapable of functioning inside the dreaded bureaucratic straitjacket. Johnson and Rich, who masterminded the most inventive spy planes in the world—the U-2 and the A-12 and SR-71 that succeeded it—made a corporate credo of small engineering teams that were left free to work unfettered by outside supervision.

The imperative for tight organizational structure and the spirit of laissez-faire science and engineering first collided head-on in early December, before the NASA-JPL union was officially consummated. It came when Glennan and Dryden paid Pickering a visit in Pasadena. "They gave me a little lecture on the fact that things were going to be different," Pickering remembered many years later, his voice edged with lingering pride and anger. "And that working with Ordnance, I was basically working with a group of colonels who didn't quite *understand* the technical business that I was involved in and [that's why] they were prepared to leave me alone. But now I was going to be working with a group of people who were primarily trained in the NACA, and they thoroughly understood the technical problems, and so on and so forth, and that Washington was going to look over my shoulder in a lot more detail than Ordnance ever did," he added. "I didn't quite say go to hell."

Glennan considered both von Braun and Pickering to be "outsiders." Yet he would call the former "persuasive and enthusiastic . . . a very able engineer . . . a born leader" who was able to command greater loyalty from his colleagues. Pickering, in Glennan's view, could not measure up to that. Years later, when Homer Newell asked him to go over the manuscript for a book on the history of space science called *Beyond the Atmosphere,* Glennan vented his frustration. When he read Chapter 15, titled "Jet Propulsion Laboratory: Outsider or Insider?" NASA's first leader couldn't resist penciling this on one of the pages: "JPL was the beneficiary of tolerance by NASA peers. . . . I suppose that the payoff of success is the final answer—but did it need to cost so much in $—in tolerance and accommodation by Newell and others? I again use the terms—able, extremely brilliant and spoiled brats."

While that may have been true, Glennan would have understood that institutional, or cultural, differences were anything but unique to headquarters and JPL. A show of solidarity to the outside world almost always obscured internal conflict. The test of wills with JPL was one of NASA's earliest and most apparent bureaucratic problems, but it was far from the only one. The early competition with the Army Ballistic Missile Agency and later, between the manned and unmanned programs, would amount to a permanent fact of life once a bead had been drawn on the Moon. So, for that matter, would competition between its own scientists and outsiders from universities and elsewhere, and between some of the centers within NASA itself. It was a cat's cradle of tension. But it is tension, after all, that holds a cat's cradle together.

Politics As Usual

Glennan, Dryden, and others among NASA's earliest leaders understood that, like the Department of Defense and other large federal institutions, their infant agency's survival prospects would be enhanced if they spread its political and economic base around the country. That meant acquiring or starting facilities in as many states—i.e., congressional districts—as possible. Nothing in that process would equal the battle with the Army over control of Huntsville or the stubborn test of wills with Pasadena.

With NACA's specialized laboratories as a goal, Glennan started the process of accumulating or building centers, a plan his successors would continue. Langley and Ames would work primarily on aeronautics, with the former extending aircraft development to spaceflight, and the latter conducting some solar system exploration in a program to be called Pioneer. JPL would specialize in lunar and planetary exploration, just as its stubborn leader wanted, in the process becoming a rival of Ames. Lewis would concentrate on researching and developing advanced propulsion and space power generation; its director, Abe Silverstein, was destined to play an important role in the manned space program. Wallops Island would continue as a sounding rocket research facility. The Flight Research Center at Edwards would test a wide variety of NASA aircraft, including the X-15 and a series of pumpkin seed–shaped "lifting bodies" whose fuselages would provide lift. The ungainly looking little craft would figure very importantly in the space program even into the twenty-first century. That much was in place when NASA congealed.

The space agency's first new installation, which had been authorized by Congress in August, was the Goddard Space Flight Center at Greenbelt, Maryland. Goddard would run most of the satellites—Earth huggers that did communication, navigation, meteorology, and remote sensing—with an initial contingent of specialists from the Naval Research Laboratory.

Eisenhower signed an executive order on March 15, 1960, turning the Army Ballistic Missile Agency into the George C. Marshall Space Flight Center. That honored his old comrade in arms, who had died the year before and whose massive plan to rescue Western Europe from the devastation of the war had made him the only professional soldier to win the Nobel Peace Prize. The Marshall Center would be responsible for designing, developing, and testing the heavy lifters that were going to carry Americans to space. Medaris and Brucker no doubt concluded that the day Ike issued the order—the Ides of March—perfectly fitted the deed.

By the summer of 1961, it would become clear that the space agency also needed a separate center to run manned spaceflight operations. On September 19, following consideration of twenty cities, James E. Webb, who by then had succeeded Glennan, announced that Houston won the competition and that a Manned Spacecraft Center would be built there to oversee the design, development, and manufacture of all manned spacecraft, select and train their crews, and run their missions. Given Vice President Johnson's interest in the space program, the selection of his state for the MSC reflected NASA's political adroitness.

Cape Canaveral, which Truman had authorized as a missile launching range in May 1949, would be run exclusively by the Department of Defense under several names until NASA occupied eighty thousand square acres north and west of the military launch sites on July 1, 1962.

The centers would be supplemented by smaller facilities: a computer center in Louisiana; a rocket-engine test facility in Mississippi; an electronics research center in Massachusetts; and a rocket test facility at White Sands, for example.

Headquarters itself scattered to a number of locations that provided more room than Mrs. Madison's place for the growing agency. More important, four new offices were established in 1961 that reflected the overall direction in which NASA was headed: Manned Space Flight, Space Sciences, Applications, and Advanced Research and Technology. By the end of that year, the space agency employed 18,454 full-time civilians. That was telling. Perhaps more telling was the fact that its growing number of contractors employed 58,000 people in 1961 and 116,000 the following year in the beginning of an expansion that would peak at 377,000 jobs in 1965. Its budget, a far cry from the $6 million to $7 million Glennan had at Case, would be $964 million in fiscal 1961 and would climb steeply. The new agency would run up a $32 billion bill during its first decade.

At the heart of the buildup was a commitment to send Americans into space and their robotic proxies to the planets as soon as possible. Eisenhower remained ambivalent about such goals to the end, insisting that there should be no race with the Russians, that the United States could easily hold its own in both war and peace, and that he had no intention of hocking the family jewels, "like Isabella," to send Americans to the Moon. (Actually, she had done no such thing; Columbus was financed through the Spanish Treasury.)

Glennan approved Project Mercury on October 7, 1958, just seven days after his agency formally came into existence, and Eisenhower signed off on it. But Ike emphatically did not sign off on Glennan's plan to send astronauts around the Moon; on squandering a sizable part of the nation's treasure on what he took to be a pointless excursion for the sake of mere propaganda. So he turned a thumb down on the Moon mission. But he was a lame duck whose frugality was widely believed to have gotten the country into a military and space deficit in the first place.

Whatever the old soldier said, however, the civilian space establishment was looking hungrily at the Moon. At the end of July 1960, 1,300 representatives of the government, the aerospace industry, and academic institutions attended the first in a series of planning sessions to chart the future of the manned program. Many came back the following month energized, and heard NASA's Space Task Group describe how von Braun's Saturn was going to send a manned spacecraft around the Moon on a flight that would directly support landings.

And Saturn's payload—the orbiter-lander—already had a name. Silverstein, the feisty dreamer who was brought in from Lewis to direct NASA's Office of Space Flight Programs in Washington (which at that point ran both the unmanned and nascent manned programs), pulled it out of a book on Greek mythology before the July meeting. "I thought the image of the god Apollo riding his chariot across the sun gave the best representation of the grand scale of the proposed program," he later recalled.

Foundation for a Global Village

The debate over whether there was any serious advantage in sending people to space was already well under way by the end of Ike's presidency. But there was no debate over the enormous potential space had to change the way people on Earth communicated with each other, dealt with the fickle and often deadly vagaries of weather and climate, and efficiently used the planet itself as a resource.

In February 1945, Arthur C. Clarke, a young and almost unknown Royal Air Force officer and scientifically literate space buff, published an article in *Wireless World* magazine explaining that a manned "space station" put into orbit 22,213 miles from Earth would keep circling the planet at the same speed as Earth's surface rotated and would therefore be parked over the same place all the time. Clarke went on to suggest that a spacecraft in such an orbit, which would come to be called "geostationary" or "geosynchronous," would be ideal for ricocheting radio messages around the world. John R. Pierce of Bell Laboratories in New Jersey soon came to the same conclusion and made the point that communicating by bouncing or relaying communication signals off satellites would be infinitely cheaper than laying endless cables across the floor of the Atlantic, which was first done in 1956 at a cost of $35 million, and have infinitely more capacity.

The first communication relay satellite was the Moon itself, which the Navy used in 1954 as a passive relay point and which was put into regular service for radio traffic between Washington and Hawaii in 1959. Passive communication satellites just bounced microwave beams from transmitters at one place on Earth to receivers on another. Active spacecraft, on the other hand, received and transmitted messages. Some were known as "delayed repeaters" because they could store messages on tape and then send them back down when they were within range of the receiving antenna. Once the rocketeers built the stairway to orbit, NASA mimicked the Moon itself by sending up a hundred-foot-diameter inflatable metallic and plastic ball named Echo 1 in August 1960 and a larger second one, Echo 2, four years later. Echo 1 was the first man-made object to allow live, two-way voice communication from space. Within a week of its launch, Bell Labs sent the first transoceanic satellite signal to the French National Communication Study Center in Paris.

The first active satellite was called Score (Signal Communication by Orbiting Relay Equipment), an Army transmitter that was shot into space on the same Atlas that broadcast Ike's Christmas message as the first "voice from space" on December 18, 1958. He almost certainly was less than merry at having his yuletide greetings sent from a modified ICBM. But the missile was lobbed into orbit, not in the spirit of the season, but in the spirit of demonstrating to Marshal Andrei Grechko and other Kremlin hard-liners that Santa's sleigh was not the only Western machine that could fly over the North Pole carrying presents. Score was soon replaced by two other military communication satellites: Courier and then Advent, a $170 million failure because more was expected of it technologically than it could give.

By the time Kennedy succeeded to the presidency in January 1961, it was taken for granted that communicating by satellites that ringed the planet would fundamentally change the world in several ways. It would educate and entertain many millions of people in Earth's most developed nations and deep inside its vast and often inaccessible backwaters. It would beam news and political messages past every checkpoint, wall, barrier, and guard post in the world. It would carry stock market and other financial information around the globe in seconds. The communication satellite would soon take its place with the printing press, the telegraph, radio, and television as a fundamental way to spread knowledge (if not virtue). Human civilization would come to depend upon this spacecraft as it would depend upon a central sensory system because, together with military reconnaissance, weather, and civilian imaging satellites, that is precisely what it was.

And it was also potentially profitable. Eisenhower understood that and worked for an international communication satellite system during his last days in office. But it was Kennedy who took decisive action to create a global union. He delivered a "Policy Statement on Communications Satellites" on July 24, 1961, that called for "all nations to participate in the communications satellite system, in the interest of world peace and closer brotherhood among peoples of the world." This led to passage of the Communications Satellite Act of 1962, which spawned the Communications Satellite Corporation, or

Comsat, the following year. Comsat was conceived as a profit-making private corporation with ownership shared by the public and communication carriers. That done, the United States and eighteen other nations formed the International Telecommunications Satellite Consortium, or Intelsat, in 1964, with Comsat representing America. By 1969, Intelsat would have sixty-eight members, most of them representing government agencies.

By the time Intelsat came into being, commercial satellite communication was already two years old. Telstar 1, developed by Bell Labs and AT&T, was shot into a relatively low elliptical orbit on July 10, 1962. It was a 170-pound faceted sphere covered with solar cells for power and wore an antenna around its waist. That day the first live television was transmitted overseas, from the United States to England and France and then back to the United States from France. Telstar quickly became so well known that its very name became synonymous with satellite communication the way Sputnik was taken to mean any satellite in the Soviet Union. Telstar was also the first significant private space venture.

Telstar 1 was followed by Telstar 2 and then by improved spacecraft called Relay and Syncom. Syncom 2 became the first communication satellite to be parked in a geostationary orbit when it was launched in July 1963, fulfilling Clarke's prediction. Soon, viewers around the world were looking at news and sports events over captions that boasted they were coming "live by satellite." Intelsat 1, or "Early Bird," which was sent up in April 1965, was the first in a long series of versatile communication satellites developed by Comsat in cooperation with Intelsat. Its successors, all squat cylinders that were also covered with solar cells, transmitted voice, data, facsimile, and telegraphy, as well as television.

The civilian spacecraft soon had established military counterparts. A series of interim communication programs followed Score and culminated with the Defense Satellite Communication System (DSCS), which was a constellation of twenty-six yard-wide satellites that became fully operational in June 1968. They in turn were followed by improved versions, as well as by a dedicated series of 4,000-pound Navy satellites that would start heading for geosynchronous orbit in 1978. The purpose of the Fleet Satellite Communications System, or Fltsatcom in militarese, was to make sure that Navy ships, planes, and submarines scattered around the world were connected to the national command authority. An Air Force Satellite Communication System (Afsatcom), designed to make certain that the far-flung bomber and missile crews also stayed in touch, would have no satellites of its own but would hitchhike on Navy and civilian spacecraft.

The Soviet Union had its own communication satellites, the most notable being Molniya (Lightning). Since the Russians were ten years behind the

United States in perfecting the technique of sending spacecraft into geosynchronous orbit, they shot Molniyas into very high elliptical orbits, where they arced as high as 25,000 miles over the USSR and as low as 300 miles over the Southern Hemisphere. That kind of orbit, which took twelve hours, would also be used by American radar ferret satellites, including one named Jumpseat. The international fraternity would honor Molniya by naming the orbit after it. The first Molniya, called Cosmos 41, left Tyuratam on August 22, 1964. It made it to orbit but would not communicate. Raduga (Rainbow) 1, the first Soviet geosynchronous communication satellite, would be orbited in December 1975 and would be followed by six other Radugas and by still other communicators called Gorizont (Horizon) and Ekran (Screen, as in TV screen).

The comsats started a true revolution. Applications Technology Satellites, for example, brought news and information to isolated communities. ATS-6, a one-ton communication satellite that was sent into geosynchronous orbit on May 30, 1974, was typical of a class of machines that were beginning to change the world profoundly. Not long after it went up, "Teacher," as it was nicknamed, began beaming evening college classes to schoolteachers in Appalachia. When night fell, its antenna was turned on the Rocky Mountains, where it broadcast vocational courses to junior high school students. With evening creeping across the West Coast, ATS-6's antenna was again moved, this time to the Pacific Northwest and Alaska, where a two-way television hookup was used to send medical assistance. When an Athapaskan Indian's face was ripped open from his mouth to his temple in a construction accident and he began to bleed to death, a health aide used "Teacher" to contact Anchorage, showed a local doctor the wound, and got immediate instructions on how to sew it up, "directing every stitch over TV," according to a former ATS program manager. "The patient did fine." The satellite went on to beam educational television programs on agriculture, health, and family planning to remote backwaters in India, Africa, South America, and elsewhere.

It was the same with weather prediction. By the mid-1960s, satellite weather maps became permanent fixtures in newspapers and on the evening television news. The weather satellites, starting with the drum-shaped Tiros (Television and Infrared Observation Satellite) in April 1960, were the first to carry tiny television cameras, and therefore human eyes, to space. They successively improved weather prediction. But they did a great deal more than that, however subtly. They also provided vivid descriptions of what was happening in faraway places—that people in Missouri were being drenched by thunderstorms, others in Arizona were baking, and still others in Stockholm were being buried in snow—and they therefore played their own part in drawing the global community a little closer together.

Meanwhile, the National Reconnaissance Office took the lead in developing the Pentagon's own weather satellites since it was understandably loath to have its KH cameras photograph blankets of impenetrable clouds (which covered Eastern Europe and parts of the USSR as much as 70 percent of the time in winter). So Charyk, who did not think the civilians could deliver what the National Reconnaissance program needed, and who was impatient with their seemingly endless review process, did what others who ran black systems were learning to do with their secret budgets. He simply set up a Defense Meteorological Satellite Program. The first NRO satellite left Vandenberg on August 23, 1962, and sent cloud cover data to a specially built receiving station in Florida the next day. Four of the spacecraft were in orbit within two years.

E Pluribus Unum

When he was the NASA administrator, Keith Glennan had "UNITED STATES" painted on his civilian agency's rockets to show that they represented the whole nation. And, of course, they did. But they came from the Army and the Air Force. Painting the space program in civilian colors was illusionary, as NASA understood. The machines that were going to space, whether to carry television cameras or people, were military at heart, regardless of which agency owned them or what was painted on the glistening launchers, lifters, boosters, or sustainers—really the missiles—that shoved them into orbit and beyond.

That fact was not lost on Bernard Schriever. "Today, as never before," he said with absolute honesty, "our military and civilian aims and actions are inseparable." Civilian and military branches were growing out of the same technological trunk, and that was the single most salient fact of the dawn of the space age.

A political clash of wills—a bitter battle over turf—would develop sometime in 1960 over control of Corona and the whole space reconnaissance system. Both sectors understood by then that space-based intelligence collection would have a direct bearing on how the enemy was gauged and therefore on how the entire defense posture of the United States was organized, particularly relative to the much-feared surprise attack.

By 1960, the Air Force correctly concluded that space reconnaissance figured to be an immensely important part of the whole national security apparatus and that control of it was properly in its own domain. So the airmen began to move into space reconnaissance aggressively. Cooke Air Force Base, which was built on a knob of land that stuck into the Pacific northwest of Santa Barbara, California, had already been selected as its satellite launch and test facility and renamed Vandenberg Air Force Base after General Hoyt S. Vandenberg. The airmen were also piecing together tracking and data retrieval stations. All of this was done with the understanding that the Strategic Air Command would become responsible for the operational control of all

WS-117L space reconnaissance and that intelligence collected by the satellites would be processed at SAC headquarters in Omaha.

The CIA thought otherwise. So did Eisenhower, who remained convinced that strategic intelligence had to be a national program, meaning that it needed to be run by and for the agencies that used it and that it had to be controlled by civilians. Knowing this, CIA Director Allen Dulles moved quickly to take over Corona with the concurrence of the White House. No sooner had Eisenhower approved Corona than Dulles chose Richard M. Bissell Jr., his special assistant, to direct it. The tall, bespectacled former Yale economist was already running the U-2 operation. Years later, he would vividly recall how Dulles had moved against the Air Force:

Dulles made it plain on that winter day that Corona would technically be comanaged by Bissell and an Air Force general, just the way the U-2 program was, Bissell said. But in reality, Dulles explained, Corona would be funded with CIA money and run by the agency. He told Bissell that he hoped the comanagement would mollify the Air Force, but whether it did or not, there was absolutely no question but that the CIA would control space reconnaissance. The Air Force, Dulles continued, would get the satellites into orbit, maneuver them and operate their photographic and other systems, and recover their film. Period. In a manner of speaking, he wanted the Central Intelligence Agency to have the primary responsibility for designing, building, and using the all-important spy craft, while the Air Force was relegated to driving the truck that carried it. That, the Princetonian and the Yalie undoubtedly reflected, was what Air Forces were supposed to do in the first place.

But there are few clear-cut victories in Washington's political environment. The Advanced Research Projects Agency was created within the Department of Defense that same February as a response to the Soviet challenge by stimulating research and development in highly advanced military technology. ARPA's responsibilities included the development and control of all military space projects, into which category the Pentagon insisted Corona fell. An ARPA directive on February 28 therefore in part salvaged the situation for the Air Force, at least for the time being, by dividing WS-117L into two parts: the CIA was to get Corona, while the Air Force was given responsibility for both Samos, the rapid readout spy satellite, and Midas, the missile warning system.

Whatever Dulles's intention, Corona would actually evolve as a CIA-ARPA program with Air Force participation. The CIA's own history of the program would note that it actually lacked central direction and a clear delineation of authority: "The vague assignments of responsibilities caused no appreciable difficulties in the early years . . . but they later (1963) became a source of severe friction between CIA and the Air Force over responsibility for running the program."

Friction was not the word for it. The battle over control of the reconnaissance satellite system and its "product" would grow so nasty that it became a part of the folklore in the space program's dark inner precincts. At the heart of the matter was the question of whether the military would or would not be permitted to choose what would be photographed and analyze the pictures that came back. From its standpoint, that was precisely the way it should have been done, since military officers were in the best position to judge what their opposite numbers were up to. But civilians, including Eisenhower, his top advisers, Dulles, and others at the CIA, worried that if the military controlled space-based intelligence collection, a whole garden of monsters would be found in the Soviet Union and elsewhere, justifying ever-higher weapons appropriations.

8

The Greatest Show on Earth

Nikita Khrushchev cared so little about the process of spaceflight that he was never known to attend a launch, let alone ponder worlds beyond his own. He was no more inclined to learn about rocket propulsion than he was to absorb details on the working of tanks and submarines. The first secretary was too good a politician for that. He knew precisely what he had to know: that tanks and submarines were necessary to protect the U.S.S.R. and had to be used for that purpose. Generals and admirals were paid to know the details.

It was the same with space. If an admiral wanted to look at octopuses or whales while under water in a submarine, that was his business provided he understood that the vessel's only real use was to fire torpedoes and missiles at the enemy. And if Sergei Korolyov and the other rocket people wanted to dream about trips to the other planets, that was perfectly acceptable, too, provided they understood that the requirements of national security had nothing to do with Jupiter. And where Korolyov, Valentin Glushko, Vladimir Chelomei, head of the powerful OKB-52 design bureau, and the others were concerned, the security of the state rested on two pillars: machines—long-range missiles, launchers, and spacecraft—that would enhance the safety of the Union of Soviet Socialist Republics; and propaganda that would enhance its appearance and strengthen its political underpinnings. That is why Khrushchev lavished his best rocketeers with funding, power, and personal amenities. And because Korolyov and his colleagues, like von Braun and his, were driven to send men and machines across the solar system because that was the ultimate challenge, they eagerly accepted the largesse by concluding, correctly, that there was no other way.

Yet there was an insidious price to pay for Khrushchev's favor. Propaganda, as all the world knew from Sputnik, came from establishing firsts, from setting records. If that was true, then little or nothing was to be derived from repetitive missions, and they were therefore relegated to second place by Korolyov and other chief designers unless there was a military necessity for them. But that was anathema to many other scientists and engineers whose grail was the testing of theories by follow-on experiments. Whether in the biology laboratory or in outer space, it was going back again, and then again, and then yet again, performing the same experiment more than once, that distilled raw data into knowledge.

Korolyov, on the other hand, was far more interested in keeping his design bureau and the program it supported vigorous and well funded. As was the case in the United States, science therefore took second place—a distant second place—to spectaculars; it was strictly a passenger, never a driver. Khrushchev was widely credited during the cold war with pushing Korolyov to perform one spectacular feat after another, often at a killing pace. That was more apparent than real, however. The chief designer was very much his own taskmaster and, in fact, managed to push Khrushchev by using the spectaculars to generate support. The two *Pravda* hacks who ghost-wrote Yuri Gagarin's shamelessly saccharine autobiography, *Road to the Stars,* quoted Korolyov as telling his cosmonauts that "Soviet space exploration was Khrushchev's favorite child." If Korolyov really said that, it was a self-serving distortion, since Khrushchev's favorite "child" was necessarily the intercontinental ballistic missile, which was needed to hold the West at bay. Besides dulling fundamental science, extravaganzas unnecessarily taxed the entire space establishment since forced, incessant innovation is by definition exhausting. The parable of the tortoise and the hare is not inappropriate where competition between the two superpowers was concerned.

Khrushchev did use his rockets to enhance accomplishments and mask problems in foreign policy. The fact that all of the rocket work had to be done anonymously would have been galling to Korolyov, who was not allowed to travel outside the country, publish in foreign journals, or even have the pleasure of entertaining Hermann Oberth at the relatively ornate two-story dacha he was given on a wooded acre in a northern Moscow neighborhood. The glory went to Khrushchev and to the cosmonauts whom Korolyov and the others worked feverishly to send to space, and to that most amorphous of entities, "the People." So tightly leashed was Korolyov, in particular, that even his occasional articles in *Pravda* were signed "Konstantinov" or "Professor K. Sergeev." A decade after Sputnik, one insider who defected would write that Korolyov shared "the same fate as H. G. Wells's *invisible man.*" Ten months after it went up, Khrushchev heaped praise on his rocket men in a speech in Bitterfeld, East Germany, saying that their names would be inscribed in gold in an obelisk so that posterity would know them. But for the time being, he added, it was imperative to "protect their safety from hostile agents who might be sent to destroy these outstanding people, our valuable cadres." But there were deeper reasons for making Korolyov and the others as classified as their designs, and they had to do with the Byzantine nature of Russian politics.

Korolyov was officially invisible. Unofficially—what came through Western intelligence services' transoms—he was portrayed as larger than life. While he was technically a nonentity, many both inside the country and elsewhere knew him to be a very definite entity. They knew him to be, or thought they knew him to be, the mysterious, ubiquitous science and engineering prodigy who fathered the Soviet space program. The Invisible Man was referred to with abiding reverence as the "Chief Designer" the way leaders of other orders were called "High Priest" or "Paramount Leader" or "First Among Equals."

There is no question that Korolyov was a highly gifted, imaginative engineer who had an instinctive, keenly developed grasp of the mechanics of both rockets and spacecraft. But the end of the cold war and the coming of perestroika shed new light on the enigmatic chief designer. A closer look shows—and it takes nothing away from him to say this—that his greatest strength was not the creation of space systems themselves but managing their creation and inspiring others to follow his vision.

Sergei Korolyov's real genius lay in his ability to conceptualize daring missions, spot and attract exceptional talent, and galvanize it to generate remarkable engineering. What was not generally known during the cold war, even among those who were highly knowledgeable about the Soviet space program, was that every design bureau had a chief designer in the same way that every Western corporation has a president. There were easily more than a hundred chief designers, and most of them routinely took personal credit for their bureaus' successes while passing blame for mistakes or failure on to their underlings or to the suppliers of components. Mikhail Tikhonravov, to take one outstanding example, was a profound innovator among Korolyov's designers who never received the credit due him. This is not to say that Korolyov was not outstanding, for he surely was. Rather—and this is borne out by a number of men who worked in the Soviet program—it is to say that he reached the pinnacle of his profession more by dint of managerial skill and charisma than by innovative engineering.*

* Sergei Khrushchev made that point in a conversation with the author. At least two other knowledgeable participants in the Soviet space program have made the same point in oral histories. Viktor V. Kazansky, a former deputy director of the Scientific Research Center for Space Documentation of the U.S.S.R., likened Korolyov to a "coach to be rallied around . . . the guiding light," but added that his engineering achievements tended to be overemphasized in biographies to the point of overshadowing the work of others. Aleksandr A. Maksimov, head of the Space Department of the Ministry of Defense, said that Korolyov "was not only a scientist and designer, but also a great organizer . . . who knew how to put himself and others to work." (Rhea, *Roads to Space,* pp. 52 and 83, respectively.)

Structure of the Soviet Space Program

As it was to evolve from the mid-1960s, the Soviet space program would be shaped like a pyramid, with supreme authority held at the top by high-level Communist Party officials. This simplified the decision-making process. It also created pervasive corruption. Laid on top of a hopelessly inefficient economic system in which there were no real performance incentives, the endless political maneuvering and institutionalized deception hobbled even the best of the Soviet rocket men.

The Soviet space program differed from the American in at least two fundamental ways. For one thing, all executive decisions came from the top of the pyramid: from joint decrees issued by the all-powerful Politburo of the Central Committee of the Communist Party of the Soviet Union and the Council of Ministers. Unlike Congress, the legislative branch—the Supreme Soviet—was anything but supreme and in fact merely rubber-stamped all decisions having to do with space and military affairs. What's more, while the civilian and military space programs were at least technically separate in the United States, there was basically no such division in the Soviet Union. That meant ministers of defense and generals would effectively control space science, which they thought was irrelevant, and the nascent manned Moon program, which they thought was tremendously wasteful. "Society was militarized," Korolyov's eventual replacement, Vasily Mishin, noted many years later, "and military interests were not disputed." As a consequence, he added, the program to send cosmonauts to the Moon was funded with "three billion wooden rubles."

Responsibility for running all space and missile programs rested with the deviously named Ministry of General Machine Building, or MOM, which was right under the party pinnacle. MOM was one of nine military-industrial ministries supervised by the Defense Division of the Central Committee and the Military Industrial Commission of the Council of Ministers. In effect it was the space program's customer, financing agency, and producer. Next, all spacecraft and launch vehicles were procured by the Ministry of Defense's Directorate of Space Means, with funds for them going directly from MOM or one of the eight other ministries to the industrial contractors. The party's control of the development and production of space systems was so tight that even the military itself got only an advisory vote (though the generals still counted for more than the civilian designers). This meant that the nation's military space forces usually had to accept whatever manufacturers could supply. And unlike in the United States, the Air Force was only marginally involved in the space program, mainly to select and train cosmonauts and rescue them after missions. Understandably, pure science—solar system exploration, for example—got short shrift from the Ministry of Defense. What the scientists did get came from the Central Committee, which tried to use them and their achievements for political advantage.

Since the party was technically infallible, there was no real independent analysis of the costs or technological consequences of whatever projects were proposed, and party directives to proceed with them were almost irreversible. That being so, there was an overwhelming emphasis on connections by special interests and, consequently, endless lobbying. The Ministry of General Machine Building itself had essentially no interest in projects that benefitted either science or the economy. Instead, it saw its own vital interest as raising its enterprises' workloads, increasing the scale of financing, and creating and maintaining as much independence as possible for itself and its industrial base. It was MOM that controlled international cooperation in space research, such as it was, and proposed the appointment of general and chief designers to the Central Committee.

And it was MOM that controlled design bureaus such as Korolyov's, pure research facilities like IKI, the Space Research Institute, which was founded in 1967, and state companies that actually produced the hardware. To further complicate matters, there were separate design bureaus for the manned Moon program (Energia) and for the unmanned (first Babakin, named after its chief designer, Georgi N. Babakin, and then Lavochkin, named after his successor, S. A. Lavochkin). The cream of the chief designers formed their own elite Council of Chief Designers.

Korolyov himself headed OKB-1, a large special design bureau that had overall responsibility for both launcher and spacecraft design. Glushko's Gas Dynamics Laboratory in Leningrad, also known as OKB-456, specialized in powerplants. Chelomei's OKB-52, which was started in Moscow in 1959, competed with OKB-1 in the development of launchers and long-range missiles. Two years after it came into existence, OKB-52 would be ordered by Khrushchev to develop a launcher and spacecraft to carry a cosmonaut around the Moon. Mikhail K. Yangel, a rising star, was given his own rocket design bureau. So was Vladimir Barmin, a veteran of the trip to Germany in 1945 and now a prestigious academician, whose OKB-586 designed ballistic missiles. The giant bureaus were in turn supplied with components—communication equipment and propellant tanks, for example—by smaller bureaus that were the equivalent of subcontractors.

All of the design bureaus were encouraged to compete in an environment that superficially mimicked capitalism. But obviously free-market philosophy had nothing to do with it. Rivalries were fostered for at least two reasons. First, it was assumed that pitting individuals and bureaus against each other would produce the best possible products. More insidiously, dispersing the creative and industrial centers, keeping their leaders anonymous, and forcing them to expend physical and psychic energy competing with each other was calculated to prevent any of them from gaining undue and potentially dangerous influence. This had been Stalin's practice, and no successor would be bold enough to change it until Mikhail Gorbachev came to power. But by then it would be too late.

The system was subtly but relentlessly oppressive. It thrived on secrecy, punished failure, and rewarded success with ubiquitous plastic medals, better amenities, and increasing intimidation. The state security apparatus permeated society on all levels, and the space program was no exception. Almost everyone had either spent time in a gulag or knew of someone who had, and they certainly remembered Stalin's and Beria's rule by terror. Colonel-General Alexander A. Maksimov, a retired head of the nation's space forces, had vivid memories of the time when "the 'Black Raven' police van made frequent late-night trips around Moscow carrying arrested people to Lubyanka and various other prisons." Many others remembered as well.

Everyone understood profoundly that existence was tenuous and that whatever privileges they enjoyed depended upon the graces of a Central Committee stuffed with scientific and technological illiterates, some of them bumpkins in suits and many as vicious as they were insecure. Privilege required no less. No wonder feuding became the norm, political maneuvering was rampant, successes were exaggerated and often snatched from the underlings who deserved credit for them, blame for mistakes was routinely placed elsewhere, and obsessive secrecy was used as a security blanket. The result was a degree of institutional disarray, of bureaucratic fraying, that could not be imagined by Westerners who believed that the iron discipline imposed by successive Soviet dictators extended throughout the system and made it efficient.

Roald Z. Sagdeev, who headed the Space Research Institute long after the events described here, explained that the iron-fisted secrecy coming from the Kremlin—a secrecy Korolyov hated because it impeded his work rather than helped it—did not have to be routinely enforced because years of repression provided the medium for it to take hold on its own. "[T]he [aerospace] authorities were reluctant to put themselves at risk of failing in the eyes of the press and the public," he would write. "This was not simply imposed by the government or by its agencies. To a very large extent, it was cultivated by the industry itself. In the event of failure, secrecy could provide protection, a kind of escape." Secrecy, the physicist added, "had very long, very deep historical roots in the deformation of the psychology of the people. Those who worked in this area told me that, to some extent, they felt even happier in such an atmosphere of secrecy."

The feuding was low-key but constant. Korolyov and Chelomei, onetime colleagues, became bitter rivals. Though the former could be as charming as the moment required, he tended to be relatively outspoken and given to showing anger or frustration in plain, idiomatic Russian. Chelomei, a very gifted scientist, was far more controlled but no less canny. He responded to the accolades and awards bestowed on Korolyov for producing the R-7, for example, by hiring Sergei Khrushchev, the son of the first secretary, who was a talented missile guidance-system engineer in his own right. The younger Khrushchev himself would later explain that differences between the two men extended across the full spectrum of their lives. Whereas Korolyov did not care much for fine clothes or furnishings, Chelomei was a bit of a dandy. "He looked like an artist, dressed well, wore natty ties," Khrushchev recalled. "His office had elegant furniture, clean carpets. He spent two months designing his own desk. . . . He was very cultured, liked having associations with painters, writers." That was emphatically not Sergei Korolyov.

Korolyov and Glushko also had a serious falling-out, at least in part over the type of fuel to be used to propel a rocket to the Moon. Mishin, who was Korolyov's deputy at OKB-1, still bristled years later at what he believed was Glushko's mind-boggling stupidity. "The most promising fuel for injecting heavy satellite payloads into orbit is liquid hydrogen," Mishin explained, echoing Tsiolkovsky and innumerable others. "Its main detractor," he added scornfully, "was Glushko, who pronounced the following immortal words: 'Liquid oxygen is not the best oxidizer by far, and liquid hydrogen will never be widely used in rocket engineering.' " Liquid hydrogen was far more powerful than the hydrazine derivatives and nitrogen tetroxide favored by Glushko, but it could be stored in a rocket for only about forty-eight hours, whereas Glushko's fuel could be stored almost indefinitely. That meant it was perfect for ICBMs. The vituperative duel sent Glushko into Chelomei's camp and both of them into the arms of the powerful Ministry of Defense, which had long-range ballistic missiles as its top priority.

Almost everyone was drawn into one intrigue or another, hampering the work of very capable men such as Yangel, Tikhonravov, Mishin, Leonid Voskresensky, Mstislav Keldysh, and many others. As Winston Churchill put it in another context, the chief designers tended to behave like dogs fighting under a carpet. Indeed, the wonder was that they were able to accomplish what they did in a system so laden with gross politics and institutionalized inefficiency. In those circumstances, what they achieved was remarkable to the point of being almost miraculous. They were preparing to send people into orbit within a year of Sputnik. And, as all of the insiders knew, the only logical place to go after orbiting Earth was to the Moon.

The Cosmonauts

Vostok, which means "east" or, more to the point, "upward flowing," as in the Sun rising in the east, was the first man-carrying spacecraft. In keeping with Korolyov's illustrious legend, he has generally been credited with designing Vostok, too. In fact, it was designed in 1958, mainly by Oleg G. Ivanovsky, one of the chief designer's anonymous engineers.* Ivanovsky created a self-contained spherical capsule—a large ball—attached to a conical equipment module that held telemetry and other communication equipment, oxygen and nitrogen gas containers, several antennas, a retro-rocket, and related hardware. A ring of round oxygen tanks circled the section where the sphere and the cone were joined. The idea was to send both of them up together, with the cosmonaut strapped to a seat in the ball and the equipment module attached behind him. When it was time to return to Earth, the equipment module would be jettisoned and the ball—now a descent module—would plunge into the atmosphere with its human cargo and eventually pop open parachutes.

Vostok and those who would ride it were developed simultaneously. The decision to train cosmonauts was made in the autumn of 1959, several months after their American rivals made their public debut, when twenty of them started training at the Star Town facility outside Moscow on March 14, 1960. They were soon reduced to twelve. The survivors of the elimination process were called the "Star Town Twelve" in an evident effort to create esprit de corps and good press. It was a star-crossed group: Yuri Gagarin, German Titov, Andrian Nikolayev, Pavel Popovich, Valeri Bykovsky, Vladimir Komarov, Pavel Balyayev, Aleksei Leonov, Boris Volyanov, Yevgeni Khrunov, Georgi Shonin, and Viktor Gorbatko. One would be the first human to orbit Earth; one would be the second; one would take the first "walk" in space; one would marry another cosmonaut (Valentina Tereshkova); two would be the first to fly in formation; and one would be the first to die on a mission.

The criteria for cosmonaut selection were basically the same as those used by NASA to pick its astronauts. The men had to be the right size for a spacecraft's cramped quarters, be comfortable flying, and possess excellent health, perfect reflexes, emotional stability, and be capable of reacting coolly and decisively in emergencies. And as potential heroes, they had to be reasonably attractive and personable. Individuals with those qualities existed throughout society, but they were most commonly found in the ranks of military pilots, particularly test pilots. Given the speed with which both nations were prepar-

* See, for example, Newkirk, *Almanac of Soviet Manned Space Flight,* p. 7. The Russian authors of *Cosmonautics: A Colorful History* gloss over Ivanovsky by reporting only that the spacecraft came out of "Korolev's bureau" (p. 30).

ing for manned spaceflight as the decade of the 1960s took hold, it was logical to go to the pool of military pilots for candidates.

Another implicit reason for using military pilots was that they were trained to be absolutely obedient. While intelligent candidates were sought, intellectuals—ponderers—were scrupulously avoided. It was one thing for a cosmonaut or an astronaut to think clearly about escaping from a dangerous situation or use his initiative to override instructions from controllers if he knew better. But it was quite another to question the nature of the enterprise itself or become distracted by such technically superfluous and even debilitating ideas as whether the space race was stupid. Both sides wanted their spacemen and -women to think. At the same time they demanded political allegiance. If racing to space was a progressive nationalistic endeavor, ruminating about one-worldism or the dangers of militarism, the inherent futility of the race itself, the real possibility that the spacefarer would not make it back alive, the environmental damage caused by rockets, the problem of siphoning off precious resources that might have been better directed to the needy, or questioning whether God wanted His creatures to head for other worlds amounted to pouring glue on the wheels of progress. Neither system would tolerate that.

Sergei Krichevsky, a cosmonaut who had yet to fly, asserted years later that he would probably never be sent to space because he thought about things that were not considered correct for men of his profession. He had published one paper on the possibility of Earth being struck by an asteroid that did cataclysmic damage and another on the eventual need for human rights in space. He even published a short volume of poetry at his own expense. A number of highly regarded astronomers and other scientists were at the time expressing real concern about the possibility of a monster rock colliding with Earth, and they remain concerned. Yet the cosmonaut maintained that such work was out of bounds for him and his comrades and that he would therefore be passed up for a flight assignment.

"Soviet and Russian officials traditionally oppose open and independent discussion by cosmonauts of the pressing problems of activity in space," Krichevsky would reflect after the cold war, adding that his American colleagues were bound by the same restrictive rules and that both groups needed to be affiliated in a professional organization. "The historic monopoly of those who run the space programs, and the absence of effective, independent expertise [by the men and women who go to space], make it almost impossible for cosmonauts to significantly influence the process of improving spaceflight techniques or scientific programs," Krichevsky complained. "[Effective] influence is possible only if the cosmonaut is not only competent, but free to state his opinions."

Valentin Lebedev, a veteran cosmonaut who spent a record 211 days on *Salyut 7* in 1982, would agree. "I was worried about how the doctors would accept my frankness about my physical condition and mood," he wrote, referring to the risks of keeping an honest diary of his experience on the space station. "I am an active cosmonaut, and such frankness could count against me later. My frankness had already gotten me in trouble before," he added without further explanation, "and here I was preparing to take the same risk again." In the end Lebedev decided to keep a true record of his experiences since, as he put it, glossing over problems would have made the journal useless.

And there was another unmentioned caveat, at least for the first Soviet citizen to reach space. Political considerations required that he be a purebred Russian proletarian. Whatever the official line about the country's union of republics and their collective internationalism, the man destined to ride the first Vostok to space had to be a full-blooded Russian—not a Ukrainian, Byelorussian, Kazakh, or any other constituent nationality, and certainly not an Armenian, Mongol, or Jew—with roots that went back at least to his grandparents. Furthermore, the successful candidate had to come from a family of workers and peasants, since Khrushchev had plans to send the cosmonaut on an international propaganda tour designed to prove, in the words of one insider, that "only the Soviet system could provide the son of a worker or a peasant with a ticket to outer space."

The winner was a fighter pilot who had been born in a plain log house in Klushino, a village in the Smolensk region, and whose father was not only a peasant farmer and carpenter but a gloriously poor one at that. "I was born into a simple family that differs in no way from millions of other working families in our land of socialism," said his ghostwritten autobiography, which was meant to be deeply inspirational. "My parents are plain Russian people for whom the Great October Socialist Revolution had opened up the way to a new and promising life." His name was Yuri Alekseyevich Gagarin. Lebedev would note two decades later that Korolyov's face brightened when he laid eyes on Gagarin. "Later he [Korolyov] said that this one with the nice Russian face, the wonderful smile, the sharp mind, the peasant origins would be a perfect candidate to be the first cosmonaut." Gagarin's backup, German Titov, was a schoolteacher's son and also a pure Russian from a small town. Titov would be second to go to space. After that, candidates would be chosen specifically from the other republics to demonstrate the solidarity of the union.

From start to finish, however, cosmonauts were carefully depicted as Slavic-socialist Frank Merriwells: "simple yet wise, possessed of a steely disposition yet kind, working in the spirit of the collective, and," according to one knowledgeable observer, "being, above all else, an ordinary Soviet citizen."

That is, even the lowliest worker with Soviet upbringing and the right attitude could be a cosmonaut: a traveler in the cosmos.

The Wings of Mercury

Meanwhile, an amorphous manned space program was coalescing in the United States, too. Technically, NASA Administrator Glennan had presidential approval only to send Americans into Earth orbit, and even that was Ike's begrudging response to the Soviet spectaculars and the public clamor they provoked. Yet no one (with the possible exception of Ike) believed that it would end there: that a massive and expensive organization would painstakingly be put in place for the singular purpose of sending lone Americans into Earth orbit and, having accomplished that, disband. No. What happened, in effect, was that a kind of guerrilla war started between the space establishment and the White House.

The insurgents tended to favor brush cuts, sideburns, horn-rimmed glasses, and J. C. Penney shirts with plastic pen and pencil holders to protect pockets from the stains that the distracted habitually inflicted upon themselves. They had engineering degrees from places such as Purdue, Minnesota, and Georgia Tech and carried slide rules, countless sets of equations on everything from specific impulses to orbital injection angles, and RAND's authoritative *Space Handbook.** And they kept refining their plans, obstinately pushing for change in increments, as they impatiently waited for a more sympathetic regime. They would not have long to wait.

The Mercury program had started coming together before there was a NASA. Its incubator was at Langley's Pilotless Aircraft Research Division, or PARD, where a gifted and earnest young engineer named Maxime A. Faget headed the Performance Aerodynamics Branch. PARD tested rocket models for aerodynamic qualities beyond the speed of sound. One day in early March 1958, Faget, Langley's deputy director, Robert R. Gilruth, and some other engineers met to discuss the direction NACA should take in designing a manned satellite. They decided to go ballistic. A satellite that went on a ballistic trajectory—that was tossed into orbit like a football—would benefit from relatively simple stabilization and control requirements and could be brought

* The book, published in 1958, was a compendium of factual material on rocketry and the space environment presented in readable form. It took for granted (as did General Boushey and others in the Air Force) not only that crews would be sent to the Moon but that establishing a lunar base was essential for the manned exploration of the solar system. "Bases, more or less permanent in character, will be required at some point in an expanding space program," it said in a matter-of-fact way, adding that the facilities would have to have artificial environments similar to that of space stations, which it also described. (Buchheim, *Space Handbook: Astronautics and Its Applications,* p. 158.)

back from orbit by firing small retro-rockets forward to slow it down for descent. Maneuvering in the vacuum of space would be accomplished by using tiny thrusters that spat high-velocity gas in short bursts. The "black rockets"—the X-15s—that were to start flying the following year were expected to demonstrate the successful use of thrusters in their own very high altitude forays. The pilot of the small cone-shaped ballistic shell would fly with his back against its wider, flat end for re-entry so he could tolerate higher gravity forces, or in the engineering argot, g forces. And a special shield would absorb the terrific heat buildup on the other side of his contoured seat.

Faget and two colleagues, Benjamin Garland and James W. Buglia, delivered an historic paper at a high-speed aerodynamics conference held at NACA's Ames Aeronautical Laboratory on March 18–20, 1958. The document, "Preliminary Studies of Manned Satellites, Wingless Configurations, Non-Lifting," laid out the Langley design as it was shortly to be incorporated in the Mercury capsule. Although it could not have been known at the time, Langley's "cannonball" approach would doom the Air Force's prized X-20 Dyna-Soar program, then just getting under way, and other early attempts to get wings into space. Another paper, written by four Ames engineers, described a so-called lifting body that looked like a pudgy, wingless pumpkin seed that would glide out of orbit under its pilot's control and land on wheels at air bases instead of floating down by parachute over the high seas. Progressively improved versions of the lifting body—named because its body created lift like some fat wing—were tested at Edwards Air Force Base. The first, called the M-1, was made of plywood and towed into the air by a souped-up Pontiac. The lifting body would give way to the ballistic capsule, at least temporarily, because capsules could be produced faster and required less of their pilots. But its bloodline would go directly to the shuttle and the space plane.

In the autumn of 1958, Mercury was still more mandate than mission, since NASA itself was an unwieldy conglomeration of transferred people, places, and projects. Glennan began to shape the program out of an amorphous bureaucratic mass as early as September with the creation of a NASA-ARPA Joint Manned Satellite Panel. Six of its eight members were from the nearly extinct NACA and only two from the Department of Defense, which reflected the White House's insistence that the Pentagon's Advanced Research Projects Agency help the space agency with the manned program, not comanage it. The separation was illusory, though, since there was already permanent technical cooperation between the soldiers and civilians.

Using the work of thousands of scientists and engineers who had shaped the technology during that decade, the panel issued a report, "Objectives and Basic Plan for the Manned Satellite Project," in late September. The objective of the program was straightforward: to orbit a man around Earth and recover him while investigating human capabilities in space. The mission was defined as using a circular orbit that was high enough to allow a twenty-four-hour flight (the actual number of orbits was not specified) and a return by parachute to land or water. The vehicle was described as Langley's ballistic capsule, "designed to withstand any combination of acceleration, heat loads, and aerodynamic forces that might occur during boost and re-entry or successful or aborted missions." The report also outlined in general the life support, attitude control, retrograde, recovery, and emergency systems, and described the guidance and tracking, instrumentation, communication, ground support, and test program requirements. The plan to send Americans to space was in place by the end of the first week in October 1958, the week NASA was born.

The men who were to make the plan work came from around the empire, but their critical mass still centered at the Langley Research Center. That month, thirty-three people from Langley, twenty-five of them engineers, met to form the nucleus of what would soon be called the Space Task Group. The STG was led by Gilruth and included Faget, who would himself figure prominently in America's space program.

Having decided what they wanted in the way of a capsule, the Space Task Group mailed its preliminary specifications to more than forty prospective firms by the end of the month. Thirty-eight companies sent representatives to a detailed briefing conducted at Langley by Faget and other members of the STG on November 7. Eleven responded with bid proposals by December 7. Eight of them, including Convair, Lockheed, Martin, McDonnell, and North American, were already doing feasibility studies for an Air Force manned satellite.

On December 15, Glennan, his staff, and representatives of the military and industry were briefed at Dolley Madison House by von Braun and two colleagues from ABMA, Ernst Stuhlinger and Heinz H. Koelle, on how the ballistic missile agency might play a role in the new space program. It soon became apparent that while the ABMA's Redstone-Jupiters could be used to get individual astronauts to orbit, the three men from Huntsville had a great deal more on their minds than Mercury. They told Glennan and the others that they wanted to create a few very powerful, versatile launchers for the purpose of landing Americans on the Moon. Having been cheated out of the glory of sending the first man-made object into space because of Eisenhower's obsessive penny-pinching, von Braun had no desire to follow the Russians to the Moon.

Koelle offered Glennan the cheese, saying that perhaps by the spring of 1967 "we will have developed a capability of putting . . . man on the Moon. And," he teased, "we still hope not to have Russian Customs there." He went on to say that Huntsville's launch vehicle plans fit perfectly into NASA's own emerging manned program and that there could be no astronauts on the Moon until manned operations in Earth orbit had been established. It was right off the pages of the *Collier's* series, with one step locked into the next. No one at that meeting doubted the technical feasibility of sending men first into orbit and then to the Moon.

Getting down to specifics and unknowingly echoing Tikhonravov, von Braun said that the multiengine system that worked for large aircraft was applicable to launchers: rocket clusters. There were five possible ways to hit the Moon, he added: direct ascent, involving a straight shot, and four kinds of rendezvouses en route. He ruled out the first because, even assuming liquid hydrogen–liquid oxygen upper stages would be used, "you would need a seven-stage vehicle which weighed no less than 13.5 million pounds." Von Braun found such a launcher forbidding, and instead suggested using fifteen smaller rockets to join up in Earth orbit like prairie schooners in a wagon train.

Not uncoincidentally, von Braun had the perfect rocket in mind: his own. "The SATURN," one ABMA report said immodestly, "is considered to be the first real space vehicle as the Douglas DC-3 was the first real airliner and durable workhorse in aeronautics." Unlike Donald Douglas's immortal Gooneybird, however, Saturn's increasingly larger and more powerful versions would be anything but classically simple. Two Huntsville engineers called it "almost impossibly complex."

Far less complex were the boosters that were supposed to get the Mercury astronauts to orbit in progressive stages. There would be four of them: Little Joe, Redstone, Jupiter, and Atlas. The first would carry dummy capsules to high altitude. The second would carry test capsules and then astronauts on ballistic "up-and-down" trajectories to get them used to orbital flight without actually going into orbit. The third was supposed to carry a test capsule on a high-speed flight to test its heat shield. Finally, the Air Force's Atlas would use its 180-ton-thrust Rocketdyne engines to lift unoccupied and then occupied capsules into orbit. Of the four, only Redstone was a fully proven entity. Atlas, which was supposed to actually carry astronauts, was itself still in development at General Dynamics in Fort Worth. One Redstone cost $1 million. An Atlas would cost two and a half times as much.

It was Silverstein who came up with the program's name, Mercury, just as he would name Apollo. The gritty Ohioan knew that Mercury was the Olympian messenger who was the grandson of Atlas and the son of Zeus. If he knew that the winged courier was also associated with archthievery, patronizing traders, and blatant commerce, he kept it to himself. On December 17, 1958—Wright Brothers Day—Glennan publicly announced that the manned satellite program would be called Mercury.

NASA informed McDonnell on January 12, 1959, that it had won the contract to produce America's first manned spacecraft. The St. Louis firm was best known for its Navy jets, including the then-new Phantom II, which was so versatile that even the Air Force agreed to use it. The contract set an estimated cost of $18,300,000 (later revised sharply upward) for what was initially a dozen identical capsules but that would within months become twenty, each adapted to a specific mission. The company's fee for producing the spacecraft was a paltry $1,150,000, representing a small fraction of its business and a minuscule outlay by the government. Yet that agreement—"Research and Development Contract for Designing and Furnishing Manned Satellite Capsule"—started what was destined to become one of the largest corporate mobilizations in U.S. peacetime history. The capsule itself was a claustrophobe's nightmare: it was only nine and a half feet long and weighed less than 4,300 pounds. That was the size and mass of a small car. It had just slightly more air volume than a large adult casket. But it would employ some 4,000 suppliers, including 596 major subcontractors from twenty-five states, and more than 1,500 smaller ones.

The Magnificent Seven

The previous November, the Space Task Group had asked representatives of the military and industry to nominate 150 males from whom a pool of 36 would be selected for physical and psychological testing for Mercury. Cooperation between the space agency and the military, at least in that regard, was excellent because the armed services saw the selection of their officers as an important way to get to space (however repugnant it was to do so through NASA). A draft announcement for applications was written three days before Christmas 1958 and stipulated that the positions were for a "Research Astronaut-Candidate" in the GS-12 to GS-15 range, meaning that salaries would range from $8,330 to $12,770 a year, depending on qualifications. Only males between the ages of twenty-five and forty, who were less than five feet eleven inches tall and who had at least a bachelor's degree, would be taken seriously.

As originally planned, candidates would have had very diverse educational backgrounds, including degrees in the physical, mathematical, biological, or psychological sciences; engineering or technical research; medicine; or anyone with a minimum of three years flying airplanes or balloons, or with time in submarines as commander, pilot, navigator, communications officer, engineer, or a comparable technical position. Furthermore, all would have had to prove that they had recently demonstrated their willingness to accept hazards such as those found in research aircraft; a capacity to tolerate rigorous and severe environmental conditions; and the ability to react well in emergency conditions. "Lunatics" who might otherwise send in crank applications would have been screened by a requirement that all candidates be sponsored by reputable organizations. Sport flying, parachuting, scuba diving, exploring the Arctic, military combat, or other physically challenging occupations would have been fine with the Space Task Group.

But they were not fine with Eisenhower. During that Christmas holiday of 1958, even as his greeting to the world was beaming down from the orbiting Atlas ICBM, Ike decided that the military test pilot fraternity would automatically have the necessary traits. Using that pool would therefore save both time and money. Furthermore, they all had security clearances. On that score, at least, the president of the United States was more of a mind with the Soviet Union's preeminent chief designer than he was with his own space experts.

Accordingly, four additional requirements were added to those of age, height, and education, making the formula quite simple. The candidates also had to be in excellent physical condition, be jet pilots, have graduated from test pilot school, and have at least 1,500 hours of flying time. Department of Defense records showed that 580 military pilots in the four services had graduated from test pilot school and that 110 of them met all seven basic requirements.

Some fighter and test pilots thought of the astronauts as being helpless passengers in the capsules: "a man in a can," as Scott Crossfield of the X-15 program and others contemptuously put it. The term was soon turned into "Spam in a can."* But the STG believed strongly that real piloting would in fact be required. George M. Low, who also came from the Lewis Research Center and who would soon head the Office of Manned Space Flight, put it this way to Glennan: "These criteria were established because of the strong feeling that the success of the mission may well depend upon the action of the pilot; either in his performance of primary functions or backup functions."

Ironically, many engineers at Huntsville and elsewhere who were designing the rockets that were to carry the astronauts into space agreed with the test pilots and fighter jocks, though for a different reason. They worried that humans at the controls of the spacecraft would endanger the missions. Joachim P. Kuettner, who held multiple doctorates and who had been one of the very few men to test-fly the jet-propelled V-1s for the Third Reich before emigrating to Alabama, drew a sharp historical distinction between the all-important role of the pilot in an airplane and his strictly secondary role in a rocket. The pilot's decisions in aircraft "are probably the greatest contributions to his own safety," Kuettner maintained, while rockets were designed to send warheads

* By sheer coincidence, four Yorkshire pigs strapped to contoured couches were used by McDonnell engineers in "pig drops" to test Mercury's crushable aluminum honeycomb energy-absorption material during high-impact landings. The reclining swine impressed their handlers by walking away from 58-g crashes with only minor internal injuries. (Swenson, Grimwood, and Alexander, *This New Ocean*, p. 143.)

long distances to destroy targets without anyone at their controls. The introduction of pilots in the rocket's control loop therefore needed to be done with great caution, he added, so they don't create more problems than they solve. Bell Lab's John Pierce, the communication satellite expert, was even more blunt: "All we need to louse things up completely is a skilled space pilot with his hands itching for the controls."

That, in fact, was a persistent concern. In his carefully researched classic, *The Right Stuff,* Tom Wolfe repeatedly described the tension that existed between those who wanted to use pilots as crewmen and those who believed that they were unnecessary except as wired guinea pigs. "Above all, of course, he would be wired with biosensors and a microphone to see how a human being responded to the stress of the flight. That would be his main function. There were psychologists who advised against using pilots at all—and this was more than a year after the famous Mercury Seven had been chosen. The pilot's, particularly the hot pilot's, main psychological bulwark under stress was his knowledge that he controlled the ship and could always *do something.* ("I've tried A! I've tried B! I've tried C!" . . .) This obsession with active control, it was argued, would only tend to cause problems on Mercury flights. What was required," Wolfe wrote, "was a man whose main talent was for *doing nothing* under stress." For many of the engineers in von Braun's shop and elsewhere, an astronaut was essentially a redundant component.

This was the central conundrum of sending people to space. There was no purpose in wiring human specimens to measure their physiological and psychological reactions to flying way out there unless it was to develop a body of knowledge that would be useful for future manned missions in orbit and beyond. But was it really necessary to send people there? That was very much in the soul of the beholder. Those who believed that astronauts were as superfluous as they were expensive maintained, as did Van Allen and Nobel laureate chemist Harold Urey, that machines could do whatever men could and do it much better. They and many others, including a number of leading academicians and journalists, dismissed the idea of sending humans to space as pure politics. Since both the Soviet and American programs were ordered by politicians with political agendas, it was hard to argue with them on that score. It was only the blatant, unnecessary competition that required heroes to be sent out to hang their and their countries' hides on the line.

Yet the dreamers were always more pragmatic than even some of their own order believed. Many of the evangelists who were convinced they were part of the beginning of a great migration that was destined to send human seeds across the solar system and beyond, from the quietly implacable denizens of the rocket societies to the dynamos such as von Braun and Korolyov, Chelomei and Silverstein, Gilruth and Tikhonravov, believed that if base politics was what it took to start the process of colonizing the new frontier, so be it. Better that their fate be in the hands of expedient benefactors than altruistic opponents. Of course putting people in rockets and shooting them into space was political. But politics could be very fine indeed if its end was not only justified but reflected manifest destiny.

Donald K. "Deke" Slayton, one of the seven astronauts who on April 27, 1959, were finally picked to fly the Mercury spacecraft, understandably took a sharply opposing view to that of Kuettner, Pierce, Van Allen, and the other naysayers. In a speech to his brethren in the Society of Experimental Test Pilots the following October, the Air Force captain counterattacked: "Objections to the pilot range from the engineer, who semi-seriously notes that all problems of Mercury would be tremendously simplified if we didn't have to worry about the bloody astronaut, to the military man who wonders whether a college-trained chimpanzee or the village idiot might not do as well in space as an experienced test pilot. . . . If you eliminate the astronaut, you can see man has no place in space," he continued. "[A]s in any scientific endeavor the individual who can collect maximum valid data in minimum time under adverse circumstances is highly desirable.

The 110 potential members of NASA's first astronaut class were whittled down to sixty-nine who passed a preliminary round of written tests and technical interviews and who submitted to medical history reviews and psychological interviews. They were offered the option of volunteering for Project Mercury, and all did. Then thirty-seven either declined or were eliminated after further interviews. The remaining thirty-two were considered to be qualified to be astronauts, both in education and experience, and were given detailed medical examinations and exposed to the extreme physical stresses that were thought to be expected in space flight. Finally, the eighteen who made it through that last torturous battery of tests were further reduced to seven.

The seven—men who demonstrated (in the words of one of their testers) "the most outstanding professional background and knowledge in relation to the job requirements"—were anointed as America's Mercury astronauts on May 28, 1959, when they were brought before an executive session of the House Committee on Science and Astronautics. Besides Slayton, they were Lieutenant Colonel John Herschel Glenn Jr. of the Marine Corps; Lieutenant Commanders Walter Marty Schirra Jr. and Alan Bartlett Shepard Jr. and Lieutenant Malcolm Scott Carpenter of the Navy; and Captains Leroy Gordon Cooper Jr. and Virgil I. "Gus" Grissom of the Air Force. Political considerations being what they were, the Navy and the Air Force were allowed three astronauts apiece. The Marine Corps, a ward of the Navy and one that fought mostly on terra firma at that, would get to send only one of its own to space. But Glenn was the senior man in age and date of rank. And he was a happily married "straight arrow" who possessed that rarest of qualities in a profession noted for laconic individuals who were notoriously introverted: a presence so winning that it bordered on the charismatic. They were qualities that would serve him well in public. Glenn's temper, which could be terrible, served him less well within the astronaut community itself.

The Americans trained hard in a number of major areas. First and foremost, in the words of Lieutenant Robert B. Voas, a Navy psychologist who worked with them, they had to learn to "operate the vehicle." The new astronauts themselves called it "flying the spacecraft." The difference between "operating" and "flying" underscored the cultural chasm between the test pilots and their handlers. But there was much more to the new profession of astronautics than piloting. They also had to study the basics of propulsion, trajectories, astronomy, and astrophysics; be exposed to the stresses of spaceflight itself, such as acceleration, weightlessness, vibration, disorientation, and noise; prepare physically for those stresses; and even learn how ground stations around the world would track, monitor, and communicate with their spacecraft. And on top of all that, they had to keep their own flying skills sharp.

The process of training military officers to perform tasks in new and stressful environments—"human factors engineering," as the astronauts' various trainers called it—was not new. Submarine crews, aviators, and high altitude balloonists had been trained to adapt to dangerous new tasks since before World War II. The only important difference went back to the fact that, owing to the extreme altitude, terrific speed, and long trajectories, the capsules would be much more the masters of the men than the other way around. For the first time engineers on the ground would not merely prepare the way for the pilots, but in a sense, they would effectively *be* the pilots, since the rockets, spacecraft, and everything else in the system were so complex.

This did not diminish the bravery of the astronauts. Glenn would only half joke later when he said that when people asked them how they felt about riding the rockets, they would answer: "How do you think *you'd* feel if you knew you were on top of that thing, built by the lowest bidder on a government contract?" But it did mean that the jaunty service cap and silk scarf—instinctive flying skill, lightning reflexes, and derring-do—had forever been made secondary to scientists, engineers, and their ubiquitous computers. Control had been shifting from the air to the ground since aviators had stopped making their own airplanes, and it would continue to do so.

As it was, the astronauts themselves had to cajole the capsules' designers at McDonnell and others into adding, as Slayton put it, a control system for steering, a window so they could see, and a hatch "that would blow off so we could get out of the damned thing." Their persistence would pay off.

Public Relations and the Media "Circus"

The space agency and the news media had a particularly tense and ambivalent relationship. Like two self-absorbed and overzealous lovers, each with a shared interest counterbalanced by a strong individual agenda, contemplation, reason, and calmness gave way to ongoing bouts of lust and loathing. The relationship was particularly obsessive because NASA had an uncommonly strong need for good public relations while the press thrived on dramatic stories, graphically illustrated, that told of life, death, and adventure.

Favorable press coverage was important because, judged by the core criteria of promoting life, liberty, and the pursuit of happiness, NASA's existence was next to impossible to justify. The Department of Defense existed because it played a role that was fundamental to the survival of the country: defending it against enemies. The Departments of Justice, Interior, and Treasury, to take only three more examples, were responsible for running the criminal justice system, safeguarding public lands, and maintaining the gold reserve and circulating money. Public safety, protecting the national wilderness, and assuring solvency were undertakings that everyone agreed were also basic to the well-being of the people of the United States. Nowhere, however, was it written that Americans would come to serious harm or die or go berserk if there were no communication or weather satellites in space or close-up pictures of Mars. No one seriously believed that society would disintegrate unless a handful of test pilots sailed over the world sealed in tiny canisters with plastic bubbles on their heads, saying "I copy that," "Do you read?" and other technojargon.

In its formative years, before much of it became woven into the nation's institutional fabric, the civilian space program was essentially a partnership between highly motivated political, industrial, and academic interests. Its populist base was fickle. Where the average American was concerned, space was both literally and figuratively too far out to take seriously unless there was a real threat to the nation's security or, at minimum, great embarrassment. Building or sending rockets to space with little or no tangible return (unless one built or fired them) was an elitist game. That was understandable. All of Palos, after all, had not turned out to wave good-bye to Columbus and wish him well. The average fifteenth-century Spaniard was as preoccupied with life's more mundane challenges as were his or her American and Russian counterparts more than four hundred years later. Persuading large numbers of people to enthusiastically support elite programs therefore required a high order of salesmanship. It required that the media of the United States and elsewhere be made to fawn over, or at the very least participate enthusiastically in, the grand enterprise.

And it required even more than that in the Union of Soviet Socialist Republics, whose inhabitants were denied many of the public and private comforts enjoyed elsewhere because they had to pay for a huge and far-flung military presence and prestigious cosmic abstractions. The Soviet lunge to space was therefore not only glorified by an obedient press, but it was emblemized with countless patriotic gewgaws: steel statues of rockets and space people in all sizes, marble sculptures and bas-reliefs, flags and banners, posters, postcards, pens and pencils, stamps, models, and millions of gaudy plastic lapel pins and other trinkets of every description that were the propagandists' currency. And underlying all the glitz was the unspoken but implicit fact that space was the shoulder upon which the national honor rested. The rockets did not hang on walls and were not imported from France. They had been the first to thunder into the sky, and they were made in the U.S.S.R. Like chess and weight lifting, the Russians had found a game they could beat the West at, and it was quietly exhilarating.

In the United States, the seven Mercury astronauts made their press debut to a "waiting and breathless world," as Glennan put it, on April 9, 1959, in his agency's otherwise little-used pressroom. The administrator introduced the pilots—all of whom wore civilian clothes—and then turned the meeting over to one of his public relations officers, who described the program to reporters and photographers eager to record the first public appearance of the nation's new knights of the sky. This was the famous occasion when, asked which of them expected to be first in space, all raised their hands. Wally Schirra raised both hands. So did Glenn, easily the most personable of the group. The Marine, who held a coast-to-coast speed record and had briefly starred on the television show *Name That Tune,* quipped that he wanted to be an astronaut "because it is the nearest to heaven I'll ever get." The press loved it.

However much NASA and the astronauts themselves wanted to separate the men from the other primates, both for the purposes of bolstering morale and creating heroes, they had to contend with superiors who somehow would not, or could not, differentiate between them. The decision to send monkeys to space before a man rankled the astronauts. "The irony of playing second fiddle to a chimpanzee was particularly galling to these highly intelligent and skilled men," Shepard and Slayton would write, adding that, for his part, Shepard hungered for a "chimp barbecue." Although Keith Glennan participated in many news conferences, only the astronauts' debut and one other would later stand out in his own memory. The second took place seven weeks after the astronauts were unveiled and starred two monkeys, Able and Baker, that had just survived a suborbital test flight on a Jupiter. Glennan noted in his diary, tellingly, that he "again had the privilege of introducing two astronauts—in this case, the monkeys." And the quick-witted John Kennedy would betray the same attitude soon after he became president, drawing howls of laughter at a news conference when he quipped that one of the flying chimps "reports that everything is perfect and working well." No wonder test pilots at Edwards and elsewhere invented little ditties like "Shepard, Grissom, and Glenn, the link between monkey and man."

The imperative that the space agency have good public relations made Glennan and his successors deeply ambivalent toward the news media. For them, journalists were more or less divided between friends and enemies, depending on whether NASA was reported favorably or not. The majority of reporters and writers who covered the agency were so caught up in the adventure of sending Americans to space that they willingly became partners in the enterprise, in the process helping to glamorize it, while overlooking problems and neglecting to ask hard questions. The others were disparaged as hatchetmen and -women who sacrificed the truth as NASA saw it for sheer sensationalism. If the administrator and others in Washington had one thing in common, it was that they simultaneously courted the news media and scorned them. As the author and veteran reporter Martin Mayer put it about those who try to get favorable coverage: "If they succeed, they are contemptuous because they have manipulated the press; if they fail, they are angry."

Glennan believed implicitly that journalists were always hoping for catastrophes because, as the hackneyed expression went, bad news sold newspapers. (If that was true, it indicated that their audience enjoyed reading about calamities and watching them on television, which was not usually mentioned by those who disparaged the media.) Following news reports of a Redstone launch failure, for example, he wrote that "great headlines are only good when one can say there has been a failure." He therefore felt particularly vulnerable because his disasters, unlike those that could happen at Justice, Interior, or Treasury, were truly spectacular and involved expensive machines that were fundamentally irrelevant to real national security.

But even close scrutiny and the raising of hard questions were often interpreted as attacks that could turn insecure bureaucrats into crybabies. Glennan noted in his diary that the editors of both *The Washington Evening Star* and *The Washington Post* helped NASA with what the agency saw as "public information problems." That in itself was a departure from accepted journalistic practice. But then he added that he needed the help because he had been "concerned about the possibility—really, the probability—that the papers would make a circus of our Project Mercury as we approach the time when a man will enter the capsule to undertake a flight." Yet a circus of sorts—an extravagantly patriotic national entertainment—is precisely what NASA was trying to stage. What Glennan and his successors wanted, as all showmen and -women want, was rave reviews. Not getting them made him and the others bitter and defensive.

"I developed a special antipathy toward the trade press . . . ," Glennan told his diary. "Two publications seemed always to get in our hair—*Aviation Daily* and *Missiles and Rockets.* These publications existed primarily on advertising, and their mission was advocacy rather than objective reporting. Their attacks on the agency were hard to stomach. Similarly, the absolute authority with which people at *Time* magazine castigated our operations was also hard to live with." Here, for example, is what *Missiles and Rockets'* James Barr wrote to raise Glennan's hackles:

> Despite precautions and improvements, Mercury continues to be a technically marginal program that could easily end in flaming tragedy. Mercury, at best, is a technical stop-gap justifiable only as an expedient. It is no substitute for what is needed sooner or later, a maneuverable spacecraft similar to the Air Force's much hampered Dyna-Soar.

In fact, the program had indeed slipped seriously by the time that article appeared in August 1960, disheartening many in the Mercury program itself. There had been capsule delivery delays, some sloppy workmanship on the capsules that had been delivered, launcher development problems, aborted test firings of the various rockets, and some accidents after launch, all causing irritating and expensive delays. At the same time, the Air Force continued to push Dyna-Soar for all it was worth, and Boeing advertised in *Missiles and Rockets* (which accounted for Glennan's insinuation that the magazine was catering to its industrial supporters by extolling Dyna-Soar as the true spacecraft of the future).

Meanwhile, Glennan, the Space Task Group, and others in the program not only worked under immense pressure in a new technological environment, but were in effect under pressure from the White House not to appear pressured. The word that came down from the Oval Office was that officially there was no race. If there was no race, there was no pressure. And to believe the White House, of course, there was also no Moon program.

But the professionals and those in the news media who covered them knew better. On September 9, 1960, George Low spoke to a United Press International editors' conference in Washington and argued passionately against what he called misconceptions that the Fourth Estate was carrying to the public. Mercury, Low said, was not "merely a stunt . . . designed only to win an important first in the space program" and should not be killed in the likely event that the Russians put a man in space before the United States. Rather, he explained with candor, Mercury was an indispensable step toward the Apollo program and one which therefore "must be carried out regardless of Russian achievement." Given the fact that the Soviets not only were ahead but were almost certainly going to increase their lead by getting a man up first, Low's strategy was a good one. It tried to insulate the manned program from the vagaries of competition by giving it a life of its own that extended into the infinite future. What the Russians did, Low maintained, had no bearing on what the United States did. That theme, which would soon become official NASA policy, was as clever as it was disingenuous. Not only did the manned program owe its existence to the Russians, but their sending the first man into space would benefit Mercury and its successors, not hurt them.

Meanwhile, Low continued, the media needed to grasp the unprecedented complexity of the first step itself. "It has been a major engineering task to design a capsule that is small enough to do the mission, light enough to do the mission, and yet has reliable subsystems to accomplish the mission safely," he told the editors.

The men and women who were creating the two entities that were Mercury—the Atlas launcher and the capsule itself—and who were trying to get them to function together perfectly worked within parameters as tight as their hardware. They had to try to stay on schedule while developing a product so perfect that it would not suffer critical failure under tremendous pressure. It was not a matter of achieving technological breakthroughs. If the technology had not mostly been in place, there would have been no Mercury program.

Yet fitting it all together quickly was a formidable challenge. There were an estimated forty thousand critical parts in the Atlas and roughly that number in the spacecraft. The challenge was to get all of them to function flawlessly together and on schedule. Barr predicted that Americans would be beaten to space by Russians. That would be a serious political defeat. But it would be infinitely worse if an American died trying to take second place. So everyone in the space community, and especially those who were trying to get men into orbit and beyond during the summer of 1960, were wrestling with "reliability engineering" and with a "meticulous" versus a "statistical" approach to testing for reliability: Did a component have to work every time to pass muster, or was 999 times out of 1,000 good enough?

Semantics, which most engineers abhorred, were becoming the litmus of what they did. They had to figure out what a "system" was and define what constituted its "failure." Many engineers, including those in the STG and at McDonnell, tended to equate reliability prediction with astrology and failure-rate tables with the zodiac. They maintained, with some justification, that a technology as complex and dynamic as theirs contained too many imponderables to permit foolproof predictions. It was a point that would be born out repeatedly as the space age progressed, and most notably in the *Challenger* disaster, which had been calculated as a hundred-thousand-to-one possibility.

If Glennan had problems with the news media's own semantics, which in his opinion were often wantonly mean-spirited, he was delighted to cooperate with both NBC and CBS for documentaries on Mercury and to put the astronauts' "stories" on the auction block for the print media. The highest bidder was archenemy *Time*'s stablemate: *Life.*

Time-Life publisher Henry Luce understood that sending Americans into orbit and then to the Moon figured to be the greatest adventure of the century. So he not only agreed to pay the seven astronauts half a million dollars, but also to have their stories ghostwritten under their bylines and submitted to NASA for approval in advance of publication. Hugh Sidey, one of *Life*'s prominent columnists, took a cavalier approach to the generally frowned upon practice of paying for stories. "Checkbook journalism has been with us for as long as we've had journalism and still is," he said. "I mean, the big and the powerful get the breaks, that's all."

Yeager, who was testing jets at Edwards at the time, took a dim view of the astronauts' supplemental salaries, and his colleagues undoubtedly agreed. "They were military test pilots. You do something because it's your job," he said later, adding that NASA's contract with *Life* "exposed the guys to a mercenary environment of taking a lot of pay for doing your job. That, to me, was wrong." The first man to fly faster than the speed of sound remained as good as his word until he left the Air Force. Then he cashed in on his own reputation by selling his name and face to a watch manufacturer, other advertisers, and a video game maker.

The popular weekly picture magazine and the space agency upon which it depended for material carefully created a succession of impossible men, or rather of boy test pilots who could have walked out of the pages of a comic book. As portrayed in *Life* and, for example, in a book called *We Seven* that was derived from the magazine articles, the new heroes were living embodiments of the Boy Scout ideal: physically fit, morally straight, and mentally awake youngsters-grown-to-men believing in God, country, and their merit badges. Their devoted wives were as proud as heck and suffered their husbands' danger by biting white knuckles in dutiful silence. Their children were happy to jog with them (Lyn and David Glenn); practice archery with them (Scotty and Jay Carpenter); ride with them in shiny sports cars (Suzanne and Marty Schirra), duck hunt with them (Scott and Mark Grissom), or just sit with them at poolside (Kent Slayton). Their dogs did not make mistakes.

NASA and the armed services that fed it hero-candidates fretted so much about their image that some were actually sent to what astronaut Michael Collins would call "charm school." Besides being astronauts, he explained, the Air Force thought they should have "certain social skills as well, so they ran a short course. I think we came back to Washington for a couple of days and they told us that you're supposed to wear socks like this," he said, showing one that covered most of his calf, "that go on forever and no one wants to see hairy legs. . . . And you had to hold your hands on your hips this way [palms down] and not that way [palms up] because people you don't want to talk about hold them the other way."

The Dark Side

As the space program geared up, Cocoa Beach turned from a sleepy hamlet straddling Route 1 along the Indian River to a boomtown crammed with motels, liquor and package stores, fireworks stands, barbecue and hamburger joints, gas stations, real estate offices, car dealerships, and bars ("lounges," in the region's genteel argot). The minions of the law were still so green, according to one journalist from up north, that they charged a homicide suspect with "premedicated murder." The space program turned Cocoa Beach and the towns nearest to it into "Spaceport U.S.A.," a tacky carnival and theme park for adults.

The arrival of the astronauts and their handmaidens, the engineers, technicians, company reps, military types, and assorted bureaucrats, also drew large numbers of young women who were attracted to the Mercury Seven and their successors the way their sisters in other places followed and embraced matadors, movie stars, and champion athletes.

In the astronauts, the ladies very often found willing partners. The vast majority of the men they pursued had two different personas. The public saw the carefully cultivated Boy Scouts, while the astronauts' intimates often saw a far more earthy side. Flying fighters was a highly individualistic and dangerous job that carried a cachet like that of the Arthurian knights. But if fighter pilots considered themselves members of an elite fraternity, those who occupied an even higher perch—the test pilots—were even more full of themselves. Many believed implicitly that the choicest bourbon, steak, women, cars, and cigars were their due; were their just reward not for being heroes, but for excelling at a game in which losing often meant violent death. For many of them—the self-coronated royalty of the military caste—a kind of divine right made every attractive woman a dish fit for a deserving king. For most of the early astronauts, handpicked heroes whose macho surpassed even that of test pilots, the heady privileges of stardom easily overwhelmed caution and restraint (Glenn was a notable exception). Furthermore, flying and testing fighters are highly individualistic occupations in which there is an unambiguous distinction between winning and losing. That made the original seven and many of their successors far more competitive with one another than the public ever knew.

The women came in considerable numbers, bringing their bikinis and tight dresses to the pools and lounges of the Starlite and Vanguard Motels and other playpens where access to the rooms did not require a walk through the lobby. According to a number of accounts, one called "Wickie" made her way to the beds of six of the original seven, keeping score the way old-time gunslingers notched the handles of their .44s.

"Basically, they were all hotshot pilots and had egos accordingly," recalled John W. King, who covered the beginning of the Mercury program for the Associated Press and who then became chief of public information for NASA at Cape Canaveral. "They acted like they did while on leave from aircraft carriers," he added, mentioning the infamous Tailhook sex scandal that gave the Navy a black eye more than thirty years later.

"There were ladies in Cocoa Beach, Florida, and probably other places around the country who liked to brag about how many or which different astronaut they had managed to slip into their bed," one astronaut recounted years later in reference to "Wickie" and her colleagues. "If they only knew that it wasn't all that difficult, then they probably wouldn't have bragged that much. . . . It was unbelievable," he added, "and no one ever said a word. It never got into the press."

CBS's Walter Cronkite, a correspondent and commentator so avuncular that he was sometimes likened to Eisenhower himself, ignored the heroes' warts and instead helped to project what Apollo astronaut Russell L. "Rusty" Schweickart would call their "Boy Scout monotone image." Years later, Cronkite admitted that "we were quite aware that the image that NASA was trying to project was not quite honest. But at the same time, there was a recognition that the nation needed new heroes."

It was taboo for any astronaut to publicly describe another's sins and for almost two decades the Fourth Estate kept the secrets as well (as it did with Kennedy's womanizing). Tom Wolfe broke out of the unspoken pact in *The Right Stuff,* however, when he described some of the astronauts' passions for hot cars and the "young juicy girls" who infested Cocoa Beach and who were "always there and ready."

In *The Right Stuff,* Wolfe described a heated exchange between Glenn and some of the others during a tour of the Convair plant in San Diego. "The next day the seven of them were in the living room of a suite that had been set aside for their use, when Glenn launched into a lecture, along the following lines: the playing around with the girls, the cookies, had gotten out of hand. He knew, and they knew, that it could blow up into something very unfortunate. They were all squarely in the public eye. They had the opportunity of a lifetime, and he was sorry but he just wasn't going to stand by and let other people compromise the whole thing because they couldn't keep their pants zipped."

But Shepard, an otherwise correct naval officer who liked to cut loose when he could, would have none of the rebuke, according to Wolfe. "Commander Al, the colonel's son, knew how to put on all the armor of military correctness, in the stern old-fashioned way. He informed Glenn that he was way out of line. He told him not to try to foist his view of morality on anybody else in the group," Wolfe reported. Shepard, he added, believed implicitly that there was nothing wrong with extramarital female company so long as it affected neither individual performance nor the program as a whole.

The heated exchange and the reason for it were kept out of the public eye by both the space agency and reporters who still drew lines between individual privacy and the public's right to be informed. But the moral conflict, in which all seven took sides with varying degrees of intensity, further divided individuals who were already locked in a professional competition that was the inevitable result of being famous. It was only natural. But it was also the stuff of nightmares for those who shaped NASA's image and who therefore labored to keep the big picture from cracking wide open.

If the astronauts and cosmonauts shared common selection and training processes, so too did they share primal urges. Sometime after he achieved celebrity status, the baby-faced Yuri Gagarin appeared with a nasty gash over his left eye that he kept for the rest of his short life. Officially, the U.S.S.R.'s preeminent hero had gotten the wound when he fell while playing with his infant daughter. In fact, the world's premier spacefarer got the scar on October 3, 1961, while vacationing with his family and some space officials at a sanatorium near the Black Sea. According to a diary entry made by Nikolai Kamanin, the cosmonauts' chief (and repeated by a Russian surgeon in a book), Gagarin seems to have forced his way into the room of one of the nurses. Suspecting that something was going on, his wife, Valentina, rapped on the door. The fighter pilot got out of the room the way he would have left a burning jet: he vaulted over the balcony, which was only about six feet above the ground. But as he jumped, his feet became tangled in grapevines, so he landed face first on a cement walk, fracturing a bone and landing him in the hospital for three weeks. He later told Kamanin that he didn't know there was a woman in the room and that he was playing hide-and-seek with his wife.

And like his American rivals, Titov and his comrades had a rambunctious streak that was common to fighter pilots. Kamanin related that Titov was supposed to leave an open car and get into a closed one as his motorcade left town during an official visit to Romania in 1961, for example. But instead, the exuberant cosmonaut decided to jump onto one of the motorcycles escorting the cars and take it for a hair-raising spin through the countryside. There were a number of other instances of "misbehavior" by Titov, Kamanin noted, most involving traffic violations and heavy consumption of vodka. And in the spring of 1963, he expelled three cosmonauts after a drunken brawl with a military patrol. A fourth had been thrown out the previous year because of blatant marital problems and being absent without official leave from the garrison.

Khrushchev Versus Khrushchev

In his campaign for the presidency, Kennedy accused the Eisenhower administration of committing a spate of dangerous mistakes, chief among them the ceding of military superiority to the Soviet Union and all but abandoning the high ground. He hammered Eisenhower throughout the summer of 1960 and into that turbulent autumn for what he insisted was a dangerous gap between America's space program and that of its Communist rival. It was a gap, JFK charged, that was the equivalent of the one he said existed between their respective bomber and missile forces.

Kennedy, whose known passions did not include rockets, had no more inherent interest in space than did Khrushchev. Yet like Lyndon Baines Johnson, his running mate, he brooded about his country's appearing to be second best to the U.S.S.R. and worried about the military balance. And like LBJ, he came to see that there were profoundly romantic qualities about intrepid men venturing into the forbidden void in the service of their country. The conquest of the new frontier was compelling precisely because it embodied the classic elements of the timeless Homeric adventures: men willfully testing their own mettle on a perilous journey and mustering the courage that was necessary to prevail. There were some, such as Tsiolkovsky and von Braun, who felt this in their souls. Others, such as Frank Malina and Lyndon Johnson, felt it in their guts. And still others made it their intellectual property without feeling anything. Khrushchev was one of them, and Kennedy was another. The intensely competitive young Irish-American politician would soon understand, albeit intellectually, that the classic quality inherent in the adventure of launching Americans into the new ocean would unite the nation under his leadership. The key element in the equation, Kennedy would come to realize, was not having the technology; it was having an adversary.

Khrushchev himself played a role worthy of Sophocles. He presided over a country that was surrounded by an immensely powerful and dangerous enemy. He knew that there were generals—Curtis LeMay and Thomas Power, for instance—who wanted to obliterate his country with nuclear weapons and had the bombers to do it. Whatever feats Korolyov and the others accomplished in space, they in no way made up for the fact that the Soviet Union was still recovering from the effects of a devastating war and needed time to build its own nuclear arsenal and a powerful enough military to deter yet another invasion from the West. That being so, Khrushchev decided, it was better to bluff than to negotiate from weakness. By the time Kennedy was elected president, Khrushchev had become convinced that excelling in space (or seeming to) not only had immense propaganda value at home and abroad, but was relatively cheap for what it bought. And what it bought was the appearance of strength: a defensive mirage. Technology was the embodiment of socialism's triumph over nature and had long been the state religion. But its most important icon was no longer the tractor, the hydroelectric plant, or even the hydrogen bomb. It was the rocket.

Yet it was a tremendously expensive mirage. The pride the average Russian felt because of the feats in space was more than offset by the knowledge that they did nothing to alleviate the "disastrous situation they were in with regard to housing, clothes, food, wages, and so forth," as one Russian writer observed. The dichotomy between the thundering rockets at Tyuratam and the pitted streets, seemingly endless lines for bread and sausage, cracker-box housing, drab, swamp-colored clothes, and flimsy shoes created a quiet but pervasive cynicism about the system.

More important, the propaganda generated by the space program should have been judged not by its appearance, but by its effect, which was virtually nil. The strategy of courting Third World countries with Sputniks and other spacecraft and of sending space exhibits to Asia, Africa, Latin America, and elsewhere to gain influence accomplished nothing practical. No proletarian uprising, no socialist revolution was ignited by the Russian space program. The thought of Fidel Castro, Che Guevara, Patrice Lumumba, or Ho Chi Minh urging their followers on and spreading the revolution by invoking the image of Soviet technological achievement, let alone triumphs in space, was ludicrous. The downtrodden were galvanized by events that took place where they lived, not in the sky.

Finally, and most significantly, Khrushchev's taunting insistence that his country's string of triumphs in orbit and beyond were a barometer of its strength and heralded a socialist age managed to arouse both Kennedy's ire and that of an exuberant space establishment eager to sprout real wings and grow into a large and powerful institution. Kennedy never believed that Soviet space spectaculars fundamentally endangered the United States. But they were immensely useful to him politically because they gave him an issue. Khrushchev's trying to stave off the United States by baiting it had exactly the opposite effect: it provided Kennedy with the ammunition he needed to show that the United States had become dangerously weak on Eisenhower's watch.

"The first man-made satellite to orbit the earth was named Sputnik," the senator from Massachusetts said during his campaign for the presidency. "The first living creature in space was Laika. The first rocket to the moon carried a red flag. The first photograph of the far side of the moon was made with a Soviet camera. If a man orbits the earth this year his name will be Ivan."*

And, JFK warned on another occasion, "If the Soviet Union was first in outer space, that is the most serious defeat the United States has suffered in many, many years. . . . Because we failed to recognize the impact that being first in outer space would have, the impression began to move around the world that the Soviet Union was on the march, that it had definite goals, that it knew how to accomplish them, that it was moving and we were standing still. That is what we have to overcome, that psychological feeling in the world that the United States has reached maturity, that maybe our high noon has passed . . . and that now we are going into the long, slow afternoon." The oblique reference to Will Kane, the courageous but outgunned marshal played by Gary Cooper in the film *High Noon,* was inspired.

Accelerating the nation's space program was crucial, the senator added, dusting off the old imperative to seize the high ground: Soviet control of space would invariably lead to its control of Earth. "[A]s in past centuries the nation that controlled the seas dominated the continents," he continued, choosing to ignore the fact that neither the Royal Navy nor the Spanish Armada had figured decisively in the control of Europe and that the Western Hemisphere had been lost to the imperialists despite the frigates and men-o'-war in the British and Spanish navies. "This does not mean that the United States desires more rights in space than any other nation. But we cannot run second in this vital race. To insure peace and freedom, we must be first." It was implicit, of course, that being first would indeed bring more "rights" (whatever that meant). But the race metaphor was beginning to stick and, contrary to Glennan's assertion, it was by no means the exclusive property of the news media.

Vice President Richard M. Nixon, JFK's opponent, countered, correctly, that the nation was not behind at all. Gamely echoing Ike, he insisted that "the United States leads the world in activities in the space field that promise real benefit to mankind." Yet Nixon himself, on the defensive, was forced to couch the international competition as a race when he maintained that America was "ahead of the U.S.S.R."

What Nixon undoubtedly did not know, and what he certainly would have kept secret had he known, was that Glennan was so attuned to Eisenhower's mellow attitude about space that he was actually trying to think up projects in which both countries could cooperate. But there was no hint of cooperation from Khrushchev, not only because he was convinced that he was ahead, but

* Although the text appeared under Kennedy's name in an issue of *Missiles and Rockets,* it was actually written by Edward O. Welsh, an aide to Lyndon Johnson and eventual member of the National Space Council. (Young, Silcock, and Dunn, "Why We Went to the Moon," p. 29.)

probably also because he believed implicitly that it would have opened the way for American spying in some capacity. It was fear of espionage that prompted him to turn down Eisenhower's "open skies" aerial observation offer. At any rate, his present to Ike on May 15, 1960—on the eve of the summit he boycotted because of the U-2 fiasco—had been the orbiting of the first Vostok. The five-ton spacecraft was technically called "Korabl Sputnik 1," or spaceship satellite test craft, and it carried the first in a series of life-size, pressure-suited dummies named Ivan Ivanovich. Instead of landing after its sixty-fourth orbit, which was the plan, a defective horizon sensor sent it into a still-higher orbit. It finally came down and crashed in 1965. But the point had been made. That shot alone seemed to leave no doubt that Moscow could one-up Washington at will. "Here we are again," Glennan complained in his diary. "The Russians are successful in launching something successful for Ike's benefit as he steps out of his plane in Paris. They really seem to have much better control of their activities in this field than we do, as yet."

Nixon, however, was still closer to the mark than Kennedy. By October 4, 1960—three years after the first Sputnik and a bare month before the presidential election—the United States had successfully launched twenty-six satellites and two space probes compared to six satellites and two probes by the Soviet Union. The American spacecraft included Discoverers 13 and 14, Echo 1, TIROS 1, Midas, and Pioneer 5, the first successful interplanetary space probe, which sent back data on solar flares and the solar wind.

Not only was the United States overwhelmingly superior to the Soviet Union in all but land forces in 1960, but few knowledgeable people in government believed otherwise, regardless of the leaks that were fed to the Alsop brothers, Walter Winchell, and other conservative columnists by self-serving Pentagonians and some congressmen. But that didn't matter during the presidential campaign. What mattered was that Khrushchev handed Kennedy an issue that he could run on. The Soviet leader's taunting and false depiction of his country's military and space prowess not only helped elect Kennedy, but it provided him with the means to simultaneously consolidate his power and unite the country by launching the greatest technological enterprise in history. So Nikita Khrushchev was fated to be the victim of his own braggadocio. He unwittingly set into motion the process that would lead inexorably to his own nation's defeat at a place called the Sea of Tranquility.

Prelude to a Showdown

By the end of Eisenhower's last summer as president, as the campaign for his successor moved relentlessly toward its climax, the rate of both U.S. launches and program starts increased dramatically. There were a number of reasons for this, all of them interactive: enhancement of life (into which category both the satellite communication and meteorological programs fit), national security, the quest of scientific knowledge, and sheer prestige. And the roar of rockets played counterpoint to the threatening rhetoric coming from both sides of the Iron Curtain. Politics, national security, and technology were coming together in a synergistic explosion not seen since the Second World War.

First, and most obviously, the two superpowers were locked in competition to decide which would be the first to reach an undefined goal and reap an equally undefined reward. But within the respective camps there were sub-competitions, equivalent to regional play-offs between the military and civilian sectors, the advocates of manned and unmanned missions, scientists and engineers, American corporations maneuvering for business, and their Soviet counterparts, the design bureaus, doing the same. Finally, there were the elite rocketeers themselves and the technocrats who handled and exploited them: men whose dreams, egos, and patriotism propelled them just as surely as the combustible potions they brewed sent their machines disappearing into the firmament. Unlike the amorphous goal in the larger race, the goals in these contests were very specific and they extended from the mesosphere to the plains of Mars.

The events of the summer of 1960 make the point. On August 4, NASA test pilot Joseph A. Walker flew the X-15 to a world speed record of 2,196 miles an hour, reaching the relatively low altitude (for an X-15) of 78,112 feet. Three years later, Walker would earn astronaut's wings by reaching nearly sixty miles. The program was "intimately associated with national prestige," as one of the rocket plane's chroniclers put it. But it was a great deal more than that. As the X-1 broke the sound barrier, succeeding X-aircraft broke the psychological barrier, proving beyond doubt that men could indeed fly into space—or at any rate to the edge of it—and return safely.

The three X-15s built by North American also collected an immense amount of data on high-speed, high-altitude aero- and astrodynamics, paving the way for Mach-busting planes such as the supremely innovative XB-70 Valkyrie bomber and Lockheed's sleek SR-71A "Black Bird" reconnaissance plane, as well as techniques for space capsule maneuvering and re-entry. The performance of the black rocket at the edge of space and a Mercury capsule flying a suborbital trajectory down the Atlantic were strikingly similar. The X-15's record would stand at 4,520 miles an hour and sixty-seven miles. Its LR-99 engine could theoretically have pushed it to one hundred miles, but the aircraft could not have safely re-entered the atmosphere from that altitude. Even a slight error in the penetration angle could have turned both plane and pilot into cinders.

On August 12, 1960, four days after Walker's record-breaking flight, the U.S. space program scored two singular triumphs.

Echo 1, the first communication satellite (after the Moon), was rocketed into a roughly thousand-mile-high orbit. A message from Eisenhower was quickly sent from Goldstone, California, to the hundred-foot-wide metallic balloon, where it ricocheted down to the Bell Telephone Laboratories at Holmdel, New Jersey. Echo was Glennan's favorite program. He not only listened to it with "great glee" before carrying a recording of it to a National Security Council meeting that morning but, borrowing a page from Khrushchev, decided to record the voices of foreign ambassadors and relay them via the satellite back to their own countries, courtesy of the United States Information Agency.

The same day, the recovery of the Discoverer 13 film capsule in the high seas off Hawaii reversed a string of failures and effectively started the crucial reconnaissance satellite program. The midair retrieval of its immediate successor only a week later cinched the matter by proving that airplanes could indeed snatch "buckets" coming down from orbit. Not counting an accident on the launchpad on January 21, 1959, fourteen Thor-Agenas had been shot out of Vandenberg in less than eighteen months, reflecting the intelligence community's sheer desperation to get information from behind the Reds' long curtain.

But the competition was by no means going all the Americans' way. The day Discoverer 14 was snatched by an Air Force Flying Boxcar over the Pacific, the Soviet Union sent up Korabl Sputnik 2 with two more mutts and other creatures. It was the second test flight of the Vostok man carrier, which Korolyov's bureau had begun designing early in 1958. This one carried a small menagerie: the dogs Belka and Strelka, two unnamed white rats, more than two dozen mice, and several hundred insects, as well as plants. The canines, dressed in orange space suits and Plexiglas helmets, were monitored by television in the mission control center at Kaliningrad, outside Moscow. They were eventually ejected out of the satellite, the way all cosmonauts would be, for a relatively soft parachute landing. Unlike their predecessor, Laika, they lived to bark again another day. The satellite itself descended under another parachute while a tape of Russian choral music played grandly on its radio. That much was publicized immediately.

What was kept secret for several years, however, was the fact that the cargo also included small pieces of skin that had been donated by doctors at the Moscow Institute of Experimental Biology. Members of the new profession of space medicine feared radiation far more than the effects of zero gravity on the men who were about to go to space, particularly after Van Allen discovered the belts that bore his name. The test flight was more proof, if any were needed, that Korolyov was aiming his cosmonauts at space. He was also aiming probes at Mars and Venus, also with the enthusiastic approval of

Khrushchev, who arrived in New York on the *Baltica* on September 19. The most memorable part of the first secretary's visit, at least as far as American television viewers were concerned, came when he took off his shoe at the United Nations and pounded it on the table during speeches he did not like. The image of the chairman of the Council of Ministers and first secretary of the Central Committee of the Communist Party of the Soviet Union acting like a rowdy peasant, a boorish buffoon, in a convocation of diplomats would stay with Khrushchev all of his life, at least in the West.

That was a most unfortunate image, however, because Khrushchev was an adroit politician and a sensitive, intelligent statesman who knew when to back out of a potentially catastrophic situation, as he did during the Cuban missile crisis in October 1962. Sergei Khrushchev would tell the story years later about the day he heard General Andrei Grechko inform his father in their dacha that the Red Army could overrun Western Europe and in five weeks he—Grechko—could be standing in Portugal and looking out at the Atlantic. "And do you know what you would see when you looked up?" Sergei quoted his father as saying. "You would see American atomic bombs coming down on your head!"

At any rate, a sailor defected from the *Baltica* and promptly told reporters that there were "spaceship models" on board that Khrushchev evidently planned to present to the United Nations after some new space triumph. That would have been reasonable, since he had proudly presented a model of Luna 2 to Eisenhower after the real thing had slammed into the Moon on September 12, 1959. But there was no triumph. There was, in fact, nothing.

Early in October, *The New York Times'* Washington bureau was fed a story by U.S. intelligence that bore on the models the sailor had described. The agents reported that the Soviets had failed in what was supposed to have been a spectacular space shot. The story appeared on October 13, the day Khrushchev finally gave up and flew home, and it was correct. The next day, the fourteenth, there was another Soviet space mishap, also monitored by the CIA. Twice Korolyov shot at Mars, and, to his chagrin, twice he failed to get there. Both shots involved three-stage boosters that were supposed to get a fourth stage out of orbit and into an interplanetary trajectory. In both cases, the turbopumps in the third stage malfunctioned, so the probes never even made it to Earth orbit. It was only the beginning of the Soviet Union's long and painful effort to reach both Mars and Venus.

Soviet Setbacks and a Disaster

The two failed Mars shots were preludes to a full-blown catastrophe that itself was an indication of how hard the Russians were pushing for spectaculars. On

October 22, 1960, eight days after the second failure to reach Mars, the first full-scale R-16 ICBM designed by Yangel developed a fuel leak that had to be welded closed. The procedure called for purging the massive fuel tanks before making repairs—"safing" the rocket, in American slang—but in a desperate effort to save time because Khrushchev wanted a propaganda triumph during his visit to the States, work went on with the missile's stages fully fueled. Thirty minutes before its scheduled launch the following morning, a technician pulled an electronic component out of the second stage, which caused a spark that led to a tremendous explosion. The exploding upper stage instantaneously touched off the dimethyl hydrazine and nitrogen tetroxide in the huge first stage. It erupted in a colossal explosion that incinerated ninety-two people beyond recognition, including Mitrofan Ivanovich Nedelin, the highly regarded commander in chief of the newly formed Strategic Rocket Forces and a Hero of the Soviet Union. Twenty-three design bureau representatives, some of them high-ranking scientists, and many skilled technicians were also lost. Some of the Russians died instantly. Others, their clothes in flames, ran from the pad in agony before they collapsed. The explosion was kept secret, of course.

The accident was taken so seriously that Leonid I. Brezhnev, the nation's newly appointed president, was made chairman of a special investigation commission and arrived at Tyuratam the next morning. He was greeted by Korolyov and Yangel, who were alive only because they had stayed in the concrete bunker. Zhores A. Medvedev, a dissident Soviet scientist, wrote years later that many top scientists had blamed Khrushchev himself for the catastrophe and that it had contributed to a worsening morale problem.

Five weeks later, two dogs and other animals were incinerated when their Korabl Sputnik plowed into the atmosphere at too steep an angle and burned to cinders. As the name "Korabl" implied, the spacecraft was another test model of the Vostok man carrier.

Nor were the animals the only ones to perish perfecting Vostok. An unlucky test pilot and parachutist named Pyotr Dolgov was killed testing the craft's ejection seat when it shot him head first into the hatch during a practice drop. Korolyov then ordered that there be a two-second delay between jettisoning the hatch and firing the ejection seat and, to make certain that no cosmonaut would be snagged while ejecting, that the hatch be made larger and the cosmonauts be selected from smaller specimens. In keeping with Kremlin practice, such disasters were routinely covered up. The official line on Nedelin's death, for example, was that he had been killed in a plane crash. Dolgov was not even worth a fabricated story, so his passing was officially ignored; just another pilot who had died in the service of the motherland.

The death toll in the race to space was and would continue to be measured in terms of the men and women who went there, or tried to, and who were killed in the attempt. But, at least where the Soviets were concerned, the contest was far more costly than that. The Nedelin disaster directly deprived their space program of many individuals who were desperately needed to sustain the appearance of superiority, however meaningless that was. And there was no doubt as 1960 gave way to 1961 that the next level of the game would have to do with actually sending humans to follow the dogs and monkeys, just as they had long ago followed the sheep, the duck, and the rooster. It was time for a man to ride the great arrow.

9

A Bridge for Galileo

"Poyekhali!"

With Korolyov, Glushko, Keldysh, and others watching through a periscope in a nearby bunker, the modified R-7 carrying Yuri Gagarin ignited and began its slow climb at a little after nine o'clock on the morning of April 12, 1961. With that word—"Here we go!"—Gagarin began feeling the big rocket's acceleration build slowly at first, then faster, as he arced out over the Kazakh steppe and into the pages of history.

The twenty-seven-year-old test pilot felt exhilarated as he sped into orbit, strapped securely onto his specially contoured couch, while he tried to stay in radio contact with the ground. For Korolyov, who was more than twice his age, the flight of the first human into space was a nerve-racking ordeal that further strained an already weakened heart. The chief designer had fallen asleep, medicated and exhausted, in a specially built wooden cabin not far from Gagarin's rocket early on the previous night. But he had awakened at 3 A.M. and spent the hours until dawn talking with a colleague who also had trouble sleeping. (Both Gagarin and Titov, his backup, had slept soundly, as would be expected of men who routinely risked their physical hides but who did not have a political stake in the mission.)

Then, during Gagarin's ascent, the telemetry abruptly broke off. The telemetry operator's steady "Five . . . five . . . five . . . ," meaning that everything was perfect, suddenly turned to "Three . . . three . . . three . . ." Korolyov, fearing that catastrophe had struck, tore into the room demanding to know what had gone wrong. Moments later, with the radio link reestablished, he was relieved to hear "Five . . . five . . . five . . ." once again. Eight minutes after launch, its E-rocket upper stage having fallen away, the silent *Vostok* streaked over Siberia and on toward the north Pacific and Alaska. The CIA and the National Security Agency, which monitored the flight from start to finish by eavesdropping on the telemetry and breaking into the television transmission sent from the *Vostok* capsule to Earth, saw that the total burn time to get the spacecraft to orbit was twelve minutes, eight seconds, and that the last stage itself fired for 423 seconds out of the total. The pressure on the cosmonaut himself as he slid into orbit, they also saw, was less than 1 g.

Gagarin, now weightless, reported his and his spacecraft's condition to Kaliningrad, practiced eating and drinking from tubes, took notes on a pad that floated in front of him, broadcast revolutionary greetings to the oppressed peoples of the world, and admired the view. Free of gravity, he experienced the physical freedom for which Tsiolkovsky had yearned. "I felt wonderful when the gravity pull began to disappear," he would recall. "I suddenly found I could do things much more easily than before. And it seemed as though my hands and legs and my whole body did not belong to me. They did not weigh anything. You neither sit nor lie, but just keep floating in the cabin. All the loose objects likewise float in the air and you watch them as in a dream."

He also watched Earth. "On the horizon I could see the sharp, contrasting change from the light surface of the earth to the inky blackness of the sky," he went on. "The earth was gay with a lavish palette of colors. It had a pale blue halo around it. Then this band gradually darkened, becoming turquoise, blue, violet and then coal-black" as he plunged into the night at nearly 18,000 miles an hour.

What Gagarin did not do was control his spaceship because the manual control override was kept under a combination lock. In yet another parallel to the Mercury program, a number of Soviet scientists feared that psychological stress would not only incapacitate cosmonauts, but could actually cause them to interfere with the spacecraft's operation. And like *their* American counterparts, the cosmonauts had objected. But unlike the Americans, it had been to no avail, even though the *Vostok*'s autopilot had failed on two tests. Finally, as a concession, it had been decided to put the combination to the lock in an envelope that was accessible to the cosmonaut. On the occasion of the first manned flight into space, however, the autopilot cooperated by working perfectly. Yuri Gagarin therefore flew the single-orbit, 108-minute flight as a man in a can.

The Russians who tracked Gagarin on land and sea were ordered to establish the fact that he was actually in orbit within ten minutes of engine cutoff for reasons that were more political than technical. In the event that the flight went well, Khrushchev and his advisers wanted to get news of it on the air twenty-five minutes after liftoff so other governments could track it and listen for themselves. No one was going to accuse the Soviet Union of fabricating the story.

The sensitivity was understandable. By the spring of 1961, stories were circulating in the West that cosmonauts were dying in space; that the Reds were sacrificing pilots to win at any cost. Colonel Oleg Penkovsky, a disillusioned Soviet soldier who provided some otherwise excellent information to the United States before he was executed, sent out one report claiming that a dozen cosmonauts had preceded Gagarin. Because it was following all Soviet launches with utmost care, the U.S. intelligence community knew that he was making it up. But the idea that the Communists were sacrificing young pilots probably soothed some Americans, knowing as they did that they had been humbled by an obsessively secretive and repressive regime. Glennan himself struggled with the issue of killing cosmonauts, noting in his diary that he didn't think "they could stand the horrified criticism of the rest of the world were they to do this. On the other hand, there is no guarantee that they have not already put a man into space and left him there. There seems to be some evidence that the May 15 [1960] shot was just such a shot." He referred to *Korabl Sputnik 1,* which had carried Ivan Ivanovich, the dummy, and which had indeed gotten marooned in space for five years.

The administrator's attitude, widely shared by those of his countrymen with fresh recollections of Stalin's renowned brutality, was understandable. But it was wrong. Khrushchev was no Stalin. More to the point, Korolyov was no fool. He was immensely protective of his "little eagles," as he called the cosmonauts, and usually took every precaution to ensure their safety. All Korabl Sputnik flights transmitted live television pictures and sent continuous radio reports, both of which were routinely intercepted by U.S. intelligence.* There was absolutely no truth to the rumor, however comforting it was to frustrated Americans, that their enemy was using its best pilots as space fodder because life under communism was cheap. Korolyov would not knowingly flirt with disaster until he tried to squeeze three men into one gutted *Vostok* that was renamed *Voskhod.*

The Kremlin was so sensitive to charges of lying that it ordered Gagarin's landing method be kept secret. The possibility of fire on the launchpad or some other accident soon after liftoff had compelled *Vostok*'s designers to make the seat on which the cosmonaut was strapped ejectable so it could be shot away from the spacecraft and come down by parachute. Furthermore, the

* The television images of Gagarin were broadcast on eighty-three megacycles and were picked up by National Security Agency antennas twenty minutes after launch as the spacecraft passed over Alaska. Fifty-eight minutes after launch, the NSA reported that a reliable real-time readout of the signals clearly showed a man who was moving. U.S. intelligence therefore confirmed that a Soviet cosmonaut was in orbit and alive before he completed his single orbit around the world. Besides gaining valuable insight into the operation, this helped U.S. intelligence prevent President Kennedy's congratulating the Soviet Union for a feat that had not happened, which was a constant consideration. (Plaster, "Snooping on Space Pictures," p. 34.)

spherical capsule itself was supposed to parachute onto Soviet territory after it released its equipment module for a hard landing that could have hurt or killed its occupant. Gagarin and his seat would therefore be ejected at an altitude of about 21,000 feet, and he would float down on his own chute. But fear that a cosmonaut coming down like that would lead to charges that there had been an emergency or that he had jumped out of an airplane, nullifying the glorious deed, convinced the leadership to avoid all mention of the bailout and maintain that he descended in his spaceship. All cosmonauts would be credited with riding their ships down in a charade that would go on for years.

Contrary to what Soviet space officials and Gagarin himself told the world, however, the "can" almost killed the man. Toward the end of the flight, the spacecraft went into a potentially deadly spin when the equipment module failed to separate from the sphere that held Gagarin. The module contained the braking rocket that was supposed to slow *Vostok* as it re-entered the atmosphere and was designed to fall away as soon as the retro-rocket turned off. But it did not jettison completely because all of the "umbilical cord" of wires and electrical cables connecting it to the sphere did not fully disconnect. With some cables and wires still holding the re-entry and equipment modules together, the dangling equipment module started to swing wildly behind Gagarin, thrashing from side to side like an out-of-control kite in a stiff wind. This quickly sent him into an end-over-end lurching spin at 17,000 miles an hour for ten terrifying minutes until the sphere and the conical pod finally broke free of each other.

"Everything was spinning around," Gagarin would note in his report the next day. "The rotation speed was about thirty degrees per second. . . . I was performing a kind of ballet: head-legs, legs-head with a very high rotation speed. . . . I had time only to hide my eyes from the sun to avoid its rays. . . . I felt that too much time passed, but there was no separation. . . . 'Descending 1' doesn't go on the display, 'Get ready for ejection' doesn't turn on. There is no separation. The ballet goes on." His commander, Colonel Yevgeni A. Karpov, frantically scrawled "Malfunction!!!" and "Don't panic" on a pad in the mission control room, though it is unclear whether he did so during the flight or immediately afterward. Even after separation, the three-ton cannonball in which Gagarin was strapped wobbled as it continued its fiery plunge back into the atmosphere. Since the capsule's heat shield was between it and the equipment module, it probably would have been useless if the module hadn't finally broken away. The cosmonaut and his *Vostok* would have been incinerated.

At 9:59 A.M. local time (1:59 A.M. Cape Canaveral time), Moscow Radio crackled onto the air with yet another historic announcement: "Today, 12 April 1961, the first cosmic ship named *Vostok,* with a man on board, was orbited around the Earth from the Soviet Union. He is an airman, Major Yuri

Gagarin." Shepard, Grissom, Glenn, and the others were awakened and given the news. Not far away the Mercury-Redstone that Shepard was scheduled to ride down the Atlantic test range in only sixteen days stood locked in the steel embrace of a gantry, grounded in the night.

Meanwhile, *Vostok 1*'s retro-rocket fired on schedule over West Africa, knocking some three hundred miles an hour off its speed and sending it into a long descent toward home. Radio contact was again broken, this time because of a normal blackout caused by ionized plasma that enveloped the spacecraft as it plunged back into the atmosphere. Minutes later Gagarin's billowing parachute dropped him, resplendent in his bright orange space suit and shiny white visored helmet, onto a collective farm near the Volga. The cosmonaut had hardly climbed out of his space suit when the premier, never one to be reticent, set Gagarin's feat in perspective. "You have made yourself immortal," proclaimed Nikita Khrushchev.

Two days later, a military transport escorted by seven MiGs landed Gagarin at Moscow's Vnukuvo Airport, where he marched on a thick orange carpet at the end of which stood a beaming Khrushchev, members of the Council of Ministers, and other notables. There were kisses and bear hugs, saluting and applause, and effusive patriotic greetings, followed by a limousine parade into the capital, where millions lined the streets and cheered. Flowers and red banners sprouted everywhere. The parents of the first human being to rocket around the world were on hand, too: he in his carpenter's cap, she in her shawl. The European Broadcasting Union carried the ceremonies live on television. There had been nothing to match the celebration since the end of the Great Patriotic War.

Under a headline that spanned four of the paper's eight columns—"RUSSIAN ORBITED THE EARTH ONCE, OBSERVING IT THROUGH PORTHOLES; SPACE FLIGHT LASTED 108 MINUTES"—*The New York Times* on April 13 ran a photo of jubilant youngsters outside the Moscow Planetarium and another of their hero hinting at a smile like a cosmic Mona Lisa in a leather flight cap. The newspaper of record (as it called itself) ran a hastily assembled spate of sidebars that described the flight with maps and diagrams, provided background on the preparations for the mission, and even a transcript of some of the hero's radio chatter with Kaliningrad. It also carried congratulations from Kennedy. Werhner von Braun offered congratulations, while Edward Teller offered only sour grapes. He blamed the Communist triumph on the same sort of "unimaginative, materialistic thinking" in his adopted country that had led to the Sputnik disaster. This new and devastating blow had occurred, in other words, because of the near-subversive combination of the lamentable Dwight Eisenhower, who had pitched insouciance, and Ed Herlihy, a narrator of newsreels, whose Horn & Hardart Automat commercials pitched "less work for mother."

Less work, indeed. A Man-in-the-News profile of Gagarin showed him beside Valentina, who was reading to their two-year-old daughter (indicating that Valentina Gagarina, for one, knew how to work). The article noted that "Gagarin" derived from "wild duck." And there was a cartoon from Baltimore's *Sun* that showed Khrushchev holding a red star in space with one hand and banging a likeness of Kennedy over the head with a shoe in the other.

The *Times* also carried a roundup of congratulations from other nations. And, notably, Kennedy was quoted as promising to increase Eisenhower's space budget by 11.8 percent and ordering an acceleration of work on von Braun's huge Saturn booster. But, the new president cautioned, the United States would continue to trail its rival for some time.

And there was the full text of the victor's statement, in which the party's Central Committee indulged in the sort of excessive self-congratulation that had long since become a parody of itself and a source of material for comedians beyond the Kremlin's reach. "The first man to penetrate space was a Soviet man, a citizen of the U.S.S.R. . . . This is an unparalleled victory of man over the forces of nature. . . . In this achievement, which will pass into history, are embodied the genius of the Soviet people and the powerful force of socialism. . . . Our country has surpassed all other states in the world and has been the first to blaze the trail into space."

There was more to the exorbitant chest pounding than politics, economics, and the dialectic of history. It had a psychological dimension as well. The Central Committee's statement reflected a cultural inferiority complex that had gnawed at the Russian soul at least since the sixteenth century, when a feudal Russia had begun to look in awe at a Europe that was developing centralized states and an urban, industrialized bourgeoisie that in turn spawned standing armies, new techniques of state administration, communication and monetary policy, advances in the arts and sciences, and merchant fleets that brought back riches from around the known world. "The Western influence gained ground as we realized our material and spiritual poverty," Russian historian V. O. Kliuchevsky noted even before the Soviet Union was born. Russia had no part in those achievements. As a result, a pervasive "moral and spiritual subordination" and a belief in Russian backwardness spread across the land, Kliuchevsky maintained, and it led directly to the wholesale (and humiliating) importation of the sort of Western culture that fills the pages of *War and Peace.*

Khrushchev, whose father had been a miner and who himself had been a shepherd as a youngster, was eminently aware of all that. He also knew that a cultural condescension persisted; so many Westerners dismissed the Russians as Asians that Russians themselves had accepted the ethnologically incorrect notion. Looking back on Sputnik a quarter century later, Simon Ramo, a founder of the aerospace giant TRW, would be explicit on that score: "We knew the Russians excelled in ballet and caviar, but when the proper time came to launch an artificial moon, we Americans expected to be the ones to do it."

"Bourgeois statesmen used to poke fun at us, saying that we Russians were running around in bark sandals and lapping up cabbage soup with those sandals," Nikita Khrushchev would tell a Polish audience two years after Gagarin's flight. "They used to make fun of our culture, the culture of a people considered, so to say, to be the last among the civilized Western countries. Then suddenly, you understand, those who they thought lapped up the cabbage soup with bark sandals got into outer space earlier than the so-called civilized ones." Like six-guns in the American West, rockets and missiles were the new equalizers, at least in the eyes of those who saw themselves as history's underdogs.

Ham in a Can

Gagarin had in fact been beaten to space by an American primate. It just happened to be a chimp. On January 2, a contingent of six "astrochimps," accompanied by twenty handlers and medical specialists, arrived at Canaveral from Holloman Air Force Base in New Mexico. Their assignment was to learn how to pull certain levers. In addition, their pulse, respiration rate, and other vital signs were to be studied while they were weightless and under high g forces before the first astronaut flew. Pulling the correct lever was rewarded with a banana pellet; pulling the wrong one triggered a mild shock. Three weeks after they arrived at Cape Canaveral, all six were bored, well-fed experts at their assigned tasks.

Just before noon on January 31, a particularly frisky chimp named Ham (for Holloman Aerospace Medical Center) that was strapped onto a special "biopack" couch blasted off in a Mercury capsule on top of a Redstone. A minute later, those who were monitoring the flight in the new Mercury Control Center saw that the rocket was climbing at least one degree higher than its intended trajectory and that the worrisome angle was increasing. The Redstone had developed a "hot engine" that sucked in all of the fuel five seconds faster than called for by the flight plan. The higher angle meant that Ham, now furiously trying to avoid shocks to his feet by pulling levers with both hands every time a white or blue light flashed, was pulling more than 7 gs. Two minutes and eighteen seconds after liftoff, the cabin pressure suddenly dropped because of a malfunctioning air inlet valve that had been loosened by vibration. Then the liquid oxygen ran out, causing the Redstone's engine to shut down.

Sensing that the flight was going awry, the Mercury's emergency abort system kicked in, firing three rockets mounted on a tower on the capsule's nose. That abruptly pulled the capsule off the Redstone. Meanwhile, the retrorockets that were supposed to have slowed the capsule as it came down dropped off prematurely, meaning that the hapless primate plunged back into the atmosphere at 5,857 miles an hour, or more than 1,400 miles an hour faster than planned. Ham therefore pulled almost 15 gs, which was three more than expected. The higher speed also meant that he overshot the recovery area, complicating efforts to retrieve him. Worse, the spacecraft's electrical system went haywire. At one point, as the ape drifted weightlessly, he pulled the right levers, but instead of being rewarded with banana pellets, he got the shock treatment.

When Ham's capsule hit the water 422 miles downrange from Canaveral, or 132 miles beyond where it was expected to come down, there was no human in sight. Furthermore, the beryllium heat shield punched two holes in the capsule's titanium bulkhead because of the impact of the landing, causing it to take on water and capsize. The helicopter crew that was first on the scene, and which hoisted the spacecraft and its occupant out of the ocean, estimated that it had taken on about eight hundred pounds of seawater.

Ham's harrowing experience took the edge off his disposition. The handlers who opened his container on board the recovery ship found a thoroughly infuriated space veteran who bared his fangs and bit anything, human or otherwise, he could reach. But Ham's survival in spite of a series of dangerous malfunctions raised confidence that a man could survive a harrowing ride as well.

Typically in the big-rocket business, Ham's wild ride had been caused by a kind of Rube Goldberg syndrome. It underscored the fact that the huge lifters were dangerously complicated, with engines that had fifteen thousand or more parts, and that a glitch in any of the more than a thousand critical ones could start a chain reaction leading to catastrophe.* In the case of that Redstone, for example, a single small control valve that had not properly regulated the flow of hydrogen peroxide to the steam generator that powered the fuel pumps forcing fuel into the engine malfunctioned. This pushed too much fuel into the engine, so it got "hot."

Von Braun, who always carried the picture of an exploding astronaut in his head, and who readily understood what *that* sight would do to public support for the program, decided another Redstone would have to be tested, however minor the problem had been, before an astronaut went up. The decision sent the impatient Shepard, who wanted to ride the next rocket, into a rage. Com-

* The space shuttle's three main engines would have 120 million such parts, many with extremely tight tolerances.

plaints to Christopher C. (for Columbus) Kraft, the Mercury flight director, went nowhere because, as Kraft put it, the last word was von Braun's. But von Braun's decision to hold back Shepard would prove to be as fateful as Eisenhower's decision to hold back von Braun. The next Redstone, which could have carried Alan Shepard on his suborbital flight, flew a perfect trajectory on March 24 and landed 307 miles downrange in the Atlantic.

However happy the flight made von Braun and the other NASA rocketeers, who now pronounced Redstone "man-rated," it made Shepard so furious that, as he himself put it, he could have committed murder. Had he been on that particular rocket, he would have beaten Gagarin and, as he also put it, led the world into space. "We had 'em by the short hairs," he would later recall bitterly, "and we gave it away." Not quite. A suborbital flight and one that circled the globe were entirely different propositions. Shepard's being first to touch space as he arced over the South Atlantic would in no way have lessened the Soviet triumph.

Camelot Under Siege

Kennedy had won the presidency not only by attacking Eisenhower's domestic and foreign policies, with defense and space prominent among them, but also on his appeal as a young, energetic, and mannered aristocrat who could replace Ike's benevolent (and allegedly misguided) patriarchy with style, vigor, and direction. But within a hundred days of settling into the White House, the new president faced a series of threats that severely tested his own mettle. There was turmoil in Alabama and elsewhere in the South as Freedom Riders and segregationists fought bloody battles over racial integration. There was a crisis in Laos, where the Communist Pathet Lao were dangerously close to toppling the pro-American government of Phoumi Nosavan, causing the Joint Chiefs of Staff to warn that it would take sixty thousand soldiers, air support, and probably tactical nuclear weapons to prevent another Korea from erupting. If Laos was gripped by communism, one theory had it, Cambodia, South Vietnam, and other countries in Southeast Asia would fall like so many dominoes.

And there was a plot, hatched by the CIA during the waning months of Eisenhower's second term, to invade Cuba with less than two thousand exiles in the belief that spontaneous uprisings would take place that would rid the hemisphere of Fidel Castro once and for all. Kennedy signed off on the operation, which was scheduled for April 17, but rejected the use of aircraft to support the invasion as being too risky politically. Outnumbered and lacking air cover, the pathetically underarmed insurgents from Florida were quickly routed, turning the expedition to the Bay of Pigs into an embarrassing fiasco for the nation that was their obvious, if halfhearted, sponsor. And exactly one week before that, there had been the epic—and galling—flight of Major Gagarin.

While Kennedy had no intrinsic interest in space, he saw it pragmatically: as a new area of conflict. He laid the long string of defeats by the Soviet Union at both Truman's and Eisenhower's door and remained convinced that neither of them had sufficiently appreciated the military importance and symbolic value of superiority in space. So Kennedy had made specific reference to a U.S. presence in space in his inaugural address, adroitly calling for cooperation with the Russians to "explore the stars" while warning his countrymen that if the Communists continued to lead the way to space, they would "occupy" it with dangerous (though unspecified) consequences. The U.S.S.R.'s widely heralded triumphs, in the words of his special counsel and confidant, were creating "a dangerous impression of unchallenged world leadership generally and scientific pre-eminence particularly. American scientists could repeat over and over that the more solid contributions of our own space research were a truer measure of national strength," Theodore C. Sorensen would write soon afterward, "but neither America nor the world paid much attention."*

Yet the Kennedy administration's immediate actions did not entirely match its rhetoric. They reflected a confusion that betrayed the election campaign's political thunder. By the beginning of January 1961, for example, several senior Eisenhower officials were working with their Kennedy-appointed replacements as part of the transition process. But not at NASA. Although Kennedy had given Johnson the responsibility of picking someone to succeed Glennan in December, Eisenhower's administrator left Washington on January 19 without being able to so much as wish his successor well, let alone brief him, because there was no successor. (General Thomas D. White of the Air Force expressed interest in the job, which undoubtedly made Eisenhower and Glennan wince.) The delay annoyed Kennedy. It stemmed from confusion as to precisely what direction the space agency was supposed to take. James E. Webb, a former congressional aide, business executive, and director of the Bureau of the Budget in the Truman administration, finally succeeded Glennan on February 14. The president's confidence in the new administrator, however, was limited.

New York Times military columnist Hanson W. Baldwin picked up on the

* Lloyd Berkner, chairman of the National Academy of Science's Space Science Board, was one of the most prominent of those who repeated the assurance. "One gets a little tired these days of reading about Russian space supremacy," the physicist complained, adding that the Soviet Union looked better mostly because of its success "in the weight-lifting contest." He went on to cite Pioneer 5's trailblazing deep-space mission, in which data from one fresh discovery after another were sent back from 20 million miles away, as just one example of American prowess. ("U.S. Held Leader in Space Science," *The New York Times,* May 28, 1961.)

situation when Webb testified before Congress immediately after Gagarin's flight that the United States remained committed to scientific achievements rather than to matching Soviet spectaculars. That attitude, he grumbled, sounded almost like a phonograph record from the Eisenhower administration. It was precisely the philosophy that had marred America's image abroad and hobbled its space program even before Sputnik, the exasperated columnist wrote, adding that space had to be given a top priority or "we face a bleak future of more Soviet triumphs." There were at least two ways the United States could regain credibility, Baldwin concluded: build a space station or send astronauts to the Moon.

As Kennedy had charged, incorrectly, that there were bomber and missile gaps; now the appearance of a growing space gap loomed. But suddenly it was he, not the gentlemanly general, who was becoming the target of restive wrath in Congress and the press. This new shortcoming, this current embarrassment, was his. The day after Gagarin's flight, with all the world seemingly in awe and with America's own soldiers, legislators, and journalists venting angst, Webb tried to relieve his president's oppression and gain some points for NASA by walking into the Oval Office and presenting him with a model of the Mercury capsule. A thoroughly disgruntled Kennedy looked at the thing and wondered whether Webb had bought it in a toy store on the way to work that morning.

Johnson, as head of the National Space Council, was working on the problem with Robert McNamara, the new Secretary of Defense. So was Sorensen. "Find out when, at what point, we can overtake the Russians," Kennedy instructed his confidant. "How long is this going to continue?" He made no mention of reversing the situation in space. Yet it was clear that a response would have to be made there.

On Friday, April 14, Sorensen met in the Cabinet Room with Wiesner, Webb, Budget Director David Bell, Dryden, who had stayed on as deputy administrator of NASA after Glennan left, Edward O. Welsh of the National Space Council, and some others. They were soon joined by the president, who was briefed on their discussion.

The gospel according to NASA more or less reflected von Braun's old blueprint: build systematically on Mercury, gradually extending missions to longer one-man flights, then two-man flights, followed by an orbiting laboratory, a fixed space station, a manned mission around the Moon, a manned landing on the Moon, manned exploration of Mars, and a spaceplane that would routinely connect the United States and its orbiting outposts.

The plan was orderly and logical. But it also virtually guaranteed that the hare would stay ahead of the tortoise in the propaganda race by continuing to stage spectaculars. If the United States was going to compete, and there was no doubt in anyone's mind in that room with the exception of Weisner that the race should in fact be run, then the options became severely limited. Given its long lead in heavy lifters—*Vostok 1* weighed more than 10,000 pounds—it seemed certain that Soviet two- and possibly three-man crews were going to reach Earth orbit first and then probably be first around the Moon. There was no point, either scientifically or politically, in keeping small groups of spacefarers in Earth orbit indefinitely. It was implicit that sooner or later they would have to reach for the Moon or close down the manned program and go back to robots. That being the case, the only way to win decisively would be to land Americans on the lunar surface before the Russians got there. And that in turn, the group agreed, would require a great deal more than heavy lifters. The engineering would have to be phenomenal. So would the budget.

Life's Hugh Sidey, whose magazine had bought exclusive rights to chronicle the Mercury Seven's journey to space, had his own privileged access to Kennedy. And it was in that capacity that he showed up late on the afternoon of the fourteenth, just after Sorensen's weary group began briefing Kennedy, to find out what the president planned to do about what was broadly seen as the space fiasco. Sorensen took the columnist into the Cabinet Room, where the president rehashed the discussion for him. "Now let's look at this," Kennedy said, impatiently. "Is there any place we can catch them? What can we do? Can we go around the Moon before them? Can we put a man on the Moon before them?" Dryden answered that there was a fifty-fifty chance of landing Americans on the Moon first, provided there was an all-out effort to do so. It would have to be another Manhattan Project, the old NACA hand added, referring to the development of the atomic bomb. And it would cost $40 billion. More discussion. Now JFK spoke again, this time beseechingly. "When we know more, I can decide whether it's worth it or not. If somebody can just tell me how to catch up." Then, Sidey would soon report, Kennedy glanced from face to face. "There's nothing more important," he said quietly.

"Have you got all your answers?" Kennedy asked Sidey as the meeting ended.

"Well, yeah," the journalist answered, "except, what are you going to do?" Sorensen would later recall that Sidey kept pressing the question, compelling the president to come up with an answer. Finally, Kennedy asked Sidey to leave the room so he could be alone with Sorensen. Together, the two men reviewed the group's earlier discussion. At last Sorensen came back out into the hall, where Sidey was going over his notes.

"We're going to the Moon," Sorensen told Sidey.

But that was less certain than it sounded on the evening of the fourteenth. The wisdom of embarking on a manned voyage to the Moon would continue to be intensively thrashed out in the days that followed, not only in the White

House, but at NASA and the Pentagon. Five days later, still mulling it over, Kennedy asked Johnson for his own suggestions. He was told that the Space Council ought to hold hearings on the matter and develop a program that would fly in Congress. Kennedy responded the next day by sending the vice president a memorandum charging him with deciding what could be done in space that offered the best chance of "beating the Soviets." Webb, Wiesner, McNamara, and others were instructed to cooperate. Von Braun and Schriever told LBJ that they, too, were for sending men to the Moon. The President's Science Advisory Committee, which was responsible for reviewing the scientific value of all federal projects, was not consulted. Wiesner, the president's otherwise valued science adviser, was present at several meetings but was rarely asked for his opinion. Science and the exploration that related to it would not only take a distant second place to politics, but they would become its instruments.

The point was that while science was a sacred icon, the foundation of the technology that built the nation, it was about as relevant to the Moon program as nutrition was relevant to dining in a three-star French restaurant. It was not, nor could it be, part of the definition of the Moon program. To the complaint by scientists that science was given a relatively low priority in that enterprise, two dissenters would answer by saying, so what? Writing in *The Atlantic* two years later, NASA physicist Robert Jastrow and Homer Newell, then the director of the agency's Office of Space Sciences, argued that "while science plays an important role in lunar exploration, it was never intended to be the primary objective of that project. The impetus of the lunar program is derived from its place in the long-range U.S. program for exploration of the solar system," they added, more loftily than was warranted. "The heart of that program is man in space, the extension of man's control over his physical environment. The science and technology of space flight are ancillary developments which support the main thrust of manned exploration." The pace of the Moon program, they added, "must be set not by the measured patterns of scientific research, but by the urgencies of the response to the national challenge." That made it blatantly political.

At that point, going for the Moon was all but assured as Kennedy's prime target because of the dramatic impact it would have. And not incidentally, it would also create jobs and feed a young, aggressive, and hungry aerospace industry. "We talked a lot about do we *have* to do this," Wiesner would later recall. "He [Kennedy] said to me, 'Well, it's your fault. If you had a scientific spectacular on this earth that would be more useful—say desalting the ocean—or something that is just as dramatic and convincing as space, then we would do it.' . . . If Kennedy could have opted out of a big space program without hurting the country in his judgment, he would have. . . . I think he became convinced that space was the symbol of the twentieth century. It was a decision he made coldbloodedly. He thought it was right for the country."

But John Kennedy was a prisoner of his own temperament and, more subtly, of an increasingly powerful space establishment. He therefore could not easily have opted out. History would credit him with making the decision to send Americans to the Moon. Technically, that was true enough. But Apollo and its precursors had by then become so clearly defined, had gathered so much momentum, were pointed so certainly in one direction, that in reality there was virtually no other way. Breaking the inertia at that point, particularly given his scrappy nature, was more unthinkable than probably even he realized.

Couturier to the Stars

Some 45 million Americans, many of them with vivid memories of Vanguard and other rockets blowing up as they fought for altitude, watched their television sets at 9:34 on the morning of May 5, when an astronaut strapped to the top of a Redstone rose from its pad at Canaveral and disappeared high over the Atlantic. Technical glitches (and the nightmarish prospect of horrified Americans and smirking Russians seeing the inaugural astronaut being broiled alive in an explosion within days of Gagarin's triumph) had caused the launch to be delayed. And the delay caused an unexpected problem: Shepard, cocooned in his space suit and strapped securely onto a contoured couch for hours, had to urinate.

Mercury capsules, like all U.S. and Soviet manned spacecraft, were carefully designed to support the lives of their occupants with attention paid to air supply and filtration, radiation shields, and other systems that made them artificial environments. But it was recognized from the beginning that spacefarers, like high-flying aviators, would also need to wear their own environments to counteract the very high g forces that would come with acceleration and deceleration and to ensure that they had a backup life-support system—redundancy, in the jargon of the business—in case the spacecraft itself sprang a leak or suffered some other horrible failure. And obviously there could be no lunar excursions without appropriate apparel. A number of real dangers, foremost among them being a lack of air and high temperatures, required that people who traveled to the Moon be appropriately attired.

Space suits did not spring full-blown out of the space program. Like almost everything else that went to space, they evolved. The concept of a wearable environment went back to the clunky suits and metal helmets worn by deep-sea divers to ward off the bends and provide air, and then to their high-flying counterparts. Military pilots began wearing full and partial pressure suits at the beginning of the jet age and advanced versions of them were fitted to the mach-busting experimental rocket pilots. But pressure suits predated even the jet age.

The father of the space suit, the Calvin Klein of astronautics, was undoubtedly Russell Colley, a B. F. Goodrich engineer and frustrated women's fashion designer who played a key role in creating the Mercury astronauts' and their successors' ensembles.

Colley invented a rubber pressure suit and an aluminum helmet for Wiley Post, the one-eyed daredevil who flew his Lockheed Vega, *Winnie Mae,* to a record 55,000 feet in 1934. Colley stitched the suit together on his wife's sewing machine while Post passed the time teaching their ten-year-old daughter, Barbara, to shoot craps. The metallic cloth and tin can–shaped helmet, complete with an off-center viewing window to accommodate Post's good eye, made him look like he belonged on the set of a Saturday morning sci-fi movie serial. In fact the celebrated aviator reported that a bystander, seeing him in the eerie getup as he walked away from an emergency landing, took him for a Martian and nearly fainted from fright.

Colley's pressure suit designs for Navy pilots in World War II evolved into the suit that Shepard wore on that fateful May morning. The outfit, a modified Navy Mark IV pressure suit, was a marvel of understated complexity. Temperature was precisely regulated, for example, while body odor was drawn off and sent through activated charcoal. Oxygen entered through the torso and left through the helmet. It was the combination of artistic imagination and technical expertise, to take only one example, that gave Colley the idea of adapting the movement of a tomato worm to solve a flexibility problem in a space suit.

Yet even Russell Colley and his colleagues had one notorious lapse. Like the architects of the Yale Bowl who were so taken with the gladiatorial splendor of their design that it did not occur to them to include restrooms until seventy thousand people showed up for the first game, the space suit designers forgot to provide Shepard and the other astronauts with a waste relief system.

Shepard Wets His Diaper

And so it was that, inevitably, nature began to call to Alan Shepard as the glitches and cautious "holds" stretched the time until liftoff. "I soon discovered that the bladder was getting relatively full," Shepard later recalled. So he asked von Braun to be let out of the capsule to go to the men's room. The answer was an emphatic no. "So, I told the folks that I was going to relieve myself. And they said, 'no, you can't do that; you'll short-circuit everything.' " Shepard suggested that the power be turned off, and after a few minutes' deliberation, it was done. He then turned Colley's insulated space suit into the world's most expensive diaper. But as Shepard began to dry out, his impatience and anger flared again, so he told Mission Control that he was willing to risk a shock and that the power therefore ought to be turned back on. Ham, he did not need reminding, had beaten him to space while managing to endure his own shocks. But there was hesitation in the blockhouse. "I'm getting tired of this," Shepard finally snapped. "Why don't you light the damned candle, 'cause I'm ready to go." Von Braun ordered the candle lit.

Shepard Goes Ballistic

As the black-and-white cylinder gained velocity, with the launch complex and the adjacent ribbon of golden sand slowly falling away to the rocket's rumble, Alan Shepard performed the first of the space liturgies that would soon become more familiar to his countrymen than those they heard in church:

"Ahhh . . . Roger. Lift-off and the clock is started. . . . Yes, sir, reading you loud and clear. This is *Freedom 7*. The fuel is go. Cabin at fourteen psi. Oxygen is go. . . . *Freedom 7* is still go."

And "go" it did. The first forty-five seconds went so smoothly that it surprised Shepard. Then everything began to shake as the Redstone hit the transonic speed zone where air turbulence builds up. *Freedom 7,* still plugged into its flaming lifter, hit the point of maximum aerodynamic pressure—the equivalent of the sound barrier—at about eighty-eight seconds, causing Shepard to bounce so hard he could not read the dials. Then the noise and buffeting abruptly stopped. As the Redstone accelerated to more than 5,100 miles an hour, the man riding it was pressed into his seat by a force more than six times his body weight, making him feel as if he had turned to stone. The engine shut off 141 seconds after liftoff, right on schedule. Thirty-eight seconds later, with the Redstone separated and starting its own long plunge into the Atlantic, *Freedom 7*'s retro-rockets fired, turning it around so that it was flying heat shield–first.

As the spacecraft approached its maximum altitude of 116.5 miles, Shepard began the most important of the tasks he had been assigned: seeing whether he could control *Freedom 7*'s attitude manually. He did, causing slight changes in all three of the spacecraft's axes. It was a perfect time to test the procedure, as it turned out, because the thrusters had misaligned the spacecraft's pitch—its up-and-down position—in the automatic mode. Peering through a periscope, he could see the west coast of Florida and the Gulf of Mexico, Lake Okeechobee in central Florida, and the Bahamas. The periscope automatically retracted when *Freedom 7* began its almost-12-g plunge back into the atmosphere. At 10,000 feet, the antenna canister blew off, pulling out the main parachute, which popped open with a reassuring jolt. Shepard would later report that *Freedom 7* hit the water with about the same force as a plane landing on the deck of a carrier. It immediately listed about sixty degrees, then quickly righted itself like the large black cork it had now become. He climbed through the hatch, hopped into the water, and slipped into the "horse collar" hoisting sling that had in the meantime been lowered by a Marine Corps helicopter hovering overhead. Shepard was winched on board and *Freedom 7,* scorched but bone-dry inside, was also pulled out of the water. The man and his can were quickly delivered to the deck of the carrier *Lake Champlain,* which waited nearby.

Kennedy, who had watched the flight on television, called Shepard to congratulate him. While still on the ship, the first American to sail through space was given a careful physical examination. He reported that he felt fine and had even found his five minutes of weightlessness pleasant. "It was painless," he told the two physicians and the psychologist who examined him. "Just a pleasant ride." He was in fact in excellent condition, both physically and mentally. This was good news because it was far from certain in the spring of 1961 that humans could survive weightlessness and high g loads without being physically or mentally impaired. There was a theory, for instance, that high gravitational forces might pop eyeballs out of their sockets.

The flight of *Freedom 7* was far more important, politically and technologically, than its quarter-hour suborbital foray indicated. Not only had a man successfully controlled his ship under the rigors of weightlessness and high g force and emerged with no adverse physical or mental problems, but both the spacecraft and its launcher had performed as designed. An American and his spacecraft had now proven that they could function together in space.

And unlike Gagarin and his *Vostok,* they had done so in full view of the world from liftoff to splashdown. That in itself had taken a degree of courage that clearly surpassed what the Soviets had brought to their effort. Reporters from the Istanbul newspaper *Millyet* who saw both the Gagarin and Shepard films, for example, told the Soviet consul general that "in the Shepard film we followed all phases of his flight, but in yours we followed only Khrushchev. Why don't you show us your space flight, too?" That kind of reaction, and accolades that came into NASA from around the world, reportedly infuriated Khrushchev, who insisted that Shepard's flight had been far inferior to Gagarin's.

Freedom 7's journey played very well at home, too, turning Alan Shepard from only a celebrity to a genuine hero. Crowds lined Pennsylvania Avenue on May 8 when the young Navy test pilot–astronaut, his six colleagues, and their wives arrived for a ceremony in the White House Rose Garden in which Kennedy pinned the NASA Distinguished Service Medal on Shepard. They also were greeted on Capitol Hill and were NASA's guests at a special dinner that night. The adulation was wasted neither on the man who occupied the Oval Office nor on those who watched from the Hill. *Freedom 7* itself was sent to be displayed at the Paris Air Show. Typewriters at Canaveral and in Washington gushed wonder, pride, and excitement at Shepard's deed. The reporters even took to saying "A-OK," an old telegraph expression that had been adopted by NASA and used by John A. "Shorty" Powers, the voice of NASA, to tell them that the flight was progressing satisfactorily. Soon the whole country was using it as insider lingo that everything was all right; for what hopelessly unimaginative engineering jargon called "nominal." Shepard himself did not use that term on his flight. But he knew it was true.

A "Fluid Front" Is Extended

If Lyndon Johnson harbored any doubts about going to the Moon, the flight of *Freedom 7* ended them. With the president's mandate to study the matter in hand (not to mention Kennedy's promise to build a space center in Texas), Johnson brought McNamara, Webb, and other powerful insiders together on the weekend after Shepard's mission to consummate an unholy alliance between the Department of Defense and the upstart agency that had moved into what the Pentagon saw as its rightful domain. The vice president told them to compose a persuasive rationale for going to the Moon and then in effect stood over them as they wrote it. Barely two months earlier, Kennedy had rejected a $182,521,000 request for additional funds by NASA, some of which would have gone into the manned program. The ex-skipper of PT-109 turned sailboater had executed a dramatic change in course.

McNamara's and Webb's tract, which both of them signed, was classified top secret, and for good reason. It opened with an obligatory reference to the importance of weather forecasting, communication, and other benefits that could be derived from going to space. But then, in starkly straightforward terms, it went on to assert that the balance of world power rested on the Moon:

> It is man, not merely machines, in space that captures the imagination of the world. All large-scale projects require the mobilization of resources on a national scale. They require the development and successful application of the most advanced technologies. Dramatic achievements in space therefore symbolize the technological power and organizing capacity of a nation. It is for reasons such as these that major achievements in space contribute to national prestige. . . .
>
> Major successes, such as orbiting a man as the Soviets have just done, lend national prestige even though the scientific, commercial or military value of the undertaking may by ordinary standards be marginal or even economically unjustified.

Then McNamara and Webb spelled out the real point of sending men to the Moon:

> Our attainments are a major element in the international competition between the Soviet system and our own. The non-military, non-commercial, non-scientific but "civilian" projects such as lunar and planetary exploration are, in this sense, part of the battle along the fluid front of the cold war.

So there it was. Where international politics and the balance of power were concerned, the military and civilian space programs were not only interchangeable, they were fundamentally inseparable. As sailors with cannons had transported missionaries to bring God to heathens, and as soldiers and bureaucrats had followed the churchmen to guarantee order and bestow civilization on their converts, so did the military and civilian rocket men come together as one entity to protect their country and extend its imperial ambitions. There was really only one space program, its core was indivisible, and it was a political weapon with which to bludgeon the opposition. The document amounted to a license to wage virtually unrestricted technological and political warfare against the Russians at any cost, in the process raising the stakes of the cold war far beyond anything Eisenhower thought was sensible or Khrushchev could match. Appropriately, the memorandum was delivered to Kennedy by Johnson on the afternoon of May 8, right after he awarded the medal to Shepard.

Two days later, the president ratified the recommendations with his senior advisers, including Sorensen, National Security Adviser McGeorge Bundy, McNamara, Webb, Dryden, Wiesner, Welsh, and two officials from the budget office, Bell and Elmer Staats. Bundy would recall that Kennedy had at that point pretty much decided to go for the Moon and was not interested in hearing arguments to the contrary.* Bundy, in fact, did argue to the contrary, saying that it was unsound scientifically and, besides, that it would be difficult, complex, and in a sense, a "grandstand play." But that is precisely what appealed to JFK, who was also a football fan. Sending Americans to the Moon would amount to the longest forward pass in history. "You don't run for President in your forties unless you have a certain moxie," Kennedy told Mac Bundy.

Destination Moon

Kennedy addressed a joint session of Congress about "urgent national needs" on May 25. He couched his speech as a second State of the Union address,

* Kennedy's economic advisers, fearing that the Moon program was neither large enough nor able to stimulate the economy sufficiently to prevent the recession they warned was coming, countersuggested that he approve a large public works program instead of shooting for the Moon. He rejected the idea outright. (Logsdon, *The Decision to Go to the Moon,* p. 127.)

made necessary because of the "extraordinary times" that had been caused by the Communist challenge in the developing world. The preamble was followed by a littany of measures he said had to be taken to counter the threat of global subversion and strengthen the United States itself. Foreign economic and military aid therefore had to be increased, Kennedy said, while the Army and the Marine Corps, the U.S. Information Agency, and even the Small Business Administration had to be strengthened. There would even be a tripling of the amount of money spent for the construction of fallout shelters as a way to discourage a Soviet nuclear attack. Then Kennedy, who had written key parts of the speech himself, wound up for the pitch:

> Finally, if we are to win the battle that is now going on around the world between freedom and tyranny, the dramatic achievements in space which occurred in recent weeks should have made clear to us all, as did the Sputnik in 1957, the impact of this adventure on the minds of men everywhere who are attempting to make a determination of which road they should take. . . .
>
> Now it is time to take longer strides—time for a great new American enterprise—time for this nation to take a clearly leading role in space achievement which, in many ways, may hold the key to our future on earth.

While Kennedy pointedly said that the competition between his country and its rival was "not merely a race," he made it clear that that was precisely what it was, first by noting that the Soviet Union's head start in large rockets would allow it to exploit its lead, and then by adding that "while we cannot guarantee that we shall one day be first, we can guarantee that any failure to make this effort will make us last."

Furthermore, the president continued, "We take an additional risk by making it in full view of the world. But as shown by the feat of astronaut Shepard, this very risk enhances our stature when we are successful." It was a brilliant riposte. In one adroit move, he simultaneously scorned Soviet secrecy and set the basis for turning any American space disaster into an act of national heroism and a validation of freedom of speech and therefore of democracy itself. Then, mindful that the United States had exactly fifteen minutes and twenty-two seconds of manned space time, Kennedy got to the point:

> I therefore ask the Congress, above and beyond the increases I have earlier requested for space activity, to provide the funds which are needed to meet the following national goals:
>
> First, I believe that this nation should commit itself to achieving the goal, before this decade is out, of landing a man on the moon and returning him safely to earth. No single space project in this period will be more impressive to mankind or more important for the long-range exploration of space. And none will be so difficult or expensive to accomplish.

> We propose to accelerate the development of the appropriate lunar spacecraft.
>
> We propose to develop alternate liquid and solid fuel boosters much larger than any now being developed, until certain which is superior.
>
> We propose additional funds for other engine development and for unmanned explorations—explorations which are particularly important for one purpose which this nation will never overlook: the survival of the man who first makes this daring flight. But in a very real sense, it will not be one man going to the moon. We make this judgment affirmatively: it will be an entire nation. For all of us must work to put him there.

As he neared the end of the forty-seven-minute speech, Kennedy challenged his audience to rise to the goal he had set for them, making it clear that the decision would be theirs. "Let it be clear that this is a judgment which the members of Congress must finally make," he said. "Let it be clear that I am asking the Congress and the country to accept a firm commitment to a new course of action—a course which will last for many years and carry very heavy costs—five hundred and thirty-one million dollars in fiscal '62 and an estimated seven billion to nine billion additional over the next five years. If we are to go only halfway, or reduce our sights in the face of difficulty," he added, baiting the legislators and their millions of constituents, "in my judgment, it would be better not to go at all." He implored his audience in an almost plaintive tone, the whole time talking to the Republican side.

Although Kennedy drew applause eighteen times, he was so worried about what he took to be a tepid response by the legislators that for the first and only time he departed from his text in a congressional address to lay special emphasis on points he felt were particularly important, while dropping those he thought distracting. "Unless we are prepared to do the work and bear the burdens to make it successful," he said, no longer reading, it made no sense to try to send a man to the Moon.

Driving back to the White House, he complained gloomily to Sorensen that the reaction to his speech had been less than enthusiastic and that Congress knew of better ways to spend $20 billion.* Kennedy himself had no doubt about the feasibility of the Apollo program, Sorensen would say later. But he would worry for the remaining thirty months of his life about the resolve of Congress and the people to see it through.

What Kennedy had no way of knowing during his address and in the hours immediately following it, though, was that the majority of the lawmakers were neither apathetic nor hostile. To the contrary, most of them were so weary of what they saw as years of presidential torpor that they had a very positive, vis-

* Other estimates were as high as $40 billion. ("A 3-Man Trip to Moon by 1967 Projected by White House Aides," *The New York Times,* May 26, 1961.)

ceral reaction to his call for action. In the words of some of those who chronicled the episode, they were in "dumb accord."

Congressional criticism came from a few Republicans who objected to what they saw as reckless spending. Representative Leslie C. Arends, an Illinois conservative, was among the bluntest of his colleagues. He criticized Kennedy's "huge spending schemes," adding that it was "indeed an extraordinary occasion for the Congress to meet in joint session to have the extraordinary experience of hearing President Kennedy say nothing extraordinary." But many other members of the GOP, who had scrawled notes, inspected their fingernails, brushed back their hair, and remained in stony silence during the first part of the speech, had warmed to it as he continued. They knew that it was indeed extraordinary.

Moonstruck

Support for the Moon program came from two basic sources, though as usual there was a great deal of overlapping. There was, as there had always been, the hard core of implacable dreamers: the unabashed zealots who shared a religious conviction that it was their race's destiny to explore other worlds and then start colonies on them. Nor is the term "religious" misplaced. What else to call people who had already painstakingly worked out deep space missions—scientists and engineers who planned and would continue to plan colonies on Mars or the terraforming of Venus, for example—with the certain knowledge that they would not live to see their objective realized? How else to define people who were driven to believe implicitly that Mars and a number of the moons in this solar system would be put to use for the benefit of a resource-depleted Earth? For them, as for the members of most other religions, purpose transcended both time and the individual and became an immortal crusade. Spaceflight provided the means to undertake the ultimate, endless process of discovery and regeneration for people who took it as an article of faith that the end of exploration would lead to the end of civilization: to mental and emotional suicide. Time was irrelevant so long as the process continued. The idea of finding new worlds (and perhaps life) and of casting their own race's seeds on the limitless "new ocean," as Kennedy would soon call it, was so emotional that it could bring tears.*

* He used the term in a speech at Rice University on September 12, 1962, when Rice donated one thousand acres of pastureland south of Houston for the Manned Spacecraft Center. Promising to make the United States "the world's leading spacefaring nation," Kennedy added that whether space "will become a force for good or ill depends on us, and only if the United States occupies a position of preeminence can we decide whether this new ocean will be a sea of peace or a new, terrifying theater of war." (Swenson, Grimwood, and Alexander, *This New Ocean,* pp. 470–71.)

The second group were the political pragmatists, many of whom were in the aerospace industry. Called to the challenge of getting into the race, they shared an amorphous fear that the Russians would somehow occupy space with dangerous (if unclear) consequences unless they were challenged and defeated. More tangibly, they worried about their country's place in a world in which it was somehow diminished and embarrassed by a technology that sprang from a godless, repressive, and politically abhorrent value system. This was the fear that gripped Kennedy and the others. Many of those individuals remembered a proud nation that had emerged from World War II as the most powerful on Earth, only to see it humbled, to see that power threatened by a dangerous enemy.

Finally, there was deep sentiment in that spring of 1961 to get the country moving again: to shake off years of benign lethargy and accomplish something grand enough to do credit to what was popularly—and proudly—called "the greatest nation on Earth." It was only natural that the "military-industrial complex," as Ike had called it in his farewell address, should want to fight the cold war with programs that were both dynamic and highly profitable.

Willis H. Shapley, who headed space and defense affairs at the Bureau of the Budget and who later became a high NASA official, would say soon after the decision to go to the Moon was made that it had come at a time when both the armed services and industry were looking for an excuse to develop truly colossal launch vehicles. The rockets that were already in the inventory were adequate for what was needed and no one had a clear idea of exactly what the Russians would want to do with their more powerful rockets. Yet Americans at that time had a kind of Cadillac mentality, a feeling that bigger was necessarily better and that truly big rockets were somehow vital for national defense. It was therefore comforting to know that Dr. von Braun's monumental lifter, the Saturn, was being developed.

For their part, the companies that formed a competitive but self-protective club known as the "defense contractors" understood that they would prosper on large federal programs or shrivel up, and perhaps disappear, without them. By the time of Kennedy's speech, 538 Boeing B-52 heavy bombers and more than a thousand of the Seattle-based company's B-47s were in the SAC inventory. The nation's ballistic missile force was moving toward completion. There were General Dynamics's Atlas and Martin Marietta's Titan, which were already in service, and Boeing's Minuteman and the submarine-launched Lockheed Polaris, which were soon to go into service. Those firms and others that led the field, plus a much wider base of major subcontractors such as Bendix, General Electric, Hughes, Raytheon, RCA, Honeywell, and Westinghouse anticipated that lean times were coming. They were eager to get their slice of the new, fat pie that was called Apollo.

The Shape of Things to Come

The general specifications for the spacecraft itself, which would include a command module, a service module, and a lunar landing module to get the astronauts on the Moon and back up again, were spelled out by Faget, Low, and others at a NASA-Industry Apollo Technical Conference that was held in Washington from July 18 to 20, 1961. The written guidelines themselves weighed more than 250 pounds. A thousand or so representatives from three hundred companies, the White House, Congress, and other government agencies attended the meeting. The prime contractor would develop and build the command and service modules, an adapter to fit the spacecraft to an orbiting laboratory (eventually called Skylab) to be built after the initial Moon landings, and ground support equipment. Requests for proposal were sent to fourteen firms whose mountainous responses were evaluated throughout the rest of the summer and into late autumn by an eleven-man Source Evaluation Board, which included Faget, and more than a hundred specialists. Candidates were rated not only on the technical merits of their proposals, but on costs and their way of doing business.

The fourteen eventually became five: Martin, General Dynamics, North American Aviation, General Electric, and McDonnell. Judged by technical approach, technical qualification, and business management, Martin came in first, with General Dynamics and North American tied for a close second. The board called the company the "outstanding" candidate. Then it was abruptly overruled by Webb, Dryden, and Associate Administrator Robert C. Seamans Jr., who picked North American. They rationalized their decision on the fact that North American had scored highest in technical qualifications, largely because of its outstanding X-15 rocket plane, the Navajo winged missile, an early cruise missile called Hound Dog, the classically beautiful, MiG-killing F-86 Sabrejet of Korean War fame, and its robust successor, the F-100 Super Sabre. The company was also excellently managed. The NASA triumvirate added that North American also had the longest and closest association with both NACA and the space agency.

But the decision reeked of influence peddling and oily deals and had to be defended on Capitol Hill. It soon turned out that Bobby Baker, secretary to the Senate majority leader, and two partners owned a vending machine company that made a million dollars a year, most of it from a deal made with North American three months after the Apollo contract was signed. Baker, in turn, was described as being "like a son" to Senator Robert S. Kerr, who not only was chairman of the Senate Committee on Aeronautical and Space Sciences but had a company called Kerr-McGee Oil, one of whose directors was none other than James Webb. It had been Kerr, in fact, who had recommended

Webb to Lyndon Johnson in the first place. There was nothing illegal about these relationships. Yet they aroused suspicion and suggested that the kind of shadowy insider wheeling and dealing over financial spoils that many Americans believed was endemic to Washington had tarnished Kennedy's Moon program even before it started.

North American, which would also build the Saturn second stage, would quadruple its payroll within a year and earn more than $6 billion by the time Americans first set foot on the Moon. "The Apollo contract was . . . rather more than routine government business," one account would note nine years later. "It was the biggest civilian contract in American history and the centerpiece of a program on which a large slice of the national prestige was riding. The contract would guarantee a long and prosperous future to any aerospace firm facing the prospect of a decline in military spending." And with thousands of contracts running into tens of billions of dollars, it was inevitable that the program developed what the account called its own political underworld. "Alongside the cool technical geniuses of NASA, profiteers and powerbrokers flourished. . . . These men, far from intruding apologetically into space as if it were a world too sterile for them to inhabit, grew to a scale commensurate with the pork barrel off which they fed."

At the bottom of the pyramid there were many thousands of honest, workaday aerospace employees—welders, fitters, electricians, technicians, draftsmen, engineers, secretaries, and scientists—who believed in their hearts that they were engaged in a supremely patriotic and historic enterprise. They also thought that it was infinitely sexier than making marmalade, fountain pens, or shoes.

Science Plays the Game

The official launching of the Apollo program was by no means greeted enthusiastically by all of Kennedy's fellow citizens. Many scientists believed that the program, like manned rocketry in general, amounted to "superficial entertainment" (in Van Allen's words) that took money away from what they considered to be legitimate research. Yet their misgivings were and would remain tempered. Most scientists understood that the manned program, not the more esoteric science missions, captured the public imagination. Van Allen maintained that there had been quite a bit of public interest in the V-2 and Viking sounding flights long before Mercury—though when they had flown, they had in effect been the space program. Yet even he came to believe, begrudgingly, that science rode on the manned program's coattails and that without it, scientific research in space would itself be greatly diminished. It was a theory shared by many in the space science community.

"A convincing argument can be made that one would not get *any* money . . . for space exploration without men in the loop; that the public funds space science because of the emotional issues involved with putting men in space," Professor David J. Helfand, who headed Columbia University's Department of Astronomy, would say years later. The American and Soviet space programs, he added, had a common fundamental requirement. "The next thing has to be much more spectacular than the last thing or it's not worth doing, because they're not fundamentally interested in doing the science; they're fundamentally interested in the technological achievement. . . . The agency is driven, from top to bottom, as far as I can see, by creating reasons for itself to exist."

But the scientists themselves shared the blame, Helfand added, because they implicitly endorsed the manned program in order to protect their own budgets. "The community is at fault because the community of scientists gets sucked into this," he said. There were very few people such as Van Allen and Cornell's Thomas Gold who would "stand up and say what they honestly feel. A large fraction of the community feels that the manned program is a waste of time.* But I constantly get letters, as does everybody else, from the leaders of the community . . . that you have to support the agency budget. You can't go in there and say, you know, cut all this manned crap and we could do some good science for a tenth the cost because that's a disaster on Capitol Hill. You're co-opted into playing the game."

Nor were scientists the only Americans who felt co-opted. Sending astronauts to the Moon did not seem to captivate most laymen, either. A Gallup Poll taken immediately after Kennedy announced his decision showed that 58 percent of them opposed it.

Liberty Bell's Toll

Gus Grissom, the first American to reach space after Kennedy's speech, provided a sour prelude to the majestic adventure envisioned by the president. Unlike Shepard's *Freedom 7,* which had had two small side ports, Grissom's *Liberty Bell 7* had been given a relatively large centerline window and a side hatch that was held in place by seventy explosive bolts. Both had been added at the insistence of the astronauts themselves, who remained sensitive to the derisive man-in-the-can cracks coming out of Edwards and elsewhere. But the hatch had not been added simply to placate the Mercury Seven. As the capsule had originally been designed, the astronauts were supposed to remove a small bulkhead in the antenna compartment at the top and climb through the open-

* Gold was a highly regarded Cornell astronomer who did pioneering work on planetary rings and who first suggested the term "magnetosphere" to describe the region around a planet that contained trapped radiation. He was also a NASA gadfly.

ing after they hit the water. But it was a difficult way to get out, especially in a space suit, and trying to do it while injured would obviously have been even harder. So the engineers put a hatch on *Liberty Bell 7* that blew off. To blow the hatch, the astronaut had to pull a lock pin out of a plunger and then press the plunger with five or six pounds of pressure. The seventy bolts would then fire the hatch about twenty-five feet. It was basically the same system used for the ejection seats in every fighter the astronauts had flown. A pin pulled out of the yellow-and-black-striped ejection handle before takeoff would arm the seat; yanking the handle itself in an emergency would blow off the canopy and send the seat and its occupant roaring out of the cockpit on top of a small solid-fuel rocket.

The Redstone carrying Gus Grissom inside *Liberty Bell 7* lifted off at 7:20 on the morning of July 21, 1961. The flight itself was nominal (just great). Grissom's only problem was having to overcome the urge to look out of the enlarged window and instead concentrate on his dials and on manually controlling the spacecraft. He radioed to Shepard that the scene outside the window, which stretched 800 miles at his 118-mile maximum altitude, was fascinating.

Going by the book, Grissom first fired the retro-rockets, slowing his spacecraft and therefore sending it into a long downward arc, and then pitched it over so that he was coming in backward. Condensation and hot smoke spilled off the glowing heat shield as he slammed back into the atmosphere at sixty miles' altitude. The main chute popped open at 12,000 feet, just as it was supposed to, and seven minutes later *Liberty Bell 7* plopped into the Atlantic, first tilting to the left, and then rolling in the turbulent sea. Grissom disconnected his helmet and went through the rest of the end-of-flight procedures, including recording cockpit panel data and unbuckling most of the harness that tied him to his seat. He then radioed two nearby helicopters to come and get him and his capsule. Carefully following procedure, he slid the pin out of the hatch-cover detonator and lay back, exhilarated at what he had accomplished.

"I was lying there, minding my own business," he would later tell a board of inquiry, "when I heard a dull thud." The thud was made by the seventy exploding bolts, which fired simultaneously, blowing the hatch off *Liberty Bell 7* and opening the way for seawater to pour in. The third man to reach space scrambled out of his flooding spacecraft and swam away. Assuming that Grissom could take care of himself (the helicopter pilots had noted that the astronauts seemed to like the water during recovery training off the Virginia beaches), the helicopter crew went after the spacecraft. But it was to no avail. The Sikorsky chopper's engine was losing its tug-of-war with gravity as it strained to pull the swamped capsule out of the ocean, so the cable had to be disconnected. Grissom's spacecraft slid ignominiously under the waves and disappeared in 2,800 fathoms of water.

And Grissom himself very nearly became the first astronaut to drown. Not only was air escaping from his space suit's open neck section, but he had forgotten to close the suit's inlet valve, so it was filling with water at the same time. Swimming in the swells was becoming more difficult by the second as the hapless astronaut, now bobbing under the waves as he desperately looked for a swimmer from one of the helicopters, became scared and angry. He was finally tossed a horse collar from a helicopter hovering over him and was winched aboard by George Cox, its copilot. Cox had performed the same service for both of Grissom's predecessors: Ham and Shepard.

Gus Grissom, in some disgrace, would tell the board that while he had some difficulty remembering exactly what he was doing the instant the hatch cover had blown, he was absolutely certain that he had not hit the detonator plunger. Whatever the truth of the matter, the flight itself had gone so well that everyone agreed it was time to abandon the Redstone shots and go to orbit on top of one of the more powerful Atlases. In fact, it did not seem far-fetched to think that if things moved quickly enough, an astronaut riding a Mercury-Atlas could get a jump on the opposition by flying three orbits.

The Empire Strikes Back

Then, on August 6, the Russians struck again. German Titov blasted off from Tyuratam in *Vostok 2* on a daylong mission that lasted twenty-five hours and eighteen minutes and flew through seventeen days and nights. General Nikolai P. Kamanin, the cosmonauts' trainer, and Vladimir I. Yazdovsky, a leading aerospace physician, had urged Korolyov to limit Titov's orbits to three because they were not completely certain how the human body would react to an entire day in space and wanted to find out gradually. But the chief designer, arguing that long-duration missions were in store for his little eagles anyway, overruled them.

The point was to start living in space as quickly as possible, and if that meant embarrassing the United States yet again, so much the better. Titov tested the *Vostok*'s attitude-control system and ate the first real meals in space on the third and sixth orbits. As doctors, physiologists, psychologists, and others (including the chief designer, who took notes) watched him on television, Titov squeezed a tube, forcing out puree like toothpaste that floated in front of his mouth. He also ate bread, liver pâté, peas, and some meat and drank black-currant juice from another tube.

On the fifth orbit, the twenty-six-year-old cosmonaut also became the first person to be afflicted by spacesickness. But after sleeping for five orbits, starting with the seventh (making him the first to sleep in space), the queasiness eased a little. Knowing that the radio link with the spacecraft was being monitored by U.S. intelligence and determined that no one on the outside was going to call the cosmonauts sick sissies, it was decided that anyone who became ill on future missions would signify it by telling the controllers that he or she was "observing thunderstorms." The afflicted spacefarer would then be brought down earlier than planned.

Titov's illness deeply concerned Korolyov and the physicians and trainers. Like their American counterparts, they knew next to nothing about how the actual rigors of spaceflight would affect those who did it. They were haunted by the notion that the weak link—perhaps the decisively weak link—in the chain that stretched from the Earth to space was not the rockets, after all, but the pathetic frailty of those who rode them. Machines could be engineered to withstand radiation, high g forces, micrometeoroid hits, disorientation, and other dangerous problems. But people were harder, perhaps impossible, to engineer.

Besides his other chores, Titov used a handheld Zritel camera to take pictures of Earth. This was important because it helped heal a rift between Korolyov and the military. When the chief designer proposed an accelerated manned program in July 1958, following the success of the Sputniks, he met resistance by military planners who preferred that funding be put into photoreconnaissance satellites.

The previous November, the Council of Chief Designers, to which even Korolyov had to report, approved a compromise calling for the priority development of a manned spacecraft, Vostok, that could be modified as a camera-carrying spy satellite without a great deal of effort. The prolific Mikhail Tikhonravov had first suggested using satellites for such practical purposes as reconnaissance in the late 1940s and early 1950s, and Korolyov had weighed in during the mid-1950s, but they had gotten nowhere until Sputnik became a reality. The result was Zenit (Zenith), which would take to the sky for the first time the following year, 1962, and begin returning usable pictures early in 1964. In place of crew quarters and life support equipment, the Zenit version of Vostok's re-entry module contained cameras, film, recovery beacons, parachutes, and an explosive charge that would blow the module to pieces if it appeared to be coming down over the wrong place. The service module was similar to Vostok's and held batteries, orientation equipment, radios, and a liquid-propelled retro-rocket to slow Zenit before its re-entry module separated and came down.

At any rate, Titov became the first person to photograph Earth from space. He went into orbit as a captain and, like Gagarin, left it as a major and a full member of the Communist Party of the Soviet Union. Khrushchev could do no less for the nation's new heroes.

But Titov's spacesickness so worried Korolyov and everyone involved in the program's human dimension that extensive physiological and psychological research was stepped up on a crash basis. Fearing that a sick cosmonaut would cause some calamity, Vostok launches were stopped for a full year while rigorous human testing was done on the ground.

An American Orbits the World

Meanwhile, like one of the rockets themselves, momentum was building on the other side of the Atlantic. On September 13, 1961, an unmanned Mercury capsule was launched on top of an Atlas as a dress rehearsal for a manned flight. It came down intact east of Bermuda an hour and fifty-five minutes later and was retrieved by the destroyer *Decatur.*

Six days later Webb announced that, as agreed, the headquarters for the manned space program—now consisting of Mercury, Gemini, and Apollo—was moving to the thousand-acre site near Houston that had been donated by Rice University. Its population at that moment consisted of a mixture of people and wildlife—eight hundred families, including assorted cowboys, and wolves, turkeys, deer, foxes, and snakes—that was going to be forcibly displaced within three years. The human residents of the area, incensed, would turn their wrath on Lyndon Johnson and the space program he embraced. But big things were in the works, and they were little people, so it was to no avail.

On November 1, the Space Task Group itself was officially given the name of the new facility: the Manned Spacecraft Center. Twenty-eight days later, the ever-conservative NASA engineers sent another chimp, Enos, on an Atlas for a successful two-orbit mission. The little guy, who was weightless for 181 minutes, performed his tasks perfectly. Unlike Ham, whose nerves had been frazzled by his freakish flight, the poised anthropoid took a news conference in stride, even responding to the popping flashbulbs with impressive composure. Gilruth and Director of Operations Walt Williams also had the situation in hand. Asked for the name of Enos's immediate successors, Gilruth named Glenn and Carpenter, with Slayton and Schirra as their respective backups.

John Glenn roared into space on the morning of February 20, 1962, with an estimated fifty thousand festive "bird watchers" cheering on the beach and another hundred million absorbed in front of their television sets. He had been told before his *Friendship 7* thundered over the Atlantic that he would get at least seven orbits. But the computers at the Goddard Space Flight Center in Maryland soon reported that the Atlas had done such a good job of getting the spacecraft "through the gates"—through every phase of the launch—that almost one hundred orbits would be possible (though von Braun would not have dreamed of taking such a senseless chance).

Cruising along weightlessly and backward at 17,544 miles an hour, the first American to reach orbit did his control and systems checks, took pictures, looked at the Atlas that tumbled end over end behind him at a steadily widening distance, and enjoyed an unobstructed view of the Canary Islands and the

African coast. He sailed into night over the Indian Ocean, first describing the brilliant orange sunset and then noting that the sky had become very black and that a thin blue band crowned the horizon. No nausea for John Glenn.

Friendship 7 glided over Australia in the dead of night and then into dawn over the Pacific. It was there, over Canton Island, that he noticed "little specks, brilliant specks, floating around outside the capsule." The "fireflies," as he called them, were fluid that was jettisoned by the environmental system, stuck to the spacecraft, and froze. The ice particles could be seen only at night, and banging on the inside of the spacecraft would loosen a flurry of them. The "frostflies," as Scott Carpenter would call them during his flight on *Aurora 7* in May, were not troublesome. But trouble did develop.

First there was an attitude-control problem. The automatic stabilization control system began allowing the spacecraft to drift slightly to the right, forcing Glenn to take manual control to get it back on course (and justifying the pilots' demand for control capability). Then a potentially deadly problem developed when an engineer reported that telemetry data seemed to show that *Friendship 7*'s heat shield and compressed landing bag were not locked into position. If true, it meant that the device that was supposed to keep Glenn from immolating on re-entry was being held in place only by the metal straps that kept the retro-rockets on. The first in-flight crisis of the U.S. manned space program was developing by the second.

Seven orbits were now out of the question. The operations team at Canaveral quickly began thinking about how to get Glenn back down. It was decided to keep the retro-rocket package on after the rockets fired to slow *Friendship 7* in the hope that the straps would hold the heat shield in place through re-entry. The engineers knew that the retropack would burn away during the early part of re-entry, but quick calculations seemed to show that aerodynamic pressure at such high speed would hold the shield in place even after the retropack disappeared. Faget, who was reached in Houston, said that the plan ought to work provided all of the rockets fired. If they did not, Mercury's chief designer warned, they would have to be jettisoned because unburned solid rocket fuel would definitely explode during re-entry, turning Glenn and his spacecraft into flaming grapeshot and then particles of debris.

A series of tracking stations that ringed the planet from Australia to Zanzibar, and that included specially configured ships with radar and communication capability, had been put in place before the Mercury program began so spacecraft could be precisely monitored and guided.* Now, an already suspicious John Glenn was being instructed by a number of them as he passed

* They were at Muchea and Woomera, Australia; Canton Island; Kauai, Hawaii; Point Arguello, California (eventually Vandenberg Air Force Base); Guaymas, Mexico; White Sands, New Mexico; Corpus Christi, Texas; Eglin Air Force Base and Cape Canaveral, Florida; Bermuda; Grand Canary Island; Kano, Nigeria; and Zanzibar.

overhead to do things—hitting the landing-bag deploy switch to see whether the bag was still there and operating, for example—that convinced him there was a serious problem. He was then told to prepare to come down after his third orbit.

And come down he did, in what was the most fearful and dangerous part of the flight. As *Friendship 7* began its long downward trajectory, with Glenn doing the flying and the retropack still on, he heard what sounded like "small things brushing against the capsule." Then came the blistering heat and high g force as man and machine plowed into the air. Glenn was instructed not to jettison the retropack until his g meter showed 1.5, meaning that his velocity had slowed to the point where the heat would no longer be dangerous.

"That's a real fireball outside," Glenn told Canaveral, a touch of anxiety in his voice. To make matters worse, he could see one of the retropack's straps fluttering like a pennant in the red glare and smoke on the other side of the window. "I thought the retropack had jettisoned and saw chunks coming off and flying by the window," he would say afterward. He worried that the chunks were parts of his disintegrating heat shield and that he was seconds away from turning into hot vapor.

Then, right after *Friendship 7* made it past the maximum g region, it began heaving so violently that Glenn could not control it manually. So he kicked in the auxiliary damping system, which automatically fired the steering thrusters until the spacecraft was relatively stable again. As Glenn wondered whether the steering systems had enough fuel to keep him stable until the drogue chute opened, the fuel ran out. Then the heaving started again. Finally, as the astronaut and the scorched, lurching cone in which he was strapped reached 28,000 feet, the drogue chute popped open. To his immense relief, the main parachute blossomed above him and the charred heat shield and impact bag dropped into place for a cushioned landing in the Atlantic. The rest of the flight was nominal.

John Glenn came out of *Friendship 7*'s inferno as the Lindbergh of his time. He was lavished with praise by a grateful (and suddenly prideful) nation and touted by the news media at home and abroad as a genuine hero. The spacecraft itself was dispatched by NASA on a round-the-world, seventeen-nation tour that was promptly called its "fourth orbit." Kennedy seized the moment to pour his own fuel on the competition with Khrushchev, saying in a Rose Garden ceremony that "we have a long way to go in this space race. But this is the new ocean, and I believe the United States must sail on it and be in a position second to none." Those who had labored under a barrage of scorn by politicians and pundits who had called them the pitiful laggards of the space age—Dryden, Gilruth, Low, Faget, Williams, Kraft, and others—were now praised for having "stuck to their guns."

Harbingers of Triumph and Tragedy

Mercury sailed on through May 1963. Three months after Glenn's historic ride, Carpenter flew essentially the same mission in *Aurora 7.* Schirra went up in *Sigma 7* on October 3, 1962, and stayed in orbit for nine hours. And finally, on May 15, 1963, Cooper flew *Faith 7* on a thirty-four-hour, nineteen-minute mission that covered 546,167 miles in twenty-two orbits. Gordon Cooper was the last astronaut to go to space alone. Deke Slayton, diagnosed with a slightly "erratic" heart rate—an idiopathic atrial fibrillation, as the cardiologists put it—was grounded but would get to space another day. All seven were toasted from coast to coast as their stories played out in *Life* and elsewhere. Meanwhile, the successor program, named Gemini, got under way.

But somewhere beneath the appearance of technological efficiency that NASA tried to spread like warm honey across the pages of the nation's newspapers and magazines, there existed what Norman Mailer would call "the psychology of machines." That is, no matter how careful the engineering or how painstaking the quality control by the technicians, the immensely complicated devices that were shot into space functioned in a mysterious and confounding dimension in which things could suddenly go horribly wrong for little or no apparent reason. No matter how careful the designers and mission planners were, no matter how heavily they fortified their papers and reports with precise mathematical calculations, they were relentlessly stalked by the one phenomenon all of them dreaded because it could not and never would be controlled: plain luck.

"The early history of rocketry reads like an account of the burning of witches," Mailer would write when he chronicled Apollo. "One sorceress did not burn at all, another died with horrible shrieks, a third left nothing but a circle of ash. . . . At the *Raketenflugplatz* outside Berlin in the early Thirties, rocket engines exploded on their stands, refused to fire, or when tested in flight lifted in one direction, then all but refired at right angles and took off along the ground. Indeed the early history of rocket design could be read as the simple desire to get the rocket to function long enough to give the opportunity to discover where the failure occurred. Most early debacles were so benighted that rocket engineers could have been forgiven for daubing the blood of a virgin goat on the orifice of the firing chamber."

Evocations of the supernatural aside, Mailer described a phenomenon that not only outlived the VfR but haunted Mercury and continues to this day. The schedule for the manned Mercury flights slipped from sixteen to thirty-six months because of "glitches" and other "anomalies." Worse, as Mailer himself pointed out, all of them had been plagued by errors, mishaps, and trouble. There had been Shepard's thruster problem, Grissom's near-deadly blown hatch, and Glenn's spate of malfunctions. The Atlas that had sent Schirra into orbit had done a heart-stopping clockwise roll after it left the pad because of misaligned steering engines. And Cooper had had several problems, including a failure of the automatic control system, forcing him to fly *Faith 7* through reentry with his own hand moving its stick. Precisely the same thing happened on every Soviet flight. "Both sides were lucky they didn't have a catastrophe early on," Valentina Ponomareva, one of the Soviet Union's first female cosmonauts, would remark years later.

This is not to say that the rockets and the machines they carried were not improving, only that there was no such thing as a risk-free ride to space and there never would be. The only element that would never change, in fact, was the most fundamental equation of all: those who chose to send their machines or themselves beyond the atmosphere for adventure, advantage, or transcendence would always be accompanied by the potential for violence, destruction, and death.

10

To Hit a Moving Target

Whatever Khrushchev and others in the Kremlin thought about sending cosmonauts to the Moon during the early 1960s—and even tentative approval to do so would not come until 1964 and final approval two years after that—Korolyov never doubted that it could be done. Like their American rivals, the Soviet Union's leading engineers were substantially ahead of their political leaders on that score, knowing full well that firing men at the Moon was perfectly feasible if technology was pushed hard but prudently.

Yet explosions and other maddening mishaps occurred all the time. The record of both superpowers during the early to mid-1960s shows both an increasingly frenzied launch rate and an appalling string of failures. While both sides were getting the hang of orbiting satellites that tracked storms, relayed radio and television signals, mapped Earth, unveiled military secrets, and told fliers and ship captains where they were, the competitors were not faring so well in the more difficult business of racing to the Moon and the inner planets. The first three Luna probes—a flyby, a hard landing, and the photographing of the far side—had been outstanding. But the Russians wrecked or lost two probes for each success. And even at that, they were doing considerably better than the Americans, who racked up eight straight failures in nineteen months. The gap between what the technology was supposed to do in order to hit the Moon and what it actually did do remained dauntingly large.

The satellites that carried the people who were supposed to go to the Moon did substantially better than the long-range robots, yet the specter of catastrophe loomed on every flight as technocrats and technicians on both sides pushed ever harder for advantage. No one was oblivious to the multiple dangers that stalked Mercury, Gemini, and Vostok. Korolyov knew that his spacecraft and those of his rival were no more than 50 percent reliable. So he tried to stave off calamity by sending up an unmanned spacecraft that did the mission that was to follow within a few days with cosmonauts. Those in the United States who closely followed the Soviet space program soon learned that such an unmanned flight was a sure indication that a manned one would soon follow. Yet relentless pressure from the top, plus his own obsessive need to stay ahead of the competitors who justified what he did, would cause him to cut corners.

Valentina Tereshkova Seagull

Knowing that a manned Moon mission would require two or more spacecraft to link up in either Earth or lunar orbit, Korolyov staged the world's first formation flight in mid-August 1962, when Andrian Nikolayev and Pavel Popovich flew *Vostok 3* and *Vostok 4* to within four miles of each other with impressively close velocities. Although the mission won pleasing headlines, that was not the point; the chief designer and his colleagues were pressing the increments they knew they needed to get to the Moon. So they repeated the mission in mid-June 1963 with *Vostok 5* and *Vostok 6.** This time, however, they added still another headline grabber: the pilot of *Vostok 6* was the first woman in space, a twenty-six-year-old former textile worker, amateur parachutist, and exemplar of the ordinary named Valentina Vladimirovna Tereshkova. And as was the case with Gagarin, "ordinary" was precisely the point. By contrast, Valentina Ponomareva, who was one of Tereshkova's backups, held a diploma in engineering from the prestigious Moscow Aeronautical Institute and flew Polikarpov and Yak trainers. "The requirements for cosmonauts were not as strict" after Gagarin and Titov, the diplomatic Ponomareva would one day recall.

The formation flight was definitely the most extraordinary "date" a man and woman ever had. "*Chayka*" (Seagull), as Tereshkova was called while in space, racked up forty-eight orbits in almost seventy-one hours, or more than all of the Mercury astronauts combined. Her colleague in *Vostok 5,* Valeri Bykovsky, went even beyond that. Besides flying to within three miles of Tereshkova, Bykovsky stayed in orbit fifty-four minutes short of five days, setting a record that would stand for more than two years. That much was heralded by TASS and on the streets of Moscow, where rapturous throngs cheered the space couple.

According to an official account, and in keeping with the new myth promoted by both sides that spacemen and women were so unemotional and self-

* By way of reemphasizing that no tricks were being played, Korolyov not only had Nikolayev and Popovich communicate over open radio channels, describing each other's spacecraft, but they made the first live television broadcasts from space, which were shown in both the U.S.S.R. and Europe. (Newkirk, *Almanac of Soviet Manned Space Flight,* p. 29.)

controlled—so cool under life-and-death pressure—that they could sleep through anything, Tereshkova "overslept so long that Ground Control began to wonder if she was still alive." It was cute. And it was certainly impressive. But it was not true. If anyone worried about whether Tereshkova was alive, it was undoubtedly the cosmonaut herself, since she too was hit by spacesickness and had to endure agonized hours alternating between nausea and fitful sleep in her tiny speeding cocoon. The veteran parachutist got so sick at one point, in fact, that her voice became almost unintelligible.

"What are you doing to her?" a suddenly alarmed Korolyov shouted to a nearby operations officer. What the man did to her—what he did to bring her out of a sobbing nightmare that incapacitated her—was switch her cabin light on and off repeatedly. And it worked. Tereshkova awoke with a start and began reflexively repeating her code name over and over again: "Seagull here, Seagull here . . ."

Clare Boothe Luce, the doyenne of the Time Inc. publishing empire, read into Tereshkova's singular odyssey precisely what Khrushchev wanted. "In entrusting a 26-year-old girl with a cosmonaut mission," Luce wrote in *Life*, "the Soviet Union has given its women unmistakable proof that it believes them to possess these same virtues [as men in carrying the nation's prestige and honor]. The flight of Valentina Tereshkova is, consequently, symbolic of the emancipation of the Communist woman. It symbolizes to Russian women that they actively share (not passively bask, like American women) in the glory of conquering space."

Tereshkova's colleagues, including the chief designer himself, would have snickered at Luce's emancipation proclamation and her disparaging remarks about NASA. It had been Nikolai Kamanin and Chief Academician Keldysh who wanted a woman to ride a rocket, not the chief architect of the space program. And they, in turn, were bending to Khrushchev's wishes. The decision to launch a woman, Aleksei Leonov would say many years later, had been made "at the highest level." Korolyov, feeling far more comfortable with men, and with test pilots in particular, had been emphatically against the idea. But he had been overruled. Seagull's plight only reinforced his belief that women in space were more trouble than they were worth and amounted to a serious liability. "I'm not going to deal with women anymore," Ponomareva quoted him as vowing after Tereshkova had floated safely home.*

Far from being unique in his abhorrence of female cosmonauts, the chief designer spoke for other professionals who felt the same way. Years later,

* Two points can be made on Tereshkova's behalf. She was apparently no sicker than was Titov. More to the point, she may have became sick in the first place precisely because she was not a pilot, let alone one who endured high-g turns and disorienting maneuvers while testing jet fighters.

chief cosmonaut Vladimir Shatalov told reporters that space was too demanding for women. "In such conditions we just had no moral right to subject the 'better half' of mankind to such loads," he said. And there was other derision. Leonov drew laughter when he, too, quipped that spaceflight was beyond women's true capability and added that "they can do other things down here." Ivan I. Spitsa, a communication specialist, remembered that soon after Tereshkova was made a cosmonaut, it was suggested that she be assigned the call sign Golden Eagle. But someone objected. "Why are you giving her the name of a predator like the golden eagle?" he wanted to know, adding that the birds were noble hunters. The implication was wasted on no one: the call sign of a bird as aggressive and respected as the golden eagle should go only to a man. Spitsa's deputy then came up with Seagull, pointing out that they too were predators: of fish. This brought howls of laughter from men who either did not know, or more likely did not care, that golden eagles catch fish too. Popovich was awarded the coveted Golden Eagle call sign, and Bykovsky became Hawk.

The attitude toward Tereshkova and her female colleagues reflected a systematic discrimination that existed throughout the nation as a whole and still does, Marxist theory and Clare Boothe Luce notwithstanding. Russian women have always complained about their inferior professional and financial status in what is a steadfastly male-oriented society. Hedrick Smith, a *New York Times* correspondent in the Soviet Union, called it "inbred unselfconscious male chauvinism" that occasionally surfaced in embarrassing ways, as when it turned out that the head of the Soviet Commission for the International Women's Year in 1975 was a man.

It was far different in the United States, but not in the space program. Jeraldine Cobb, one of several American women who yearned to go to space, was no passive basker. An unmarried thirty-two-year-old with ten thousand hours of flying time and four world flying records, Jerrie Cobb made an impassioned plea to a congressional committee that she and the others be allowed to compete to be astronauts. "I find it a little ridiculous when I read in a newspaper that there is a place called Chimp College in New Mexico where they are training fifty chimpanzees for space flight, one a female named Glenda," the peppery pilot complained, adding that she would even have been willing to "substitute for a female chimpanzee" if called upon to do so. Cobb finally won the right to take the same rigorous examinations as Shepard, Glenn, Grissom, and the others and passed them successfully. Then she was told that her lack of experience as a test pilot disqualified her. Jane Briggs Hart, the wife of Senator Phillip A. Hart of Michigan and mother of eight, fared no better even though she held single- and multiengine ratings with thousands of hours, had raced in the Powder Puff Derby, and was the first woman in her state to be licensed to fly helicopters.

Seven years after her acclaimed feat, Tereshkova herself would write that no work, however vigorous and demanding for a woman, "can enter into conflict with her ancient 'wonderful mission'—to love, to be loved—and with her craving for the bliss of motherhood."

On November 3, 1963, Tereshkova and Andrian Nikolayev were married, making theirs the first wedding of a spaceman and a spacewoman. Khrushchev, who was widely called the couple's "godfather" because of his publicist's penchant for such conspicuous events, was at the party, as were Korolyov and Glushko, who were allowed to make rare public appearances, and led the toasts and singing. The couple had the first space baby, a girl named Yelena, in June 1964. The blessed event was celebrated not only by Yelena's parents but also by a biologist named Aleksandr Neyfach, who carefully tested the unique infant for signs of any cellular abnormality that could have developed because of her parents' exposure to cosmic radiation. There was none. Yelena would grow to adulthood carefully insulated by her mother from publicity. Valentina's and Andrian's nuptial bliss eventually ended, and they became the first divorced spacefarers.

Gemini: "Twins" in Space

By the autumn of 1964, a series of requirements that would have to be met in order to get men onto the Moon was being articulated and solved by engineers and mission designers on both sides. Obviously, such an arduous and complicated voyage would require more than one man, meaning that crews would have to practice working as teams in a process that was stupendously complicated.

Since it was likely that two spacecraft were going to have to link in some capacity on the way to the Moon, American Gemini spacecraft (as in Gemini Twins) were designed to carry two astronauts and test a rendezvous and docking maneuver with Atlas-launched Agena spacecraft (the same basic satellite upper stage that carried the Corona reconnaissance program's cameras). Furthermore, landing men on the Moon would mean that at least one of them would have to get out of the spacecraft, perhaps on the way there in order to transfer from the carrier to the lander, but certainly on the lunar surface no matter how it was reached. Leaving the safety of the spaceship raised the prospect of being doused by lethal radiation, struck by micrometeoroids or other particles, fatally encumbered by the space suits themselves, or being afflicted by some other terrible and unforeseen menace.

A committee chaired by Ames's Harry J. Goett had therefore considered a post-Mercury team approach as early as May 1959 and the Space Task Group gave the go-ahead three months later. The spacecraft were at first called Mercury Mark II, but on January 3, 1962, they were renamed Gemini because they were designed to carry two-man crews. Far from being a mere scaled-up version of its predecessor, Gemini was much more complicated, since it was designed to fly missions of up to two weeks, be able to maneuver for rendezvous and docking in orbit, and have a hatch and pressure system that allowed astronauts to do extravehicular activity, or EVA in space slang.

While Mercury was only a conical capsule, Gemini represented a true link between Mercury and Apollo, since it was composed of three distinct modules: a crew compartment that superficially looked like an enlarged version of Mercury except for a cylindrical rendezvous and recovery section up front; an equipment module holding the electrical power system, propellant tanks, communication and instrumentation equipment, drinking water, coolant pumps, and attitude-control thrusters; and an adapter ring that connected both modules and which had its own maneuvering thrusters. In that regard, in fact, it quickly evolved less as a successor to Mercury and more as precursor to Apollo itself. The whole contraption was almost nineteen feet long and weighted more than seven thousand pounds.

Gemini was shot into orbit by a ninety-foot-long Titan 2 ICBM showing, yet again, that the civilian and military programs were as synergistic as they were interdependent. "The Gemini is an excellent example of a joint NASA/Air Force cooperative space venture," bragged an Air Force lieutenant colonel named Jay R. Brill.

Voskhod: Sardines in a Can

As Gemini was a follow-on to Mercury, Soyuz (Union) was a logical step beyond Vostok for the Soviets. It was an intelligently designed, versatile spacecraft that could carry up to three crewmen in relative safety and comfort. But Soyuz would not be ready to fly until at least 1965, and Khrushchev wanted to trump Gemini as soon as possible. That meant sending three men to space.

In October 1964 the Russians launched two Vostoks that were modified to carry three crewmen. The first, an unmanned test model carrying the deliberately ambiguous name Cosmos 47, went up on the sixth and came down for a soft landing the next day. The second, *Voskhod 1,* carried three cosmonauts into orbit on the twelfth and soft-landed them on the thirteenth. The trio consisted of Vladimir Komarov, a veteran of the original cosmonaut class; Boris Yegorov, the first physician in space; and Konstantin Feoktistov, a leading scientist-engineer from Korolyov's OKB-1. Having made their point about glorifying ordinary folk, the Russians were now aiming quite a bit higher. The three men who flew in *Voskhod 1* were first-rate specialists who were supposed to test the three-man concept for space stations and a trip to the Moon. As much could not be said about their spacecraft.

Leonid Vladimirov, a journalist who covered the space program before his defection, made that point in his book *The Russian Space Bluff.* "Khrushchev asked whether Korolyov was aware that the Americans were planning to launch two men into space in a single capsule. (This aspect of the project was the only one that interested the Soviet leader.) When Korolyov answered in the affirmative, Khrushchev simply gave him his instructions: 'All right then—by the next Revolutionary anniversary [7 November 1964] we must launch not two but three people into space at once.' " Whether this is accurate or not, the prospect of racing the United States to the Moon was becoming a reality, so Korolyov knew that three cosmonauts would probably be needed for the trip. He also knew that there was only one way to meet the tight schedule: squeeze three people into a Vostok. The chief designer is said to have warned his political patron that such a flight would be very risky, and Kamanin agreed, but they were apparently summarily overruled.

The spacecraft itself was a dangerous half measure that was judged even by its creators to be less than 50 percent reliable. The only way to get three men into a Vostok was to clothe them in lightweight flight suits, not insulated space suits, place them on custom-made couches arranged so that one was both between and slightly in front of the other two, remove all nonvital electrical equipment, and provide food for only twenty-four hours instead of ten days. The standard nose-mounted rocket that was supposed to pull the spacecraft off its booster in case of an emergency on the pad took twenty seconds to kick in after launch because it was mounted under the shroud that protected the spacecraft, not over it. Had there been an explosion on the pad or some other calamity during the first twenty seconds, Komarov, Yegorov, Feoktistov, and their two successors, Pavel Belyayev and Aleksei Leonov, would have been killed. And, of course, they knew it.

Squeezed in shoulder to shoulder, Komarov, Yegorov, and Feoktistov rode a more powerful version of the venerable R-7 to orbit on the morning of October 12, 1964, and floated back down almost exactly twenty-four hours later. *Voskhod 1*'s occupants were duly televised and had the usual chat with Khrushchev, who was vacationing with Anastas Mikoyan, his trusted supporter, on the Black Sea.*

"Anastas Ivanovich is pulling the receiver out of my hand," a delighted Khrushchev told his cosmonauts, playfully, while millions of their countrymen listened. It was Khrushchev's last public statement. He was called to Moscow on the fourteenth and driven straight to the offices of the Central

* Mikoyan's younger brother, Artem, was an Air Force major general and partner of M. I. Gurevich. Together the men designed their country's most famous fighters, whose prefix was a contraction of both of their names: MiG.

Committee, where he was labeled an "irresponsible voluntarist" (dangerously headstrong) and removed from political and party office on the spot. Meanwhile, the three cosmonauts were left to cool their heels for a week while Leonid Brezhnev consolidated his grip on the Kremlin.

When they were finally summoned to a reception in the capital, they were praised by Brezhnev and Aleksei Kosygin, not by the man who had sent them to space and then vanished. But the official praise was relatively faint. It was so faint, in fact, that it portended a sea change in the fortunes of Korolyov and others in the space program. Brezhnev could no more stop his country's manned program in its tracks than Kennedy's successors could stop their own: the inertia was already too great. But Brezhnev did not like the Moon program. Too many of its planners would have lingering loyalties to his predecessor. Moreover, there was only so much money, and the military establishment on which he depended wanted submarines and missiles, not Moon rocks.

The flight of *Voskhod 1* was publicly celebrated not only in the Soviet Union, but in the West as well by writers and others who romanticized what was in fact a cynical gamble that risked lives for political expediency. The desperation that forced Korolyov to dispense with space suits, for instance, was interpreted as evidence of the ingenuity of Voskhod's design. It "was a vastly improved, flexible, commodious, and comfortable spaceship," one admiring New York journalist who visited the Soviet Union reported when he returned home. "The unsinkable and hermetically sealed craft and its relatively low oxygen-gas system also permitted a shirtsleeve environment."* What this showed, he added in awe, was a "determination to explore the moon and inner planets . . . [that] is deeply rooted in its national consciousness." What it showed was that Tsiolkovsky's and Tsander's pioneering work notwithstanding, the Soviet space program was now rooted in defense requirements and political expediency. And its roots were starting to tangle.

Leonov Takes a Historic Walk

The second and last Voskhod mission on March 18 and 19, 1965, was the one in which a man stepped out of a spacecraft's protective environment to "swim in open space" for the first time, as a Russian account put it. In what could be called Nikita Khrushchev's last spectacular, Leonov spent twelve minutes at the end of a tether trying to photograph *Voskhod 2,* five times banging into it with such force that Belyayev, the mission commander, felt the capsule shud-

* That was wrong. The atmosphere was just like Earth's at sea level: roughly 78 percent nitrogen and 21 percent oxygen.

der the way a canoe does with a swimmer clinging to its side. That first EVA—the first space walk—also fit into the scheme of the manned Moon program, then gaining momentum in the design bureaus.

Leonov's historic venture into the void was very nearly his last anywhere. He found while floating 128 miles above Earth that no matter how hard he tried to bend his legs, he could not squeeze back into the flexible eighty-inch-long air lock that would repressurize after he closed its hatch and therefore allow him to get back into *Voskhod 2*'s cabin. Finally, the increasingly worried cosmonaut let some of the air out of his suit, reducing the pressure to shrink it a bit, which was very dangerous. Seeing that even then he could not climb back into the air lock legs first, he turned over and climbed in headfirst, squirming and twisting desperately once he was inside so he could close the outer hatch. Leonov said afterward that he had lost twelve pounds on the flight and been on the verge of heatstroke by the time he was safely back inside.

But the ordeal did not end there. He and Belyayev discovered while getting ready to come down that *Voskhod 2*'s automatic guidance system had failed. This forced them to navigate by eye. Furthermore, they used their liquid-propellant rocket to slow them down, but it was fired when the spacecraft was at the wrong angle. The combination of problems sent *Voskhod 2* skimming two thousand miles past its recovery point. It landed in a snow-covered forest near the northern city of Perm. To make matters even worse, the spacecraft's radio-beacon antenna, which was supposed to transmit its location to the recovery team, snapped off as the capsule came crashing through the conifers trailing its orange and white chute. Then hungry timber wolves showed up. The utterly miserable cosmonauts spent the night inside their dark capsule with their cumbersome space suits on and with the hatch open just enough to let air in and keep the wolves out. A fire was out of the question and so was using their flashlights, which they decided to save in case they had to signal rescuers at night. They were picked up the next day and returned to Star Town.

A Fatally Flawed System

The saga of *Voskhod 2* was an ironic, and probably appropriate, finale to Khrushchev's penchant for appearance over substance. The first swim in space, like the flights of the first man and woman, masked a grim reality that Sergei Korolyov and at least some of the most thoughtful of the other chief designers would have known. Eisenhower had really been right all along. Down on the ground, where the essential components of future missions had to come together with a synchrony of political will, efficient management, abundant resources, outstanding technology, and high morale, the capitalist behemoth was moving relentlessly ahead of its socialist rival. And the average Russian could have surmised as much, not by looking up at the sky but by looking down at his or her shoes.

What was not generally understood in the West in the throes of Sputnik, the flights of Gagarin, Titov, Tereshkova, and Leonov, and other dazzling deeds (including the impressive expeditions to the Moon with robots) was that the Russians were decisively hobbled by their political system. Marxism-Leninism led to institutional self-destruction no less in the space program than in society as a whole.

The basic problem was not, and never had been, a shortage of skilled scientists and engineers. Russia abounded with them. But starting in the late nineteenth century and continuing all the way through the socialist era, the men and women who practiced science and technology (as well as medicine) worked in a society that held the best of them and much of the rest of the intellectuals in contempt. To be sure, Lenin claimed to want a Communist intelligentsia, but that would have been a contradiction in terms in a system that deified workers and peasants while isolating, imprisoning, or murdering independent thinkers because they were potentially threatening to the system. The most probing minds in the country, artistically as well as scientifically, were therefore permanently alienated.

And if a true intelligentsia was anathema to Marxists, then scientists and engineers were taken to be the most potentially dangerous elements of all. This is because revolution depends by definition upon discontinuity, while the very process of science rests on continuity: on a progression of experiments, one building upon another, that are conducted in an unpoliticized and absolutely free environment. The so-called scientific method depends entirely upon the scientist being able to pursue whatever interests him or her; that is, to think freely. Art was easy enough to corrupt (although many individuals such as Boris Pasternak, Mikhail Bulgakov, and Aleksandr Solzhenitsyn fought back with stunning courage), but science was quite another matter.

Both science and engineering had a predictive power that the party's ideologues could only envy. It was science, not Marxism, that could predict an eclipse a thousand years in advance, reduce or eliminate dread diseases, and propel people under the oceans and into space. It was engineering, not Leninism, that built the bridges, moved the rivers, and calculated orbital velocities and lunar trajectories. The theoreticians and party hacks on all levels therefore saw science and engineering—but particularly pure science—as dangerous competitors that threatened them with laws that were better than theirs; laws that were truly immutable.

They reacted in two ways. First, they co-opted the word "science" for themselves, as in "scientific objectivism," to give their dubious theories the appearance of unchallengeable validity. Marx, who had infamously predicted that the first workers' revolt would come in an industrialized country such as Germany, not an agrarian one such as Russia, thereby got to wear Pasteur's intellectual mantle. And second, they either appropriated scientists such as Trofim Lysenko, the undereducated agronomist who invented Marxist biology, crippling much of the other kind in the process, isolated and carefully controlled the best scientists they had (Andrei Sakharov and Pyotr Kapitsa being but two), or simply liquidated those who were uncontrollable. Pyotr Palchinsky, a rebellious engineer who sounded the alarm over huge and grotesque civil engineering errors, was ordered shot by Stalin for doing so. (Kapitsa, who intervened with Stalin to get scientists out of the gulags, was himself marked for assassination by Beria, but it somehow never happened.)

And there was another problem: science and technology do not, and cannot, function in a political void; they are inexorably linked. A system that held scientists and other independent thinkers in contempt treated like-minded politicians the same way. As a result, the political system itself was infested with functionaries and apparatchiks of all descriptions, some of them exceedingly wily or intelligent, but none either freethinking or freely elected. To the contrary, according to veteran space journalist Grigori S. Khozin, the men in the Kremlin were "afraid of independent thinkers." As a result, the party stalwarts invariably held their highly tenuous jobs because of absolute loyalty to the system, relentless maneuvering and intrigue, and the incessant undermining of anyone who posed a real or perceived threat. An old Russian joke has it that the way to deal with a neighbor whose cattle are better than yours is to burn down his barn.

The intrigue, which crowned an economy always severely strained by the preeminent requirements of the military, defeated even the best efforts of the best minds in the space program. Ustinov continuously tried to check Korolyov because the chief designer was not only an important personality in his own right, and therefore potentially dangerous in some undefined way, but also because he talked directly to Khrushchev. Ustinov, whose roles at one time or another included defense minister, armaments minister, and secretary of the party Central Committee, therefore tried to foil Korolyov—"press him down," as one insider put it—at every opportunity. Years later, he would take away the telephone, called *kremlevka,* that linked the Space Science Institute to the leaders of the space industry, evidently also to keep power diffused. When Roald Sagdeev, the head of the institute, complained, Ustinov pulled his *kremlevka,* too.

When Ustinov was Central Committee secretary, he and Defense Minister Andrei A. Grechko had their own series of notorious battles over ballistic missile development with each using a chief designer as a proxy: Yangel and Chelomei, respectively. Pitting design bureaus against one another in internecine warfare over funding and prestige was done for at least two reasons, both of them as utterly cynical as they were stupid: to make certain that no chief designer grew inordinately strong (as if Korolyov or Glushko aspired to the Supreme Soviet), and also on the theory that competition would get the most out of the adversaries. So when Khrushchev decided just before he was toppled to race the Americans to the Moon, and Korolyov voiced caution about such a course, explaining that while it could be done it would be horrendously complicated, the premier responded by setting Chelomei to work on a Moon rocket in direct competition with Korolyov. The old man's ire was brought home to Korolyov when his son, Sergei, went to work for Chelomei designing cruise missile guidance systems. On the other hand, Khrushchev and Korolyov were like quarreling lovers, each bringing a whole bag of often conflicting needs and emotions to what was a surprisingly synergistic relationship.

Far from being confined to the space program, what happened to Korolyov and his colleagues was only symptomatic of a systemic cynicism and ferocity so deeply embedded that they can accurately be called traditional. The dreaded Rasputin was a supreme maneuverer. Stalin was notorious for vindictively playing his subordinates off against each other. So would Boris Yeltsin, who, as one observer put it, "stretched the meaning of creative tension." "It is our history and it is very much Yeltsin's style," a former Yeltsin aide would say about his boss's divide-and-conquer strategy. "He is creating a dynamically tense complex of power where each person is capable of something important but none has the independence needed to act alone."

Alessandra Stanley, a *New York Times* Moscow correspondent who covered Yeltsin's electoral triumph in the nation's first real election, would speak eloquently to the phenomenon when she observed, "In Soviet times, the closest thing to a festive inauguration ceremony for a new leader was the funeral of his predecessor."

The competition forced upon the chief designers led to Chelomei's producing what would become the powerful and reliable Proton booster, while Korolyov developed the mammoth, but unlucky, N-1 Moon rocket. The point, however, is that political expediency created parallel programs using redundant technology that was ruinously expensive and therefore counterproductive in an economy as fragile as the Soviet Union's. Better management, which was inherently impossible in such a divisive environment, would not have allowed that to happen. So the system carried the seeds of its own failure. It is a tribute to Korolyov's and his colleagues' genius and tenacity, in fact, that they were able to accomplish what they did.

To their lasting credit, Soviet rocket designers developed simple, reliable, and relatively cheap engines and launch vehicles that were models of innovation. The idea was not to push technology to its limit but to be clever. Unlike American engines, for example, Soviet engines used a single driveshaft to spin both fuel and oxidizer pumps, reducing the number of moving parts, and they even used the one driveshaft to feed as many as four combustion chambers. The designers pioneered wrapping kerosene pipes around hot rocket nozzles to cool them and increase efficiency. They relied far more on kerosene, which is cheaper, safer, and far easier to handle than liquid hydrogen, by increasing combustion chamber pressures and therefore thrust. And unlike their Western competitors, who used so-called open-cycle engines that spat out some unburned fuel, they took advantage of every ounce of energy by using a closed-cycle system that preburned some propellant to run pumps before it shot into the combustion chamber, increasing heat and therefore performance.

All of that and more went into building a fleet of powerful heavy lifters. The pity of it was that the engineering could never be good enough to overcome political gravity: a fatally flawed economic system and pervasive, institutionalized deceit and paranoia. The American program had no such afflictions. Certainly the economy was healthy enough to get to the Moon. And while American democratic politics could certainly be rough, it had no dark side that even approached the Soviet Union's. The differences were sharp and they were slowly but relentlessly proving to be decisive. The first real indication that the test of national wills was shifting in America's favor, at least in space, came with Gemini.

The "Twins" Fly

Mercury-Redstones and Mercury-Atlases were called MRs and MAs, respectively, and Gemini-Titans were known as GTs. GT-1, which flew without a crew, was sent into orbit with its Titan booster attached on a shakedown flight on April 8, 1964. Air Force and NASA officials pronounced it a "storybook launch." That may have been true, but GT-1 was also eleven months late because of management and engineering problems, schedule slips, skyrocketing cost overruns, and development problems with the Titan 2. Although that kind of slippage would not have been especially serious in some other programs, it was deeply worrisome to NASA, which all the world expected to see land men on the Moon within only five years.*

Storybook analogies notwithstanding, the days were filled with as much danger for the Americans as for the Soviets. The Atlas and the Titan had been designed as ballistic missiles with no expectation of perfect launch rates. That was acceptable where carrying warheads was concerned, since it was cheaper

* McDonnell increased the cost of the spacecraft alone from $240.5 million to $391.6 million to $498.8 million—a jump of 107 percent—in 1962 alone. By the following March, the projected cost of the overall program had jumped to more than a billion dollars, raising hackles at headquarters. (Baker, *The History of Manned Space Flight,* p. 182.)

to buy extra missiles than it was to design them for nearly total reliability. "We knew we were going to have a lot of failures," one NASA official later explained. "I used to refer to our attempts to launch as random successes." But carrying astronauts was a far different matter. Failure, as the hackneyed cliché had it, was not an option with lives at stake. There was no such thing as an acceptable loss rate in the manned program, and that put enormous pressure on the engineers to design out catastrophic failure. Gemini had no catastrophes. But there were some heart stoppers.

Following a second successful unmanned test in January 1965, Gus Grissom and John W. Young flew GT-3 in a basically uneventful mission two months later. They evaluated the spacecraft, did some maneuvering, and performed three experiments. Grissom, showing either almost unbelievable insensitivity or very dark humor, reportedly wanted to name the craft *Titanic.* That appalled officials who had vivid memories of his nearly drowning as *Liberty Bell 7,* its hatch wide open, slipped under the waves like the ocean liner that had hit the iceberg. They vetoed the idea outright. So the plucky astronaut countered with *Molly Brown,* whose namesake, after all, had at least been unsinkable on Broadway. Grissom was allowed to use the name, but the ever publicity-conscious people at headquarters decided that GT-3 would be the last manned spacecraft to carry a name, at least for some time. (Cooper had trouble selling *Faith 7* to his superiors, who dreaded an accident that would almost inevitably be followed by headlines saying: "The United States Lost Faith Today.")

At the end of their first orbit, Grissom and Young used their relatively sophisticated Orbit Attitude and Maneuvering System for a seemingly mundane operation that actually amounted to an historic first. They fired the OAMS to alter their flight path—change their orbit—and therefore proved that a rendezvous could be done. Fortunately, *Molly Brown* lived up to her name.

Gemini 4, which went up on June 3, 1965, with James A. McDivitt and Edward H. White, marked a true transition from Mercury's up-and-down operation to landing Apollo on the Moon. In sharp contrast to the showy flights of the Voskhods, this four-day mission—the time it would take to reach the Moon—truly extended the Americans' reach. In technical terms, the flight was supposed to demonstrate and evaluate spacecraft systems and the effects of "prolonged exposure of crew to [the] space environment." This included turning the capsule around and trying to rendezvous with the Titan 2's second stage, which was the kind of docking maneuver that would have to be done when Apollo set out for the Moon. It was the first time anyone tried to actually link up with another object in space. But the mission was abandoned when the upper stage, which was two hundred yards to the right of them, started to tumble.

The first American EVA, on the other hand, was an unmitigated success. The hatch cover opened four hours and eighteen minutes into *Gemini 4*'s flight, and twelve minutes later White appeared holding a compressed-air maneuvering gun while remaining firmly secured to the spaceship by a tether. Pointing the gun at the capsule and gently squeezing its trigger propelled him out of the hatch and into free space. Soon the gun's air was gone, leaving the exhilarated astronaut to cavort on his tether and enjoy the mind-boggling view.

"It was very easy to maneuver with the gun. The only problem I have is I haven't got enough fuel," White reported. "This is the greatest experience; it's just tremendous. . . . Right now, I'm standing on my head and I'm looking right down and looks like we're coming up on the coast of California . . . as I go on a slow rotation to the right. There is absolutely no disorientation associated with it." Edward H. White was flying upside down at about 17,500 miles an hour, a vast expanse of blue, green, and brown just on the other side of his visor, and feeling just great.

The space walk lasted twenty minutes—"the longest extravehicular activity on record," as briefers would later tell the news media—and ended with White's climbing back into his craft without experiencing the frightening ordeal that Leonov had endured. Besides the romp in space, eleven experiments were performed.

More prosaic, though necessary, operations were also accomplished. There was a bungee cord with a handle on one end and a foot strap on the other to be used for exercise in the cramped space and to tone up the astronauts' hearts before re-entry.

Longer flights also meant that food had to be considered, as did its elimination. McDivitt and White had four meals a day—breakfast, lunch, dinner, and supper—that were marked by day and meal and connected by a nylon cord in the order they were to be eaten. The food itself was freeze-dried, dehydrated in powder form, or compressed into bite-sized chunks. A specially designed water gun was used to reconstitute the freeze-dried and dehydrated morsels. A typical breakfast included apricot cereal bars, reconstituted coffee, ham and applesauce, and cinnamon toast. Chicken salad, beef sandwiches, peaches, and banana pudding were a typical dinner, while supper consisted of potato soup, chicken and gravy, toast, peanut cubes, and reconstituted tea. The nutritionists figured that the four low-residue meals combined averaged 2,500 calories.

And all of it had to come out. Eliminating waste had not been a problem on the short Mercury flights (with the notable exception of Shepard's delay), but Gemini, Apollo, and other relatively long duration missions required an efficient and tidy waste elimination system, particularly since the astronauts had to function in an area the size of two phone booths. Urine was passed through a relatively simple hose and into a container from which it was flushed overboard. Getting rid of excrement, however, was a far different matter because of the cramped quarters and weightlessness. An astronaut who felt nature's call had to pull off his space suit and attach a small bag with an adhesive surface to his buttocks. Since there was no gravity, he then had to squeeze his excrement to the bottom of the bag and then thoroughly knead it like dough so it would mix with a chemical that killed bacteria. It was not a process NASA's public information people went out of their way to talk about.

There were eight more Gemini flights, the last one flown on November 11, 1966, by James A. Lovell Jr. and Edwin E. "Buzz" Aldrin Jr., both of whom were destined to go to the Moon, though under vastly different circumstances. Combined, they increased U.S. experience in manned spaceflight by 540 orbits, 865 hours, and just over 24 million miles, or the equivalent of fifty-one round trips to the Moon. The astronauts performed fifty-four experiments, including in-flight sleep analysis, cardiovascular conditioning, measuring the effects of weightlessness and radiation on blood, communicating during re-entry, collecting micrometeorites, navigating by starlight, and measuring surface features. In addition, astronauts racked up eleven more hours floating in space, six of them working in the open hatchways of their spacecraft.

Largely unnoticed was the fact that an important transition had begun. The difference between Mercury and Gemini was the difference between taking short excursions to space and living there.

Yet Gemini, like Mercury, was menaced by a series of glitches—one of them nearly fatal—that underscored how fragile manned space operations really were. GT-4 was unable to rendezvous with the target spacecraft because it ran low on maneuvering fuel and then had to be landed by hand because its computer forgot the automatic re-entry sequence; GT-5 also missed its rendezvous, this time because its special fuel cell, designed to provide long duration electrical power, broke. *Gemini 6*'s launch was delayed because of a problem with the Titan 2, while *Gemini 7* not only had problems with its fuel cell, but with two failed attitude thrusters. Despite that, however, they had a beautiful rendezvous that came within a foot of each other in identical orbits. Neil A. Armstrong and David R. Scott were nearly killed when *Gemini 8* spun wildly out of control during the program's first docking attempt. They regained control by using the re-entry control system, but that forced them to come down early. *Gemini 9* was launched late because of a guidance system computer problem and, once up, Eugene A. Cernan was unable to test a new Air Force maneuvering unit during his own EVA because his visor fogged up.* *Gemini 11*'s launch was postponed twice because of problems with the

* Cernan and Thomas P. Stafford became GT-9's crew in the first place only because its original crew, Elliott M. See and Charles Bassett, had been killed on a training flight three months earlier.

Titan 2 (one of them a dangerous oxidizer leak), and its successor, the last Gemini that went to space, was afflicted by problems with its fuel cell and attitude thrusters.

Slaves in Space

While people were beginning their first, tenuous forays into low orbit, their proxies were multiplying there and far beyond. Like the supporting cast in a drama of majestic proportion, a class of robotic spacecraft served the humans who were being shot into space and readied for the Moon. Where no man had gone before (to borrow a phrase from a then-new television series called *Star Trek*), robots would go first.

By the mid-1960s, the United States and the Soviet Union had developed two basic kinds of robotic spacecraft: the Earth huggers and far-ranging explorers. The explorers were sent to the Moon and to Earth's nearest planetary neighbors, Venus and Mars, for the sake of both knowledge and politics. The trips to the Moon were reconnoitering expeditions in advance of the arrival of humans.

Having been upstaged by the Army's Explorer 1 on January 31, 1958, the Air Force decided to strike back by using an Atlas to launch a probe to the Moon that August. Fearing that the GIs would get wind of it, the airmen took the precaution of classifying it top secret. And since they needed first-rate tracking for the mission, they decided to go to Sir Bernard Lovell at Jodrell Bank in England. The astronomer first heard about the Air Force plan when a colonel showed up in Manchester just before Easter. After making certain that the windows in his office were closed and the door was locked, the American visitor asked Lovell in a near whisper whether he would be willing to use his telescope to track an Atlas-launched Moon shot. Lovell agreed to do it and also to keep the secret. But the secrecy soon evaporated.

As it turned out, the Air Force tried to reach the Moon with a Thor IRBM and a specially designed upper stage called Able, both of which disappeared on August 17, along with a tiny hitchhiker named Pioneer, in a well-publicized explosion shortly after being launched from Cape Canaveral. A second try with a successor named Pioneer 1 failed on October 11, though the midget spacecraft went high enough to send back forty-three hours of data before it dropped into the Pacific like a scorched rock. Pioneer 2 also fell back down a little less than a month later. That was it for the Air Force. The old Jet Propulsion Laboratory–Army Ballistic Missile Agency team led by Pickering and von Braun, in effect shaking their collective head on the sideline, saw the Vanguard fiasco repeating itself and prepared to rescue the nation's honor yet again.

The Russians, meanwhile, were also aiming beyond Earth orbit at a pace that reflected their own desperation. Knowing that the R-7 lacked the muscle to get even a tiny payload to the Moon, Korolyov invented an upper stage whose RO-5 engine added another five tons of thrust, and took his own shot from Tyuratam on September 23. The *Semyorka,* carrying a small probe named Luna, shook itself to pieces and blew up ninety-two seconds after ignition. On October 12, the day after Pioneer 1 was lost, a second R-7 disintegrated a hundred seconds after launch. A third try, on December 4, lasted a little more than four minutes before engine failure struck.

Two days later, just as that month's launch window was closing, JPL and the ABMA launched the twenty-eight-pound Pioneer 3 on top of a souped-up Jupiter. There would be no gloating this time, however, because the booster's engine cut off early, sending the little spacecraft into an arc that stretched more than 63,000 miles before plowing back into the air blanket and disintegrating in its own fire.

The Russians opened 1959 with a faultless launch on January 2 that sent a 796-pound probe named Luna 1 toward the Moon. "Lunik," as it was called by its creators, was a fully pressurized and temperature-controlled aluminum sphere that drew its energy from a battery and that measured the Earth's and Moon's magnetic fields, collected data on solar radiation, and more. The probe missed the Moon by 3,700 miles and instead sailed into an orbit around the Sun. "Mankind's age-old dream of flight to the Moon was coming true—which is why Luna 1 was renamed 'Mechta' (Dream) when it became the first manmade planet," one sympathetic account would later explain. The Russians had indeed produced the first machine to fully escape Earth's gravity and enter deep space. The American space community's inner circle knew full well that it was the Moon the Soviets were aiming at, but that was only acknowledged years later.

On it went, well into 1959 and then into the new decade, with the rockets improving all the time and the contestants hammering at each other relentlessly. Pioneer 4, launched two months after Lunik, was also aimed at the Moon but came no closer than 37,000 miles before it became the second craft from Earth to swing into orbit around the Sun. JPL and the Huntsville rocketeers could take some consolation in knowing that they had at least accomplished what their Air Force rivals had not been able to do. But it was small satisfaction, since Korolyov scored yet another victory on September 14, 1959, when Luna 2 became the first man-made object to actually hit the Moon.

In keeping with their perchant for anniversaries, the Russians launched another Moon probe on October 4, 1959, or exactly two years after Sputnik 1 went up. On October 7, Luna 2's achievement was topped by Luna 3's sensational orbit of the Moon and the return of the first photographs—twenty-nine of them—showing 70 percent of its far side. To accomplish that stunning feat, the spacecraft had to fly between the Sun and the Moon, stop its rotation, point and shoot its two cameras for forty minutes while the attitude-control system kept it steady, and then head back toward Earth so the imagery could be scanned electronically and transmitted to ground stations by television as the spacecraft swung by from 25,000 miles out. In the view of Nicholas L. Johnson, an expert on the Soviet space program, the mission amounted to "one of the most astounding technological achievements of any nation, considering the state of the art at the time." Pickering, von Braun, Schriever, Medaris, Killian, and others were soon treated to pictures of the Sea of Moscow, the Gulf of Cosmonauts, the Sovietsky Mountain Range, Tsiolkovsky Crater, and the Sea of Dreams.

But now the momentum moved back to the United States. Unlike its predecessors, Pioneer 5 was sent to explore interplanetary space, not the Moon. And also unlike the others, it did what it was told to do. Launched on March 11, 1960, and run by the Goddard Space Flight Center, the ninety-five-pound sphere carried four scientific instruments into orbit around the Sun and returned a great deal of data about what was happening out there: specifically, about radiation, magnetic fields, cosmic rays, and solar activity. Pioneer 5 was the first spacecraft to track the magnetopause, which is the boundary where Earth's magnetic field gives way to the solar wind. Whatever it did for the fledgling field of space science, it gave the United States a badly needed boost in both senses of the word.

Within three years of the start of the space age, exploring machines were being developed for missions both to the Moon and to other planets. Three kinds of missions, each in logical order and of increasing complexity and sophistication, would be worked out for the new explorers during the 1960s: flyby, orbit, and landing.

First and most basically, spacecraft would shoot past the planet or moon and collect preliminary data before either being pulled into an eternal orbit around the Sun or heading out of the solar system. The information, combined with what astronomers turned up with their telescopes, would define the next phase: orbiting. As the term implies, spacecraft following the ones that flew by would drop into orbit around the planet. The obvious advantage of circling for months or years rather than flying by was a phenomenal increase in data because they were collected continuously. And that wealth of new information would in turn tell mission planners where to put the landers down for the final, touch-and-feel phase of robotic exploration. Then, according to plan, people would follow.

Like most of his colleagues, Oran W. Nicks, an aeronautical engineer who ran NASA's Lunar and Planetary programs during their formative years, liked to give the robot explorers the qualities of living creatures. A "far traveler," as he lovingly called lunar and planetary spacecraft (more than thirty of which he helped send to the Moon, Venus, and Mars), had to have lifelike qualities and be able to get along on its own very far from home. First, and most obviously, Nicks explained, it had to have a rigid structure that held it together just as a skeleton defines the human shape. It also had to see with its cameras, speak with its radio, listen with its antennas, sniff with its other sensors, keep its balance with gyroscopes and optical sensors, and remember with its tape recorder. And it had to be able to eat and digest sunshine for energy. "Incoming solar energy had to be assimilated to sustain it," Nicks explained. "Attitude orientation was required to obtain power, to maintain communications, for pointing sensors, and for thermal control. A method of knowing where it was headed was required; thrusters were needed to serve as 'muscles' for attitude and course corrections. It had to have some memory and a time sense, plus an ability to interpret and act on commands and to communicate its state of health and its findings." In the same vein, Donna Shirley, who directed JPL's Mars exploration program three decades later, would talk about the robots that were sent to crawl over the Red Planet using artificial intelligence as being as "smart" as bugs. And they were.

Lone Rangers

While Mercury, Gemini, and Apollo progressed, a series of three kinds of robots were developed to successively scout the Moon. The first were called Rangers and were designed and built by JPL. The plan called for sending the first two flying over the Moon in highly elliptical orbits so they could measure radiation, magnetic fields, and other properties at close range. Then three others were supposed to crash onto the lunar surface, taking detailed television pictures and collecting data on the electromagnetic field, particles, and other things before they were smashed to splinters. The pictures would be used to calculate where a series of follow-on spacecraft, named Lunar Orbiters and Surveyors, would circle and set down, respectively. The orbiters were supposed to send back high-resolution pictures of likely landing sites, spotting boulders, depressions, and other potential hazards. The lander would hopefully prove beyond doubt that a manned spaceship coming to rest on the lunar surface wouldn't sink into cosmic quicksand. The best geologists in the country thought the lunar surface was indeed firm enough to support the Apollo landing. But given the catastrophic consequences of being wrong, educated guesses were not good enough. Like Ranger, Surveyor was developed by the Jet Propulsion Laboratory. Lunar Orbiter was awarded to Langley.

Ranger itself was designed in close conjunction with an interplanetary explorer named Mariner (the family resemblance was unmistakable) and, unlike its far-ranging cousin, very nearly caused the demise of JPL.

Rangers 1 and 2, launched on August 22 and November 18, 1961, were done in by Agena upper stages that failed to restart, leaving both spacecraft stranded in the parking orbits they were supposed to use before heading toward the Moon. Ranger 3 went up on January 26, 1962. This time the Atlas booster went deaf. Unable to receive a command from Canaveral to shut down at the right instant, the big rocket's engines fired for too long, sending its passenger hurling ahead of and below the Moon at an uncorrectable distance of 20,000 miles. That much was not JPL's fault.

Ranger 4 actually made it to the Moon on April 26 but crashed on its far side without sending back so much as a single picture because its timing sequencer did not tell the antenna and solar panels to deploy. It therefore virtually ran out of energy and slammed into the lunar surface like a stone.

Webb nevertheless proclaimed the flight to be a resounding triumph. An "outstanding American achievement" is what he made of the embarrassing mishap because, he explained to newsmen with an absolutely straight face, it was the first American spacecraft to reach the Moon. As such, it contributed to what he called the "long strides forward in space" made by the United States. And, NASA's administrator could not resist adding, Ranger 4 was far more sophisticated than Luna 2. What Webb neglected to mention was that small spheres made from metal pennants and embossed with hammers and sickles rested on the lunar surface as he spoke. They had been delivered there by Luna 2 despite its relative unsophistication (or, more likely, because of it).

"The Americans have tried several times to hit the Moon with their rockets," Khrushchev retorted at his own impromptu news conference. "They have proclaimed for all the world to hear that they launched rockets to the Moon, but they missed every time." The pennants, he added gleefully, were getting lonesome waiting for an American companion.

"Ranger 4 was tracked by the Goldstone receiver as it passed the leading edge of the Moon," JPL's infuriated director shot back before giving the coordinates of the crash site. And if the Russians wanted to confirm that fact, William Pickering added with unbridled anger, they could send one of their cosmonauts to the scene to see for themselves. But he was far from sanguine. The nature of the Ranger program itself was now being questioned by Apollo managers and engineers who needed reliable data on the Moon's radiation environment and the characteristics of its terrain so they could pick landing sites for the astronauts.

Nor did matters improve on July 22, when Ranger's close cousin, Mariner 1, was launched on an incredibly short and ill-fated trip to Venus to answer four frantic Russian attempts to get there and to Mars ahead of the Americans in October 1960 and February 1961. The galling thing about losing Mariner 1 was not that it was kidnapped by an errant Atlas, or even that it went brain dead or became hard of hearing, but that it met its untimely end because of an error in punctuation: a minus sign was left out of the booster's steering code, causing it to veer left and nose down. Following procedure, the range safety officer hit the button that set off the explosives that ignominiously ended America's inaugural planetary mission.

Ranger 5 blasted off from Canaveral's Launch Complex 12 and into a leaden sky on October 18 while its creators kept their fingers crossed. But that did not help. An hour and thirteen minutes after liftoff, an apparent short circuit knocked out electrical power coming from its solar panels, which automatically turned on Ranger 5's battery. Since the battery had only a couple of hours of electricity, project manager James Burke decided to send the crippled spacecraft crashing into the Moon in what would euphemistically be called an engineering test. But Ranger's telemetry came back so garbled that no one knew precisely where it was. Three days after it was launched, Ranger 5 cartwheeled 450 miles over the trailing edge of the Moon and joined some of its predecessors in solar orbit.

This fifth failure set off a political explosion. Boards of inquiry were set up both at JPL and in Washington. The latter, chaired by a Navy scientist named Albert J. Kelly, handed in a report that called for an end to sterilizing the spacecraft with heat, which was done to keep Earth germs off the Moon, but which degraded components. It also called for a number of other changes, including a management shake-up, better prelaunch inspections of the rockets, the use of an industrial contractor to build Ranger 10 and subsequent spacecraft (the number had since grown from the original six), a restatement of the program's objectives, and elimination of such extraneous work as the simultaneous development of technology for planetary missions. JPL, in other words, was being distracted by Mars and that had to stop. Furthermore, all space science experiments were scrapped in favor of the remaining Rangers carrying only a television camera. "A few TV pictures of the moon, better than those taken from earth, is the only mission objective," ordered Harris M. "Bud" Schurmeier, who replaced Burke as project manager. That angered the irascible Harold Urey, who complained to Newell that science was once again being given short shrift by NASA. But so it had to be.

Ranger 6's flight to the Moon began as the space age's first love-in. Having passed meticulous inspections by JPL and Air Force technicians, the spacecraft, bolted securely inside its Atlas-Agena, thundered away from Canaveral on the morning of January 30, 1964, and headed for the lunar surface with no apparent problem. Two nights later von Karman Auditorium at JPL was filled with expectant print and broadcast reporters. JPL's own television camera took pictures of the television cameras that were there to take pictures of the pictures Ranger 6's television cameras took of the Moon during its final nineteen minutes. "As we swing farther forward there are some members of the press now coming into view with the stage," a narrator told everyone else at JPL who had his or her closed-circuit set switched on. "At one o'clock this morning," the soothingly authoritative space voice continued, "some twenty-two or twenty-three minutes from now, the spacecraft will be 237,270 miles away from Earth." That kind of precision was impressive and reassuring. JPL seemed to know what it was doing.

Over in the Space Flight Operations Facility—the control center—Pickering and Robert Parks, JPL's lunar and planetary program director, were hosting Homer Newell, Edgar Cortright, who was Parks's opposite number at headquarters, and William Cunningham, who ran the Ranger program from Washington. Newell, Cortright, and Cunningham had flown in for the grand occasion. In the capital, NASA's own auditorium was also packed with reporters and technocrats, all of them desperately hoping that they were about to witness an event of historic proportion. They were indeed.

With nineteen minutes to go before impact, Ranger 6 reported to Earth that the first of its two camera systems was switching on and warming up, as planned. A few minutes later, still following the script, its other system went on. Then, nothing. The time when the cameras were supposed to start sending back pictures of the crater-pocked Sea of Tranquility came and went without a single image brightening the television screens.

"The signal level is steady but there is still no indication of full-power video," said Walter Downhower, the chief of the Systems Design Section, who had volunteered to be the temporary voice of mission control. Over in von Karman, the members of the Fourth Estate had stopped murmuring and, sensing a looming catastrophe, had fallen absolutely silent. The red hand on every clock at JPL, every clock at headquarters, every clock at Canaveral and Houston and everywhere else in NASA's hushed and expectant empire was a wand that was transforming the elusive specter of triumph into a nightmare of grotesque proportion with every sweep. Pickering sat with Parks and their guests in a balcony that looked over the mission controllers who worked at their consoles. He watched the activity—or lack of it—on the floor below, and as he watched, he began to squirm.

"We are coming up on impact minus two minutes. We have no indication of video. Repeat, there is still no indication of full-power video. . . . Still no video. . . . Coming up on impact minus ten seconds. No indication of full-power video. . . . DSIF reports loss of signal." Loss of signal. The Deep Space Instrumentation Facility, which operated the network of bowl-shaped antennas and other equipment that tracked Ranger 6, did exactly what its counterpart in hospital emergency rooms did with humans: it reported that the spacecraft's vital signs had stopped.* The patient had plowed into the lunar landscape and expired in a heap of twisted and smashed rubble without having opened its eyes. Not once.

Cunningham stalked out of the building, staring straight ahead and mumbling over and over again, "I don't believe what's happened. I don't believe what's happened." An ashen Pickering, who had walked onto the floor right before impact, looked hard at those nearest to him. "I never want to go through an experience like this again," he told his subordinates. *"Never."* Back in Washington, many of those gathered in the headquarters auditorium winced when they were treated to this additional message, delivered over the intercom because a technician accidentally crossed lines: "Spray on Avon Cologne Mist and walk in fragrant beauty. Or splash on an Avon . . ."

Newell, Pickering, and Schurmeier, the Ranger project manager, had to meet the press while the dust was still settling on the Sea of Tranquility.† Here is what they hastily agreed to tell the world: it was too soon to tell what had gone wrong; the lack of imagery was disappointing; JPL and NASA would try to do better next time; Ranger's trajectory and impact point were virtually perfect: within only twenty miles of the aim point. It was, Schurmeier said resolutely, "a very significant achievement in command and control of a space vehicle."

An in-house postmortem completed at JPL on February 14 put the blame on a radio channel that was supposed to monitor the television systems during the cruise part of the flight but that had accidently turned on within minutes of launch. It, in turn, caused a short circuit that knocked out the power supply to the television cameras. That was true as far as it went. But headquarters wanted to step quite a bit farther back. In mid-March, NASA's own board of inquiry took a politically charged shot at the snotty Pasadenians. The board

* The DSIF, soon renamed the Deep Space Network, was a worldwide string of antennas and communication facilities developed and operated by JPL that was responsible for tracking and providing communication links with NASA's lunar and planetary spacecraft, and occasionally with military spacecraft as well. The network is referred to as DSN.

† There was and remains a distinction between projects and programs. Projects such as Ranger, Mariner, and Pioneer were run by project offices at JPL, Ames, and other centers that were responsible for developing spacecraft or rockets or testing aircraft such as lifting bodies and the X-15. They were concerned primarily with the hardware and how it functioned. Every project was in turn part of a larger program with the same name that was controlled at headquarters. The program office was responsible for coordinating and integrating every aspect of the mission, including, for example, making certain that the spacecraft and the booster were compatible, the launch facility was ready for them, and the tracking system was in place and programmed. Technically, the program manager was therefore the project manager's superior. Beginning in the mid-1990s, a few whole programs, such as JPL's New Millennium, would be run at the center, not in Washington.

charged that Ranger was too complicated and that its design and testing procedures were seriously flawed. It suggested that headquarters should "closely monitor" (i.e., rein in) work on Ranger 7, which had been scheduled for launch that month. And contrary to what he said in public, Webb was so angry at what he took to be JPL's stubborn refusal to accept direction that he sent letters summarizing the board's criticism to Clinton P. Anderson, chairman of the Senate Committee on Aeronautical and Space Sciences, and George P. Miller, chairman of the House Committee on Science and Astronautics. It was like one parent of an unruly child saying to the other parent: "I can't do anything with him; *you* talk to him."

And talk they did. Webb, Pickering, Schurmeier, Cunningham, Cortright, and Newell were summoned to congressional hearings in late April, along with Parks and Nicks. All, including the space agency's irked administrator, tried to put the best face on a very troubling situation. The committee, chaired by Representative Joseph E. Karth of Minnesota, did not excoriate JPL (leading *Missiles and Rockets* to call the four-day inquiry a virtual whitewash). But it did issue four recommendations, three of which boiled down to forcing JPL to be more responsive to direction from Washington.

What really haunted NASA, expressed by both Anderson and Karth right after the Ranger 6 fiasco, was not Ranger's performance per se. The real problem was that the program's setbacks looked like they were going to delay the landing of astronauts on the Moon. Anderson fretted in public that NASA was trying to compress all of the elements in the Moon program in too short a period of time, partly because of the mess in the Ranger program. "Congressional critics," *The New York Times* reported, "are becoming worried over . . . how the space agency can prudently proceed without more photographic and scientific evidence on the nature of the lunar surface, to design the lunar excursion module (LEM) that will carry two men to the moon."

In fact, the whole immensely complicated Apollo program was gathering momentum without Ranger. Even before Ranger 6 plowed into the Moon with its eyes closed, its managers had gone to Houston to tell their counterparts in the Manned Spacecraft Center where they planned to take pictures, but they were rebuffed. JPL was informed, as one of them later recalled, "that the design of Apollo's landing gear was already frozen" and that the engineers in Houston "frankly didn't care where we put Ranger 6."

Moon à la Mode

If one technical decision can be said to be the single most important in the manned lunar program, it was the choice of mode, or exactly what route the spaceship was going to take to the Moon. On that decision rested the design of the spacecraft, its launcher, support facilities, costs, and schedules. There were two apparent possibilities, at least at the outset, which had been hotly debated at least since 1959: direct ascent and Earth orbit rendezvous.

Direct ascent involved a straight shot from Earth, more or less leading the Moon as a skeet shooter leads a clay pigeon and firing so that both objects—rocket and Moon—would arrive at the same place at the same time. This had the benefit of being relatively simple. The problem was that getting almost five thousand metric tons of rocket, spacecraft, and men off the ground and headed into orbit before only part of it—a spacecraft containing the crew, landing equipment, and fuel to get down to the lunar surface, back up again, and return home—was "injected" into a lunar trajectory would involve a staggering propulsion requirement. The answer, at least from 1959 to 1964, seemed to be the development of a colossal rocket called Nova. But the "super booster," whose first stage alone would have to have ten powerful engines, would be both very costly and difficult to develop.

One direct ascent method, briefly raised in 1961, would in fact not have required a super booster at all because it involved a one-way trip. The idea, first proposed by E. J. Daniels of Lockheed Aircraft and then by two Bell Aerosystems engineers, was to deliberately strand one astronaut on the Moon (the way poor Joe had been left there in *Destination Moon* in 1951). The space-age Robinson Crusoe would be resupplied by rockets for years, during which time he would perform "valuable scientific experiments" while people back home tried to come up with a way to retrieve him. The Bell engineers, John N. Cord and Leonard M. Seale, acknowledged that the mission would be very hazardous, but they explained in a paper they presented in Los Angeles in June 1962 that it seemed to be the only way to beat the Russians to the Moon. NASA mission planners took the idea for what it was: lunacy.*

The other serious possibility was Earth orbit rendezvous, or EOR. That would have involved separately sending up the spacecraft, its equipment module, a lunar lander, and whatever else was needed for the journey into Earth orbit, forming them up like a high-tech mule train, and then heading to the Moon. The advantage was that orbiting the components separately would require far less energy for each launch than the amount needed for direct ascent. The problem, at least in the eyes of many engineers, was that EOR was impractical, complicated, and therefore very dangerous.

In the end it would be a third mode, lunar orbit rendezvous, or LOR, that was selected. This mission profile was first suggested by Langley's William Michael in 1960 and then championed by his boss, John Houbolt, against stiff

*Seeing that his country was losing the race to the Moon a few years later, a cosmonaut named Mikhail Nikolayevich Bordeyev is said to have volunteered in a similar mission to Mars. He was taken no more seriously than were the American engineers. (Author interview with Boris Kantemirov and Valentina Ponomareva.)

opposition. It required that the Apollo return capsule—the command and service module, as it would be called—be "parked" in an orbit around the Moon while a lander went down to the surface with the crew. Critics of LOR, at first including von Braun, said that rendezvousing in Earth orbit was dangerous enough, but at least there was an abort capability so that astronauts in trouble could be brought down relatively quickly. A missed rendezvous in lunar orbit, on the other hand, would spell almost certain death for the astronauts, who would be condemned to circle the Moon for years in an orbiting coffin that had been made in America.

Yet there were overriding pluses to lunar orbit rendezvous. The most important of them had to do with the size of the spaceship that was supposed to carry the astronauts to the lunar surface. Both direct ascent and Earth orbit rendezvous would require a "big propulsion module" to land on the Moon and get off again: a rocket and life-support system as high as a six-story building that would be phenomenally expensive to produce and get to the Moon and that, in addition, would be dangerously complicated. Having a command module collect astronauts who came up to rendezvous with it in an ascent module, on the other hand, would permit the use of a much smaller, lighter, and more efficient lander, since part of it could be left on the Moon when the ascent module shot up to meet the orbiting command module. And once the astronauts were safely back inside the command module and heading home, the ascent module itself could be jettisoned. Unlike direct ascent, that operation would not require a monster like Nova, which was then under early development along with Saturn, von Braun's mammoth project. This meant that LOR would cost 10 to 15 percent less than direct ascent (calculated at $10.6 billion) or Earth orbit rendezvous ($9.2 billion).

Von Braun himself eventually came out in favor of the lunar orbit plan, mostly because it was the easiest to do from a managerial standpoint and in fact was the only way to get to the Moon within the time limit set by Kennedy. NASA's Manned Space Flight Management Council—the agency's high priesthood—endorsed LOR on June 22, followed by a formal announcement on July 11, 1962, seven and a half years before the deadline. The president had in the meantime put Apollo in the "DX" procurement category, giving it the highest national spending priority. Apollo's overall budget that year was a relatively paltry $75.6 million, but it would pass the $1 billion level the following year and keep climbing until 1966, when it peaked at almost $3 billion.

Moving the Pillars of Hercules

All of the technical stuff—conditioning and training the men who were headed to the Moon; developing and launching the reconnaissance craft that had to go before them; conceiving, designing, coordinating, and building the machines to take them there; and putting into place a worldwide support system to communicate with and track them—rested squarely on a management process that was the equivalent of the one that built the pyramids. The astronauts and their rockets were only the uppermost elements on a massive structure so vast and complex that it virtually defied close description.

Arnold S. Levine, the author of the classic work on the subject, *Managing NASA in the Apollo Era,* put it this way:

> To understand what NASA did, one must begin by considering it as an institution coordinated to achieve certain goals that were neither fixed nor always precisely determined. Coordination had to be achieved on different levels: within the agency among the substantive program offices, the several field installations, and the central functional offices; between NASA and the Executive Office of the President, which determined the funding levels of each item in the NASA budget before congressional review; between NASA and the congressional committees that authorized its programs, allocated its funds, and provided continuous oversight; between NASA and the scientific community, which was client, critic, and not-so-loyal opposition; finally between NASA and other Federal agencies, which might be partners . . . rivals . . . or symbiotic."

Outside the government itself, there were dozens of prime contractors and thousands of subcontractors, all of which had to work effectively with each other and with the space agency.

On November 28, 1961 (after Houbolt had made his pitch for lunar orbit rendezvous but before it had been selected), NASA picked North American Aviation as the prime contractor to design and build what was already envisioned as a two-section Moon ship. Like Gemini, it would consist of a cone-shaped command module attached to a cylindrical service module. Together, they would be called the command and service module, or CSM. The ungainly contraption that was going to get astronauts onto the Moon and off again—the lunar module (pronounced "lem" and therefore sometimes called the lunar excursion module)—would be built by Long Island's Grumman Aircraft Corporation, which beat out eight other firms for the contract on November 7, 1962. Both companies, as well as others that would be responsible for developing everything from onboard computers (IBM) to the stabilization and attitude-control system (Honeywell), the CSM's instruments (Bendix), the lunar module's all-important ascent stage (Bell Aerospace), the guidance and navigation system (MIT), and even the space suits (International Latex), would be coordinated by an Apollo Mission Planning Panel that was set up in February 1963.

Meanwhile, the von Braun team (as it was called at the Marshall Space Flight Center in Huntsville), working closely with the Lewis Research Center, was straining to perfect Saturn. In 1962, its leader attributed the expatriate German team's success to its being "a fluid, living organization" that responded to external forces, and by never losing sight of its only real purpose: to push "the continuous evolution of space flight." The forces to which the rocket team responded were invariably military and the spaceflight to which von Braun referred evolved in the form of missiles: V-2s, Redstones, and Jupiters. While the heavy lifter (at first unofficially called Juno 5) was never intended to be an ICBM, it was conceived as a huge rocket powerful enough to fling very heavy Department of Defense payloads such as Dyna-Soar into orbit.

NASA took over the heavy lifter project in November 1959. By that time, Rocketdyne (soon to be acquired by North American) had designed three engines, the H-1, J-2, and F-1, that could be clustered like the Russian RD-107s and 108s to maximize thrust. Throughout most of the 1960s, these powerful engines—the H-1 and J-2 developed 200,000 pounds of thrust and the F-1, an incredible 1.5 million pounds—were mounted in stage combinations on a family of progressively larger and more powerful Saturn boosters: the two-stage Saturn 1 and Saturn 1B, which would test the engines and carry astronauts to Earth orbit; and the staggeringly large, three-stage Saturn 5, which would fling them at the Moon. Saturn 5 would stand 364 feet high and weigh nearly 5.8 million pounds fully loaded. Its first stage would use five F-1s that ran on kerosene and liquid oxygen; its second would use five J-2s that were fed liquid oxygen and liquid hydrogen; and its third would use a single LOX and liquid hydrogen–fed J-2. Combined, the eleven engines would develop 8.7 million pounds of thrust.* That kind of thrust could send 124 tons into Earth orbit or more than 50 tons to the Moon at up to 25,000 miles an hour. It could also bite its creators. A fully fueled Saturn 5 was a bomb that could turn into a fireball a half mile wide. The man chosen to head the bomb's design team was none other than Arthur Rudolph, the production manager at the Dora concentration camp.

While Soviet rockets were moved to their launchpads lying horizontally on flatcars and were then erected, fueled, and fired, the Saturns would be assembled upright at Canaveral in a colossal box called the Vehicle Assembly Building (originally named the Vertical Assembly Building) and then carried to the launchpad locked in the embrace of a vertical mobile launcher and on top of one of two specially built crawler-transporters. The crawlers would themselves be huge diesel-powered versions of self-propelled strip mining shovels. And

* Its first stage, called an S-1C, had five F-1s that produced 7.5 million pounds of thrust; its second stage, or S-2, had five J-2s that accounted for another million pounds, and its upper stage, an S-4B, gave the final kick with 200,000 more pounds of push from a single J-2. Total thrust is irrelevant, however. What counts is what each stage does independently.

like their smaller distant relatives, the squat, flat-topped crawlers would inch between the VAB and Launch Complex 39's two pads at a snail-like mile an hour. The VAB itself, which would be the heart of Launch Complex 39, would amount to one of the wonders of the world: a cavernous structure that towered a dizzying tenth of a mile high and dwarfed even the Saturn 5. Boosters would be fitted together like gigantic toys in the VAB and then their spacecraft would be hoisted high into the air and bolted onto them. The structure boasted seventy lifting mechanisms, including a pair of 250-ton bridge cranes that one day would pluck space shuttle orbiters off the floor like stuffed birds. Many people become disoriented by being high off the ground, but the VAB was so cavernous that it would even disorient those standing on its floor and looking up. Its counterpart at Tyuratam was lower, since Soviet launchers and missiles were stacked horizontally, but it was a full quarter of a mile long and, in terms of sheer volume, was the largest building in the world. The two would be the cathedrals of the space age.

On Thanksgiving Day 1963, less than a week after Kennedy was murdered in Dallas, Lyndon Johnson announced in a televised speech that the Pentagon's Atlantic Missile Range and NASA's Florida Launch Operations Center, which shared the facility, was to be renamed the John F. Kennedy Space Center as a memorial to the man who had pointed the nation toward the Moon.

The Mission Profile

There were two parts to the Moon problem, each of them horrendously complicated: engineering and navigation.

The plan was to launch a Saturn 5 carrying the Apollo "stack" from the Kennedy Space Center. The stack would consist of the command and service modules attached together and, stowed behind them, the lunar module in a long adapter section. The LM itself would be designed in two sections: a descent stage and an ascent stage. The flight to the Moon and back would be a meticulously choreographed ballet in what was unquestionably the most impressive engineering feat in history.

Early in the flight, after the Saturn 5's first and second stages had done their job and been jettisoned, the third stage, carrying the men and their Moon craft, would fire itself into Earth orbit. A second firing, done at precisely the right instant, would swing stage three out of Earth orbit and on a rendezvous course with the Moon. Then the command and service modules, locked together, would separate from the third stage, move a few yards ahead of it, and do a 180-degree turn. The third-stage adapter section's outer panels would be blown away. Next, the command and service modules, still attached to each other, would dock nose-first with the lunar module inside stage three's adapter section and pull it out. All three would then fly to the Moon locked together like mating insects (appropriately, Jim McDivitt's *Apollo 9* crew would name the ungainly LM "Spider" because of its thin legs and buglike body).

After swinging into lunar orbit, two astronauts would take the whole LM, both descent and ascent stages, down to the surface for a landing while the third astronaut continued to orbit in the CSM, waiting for them to blast off in the lunar module's ascent stage. Leaving the now-useless descent stage on the Moon, the ascent stage would take off and rendezvous with the CSM. After both spacecraft docked securely, the lunar explorers would climb inside the CSM, and the ascent stage itself would be discarded. The three astronauts would then head back to Earth. On the return the service module would also be abandoned, leaving the conical command module to plunge into the atmosphere backward at a blistering 36,000 feet a second and then parachute down to the usual watery landing. It was an extraordinarily complex and audacious plan in which a frighteningly long list of things could go catastrophically wrong. But there was no other way.

"Moving Targets"

Virtually every NASA engineer and technocrat who had major responsibilities for sending Americans to the Moon, and most of their counterparts in the contracting companies, would remember the years of the Apollo program as a bittersweet time. They were years crowded with exhilarating, unparalleled challenges in what many deeply believed was a cause so worthy that it amounted to a crusade: not to beat the Russians but to accomplish the most daring feat of exploration of all time. Yet the intensity of the mission—planning this greatest of technological enterprises—necessarily caused exhaustion and frustrations so pervasive that they could, and often did, let loose their own demons.

The challenges crossed the entire engineering spectrum, from the almost incomprehensible to the mundane. Getting a spacecraft from Earth to the Moon and back again, for example, was less a propulsion than a navigation problem. While the relationship of two objects in space relative to each other had been understood for centuries (allowing solar eclipses to be predicted), the so-called Problem of Three Bodies was an entirely different matter. Here, the relative positions of three bodies, each influenced by the other's gravitation, had to be predicted with absolute precision. This would be like planning ahead of time that a tennis ball could be tossed into a coffee can from a distance of three hundred miles while the thrower, the ball, and the can were all moving. The nonlinear differential equations necessary to solve the problem were monstrous and had literally defied comprehension for centuries. One eminent mathematician claimed in 1944 that more than eight hundred of his predecessors and colleagues, many of them distinguished, had tried for two hundred years to come up with a formula that solved the problem but that none had succeeded. "The problem of three bodies," he concluded, "cannot be solved in finite terms by means of any of the functions at present known to analysis." The august *Encyclopaedia Britannica* was even more emphatic, stating that "no general solution of this problem is possible."

There were specific solutions that could be found on a mission-by-mission basis, as it turned out, but it took computers to find them. The computers would be set to work doing the painstaking refinement—the massaging—of lunar and planetary ephemerides (exactly where the planets and moons were and were going to be at any given moment) for what aerodynamicists and astronomers call differential corrections. High-speed computers allowed so many minor adjustments, so many cosmic corrections, to be calculated about the direction and velocity of planets and moons and how their gravitational pull, plus the Sun's gravity, affected each other that selective prediction ultimately became possible. And with the ability to predict where Earth, Moon, and spacecraft would be at any given moment relative to one another, so too did the voyage itself become possible.

Professor Stark Draper, an engineer and navigation guru who founded MIT's renowned Instrumentation Laboratory, assured Webb in late 1961 that he could navigate Apollo to the Moon by using his computers to make the necessary corrections. "If you'll guarantee the propulsion," he told the administrator, "I'll guarantee the guidance and navigation." He even volunteered to go along on the first flight. And on the other side of the continent that same year a UCLA doctoral candidate in mathematics and physics named Michael A. Minovitch was coming to the same conclusion independently. Minovitch did vector analysis—refining lunar and planetary movement—for fun and used JPL's and then UCLA's computers to get his own pioneering grip on the three body problem.

But the Earth, the Moon, and the spacecraft were not the only objects in motion. The program itself was a so-called moving target, meaning that everything was constantly changing. Whole design departments groaned under the weight of requirements that seemed to change as much as Minovitch's constantly refined vectors. North American Aviation's Harrison Storms, an engineering genius on the order of Lockheed's legendary Clarence L. "Kelly" Johnson and a handful of others, virtually commuted by air between Los Angeles and Houston with his top engineers as the pace of Apollo design work increased and became hideously complicated. It was so complicated, in fact, that even the computer that was trying to keep track of thirty thousand separate activities at North American was swamped.

Two years into the design of the command and service module, to take one example, NASA still had not decided whether the command module should parachute onto land or water. The Russians' coming down on the great steppe looked impressive (no one in the United States then knew for sure that the cosmonauts had really bailed out of their Vostoks before they hit terra firma), and NASA wanted to match that if possible. But what land to use for the landing? Lyndon Johnson, who forgot neither his grass roots nor his place in history as the space program's foremost political booster, graciously offered to turn his own ranch in Texas into the Apollo landing zone. Even assuming that such a politically loaded and potentially ludicrous idea was accepted, there would be unacceptable physical problems. Landing on a flat, sprawling cattle ranch was one thing. But the mission planners worried that if the astronauts overshot the LBJ Ranch and came down in the Rockies, they might be killed. They could get killed coming down in a lot of places, in fact, so North American finally told NASA that it could not guarantee a safe landing unless the capsule came down at sea. The space agency bought the argument.

The Apollo program was like a huge amorphous ball; it was moving in a definite direction but changing shape all the time. At one point a North American vice president of engineering named John McCarthy complained angrily that changes ordered by the space agency had created a "complete mess. The design is changing so fast that the changes can't catch up. You go down to the factory and watch the thing being built and it has nothing to do with the current configuration."

Not all of the problems were created by NASA, however. The S-2—the second of the Saturn 5's three stages—was also contracted to North American. It had its own setbacks, each of which frustrated von Braun, Arthur Rudolph, and the others at the Marshall Space Flight Center who were trying to fit the entire rocket together on schedule.* The S-2 was so big that three railroad tank cars stacked end to end would have fit into it with enough room on top for a caboose lying sideways. Yet it weighed only 95,000 pounds because its skin was so thin it amounted to the technological equivalent of an eggshell. Putting sections of that skin together in turn required welded joints that had to be surgically clean, flawless, and accurate to within a third of a millimeter, or about the thickness of this page. That would have been tough enough. But it was made even worse because the aluminum alloy that North American picked for the skin, a type called 2014 T6, which actually strengthened when it was near

* The first stage was contracted to Boeing, the third to Douglas, and the rocket escape tower on top of the command module to Lockheed. Spreading the contracts was done in part to prevent any one contractor from becoming overloaded, but also to widen the space agency's political support. The Department of Defense routinely did the same thing with weapons systems.

supercold propellants such as liquid oxygen and liquid hydrogen, was brutally difficult to weld because it weakened under heat.

On top of that, the aluminum sheets varied in size, shape, and thickness, which added to the welding problem. The shifting thickness, in the understated words of one NASA historian, "frequently made temperamental men of normally even-tempered welding engineers; weld speeds, arc voltages, and other regimes had to be tailored for each variance during the welding pass." The extraordinary precision in joining the metal was vital because the design teams knew that even a hairline crack, moisture, the tiniest foreign object, or the seemingly most innocuous imperfection could cause a propellant leak under the immense pressure of flight that could have catastrophic consequences. Huntsville and Houston understood on a cognitive level that North American and the other contractors faced daunting problems. But the primal part of NASA's collective brain showed contempt as the companies hit one snag after another.

In late September 1965, for example, the first finished and flightworthy version of the S-2 second stage was ready for pressure testing. This would be done by filling its tanks with water under increasing pressure and then shaking the stage and otherwise stressing it as if it were really lifting off. Since the whole Saturn 5 was supposed to be designed to withstand one and a half times the maximum calculated stress load of a real flight, Huntsville ordered that the S-2 be stressed right up to that point. But what Rudolph and the others at Marshall apparently did not take into consideration, and what filled Storms and McCarthy with dread and anger, was the fact that water weighs fifteen times more than liquid hydrogen. Furthermore, liquid hydrogen would cool the meticulously shaped and welded 2014 T6 to 400 degrees below zero, increasing its strength. Water would do no such thing.

On the night of September 29 the S-2, strung on an open-air gantry at North American's Seal Beach plant, south of Los Angeles, creaked ominously as the pressure inside hit 144 percent of design capacity. Then there was an even more ominous groan. Finally, stressed beyond its limit, the S-2 gave up with a loud metallic belch and exploded, dumping a fifty-ton torrent of water onto the floor of the test facility.

Brigadier General Edmund F. L. O'Connor, director of Industrial Operations at the Marshall Space Flight Center, responded to the deluge by sending a memorandum to von Braun saying that the S-2 program was "out of control." Eberhard Rees, the deputy director of von Braun's Development Operations Division, seethed. He wrote his own scathing memorandum in early December in which he threatened to use draconian measures to get North American to change what he saw as a dangerously unwieldy and unresponsive operation. Rees, who had been deputy for in-house manufacturing at Peenemünde and who had therefore seen production schedules slip in a previous life, was so frazzled that he made his point in italics: *"It is not entirely impossible that the first manned lunar landing may slip out of this decade considering, for instance, the present status of the S-II program."* Like everyone else in Apollo, Rees knew that the time limit was in fact about the only thing in the program that was cut in stone. A "tiger team"—NASA's name for ad hoc troubleshooters—was dispatched to California to get a handle on the problem.

Storms reacted by having a heart attack (from which he recovered). So did Norman Ryker, who supervised the engineering group that put together the winning CSM proposal. Others, including the chief project engineer, the man in charge of material, and someone in guidance and stabilization, simply collapsed of exhaustion. Joseph Goss, a rocket engineer who worked on the S-2, spent weekends in the office and took work home with him. One day his son, a smart second-grader named Matthew, drew on his own experience with less gifted classmates who had to stay after school or do extra homework. "Dad," Matthew asked his father, "are you in the slow group?"

Nor was the situation much better where the command and service module was concerned. Eleventh-hour problems with the water glycol pump in the command module's environmental control unit, to take only one example, delayed the arrival of the first spacecraft to Cape Kennedy by nearly a month. When it did arrive, the Apollo program manager at the scene complained about the amount of engineering work that still had to be done, adding that more than half of it ought to have been completed before the spacecraft left California.

The disgust soon spread to the astronauts themselves. Grissom, White, and Roger Chaffee, who were scheduled to fly the first manned Apollo flight in an orbital test, had to rely on a simulator much more than they had to in Gemini. But the command module was being changed so quickly—more than a hundred changes at one time—that the simulator could not keep up. The fact that the trainer was always outmoded deeply angered Grissom, who expressed his irritation the way most test pilots would: he hung a lemon on it.

Estranged Bedfellows

Grissom was not the only one hanging lemons. So were journalists, scholars, artists, and others who, like many scientists, were not persuaded that sending men to the Moon would accomplish anything worthwhile. They believed, to the contrary, that its staggering cost—$19.5 billion through 1966—and circus-stunt aura would bleed money from worthier causes and distract the nation from serious and pervasive problems on Earth.* Or else they thought it was

* That figure does not include Mercury, Gemini, or the three robotic missions that were being flown in direct support of Apollo: Ranger, Lunar Orbiter, and Surveyor. Together, the five programs added nearly another $2.5 billion. (Levine, *Managing NASA in the Apollo Era,* p. 155.)

just plain stupid. "It means nothing to me," said Pablo Picasso. Novelist Arthur Koestler complained about humanity's "unprecedented spiritual vacuum," declaring that "Prometheus is reaching out to the stars with an empty grin on his face."

One Gallup Poll taken in July 1965 underscored that point, with slightly more than twice as many Americans saying they wanted to see spending on what was called "space exploration" decreased as those who wanted to see it increased (though the largest number said they wanted it to stay the same). Asked which country they thought was farthest ahead in "space research," 47 percent said they believed it was the United States; 24 percent said they thought it was the Soviet Union; and 29 percent claimed to have no opinion, which itself indicated a lack of interest.

Beyond those who were polled, there were countless numbers of citizens who were deeply angry with both science and engineering. Great feats had been accomplished during the hundred years that spanned the mid-nineteenth and mid-twentieth centuries. There had been the construction of world-class bridges and skyscrapers, continental highways and railroads, and amazing advances in everything from pharmaceuticals to commercial aviation: advances that had turned their country into an "arsenal of democracy" that outproduced the Axis and drowned them in mass-produced technology on an unprecedented scale.

But there was a sea change between 1950 and 1965. The glorious technology also visited on them atomic and then hydrogen bombs, fallout shelters, a Bikini Atoll atmospheric nuclear test that drenched a boatload of Japanese fishermen with radioactive fallout, and a wave of stories about how Earth was being surreptitiously destroyed by Promethians bearing split atoms. Rachel Carson's *Silent Spring* warned that DDT and other man-made chemicals were causing sickness, madness, and death not only in adults but in their children and generations to come. Vance Packard's *The Waste Makers* made a convincing case that industry was selling products cynically designed for early obsolescence and in the process wasting limited resources. Ralph Nader's widely read *Unsafe at Any Speed* accused the automobile industry of deliberately cutting corners on safety to increase profits. And the cars and trucks themselves, plus the factories that produced them and other goods, were suddenly seen to be spewing toxins that made the very air people breathed deadly. Tom Lehrer lamented at the piano that "You can use the latest toothpaste, and then rinse your mouth with industrial waste." By the mid-sixties, all this and more was laid at the door of the ubiquitous "scienceandtechnology," which had turned from savior to arch villain where large numbers of Americans were concerned.

Edwin Diamond, an irreverent veteran reporter who became the science editor of *Newsweek* in 1957 and then took over as the magazine's general editor, covered the space program virtually from its beginning. The result was a book that came out in 1964 that lambasted the notion of racing the U.S.S.R. to the Moon. In the process, Diamond ridiculed the very idea of a race and asserted, correctly, as it would turn out, that there was less to the Soviet space program than appearances indicated. He acerbically likened the race to the potlatch ceremony conducted by the Kwakiutl Indians of North America, in which the chiefs of clans tried to glorify themselves and humiliate their opponents by tossing their most valuable possessions into a fire. Diamond used potlatch as a metaphor for what he considered to be the senseless and ultimately destructive pouring of billions of dollars and rubles into the space race. "The cumulative effect of the space race psychology," he added sourly, "was to elevate the Kwakiutl *potlatch* rite into contemporary national policy."

The race, he went on, also had a number of unfortunate consequences. One was creating the "unwarranted impression that rockets and 'hardware' . . . can be made the pre-eminent standard to judge national achievement. . . . The most exciting and, perhaps, important science being done today, for example, is in the field of molecular biology and biochemistry; but to date space activities provide little measure of these achievements." Furthermore, he wrote, the race did not reflect where both societies stood on such immeasurable qualities as civil rights and human freedom in general.

Diamond, a dedicated liberal and First Amendment fundamentalist, declared that turning the competition into a full-blown race against both time and the Communists tended to shut off discussion about the space program's goals and the techniques for attaining them. "The normal governmental procedures for funding and programming tend to give way to wartime-style 'crash programs' and a doctrine of 'concurrency,' in which design may be only half a step ahead of construction, and construction only half a step ahead of procurement. At the same time," he added, "an increased degree of secrecy also becomes necessary 'to prevent the opposition from knowing too much.' Thus momentous decisions may be made under pressure or behind closed doors, and to question them too closely becomes somehow un-American."

In a book called *The Moon-Doggle* that also appeared in 1964, sociologist Amitai Etzioni attacked the Apollo program, too, charging that it was being conducted by an elitist conglomerate of self-serving government bureaucrats and industrialists while little or nothing was being done to reduce the threat of nuclear war, alleviate the suffering of the 20 million or more Americans who languished in poverty, improve education, or address all of mankind's other earthly ills.

Etzioni took the Kennedy White House to task for abandoning a promising start at reversing years of social and economic inactivity and complacency in favor of relatively easy, ultimately pointless, mind-numbing, and horrendously expensive rocket glitz. "[T]he enthusiasm of the New Frontier ran into the same barriers that had for years frustrated any effort to make America confront the problems brought about by the modern age," he wrote, scathingly. "Then the moon reflected in the water of the Potomac."

Even beyond the profit motive and the unwillingness to come to grips with notoriously difficult social problems, Etzioni thought he saw an imperialist sour grape of enormous dimension. "Basically, the prestige race to the moon expresses the refusal of the West to accept the Soviet Union as a first-rate power and full member of the global community," he asserted. "The United States was for many years a *superior* power. No nation invaded it, and after the turn of the century, whenever the Europeans got into a fight, the conflict was always decided when the United States mounted its war horse, and it was always won by those Europeans whose side the United States chose to be on." America, the sociologist added disdainfully, had in effect been the world's town marshal since 1917 and simply did not like the fact that there was now another fast gun in the neighborhood.

"The Soviet Union has developed intercontinental missiles and a nuclear arsenal that do not match the American stockpile in numbers, but are strategically sufficient to deter a U.S. attack," Etzioni wrote, adding that the weapons put the U.S.S.R. "in the same class" as the United States, making it "another first-rate power. A large part of the American people and government have a hard time accepting this fact and openly facing its consequences." (Lewis Mumford, the urbanologist and historian, was solidly in Etzioni's camp. He called landing on the Moon "a symbolic act of war" and predicted that the very triumphs of technology would soon turn the planet into either a lunatic asylum or a crematorium.)

"In the realm of prestige," Etzioni continued, "the moon is the psychological parallel of the missile race." Then he said something that would have drawn a smile from Khrushchev. The United States had spent years insisting that the Communist system could not work and calling the Russians "a backward, barbarian people," he wrote, adding that Americans found a convenient rationale for every Russian achievement, from the use of abducted scientists to spying to simply copying technology. "Scientific exploration of space ought to be continued—the moon included—but the multibillion-dollar haste and waste that is perpetuated so that an American can visit the moon in person is predominantly a publicity stunt."

Sure it was. But what a stunt! It was a stunt that would have many repercussions, some of which were very difficult to fathom as the race to the Moon wore on. The most important and longest lasting of them, ironically, would become the bane of both competitors, but especially of the United States: the institutionalization of the stunt itself. The selling of the space program in

America in the aftermath of Apollo would depend increasingly on appearance over substance. Public relations would become as important as the rockets themselves. That would have consequences that ranged from beneficial to catastrophic.

Yet the stunt was substantial, not trivial. A stunt performed on a high wire without a safety net may not be socially useful in the strict sense of the term. But it embodies two qualities, courage and expertise, which can infect the audience and which are useful for achieving more earthly goals. Whatever the motive, trying to land men on the Moon, particularly in public and with a severe time constraint, would be an act of Herculean proportion. It would show that the United States had the resources, the resilience, and the will to send its citizens to another world for whatever reason it chose and then bring them safely home again: six times.

The main sticking point about Apollo would be this: after all the arguments about wasted resources, impoverished multitudes, national insecurities, base political motives, industrial greed, presidential machismo, and more—many of them valid—had been made, the voyages of men to the Moon would still reflect unprecedented managerial and technological prowess within a time frame that was incredible. And the result of that effort was not the conquest of another country or the creation of an awesome weapon, but the greatest human voyage in history; the *Odyssey* of the new age. The question that would haunt the United States had to do with deciding just how important that was.

What confounded both the pragmatists in the manned space program and the dreamers who believed implicitly that the human exploration of space was manifested by destiny, was not limited by imagination, but by the budget. However they tried, they could not justify going to the Moon, any more than they would be able to defend its "logical" next steps—a space shuttle, an orbiting station, and an expedition to Mars—on grounds of cost effectiveness. There was no apparent way to answer the frequently made assertion that sending people to space was "unnecessary" because machines could do a better job far more cheaply.

Yet the problem, at least in the view of Hans Mark, was in the equation itself. Mark, a nuclear physicist who directed the Ames Research Center and was both an undersecretary of the Air Force (and head of the National Reconnaissance Office) and a NASA deputy administrator, would have his own epiphany after the last Apollo mission in 1972. One day while sailing with von Braun on San Francisco Bay, he heard the master rocketeer say that Antarctica had gone unvisited after its discovery until "enabling technology"—the airplane—had made revisits practical and relatively easy. The space station, von Braun said, would do the same for the Moon. Mark was won over and soon decided that it was pointless to debate critics of manned spaceflight on their own terms because their equation—balancing costs with clearly defined benefits—was inherently inappropriate. "I do not know whether [sending people to space] is right or not because I do not know how to quantify the value of putting people in space," Mark reflected. "What I do know is this: The argument is almost certainly beside the point. People will go into space for reasons that have nothing to do with the cost-benefit analysis." He was talking about the compulsion that drives the human spirit to exploration and settlement. It is a spirit, Mark and his soul mates believed, that is beyond the judgment of accountants, managers, and others who count beans.

PART TWO
The Infinite Voyage

NASA

A few million years ago there were no humans. Who will be here a few million years hence? In all the 4.6-billion-year history of our planet, nothing much ever left it. But now, tiny unmanned exploratory spacecraft from Earth are moving, glistening and elegant, through the solar system. We have made a preliminary reconnaissance of twenty worlds, among them all of the planets visible to the naked eye, all those wandering nocturnal lights that stirred our ancestors toward understanding and ecstasy. If we survive, our time will be famous for two reasons: that at this dangerous moment of technological adolescence we managed to avoid self-destruction; and because this is the epoch in which we began our journey to the stars.

—CARL SAGAN, *Cosmos*

11

From the Earth to the Moon

Two space races were under way by the summer of 1964. The one that grabbed most of the headlines had to do with people being catapulted into orbit for the glory of their countries. Tereshkova had become the first woman to go to space the previous June and Korolyov was hurriedly converting the basic Vostok design to the three-man Voskhod that would fly within a matter of weeks. Gordon Cooper had wrapped up the Mercury program the month before Tereshkova made the last flight in a Vostok, and an unmanned boiler-plate Gemini had gone up on top of a Titan in April. Max Faget's engineering division had 1,400 engineers working on the Apollo program that summer, while North American had 4,000 more working on the command and service modules alone.

But the robots were doing their part, too. Ten years after the Technological Capabilities Panel turned in its report to Eisenhower, the possibility of nuclear attack against the United States had grown in proportion to the Soviet Union's ability to launch it. Before they were succeeded by more advanced spy satellites in 1972 and again in 1976, the ubiquitous KH-4s that operated in the Corona program would drop close to 800,000 pictures out of orbit. Photointerpreters scrutinizing the imagery under their stereoscopes in 1964 counted 220 Soviet heavy bombers and tankers and 78 hardened ICBM sites, with as many as 250 sites projected by the end of 1966. They were no match for SAC's 626 B-52s, more than that number of KC-135 tankers, 930 Atlases, Titans, and Minutemen, and 500 Hound Dog standoff missiles that could be launched by the B-52s hundreds of miles from their targets. Two points can be made about those numbers. One, given the destructive power of nuclear weapons, the Soviet strategic air and rocket forces did not have to match their American counterparts plane for plane and missile for missile. They could have clobbered the United States. But two, the huge American advantage gave perspective to the Soviets' fears of a devastating first strike against them.

Nor were the photoreconnaissance satellites the only ones watching Earth. Four satellites in a program code-named "Vela Hotel," each honeycombed with infrared sensors, circled the planet from more than 60,000 miles out and watched for nuclear test explosions that would violate the Limited Test Ban Treaty signed the year before, which prohibited explosions in the atmosphere, underwater, or in space. The eighteen X-ray, gamma-ray, and neutron emission detectors on each satellite could have detected a nuclear blast on Mars.

Still other robots were in the fourth year of creating a revolution in weather forecasting. These were the TIROS series, which provided twenty-four-hour television monitoring of millions of square miles of Earth's landmass and oceans. The first Nimbus satellite, a successor to TIROS, went up late in that August 1964. And other revolutions were in the making. Telstar, the first commercial spacecraft, was already making instantaneous round-the-world communication routine and would soon be joined by Intelsat. Meanwhile, a slew of signals intelligence collectors were eavesdropping on foreign communication traffic, including America's allies, and ferreting out radars' pulses. A dedicated military communication satellite system soon to be called the Defense Satellite Communications System (DSCS, pronounced "discus") was on the drawing boards and would be launched within two years.

There was action elsewhere in the solar system, too. Right after Mariner 1 plopped into the Atlantic because of the typographical error, its sister became the first machine from Earth to reach another planet when it reached Venus on December 14, 1962. Mariner 2 sent back a stream of data on interplanetary space even before it streaked past its fog-shrouded target: information about the interplanetary magnetic field, for example, and solar plasma. The 446-pound spacecraft sent back few surprises, but it did confirm what until then had been only theories. It proved, for example, that there really was a solar wind radiating from the Sun. And during its forty-two-minute encounter with Venus—the first close look at another planet by a machine from Earth—it found that the planet's high temperature was roughly the same everywhere, indicating that its thick, carbon dioxide–soaked clouds spread the heat evenly. Not being able to see through those clouds was the most frustrating part of the mission for the scientists at JPL and elsewhere who carefully picked through the mass of data it returned.

Ranger Hits the Moon

But it was Apollo and its robotic support programs that most engaged NASA. The robots, as usual, were the mechanical canaries that would precede humans into the void. The first scouting missions that were needed to get Americans to the Moon—Rangers 1 through 6—had been unmitigated disasters. But in midsummer 1964, a successor dramatically redeemed its creators' honor.

Ranger 7, looking like a mechanical dunce cap with flippers, bored in on the Sea of Clouds early on the morning of July 31. At 6:07, George Nichols, the spacecraft's new "voice," announced to the reporters who already filled von Karman Auditorium and JPL employees elsewhere that the command to warm up its two cameras had just been given.

Not a great deal was riding on Ranger 7. And everything was riding on it. Having lived through six successive failures in the program, Apollo's designers, who were themselves under enormous pressure, had vowed in disgust to press on with their work despite the notable absence of help from Pasadena. As far as they were concerned, Ranger was all but superfluous, and so was the Jet Propulsion Laboratory.

Yet a great deal was riding on Ranger 7 for the lab itself. The pressure to succeed, in the words of at least one member of the Ranger team, was "unbelievable." During the half year since the Ranger 6 calamity, three top-to-bottom reports that were critical of JPL's handling of the program were issued. RCA's Astro-Electronics Division, whose television cameras had been performing flawlessly on eight experimental TIROS weather satellites since April 1960, did its own investigation (in the process finding a bag containing fourteen screws and a lock washer in one of Ranger 7's electronic modules that led to unfounded suspicions of sabotage). Meanwhile, the system that had somehow short-circuited was modified, simplified, and thoroughly checked out. And science instruments were still mainly kept off the spacecraft to keep the mission as simple as possible.

Yet the specter of disaster chased Ranger 7 during the four days it took to reach the Moon. JPL's long and contentious relationship with headquarters would almost definitely result in what was euphemistically called "personnel changes" if no pictures were sent back, and there was a distinct possibility that the lab, itself, would be severed from NASA or shut down altogether. There had been fourteen failures in the lunar and planetary programs since the Thor-Able had blown up in August 1958, against one notable success: Mariner 2's Venus flyby four years later. Eleven of the losers (and that one winner) came from Pasadena.

Ninety seconds later Nichols sent a ripple of excitement, then applause, through von Karman when he announced that Ranger 7's full-scan cameras were on full power. There was imagery. Three minutes after that, he relayed word from Goldstone that there was also full power from Ranger's four telephoto-equipped partial-scan cameras. The applause turned to cheers.

"Ten minutes . . . no interruption of excellent video signals. All cameras appear to be functioning . . . all recorders at Goldstone are 'go.' . . . Seven minutes . . . all cameras continue to send excellent signal. . . . Five minutes from impact . . . video signals still continue excellent," an excited George Nichols reported as the first successful U.S. lunar reconnaissance mission unfolded with breathtaking speed. "Three minutes . . . no interruption, no trouble. . . . Preliminary analysis shows pictures being received at Goldstone. . . . One minute to impact. . . . Excellent. . . . Excellent. . . . Signals to the end." Ranger 7's last picture, taken from 1,500 feet, captured a section of moonscape about a hundred feet across that clearly showed craters only three feet wide. Then: "IMPACT!" The stream of telemetry turned into the hiss of static at exactly 6:25, but it was drowned out by thunderous cheers, paper flung into the air, and, in some places, the discreet shedding of tears. Schurmeier broke open several cases of champagne he had bought for Ranger 6. Pickering, Newell, and Schurmeier, the goats of winter, were now the lions of summer. They walked back down to von Karman again, this time to a standing ovation from an unabashedly delighted press.

Pickering introduced the Ranger experimenters at a nationally televised news conference at 9 P.M. Gerard Kuiper, Ranger's principal investigator, Carl Sagan's teacher, and the most renowned American astronomer since Edwin Hubble, led off. "This is a great day for science, and this is a great day for the United States," he declared. Whatever it was for science, which was questionable, it was certainly a great day for the United States.

Ranger 7 returned 4,316 of the first close-up pictures of the lunar surface on videotape and 35-millimeter film. They were followed in mid-February 1965 by 7,137 more from Ranger 8, and a month after that by 5,814 more from Ranger 9, the last of the series. With Ranger's detailed pictures in, landing places could be chosen for the Surveyor landers based on hard information, not educated guesses. That much was apparent. More important, though not yet obvious, was that Ranger's ultimate success, and therefore America's, was another signal of the beginning of a change relative to the faltering Russians. Even beyond supporting Apollo, the lunar reconnaissance program would provide NASA and some of its key contractors with mission experience that would simultaneously get astronauts to the Moon and open the way to the planets in a burst of exploration without parallel in history.

A Quagmire Called Surveyor

While the Mariner-Venus science and engineering teams and Ranger's team reveled in their hard-won victories, their colleagues in the Surveyor Moon lander program faced their own ordeal. Driven by Apollo's needs, Lunar Orbiter and Surveyor had begun to evolve as separate programs in 1962–63, with the Langley Research Center being given the orbiter and JPL keeping Surveyor, the lander. Lunar Orbiter was Langley's first major space project and it relished the opportunity to extend its work from air-breathing planes to spacecraft. In May 1964 a contract was awarded to the Boeing Company to build the orbiter.

Surveyor was an instrument-loaded, six-hundred-pound insect that was supposed to act like a miniature lunar module by making a soft landing on the Moon's surface with little padded feet. In January 1961 the Hughes Aircraft Company was picked to build seven of the landers, the first of which was supposed to touch the lunar surface in August 1963. But no one at JPL, Hughes, or headquarters could have foreseen on the day the contract was signed that the Moon bug would run into two enormous complications: Ranger and Centaur.

The setbacks with Ranger turned Surveyor into a stepchild. Pickering astounded Newell and Cortright in July 1964, at the time Ranger 7 hit the Moon, by telling them point-blank that he considered Surveyor a "back-burner" project that could be left to Hughes to worry about. The remark set off yet another explosion at headquarters. An enraged Newell fired off two letters to Pasadena saying that Surveyor was one of NASA's highest-priority projects and warning JPL to tighten its management practices or else. "Or else" included demanding that Caltech fire Pickering, once and for all. It was a tempting idea that was soon dropped because of his expertise and so, too, was the notion of turning the laboratory into a civil service operation. "The fierce pride" that JPL took from its heritage as part of Caltech, Newell would later write, "left grave doubts as to whether the laboratory could be converted without seriously disrupting the ongoing program."

Centaur Is Stalled

By then, it had long since been decided that the relatively heavy Surveyors would have to be shoved toward the Moon not by Agenas, which propelled the eight-hundred-pound Rangers, but by more powerful upper stages called Centaurs. They were more powerful because they ran on liquid oxygen and liquid hydrogen. At least they were supposed to be more powerful. The problem was they weren't running at all.

Krafft Ehricke, the brilliant astrodynamicist who knew his Tsiolkovsky and Oberth as well as the next rocketeer, and who by then was with General Dynamics, was convinced that only an oxygen-hydrogen upper stage could heft Pentagon and other massive payloads to both low orbit and deep space.

The Pentagon's Advanced Research Projects Agency initially wanted to use Centaur on top of the Atlas to launch very heavy spacecraft, including reconnaissance satellites. But the upper stage suffered from development problems right from the start. Then, in July 1959, it had been transferred to NASA over strong ARPA and Air Force objections and been given to von Braun's group at Marshall on the theory that working on several oxygen-hydrogen engines—the others were for Saturn—under one roof would be efficient. But it was not. Saturn so preoccupied von Braun's engineers that work on Centaur slipped badly. It was the direct equivalent of Surveyor's place at JPL. The fact that two such vital programs were relegated to a distant second place showed how NASA was straining to stay on Kennedy's schedule. Keith Glennan described the situation at Huntsville cogently: "Saturn was a dream; Centaur was a job." So the orphaned upper stage was transferred to the Lewis Research Center in Cleveland where work continued with heavy ARPA and Air Force participation (showing, yet again, that space technology was indivisible).

The decision to use Centaur to get Surveyor to the Moon was necessary given the lander's weight. But it was also appallingly dangerous because it meant that both the spacecraft and its upper stage would have to be developed simultaneously. It did not take a statistician to figure out that a single mission dependent on two complicated and unproven devices—Surveyor needed a sophisticated terminal-descent guidance system, and Centaur required the equally sophisticated hydrogen-oxygen-fed RL-10 engine—had a lower probability of success than if one had already been a proven article. "When an open-ended project, such as Surveyor, was assigned to an open-ended launch vehicle," one administrative expert would note with remarkable understatement, "troubles were created for both." Indeed, during the first half of the Surveyor program, there were serious doubts both at Hughes and JPL about whether Centaur would ever fly.

Meanwhile, at Houston, Apollo's father had become so fed up with what he saw as bungling by JPL in both the Ranger and Surveyor programs that he decided to plan his own scouting mission on the assumption that neither of the others would send back anything useful. Max Faget's utter exasperation had come to a head during a lunch with Joseph Karth in 1962, the year three Rangers went belly-up. The congressman came to the Manned Spacecraft Center to talk about Surveyor appropriations in light of its importance to Apollo.

"What kind of a problem would it amount to if the Surveyor program failed?" Karth asked. Faget's answer was driven by fury. He told Karth that Surveyor would not matter because the Apollo program was going to conduct its own lunar reconnaissance before the first manned lunar module got there. "We can do it without those guys," he told Karth. "We've got a great big wide landing gear and we just can't afford to be vulnerable to the loss of that [Surveyor] program. We'd go ahead anyway." What Faget had in mind was a closely guarded secret in Houston: plans for a manned Lunar Survey Module—"a fairly large diameter can"—that could be sent into lunar orbit with astronauts ahead of the manned lander for a week's detailed radar imaging of the Moon's surface. The Apollo program would be doing its own landing site selection and would therefore be in total control of its destiny. No more JPL. It would be wonderful.

But it did not happen. With the triumph of the Rangers that began on the Sea of Clouds during that eventful summer of 1964, JPL quickly began a transfusion of energy into its own anemic lander project. The result was the bringing together of an upper stage that was not yet as powerful as its creators wanted and a lunar probe with reduced capability. Not only was Surveyor three years behind schedule, but it carried only 114 pounds of science instruments instead of the 343 pounds that had been planned. The spacecraft itself was an ungainly triangle of pipes and tubes that supported a vertical mast which itself was topped by antennas and a solar panel that looked like a folded newspaper being used to shield it from the Sun. It had three legs, each of which had a shock absorber and a footpad. Surveyor was expected to make a gentle landing by using a retro-rocket and three vernier engines for maneuvering.

Surveying the Moon

Surveyor 1, which was Atlas-Centaur-propelled, touched down on the Sea of Storms on June 2, 1966. It returned more than eleven thousand high-quality pictures, together with seismological and other data, in its six-week lifetime. And it proved, if proof was still necessary, that a machine could land on the lunar surface without sinking out of sight. Having been rebuked by Gilruth, who had been rebuked by Webb for the remarks he had made to Karth, a contrite Faget called headquarters and, "eating crow," congratulated lunar and planetary chief Oran Nicks.

Life celebrated the event with a cover story called "The True Color of the Moon" that ran four weeks later. The spread showed a grainy moonscape of "airless beauty" while omitting any reference to the turmoil from which the six pages of pictures emerged. The magazine reported what anyone looking at the Moon through a telescope could have seen: it was gray. Yet plain grayness would not suffice. The Moon was "a rich and varied gray," America's favorite picture magazine explained. "The color photographs constitute a crowning achievement in Surveyor's success. From the moment the . . . vehicle . . . had set down on the lunar surface, it carried out its trail-blazing tasks so efficiently that even its most optimistic designers were dumbfounded," according to a rhapsodic caption writer who, like others in Henry Luce's powerful empire, used superlatives the way movie theaters used buttered popcorn.

But the hype was not necessary. The pictures, the first taken on the surface of another world by an American spacecraft, were spectacular enough. Only a geologist would have found the barren, rock-strewn panorama beautiful, however. The truly important thing about the pictures was not what they showed but that they had been taken at all. Surveyor 1 had been beaten to the Moon by Luna 9, which had landed on February 3, opened its petal doors, and televised the neighborhood. But the accomplishment had been costly. At least five previous attempts by the Russians to make soft landings on the Moon between April 1963 and December 1965 had ended disastrously, with two spacecraft straying into solar orbit and three others coming in so fast they smashed to pieces.

Surveyor 2 crashed because of trouble with its small vernier steering rockets. But Surveyor 3 made a perfect landing on the edge of a shallow crater in the Sea of Storms on April 19, 1967. In the days that followed, it sent back not only 6,315 detailed photographs of the surrounding terrain but data on the composition and surface strength of the Moon's crust by digging trenches with a "surface sampler mechanism"—a claw—and televising the operation as it progressed. The direct consequence of this would be to provide a landing place for Apollo 12 thirty-one months later. Surveyor 4 ran into a catastrophic problem of some sort, possibly an explosion, two and a half minutes before it was supposed to land and instead crashed. The three remaining Surveyors, launched between September 8, 1967, and January 7, 1968, performed with distinction.

Rings Around the Moon

Five Lunar Orbiters were launched in the year between August 10, 1966, and August 1, 1967, all of them successful. Hundreds of medium- and high-resolution pictures were sent back, including some taken of the far side of the Moon, as the sensational Luna 3 had done. They were applied to the final selection of Apollo landing sites. NASA itself called the imagery "orders of magnitude" better than Ranger's. Langley's orbiting spacecraft was so effective, in fact, that all of the Apollo program's requirements would be met by the end of the third orbital mission in February 1967. The lunar module's designers at Grumman's Bethpage, Long Island, plant were fed piles of data on radiation on and around the Moon, on gravity, micrometeorites, and a great deal more.

In the end, and after an expenditure of nearly $905 million, sizing up the Moon in preparation for Apollo proved to be a mixed bag. Ranger, Surveyor, and Lunar Orbiter did not send back much in the way of real science data, let alone provide clues about the origin of the Moon itself, and that disappointed

Harold Urey and others. But they weren't really supposed to do that. What they were supposed to do was help ensure that men would be able to land on the Moon without loss of life. Measured by that standard, the three scouting programs were unquestionably successful.

Beyond that, they started a revolution in lunar and planetary mapping by capturing in great detail what Earth-bound telescopes could not resolve. Galileo would have been pleased by that, as well as by the equally important role they played in accumulating design and operational information that would go into planetary explorers such as Mariner, Pioneer, and their successors.

Meanwhile, Lunar Orbiter whetted Langley's appetite for other space missions, the first of which it had done with distinction. And Ranger and Surveyor went far in resolving the troubled relationship between JPL and headquarters. The Ranger failures, in particular, forced Pasadena to come to terms with a smugness and self-satisfaction that were seriously compromising its enormous potential. Similarly, the eventual triumph of both Ranger and Surveyor showed headquarters that technological excellence was not the exclusive preserve of the civil service or of people who lived inside the Washington Beltway. Most important, six and a half years of actual lunar reconnaissance helped headquarters and several of the centers, including those in Alabama and Texas, get each other's measure while establishing management techniques that would stress cooperation to achieve specific goals. It was a vital ingredient and one that was eluding their Soviet competitors at immense cost.

Military Insecurity

No sooner had the Ranger program gotten off to its woeful start than the question was raised of how much data to release. Glennan and Dryden had stressed absolute openness from the day NASA came into being and Kennedy had emphasized that very point in his first, and most famous, public reference to sending men to the Moon. The Space Act itself called for "the widest practicable and appropriate dissemination of information . . . for the benefit of all mankind." Yet when the imagery from the Rangers came in, followed by high-resolution pictures from Lunar Orbiter and Surveyor, a number of people in NASA itself who had military backgrounds questioned whether the raw data ought to be shared with the Soviet Union. Helping America's bitter rival, they said, was not the way to win the race. They therefore suggested in discussions behind closed doors that there be either a selective release of photographs or their delayed publication. In the end, though, it was decided that the surest way to gain the respect of the world was to follow JFK's path. So all lunar imagery was made available to libraries, observatories, and technical information centers throughout the world as it became ready for distribution. And so the Russians used the fine Lunar Orbiter imagery to pick landing sites for their own spacecraft. But the pictures would not be enough to get them there.

The Russians Aim for the Moon

Given the propaganda triumphs that the Soviet Union had racked up with rockets starting with Sputnik, it was literally inconceivable that Khrushchev could see Kennedy point Americans to the Moon in 1961 and cede the feat to his country's rivals. He therefore encouraged Korolyov and others who were straining for both the Moon and Mars. And encouraged—not "drove"—is the right word. The rocket men did not need to be driven. What they needed was a high degree of coordination, but it was not there.

Korolyov pursued three separate programs to get Russians to the Moon that paralleled their American counterparts. The first, Vostok, carried cosmonauts into space the way Mercury and Gemini carried astronauts. The second, Luna, was the counterpart of the three U.S. lunar reconnaissance programs. And the last, the so-called N-program, was supposed to develop a launcher similar to Saturn. Like Saturn, it was not explicitly dedicated to propelling men to the Moon when initial studies started in 1960–61, but most insiders knew that was where it was headed.

While Korolyov's OKB-1 was given permission to do only engineering studies, his archrival, Chelomei, was awarded the assignment of developing both a rocket and a spacecraft that could orbit at least one cosmonaut around the Moon. The orbiter would be called LK-1. The huge launch vehicle would be a departure from the incrementally improved rockets that had grown out of the venerable Semyorka—lifters named after the spacecraft they carried: Luna, Vostok, and Voskhod—and instead be a thoroughly new three-stage design that burned toxic but easily storable nitrogen tetroxide and asymmetrical dimethylhydrazine. It would eventually be called UR-500K, or simply Proton, and it would be destined to carry heavy loads to space throughout the rest of the century. Glushko was given responsibility for developing Proton's first-stage engine, called the RD-253, and Simon Kosberg, who had designed the problem-plagued upper stage of the Yangel rocket that had killed Nedelin and the others in October 1960, was made responsible for the two upper stages. To make matters even more complicated and fractious, Yangel's bureau in Ukraine had its own heavy launcher, called the R-56, which also was propelled by Glushko's ubiquitous engines.

Forcing three bureaus to compete to produce a huge rocket-spacecraft combination like cocks in a ring was typically cynical and squandered both funds and people. Furthermore, Glushko and Korolyov were embroiled in their own ancient, bruising feud. The preeminent engine designer therefore decided to throw in his lot with Chelomei, no doubt because he considered his enemy's enemy to be his friend, and also because Sergei Khrushchev remained in Chelomei's bureau.

The powerful engine Glushko designed for Proton used the storable propellant, which was understandable since he also courted the Ministry of Defense and the military needed ballistic missiles that used storable propellants. Suffice it to say that Glushko adamantly refused to work on Korolyov's Moon monster, which was called the N-1. Nor did Korolyov want him. The decision reflected not only a profound disagreement about propellants—Korolyov favored the more powerful but dangerous and notoriously unwieldy oxygen-hydrogen combination—but a mutual hatred.

The feuding giants in turn presided over fiefdoms that were themselves hopelessly inefficient and ensnared in a political system that was built on flattery of superiors, treachery, low-key but ferocious competition, and paranoia. The system created by Stalin and lackeys such as Beria had bred mid- and high-level prima donnas, suspicious and wary as foxes, who suspected virtually everyone and trusted almost no one. Many years later and in another context, Aleksandr I. Lebed, the outspoken general turned politician, would say that such people were "like a dog guarding a haystack. They can't eat the hay themselves, but they won't let anyone else eat it either."

Weakened as it was by jealousy and bickering, the space program was still the country's most impressive undertaking. But it, like the nation as a whole, was bogged down by factory managers who were chronically lacking in production know-how, by poor quality control, badly organized plants, hopelessly inadequate cost accounting, and by microelectronics and computers that were a decade or more behind the West. Since the problem was intrinsic to the political system itself, and therefore could not be changed by reform, the Council of Ministers and the Central Committee would be forced to resort to espionage to redress the problem. In 1970, they would create a special unit within the KGB—Directorate T—solely to collect information on the research and technology programs of Western economies. But stealing rocket secrets was reactive, not creative, so it could never compensate for the debilitating institutional problems that weighed down Korolyov and the other rocket men.

The inherently flawed economy had struggled since World War II to maintain the largest standing army and occupation force in the world while trying to provide enough consumer goods to keep a pervasively cynical proletariat in line. Khrushchev had to use a stick (or, more accurately, the appearance of one) to offset the American military threat. But Brezhnev preferred a carrot—détente—to hold his society together for the indefinite future. "We Communists have to string along with the capitalists for a while," he would tell the

Politburo at the start of détente in 1971. "We need their credits, their agriculture, and their technology. But we are going to continue massive military programs and by the middle 1980s we will be in a position to return to a much more aggressive foreign policy designed to gain the upper hand in our relationship with the West."

More insidiously, the monolith that the rest of the world took to be the Soviet Union was more apparent than real when it came to adapting the theory of party discipline and dedication to actual practice. The Council of Ministers and the party should have been able to point the space leadership and its vast support structure at the Moon and then see cosmonauts land there reasonably quickly. That was how it was supposed to work. But the impressive-looking command structure itself was so corroded that it was essentially ineffective. The farther down the chain of command orders went, the more they unraveled.

The various ministries—General Machine Building, which was responsible for both the manned and unmanned programs, and Defense, for example—amounted to Communist baronies in which there was no effective control from the top and where funds were routinely squandered. On organization charts, the Commission on Military Industry was directly responsible for the day-to-day running of the space program. In fact, it presided over twenty-six departments and about five hundred "enterprises"—Communist parts-supplying companies—that in reality did not answer to the commission at all and that were, in addition, models of lethargic inefficiency. Korolyov, Chelomei, Glushko, Yangel, Chertok, Tikhonravov, and the other top designers were therefore plagued by late shipments, wrong components, and an unresponsiveness to requirements that was crippling. "We had no control levers to influence our suppliers," Vasily Mishin would one day complain in despair.

Driven to near desperation by Glushko's animosity, Korolyov himself went to Nikolai D. Kuznetsov, a general designer of aircraft engines who had helped with an earlier rocket engine, and won a pledge of support. Although the oxygen-hydrogen combination was far and away the most powerful propellant combination (followed by oxygen-kerosene) it came with a paradox that frustrated engine designers on both sides of the Iron Curtain. Since the molecular structures of oxygen and hydrogen were far less compact than other propellants, they required much larger tanks, which required much larger launch vehicles, which added so much weight that the advantage of using them in the first place could be greatly diminished. Kuznetsov therefore chose many engines of moderate thrust over a few that were very powerful and, regardless of what Korolyov preferred, they burned liquid oxygen and kerosene. Korolyov's own spacecraft, called the L-1, was designed to have both a crew compartment and a re-entry module. It was the direct product of the competition with Chelomei.

Korolyov's initial concept for a circumlunar mission, which was set down in March 1962, envisioned having cosmonauts go up in a Vostok and then put together three rocket stages in Earth orbit after they had been launched independently. They would then be joined by as many as three cosmonauts in an L-1, which would be fastened onto the assembled rockets and sent to the Moon by successive firings. That was why "spacewalks"—EVAs—and rendezvous operations were emphasized on the Vostok missions. So were increasingly longer duration missions. The dual *Vostok 3* and *Vostok 4* flights in August 1962 were important because Andrian Nikolayev's completion of sixty-four orbits in nearly four days without ill effect showed that a seven-day round-trip to the Moon was feasible.

A year later, however, the plan was revised to eliminate the Vostok. Instead, a new spacecraft called Soyuz (Union) was supposed to join a rocket that had previously been fueled in orbit and both would go to the Moon. In the end, though, delays in developing Soyuz and the complexity of the mission killed Earth orbit rendezvous altogether. Meanwhile, Korolyov's Moon rocket design was reviewed by a special commission headed by Keldysh that recommended a series of changes that, if anything, increased the size and weight of the original design.

The launcher that evolved from Keldysh's recommendations would be a three-stage behemoth that looked like a 2,700-ton, elongated white cone. The basic launcher was topped by two additional stages—numbers four and five—that were supposed to push a pair of cosmonauts on the last leg of the trip to the Moon. The whole thing stood 370 feet high. Standing next to a foot-high model of the N-1 configured for the Moon voyage, a person reduced to scale would be the size of a small ant. The N-1 and the Saturn 5 were therefore monsters of nearly the same size.

But size is where the similarity ended. Unlike Saturn's first stage, the bottom of the Russian launcher had thirty liquid oxygen–kerosene engines, each capable of producing 338,800 pounds of thrust. They were arranged with twenty-four mounted around the first stage's circumference and six in its center. Together, they would produce 10.1 million pounds of thrust, or 25 percent more than Saturn's five Rocketdyne F-1s. The second stage had eight more engines that developed another 3.1 million pounds of thrust, followed by a third stage propelled by four engines that developed a total of 360,000 thousand pounds. Both the second- and third-stage powerplants also used oxygen and kerosene. All three massive bottom stages, called A, B, and V, therefore would end up having forty-two engines that produced 13.5 million pounds of thrust, which was well over twice the total thrust of Saturn 5. The two special lunar stages had an engine apiece that developed another 90,200 pounds and 19,200 pounds, respectively.

But the whole rocket package was less impressive than its individual parts. The engineer's eternal conundrum is that while more engines increase propulsion, they also increase the possibility of failure. That would be the case with the N-1. And the fact that none of the engines used the oxygen-hydrogen combination spoke to the nagging technical difficulties that stalked the program in part because of the deep enmity between Korolyov and Glushko.

Following a presentation made by Korolyov to the Presidium of the Interdepartmental Council on Space on the N-1's potential to not only land men on the Moon but to support the construction of a space station, the all-powerful Communist Party Central Committee tentatively approved plans for a lunar landing by the end of the decade. It issued a top-secret decree, 655-268, on August 3, 1964, setting 1967–68 as the target date for landing a single cosmonaut on the Moon, obviously to get there before astronauts arrived.

Korolyov could bask in his triumph. But not for long. He was confounded to learn shortly afterward that the Council of Ministers had selected Chelomei to send his LK-1 into lunar orbit. And if that were not bad enough, he lost his most powerful ally, Khrushchev, two months later. The wily old premier's penchant for spectacular feats in space and the chief designer's obsessive ambition created a special synergism that simply evaporated when Khrushchev was removed from office. Korolyov signed off on the Moon landing program, known as N-1/L-3, on December 25.

By mid-March 1965, with approval to at least begin a program to go to the Moon in hand and Leonov gamboling precariously at the end of his rope outside of *Voskhod 2,* an exhausted Korolyov was hurriedly putting into place a highly detailed and comprehensive plan for reaching the Moon within the time period specified by the Central Committee. It is no exaggeration to say that the fact that all the pieces were being conceptualized in detail by a single individual was extraordinary. And, like every other aspect of the space program, it was highly classified. Here is what that individual imagined:

• The development of Soyuz, a flexible successor to Vostok, that would safely hold as many as three cosmonauts in an orbital module, behind which would be a descent module and then a service module that not only held an electronic power supply, thermal regulation, long-range radio and telemetry equipment, and orientation and control gear, but two liquid-fuel engines (primary and backup) for maneuvering in orbit and coming back down. Most important where reaching the Moon and eventually building a space station were concerned, the spacecraft would also have an automatic docking system that could be operated either hands-on by the ship's commander or automatically. The latter maneuver would use radar to feed data on distance, relative velocity, angular velocity, and relative angles of the two spacecraft to a computer that would maneuver accordingly. Meanwhile, the target spacecraft would use its own attitude-control rockets to stay aligned with the Soyuz closing in on it, rolling when the Soyuz rolled, as the Soyuz extended its probe into the target craft's docking collar. Soyuz was destined for a life as long as Proton's and would constitute the structural foundation for Korolyov's three lunar vehicles.

- The L-1 itself, basically a stripped-down Soyuz, that would send a single cosmonaut into lunar orbit. Although Chelomei set a milestone in his country's space program by successfully launching the first Proton on July 16 with a record-breaking thirteen-ton-plus scientific satellite, he was running into trouble with his LK-1 manned lunar orbiter. On December 15, 1965, Korolyov snatched the circumlunar mission away from Chelomei by convincing members of the Council of Ministers that the L-1 pushed by one of the N-1's upper stages could do the job faster and better than his rival's LK-1.
- The L-2, a beefed-up Soyuz, would be called the lunar orbit module and would be the counterpart of the Apollo command and service module that was designed to carry a crew to the Moon for a landing.
- The L-3, or lander, would be the equivalent of Grumman's lunar module and would have to get a cosmonaut onto the lunar surface and back up again. Although strictly speaking L-3 referred only to the lander, in practice it was used to mean the entire spacecraft, including the L-2, that would carry a crew to the Moon. The whole program was therefore known as the N-1/L-3.
- The N-1 itself, with all of its own staggeringly complicated permutations.

The idea was to propel two cosmonauts and the whole L-3 "stack" to the Moon using the first four N-1 stages. Then the fifth stage—the D with its 19,200-pound-thrust engine—would brake the stack's speed so it could swing into lunar orbit. One cosmonaut would then climb out of the L-2 and make his way back to the L-3 using a mechanical arm. He was then supposed to climb into the lander, after which it and the D rocket stage would separate from the orbiter and the re-entry and service modules. Like the Apollo command module, they would continue on in lunar orbit. Meanwhile, the D rocket would be relighted and, with steering thrusters on, angle the lander for a descent and bring it down to within a mile of the lunar surface. At that point, the rocket was supposed to separate from the lander and crash onto the Moon not far from where the lander touched down.

The lander was then supposed to continue its descent by using its own small engine and thrusters to make a soft landing. Scientists at IKI, including Aleksandr Basilevsky, a gifted geochemist; Natasha Bovina, the chief cartographer; and others were set to work in 1968 to pick a suitable landing site and tell the rocket designers about the composition and topography (including the location of slopes greater than thirty degrees) of the lunar surface. To accomplish that, they mostly used the Lunar Orbiter pictures, which were superior to anything returned by their own spacecraft. Theoretically, the lander would be able to support life for two days, but a maximum of four hours would be enough for the first time out.

With the Moon walk completed, the descent engine would be refired, lifting the lander out of its legs and back into orbit where, like Apollo, it would rendezvous with the orbiter and head home.

During the cosmonaut's brief visit to the Moon, he would be expected to plant the hammer-and-sickle, take photographs, collect soil samples, and leave token scientific instruments. The work would not vary appreciably from what his American competitors were supposed to do. Yet there would be one fundamental, and potentially hazardous, difference between the Russian operation and the American. Given the fact that the cosmonaut would be alone and wearing a space suit that gave him about as much mobility as the puffy Pillsbury Doughboy, a fall backward could have turned him into a belly-up space turtle. There was a hoop attached to the suit that was supposed to have helped him turn over. But were that to fail, the unlucky Russian could have flailed helplessly in the Moon dust until he expired.

The Death of Korolyov

In November 1959, the month after Luna 3's sensational photo-run over the far side of the Moon, the chief designer had been awarded a two-story wooden frame house on Ostankinsky Lane in a pleasant neighborhood in northwest Moscow. Although it was a modest dwelling by the standards of the nation with which he competed, it was a "mansion" by Soviet standards, boasting a high-ceilinged art deco living room with a marble fireplace, French period furniture, and potted palms. A sharply angled wooden staircase, partly enclosing a sculpture of a godly creature flinging a rocket to space, led to the Korolyovs' bedroom and to his study. The centerpiece of the room in which the chief designer worked was a large, ornately carved wooden desk piled with journals and reports and separated from the windows by more palms. When Korolyov sat at that desk he did so under the intensely serious gaze of a young and dignified (not haggard) Tsiolkovsky, complete with rimless glasses and a jacket and tie that almost gave him the air of a dandy. Tsander, contained in a frame a quarter the size of Tsiolkovsky's, was where he belonged: on the same wall as the father of rocketry but beneath him.

Korolyov spent four days in December 1965, the fourteenth through the seventeenth, undergoing what he was told was a routine medical examination in the Kremlin clinic. Certainly its length suggested a thoroughness indicating that the medical community, not to mention the political leadership, appreciated the patient's importance. The only discovery that even mildly concerned his examiners were some polyps in his intestine, one of which was bleeding. Since such growths were often precancerous, the Korolyovs were told, he was best advised to have them taken out. But they were so small, the patient was assured, that they could be plucked out without even using a scalpel. The chief designer could cheerfully see in the New Year, the doctors added, and return afterward to have the simple operation performed.

The yellow house at 6 Ostankinsky Lane, protected by an acre of dense trees and shrubs, provided Korolyov with a refuge from the Kremlin's treacherous currents and from his enemies in the other design bureaus. It was in the refuge on that secluded tract, in front of a fire, that he and two of the very few people he completely trusted—Gagarin and Leonov—stayed up until 4 A.M. one night at Christmastime 1965 while he shared intimate recollections of the high and low points of his life. The nadir, he told his pet eagles, had been in the gulag; the zenith had been their flights. And the constant scourge throughout it all was what he called politics and betrayal.

He was out of his house and back in the clinic by January 13. That day—the day after his fifty-ninth birthday—he was again told as his second wife, Nina, looked on that there was nothing to worry about. It was the best possible birthday present.

"Doctor," Sergei Korolyov said, placing his hand over his heart and smiling with relief, "you are our friend. How much longer will I be able to continue my work?"

"About twenty years," answered Dr. Boris Petrovsky.

"Ten will be enough."

But there would not be ten days, let alone ten years. Whatever his stature on the outside, however his designers might tremble when they faced his wrath or competitors became enraged by his audacious genius, the clinic was a leveler. Lying in bed there, pinned under sheets within the thick walls of a fortress in the heart of a vast gray city in the throes of winter, the chief designer was as helpless as one of his rockets locked in the grip of an erector. Like everyone else who went to the clinic, he had to give himself over, entrust his existence, to men and women who might as well have been shamans for all he understood about what they did. He had to have faith in people who neither knew nor cared about specific impulse, propellant pressure, or the other vital signs that defined his world. Sergei Korolyov's vital signs were not *their* vital signs.

Petrovsky, the minister of health of the Soviet Union, did not know an accelerometer from a combustion chamber. But he did know that Korolyov was an important patient and one who had the ears of the Council of Ministers and the Central Committee. He therefore decided to impress his celebrated patient with his own knowledge—in the process neglecting to get a second opinion—

and take it upon himself to perform the operation. Petrovsky had once been an "outstanding" surgeon, according to Zhores A. Medvedev, the dissident biologist, but had had no opportunity to practice that art since becoming a bureaucrat. Yet given what he took to be a straightforward hour's work, he would easily have felt up to the task. It would be so routine, Petrovsky evidently thought, that he scheduled another operation after Korolyov's.

Petrovsky opened the chief designer at 8 A.M. on the morning of the fourteenth. Plucking out the polyps ruptured a blood vessel that caused severe hemorrhaging. Korolyov was now bleeding so severely, in fact, that Petrovsky decided to cut into the abdomen to stop it. There, the startled surgeon found a large malignant and rapidly spreading tumor that had not been visible earlier. Petrovsky was now making his way through a professional minefield. During the next seven hours, he removed parts of Korolyov's rectum in a frantic effort to take out the tumor. Since Korolyov could not be intubated because of the broken jaw he had gotten in the gulag, his face had to be covered with an oxygen mask for the whole time he was on the operating table. And he continued to lose blood. Knowing that his patient had a history of heart problems and that there was an inadequate supply of blood in the clinic, and now seeing that the man under his knife was developing heart spasms, Petrovsky finally panicked and sent for the best surgeon on his staff, Aleksandr Vishnevsky, to help. Vishnevsky was reached at a nearby resort and eventually rushed into the operating room and up to the gutted patient. He took one look at Korolyov and announced that he did not operate "on dead men." Then he stalked out of the room.

It was indeed the end of Sergei Korolyov. His death on the operating table was publicly attributed to "a long and fatal illness" in a long obituary that was the first reference most Russians saw to their chief designer. Within the confines of officialdom, his death was blamed on a heart attack. He was cremated and interred in the Kremlin wall, where he would eventually be joined by Gagarin, Komarov, and others in their country's pantheon of space pioneers.

Korolyov's and Khrushchev's synergy had kept the space program moving through a storm of institutionalized arrogance, incompetence, jealousy, and misappropriated resources, human and financial. Now they were both gone. With the death of Korolyov, one technical specialist would tell an American journalist, "we could not compete with you Americans. There was confusion, disorganization. Too many 'chiefs' and not one boss. You can't get anywhere without a big man in charge." Korolyov was such a man. He unabashedly used ministers and generals to advance his work but he was never afraid to stand up to them. His deputy, Vasily P. Mishin, became the new head of OKB-1 (which was soon renamed TsKBEM, for Central Design Bureau of Experimental Machine Building). Chertok, who had led the team to Germany in 1945 and who had designed a Mars probe by 1962, would have been a better choice. But he had two serious strikes against him: he had been born in Poland, and he was Jewish. Mishin would always try to give the ministers and generals what they wanted, however dangerous it was or however stressed the system. He was no Korolyov.

The Soviet manned lunar program would continue moving forward and, initially, scoring notable successes. Luna 10 went into orbit around the Moon in early April 1966, beating Lunar Orbiter by four months. It was followed by three more successful missions, two of them placing cameras and scientific instruments in orbit and a third making another perfect soft landing. These encouraging developments persuaded the Keldysh commission to issue final approval on November 16 for the N-1/L-3 mission with the understanding that a cosmonaut would stand on the Moon within two years. What was not yet apparent, however, was that Khrushchev's banishment and Korolyov's death had decapitated the program.

As the chief theoretician of cosmonautics and his commission went through the last hours of deliberation before giving Mishin permission to go for the Moon, Buzz Aldrin flew overhead tethered to *Gemini 12*.

Apollo Goes "All Up"

Back in September 1963, George Mueller, NASA's newly appointed associate administrator for manned space flight, had asked two subordinates to calculate the chances of Apollo getting to the Moon within the time limit set by Kennedy. After surveying the entire array of problems, from horrendously tangled logistics to the Ranger mess to Centaur's birth pains to Saturn's encyclopedic complications, they concluded that the odds were one in ten.

If most of the engineers in the world shared one credo, it was this: incremental advance. The notion of consolidated testing, of making large leaps, perhaps even of making what journalists and other outsiders liked to call "breakthroughs," was anathema. Instead, the idea was to advance in carefully measured steps, one building on another. If the engineers, technology's tortoises, had a bible, it would have been called *Increments.* Slide rules slid, after all; they did not leap or jump.

But Mueller, who had a Ph.D. in physics from Ohio State and more than two decades of academic and industrial experience, much of it in ballistic missiles, became so alarmed that he began searching for a way to drastically shorten Apollo's testing process and save precious money at the same time. He finally decided to abandon incremental testing—the step-by-step firing of individual stages of a rocket before sending all of them up together as had been done by, for example, von Braun—and instead test several components in their final configuration at the same time.

Mueller reasoned that there was no statistical advantage to testing an individual rocket stage eight times as opposed to six when it was only going to have to be mated with other stages anyway; that incremental buildups were time-consuming and wasted hardware; and that they amounted to an approach that assumed there would be failure rather than success. He therefore radically changed Saturn's test procedure by ordering that the rockets be fired, not incrementally, but as whole systems the way they would go into space. This was called "all-up" testing.

When John Disher, one of the two men who had come up with the heavy odds against Apollo's making it to the Moon on time, briefed the men at Marshall about the new procedure his audience became incredulous. Instead of testing Saturn 5's three stages individually, he told Willy Mrazek, who ran von Braun's Structures and Propulsion Laboratory, all three would have to go up connected to each other on the first flight. Mrazek, knowing that the F-1 that propelled the first stage was having combustion problems and that the second stage was inherently difficult because it ran on hydrogen, explained that testing both of them and the third stage at the same time would be impossible. "And how many launches are you going to have before you put a man on?" he asked sarcastically. "One successful," Disher answered. "You're out of your mind," said Mrazek.

Arthur Rudolph had also been dubious. So had von Braun. But none of them quibbled with Mueller's contention that there was no hope of meeting the deadline unless all-up testing was done. In any case, Mueller was a strong-headed individual who had issued an order, not asked for a consensus. The all-up memorandum had gone out on November 1, 1963. Von Braun himself would later write that "Mueller's reasoning was impeccable" and that the manned lunar landing would not have occurred on time without the new test procedure. The Soviets would adopt the strategy, too, but it would not help.

The First Casualties

Cabin atmosphere—what the spacemen breathed—was another source of contention in both the American and Soviet programs. Pure oxygen was considered simple to use, since it came in bottles and was easily measured and kept at the right pressure. Plain air, on the other hand, was difficult to use because its composition—78 percent nitrogen and 21 percent oxygen—was difficult to keep balanced in the right pressure. It required an elaborate sensing device to maintain the right mixture. Faget and others in NASA had always been vehemently opposed to a two-gas system precisely because keeping the mixture balanced was a headache that could turn into a nightmare. They were afraid that the sensors controlling the mixture, in Faget's words, "would get confused and would put too much nitrogen in the cabin, a very insidious thing because there was no way to detect [it]." The astronauts would just get sleepy and die. What plain air had going for it, though, was the fact that it would not burn, while pure oxygen would burn dangerously at a half pound of pressure. Under pressure of five pounds a square inch of oxygen, a piece of paper would burn like a torch; under sixteen pounds, it would vanish in an explosion of flame. Saunders B. Kramer, who participated in oxygen experiments at Lockheed, called a pure oxygen atmosphere at sixteen pounds of pressure "a time bomb (worse than gasoline vapor) waiting to go off."

But Faget, Gilruth, and others who ran the manned program considered pure oxygen's advantage to be worth its risk. Mercury and Gemini crews breathed it without incident. The Gemini oxygen supply, for example, was carried in liquid form in a 104-pound spherical container and pumped into the cabin at a pressure of 5.3 pounds an inch. Not to take any chances, leftover oxygen was jettisoned before the spacecraft made their fiery entries back into the atmosphere. North American's design team, on the other hand, strongly objected to using pure oxygen in Apollo's command module precisely because it feared a fire and had therefore proposed the two-gas system during contract bidding. But in what was to prove to be one of two fateful decisions, North American was overruled and forced to go with pure oxygen.

The other decision had to do with providing the command modules with the same kind of explosive bolts that were also built into the Mercury and Gemini spacecraft. The idea was to be able to get the astronauts out quickly in an emergency, a technique that North American engineers, who had used ejection seats in their famous fighters since the first F-86 Sabre flew in 1948, strongly favored. But again NASA disagreed. If the hatch blew while the astronauts were out of their suits, Faget worried, their blood would boil instantly. The only time a hatch had blown accidentally was after Grissom had hit the water in *Liberty Bell 7*. No one at North American seriously believed that the bolts had blown by themselves, as Grissom insisted, but rather that he had accidentally set them off in the confusion of the moment. But the space agency, which officially bought Grissom's version, insisted that the command module have a hatch that could be opened only manually and pulled inward, not out. Pressure inside the spacecraft would help keep it tightly shut. A year later Gus Grissom, who had originally wanted explosive bolts on Mercury but who remained humiliated over that notorious episode, came out emphatically against using them in the Apollo capsule. That second decision would prove to be fateful indeed.

Five years later, on the evening of Friday, January 27, 1967, Grissom, Chaffee, and White—the crew that was supposed to test the first Apollo spacecraft by flying it around Earth the following month—were strapped onto their couches inside the command module. The module itself, number

012, stood twenty stories above Launch Complex 34's concrete pad and on top of a Saturn 1B. When it was permanently mated with its rocket, they would be known together as AS 204, for Apollo-Saturn. As was the custom, 012 was their particular spacecraft, and they had stayed with it from the production line, through its test program, and on to the launchpad, learning its peculiar idiosyncrasies and the location and use of the hundreds of switches that controlled its eighty-eight subsystems. Now they worked their way through the long prelaunch checklist, then waited patiently while yet another of the seemingly interminable communication problems with those who monitored them from the outside was being fixed. By then, all traces of sea-level atmosphere were purged from 012 and replaced by pure oxygen set at a pressure of 16.7 pounds, or two pounds higher than the normal atmospheric pressure on Earth.

At 6:31, with the Sun sinking fast, Grissom was heard to say, "Hey!" Then Chaffee came on, reporting, "There's a fire in here!" Donald O. Babbitt, North American's crew chief, was sitting in a closet-sized control room on top of the gantry outside and heard them over the radio. "Get 'em out of there!" Babbitt shouted to nearby technicians as a sheet of flame could be seen on the other side of the hatch's window. Panicking, they started to race across the gantry's swing-arm tower, but then turned and went back in a frantic search for gas masks and fire extinguishers. Babbitt also managed to call for firemen and ambulances.

An RCA employee looking at 012 on a television monitor on the floor of the launch complex saw a bright yellow glow inside the spacecraft. As he watched in horror during the next three minutes, silvery arms on the other side of the window fumbled to open the hatch. "Blow the hatch," he cried, "why don't they blow the hatch?" He learned only afterward that there had been no way to blow it. Even prying it open by hand could have taken up to a minute and a half.

Now it seemed to be taking forever. In the small clean room through which the astronauts had to pass to get into their spacecraft, Babbitt and five colleagues battled with 012, alternately moving back to gasp for air and trying to pry loose the blistering hatch. They finally got it open five and a half minutes after the alarm had sounded. By the time the first firemen arrived, the smoke had cleared enough to see the astronauts quite clearly. Chaffee was still belted to his seat. But Grissom and White were so tangled below the hatch sill that it was hard to tell them apart. Even pulling them out of the charred wreckage was difficult because their nylon suits had melted and glued them together. Autopsies performed that night showed that their deaths had been caused by the inhalation of toxic gases, not by the victims' second- and third-degree burns. Hatches had been Gus Grissom's nemesis from beginning to end. He and Chaffee were buried at Arlington National Cemetery; White was interred at West Point, his alma mater.

The deaths of the three astronauts as they prepared for the inaugural Apollo test sent the space establishment, those who covered it, and those who supported it politically into an orgy of accusations. The stench of betrayal hung in the air around Pad 34 and radiated in all directions.

Life, which had planned to cosponsor a "fiesta" in Houston the day after the fire (all three victims were supposed to have been there) tried to take the high road. It delivered a campily purple eulogy seven days after the accident and behind a cover that showed the immortal trio, eyes closed and brows furrowed, in a purported planning session deep in the Hollywood Hills that looked more like a seance. "Grissom . . . White . . . Chaffee . . . They bought the farm right on the pad, cooked in the silvery furnaces of their spacesuits, killed in a practice run before they could ever know the surge of their great Apollo craft driving upward to orbit. But put these astronauts high on the list of the men who really count."

But many of Apollo's cheerleaders in the Fourth Estate turned on the space agency for the way it "managed" news of the disaster during the first hours and, later, for the blunders that caused the tragedy in the first place. The American Society of Newspaper Editors castigated NASA for deliberately withholding information. "Although the agency knew within five minutes, it took two hours to learn that all three astronauts were dead. And this came round about from Houston Manned Spacecraft Center rather than from Cape Kennedy," an ASNE committee would report three months later. Others accused NASA of reflexively lying (as it did when it announced that Grissom, Chaffee, and White had apparently died instantly and in their couches) or of plain stonewalling (as happened when a man from NBC was told on the night of the fire that "nothing has happened; go on home.") Lying, of course, was inexcusable. But it was also unfair to demand that stunned and deeply distracted technicians, engineers, and middle managers on the scene of a sudden, horrendous, and intensely emotional accident react with clearheaded immediacy to the requirements of the news collection process. There is a natural tendency in such circumstances to protect the situation; to freeze the action until some sense can be made of what happened.

Taking it a step further, *The Boston Globe* argued in an editorial on April 12 that the men died because of the program's "many deficiencies in design and engineering, manufacture and quality control." The newspaper added that speed—the race with the Russians—had been another key factor. "Space exploration is not the same as, say, an international boat race on the Charles. The ultimate in precautions is a first requisite. The precautions have obviously been lacking. If the lack is not downright criminal, it is little short of it."

The day after the fire, Webb and Seamans asked Floyd L. Thompson, director of the Langley Research Center, to head a board of inquiry.* Anticipating public demands for answers and reforms, and dreading a call to postpone or even stop Apollo altogether, they also asked Congress to hold off the inevitable investigation until Thompson's committee presented its findings.

The findings—almost three thousand pages divided into fourteen booklets—were issued early in April. They concluded that it was impossible to fix the exact cause of the fire but that conditions had certainly been ripe for one, specifically because of the oxygen, the presence of combustible materials, including velcro, "vulnerable" wiring and plumbing, and inadequate provisions for escape and rescue. Yet the report skirted the heart of the matter, at least technologically: that the fire had happened as a direct consequence of NASA's insistence on using pure oxygen and on the fact that the gas had been so heavily pressurized.

Ultimately it was politics, not engineering, that killed the astronauts. The Russians were known to be aiming their own cosmonauts at the Moon. Not landing on it by Kennedy's deadline would have been embarrassing enough. Getting there after the Soviet Union was absolutely unthinkable. So the technocrats and engineers, and the astronauts, too, were driven like dogs. "Insane" was the way Deke Slayton described the work schedule. "We got in too much of a God-damned hurry," Chris Kraft, the manned flight director, would say when he reflected on the catastrophe. "We were willing to put up with a lot of poor hardware and poor preparation in order to try to get on with the job, and a lot of us knew we were doing that."

When they were safely in the shadows beyond the glare of the public spotlight, at least for a moment, Slayton, Faget, and Frank Borman, another astronaut, reacted to the tragedy not like Boy Scout–martinets but like the real people they were beneath all the hype and glitz. They grieved for their dead friends and colleagues, for their program, and for themselves. And they got roaring drunk. While the investigation was starting at the Cape, Borman would recall years later, "We went out one night and got bombed. I'm not proud to say it; I don't drink anymore. But we got bombed that night. We ended up throwing glasses, like in a scene out of an old World War I movie." It was definitely not a scene that would appear in the "*Life* Goes to a Party" series. Borman, in particular, had reason to get loaded. While Grissom and the others were undergoing the oxygen test at Canaveral, he and Thomas Stafford were doing exactly the same thing in a command module at North American.

* It included Faget, astronaut Frank Borman, a representative of North American Aviation, another from the Air Force Inspector General's office, an explosives expert, and three NASA officials, one of whom came from the Kennedy Space Center itself. (Brooks, Grimwood, and Swenson, *Chariots for Apollo*, p. 219.)

Then a hailstorm developed. A writer named Erik Bergaust began research on a book titled *Murder on Pad 34* that appeared the following year, charging NASA with killing the astronauts by emphasizing speed over safety in the great race. "NASA's overall manned spaceflight effort should be devoted to *refinement of the present Apollo,*" Bergaust wrote, adding that "some experts say it should make Apollo flights in the next couple of decades as safe as airplane flights across the American continent." He not only ignored the fact that the situation that created the Apollo program in the first place necessarily precluded its being decades long, but also that airline mishaps still occurred with far greater loss of life than that of 012.

Meanwhile, Mueller was admitting to a bipartisan executive session of Congress that his agency's approach to fire prevention had been wrong despite six years of safe spaceflight and that corrective measures were already in the works. And Webb was also fending off congressional attacks, some of them ferocious. "The level of incompetence and carelessness we've seen here is just unimaginable," one congressman told the head of the agency that had launched sixteen safe manned missions in a row on the most dangerous vehicle ever conceived: the liquid-fueled rocket. Fully fueled, the NASA administrator answered, the Saturn 5 "has the rough equivalent power of an atomic bomb." And in what was probably the greatest understatement of his career, Webb added that "this is not a light undertaking."

Although it was not articulated, there was a sense of embarrassment, not just betrayal, about the accident: embarrassment that the first fatalities in a space program had apparently occurred in the United States. This was understandable because the image makers had carefully portrayed the manned program as being dangerous enough to confer hero status on the astronauts. But no one who valued his job would have dreamed of saying in public that the chances were better than even that people would actually get killed trying to get to the Moon and that the nation would have to accept that fact and press on anyway. But the humiliation that came with being first in that grim game was misplaced. Aleksei Leonov would admit many years later that a cosmonaut, Valentin Bondarenko, who had "never [been] on the short list" for a mission, had perished four hours after being pulled out of a similar oxygen fire at Star Town in 1961. Sharing that information would have benefited the Apollo program, the former cosmonaut said, and it had therefore been out of the question.

The Thompson committee report angered North American, which had thick files of memos warning about the potential for a fire in an oxygen atmosphere and which briefly considered using them to clear its besmirched name. But doing so in public would torpedo NASA, perhaps fatally, and it was not in the company's interest to either sink the space agency or make it vengeful. North American would therefore take its lumps, the chief one being the sacrificial purging of Harrison Storms, who was removed from the program as a token gesture to America's space gods. North American would also be sued for negligence by the three widows, who collected a total of $650,000 in an out-of-court settlement in 1972. Meanwhile the necessary improvements were made to Apollo.

Buried beneath the surface, however—far below the conscious assignment of blame leveled in the reports, committee minutes, editorials, and political blustering—there lurked a feeling that whatever the technical reason for the fire on Pad 34, it had a deeper meaning. There was a new and uneasy feeling that high technology, America's self-celebrated gift to the world, was abidingly ambivalent about its creators. Technology, certainly including computers, was meant to be the slave of mankind. But the master had come to worship the slave. Furthermore, those who had created the slave machines saw in the deadly blaze at Canaveral that the slaves could turn on their master with spectacular viciousness. The deaths of Grissom, Chaffee, and White showed that the technology that was supposed to serve its creators by carrying them to the stars could just as easily and without warning become a dangerous, malevolent, unpredictable monster that defied orders with terrible results.

It was a theme that went directly from the Faustian legend to Frankenstein to *2001: A Space Odyssey,* the first large-scale, big-budget space fiction film, which was in production at the time of the accident and which premiered only fifteen months later. Based on a story by Arthur C. Clarke, *2001* was about a secret mission to Jupiter in search of evidence that other intelligent life existed. During the journey, the computer running the nuclear-powered spacecraft—HAL, for HAL 9000 series—decided that it could handle the mission better than the humans on board, so it turned on them, murdering four (including three hibernating scientist-astronauts) and nearly killing another before he pulled the plug on its key intelligence functions, turning it into a mechanical vegetable. If HAL really were "foolproof and incapable of error," as was claimed, its creation said something about mankind's willingness to cede life and death power to machines that would commit murder with no consideration other than logic. *2001* became a classic because audiences were lifted by its sheer imagination and technical brilliance and at the same time were haunted by a machine that turned from compliant servant to diabolical man killer on its own.

Apollo 1, as it would soon be called, left an indelible mark on NASA's soul. "The fire did give us a baptism," said one launch official. "We knew from then on there would be no forgiveness." The Apollo fire showed the space agency and the wider community that there would be no tolerance for catastrophic failure caused by human error. The constant wariness would haunt the Apollo program until its end. "There is a list of all the times we went to the Moon," one NASA engineer would recount years later. "Finally, I said, look, this is the last time. We are not going to go again, because you are just asking for trouble. It's a very, very risky thing to do."

The Death of a Cosmonaut

At no time during the race to space, and certainly not during the days of Apollo, was the codependency of the nations in the race ever acknowledged publicly. Nowhere does the record show any official admitting that his livelihood and his agency's very existence depended in large part on the competition's staying in the race. In private, however, the competitors could and did acknowledge that they owed each other a great deal. In one of the least appreciated or publicized, but most telling, exchanges of the period, American space officials at a meeting in Madrid a few months before the Apollo fire confided to their Soviet counterparts that their budget had peaked (at $2.9 billion in fiscal 1966) and that they were concerned about cutbacks. "Don't worry," they were assured by one of the Russians, "we'll help you out with that shortly after the first of the year."

The reason for the encouragement soon became apparent. On April 23, 1967—within three months of the Apollo disaster—*Soyuz 1* roared off a pad at Tyuratam carrying Colonel Vladimir Komarov, the commander of the first Voskhod mission. While U.S. intelligence carefully tracked the spacecraft and eavesdropped on its communications, expecting it to be joined by a second vehicle for a docking and cosmonaut exchange, East Berlin radio crowed that the two-year hiatus in Soviet manned launches was about to be followed by an orbital feat that would accomplish at one time what it took a dozen Gemini flights to do. An AP story out of Moscow described the mission as "the most spectacular Soviet space venture in history."

But it was not to be. There was no second launch. Instead, U.S. radar tracked the spacecraft as it tumbled during its fifteenth orbit, indicating an attitude-control problem. Later analysis would show that there had probably been multiple failures, including a solar panel array that did not deploy properly. That, in turn, caused a cooling problem in the craft's electronic package that led to the partial crippling of the attitude-control system. *Soyuz 1*'s retrorocket was therefore fired earlier than planned in the eighteenth orbit, while it was spinning, so the spacecraft started to come down much faster than it should have. Its wild gyrations caused its main chute to tangle, and when Komarov opened the emergency chute that was intended for aborts, it became tangled with the larger one. *Soyuz 1,* its two twisted parachutes fluttering madly behind it, hit the ground like a cannonball going 400 miles an hour. And Komarov would have seen it coming.

The Soviets attributed the first fatality on an actual mission to the tangled shroud. Mishin, who by then had replaced Korolyov, would say later that there were also "some design imperfections." Left unsaid was the fact that Tyuratam's ego-driven rocketeers, like their American competitors, continued to push their men and machines into the jaws of danger while their otherwise distracted political leaders acquiesced and hoped for the best. Only one cosmonaut would fly until *Soyuz 4* and *Soyuz 5* linked up in mid-January 1969 for yet another record: the historic first docking of two manned spacecraft.

Apollo Goes to Space

Pressure for a "nominal" Apollo test after the deaths of the astronauts became phenomenal. During the nine months after the fire, North American, its subcontractors, and the space agency bent to the task of cleaning up any conceivable problems in the command and service module while von Braun's men at Marshall fine-tuned the Saturn 5. At one second after seven on the morning of November 9, 1967, four years after Mueller's all-up manifesto, a Saturn 5 riding a torrent of fire lifted the unmanned *Apollo 4* command and lunar modules off Launch Complex 39A and into a trajectory that could have sent them to the Moon. It was the first all-up test of the whole system, including the rocket itself, and it was spectacular even for veterans of the space program and those who covered it. The booster created its own earthquake. It was not a question of whether the rocket had risen, someone remarked later, but whether Florida had sunk.

Besides testing the structural integrity of the spacecraft and its launch vehicle and a thousand other things, including the ground support system, NASA intended to toss the command and service modules into a 115-mile-high parking orbit, then send them out to 11,185 miles on the end of the Saturn's third stage, and finally slam the command module back into the atmosphere as if it were returning from the Moon. *Apollo 4,* blunt end forward, returned on the button at almost 25,000 miles an hour and then floated down only ten miles from the recovery ship. "No single event since the formation of the Marshall Center in 1960 equals today's launch in significance," von Braun chirped at a postlaunch news conference. "I regard this happy day as one of the three or four highlights of my professional life—to be surpassed only by the manned lunar landing."

Indeed, *Apollo 4* was virtually flawless. The reporters who came to Canaveral, mindful of what had happened in January, nevertheless chased the story with their old energy and enthusiasm. They knew that they were living through a profound moment there on the shimmering Florida coast, where fire and steel were being mated to carry members of their race to another world for the first time. The journalists, decked out in bush jackets, sunglasses, and the ubiquitous laminated press cards hanging from their necks on beaded chains or clipped to pockets, prowled beneath the huge gray gantries and inside the Vehicle Assembly Building, where the rockets came together, and searched for ways to describe the moment.

The size and power, the sheer immensity and brute force of the machines, lent themselves to the kind of simple comparisons and symbols that oil the public's digestive process. The Saturn 5's first stage was so wide—thirty-three feet—that three moving vans could drive through it side by side; the LOX tank held enough liquid oxygen to fill thirty-four railroad cars (or fifty-four, depending on the handout); the pumps in the first stage worked with the force of thirty diesel locomotives; the whole Saturn 5 was as high as a thirty-six-story building, towered well above the Statue of Liberty, and weighed a lot more than a Navy destroyer. The U.N. Secretariat building could roll through the VAB's doorway. A writer for *Fortune* figured that von Braun's behemoth could lift 1,500 Sputniks on a single launch, or 9,000 Explorer 1s, or 42 Geminis with astronauts inside.

But beyond the metaphors and analogies, beyond the hype and hokum, there was this plain fact, and it needed no embellishing: the United States of America was actually going to send men to the Moon. *To the Moon!* It was really going to happen! *Apollo 4* was important because it brought all of the diverse pieces of the program together for the first time—not the least of them being managerial—and demonstrated that they could work together. The Great Pyramid now looked like what it was supposed to look like. And it was pointing at the Moon.

The first working lunar module, three months late because of changes that had resulted from the fire, went to space as part of *Apollo 5* on January 22, 1968. In orbit, the LM's attitude control engines pulled it away from the Saturn upper stage just the way they were supposed to do on the way to the Moon and both its descent and ascent engines were tested. That accomplished, the spider's fiery carcass crashed into the waters southwest of Guam on February 12. With a few minor exceptions such as slight instability in its Bell engines, this flight, too, was an outstanding success.

The Death of Gagarin

Meanwhile, fate delivered another blow to the Russians, this one emotionally devastating. On the morning of March 27, Yuri Gagarin and Vladimir Seregin, an instructor, took off on a routine proficiency flight in a MiG-15 trainer and never returned. Their air traffic controller radioed that there was cloud cover at about 13,000 feet. In fact, the clouds began at 4,000 feet. Furthermore, the old fighter's altimeter had probably been set too high. So as 625—Gagarin's call sign—flew into the clouds northwest of Moscow, he apparently thought he was higher than he really was and that the clouds were higher, too. He therefore felt he could push the MiG's stick forward. But that put it into a steep dive that sent it slamming into the earth before either he or the man behind him had time to eject. Both were so pulverized that there were no recognizable remains. Ironically, Gagarin had been Komarov's backup, and now both were dead because of tragic accidents.

Yuri Gagarin was undoubtedly the most beloved individual in the country. Whether the average Russian cared about the space program or not—and a majority emphatically did not care—Yura or Yurka or Yurochka was a national emblem of transcendental magnitude: a soft-spoken, baby-faced, shy, virginal-appearing hero who, like the legendary Aleksandr Nevsky, had carried Russia's fragile honor to the citadel of history with courage, dignity, and humility. "Yuri Gagarin is not with us," one journalist who was not given to cultural or political bilge wrote as a eulogy two decades later. "He who was the first to realize the eternal dream of mankind, who stepped into the ocean of stars, flew over the planet, saw it from the side. A tragic, stupid, unfillable and still enigmatic loss. Our pain and grief. Forever!"

The nation truly grieved for the world's premier spacefarer. His ashes were carried through Moscow by nine cosmonauts. Then, with his widow sobbing and his portrait and decorations laid out on display, he was buried in the Kremlin wall with Korolyov and Komarov. Gagarin's likeness remains carved in marble, cast in bronze, and depicted on everything from posters to postage stamps throughout the country. On Leninsky Prospekt, a main thoroughfare leading to the heart of Moscow, a towering monument to the national idol showed a stainless-steel socialist superman clad in what amounted to light armor standing absolutely erect, his arms spread like an Olympic high diver, poised to plunge into space. A model of Earth, set at the foot of the hero's pillar, memorialized the flight of April 12, 1961. Star Town's official name was changed to the Y. A. Gagarin Memorial Cosmonaut Training Center. Gagarin was for years the most incongruous of all ghosts: a saint in a steadfastly secular society. Korolyov and Gagarin, the two guiding lights of the space program, were now dead, and their patron, Khrushchev, was banished to limbo. He would die three years later.

Apollo Moves into High Gear

Meanwhile, the pace of their rival's Moon program accelerated as the pieces of Apollo continued to come together. *Apollo 6,* also unmanned, was launched just after dawn on April 4. This time, however, an old and troublesome phenomenon called the "pogo effect" reappeared. The thrust of the first stage's

F-1s began to fluctuate in a series of spasms or pressure surges known as low-frequency modulations, or pogoing, which caused the entire Saturn to bounce for thirty seconds right after liftoff. Then two of the second stage's five J-2s conked out, forcing its controllers to increase the burn time of the other three, which caused them to run out of fuel before the huge rocket reached the required speed and altitude. So the third stage had to be fired for longer than planned in order to make up for the low velocity. This put *Apollo 6,* still attached to the now-dead third stage, into an elliptical orbit that was higher than the circular one in the flight plan. To make matters even worse, the third stage would not restart. This was anything but nominal.

While *Apollo 6* was taking a series of spectacular color stereo pictures around the world, the controllers decided to salvage the mission by firing the service module's own engine long enough to send the spacecraft out to 13,800 miles, where it mimicked the translunar injection maneuver that would get its successors headed to the Moon. Ten hours after it was launched, *Apollo 6* splashed down in the Pacific and was brought home aboard the carrier *Okinawa.* A "pogo task force" heading some thousand engineers at Rocketdyne, Marshall, and elsewhere eventually solved the problem by using helium to absorb surges in pressure in the liquid-oxygen feed system.

Apollo 7 blasted off on October 11, 1968, carrying the crew that was supposed to follow Grissom's: Schirra, Don F. Eisele, and R. Walter Cunningham. During almost eleven days in orbit, the astronauts rehearsed virtually every operation that would get their successors to the Moon. They also took television pictures of themselves to show taxpayers at home, nonaligned Third Worlders, and other earthlings how they operated the spacecraft: the "magnificent flying machine," as Cunningham called the capsule. The public relations people wanted the audience to see how the nation's heroes lived and frolicked in their exotic new environment, and swimming topsy-turvy in a weightless world was a guaranteed grabber. "Hello from the Lovely Apollo Room High Atop Everything," said one sign the clowning spacemen held for a camera.

What the television audience missed, however, was the first real space rebellion. *Apollo 7*'s crew went up knowing that they were on the first manned flight since the fire that had killed Grissom, Chaffee, and White, and that made them tense; they felt the future of the now-beleaguered program rested squarely on their performance. And their performance, they now saw, was being turned into a circus by the public affairs people.

"I had fun with Mercury. I had fun with Gemini," Wally Schirra would explain later. But then, "I lose a buddy, my next-door neighbor, Gus, one of our seven; I lose two other guys I thought the world of. I began to realize this was no longer fun. I was assigned a mission where I had to put it back on track like

Humpty-Dumpty." The pressure to do that was intense enough. But Schirra thought that having millions of people looking over their shoulders while they did it just to get publicity was contemptible. "They decided on Saturday morning—we launched on Friday—to turn on television. . . . The NASA PR guys want to have some television today because you've got some quiet time. We can get national media," he remembered Mission Control telling him. "We've got a new vehicle up here and the TV will be delayed without further discussion until after the rendezvous," an angry Schirra snapped back. Glynn Lunney, the mission's flight director, said later that the *Apollo 7* crew was "openly difficult to deal with and hostile to some of the things we were trying to do." That's because, as Schirra growled over the open mike, he thought several unscheduled chores had been invented by an "idiot." In his novel *Space,* which commemorated the Apollo program, James A. Michener likened such performances to "dancing bears" at the circus. Schirra would have agreed.

To make matters worse, Schirra probably took the first head cold to space and spread it to his companions in the command module's cramped quarters. Since there was no gravity, their mucus accumulated in their nasal passages, forcing them to blow extra hard to clear them. That built painful pressure in their ears. When it came time to come down, Schirra told Houston that he wanted the crew to take off their helmets so they could squeeze their nostrils with their mouths shut while blowing as hard as they could. The technique, known as the Valsalva maneuver, reduces pressure in the middle ear. But the men in Houston thought that taking a helmet off during re-entry was more dangerous than playing football without one, and they therefore refused the request. Schirra lost his temper again. He told Mission Control that he was *Apollo 7*'s commander—that the lore of air and space gave the commander the last word—and that he would therefore use his own judgment. Chris Kraft reacted by calling Schirra "paranoid."

And the reality of spaceflight brought another problem. Although crumbled food floating around the cabin looked like fun in a place where gravity held eggs on plates and soup in bowls, it quickly became annoying to those who had to catch it and try to get it down. Schirra and the others came to hate flying food. The combination of afflictions made Wally Schirra really testy. Deke Slayton, who would deliver a tongue-lashing to Schirra after he was safely back on Earth, called it "the first space war." It would not be the last. After logging more than 260 hours in space, Schirra, Eisele, and Cunningham hit the Atlantic off Bermuda a little more than a mile from where they were supposed to land. It was time to go to the Moon.

The final decision to send astronauts around the Moon—to go for it—was made in August 1968, three months before *Apollo 7* flew. It came after intelligence reported that the Soviets were getting ready to put five tankers into

Earth orbit and then use a manned Soyuz to herd them to the Moon like a dog herding sheep. And all that to land just one cosmonaut. The fact that Soyuz was being pushed so hard that it had killed Komarov the year before indicated what seemed to be an all-out drive to get a man on the Moon first. And if there was any lingering doubt about a manned Soviet Moon shot, Zond 5's becoming the first spacecraft to orbit the Moon and return to Earth in mid-September 1968 dispelled it. Whether the U.S.S.R. got a man onto the surface or not, the CIA warned NASA, Zond showed that it was straining to at least circle it with a man before Americans did. And Zond 5, really a lighter version of Soyuz without an orbital module that was designed to be launched by a Proton, carried passengers: plants, turtles, flies, and worms to test radiation effects.

The decision to send men around the Moon was an agonizing one because it was potentially perilous. No one had to be reminded during the summer of 1968 that the Saturn carrying *Apollo 6* had pogoed and lost three engines. Nor did anybody forget that no one had actually flown an Apollo mission. "Remember now," Deke Slayton would later recall, "we had not yet flown *Apollo 7,* had not flown a manned command module, and the last time we had flown the Saturn 5 it had almost come unglued because of the vibration problem. Shit, we didn't even have the software to fly Apollo in Earth orbit, much less to the Moon." Webb was a doubter who soon came around. But then he was replaced by former General Electric research manager and Deputy Administrator Thomas O. Paine because Richard Nixon won the election. Paine came around, too, on the first manned Moon shot after von Braun, Slayton, Gilruth, Kraft, Low, and others ganged up on him. Slayton was particularly persuasive: "It is the only chance to get to the Moon before 1969," he told NASA's new administrator.

Santa Circles the Moon

Apollo 8 roared off Pad 39A just before eight o'clock on the morning of December 21, 1968, carrying Borman, James Lovell, and rookie William Anders on what Shepard and Slayton would one day call "the single greatest gamble in space flight then, and since."

Even as the spaceship neared the Moon on Christmas Eve, no firm decision had been made as to whether it would actually orbit or simply swing around and head back home. If everything looked good—was "in the green"—then Borman would be given permission to circle. But *Apollo 8* was riding the sharp edge between triumph and catastrophe. In order to swing into lunar orbit, the trajectory guys calculated, the command and service module's engine would have to burn for precisely 247 seconds. A miscalculation or an engine problem could send it back to Earth too soon, or into an elliptical orbit around the Moon, or, with too long a burn, it would dig a new crater on the Moon.

Finally, the historic command went out. With all the indicators weighed and *Apollo 8* about to slip behind the Moon for twenty minutes during which it would be out of radio contact, Houston gave the historic order: "You are go all the way."

"We'll see you on the other side," Lovell answered as his spacecraft disappeared, locked in lunar orbit a little more than sixty-nine hours after it left home. Twenty minutes later, with Mission Control repeating *"Apollo 8 . . . Apollo 8 . . . Apollo 8"* in a monotone that gave no hint of the tension in the room, the silence from space was finally broken. "Go ahead, Houston," Lovell answered with equal reserve. Those three words created pandemonium in the Manned Spacecraft Center, with wild cheering, shouting, and applause.

"The Moon is essentially gray, no color," Lovell, now the first lunar guide, reported. "It looks like plaster of Paris, like dirty beach sand with lots of footprints in it." But the first men to see the Moon up close were transfixed, absolutely captivated, by the starkness over which they flew, as well as by the splendor of the distant blue, green, and white globe that was their precious home. "It makes us realize what you have back on Earth," Lovell said. Indeed, some of the most important photographs in history, including an earthrise and others showing Earth against the dark void—pictures destined to become the icons of the environmental movement—were taken by the *Apollo 8* crew. But there was more practical work to do. Five potential landing sites that were under final consideration, including the Sea of Tranquility, were also photographed up close. One of the Earth pictures, showing the New World and its deep blue oceans beneath swirling clouds—a disc surging with life and framed in absolute blackness—would make the cover of a special issue of *Life* the following January that celebrated "The Incredible Year '68."

It being Christmas Eve, Anders and the others came up with an appropriate greeting. "For all the people on Earth," he said, "the crew of *Apollo 8* has a message we would like to send you." He then stunned those who listened by reading from Genesis: "In the beginning, God created the heaven and the earth." Lovell and Borman read from it, too, in a blatant back of the hand to the godless Commies who had not made it. If, as Khrushchev had said, some American planetary probes failed because they were made by capitalists, then it was fair to share the glory of the moment with the God he also disdained. Borman gave his and the others' own greeting: "And from the crew of *Apollo 8,* we close with good night, good luck, a Merry Christmas, and God bless all of you—all of you on the good Earth."

Apollo 8's engine kicked in for the required 304 seconds during the last of its ten orbits on Christmas Day, easing the spacecraft out of its circular flight path and pointing it home. A relieved Lovell took the occasion to radio Houston: "Please be informed there is a Santa Claus. The burn was good." They hit the water late on the morning of the twenty-seventh after traveling farther than anyone else in history, 580,000 miles, and were given a tumultuous welcome home.

A Bear Falters

There was not cheering everywhere. The day *Apollo 8* left for the Moon, Lev Kamanin, the son of the cosmonauts' own leader and himself a Kremlin space expert, penned this melancholy thought in his diary: "For us this [day] is darkened with the realization of lost opportunities and with sadness that today the men flying to the moon are named Borman, Lovell, and Anders, and not Bykovsky, Popovich, or Leonov."

And worse was to come. On February 21, 1969, the first N-1/L-3 test launch ended disastrously after two of its thirty first-stage engines were accidently stopped by the emergency shutdown system, followed seconds later by a ruptured oxidizer line that leaked liquid oxygen, which started a fire. Doing as it was programmed to do, the shutdown system immediately stopped the other twenty-eight engines. The towering white rocket and its shrouded Moon craft came crashing back to Earth like a mastodon with a hot foot.

Four months later a Proton carrying a soft lander that was supposed to collect Moon soil and bring it home malfunctioned before it could reach orbit and fell into the Pacific.

Then, in the beginning of July a second N-1 carrying an L-3 blew to pieces after a metallic object fell into the number eight engine's oxidizer pump, causing an explosion that ripped up cables and severely damaged the engines around it. The world's second biggest rocket collapsed onto its pad and disappeared in a hideous eruption of fire and smoke. Years of political ambivalence, institutional fear and scorning of science, competition with the military, internecine treachery, a chronically hemorrhaging economy, wasted resources, obsessive secrecy, management debacles, and an unwieldy and often-unresponsive support system had finally defeated the best and brightest of Tsiolkovsky's children. It was the Fourth of July 1969 at Tyuratam and the race to the Moon was for all intents and purposes over.

Notables Weigh In on *Apollo 11*

Meanwhile, an unprecedented pageant was building on the Florida coast, a gaudy celebration of technology and resolve and patriotism that was as close as America has ever come to throwing a spectacular medieval tourney and national circus. Even those who faulted the purpose of the great journey—and they were many and often eloquent—did not doubt that it was really going to happen.

McDivitt, Scott, and Rusty Schweikart had taken *Apollo 9* on a ten-day mission in March that had fully tested the command and service module–lunar module separation maneuvers. Two months later, Stafford, John W. Young, and Gene Cernan had gone back to the Moon itself in *Apollo 10* so they could take a command module named *Charlie Brown* and LM-4, which they called *Snoopy,* for a dry landing run. Snoopy got to within 50,000 feet of the surface, which is where its successor would be when it began its descent. All objectives of both missions, as NASA put it, were achieved. Now it was time for *Apollo 11* to land Americans on another world.

Knowing that an event of truly historic significance was unfolding (whatever the reason), many of the century's most famous scribes and philosophers, soldiers and merchants, jesters and acrobats, priests and kings turned out to see what the alchemists and their knights could do. The notables came to see, but in many cases also to be seen, as if their presence would add a mystical synergy to the grandest feat of exploration since Columbus challenged the Atlantic. An event as spectacular as this—and it figured to be spectacular regardless of whether the three astronauts actually went to the Moon or were cooked on the spot by the combusting oxygen and hydrogen—whetted the sense of being alive; of being part of a truly epic enterprise. Their species was pushing off for another world for the first time and their presence at the Cape at that particular moment meant that they were somehow going too. The celebrities saw Apollo through the prism of their senses. What came out took the form of solemn pronouncements for the edification of the masses or wisdom to be chiseled onto history's tablet. They were at a momentous event and that energized their psyches. It made them momentous, too.

Norman Mailer made a pilgrimage to Florida to write about the voyage of the century in a book called *Of a Fire on the Moon.* On its pages, he himself would actively participate as a chronicler named Aquarius, in the process helping Apollo to reach its destination. It was Heisenberg's Uncertainty Principle brought to the exploration of outer space rather than to subatomic particles: Mailer would somehow influence the story by dint of climbing into it. Part of his soul would stow away on *Apollo 11.* The Saltine Warrior would have understood.

"Yes, he had come to believe by the end of this long summer that probably we had to explore in outer space," Aquarius meditated, "for technology had penetrated the modern mind to such a depth that voyages in space might have become the last way to discover the metaphysical pits of that world of technique which choked the pores of modern consciousness—yes, we might have to go out into space until the mystery of new discovery would force us to regard the world once again as poets."

But Kurt Vonnegut Jr. was not buying it. The author of *Slaughterhouse-Five* articulated the doctrine of the leading naysayers in a piece that appeared in *The New York Times Magazine* three days before *Apollo 11* departed. "We have spent something like $33 billion on space so far. We should have spent it on cleaning up our filthy colonies here on earth." Nor was Vonnegut impressed by the view of Earth from space, which he said obscured the very problems that the Apollo funding could have gone to eliminate. "Earth is such a pretty blue and pink and white pearl in the pictures NASA sent me. It looks so clean. You can't see all the hungry, angry earthlings down there—and the smoke and the sewage and trash and sophisticated weaponry."

Arnold J. Toynbee, afflicted with a head cold, ruminated in a London flat thickly insulated by books that Apollo represented an unconscionable gap between technology and morals and smacked of grossly misplaced priorities. "In a sense, going to the Moon is like building the pyramids or Louis XIV's palace at Versailles," the distinguished British historian told a *Wall Street Journal* reporter. "It's rather scandalous, when human beings are going short of necessities, to do this." Reinhold Niebuhr, a respected theologian, agreed. Priorities were upside down, the failing old man wheezed, before adding that he was delighted there was no life on the Moon because finding it would surely "enlarge the realm of antagonism" in the world. Mark Van Doren, Columbia University's beloved poet and teacher, sat on the front steps of his modest home in Cornwall, Connecticut, and said with gentle conviction that he just wished people would leave the Moon alone. The old white-maned professor, now grown frail, was defiantly ignorant about translunar trajectories and T-minus-anything and wished to remain so. "Man's true business," Van Doren mused, "is to live quietly, sweetly and generously."

But over in Cambridge, on the banks of the Charles, Isaac Asimov rhapsodized about Apollo. The muttonchopped master of s-f, then approaching the completion of his hundredth book, took strong exception to the argument that the nation's priorities were backward. Money to help the poor and wretched could be rung out of the Pentagon with no appreciable effect on the country's ability to defend itself, the prolific storyteller maintained, his voice rising dramatically. Furthermore, mankind needed adventure. "The world is being Americanized and technologized to its limits, and that makes it dull for some people," Asimov went on. "Reaching the Moon restores the frontier and gives us the lands beyond."

At the American Museum of Natural History in New York, where Bernard Smith and the other dreamers had met almost forty years earlier to rocket themselves out of the Depression, Margaret Mead weighed in with Asimov.

The feisty anthropologist, famed for her pioneering work in the South Pacific, said flatly that men would "despise" themselves as "unadventurous stay-at-homes" if they stopped short of landing on the Moon.

Nor did she sympathize with the hippies and other dropouts who were reacting to the futile and frustrating war in Vietnam; riots in Chicago, Detroit, Los Angeles, New York, and elsewhere; the murders of the Reverend Martin Luther King Jr. and Senator Robert F. Kennedy; and other scourges of American society, most of them blamed on technology gone out of control. "We're not going to come out of our present problems by going backward. We can only come out of them by going forward, and part of that thrust forward will be to push exploration of space, exploration under the sea, exploration into the infinitely small and infinitely large," Mead said angrily. "There is no hope of saving the number of people we have on this planet now by primitive technology and the village green. . . . The only way to get out of a badly built building . . . is to build a better building, not go back and sort of burrow in the sand."

Philip Morrison, an eminent MIT physicist who had worked at Alamogordo on the Manhattan Project to build the atomic bomb, took the position that scientists were the wrong people to be called upon to judge the mission to the Moon because they had long anticipated it and therefore took it for granted. It was the ordinary people and the poets who had been impressed by Magellan's voyage (which he likened to Apollo), not the grandees and scientists. The latter knew perfectly well that the planet was round, Morrison explained, and they also understood that one of the primary objectives of Magellan's trip was to show his patron's flag. It was the same now, the physicist said, adding that the flight to the Moon was motivated by a number of inseparable factors, including adventure, patriotism, and the need to keep skilled people employed and the aerospace business healthy in southern California and along Route 128 outside of Boston.

In any case, the quick-talking scientist added, Sputnik was more important than Apollo because it refuted Aristotle in a tangible way for ordinary people. The fact that men could fling an object they had made—and a slightly imperfect one at that—into the space beyond Earth proved once and for all to everyone on the planet that Aristotle's two-dimensional universe and the perfection of everything in it was fundamentally and demonstrably wrong.

Marshall McLuhan, the Canadian media guru, was not the slightest bit impressed by what the other notables said, nor by the event itself. Apollo, he cackled, was really just "a great big spectacle" that would turn the Moon into a junkyard and inebriate the world. That's right: *inebriate.* The Apollo program's budget was $24 billion, McLuhan added, or "exactly the same as the budget for booze in the U.S. per annum." (Its cost actually varied from $350 million to $30 billion, depending on what was counted. NASA put the final cost at $25 billion for Apollo alone, plus another $4.5 billion with Mercury, Gemini, and Apollo follow-up programs thrown in. The money covered salaries, research and development, facilities, operations, and hardware.)

As the excitement continued to build, someone thought to interview Esther Goddard, who had moved back to Worcester. "It's been such a long time coming," she told a reporter from the Associated Press. "My husband would have been beside himself with delight." Then she quoted from his diary. "When old dreams die, new ones come to take their place. God pity a one-dream man."

The Great Escape

The morning of the launch was hot, sunny, and humid. Special stands had been set up near the Vehicle Assembly Building to hold about five hundred of the seven thousand VIPs who had been invited. By 8:30, an hour and two minutes before launch, the bleachers were nearly full. Three and a half miles away, just off the white sandy beach, the celebrities could see the gleaming Saturn 5 poised inside 39A's bright orange launch tower. It had stood there through the night, bathed in light like the national monument it was. Half of Congress had accepted the invitation and so had nearly half of the nation's governors. There were 3,493 accredited journalists from the United States and fifty-five other countries. The dignitaries included foreign ministers, diplomats, and diplomats' wives. ("The State Department warned us we'd need hats," the Marquesa Merry del Val, wife of the Spanish ambassador, confided to Charlotte Curtis of *The New York Times*. "I brought a scarf instead.") Mrs. John Lacy, whose husband worked for NASA, tried to make an apparent newcomer feel at home. "Have you ever seen a shot before?" she asked a woman sitting nearby. "I am Mrs. Wernher von Braun," the woman answered.

Lyndon and Lady Bird Johnson were there, and so were Mayor Walter Washington of Washington, D.C., the Webbs, Daniel Patrick Moynihan, Sargent Shriver, Fiat's Gianni Agnelli, French supersonic test pilot André Turcat, Senator Jacob Javits, Jack Benny, Johnny Carson, Mrs. John N. Converse (a Southhampton neighbor of William F. Buckley who had taken the precaution of bringing cold cranberry juice and fried chicken from New York because she had been warned of a food shortage in the area), Barry Goldwater, Terence Cardinal Cooke, Leon Schachter of the Amalgamated Meat Cutters and Butchers, and Prince Napoléon of Paris, a direct descendent of the emperor. The Honorable Spiro T. Agnew was there, too. A former governor of Maryland and then chairman of the Space Task Group, he would eventually resign the vice presidency in disgrace instead of facing a trial and impeachment for tax evasion and for accepting $147,500 in bribes in unmarked envelopes, $17,500 while vice president of the United States.

The Space Task Group had been set up within six weeks of Nixon's taking office with a mandate to recommend where the space program should go after Apollo. Even before Kennedy's and Johnson's astronauts left for the Moon, Nixon had prudently looked for a way to leave his own mark on the space program. Making Agnew the group's nominal leader seemed to show that the White House cared about space, though the options were shaped by Secretary of the Air Force Robert Seamans, Presidential Science Adviser Lee DuBridge, and Paine. That September the group, noting that both the public and the science community was increasingly "vocal" about Apollo's high cost, would offer a mix of options that balanced manned and robotic missions with an eye to keeping costs down. Out of them would emerge a space transportation system, to be followed by a space station, just as von Braun had described on the pages of *Collier's* a quarter of a century earlier. But now, in July, the vice president used the occasion to call for a manned mission to Mars before the end of the century. However implausible and horrifically expensive, especially to a Republican administration, raising the prospect of sending Americans on to the Red Planet seemed to show that Nixon had just as much vision as his glamorous and martyred predecessor. Talk was cheap enough.

The president himself, who watched the launch on television in Washington, proclaimed the following Monday a national holiday to mark the "moment of transcendent drama." It had been 2,974 days since Kennedy had first pointed the way to Moon. Now the crew of *Apollo 11* was poised to carry a commemorative plaque with his—Nixon's—name and theirs to the Sea of Tranquility as a permanent memorial. It would say:

> HERE MEN FROM THE PLANET EARTH
> FIRST SET FOOT UPON THE MOON
> JULY 1969, A.D.
> WE CAME IN PEACE FOR ALL MANKIND

Whatever the plaque said about representing the human race, the flag the astronauts would plant on the Moon would bear stars and stripes, not a representation of the whole planet wreathed in white olive branches on a field of light blue.*

Mailer contemptuously surveyed the other notables, the newcomers to the

* This was the clear will of Congress. As amended four months after the *Apollo 11* landing, the National Aeronautics and Space Act stipulated that: "The flag of the United States, and no other flag, shall be implanted . . . on the surface of the moon, or on the surface of any planet" by an American crew in the Apollo or any subsequent program if the United States paid for the mission. "This act is intended as a symbolic gesture of national pride in achievement," the legislation was careful to explain, "and is not to be construed as a declaration of national appropriation by claim of sovereignty." (*National Aeronautics and Space Act of 1958, As Amended, and Related Legislation,* December 1978, p. 44.)

preserve he had staked out for himself months before so he could do his art, and decided to stay with "his own sweaty grubs, the Press and photographers." Aquarius "dislikes the VIPs, dislikes most of them taken one by one, and certainly dislikes them as a gang, a Mafia of celebrity, a hierarchical hive," the author of *The Naked and the Dead* decided ruefully. If Saturn had a perfect launch, he concluded, "it will not be the fault of the guests. No, some of the world's clowns, handmaidens, and sycophants and some of the most ambitious and some of the very worst people in the world had gotten together at the dignitaries' stand. If this display of greed, guilt, wickedness, and hoarded psychic gold could not keep Saturn V off its course, then wickedness was weak today."

More than half a million people, many of them having arrived during the night, clogged roads for ten miles outside the space center and dotted the beach with twinkling fires. Cars, campers, pickup trucks, and motorcycles, tents of every size and description, sleeping bags, and plain blankets turned the area into a vast campground. Radios, television sets, cameras, and binoculars were everywhere, with people packing jetties, wading into the surf, and sitting on the tops of their vehicles for a look at the historic spectacle. The air smelled of grilled meat and vibrated to the pounding cadence of police and military helicopters. An armada of boats with names such as *Tinker Toy, Knot Again, Bobbin Robbin II, Rum Bum, Insanity,* and *Miss Grapefruit,* some of them anchored for days, clogged the Banana River. Near Highway 1, about 150 members of the Reverend Ralph Abernathy's poor people's campaign marched with mules to call attention to the nation's hungry. They were heckled but not hassled.

Twenty-five million television viewers watched from beyond the horizon and countless others heard it on home or car radios. The three major commercial networks, working in a pool that cost a lavish $750,000 and used 150 people at the Cape alone, would provide virtually uninterrupted coverage on Sunday and Monday, the twentieth and twenty-first, from before touchdown on the lunar surface to after the astronauts left it. A live television broadcast was beamed to thirty-three countries on six continents.

Mishin, Chelomei, Glushko, Tikhonravov, Chertok, Yangel, Kamanin, and their cosmonauts would watch a taped summary on their own evening news (it ran fifth, well behind a story saluting Soviet metalworkers and another about Polish Liberation Day). Meanwhile Luna 15, a sample-return mission, raced ahead of Apollo in a last, desperate attempt to upstage the Americans by retrieving an infinitesimal part of the Moon and bringing it home. It would crash into Mare Crisium—appropriately, the Sea of Crisis, where the lone cosmonaut was to have landed—the day after the astronauts arrived at Tranquility. Typically, a commission of inquiry grilled IKI scientists in an apparent effort to cast the blame on them for selecting a dangerous landing site, but it was soon shown that it had probably been a faulty ground-sensing instrument that caused the ignominious accident. Just as typically, TASS issued a terse report saying only that Luna 15's work "had ended." In September 1970, Luna 16 would manage to return a lunar soil sample to Earth and two months later Luna 17 would deposit the first of two television-equipped Lunokhod rovers—contraptions that looked like mechanized baby carriages—on the Moon. They were clever, but they would in effect amount to toys that crawled in the shadow of astronauts.

Apollo 11 left for the Moon at 9:32 Eastern Daylight Time on the morning of the sixteenth. As the throng at the Cape cheered or watched in quiet awe, Armstrong, Aldrin, and Michael Collins rose slowly past the top of 39A's tower while the Saturn began its programmed turn. Amber lights blinked on the instrument panel. Aldrin felt as if he was on top of a long swaying pole. The F-1s, spewing fire and smoke and sending tremors through earth and water, sounded to him like a freight train rumbling far away in the night. But it was different outside. There was only one man-made noise that was louder than the Saturn 5's first stage: a nuclear explosion. Twelve seconds after liftoff, control of *Apollo 11* passed from Canaveral to Houston. "You are go for staging," Houston told the men in the command module, which they had named *Columbia.* The lunar module was called *Eagle.* "Staging and ignition," Armstrong answered. Two minutes and forty-two seconds after it lifted the two upper stages and the Apollo stack off the pad, the first stage separated and fell forty-five miles into the Atlantic.

The distance between Earth and the Moon was bridged in four days, as had been done by Verne's large cannon shell. On July 20, after a number of orbits around the Moon, Armstrong and Aldrin, now inside *Eagle,* left Collins still in orbit and began their descent to the lunar surface. Seeing that they were headed toward a forty-foot-wide crater surrounded by boulders, Armstrong overrode the computer and steered the lunar module to a clear spot a few miles away. Tingling with excitement, and after a series of alarm warnings from an overworked computer that would have caused an abort had they not been ignored by a NASA controller who suspected they were insignificant, the astronauts kicked up Moon dust as they slowly settled on the Sea of Tranquility. Armstrong came in so slowly that *Eagle*'s descent engine had just six seconds of propellant left when it came to rest at 4:17 EDT that afternoon. It was 102 hours and 45 minutes since they had left Earth. Armstrong called home:

"Houston, Tranquility Base here. The Eagle has landed."

"Roger, Tranquility," answered a relieved Charles Duke, the capsule communicator in Houston. "We copy you on the ground. You got a bunch of guys about to turn blue. We are breathing again. Thanks a lot."

The sudden stillness felt strange to Aldrin. Spaceflight—flying—had always involved movement. But now, suddenly, he and Armstrong were absolutely still on a ghostly world. It was, Buzz Aldrin thought, as if *Eagle* had been sitting there since time began. He reached over and gave Armstrong a hard handshake as the Sun rose behind them like a huge spotlight. Pulling out a small silver chalice and a vial of wine, Aldrin asked "every person listening in, whoever and wherever they may be, to pause for a moment and contemplate the events of the past few hours, and to give thanks in his or her own way."

Six and a half hours later Neil Armstrong backed slowly out of the LM's hatch and, with Aldrin guiding him, carefully worked his way down a ladder attached to the spacecraft's forward landing leg.

"I'm at the foot of the ladder," Armstrong told Houston. "The LM footpads are only depressed in the surface about one or two inches. I'm going to step off the LM now."

With Aldrin watching through one of *Eagle*'s windows and a television camera attached to another of the lander's legs recording the scene for instant transmission home, Neil Armstrong's blue lunar overshoe touched the Moon's gray powder.

"That's one small step for . . . man," Neil Armstrong told the world, "one giant leap for mankind."

What he said to Aldrin, however, was a little less weighty: "Isn't it fun?"

"I was grinning ear to ear, even though the gold visor hid my face," Buzz Aldrin would later recall. "Neil and I were standing together on the Moon."

From northernmost Norway, where Lapps herded reindeer with transistor radios pressed to their ears, to Wollongong, Australia, where a local judge tried cases with his television on, people all over Earth followed the event. An estimated 600 million of them—a fifth of the planet's population—watched the live transmission of Armstrong setting foot on the other world, followed thirty minutes later by Aldrin. The audience was also treated to Orson Welles recounting his infamous *War of the Worlds* broadcast, James Earl Jones and Julie Harris reading from the letters and journals of other explorers, Steve Allen discussing the Moon's role in popular songs, and Duke Ellington playing and singing his original "Moon Maiden." It was by far the biggest telefest in history. A tape of the Moon walk was shown three times on Moscow television and played on page one of both *Pravda* and *Izvestia.* Konstantin Feoktistov, the scientist who had flown with two other cosmonauts on *Voskhod 1,* was effusive in praising his foreign brethren, saying that he and his countrymen "rejoice at the success of the American astronauts."

As the world looked on, the plaque was unsheathed on the descent stage and read aloud. Then Aldrin, who had always been fascinated by explorers planting flags on exotic shores, pushed his own nation's banner a couple of inches into the most exotic place of all and saluted it.

"I looked high above the dome of the LM," Buzz Aldrin would later write. "Earth hung in the black sky, a disc cut in half by the day-night terminator. It was mostly blue, with swirling white clouds, and I could make out a brown landmass, North Africa and the Middle East. Glancing down at my boots, I realized that the soil Neil and I had stomped through had been here longer than any of those brown continents.* Earth was a dynamic planet of tectonic plates, churning oceans, and a changing atmosphere. The moon was dead, a relic of the early solar system."

The astronauts collected forty-six pounds of rocks and soil, left a pouch containing goodwill messages from the leaders of seventy-three nations, an *Apollo 11* patch, a small gold olive branch, and other mementos, and quickly set up science instruments, including a seismometer to detect "moonquakes" and a solar wind collector. Armstrong and Aldrin were back inside the lunar module two hours and thirty-one minutes after they had left it. They blasted off twelve hours later, at 1:54 P.M. on the twenty-first, and docked with Collins less than four hours after that. With all three safely back inside the command module, they jettisoned the ascent stage and set course for home. A charred and peeling *Columbia* splashed down 950 miles southwest of Hawaii at 12:50 EDT on the afternoon of the twenty-fourth.

After being plucked out of the water and taken to the aircraft carrier *Hornet* in isolation garments that made them look like alien creatures, the three astronauts were put in a small mobile quarantine facility aboard the ship. Armstrong, Aldrin, and Collins would be treated like heroes, but only after they had been treated like Typhoid Marys because some scientists were afraid they would return with deadly organisms and touch off a plague.† "This is the greatest week in the history of the world since the Creation," Nixon told the astronauts from the other side of the glass. It could have been the worst if Aldrin and Armstrong had indeed picked up some monstrous microbes, since they and Collins walked across the carrier deck, packed with sailors, in the less than perfectly sealed protective clothes. They would spend three weeks in

* It also occurred to him that he and Armstrong would inevitably breathe some of the same dusty soil after they removed their helmets inside *Eagle*. "If strange microbes *were* in this soil," he reflected, "Neil and I would be the first guinea pigs to test their effects." (Aldrin, *Men From Earth*, p. 242.)

† The possibility of deadly organic life on the Moon was taken very seriously before the Apollo landings because of sheer ignorance. Most scientists doubted that a catastrophe would happen, but many felt that it was wise to be conservative. A Space Science Board report in 1962 had warned that "the introduction into the Earth's biosphere of destructive alien organisms could be a disaster." So had a Harvard medical student named Michael Crichton, whose first novel, *The Andromeda Strain*, made the best-seller list only months before *Apollo 11* went to the Moon. The book described how a virus from space nearly wiped out Earth. (Duff, "The Great Lunar Quarantine," p. 39.) The notion that viruses and bacteria responsible for plagues came from space was given great credence by Sir Fred Hoyle, the eminent British astrophysicist, in a book that was published a decade later. (Hoyle and Wickramasinghe, *Diseases from Space*.)

isolation and then go on an unprecedented public relations tour. Nixon himself, who went from the *Hornet* to Jakarta, announced that he intended to send pieces of the Moon to the world's leaders after scientists had finished analyzing them.

About 750 pounds of the Moon—rocks and soil samples taken from six sites—would be brought back through December 1972, when *Apollo 17* made the last foray. They would prove to be valuable for understanding the origin of the Moon, and also of Mars and other bodies in the solar system. Several years of study would show that the Moon had formed when the Earth did—roughly 4.5 billion years ago—and that it had been geologically active until about 2 billion years ago. And most surprisingly, the samples would eventually convince scientists that all three hypotheses of how the Moon had come to be were wrong. In 1984, a consensus developed that the "giant impact theory," first proposed in a 1946 paper and then roundly ignored, was most likely right: that a collision between Earth and a large "projectile" (now believed to be a planet-sized body) had blasted debris into orbit that coalesced and became Earth's sole natural satellite.

Meanwhile, Pan American World Airways announced that it had received more than two hundred applications from people who wanted reservations for the first commercial flight to the Moon, scheduled for the year 2000 (eventually there would be ninety thousand, one of whom was Ronald Reagan). The airline whose logo adorned the spaceliner that ferried Dr. Heywood Floyd on the first leg of his secret mission in *2001* had run a clever commercial months earlier. "Who ever heard of an airline with a waiting list for the Moon?" the voice-over asked. "We like to think of ourselves as pioneers," a glib Pan Am spokesman explained while recalling that the airline had been first across the Pacific and elsewhere. He did admit, however, that the high cost of rocket fuel, passenger comfort during blastoff, and a scarcity of accommodations on the Moon were serious obstacles. It was a PR gimmick, of course, but the response showed real enthusiasm for space travel. Pan Am eventually went out of business.

"By the year 2000," an expansively optimistic Wernher von Braun predicted, "we will undoubtedly have a sizable operation on the Moon, we will have achieved a manned Mars landing and it's entirely possible we will have flown with men to the outer planets."

Barbara Marion Hopkins Day was not on Pan Am's Moon manifest, definitely did not care to be on it, and cared not at all about operations on the lunar surface, sizable or otherwise. It was not as if the eighty-one-year-old lady, who had run a boardinghouse in McGehee, Arkansas, since 1926, was against technology. She had ridden on the first railroad train to push through northern Louisiana before the turn of the century, touched the first automobile in North Little Rock in 1910, listened to the first radio in McGehee, and had seen her woodstove replaced by one that used natural gas. She even fondly remembered the day Lindbergh had landed the first mail plane to come to the region.

But the Moon was something else. Sitting on a swing on her porch as Armstrong, Aldrin, and Collins had borne down on Earth at blistering speed, the lady told a visitor that she did not believe for a moment that anyone had really walked on the Moon, that she had slept through the alleged Moon walk, and that she had not turned on her television once during the entire voyage. "I don't believe it. I don't believe they've ever been there," said Mrs. Day. "If God had intended for us to go to the Moon," she added firmly, "He would have built a ladder up there." So she emphatically had no intention of setting out herself. What she did want was her old stove back. "I'd give anything to have a good woodstove right now," said Barbara Day. "They make the food taste better."

A Last Dig by the Russians

More than a year later, the Soviets decided to make a virtue out of necessity. On September 12, 1970, following a year's careful preparation, they sent Luna 16 to the Sea of Fertility to collect soil samples and bring them home. The spacecraft became the first robot to return to Earth with samples from another world when it landed twelve days later. It was a significant achievement. But it was marred by commentators in the press who explained that using automated spacecraft for such missions was preferable to using men because it did not put lives at risk.

A Bittersweet Legacy

Looking back on Apollo twenty-five years after men first set foot on the Moon, Tom Wolfe would call it an "astonishing accomplishment. It was staggering that we happened to be alive at the moment."

John Logsdon, director of the Space Policy Institute at George Washington University, would not quibble with that. But he would assert—also a quarter century after the fact—that the political nature of the decision to go to the Moon made it an "excellent example of how not to develop a sustainable program of lunar exploration and development." The exploration of space, Logsdon would reason, should not be held hostage to the vagaries of politics. Within two years of Kennedy's ordering Americans to the Moon in a near-desperate effort to trump the Soviet Union, he faced Khrushchev down in the Cuban missile crisis and signed the Limited Test Ban Treaty, both of which improved the position of the United States and proportionately undermined Apollo, at least as far as the president was concerned. "When the political base of support for the program disappeared, there was insufficient scientific, or any other, rationale for continuing beyond the initial few sorties to the moon," Logsdon would write. It was true. The greatest feat of human exploration to that time had been undertaken for exactly the wrong reason. Yet it remained the greatest human adventure; the *Odyssey* of the millennium.

12

Destination Mars

However the two superpowers used the Moon to score political points at home and abroad, it had really always been a way station for the best of the rocketeers and the scientists who wanted to hitch rides on their lifters: a place to fill their programs' tanks on the way to the planets.

Where the public was concerned—even the public that did not much care about sending men to the Moon or that opposed it for squandering precious resources on unmitigated hubris—Apollo was a technologically formidable accomplishment.

The inner circle, on the other hand, shared a secret that none dared talk about openly while Apollo's momentum (and funding) was increasing. The Moon appeared to be stone dead from both a biological and a geological standpoint. More important, in the words of Philip Morrison, it had been such an obvious target from the start that something like the Apollo program was anticipated and accepted long before it existed. Sending people there had been taken for granted since Galileo. It took nothing away from Apollo to say that, where the fraternity was concerned, the Moon was too close and it was therefore too easy. The essence of exploration, the noblest of human qualities after love, is discovery. Whether it is done within the human body or beyond Earth, the unending wonder of it all, at least for Earth-bound explorers, is what Merton Davies has described as finding answers for which there are not yet questions. That meant voyages of truly Homeric proportion. By the time of Apollo, the dreamers understood that Tsiolkovsky's and the others' gift—the chemical rocket and its eventual successors—provided them with the means to extend the *Odyssey* to infinity. That was their real target.

The universe beyond Earth and the satellite that faithfully attended it throbbed with mystery, vitality, and energy of stunning proportion. Galileo's descendants peered into their mechanized optick sticks and marveled at a galactic circus of colossal turbulence on Jupiter, shimmering rings around Saturn, the blotched deserts and icy poles of Mars, fiery comets like the majestic Halley cleaving across the night sky, retinues of stony worlds circling the major planets like fawning servants and strung out as a belt of asteroids, and the huge nuclear furnace at the center of the solar system itself. And well beyond the edge of the Sun's influence, other stars were plainly dying and being born in massive convulsions of pure energy. The universe was as dynamic as it was endless. It would take a speeding spacecraft four and a half years at light speed to reach Alpha Centauri, the nearest star after the Sun, but that was cause for some to want to get under way, not give up.

The *Encyclopaedia Britannica,* trusted tutor to generations of schoolchildren, maintained that there was probably life under the fog that enveloped Venus. Reports on Venus by a generation of youngsters would have faithfully reflected that fact. And if there was probably life of some kind on Venus, there was an even better chance that it existed on Mars. The classic *Physics of the Planet Mars* by Gérard de Vaucouleurs, published in 1954, interpreted the bluish or greenish blotches seen on the Red Planet as being some sort of vegetation and stated with no apparent second thought that "the polar caps are without the slightest doubt layers of crystallized water—probably more like white frost than solid ice or snow—which condenses during the cold season." The respected book also claimed that there were three different kinds of clouds on Mars, one of which was attributed to dust, and the other two to some kind of condensation like water or carbon dioxide. It—and much more—was all plausible. But nobody knew for sure. The RAND Corporation's own *Space Handbook,* which came out in 1958, cited Soviet sources as claiming that the dark areas "are really Martian vegetable life." That much seemed perfectly obvious.

Knowledge for its own sake, certainly including the possibility of finding life beyond Earth, helped to drive scientists to want to explore the solar system and beyond. But even that was defined and tempered by politics and competition on every level. It was the pragmatists—the managers, technocrats, and engineers in corporations and in the space agency and the politicians supporting them—who provided the mechanism that carried exploration. It was they who would weave the carpet upon which the Aladdins would fly. The scientists whose instruments rode on the rockets would lend nobility to an endeavor that was basically yet another theater of the cold war. Exploration was always done for the wrong reason. But it was done.

The single most important factor in the politics of space, around Earth and beyond it, was absolutely implicit, though it was almost never mentioned explicitly. It was the imperative to keep extending the missions. Whether that meant ever-longer duration flights in Earth-orbiting stations or sending probes and then people across the solar system, flights needed to be elaborated and lengthened or the technology and then the spirit behind them would wither and disappear, leaving the kind of spiritual vacuum that Margaret Mead deplored when she mentioned stay-at-homes tending their cabbage patches. Like the sharks that had prowled under the keels of Columbus, Magellan, and Henry the Navigator, the spacemen had to in effect keep moving or die. The politics of spaceflight, no less than its rockets, depended on constant forward momentum. That is why there had never been any real doubt that Mercury and Vostok would inevitably lead to longer and more complex manned missions, nor that the archers who sent their arrows to reconnoiter the Moon would want to propel others well beyond it. How many times could individuals in Mercury capsules orbit Earth or land on the Moon before the collective attention span of those who supported them drifted off?

Space Science Comes of Age

Space science spawned out of astronomy's womb in the 1960s and into the 1970s; its servant was the rocket-propelled, instrument-bearing space probe. A succession of Mariners and other planetary explorers would carry laboratories that had once been stuck on Earth to distant worlds for highly detailed observations that had been impossible only a few years before. The rocket was the engineers' gift to Galileo's heirs, and they came to use it in large numbers. By the end of the 1960s, the space explorers would coalesce into their own science and engineering culture, complete with "teams" that specialized in imaging, plasma waves, infrared and ultraviolet spectroscopy, magnetic fields, cosmic rays, charged particles, radio astronomy, and more. "Space science" itself was a misnomer, in fact, since its practitioners were really regular scientists who applied what they knew about the physical world on Earth to other places. Caltech's Bruce Murray, for example, was an eclectic geologist who liked to work in volcanoes and who therefore knew one when he saw it, even on Mars. For Murray and many others, getting there was a lot more challenging than taking a plane to Hawaii to study Mauna Loa or to Sicily to work at Mount Etna.

The culture would take root throughout the United States and the Soviet Union—where the rockets were—and, by invitation, elsewhere, attracting men and women who were drawn by the excitement of effectively being able to get off Earth and use their knowledge to investigate other worlds up close for the first time. And like other cultures, space science soon took hold at particular institutions and developed its own star system.

In the United States, the preeminent NASA centers were JPL and the Ames Research Center, just south of San Francisco, followed by Langley in Virginia's Tidewater region. They and everyone else in science and exploration originally reported to the Office of Space Flight Development at headquarters, which was directed by Abe Silverstein, and which turned into the Office of Space Sciences under Homer Newell in 1961 and finally into the Office of

Space Science and Applications in 1963. Space Science and Applications would revert back to Space Science in 1971 as the various near-Earth and solar system exploration programs multiplied. Whatever they were called, the successive head offices spawned subdirectorates for areas like geophysics and astronomy, lunar and planetary programs, bioscience, and launch vehicle and propulsion programs. Beneath them, on the charts, there were program chiefs to manage successive Mariner, Pioneer, and other exploration projects.

The budget for space science and applications alone climbed steadily from a little more than $87 million in 1959 to $759 million in 1966 and then rolled downward like a roller coaster, along with the budget for Apollo. The average annual budget during the 1960s was $385 million and would increase during the 1970s to an average of $550 million, a figure that would also reflect several new programs (and which, because of inflation and rising costs, was not as big a jump as it appeared). Significantly, the space agency allocated roughly 17 percent of its overall budget to space science during its first twenty years. This, of course, fell far short of what the scientists wanted. But they learned early on to play the game for fear of losing even that much of the pie.

Since NASA had monopoly control of both civilian launchers and spacecraft, the vast majority of scientists saw no alternative to playing the game and, as the Columbia astrophysicist David Helfand said, not raising the space agency's ire by venting their real attitude: that the manned program was a voracious circus act. The scientists' getting a more or less fixed percentage of the budget had an obvious Faustian implication: when the whole budget increased, so did their cut. Obviously, and all other things being equal, a shrunken budget (and a vindictive NASA) would only hurt their work and careers.

The scientists themselves were encouraged to respond to NASA's announcements of opportunity and requests for proposals—commonly known as AOs and RFPs—with ideas for experiments to put on the solar system explorers. Proposals would come from throughout the United States and from some European universities and be judged by leading outside scientists, most in academe. As solar system exploration picked up during the Apollo years and in that program's wake, as the funding flowed freely, some universities developed particularly strong space science departments. Arizona (which probably had the best astronomy department in the country), Colorado, and Maryland were notable and, to a lesser extent, so were Brown, Cornell, and Michigan.

Within their walls, as well as at JPL, Ames, and elsewhere, a galaxy of scientific and administrative luminaries emerged that would leave a mark on solar system exploration for decades. There were Bradford Smith, Carolyn Porco, and Thomas Gehrels at Arizona; Tobias Owen at the State University of New York; Carl Sagan and Thomas Gold at Cornell; Lawrence Soderblom, Harold Masursky, and Eugene Shoemaker at the U.S. Geological Survey;

Thomas Donahue at Michigan; James Warwick and Lawrence Esposito at Colorado; Louis Lanzerotti at Bell Labs; Bruce Murray and Edward Stone at Caltech; Mert Davies at RAND; and David Morrison and Charles Hall at Ames, among scores of others. And they had counterparts in the Soviet Union, including Roald Sagdeev at IKI, Aleksandr Basilevsky at the Vernadsky Institute, Natasha Bovina, Kirill Florensky, Aleksandr Gurshtein, Aleksandr Vinogradov, and others.

However their respective governments and space organizations competed for prestige, the scientists themselves kept up a thriving discourse and enjoyed real camaraderie throughout the cold war. They were hardly oblivious to the political landscapes on which they worked and they certainly understood that much of their funding came from the fact that competition existed between their societies. But that was okay. Like von Braun, Korolyov, and the space world's other heavyweights, they were delighted to reap the benefits of the competition rather than see their programs weakened or killed.

And there was this: throughout the cold war there was a mutual respect among individuals who carried a passion that transgressed frontiers and ideologies. The members of the international space science community always understood that they had far more in common with one another than they had with those in their own manned programs, let alone with soldiers and politicians. While politicians from Khrushchev to Reagan hurled barbs and rattled rockets, the physicists and chemists, exobiologists and geologists, astronomers and engineers attended international meetings together, studied at one anothers' institutes and universities, worked on joint missions, exchanged papers and journals, and freely swapped everything from research data to bottles of bourbon and vodka. Murray, the geologist, was friends with Sasha Basilevsky, for example, and with a noted geophysicist named Vitali Adushkin.

So Bruce Murray savored pickled mushrooms from Kazakhstan in Adushkin's kitchen on Leninsky Prospekt and would participate in an experiment on one of the unlucky Phobos spacecraft that went to Mars in 1988. And Basilevsky, a star geologist, had a score of friends from Brown to JPL. One of them, Ellen Stofan, was a future JPL project scientist who studied with him in Moscow and made the pilgrimage to Tsiolkovsky's home in Kaluga. It was Sasha Basilevsky who used the Lunar Orbiter imagery to pick landing sites for the unmanned Lunokhod rovers that successfully rolled over the lunar surface taking pictures and collecting soil samples in 1970 and 1973.* He was also on the Voyager imaging team.

* Lunokhod was conceived by Korolyov's design bureau to carry cosmonauts much the way the Lunar Rovers carried astronauts on later Apollo missions. With the manned program scraped, they were turned into robots. (Basilevsky interview and Rhea, *Roads to Space,* pp. 123–25.)

There was never satisfaction when the other side's lunar or planetary explorer disappeared, went silent, or ended an encounter as a pile of junk in some crater without sending back so much as an exclamation point. Everyone in the extended family understood how punishing "single point" failure was. And everybody, or almost everybody, also understood that they had to live with their manned programs whether they liked it or not.

This did not mean, however, that they were above a little gloating on occasion. "I had this image of a bald-headed Russian general sitting in their tracking center just beside himself that we had succeeded," Murray thought when a Proton carrying a Soviet Mars probe blew up just after its American counterpart succeeded. "I'm not proud of that feeling now," he said years later, "but that's exactly how I felt at the time."

Shuttling to Space with Spiro Agnew

And where the scientists were concerned, living with the manned programs brought a galling irony. While there were endless things to discover in space science and planetary exploration—truly worthwhile missions of all sorts to fill thousands of lifetimes—the science and exploration budgets were relatively small compared to the manned programs. Those who ran the competing space establishments treated the manned programs like the dumb and disinterested scions of a rich family: they kept stuffing the kids' pockets with money while trying to come up with meaningful things for them to do. Meanwhile, the whiz kids were forever begging for enough allowance to buy new erector and chemistry sets.

The question of what course to take after Apollo was a case in point. With the flight of *Apollo 8* around the Moon at Christmas 1968, the Moon landing to follow became a virtually accomplished fact. And that fact was haunted by this question: What could the manned program do as a follow-up to the greatest technological stunt in history? Neither President Nixon nor Vice President Agnew had any particular interest in the space program except perhaps as it fed a large industry, somehow contributed to national security, and could be milked for political advantage at home and abroad. In other words, space was making the transition from luxury to necessity, and it therefore could not be allowed to languish. So new goals had to be found.

Accordingly, on February 13, 1969, Nixon ordered that his own administration's Space Task Group be formed to chart the nation's course after the lunar landings. It was chaired by Agnew, which indicated that the White House took space seriously, but its work was done by Secretary of the Air Force Robert C. Seamans Jr., NASA Administrator Thomas O. Paine, Presidential Science Adviser Lee A. DuBridge, and their respective staffs, plus observers who included Undersecretary of State for Political Affairs U. Alexis Johnson and Atomic Energy Commission Chairman Glenn T. Seaborg.

The STG's report, *The Post-Apollo Space Program: Directions for the Future,* was issued that September. Whatever it said about new initiatives, its heart was buried on page five: "The manned flight program permits vicarious participation by the man-in-the-street in exciting, challenging, and *dangerous* activity. Sustained high interest, judged in the light of current experience, however, is related to availability of new tasks and new mission activity—new challenges for man in space" (italics added). This was code. It meant that human beings had to be kept in space performing heroic acts or else public support would wither and evaporate. That idea in turn contained three assumptions. The first was that only people who faced physical danger could be heroic, which automatically eliminated every scientist and engineer who stayed on the ground to invent ways to explore other worlds by remote control, no matter how ingeniously. The second was that dangerous activities were not only good, but were necessary, for a thriving space program. And finally, there was the assumption that the activities had to keep getting more dangerous, or at any rate, more spectacular. At the same time, the report continued, Americans were becoming cranky about the cost of the space bill so missions would have to be picked prudently. That part was certainly true enough.

What the group came up with was a contradiction in terms; an impossible dream. It called for a program that was supposed to keep Americans in space and even get them to Mars within fifteen years, while continuing unmanned exploration, enhancing national security, and contributing to the quality of life on Earth, all of them done cheaply. And the cheapest way to keep Americans in space, at least in the view of the group, was by designing "low-cost, flexible, long-lived, highly reliable, operational space systems with a high degree of commonality and reusability." What the Space Task Group had in mind was the creation of spacecraft that would shuttle astronauts to and from orbit, followed by the construction of a permanently manned station, and then by a manned expedition to Mars. This, of course, validated von Braun's dream. Sending the president the *Collier's* series, in fact, would have saved a great deal of time and effort. The shuttle, or Space Transportation System, was beginning to make the transition from a theory to a program.

A month after the STG turned in its report, von Braun himself weighed in with an article for *Reader's Digest* that also mentioned the shuttle. But it did more than that. It described in extravagant terms how the vehicle would "drastically slash" the cost of getting people and equipment to orbit. "Present cost of orbiting one pound of payload is $500 or more. Completely reusable orbital space shuttle vehicles will slash this cost to about $50 or lower per pound," he predicted. And because the reusable orbiter would obviously be cheaper than expendables, he continued, it would eventually replace the expendables, even to launch unmanned spacecraft into orbit.

Meanwhile, George Mueller, the space agency's associate administrator for manned space flight, was calling for a shuttle that could carry as much as 50,000 pounds to orbit for five dollars a pound or less. NASA paid $600,000 for a study by a Princeton think tank called Mathematica, Inc., that showed that a fully reusable shuttle would save only a marginal $100 million during its write-off period from 1978 to 1990 at an investment of nearly $13 billion. That, in NASA's estimation, would not do. So Mathematica was instructed to recalculate the savings based on an almost mind-boggling 714 flights over that twelve-year period, or a little more than a flight a week with each flight carrying a 65,000-pound payload. The numbers were being bent with desperate abandon.

The point was to keep the space industry working while giving the Nixon White House its own benchmark program. Meanwhile, the president carefully distanced himself from Apollo; any calamity in that program would be Kennedy's, not his.

Appended to the Space Task Group's report was a separate one by NASA. Besides endorsing the shuttle (billed as a "low-cost earth-to-space and earth-return transportation system which will be designed to operate with minimum ground support"), it came out strongly for solar system exploration, which was already under way, and specifically for visits to all of the known planets; a search for life elsewhere; and an understanding of Earth based on "comparative studies with the other planets." So, as usual, there was something for the scientists as well.

A Laboratory in the Sky

NASA's grandiose plan, along with those of its contractors and scientists, had run headlong into a severely reduced budget after most of the Apollo hardware had been put into place by 1966. While the military space budget, and particularly intelligence collection, climbed steadily during the 1960s and into the 1970s, funds for civilian programs steadily declined. With Vietnam and other military commitments around the world sapping enormous energy and resources, and with turmoil at home over both the war and civil rights growing to ugly levels that included city-battering riots from Harlem to Watts, civilian space was increasingly seen as being mostly irrelevant to the nation's basic needs.

Accordingly, NASA's budget went into a long nosedive beginning in fiscal year 1966, when it slipped from $5.25 billion to $5.17 billion, and then progressively down to $3.3 billion for fiscal year 1972. The fiscal year 1974 request, which was fixed in 1972, came to just 37 percent of the 1966 budget with inflation during the intervening years factored in. It would drop to 30 percent by the end of the decade.

Meanwhile the Apollo program, mimicking re-entry's long arc, continued onward and downward. There were five more landings through *Apollo 17,* which came down at the foot of the Taurus Mountains on December 11, 1972. One of its crew members, Harrison Schmitt, had a doctorate in geology and was therefore the only scientist to go on an Apollo mission.

However well orchestrated they were made to appear, the Apollo landings on the Moon were incredibly brave and productive adventures. *Apollo 12*'s crew touched down only two hundred yards from Surveyor 3, removed its camera, set up an elaborate Apollo Lunar Surface Experiments Package, and collected seventy-five pounds of rock and soil samples.* Their successors planted other experiment packages at different locations, including laser reflectors that would help astronomers on Earth to fix the Moon's precise position, photographed a number of landing sites, returned more samples, made three forays in Lunar Roving Vehicles (Cernan and Harrison Schmitt drove a record twenty-one miles and had the dubious distinction of causing the first fender-bender on another world), and did additional surveys. In the process, they beat down the remote possibility that there was life in some form on the Moon, an idea that was taken seriously by some at the time, and found no real evidence that any kind of dynamic process, perhaps caused by a molten core that produced heat, was taking place (though it had 2 billion years earlier).†

The six successful round-trips to the lunar surface left the space agency caught in a conundrum that would haunt it for years. The wonder of *Apollo 11,* including its unprecedented complication and aura of imminent danger, gave way to a seeming mastery of the technique of getting to the Moon that did indeed vaguely suggest airline service. Airlines were certainly useful—particu-

* The so-called ALSEPs were made by the Bendix Aerospace Systems Division and consisted of several very sensitive seismometers that would record vibrations caused by moonquakes (indicating that the Moon was geologically alive) and meteoroid impacts, a gauge that would measure the atmosphere for any gas molecules coming from volcanoes (another indicator that something was going on under the lunar surface), a magnetometer that would measure the strength of any magnetic field that existed, a solar wind spectrometer that measured the energy and direction of charged particles coming from the Sun and carried on the solar wind, a dust detector, and an ion detector to measure charged particles formed from the lunar modules' exhaust gases. ALSEPs were powered by foot-and-a-half-long nuclear electric power plants developed by the Atomic Energy Commission. The recorded data were transmitted directly to Houston. (Lewis, *The Voyages of Apollo,* pp. 92–93.)

† See, for example: V. A. Firsoff, *The Strange World of the Moon* (New York: Basic Books, 1959). Firsoff was a Fellow of the Royal Astronomical Society who raised the possibility that there could be some life-form on the Moon.

larly for the millions of ordinary folk who used them—but their very safety and predictability made people take them for granted.

Ironically, the flight of *Apollo 13,* in which James A. Lovell, John L. Swigart Jr., and Fred W. Haise Jr. were nearly killed after an oxygen tank exploded on the way to the Moon grabbed public interest precisely because it did not go "nominally." The explosion ripped open part of the service module, creating an electrical emergency and dumping all of the oxygen, leading to a harrowing, near catastrophic, return to Earth. In an account of the flight published twenty-four years after it happened, Lovell likened Apollo spacecraft to the Edsel, the Ford Motor Company's notoriously failed sedan. "Actually, among the astronauts it was thought of as worse than an Edsel. An Edsel is a clunker, but an essentially harmless clunker. Apollo was downright dangerous," said Lovell. In fact, each successive, immensely complicated mission defied its own perilous odds.

In a perverse kind of way, the *Apollo 13* crew's brush with death vividly brought home the Agnew committee's point about mass fascination with danger, and it made good copy: pictures in *Life* showed an agonized Mary Haise and Marilyn Lovell keeping "an exhausting vigil" as they waited for word of their husbands' fate. Rumors that the astronauts carried poison pills so they could end their lives quickly if it looked as though they were going to die a long, horrible death marooned in space were absolutely wrong, according to Lovell. "As the atmosphere inside rushed out and the vacuum outside rushed in, whatever air was left in your lungs would explode out in an angry rush, your blood would instantly—and literally—boil, your brain and body tissues would scream for oxygen, and your traumatized system would simply shut up shop. The whole thing would be over in just a few seconds."

With the euphoria of *Apollo 11* a memory and the follow-ons being taken as expensive distractions, flights 18, 19, and 20 were quietly scratched. Instead, the manned program was refocused in two distinct directions. One used some of the remaining Apollo modules and Saturn boosters on strictly uncompetitive projects designed to show that the winner of the great race could be magnanimous. It in turn divided into two separate projects that were actually the first space stations: *Skylab,* which was designed to be used for scores of biological and physical experiments, and the Apollo-Soyuz Test Project, which was supposed to be a cooperative joint effort by astronauts and cosmonauts to link up in orbit. Both, like Apollo itself, had been planned for years and would be closed-ended. Other hardware was in effect thrown away. Two fully usable Saturn 5's, for example, were laid on their sides for display at Canaveral and Houston. The other major direction was supposed to lead to the shuttle, then to a permanent station, and then to the manned mission to Mars. It would therefore be very open-ended indeed.

The idea of using Apollo hardware to build a space station–laboratory went back to 1963, when engineers at the Manned Spacecraft Center looked ahead to the end of Apollo and suggested using leftover parts to fly a station that would hold eighteen people. NASA established an Apollo/Saturn Applications Office two years later to modify the modules and boosters so they could be used as orbiting laboratories as part of an Apollo Applications Program. At one point the Military Operations Subcommittee of the House Committee on Government Operations recommended combining the applications program with the Air Force's Manned Orbiting Laboratory, but that evaporated when MOL was killed.

Skylab was basically a converted Saturn 4B booster that was heaved into orbit by the first two Saturn 5 stages. Once in orbit, it was joined by a crew in one of the Apollo command and service modules. Two solar arrays that made it look like a windmill lying on its side were attached. Its purpose, as explained by the space agency, was reminiscent of MOL and suitably impressive: to determine man's ability to live and work in space for extended periods; to evaluate his physiological responses both during flight and afterward; and to conduct science and solar astronomy from above Earth's atmosphere and to improve Earth resources monitoring techniques. The huge spaceship was sent up from the Kennedy Space Center's historic Pad 39A on May 14, 1973. At forty-nine feet long and twenty-two across, the *Orbital Workshop,* as the Saturn 4B was called, was like a barn compared to the snug Apollo capsules. Being a true station in space, *Skylab* became the first American structure in orbit to be visited by temporary crews that were sent up in the command modules and came down the same way.

Having been beaten to the Moon, the Russians decided to make the best out of the loss and lead the way in space stations, which they insisted, with considerable justification, were far more useful than having people hop around on the lunar surface like rabbits.

On April 19, 1971—one week after the tenth anniversary of Gagarin's flight and almost two years before *Skylab* went up—the world's first dedicated station, *Salyut 1* ("Salute" to Gagarin) was rocketed to orbit. After a crew in *Soyuz 10* failed to get into the thing on April 23 because their spacecraft had a new and balky docking system, three other cosmonauts were sent up on June 6 and did manage to dock and board the station. It was tended by Georgi Dobrovolsky, Vladislav Volkov, and Viktor Patsayev for twenty-three days in June, during which they did biomedical and other experiments and set a record for time in orbit.

Work on board the station went well enough. But then disaster struck on June 29 when a pressure release valve in *Soyuz 11* opened prematurely as the three made their way back down. With the cabin quickly losing its pressure and turning into a dreaded vacuum, Patsayev apparently tried frantically to close the valve, which was right over his head, but the straps that held him to his seat prevented him from reaching it. All three cosmonauts died almost instantly. When the rescue crew opened *Soyuz 11,* the three looked as if they were asleep. Patsayev's forehead bore an ugly black-and-blue mark. The most worrisome immediate possibility, which would have been quickly dispelled, was that three weeks in space had somehow killed them; that weightlessness for such a long period had so weakened them that they had simply expired or had not been able take the stresses of re-entry. The fact that they had died for lack of atmospheric pressure caused by an accident would have been almost a relief compared to that other possibility.

The reaction of the Soviet public, which was told of the deaths fairly quickly, was compared by foreigners in Moscow at the time as being similar to the grief in the United States when Kennedy was assassinated. Typically, a scapegoat was needed, so Nikolai Kamanin was removed as chief of cosmonaut training and replaced by Vladimir Shatalov, an experienced cosmonaut. Since détente had by then come to both Earth and the civilian space programs, Tom Stafford was one of the pallbearers.

Skylab's first crew, Pete Conrad, Joseph P. Kerwin, and Paul J. Weitz, went up and rendezvoused with the "windmill" on May 25, 1973. Twenty-eight days, 404 orbits, and 11.5 million miles later, they splashed down in the Pacific and were picked up by the USS *Ticonderoga.* The second crew—amateur artist Alan L. Bean, Owen K. Garriott, and Jack R. Lousma—went up on July 28 and returned fifty-nine days later after 858 trips around the world. The last crew, Gerald P. Carr, Edward G. "Hoot" Gibson (after one of Hollywood's original cowboys), and William R. Pogue were sent up on November 16 and splashed down on February 8, 1974, after a record-breaking eighty-four days and 1,214 orbits. America's first space station would drop out of orbit on July 11, 1979, with a spectacular flourish, trailing a fiery plume that made it look like a shooting star. With seventy-nine successful experiments, from metabolic activity to bone mineral and body mass measurement to ultraviolet stellar astronomy to using a multispectral scanner pointed at Earth, *Skylab* was an unmitigated success.

Comrades in Arms

The Apollo-Soyuz Test Project, on the other hand, would be decisively mitigated.

"Achievement of the Apollo goal resulted in a new feeling of 'oneness' among men everywhere," NASA Administrator Tom Paine had read in the Agnew report. "Now with the success of Apollo, of the Mariner 6 and 7 Mars flybys, of communication and meteorological applications, the U.S. is at the peak of its prestige and accomplishments in space. For the short term, the race with the Soviets has been won. In reaching our present position, one of the great strengths of the U.S. space program has been its open nature." While the reference to "oneness" reflected more wishful thinking than reality, it seemed to Paine to provide a philosophical basis for bridging at least some fraction of the cold war animus. Détente was in the air. More to the point, any spectacle that could lend nobility to a nation that was being bloodied and bowed in Vietnam and embarrassed by anti-war demonstrators at home, had to be helpful.

The idea of some sort of cooperation in space had its seeds in the International Geophysical Year. With typical swagger and contempt for nature, Khrushchev had suggested that the Soviet Union and the United States pool their resources to "master the universe" when he sent congratulations after Glenn's flight in *Friendship 7* in February 1962. The subject had come up again in specific exchanges between Kennedy and Khrushchev the following month. Answering a letter from Kennedy that listed a number of cooperative measures, including ones involving a world weather satellite system, the exchange of tracking services, and mutual mapping of Earth's magnetic field, Khrushchev had added working together on both the expanding area of space law and rescuing spacecraft (he meant those that crash-landed, not orbital derring-do). This had been followed by long discussions on various levels, all of them tentative, and even by a U.N. address in which Kennedy had proposed a joint expedition to the Moon on the theory that even if the Russians accepted the offer, it would indicate that they acknowledged that their rival was ahead. The Kremlin had responded by not responding.

Having seen his country play catch-up with the U.S.S.R. for a decade, from Sputnik-Explorer to Vostok-Mercury to Gagarin-Glenn, Paine was overcome with largesse after his country's flag had been dramatically and incontrovertibly pounded into the Sea of Tranquility. Expansive gestures being the prerogative of winners, the NASA administrator began mulling over the possibility of cooperation in some form with the Soviets in space. Nixon, who had a stake in détente, reacted by convening an interagency committee in the autumn of 1969 to consider the idea. With the exception of the Pentagonians, who abhorred the idea of Russians poking around in an American spacecraft, everyone favored it. Paine was encouraged by Nixon to pursue the matter seriously, and he did so by mail with Mstislav Keldysh, the head of the U.S.S.R. Academy of Sciences.

Philip Handler, the new president of the National Academy of Sciences, was on the same path. One of Handler's goals as head of the nation's preeminent science body was to encourage closer cooperation between its members and their counterparts in the Soviet Academy, and that certainly included space. If the venerable scientist and academician needed further convincing, however, it came in a movie called *Marooned* that was based on a 1964 potboiler by the prolific aviation and space writer Martin Caidin. During a special screening of the film, Handler watched, transfixed, as two American astronauts who were stuck in orbit and facing almost certain death were rescued by a colleague who helped his stricken counterparts by pulling close enough to their spacecraft so he could deliver oxygen during a dangerous space walk. The message was that outer space was a mighty dangerous and unforgiving place and those who risked their lives by venturing into it had better stick together no matter what insignias they wore. Up there, they were all just earthlings, sailing through the great icy, unforgiving void and facing a common danger as brethren. Space, Caidin averred, was unforgiving. It knew no nationality, no political philosophy, and no language difference. Men who dared to go to that unforgiving place had to protect each other because their humanity was unique and indivisible. Space reduced them to common creatures. Swim together or sink out of sight. Handler liked the message, if not the medium, though he was a forgiving sort.

The negotiations that followed the Paine-Keldysh correspondence in 1969 dragged on almost endlessly, as engineers and administrators in six working groups meeting in Moscow and Houston thrashed out problems that inevitably would come with a combined mission. Joining two spacecraft in orbit, to take only one obvious example, meant that they would have to have compatible docking systems. That turned into an engineering nightmare.

George Low led an American delegation to Moscow in January 1971, but only after being warned by Henry Kissinger, Nixon's national security adviser, to stick to space matters and not so much as hint (as his naive astronauts were always tempted to do) that technical agreements could also solve political problems. And once in Moscow, Low and Keldysh had a brief interchange that again underscored not the spirit of cooperation, but to the contrary, the need for competition. Both of them readily agreed from the start that it was imperative for their countries to compete as well as cooperate because it was competition that spurred on their respective programs. That, again, was the inner circle's most closely guarded secret. Full-blown cooperation across the whole spectrum would almost undoubtedly have spelled disaster.

The January meeting, followed by another in November, led to an agreement on a test mission in 1975 that involved Apollo and Salyut (later Soyuz) spacecraft. Nixon and Premier Aleksei N. Kosygin signed a five-year "Agreement Concerning Cooperation in the Exploration and Use of Outer Space for Peaceful Purposes" on May 24, 1972. This was followed by three years of intensive planning in Houston and Moscow. While astronauts and cosmonauts were gradually brought together to practice joint exercises (and Russian and English, respectively), the working groups pounded out a unified system of mission operations, common life-support systems, flight control procedures, communication and joint tracking techniques, safety and crew operations, and a great deal more. In addition, two very different record-keeping procedures had to be integrated and the respective spacecraft themselves had to be modified for what came to be called the Apollo-Soyuz Test Project, or ASTP.

The thorniest engineering problem was unquestionably the design of a unique and indispensable piece of hardware: an international "androgynous" docking module, or thick collar, which had to make an airtight seal between two very different spacecraft so their occupants could get back and forth between them.

Tyuratam Versus Baikonur and other PR Problems

The thorniest public relations problem had to do with coordinating two distinctively different news media with very different traditions. NASA and the Soviet Academy had their own protracted negotiations for live television signals, which had to be compatible not only with their own countries' but also with those of the European Broadcast Union. Then there were the matters of accrediting newsmen and -women, creating press kits and press centers, and providing access to Mission Control at Kaliningrad and to the launch site by reporters and camera crews from the West. The ordinarily secretive Russians understood that the most important aspect of the Apollo-Soyuz Test Program was political and that unprecedented coverage inside their country was therefore necessary. So while the Soviet Academy reserved the right to bar any reporter from Tyuratam, it did agree to live coverage of liftoff by the Western correspondents (a point NASA Administrator James Fletcher made in the strongest way when *Aviation Week* accused his agency of caving in to the Academy's wanting to keep them out). There was no denying, however, that NASA's representatives were flown into and out of Tyuratam only at night and that its coordinates as supplied by the Soviets were invariably wrong. The ever-suspicious Russians understandably believed that U.S. intelligence would be sorely tempted to collect sensitive information at their open facilities. It was standard procedure on both sides at every opportunity.

Even the name of the launch site became a source of contention, though not between the Americans and the Soviets, because of Soviet security concerns. At one press session with the astronauts, ABC News' notoriously combative Jules Bergman exploded at Stafford, one of the three astronauts scheduled to fly the Apollo part of the ASTP, after Stafford described the huge facility at "Baikonur" as appearing to be much larger than the Kennedy Space Center.

What followed was vintage Bergman: "Baikonur, if you'll look on the coordinates, is a hundred and thirty-five miles away or something. Tyuratam may only be a railhead, but it is the Tyuratam Launch Complex. They call it Baikonur, I know. . . . I'm going to call it Tyuratam. ABC is going to call it Tyuratam. SAC calls it Tyuratam. Can we once and for all straighten that out and arrive at a . . . name for it, Tom?"*

Stafford tried to gloss over the discrepancy by calling Tyuratam "a little bitty old city," adding that the Russians themselves referred to the entire region as Baikonur. But Bergman and many of his colleagues would not buy it, any more than Dino Brugioni had in the late fifties, when he had christened the place Tyuratam after referring to an old German Army map. The reporters believed that the facility had deliberately been misnamed to conceal its real location. They were right.

The Cosmic Ballet

With ballistic missile early warning and a variety of military reconnaissance satellites and ground stations looking and listening to them, Valeri N. Kubasov and Aleksei Leonov blasted off in *Soyuz 19* on the afternoon of July 15, 1975. Kubasov was a brilliant flight engineer who had designed spacecraft for years. Leonov, who had made the first space walk from *Vostok 2,* was already a legend second only to the deified Gagarin. The stocky, muscular flier was a serious artist, an outstanding swimmer, fencer, volleyball player, cyclist, and yachtsman who had not only graduated from military flight school and was an accomplished fighter pilot who had once landed a burning jet rather than abandon it, but was an Air Force parachute instructor with a hundred jumps. (He was also a lucky man, since he had been on the *Soyuz 11* backup crew.) Now Kubasov and Leonov were setting a new record: theirs was the first launch from Tyuratam to be carried live on television. They were followed seven and a half hours later by Stafford, Slayton, and Vance D. Brand in an Apollo command and service module, which went up on a Saturn 1B from Pad 39B.

Two days later, Stafford inched *Apollo* to a perfect docking with *Soyuz.* With millions watching on television (including Anatoly Dobrynin, the Soviet ambassador to the United States, who followed the flight "in a state of nervous excitement," as he put it, in the Department of State conference hall) Stafford

* This was a thinly disguised reference to the fact that the Strategic Air Command's Strategic Library Bombing Index listed all targets in the Soviet Union, derived from satellite reconnaissance, with considerable precision and that it therefore was not fooled by "Baikonur."

The missing word in "and arrive at a . . . name for it," which appeared in an official NASA history of the ASTP, almost undoubtedly was a seven-letter expletive.

opened the hatch that led into the *Soyuz* orbital module. "Aleksei," Stafford said to Leonov, "our viewers are here. Come over here, please." With a hand-lettered sign in the background reading "Welcome Aboard Soyuz," the two commanders shook hands in the docking module—neutral territory—as they passed over the French city of Metz at five miles a second. Leonov gave Stafford and Slayton the traditional bear hug, and then, floating in the cabin like suspended ballet dancers, they exchanged flags and plaques. Leonov also presented the Americans with sketches he had made of them during training.

Two dockings, twenty-nine experiments, and thousands of photographs later, *Soyuz* and *Apollo* parted and returned to Earth, with the cosmonauts floating down to the steppe on July 21 and the astronauts landing in the Pacific three days later. It was the last time a manned American spacecraft would splash down.

It was also the end of Apollo. And it was the end of détente. With Cambodia and South Vietnam disintegrating, and with the Ford administration being accused of pandering to the Communists (Washington's allegedly "submissive" attitude toward Moscow in the Strategic Arms Limitation Talks being only one example), the lines of the cold war began to harden again. *Apollo-Soyuz* not only did little or nothing to stop the deterioration in relations, but to the extent that it represented cooperation between the two superpowers, it provided fuel for détente's implacable enemies in both camps.

Yet the arguments of cynics aside, the spirit of *Apollo-Soyuz* would outlive not only the end of détente but the cold war itself. Two decades later, other space travelers would join forces in the Russian *Mir* (Peace) station and the American shuttle. That meeting, like that of *Apollo-Soyuz,* would be highly publicized as an example of the possibilities of cooperation in space. But, also like *Apollo-Soyuz,* it would be only the most apparent example of a broader, more pervasive, cooperation. The other was far less obvious, yet more important, because it had to do with relatively large numbers of scientists and engineers and was therefore an inherent part of the fabric of both space programs. Political conservatives on both sides stayed well clear of the scientists because they did not understand them and, in any case, did not want to become tainted by what they saw as their misguided, if not alarmingly naive, ways. The scientists thought that insulation from the reactionaries was wonderful.

The Imperative to Explore

The NASA report that accompanied the Space Task Group's 1969 study was arranged according to the space agency's own priorities, with the manned program coming first, followed by the exploration of Earth and the solar system. As was the case with the manned program, solar system exploration was internationally competitive on a number of levels, most notably propagandistic. But both exploration programs were also internally competitive.

The process of selecting experiments to put on a spacecraft was like the eliminations in sports play-offs. Before scientists were allowed to represent their own countries in the race to scout distant worlds, they had to survive competition at home, some of it political, some scientific. In the United States, NASA would publish its announcements of opportunity for scientists who wanted to win precious space for their experiments on the probes, and the scientists would respond by submitting detailed proposals for space agency evaluation. Given the fact that a maximum of only ten or eleven instrument packages could go on a spacecraft, the competition among scores or hundreds of scientists was often intense.

The popular image of the individual, selfless scientist, a politically oblivious Louis Pasteur, Walter Reed, Marie Curie, or Ernest Rutherford laboring in the forlorn isolation of their laboratories and worrying only about the good of mankind and the pursuit of knowledge was a myth that had been shoved aside by reality in the second half of the twentieth century. The outstanding scientific inventions that had come out of the pressure cooker of World War II—nuclear weapons, radar, infrared photography, and the ballistic missile, to take only four examples—had been produced by groups of individuals working together. The loners had mostly given way to coordinated teams that did research in laboratories whose lifeblood was the competitive government grant. If von Braun and Korolyov and their respective rocket establishments were on the government teat, so were thousands of scientists who, like Van Allen, could go nowhere else for the amount of money they needed, let alone to private rocket launchers.

Individually, the men and women with doctorates in the hard sciences inhabited a world that was grossly underpaid by corporate standards, but which certainly was as competitive. Their bottom line was neither stock options nor megabuck raises. At its heart it had to do with trying to understand how the natural world worked through a painstaking process that required a solid grounding in the knowns, formulating good questions about the unknowns, establishing parameters that would show what was true and what was not, designing experiments that could produce answers, and then getting those experiments—measuring instruments—on board spacecraft. The scientific community kept and continues to keep an internal score no less than its counterpart on Wall Street. Reputation, promotion, and tenure rested squarely on the publication of significant discoveries, pulling in grants, and by being quoted correctly (but not too often) by the news media. Carl Sagan, a first-rate astronomer, for example, would become too visible by writing eclectic, popular books and starring in a dramatic television series called *Cosmos*. For a time he would therefore be shunned by fellow scientists whose jealousy was camouflaged by apparent distaste for his "popularizing" their work. But that would change.

In the Soviet Union, there was fierce competition between the design bureaus and the state companies that manufactured the spacecraft and their components, all of it cloaked in secrecy and deception. And there was competition at IKI, the Space Science Institute, with consequences that mirrored the Moon program's multiple afflictions; a competition whose destructiveness managed to overwhelm the brilliance of its participants. Sagdeev knew that there was a natural tendency for competing scientists to challenge each other in any institution in any country. But, he explained, the divisiveness could be overcome and positively directed by a few bright, strong leaders. Not so in the Soviet program. The physicist who would head IKI during the last flushes of the cold war would complain bitterly in his memoir that "barricades" separated the space industry (the design bureaus) from the space science community and, perhaps more perniciously, divided the community itself. "Deep divisions existed within the institute," he would note, "which split into a number of small strongholds—in the hands of different scientific clans."

Cooperation between the design bureaus and IKI, and between the institute's laboratories and divisions, was held to an absolute minimum. Astronomers, for example, were openly contemptuous of their colleagues who studied Earth's upper atmosphere or even the solar wind because those disciplines were more practical, which is to say, useful to a degree that was distasteful. "Despite the fact that early successes in space science were associated with this very discipline, the leader of IKI's space astronomers . . . was sure that local 'environmental' science would be nothing but a *'caliph'* for an hour." Translation: studying Earth's relatively mundane problems could not compare to delving into the grandeur that came with astronomy's infinitely more important concern with time and space: with what the science writer Timothy Ferris would later call "The Whole Shebang."

"It's not a scientific institute," Sagdeev would quote a colleague named Lev Artsimovich as saying in frustration, "it's a travel agency for space science."

Exploration's Engine

Still, the motives of even science's jackals were as pure as Dr. Faust's: to know the world beyond their grasp, and in doing so, to extend themselves beyond what was required to merely survive. Murray made that point by challenging the notion that research on diseases, and even ending hunger, had a higher claim on the U.S. science budget than did space exploration. A civilization is ultimately measured, he said, not by what it has to do, but by what it wants to do. "Space exploration is forward looking, intangible, and appeals to the imagination and a sense of adventure, as distinguished from AIDS and cancer [research], which are basically utilitarian. . . . They're important. But the reason we work on that is because we're worried about dying." Trying to survive—a trait humans share with viruses—is perfectly understandable, Murray added. But it is not particularly admirable.

In going to Mars, the former JPL director explained one sunny afternoon in his Caltech office, "we're taking some of our valued treasure and spending it on something that doesn't pertain to our territory, to our life and death. It's an intellectual, spiritual, cultural endeavor. . . . Suppose the Russians could get the solution to AIDS," Murray pondered aloud, "and we could have the planets?"

It was Venus and Mars and the great gas balls, Jupiter and Saturn, with their own retinues of orbiting worlds, that had captivated Galileo, Schiaparelli, Tsiolkovsky, Hubble, and all of their descendants from Tsander to Oberth, Goddard, Korolyov, von Braun, Pickering, and the great science and engineering establishments. And it was the whole solar system and the galaxies beyond that intrigued the dedicated amateurs who faithfully pored over the journals and kept long nocturnal vigils squinting through telescopes (many of them homemade) in hopes of spotting a new comet or just savoring the timeless magnificence of Saturn's rings.

Von Braun had earned his living in 1948 by turning the V-2 into an advanced ballistic missile so his new masters could grow mushroom clouds on enemy territory if that was considered necessary. Shades of Dornberger and Korolyov. But he had entertained himself in his spare time that year by using a slide rule to plan in considerable detail an expedition to Mars that involved seventy astronauts and ten spacecraft, some carrying passengers, some cargo. What von Braun wanted to do was to end once and for all what he considered the myth that a boy, a girl, and a dog, or two intrepid brothers, or a guy named Flash Gordon, or even a few adventurous astronauts could fly to Mars almost as a lark. The point, as he put it, was to "explode once and for all the theory of the solitary space rocket and its little band of bold interplanetary adventurers. No such lonesome, extra-orbital thermos bottle will ever escape earth's gravity and drift toward Mars."

Just as Columbus had used three ships to explore the far reaches of the Atlantic and the pioneers had assembled wagon trains in long lines to move westward rather than risk everything by depending on only one ship or a single prairie schooner, an expedition to Mars would require a large number of people traveling in a convoy, von Braun had prophesied. He had gone on to describe the necessary propulsion systems, the trajectories, the winged "landing boats" that would be used to get down to the Martian surface and back up again and a great deal more in dense detail and with a generous use of equations. The book, which he called *The Mars Project,* was definitely not beach reading unless the reader was entertained by calculations for specific impulse and exhaust velocities or the wing areas and speeds of the landing boats and how hot their skins would become as they plunged into the atmosphere. It was published in German in 1952 and in English the following year, after his name was well enough known to help guarantee at least a break-even sale.

Until the 1950s, professors had actively discouraged students from careers that concentrated on the solar system because, they told them, most of the important information that could be collected was already in hand. Worse, most of the astronomers believed that optical telescopes had about reached their theoretical limit. But ten years after von Braun wrote his book, Sputniks and Explorers gave astronomers and other scientists cause to hope that far-ranging robots could actually reach well beyond the Moon to collect both data and samples. They were thrilled by the possibility of sending proxies to get up close.

One scientist at the University of California had a long wish list. Otto Struve anticipated Luna 3 by noting that he wanted to see the far side of the Moon out of "idle curiosity," have another spacecraft plant an explosive charge inside the icy nucleus of a comet, and have still others collect lunar soil and bring it home, study the atmospheres of Venus, Mars, and Jupiter, and snap up and retrieve fragments of Saturn's rings. "[T]he year 1957 will be remembered in the history of astronomical exploration as the year 1492 is remembered in the history of geographical exploration," Struve told an audience of scientists in Eugene, Oregon, in 1958.

"There will never again be a lecture on the solar system which does not in some way recognize the achievement of the Russian scientists" in developing the Sputniks, Struve said before quoting two other noted astronomers, Fred Whipple and J. Allen Hynek: "In his millennia of looking at the stars, man has never found so exciting a challenge as the year 1957 has suddenly thrust upon him."

The published version of Struve's paper was illustrated with the best pictures available at the time of Mars, Jupiter, and Saturn, all of them taken by the two-hundred-inch Palomar telescope in California. They were maddeningly blurry blobs wholly lacking in real definition. The telescopes could not get close enough. And even if they could, they would not be able to touch what they saw.

The distances were colossal, beyond the grasp of most imaginations. They were so great that they invariably had to be reduced to basketballs and miles so the average mind could contain them. If the Sun was a house in New York and Jupiter was a basketball, Jupiter would be in Chicago. If Earth was a basketball and the Moon was a tennis ball, Harvard's Owen Gingerich, a professor of astronomy and the history of science, explained one day in 1969, the

Sun would be a house in Harvard Square, several blocks away. And, he added, the nearest stars—the triple star system called Centuri—would be houses on the real Moon. At some point the imagination that tried to leap such distances fell so far short that it had no choice but to shut down.

As both superpowers were finding out, reaching Venus and Mars, the nearest basketballs, was challenging enough for flights in which any number of otherwise minor glitches could ruin a mission. But at least the two planets themselves could be approached without a forbidding amount of velocity. Visiting the worlds beyond them, however, was an entirely different matter.

The Limits of Self-Propulsion

Conventional wisdom up to 1961, when the exploration race had begun to gather real momentum, said that the distance problem could be overcome only by increasing the sheer velocity of the rocket and using the old Hohmann Trajectory, or Hohmann Transfer.

The people who pondered solar system exploration going back to the pioneers thought that it depended squarely on very high velocities and precise trajectories: on brute power and on knowing exactly where to go and how to get there. Everyone believed that increasing velocity was the key to exploring the solar system, so as the 1960s got under way, mathematicians, physicists, engineers, and others considered ways to squeeze more velocity out of the rockets. Between 1959 and 1962, while the world watched the duel taking place between Gagarin, Titov, Shepard, and Glenn, while generals and their staffs were fantasizing about lunar bases, orbital bombardment systems, space fighters, and real-time reconnaissance, and while the race to the Moon itself was gathering momentum, the scientists and engineers who were thinking about sending spacecraft on really long-distance missions were busily working the problem. Technical papers, journal articles, magazine pieces, and even chapters in anthologies that tried to come to terms with velocity proliferated.

Some considered, and invariably rejected, liquid-propelled chemical rockets of gargantuan proportion as still being inadequate. One, at 660 feet high, the length of a battleship, was named Sea Dragon and would have towered 125 feet above the Washington Monument. It would have weighed 100 million pounds and developed 130 million pounds of thrust. But even that monster would not have been able to carry a serious payload to the outer planets and beyond.

There was also a long-standing flirtation with atomic energy. As early as 1953, two British authors with technical backgrounds wrote a book entitled *Space Travel* in which they stated categorically that "nothing short of an applied form of atomic energy will ever be adequate" to propel men to the

Moon. A few years later, Theodore B. Taylor and Marshall N. Rosenbluth of the General Dynamics Corporation came up with a concept called Orion in which a very large spaceship would be pushed across interplanetary space at high speed by a series of nuclear explosions. Freeman Dyson, the Princeton University physicist who was renowned for his imaginative speculation about the future, also worked on the project. He explained years later that Orion's creators wanted to build a ship that would carry mankind peacefully from one end of the solar system to the other. But the project was funded by the Air Force. As recounted by Dyson, one Air Force captain became fixated by the notion of using fleets of Orion spacecraft as battleships that would stay on constant patrol, "cruising majestically" beyond the Moon, in a Deep Space Bombardment Force that would, as the physicist derisively put it, "stand between the tyrants of the Kremlin and dominion of the world." The captain's superiors thought they had better things to do with their money, however, so Orion soon sank out of sight.

Other plans described in intricate detail at least three kinds of nuclear engines that would "burn" fissionable materials, or use fusion, or decaying isotopes to generate heat from alpha, beta, or gamma particles that would heat hydrogen. And still others suggested three kinds of so-called ion, or electric, engines, one of which would strip electrons off a gas such as xenon and shoot them through an electrical field, which would make them accelerate. Ion propulsion was and remains one of the most feasible and attractive energy sources because the engine weighs only about a hundred pounds and could reach speeds as high as 300,000 miles an hour. The problem was that it would have to be carried to orbit, initial thrust would be measured not in pounds but in millipounds, and it would take ten years for the little beast to work its way up to top speed. Even solar sails, which would coast on the solar wind—Kepler's "heavenly breezes"—were considered.

It was widely accepted that only the spacecraft itself could increase its velocity and that the most efficient way to get from one planet to another was through the classic "minimum-energy" Hohmann Transfer or Trajectory, first articulated in 1925 by architect and VfR member Walter Hohmann. This involved a spacecraft leaving Earth and catching up with another planet by flying a long, flat ellipse until it and the target planet converged. Three Grumman engineers maintained in 1960 that the Hohmann Trajectory was absolutely the "optimal . . . least fuel path" for getting around the solar system. No one disagreed. Yet the farther out a planet was, the longer the ellipse, until it translated into more than thirty years for a flight to Neptune. During that time some of the mission scientists (not to mention their journal editors, publishers, and colleagues on promotion and tenure committees) would inevitably also leave this world, though not for Neptune. Indeed, it would take Rip Van Winkles to run such missions.

And if the object of Hohmann's nice clean elliptical flight path was a smooth, graceful, and efficient voyage, then planetary "perturbations"—the tugging and pulling of the heavenly orbs—amounted to an annoying hindrance that had to be canceled out by using the spacecraft's own engine.

Free Lunch

All of that was turned upside down in 1961 by a precocious twenty-five-year-old doctoral candidate named Michael A. Minovitch, who was studying mathematics and physics at UCLA and whose idea of fun was doing vector analysis. New to celestial mechanics, vector analysis had to do with developing equations to calculate orbits in three dimensions under all sorts of conditions. What Minovitch had done for fun as a youngster, besides read and study science voraciously, was watch *Destination Moon* and build model airplanes. Now it was all leading somewhere.

He landed a job that summer with JPL's Trajectory Group, which was trying to provide Mariner mission planners with one-way routes to Venus and Mars. They were doing that with their pride and joy, a new IBM 7090 mainframe computer that churned out pounds of calculations, and with their knowledge of the Hohmann Trajectory. The work was immensely complicated and tedious and involved using complex numbers to integrate the launch location (Cape Canaveral) with every phase of the flight. This would enable the navigators to set up simulated missions that told them where thrust vectors had to be pointed in order to get the rocket into a so-called parking orbit around Earth and then off to intercept Venus or Mars. Spacecraft were parked in parking orbits until the perfect moment came for them to strike out in a new direction.

"It's a killer," Minovitch would explain later. "I mean, talking about numerical headaches. It's like a labyrinth . . . very detailed number crunching. . . . But it's the type of work that has to be done to translate Oberth's and Tsiolkovsky's theoretical ideas of space travel into engineering realities. This was the nuts and bolts of preparing for rocket flight," he said, and it was happening under the pressure of beating the Soviet Union to the planets. "The federal government was throwing millions and millions of dollars at JPL like there was no end to the water spigot," Minovitch would recall. "It was like a miniature Manhattan Project," he added, still savoring the government's largess.

What Minovitch wanted to do was solve the endlessly frustrating Problem of Three Bodies. He therefore avoided Hohmann's elliptical approach to calculating trajectories. Instead, he applied his beloved vector analysis to three-body trajectories—Earth, spacecraft, and Planet X—and specifically to differential corrections. "Differential corrections" is jargon for using a computer to pick the best possible trajectory for a spacecraft and then gradually refine the model, making small corrections, until the combined effect of all three bodies on each other is precisely established.

Minovitch, feverishly working alone through the night at JPL's computer or at home, was trying to use vector analysis to get a spacecraft to a given destination by knowing where it, the planet it came from, and the target planet were at all times. The secret was to start with the best possible trajectory and then keep refining it: by using a high-speed computer to reduce the three-body picture into a series of Problems of Two Bodies and then combine all of them.

And that was not all. The young physicist-mathematician had taken the Problem of Three Bodies out of mathematics and put it into physics. It was then, while working on the problem from a physical perspective, that he came to see that the amount of energy, expressed as gravitational pull—the hated perturbation effect—could actually be used to fling a spacecraft toward another planet or a moon without using the spacecraft's own engine. It dawned on the driven young scientist that a spacecraft swinging around a planet would pick up some of the huge body's own energy as gravity pulled it toward the planet. It would then leave the encounter flying much faster relative to the Sun than it had been when it arrived.

This meant that once a spacecraft reached Venus, for example, with minimal velocity from its own rocket, the planet itself would fling it in any direction its navigators on Earth chose, provided it skimmed past the planet at just the right distance. If it was too far away to be fully affected by the planet's gravity, it would race out to infinity; too close, and it would be pulled in by an embrace that ended in a crash. Since the planet itself would be speeding along when the spacecraft caught up to it, the machine from Earth would gain speed, perhaps doubling it, as it was hurled out in a new direction.

But, he realized, it got still better. If this worked once, it would work repeatedly. In fact, it would work infinitely. Every time a gravity-propelled spacecraft reached a new planet and was steered into the same maneuver, it would get another massive shove through no effort of its own. Theoretically, then, a robotic explorer could fly around the solar system indefinitely like a billiard ball bouncing off the cosmic cushion with absolutely no further assistance from the engine that got it off Earth in the first place. Gravity would provide an infinite free lunch.

Minovitch was not the first to think about using planetary perturbations to help propel spacecraft. Tsander had done so, as had New Zealand's Derek F. Lawden, one of the century's outstanding astrodynamicists, and both UCLA's Samuel Herrick and Krafft Ehricke at General Dynamics. All of them except Lawden, however, either disregarded gravity propulsion or wanted to use it with the Hohmann Trajectory. But using Hohmann's Trajectory necessarily also meant using the spacecraft's own engine the way a swimmer pushes off from a boat or the wall of a pool. Lawden had briefly mentioned using a large moving body to increase velocity "without expenditure of fuel" in an article in the *Journal of the British Interplanetary Society* in 1954, but it had gotten lost in the literature and been all but forgotten.

Minovitch turned his research into a forty-seven-page paper, loaded with equations and charts, called "A Method for Determining Interplanetary Free-Fall Reconnaissance Trajectories." The document, dated August 23, 1961, would shrink the distance separating Earth from the outer planets and in the process create a revolution in solar system exploration. It would come to be called "gravity propulsion" or "gravity assist," and the credit for discovering it would be controversial for years.*

The Soviet Union, which probably had the best mathematicians in the world, was nowhere near nailing down gravity propulsion despite the head start bequeathed by Tsander. This is undoubtedly because it trailed the United States in computer development. The Russians only sent probes to Venus and Mars, which did not require gravity-assisted trajectories. But there was ample evidence, certainly in the papers they presented at international meetings, that they considered all planetary missions to be straight, direct transfer shots without intermediate detours. Two prestigious Soviet scientists, G. N. Duboshin and Dmitriy Y. Okhotsimsky, made that point in a paper they presented at the Fourteenth International Astronautical Congress in Paris in 1963.

Meanwhile, Minovitch continued to work on gravity propulsion and related areas sporadically at JPL, but mostly back at UCLA, during the next few years. He soon coaxed the computers to lay out the basis for a long gravity-assisted deep-space mission. Given the fact that Jupiter was by far the most

* There was a persistent unwillingness at JPL to credit Minovitch, possibly because of an institutional unwillingness to admit that a twenty-five-year-old summer employee, and a newcomer to the subject at that, could accomplish what the best minds in trajectory analysis there and throughout NASA and beyond were not able to do. Minovitch's boss that summer, Victor C. Clarke Jr., was credited with the discovery nine years later. Clarke was cited in a NASA press release for a prize-winning essay on applying gravity assist to space flight, for example. The release stated that "he first demonstrated in 1961 the possibility of bouncing a spacecraft from planet to planet." The use of a planet's gravitational field to change spacecraft speed and direction "can significantly shorten flight time," it added. Minovitch was given credit at the end for helping. ("Mariner-Venus '73 Flight Genesis," release No. 70-112, July 5, 1970.) Nineteen years later, Minovitch's discovering gravity assist was acknowledged by JPL in its *Voyager Neptune Travel Guide* (pp. 104–5). In his own fine work, *Mapping the Next Millennium,* Stephen S. Hall gave credit to Gary A. Flandro, another JPL scientist, though he acknowledged a "rivalry for credit" between the two. Hall cited a "key" 1966 paper by Flandro, which appeared five years after Minovitch's. (Hall, *Mapping the Next Millennium,* pp. 40 and 406.) And in 1994, a writer in *Air & Space* advanced the sophistic argument that, in any case, "it was just a matter of time" before someone else solved the problem because gravity assist was an evolutionary development. (Reichhardt, "Gravity's Overdrive," p. 78.) That discounts the inventive process. By such reasoning, it was just a matter of time before someone else invented the telephone, penicillin, and cubism, so Bell, Fleming, and Picasso should receive no special credit for doing so first.

massive of the planets—the solar system's most powerful sphere of influence after the Sun—it was a logical candidate to get probes to the other outer planets off to the fastest possible start. In August 1964, three years after the birth of gravity propulsion, Minovitch got a computer to tell him that there would be a perfect window to Jupiter in 1977 that would allow a spacecraft to skip like a rock on water from it to Saturn, Uranus, and Neptune, all three of which would be strung out in the right positions for successive encounters. That same year, 1964, Minovitch's gravity propulsion work was included in a JPL mission design paper. The following year, Gary A. Flandro designed his own set of outer-planet missions based on gravity propulsion.

The Great Leap Outward

Mission planners in the Soviet Union began thinking seriously about sending spacecraft to Venus and Mars almost as soon as Sputnik's flight was a reality, and their American counterparts were close behind. The drive to reach Earth's closest planetary neighbors was a complex brew of scientific curiosity and nationalism that was not easily untangled, though there is no doubt that the former far outweighed the latter as far as the scientists and the engineers who were supposed to get the spacecraft to the planets were concerned. The problem was that science, as always, was captive to politics. "There was always a sense of competition," Soviet physicist Roald Sagdeev explained about the exploration of Venus. "The Soviet government always was supportive of such missions from the point of view of propaganda and flexing muscles. Venus was considered by them as one of the battlegrounds of the cold war." For both Soviet and American scientists, it was a battleground where the flower of knowledge blossomed. But as usual, the politicians cared more for prestige, and they were the ones who provided the funding. So Korolyov, supported by Keldysh, needed to protect his lead against the United States by being first to reach Earth's nearest neighbors.

But steady, relentless progress in the Gemini and Apollo programs in the early and mid-1960s and the faltering Soviet manned Moon program was eroding the lead. And as it slipped away, desperation set in. Pressure on Korolyov and his colleagues to leapfrog the American manned program by going to the planets became intense. The record of Soviet shots at Venus and Mars between October 1960, when the first was attempted, and October 1967, when an "Automatic Interplanetary Station" named Venera 4 actually returned data from the Venusian atmosphere, indicates sheer frenzy. And the results, at least to begin with, were appalling.

There were at least twenty consecutive failures, several of which were neither named nor announced, in which the spacecraft either blew up on the pad, failed to make it out of Earth orbit, or had a catastrophic communication problem once they were headed in the right direction. But the circle of listening and tracking stations whose eyes and ears soaked up every launch at Tyuratam, Kapustin Yar, and Plesetsk from West Germany to Iran to Alaska made the secrecy all but pointless except where propaganda was concerned.

The perfect time to send a spacecraft to Mars, for example, comes during only a limited number of minutes or hours during a limited number of days every twenty-five months. Such "windows" are governed by the laws of physics, not politics, and have to do with Earth's and the other planet's positions relative to each other based on a number of factors, including the planets' velocities, where the launch site is, and most obviously, which side of the Sun they are both on. When a large and powerful rocket is launched on a trajectory that would take it to space during a window that was choicest for a shot at the Moon, Mars, Venus, or another body, it was obvious to U.S. analysts that such a shot had indeed been tried. They therefore knew that successive Soviet attempts to reach Venus and Mars—one painful try after another—became a litany of frustration, despair, and punishment: "Failed to leave Earth orbit. . . . Failed to leave Earth orbit. . . . Failed to leave Earth orbit . . . ," their reports noted in dismal counterpoint to the mayhem at Tyuratam. "Apparent launch failure. . . . Communications failed. . . . Failed to leave Earth orbit. . . . Lander crashed. . . . Engine malfunction. . . . Apparent launch failure. . . ."

It was not until early in 1966 that Venera 2 and 3, both of them developed by Korolyov's design bureau, actually reached Venus. One of them shot past the cloud-shrouded hothouse at an altitude of 15,000 miles and the other, carrying a Communist souvenir, crashed into it. Before they got there, the two spacecraft sent back data on interplanetary magnetic fields, cosmic rays, solar radiation, and more. But not a single bit of information came back from Venus itself because the radios conked out before they could transmit anything from the planet. It was the same at Mars. Successive spacecraft named Mars and Zond sent to the Red Planet between 1962 and 1964 either suffered communication failures short of the goal or were afflicted with other disastrous problems. Those who sent them quietly maintained that they had, after all, gotten to both planets first. They still do. But the claim was lame, since the reason for the missions had been to collect data about the planet, but none had come back.

Mariners Uncover Venus

On the most primal level—far beneath the engineering and logistics—Soviets and Americans conceived of planetary missions in different ways. The Soviets focused on the objective and therefore tended to name their spacecraft after their targets: Venera, Mars, and Phobos. The Americans, on the other hand, delighted in the romance of voyages themselves; in the challenge of traversing immense distances. NASA therefore chose metaphors that suggested great journeys when it named its explorers. Ames's would be Pioneer, which suggested the wagon trains that had relentlessly opened the Western frontier. JPL's collective imagination tended to go to sea with Mariner, Viking, and Magellan, though it also selected the universal Voyager as the name for spacecraft that would soon perform the greatest feat of exploration in history.

The first successful planetary encounter was made by JPL's Mariner 2, and the inaugural target was Venus. While American scientists itched to get a close look at Venus's mysterious, "cloud-laden atmosphere" and perhaps even peek under it, it was Korolyov and his colleagues who dictated the choice of the morning and evening "star" as America's premier planetary target, not NASA. Venus was picked because the space agency wanted to reach a planet before the Russians. It would take only about three months to reach Venus, while getting to Mars could take three times that long. Understandably, Korolyov picked Venus for the same reason.

Following its twin's ignominious disappearance into the Atlantic because of the minus sign, Mariner 2 sped past Venus on December 14, 1962. The 446-pound spacecraft, whose ungainly family resemblance to Ranger was unmistakable, carried a magnetometer, ultraviolet recorder, and other instruments. Like Venera 2, it sent home a stream of data about the interplanetary magnetic field, cosmic dust, and charged particles, and also bore out the fact that there was a vast wind that radiated from the Sun and moved across the solar system, just as Kepler had imagined. But Mariner 2 also provided Earth its first close-up look at Venus. The long-held idea that Venus was Earth's "twin" because both planets had similar masses and orbits evaporated under the spacecraft's scrutiny. During its forty-two-minute encounter, Mariner 2 skimmed to within 22,000 miles of Venus and sent data to Pasadena showing that the surface temperature of the cloud-shrouded planet was some 800 degrees Fahrenheit, or twice as hot as had been predicted, and that the clouds themselves were relatively cold. If Venus had radiation belts, they were nowhere near as strong as Van Allen's. But like Earth, Venus showed the effects of the Sun's plasma flow near its surface. That was indeed something for the budding planetary science community to build on. The intense heat and lack of a protective magnetic field, for instance, just about ended the notion that there could be any kind of carbon-based life on Venus.

A Rembrandt from Mars

It was not by coincidence that scientists anticipated the Mariner-Mars mission by inventing a new science: exobiology, or the study of life beyond Earth. All the years of speculation about life on Mars, going back to Lowell, had whetted the appetites of Joshua Lederberg, who had won a Nobel Prize for his work on the genetics of bacteria, Wolf Vishniac of the Yale University School of Medicine, and others. As the low-Earth orbiters had finally gotten Van Allen close to his radiation belts, so too would the planetary probes finally get Lederberg, Vishniac, Sagan, and their fanatically curious colleagues close enough to Mars so that the life theory could be settled once and for all. That, at least, was what they hoped.

The big question—whether there was life out there—had driven fiction writers, astronomers, UFO cultists, filmmakers, Orson Welles, hardheaded scientists who trained radio telescopes on other galaxies hoping to pick up signals that would show signs of intellect, and others. Certainly Lederberg, Vishniac, and Sagan did not think there were little green men, or intelligent life in any form, on Mars. But they did believe implicitly that even finding the remains of microbes or any other trace of life on another world would be the greatest discovery in history and would have profound implications for the human race. As physicist Philip Morrison (a founder of the Search for Extra-Terrestrial Intelligence radio-telescope project) said, it would change life from being a miracle to being a statistic.

Beyond the science fraternity, there remained the news media, which knew (or thought they knew) a hot story when they saw one. And the story was the life angle.

From the beginning of the mission, JPL and NASA carefully explained that the quality of the pictures they hoped would come back from Mars would not have a high enough resolution to spot vegetation or animal life. The reporters did say that the quality of the imagery was not expected to answer the question of whether there was life on Mars. But some of them could not resist a little hype, so many of the stories that appeared while the first visitor from Earth neared Mars milked the life angle by reporting that the pictures might well "answer long-standing questions about the canals."

Mariners 3 and 4 were dispatched to Mars during the window that opened in November 1964. Mariner 3 left on the afternoon of the fifth. But within an hour, it became apparent that no sunlight was shining on its solar panels, and for very good reason: the fiberglass shroud on the Agena upper stage carrying the planetary probe never opened. Its four long panels remained pinned down like arms pressed against hips. No panels, no power; no power, no mission.

Mariner 4 headed for Mars on November 28 and sped past it without a hitch on July 14, 1965, all sensors going. The engineers at JPL racked up another record: theirs was the first visitor from Earth to reach the Red Planet. By the time Mariner 4 sped past Mars, many Americans and others were waiting for the big question to be answered. Any real clue, any clue at all, would be delicious. The spacecraft took pictures of 1 percent of the Martian surface from a distance of a little more than 6,000 miles during its twenty-five-minute closest encounter. Then it sent them home. The imagery, which was pulled in by the giant antenna at Goldstone at an agonizingly slow rate of eight bits a second—the state of the art that year—was then relayed to JPL. The twenty-one and a half fuzzy pictures, each of which took ten and a half hours to materialize, came in as digital data on thin strips of paper, tiny numbers that represented the Martian landscape's features in varying hues. Then the strips were glued to a plywood board and hand-colored with colored pencils in shades of yellow, orange, and brown that matched the three-digit numbers.

What slowly emerged was good stuff. And it was terrible. Mars was no longer an elusive orange blur with whitish poles and alluring dark blotches. It had been transformed into a place that had recognizable features with which earthlings could identify. And what a place. Gone were the canals or anything else that could have been purposely dug or built. Gone were the oases holding precious supplies of water. Gone were creatures in any form. Gone, too, were ocean basins, vegetation, or a landscape that even remotely looked like Earth. In fact, there did not even seem to be such Earth-like features as mountain chains, great valleys, or continental plates. What the pictures showed at two-mile resolution (which was still twenty times better than the best telescope imagery) was a desolate scene pockmarked by scores of craters, some of them with walls that seemed to rise as high as three hundred feet while others appeared to be several hundred feet deep. Mariner 4 also reported that there was no magnetic field to speak of, so the planet was exposed to cosmic-ray bombardment, and that the atmosphere was very thin compared to the one it had left on Earth. The most important news that came from Mars during that first encounter, in fact, was that the place seemed to be like the Moon. And the Moon appeared to be stone dead. Bummer.

The first of the Mariner 4 pictures, the first image taken of Mars by a machine from Earth, hangs at JPL today and is like a "priceless Rembrandt," according to Jurrie van der Woude, a venerated member of the lab's public affairs department and a former countryman of the great Dutch master.

Science had searched for the truth and seemed to have found it. But that truth, or what seemed to be the truth, was a stunning blow to the exobiologists. Not only was there no sign of life in any form, but the craters indicated that there had been no volcanic or other internal activity. And there was no evidence of oceans. No oceans—no water—meant no life. Ever. "We were all shocked by seeing such large lunar-like craters," Murray would recall. "It meant that Mars had not recycled its surface the way the Earth does. There must have been no rainfall, no weathering, in any way comparable to Earth's for billions of years, in order for Mars to resemble the Moon." But as the military intelligence people had hated not knowing what was hidden deep inside the Soviet Union before spy satellites provided virtually unlimited access to that vast country, Murray and all the others in the new fraternity of space explorers hated not knowing what lay beyond Mariner 4's myopic camera. Furthermore, the planets beyond both Venus and Mars beckoned, and so did the space between all of the planets and the staggering distances beyond them.

There were old questions and an explosion of new ones. There were questions about the nature and creation of Saturn's rings (and indeed why it was the only planet that seemed to have rings). There were questions about the exact number of planets in the solar system, about why some were solid and others were gaseous, and about how many moons really existed and how they themselves had been formed. There were questions about why planets and moons were different at all, and still others about the nature of asteroids and comets, solar flares and magnetospheres, and the stuff that moved unseen through the immense void. Even the questions—*especially* the questions—could be delicious. Why, for instance, was Io, one of the four Jovian satellites first spotted by Galileo, seen through telescopes to be spewing sodium all over its neighborhood, and why did it seem to be generating a million amperes of electrical current? Pretty lively work for a supposedly dead world.

Mariner 5 blasted off for Venus on June 14, 1967, and returned more detailed data on the planet's atmosphere, magnetic field, and mass. Two years later, Mariners 6 and 7 (or Mariner-Mars 69, as they were collectively called) were launched on the powerful hydrogen-oxygen Centaur upper stages. They brought the Red Planet still closer, imaging its full disc and about 10 percent of its surface as they sailed past it that summer. Some of the 1,177 pictures they sent home, taken from as close as 2,000 miles, showed terrain that was heavily pockmarked with craters, other areas that were called "chaotic" because of their valleys and angular geological formations, and featureless areas that could not be resolved. Taken together, Mariner-Mars 69's imagery delivered a coup de grâce to Lowell's canal theory once and for all.

By the time Mariners 6 and 7 had their encounters with Mars, two salient facts of life had become as clear to American space scientists and technocrats back on Earth as they had to their colleagues and rivals in the manned program. Whatever the results of ultraviolet and infrared spectrometer experiments that measured the composition of the atmosphere, or the infrared radiometer that collected data on surface temperature—and the information had an important bearing on the possibility of life on Mars—it was pictures that captivated the public because they were both dramatic and relatively easy for lay people to understand. And the route to the public (and therefore to congressional budget makers) was controlled by the gatekeepers in the Fourth Estate, many of whom had long since hatched specialties as science and even aerospace writers. Planetary encounters were therefore handled by JPL and Ames the way the Ranger, Surveyor, and Lunar Orbiter missions had been handled: by news conferences, science briefings and the lavish use of printed news releases, background material, and photographs.

"Before the space age, Mars was thought to be like the Earth, polar caps, seasons, . . . rotates in twenty-four hours, etc.," Robert B. Leighton, the imaging team's principal investigator, or leader, told reporters at a news conference on September 11. That view of Mars, he continued, "was largely the legacy of Percival Lowell, who popularized the idea of reclamation projects to get the water supposedly from the polar caps down to the equator, where the farmers were." Although that idea had been rejected by most scientists long before Mariner 4, they had not been prepared for the stark, lunarlike vistas that were recorded by Mariners 6 and 7. Mars, the Caltech physicist added, was "like Mars." That is, it was now seen to have its own characteristic features, "some of them unknown and unrecognized elsewhere in the solar system."

Indeed, a lack of the kind of tectonics that came from tremendous internal pressure, and which caused the formation of continents on Earth, combined with very old impact craters, indicated that the Martian interior was far less active than Earth's.

The Hazards of Exploration

The stream of data that came back from Mariner-Mars 69 belied the fact that Mariner 7 was afflicted with a succession of unexplained problems. It would suddenly go dead, for example, sending back not a scrap of telemetry for hours. Then, just as mysteriously, it would come back on in a halting whisper and on a slightly different course. It was spooky. On May 8, 1971, during the next window, Mariner 8 was dumped into the ocean by a Centaur that could not get it into orbit. Two days later, one of the big new Soviet Protons managed to get Cosmos 419 into Earth orbit, but the injection stage did not ignite, so it never left the precincts of home. The accident was caused by "a most gross and unforgivable mistake" in communication with the launch vehicle's computer, according to a Soviet source. It would not be the last communication problem to ruin a Soviet deep-space mission.

Two Soviet orbiter-landers, Mars 2 and 3, were launched on May 19 and 28, followed by Mariner 9, which was identical to Mariner 8, on the thirtieth. Both Soviet spacecraft happened to arrive at Mars as the worst dust storm in its recorded history was building. The state of Soviet computers and communication being what they were, both probes had been programmed to do the mission sequence before they left Earth. There was no possibility of modifying what they did. So Mars 2's lander separated from its orbiter on November

27, exactly as it had been told to do, and then disappeared into a vast cloud of ocher dust. It slammed into the Martian surface without returning a single scrap of useful data. Mars 3's lander followed five days later, entering the atmosphere at a blistering 13,320 miles an hour, popping its braking and main chutes on the button, discarding its heat shield, kicking in its retro-rockets at the right instant, and touching down as planned. Next, the sterilized spacecraft obediently unfurled its petal-like covers as if it were a blooming flower and began to transmit imagery up to the orbiter for relay to Earth. Then disaster struck again. Twenty seconds after it started beaming signals, they abruptly stopped, leaving its stunned team with one darkened partial panorama that showed virtually nothing. Some analysts claimed that the orbiter had failed to relay the signals. More likely the lander was blown over, perhaps literally going belly-up, and expired on the spot. Being locked into a flight program that could not be changed in any circumstance meant that there was no way to react to unforeseen developments such as the dust storm.

"They can't modify what the spacecraft is really going to do," Nicholas L. Johnson noted later. "They can't reprogram very well from the ground. They wanted a dead-shot trajectory for the planet; it was then or never, so they went down into some of the worst storms we know about, and neither one of them survived. They just don't have the onboard processing." The loss of both probes reflected a vulnerability that increased in direct proportion to distance: a fundamental weakness that would force the Soviets to concentrate on Earth-hugging space stations and cede Jupiter, Saturn, and the other outer planets to the Americans.

If the essence of exploration is discovery, a mission that could not be altered made the reason for the trip almost pointless. As Mert Davies has said, if the purpose of exploration is to find and describe the unknown, then explorers "must be prepared to respond to new things." That is, they have to have enough flexibility so they not only can search in unchartered places but can either independently study what they turn up or be told to do so from Earth. A lack of flexibility implies that those who program the spacecraft pretty much know what it is going to find. But that is not exploration; it is tourism.

The Ghoul

The spate of calamities, several of them technologically inexplicable, led inevitably to speculation that there was a mysterious, perhaps evil, presence out there; a kind of galactic version of the dangerous dragons that were once thought to swallow whole ships without leaving so much as a barrel floating on the water. John Casani, the Mariner-Mars project engineer, was asked by reporters whether there was any particular physical danger in deep space that lurked in wait for visitors from Earth. When he told them that some kind of creepy danger zone did not exist, *Time* magazine's Donald Neff quipped that the problems were really being caused by a force that harbored nothing but ill will for the emissaries from Earth and therefore kept trying to do them in. Neff whimsically named his remorselessly insatiable beast the Great Galactic Ghoul. It stuck. The Ghoul became the bane of astronautical engineers everywhere. It seemed to prefer to chomp on any Soviet spacecraft that unwittingly got too close to its lair, which was in the vicinity of Mars. The Ghoul was fondly embraced by the American space exploration community, and especially by JPL, as a monster that no amount of perfect engineering could overcome once he decided to strike. One contractor's artist depicted an immense androgynous creature that was invisible to radar—making it the first stealth monster—floating in space and happily clutching Mariner 7 the way King Kong had clutched the biplane on the top of the Empire State Building. The Ghoul, however, clearly intended to turn the hapless spacecraft into an hors d'oeuvre. The portrait hung at JPL for many years.

Triumph and Trauma at Mars

But even the Ghoul could not be everywhere. While it was busy scuttling the Soviets' Mars 2 and 3, Mariner 9, its onboard processor working perfectly, was ordered to stay in orbit until the storm blew over. That happened by late February 1972, leaving the Martian atmosphere clear enough for the spacecraft to map 85 percent of the planet's surface during 349 Earth days of operation. The orbiter sent back 7,329 pictures and 54 billion data bits: twenty-seven times as many as the three previous Mariners combined.

The return was stupefying. There were the first pictures of Phobos and Deimos, the Martian moons; Valles Marineris, the 2,800-mile-long canyon that stretched a quarter of the way around the planet (and that was named after the robot that took its picture); the towering, fifteen-mile-high Olympus Mons volcano, the highest known mountain in the solar system; surfaces that were either ancient and covered with craters or young and smooth; tantalizing fractures and faults; and five different kinds of channels, many of them appearing to be old riverbeds—*riverbeds* where water would have flowed!—and meteorological phenomena that included not only dust storms but weather fronts as well. In addition, the Martian gravity field was measured more accurately, the planet's ephemeris was markedly improved, and piles of data came back on the chemical composition, temperature, and pressure of the atmosphere and of the composition and temperature of the terrain. What it all added up to, in the words of two scientists who put the findings into a book, was "a knowledge explosion."

An explosion indeed. Bruce Murray and Eric Burgess, a prolific chronicler of the space age, found the mound of knowledge utterly "stupendous," thanks to the rocket that bridged the gap between the home planet and the red one. All of the useful photographs of Mars taken through telescopes during the fifty years before Mariner 4's flyby would barely equal the information its twenty-one radio-relayed images delivered, they noted. Four years later, Mariners 6 and 7 returned a hundred times as much data as Mariner 4, and then Mariner 9 sent back a hundred times more than its two immediate predecessors. "The total photographic and other information about Mars thus increased 10,000-fold between 1965 and 1972! No wonder there has been a revolution in ideas about the nature and history of Mars."

And that revolution, with its evidence of a watery past, geological activity, active polar ice caps, layered terrain, and more, rekindled the notion, made temporarily dormant by Mariner 4, that there could indeed be life on Mars: that it was not, nor had it been, dead. The old siren beckoned once more. "If Mars is empty . . . we will fill it," wrote science fiction writer Ray Bradbury after Mariner 9 went silent in October 1972 for lack of attitude-control gas. "But still the voice of Mr. [Edgar Rice] Burroughs calls out on nights when we pace our lawns and eye the Red Planet: 'All the evidence is not in! Maybe.' "

Vikings to Utopia

So audaciously crafted orbiter-landers were dispatched after Mariner 9 to press the search for life. The mission, which was conceived in 1962 and first called Voyager, had been battered and all but killed in the cuts that came with the diving budget curve in 1966 and afterward. But by the end of February 1969, a scaled-down version of the project, by then renamed Viking, had been started in earnest. The idea was to send two orbiters to Mars, each attached to a lander that in turn was stuffed with television cameras, three instruments designed to search for life, and other experiments.

Although there were thirteen science teams on the mission, including those specializing in water vapor and thermal mapping, magnetic properties, and meteorology, the main purpose of the Viking mission was to press the search for life. And that heavily publicized fact contained a nasty deal which, like so many others in the space program, smacked of Faust. A return to Mars could be sold to Congress only by promoting the tantalizing notion that life could be found there. If Viking did indeed find life (or evidence of it), the United States would have racked up the biggest scientific discovery of all time, in the process scoring a propaganda coup of epic proportion. But if the search came up empty, NASA could probably expect to face the wrath of politicians and many of their constituents who would feel duped. That would not be good for the space agency, especially coming so soon after the post-Apollo letdown.

JPL built and managed the orbiters (which bore an obvious family resemblance to the Mariners), while the Langley Research Center managed the landers and the project as a whole. The mission's cost would go through the ceiling, partly because of an especially complicated procurement system, and also because of a two-year launch delay that ran up an extra $141 million. TRW originally estimated that its biology instrument, the heart of the lander's science system, would cost $13.7 million. The final cost was actually $59 million, bringing the program's total cost to $967 million.

The delay was as pernicious as the soaring contractor charges and laid bare a problem that would haunt those who explored deep space for years to come. Since what they did was not considered important for either domestic well-being or the nation's security, and was neither comprehensible nor enduringly interesting to most people, it was always especially vulnerable to budget cutting. There were no constituents on Mars. At least not yet.

The orbiter-lander's combined weight was 7,738 pounds at launch, which required the services of massive Titan 3Es and Centaurs. Viking 1 left the Kennedy Space Center on August 20, 1975, followed by Viking 2 on September 9. NASA wanted Viking 1 to land on Mars on the Fourth of July 1976 to commemorate the nation's bicentennial. But soon after it slid into orbit on June 21, pictures of the terrain below taken by the orbiter showed that the intended landing site was dangerous, as were several others, so the spacecraft continued to circle until a better one was found. The same thing happened to Viking 2. The computers and communication link, which provided flexibility, proved their mettle again.

The Viking 1 lander finally came down on Chryse Planitia on July 20, followed two weeks later by Viking 2, which landed smoothly on Utopia Planitia. The sites were on roughly opposite sides of the planet and on different latitudes in its northern hemisphere, since far-flung locations offered the best odds for turning up life.

Thomas A. Mutch, a geologist who headed the lander imaging team, would never forget the thrill of Viking 1's approach and landing. He spoke of it, and of all far encounters, with a simple elegance that reflected what drove him and his American and Russian colleagues to explore distant worlds:

> At 1:51 a.m. the Lander separates from the Orbiter and begins its descent to the martian surface. Approximately at the same time I drive through the cool California night to the Jet Propulsion Laboratory. The windows of the tall buildings sparkle with lights. The parking lots are full. People hurry past in the darkness. I walk quickly to the building where the Lander Imaging Team is housed. Many of my colleagues, scientists and engineers, are there. For all of us there is only waiting, and I realize that I would rather wait alone, away from

forced conversation. I walk to a nearby building and take my assigned position in the "Blue Room," a broadcast area where the first pictures will be received and transmitted to the news media assembled in an auditorium.

5 a.m. The final descent begins. Conversation stops—an overwhelming silence. We listen to the mission controllers as they call out each event. After years of waiting, hoping, and guessing, the end rushes toward us—too fast to reflect, too fast to understand.

5:05 a.m. "400,000 feet"

5:09 a.m. "74,000 feet"

5:11:43 a.m. 2,600 feet"

5:12:07 a.m. "Touchdown. We have touchdown."

It worked! Amazingly, it worked. Everywhere people are cheering, shaking hands, embracing. I decide not to join the celebration. It is too soon. Forty minutes more remain before the first picture from the surface of a far planet will assemble on the television screen.

5:54 a.m. I study the blackness of the television screen, waiting for the narrow strip of light that will signal the first few lines of the first picture. *And it appears.* A sliver of electronic magic. Areas of brightness and darkness. The picture begins to fill the screen. Rocks and sand are visible and—finally, at the far right—one of the spacecraft footpads, a symbolic artifact that stamps our accomplishment with the sign of reality.*

Viking 1 began to scan twenty-five seconds after touchdown. Like that of an embarrassed or awkward intruder, the first picture sent home did, indeed, look downward, showing small rocks and one of its own feet: a footpad, complete with tiny rivets.

Those that followed, as well as others taken by the orbiters as they approached Mars and then went into orbit around it, forever changed the way humans saw the Red Planet. The orbital pictures, painstakingly arranged in mosaics at JPL, were clear enough so that the entire planet's surface could be mapped in extraordinary detail. The resolution was so good, in fact, that it would eventually be put into a *Mars Landing Site Catalog* that listed seventy-eight places to land rovers, sample return spacecraft, and instruments on balloons, in addition to other places that were likely targets for harpoonlike penetrators or weather stations. The landers sent back thousands of panoramic views of a rock-strewn ocher desert that was reflected in a disconcertingly pink sky. The pictures were so clear that shallow trenches caused by wind erosion, small pockmarked rocks, and even pebbles could plainly be seen. The Martian blur was gone forever.

* Mutch was killed in a climbing accident in the Himalayas in 1981. The Viking 1 lander was named the Thomas A. Mutch Station as a memorial to him. A plate commemorating his death is attached to a model of Viking at the National Air and Space Museum.

It was the same with the science experiments. So much data came back on the Martian climate and how it changes, on details of its topography—including impact cratering, the canyon system, volcanoes, tectonics, and ice formation—and on surface chemistry, geology, and the atmosphere that they were analyzed for years and together amounted to a detailed profile of the planet. Studying the volcanoes proved not only that Mars had an active interior but that a great deal of water was released during the planet's first 2 billion years. The presence of water was inferred by the volatile nature of the lava that poured out of the volcanoes, from the shape of basins, and then from the position of rocks. The data were so voluminous that conclusions would still be squeezed out of them two decades later. The highlights of the Viking data were published in a richly detailed tome called *Mars,* which was published as part of a prestigious space science series in 1992 by the University of Arizona Press. It was 1,498 pages long and weighed four pounds, reflecting a bonanza of knowledge. That was the good news.

The bad news was that no sign of life turned up. The Vikings used mechanical claws to scoop up small soil samples and deliver them to the biology instrument for analysis. Three experiments designed to test for life, two of them by exposing any microorganisms in the soil to a rich broth, produced nothing.

The first and most obvious rationale was that the landers had sampled only two places on the entire planet and that life or signs of it could very well lie elsewhere. A second had it that the tests were more suited to searching for life in a place like Earth than in Mars's totally different environment. But Norman H. Horowitz, the chief of the Mariner and Viking bioscience teams, would have none of it. "With the return of the first set of GCMS [gas chromatograph mass spectrometer, one of the three experiments] data from the Chryse landing site, unambiguously negative, it became clear that if subsequent analyses continued to show no organic matter in the Martian soil, then convincing evidence for life in the soil would be unattainable, regardless of any other findings," he would write. And, he added, the two landing spots were in fact probably representative of every place else.

Most other scientists believed that the Viking results remained ambiguous and, predictably, that follow-on missions covering far more territory would be necessary to answer the life question once and for all. Murray, a colleague of Horowitz's at Caltech, maintained that the results were bound to be ambiguous. As was the case with Apollo, Murray believed that political expedience had pushed Viking into an untenable corner; that as men should have gone to space more gradually and systematically than Apollo's high drama to avoid the need for ever-greater stunts, so the search for life on Mars should have taken its place after other, more basic, environmental knowledge had been gained. In any case, Murray agreed with Horowitz that Mars is self-sterilizing.

"For this reason," he concluded, "the absence of organic compounds in the Viking tests is a strong indication of the absence of life over all of the desolate planet's surface."

Nor was Horowitz disappointed because the evidence showed that Mars was lifeless. For him, that only added to the wonder and uniqueness of his own planet. "The Viking mission completed the deLowellization process started in 1963," the biologist said at a Mars conference at the National Academy of Sciences in July 1986. ". . . Viking brought to an end not only the age of Lowell but also the long history of the so-called Copernican world view—that unwarranted extension of the Copernican revolution that held the Earth occupies no special place in the solar system. . . . The failure to find life on Mars was a disappointment, but it was also a revelation. It now seems certain that the Earth is the only inhabited planet in the solar system. We have come to the end of the dream. We are alone—we and the other species that share our planet with us. Let us hope that the Viking findings will make us realize the uniqueness of Earth and thereby increase our determination to preserve it."

13

A Grand Tour: The Majestic Adventure

American planetary scientists and the engineers who carried their dreams had begun looking past Mars at least a decade before the Viking missions. In 1965, with Gemini flying, Apollo gearing up, and the Soviets shooting frantically at the Moon and the inner planets, Minovitch and Flandro published separate papers explaining how gravity propulsion could be used to fling robotic explorers around the outer solar system, starting at Jupiter. Although neither would have put it this way, particularly in a technical document, they were calling for missions with ranges that were beyond the capability of the Russians.

By the end of 1966, Flandro and others calculated that a window would open to all of the outer planets in 1977 and 1978: that they would be strung out in a staggered formation that would allow spacecraft to fly from one to the next, starting at Jupiter and ending at Neptune. The mission was called the "Grand Tour."* That December, Homer Joe Stewart, the venerated engineer who managed JPL's Advanced Studies Office and who was one of the lab's true stalwarts, published an article in *Astronautics & Aeronautics* calling for both electric and gravity propulsion to be used on a "Grand Tour of the outer planets, passing Jupiter, Saturn, Uranus, and Neptune" with "a drastic reduction in trip time."

* The popular term was used at least from the eighteenth century to describe an international jaunt taken by young English gentlemen (and sometimes chaperoned young ladies) in search of education and adventure. Such journeys were often considered to be an essential part of education. For example, John Bacon Sawrey Morritt, a squire from Yorkshire who was in his early twenties, wrote letters and kept diaries during a pilgrimage through Hungary, Austria, Italy, Greece, and Asia Minor at the end of the eighteenth century, which were published in 1914 as *A Grand Tour: Letters and Journeys, 1794–96.*

But first gravity assist would be practiced on the inner planets. In 1962, Minovitch's discovery led to the realization that a spacecraft launched at Venus during the 1970 or 1973 windows would in effect get a free ride to Mercury. The following year, the peripatetic mathematician-physicist described a complicated round-trip mission involving multiple flybys of Earth, Venus, and Mars (not Mercury) in a gigantic figure-eight. Three years later, in 1966, Flandro and Roger D. Bourke, a colleague at JPL, responded to a proposed Grand Tour mission study by suggesting that gravity propulsion be tested on a flight from Earth to Venus to Mercury during the 1973 window. The Space Science Board of the National Academy of Sciences endorsed the mission in 1968, and in September 1969 NASA formed a Science Steering Group to define the mission requirements. The agency's Office of Space Science and Applications signed off on it by requesting $3 million from Congress for seed money funding in fiscal year 1970.

The appropriation was approved by the Senate in November. The senators tossed in an additional $10 million, but only after the House Committee on Space Science and Applications voted to "defer" the mission in favor of others that were closer to home. It was another of the space program's debilitating tugs of war. Apollo was still flying, Skylab was in the works, and thoughts were turning to the shuttle. Although the Space Transportation System would not be approved by Nixon until early in 1972, it was already mandated by an industry whose Ph.D.s were losing jobs in formidable numbers in the aftermath of the Apollo wind-down. The story, which was largely apocryphal, had it that they were driving taxis. Whereas more than 77,000 scientists and engineers worked for companies that contracted with the space agency during Apollo's high-water mark in 1966, for example, only 51,350 remained two years later.

In addition, a number of innovative science missions were being planned, some of them involving the long-lived Explorer series. Other satellites, called observatories, were being created to do novel experiments in high-energy astronomy, solar observation, physics, and geophysics, all to the delight of Van Allen, who had been present at the creation, and many of his unstarstruck colleagues around the country. And the space agency was embarked on ambitious Earth monitoring programs involving both aircraft such as U-2s and, starting with a launch on July 23, 1972, Earth Resources Technology Satellite (ERTS) spacecraft, which were designed to collect data for environmental and land-use management, large civil engineering projects, resource monitoring, pollution, ice floes, and other peaceful uses for which remote sensing was ideal. ERTS—later renamed Landsat—was built by General Electric, which had experience in other remote sensing programs, notably the Nimbus weather satellites and the Discoverer capsules in the top-secret

Corona reconnaissance program. Since weather and resource satellites were "practical," they would always have an edge over their esoteric cousins that scouted the solar system.

Nor was Mariner 10, as the Earth-Venus-Mercury spacecraft came to be called, without competition in deep space. In 1962, the Ames Research Center, working in conjunction with TRW, had done feasibility studies on solar probes that so impressed headquarters it was given permission to start them up as a second Pioneer program. Whatever the validity of the engineering—and it would prove to be very valid indeed—awarding Pioneer to Ames had also been blatantly political. Charles F. Hall, the project's manager, would recall years later that Edgar Cortright had wanted to foster a degree of competition in order to keep the "spoiled brats" at JPL in line. The same tactic, deeply institutionalized on every level, was also taking place in the Soviet space program and was slowly but relentlessly helping to erode it. Not so in America's more robust economy.

Under Charlie Hall's leadership, Ames sent four of its TRW-built explorers, named Pioneer A through D, or 6 through 9, out to collect data on how the Sun affects interplanetary space, including cosmic ray and electric field information, celestial mechanics, and more, while orbiting this system's star.* They returned a stream of data on the solar wind, energetic particles, and magnetic and electric fields radiating from the Sun. In the process, they also provided valuable data on solar disturbances, commonly called flares, which can break up communication on Earth and elsewhere in the solar system and which directly affect the climate. Pioneer E, which would have turned into number 10 had it lived, was demolished on August 29, 1969, after the Thor-Delta carrying it suffered a disastrous hydraulic failure minutes after it lifted off at Canaveral.

Meanwhile, Goddard had come up with an ambitious proposal of its own. This was for "Galactic Jupiter Probes" that would make the first visits to the colossal outer planet and beyond during 1972 and 1973, when there would be open windows only thirteen months apart. Headquarters approved the Jupiter mission in February 1969, but it confounded Goddard by handing the program to Ames. That left the Maryland facility in sole possession of such Earth orbiting missions as communication, weather, the various science satellites, and, starting in 1972, the ambitious ERTS program. That was plenty for Goddard to handle, at least from headquarters' standpoint. Ames's planetary explorers would be called Pioneer 10 and 11. They would make history, not only in their own right as the first visitors from Earth to reach Jupiter and then Saturn by

* The first were the four Army and Air Force Pioneers that had made unsuccessful attempts to reach the Moon in 1958 and 1959. Pioneer 5, with instruments built at Goddard, had managed to orbit the Sun between Earth and Venus in 1960 and sent back first-rate data. (Ezell, *NASA Historical Data Book,* Vol. 2, pp. 305–07.)

gravity assist, but as trailblazers for a pair of JPL deep space explorers named Voyager 1 and 2.

The Grand Tour Comes Together

By the summer of 1969, even as *Apollo 11* was being readied for its own historic flight, the Jet Propulsion Laboratory was laying out plans for the Voyagers and the ambitious, unprecedently long mission they were supposed to fly. "After Mars and Venus come Jupiter, Saturn, Uranus, Neptune and Pluto in the most far-reaching space missions yet conceived by man," trumpeted an optimistic NASA news release issued on June 2. It heralded the birth of what JPL immodestly called a "Grand Tour" of Planets. "The best outer planet alignment in 179 years, occurring in the 1976 to 1980 time period, opens the outer planets to exploration in an effective and timely manner," James E. Long of JPL's Advanced Studies Office was quoted as saying. Well, yes; that was the way Flandro had worked it out. Pasadena's publicists and their colleagues in Washington made a point of noting that the last best time to do the Grand Tour had been during the Adams presidency and the next one would be around the middle of the twenty-second century. Strictly speaking, that was true, but only if planetary alignment was taken into account. But "best" did not mean "only." It did not take into account the flexibility that gravity assist gave the spacecraft. Even if the outer planets were not perfectly aligned, the enormous boost in propulsion opened the window wider than Long and his enthusiastic colleagues seemed willing to admit.

Exploration, by definition, meant going to new places, all of them as inherently dangerous—at least for a robot—as they were seductive. So planning required a mix of conservatism and daring in everything from conception to engineering to the selection of instruments and trajectories to flying the missions, since distances on the new ocean were immense and unchartered. The cost of trying to get to Jupiter, Saturn, and beyond by steering through the asteroids, for example—the ill-named "little stars" that were actually a ring of racing rocks on the other side of Mars about which little was known—could be obliteration: a multimillion-dollar machine, years of planning, the possibility of learning enough to make a contribution to one's discipline, and the tenuous support of a fickle public, all blown to hell by a dirty stone potato zinging along at 30,000 miles an hour. (The likelihood of a collision with a tiny asteroid was known to be far greater than running into one a mile or more wide since there were far more little ones.)

But scientists and mission planners in those days were haunted by more than rocks flying in the night. What was known for certain was that any spacecraft approaching Jupiter would run into what Bruce Murray called "murderous radiation." (Gustav Holst, who celebrated the solar system by composing his masterpiece, *The Planets,* called Jupiter the "Bringer of Jollity." But the great English composer didn't have a radio telescope that picked up Jupiter's intense and deadly radiation on three wavelength bands. "Radiator of Poison" would have been more accurate, if less fanciful.) The spacecraft's computer "brain" would therefore have to be carefully protected. "That," Murray said, referring to the deadly radiation, "was the biggest problem, the hidden surcharge for the free ride to Saturn and on to Uranus and to Neptune."

So there had to be what the engineers called a high degree of redundancy. Two centers, Ames and JPL, would run the mission and each would send out its own spacecraft. Ames would provide a pair of pathfinders to be launched a year apart. The first would head for Jupiter. What was learned navigating it through the asteroid belt (if indeed it made it through the belt) and overcoming the radiation and other problems would be applied to the second, which would fly by Jupiter and go on to Saturn, cautiously extending the reach.

Then JPL would send out four deluxe explorers named Thermoelectric Outer Planet Spacecraft, or TOPS, which would run on plutonium 238 that decayed in Radioisotope Thermoelectric Generators. The RTGs would be necessary because the spacecraft would be too far from the Sun to use solar panels. During the spring of 1969, the engineers in Pasadena began conceptualizing TOPS as a super-Mariner: a 1,200-to-1,400-pound, long-range caravel fitted with two highly advanced computing systems, a fourteen-foot-long deployable high-gain antenna for very long range communication, and a scan platform holding two television cameras, ultraviolet and infrared detectors, and other science instruments. By the standard of the time, TOPS was to be the ultimate exploring machine.

On October 15, 1970, John E. Naugle, NASA's associate administrator for space science and applications, issued a four-page invitation for scientists to define the Grand Tour missions: that is, what they were supposed to accomplish. Suggestions for specific experiments were supposed to come later. But first, the Grand Tour had to be defined to the point where funding could be requested. "Although these missions have not yet been authorized," he noted, "this invitation is being issued to assure scientific participation in the mission definition phase."

It was a siren call for scientists who yearned to pull precious data out of the then-unreachable outer planets and their gaggle of mysterious moons. Within six months, 108 scientists from the United States and six other nations had been chosen from some 500 who submitted proposals. They were in turn divided into thirteen science teams, many of whose members were already planetary exploration's all-stars: Murray, Van Allen, Sagan, Mert Davies, Verner Suomi, Tom Gehrels, William Hartmann, Toby Owen, Irwin Shapiro, Conway

Snyder, Tom Donahue, Lyle Broadfoot, James Warwick, Moustafa Chahine, Ellis Miner, Gerry Neugebauer, and several others.

The idea was to send four TOPS across the solar system: one pair to Jupiter, Saturn, and Pluto; the other to Jupiter, Uranus, and Neptune. They were supposed to explore farther out and send more information home than all other flybys, American and Russian, combined during eleven years of continuous operation. They were supposed to constitute a stunning finale to what was predicted to be the golden age of solar system exploration. They were supposed to do all that and more. They were supposed to but did not, because they never flew. TOPS died on the draftsman's drawing board, along with three Earth huggers in the Orbiting Solar Observatory series that were supposed to scrutinize the Sun in unprecedented detail. All seven were killed on the shuttle's sacrificial altar.

The Grand Tour Comes Apart

The frugal Richard Nixon, who was anything but a space buff, publicly signed off on the shuttle on January 5, 1972, with results that rippled ominously through a space science and exploration community whose budget was slipping into a free fall. Overall funding dropped steadily through fiscal year 1972, when it hit $3.3 billion. While only $100 million of that budget was earmarked for the Space Transportation System, it was clear that an ongoing program of such immensity would easily devour the single largest portion of NASA's funding for many years to come. Other long-range programs therefore had to be judged, not as temporary aberrations, but as long-range competitors to the shuttle. Space science ran head-on into the massive, humans-in-space, bread-and-butter, long-term engineering project that had been laid out in *Collier's* almost two decades earlier. The eleven-year outer-planet mission disintegrated on impact, and so did other science missions.

"The impending action on OSO [Orbiting Solar Observatory] is already widely known, but as far as we know even the Jet Propulsion Laboratory people directly affected are not yet aware of the TOPS decision," NASA Administrator James C. Fletcher wrote to Caspar W. Weinberger, the deputy director of the Office of Management and Budget. The tone of the memorandum was that of a close relative breaking the news of the death of a child to its parents. "We feel we can handle this situation better if JPL and the California Institute of Technology hear the decision directly from us." It was three days before Christmas 1971. With an election barely ten months away and the economy drained by the unpopular war in Vietnam, what the White House and Congress needed from space was not close-up pictures of Jupiter's Great Red Spot and Saturn's rings, but fifty thousand jobs that would create a device to carry more heroes into orbit and Republicans back to office. Planetary exploration was therefore seen by the pragmatists inside the Beltway as being a frivolous indulgence. Besides, the astronomers, the physicists, and the rest of them already had a major mission in the works—sending Viking to Mars—and that would have to do.

JPL got the bad news on January 11, 1972. It responded immediately by erasing Uranus, Neptune, and Pluto from its institutional blackboard and laying out what was in effect a large, souped-up Mariner. It gave the mission a conservative name: Mariner Jupiter Saturn, or MJS. Murray, who had recently succeeded Pickering as JPL's director, was now faced with a daunting challenge. He had to convince the Space Science Board that MJS would not threaten other space research projects, including an array of highly specialized science satellites.* He also had to show the White House that the mission was valuable for the nation as a whole; that it would be good PR.

"It could be incredibly visual and popular," the politically adroit geologist told an old friend who was plugged into the White House in February. "And once the Uranus, Neptune, and Pluto requirements are dropped, the mission is more within reach technically, and much cheaper." And, he added, shoving the hook all the way in, "It's certainly the most cost-effective space competition with the Soviets imaginable."

The game was played with exquisite subtlety. In the first place, Murray's friend and everyone else knew that the outer planets were hopelessly out of reach of the Russians. (Timur M. Eneyev, a corresponding member of the Soviet Academy, had suggested the year before that suspending instrument packages from balloons in the outer planets' atmospheres would be more productive than flybys, but he had not claimed that his country was ready to try such missions.) Furthermore, JPL reasoned that it was better to get a potentially open-ended mission started on a conservative note than stick with a grandiose plan that definitely was not going to fly (in either sense of the word). Put another way, there would theoretically be an option to keep going once MJS arrived at Saturn. There would be no option if it never even got off the ground.

What had to be done, then, was eliminate all mention of exploring planets beyond Saturn, at least officially. There was no sense in provoking the White House and the Hill. Yet stopping short of Uranus and Neptune if reaching

* These included still more Explorers (Explorer 45, a Meteor Technology Satellite, was at the moment being prepared for a July launch); the High Energy Astronomy Observatory series to collect high-quality X-ray, gamma-ray, and cosmic-ray data from the universe; the Orbiting Astronomical Observatory; the Orbiting Geophysical Observatory; the Biosatellite, or Biosat, program that collected data for the manned program; and Landsat (which was really an applications platform, not a purely scientific one). In addition, the science community participated in a number of European programs at the time, notably British, Dutch, and West German, and later with Canada, Italy, Spain, and the European Space Agency (ESA). (Ezell, *NASA Historical Data Book,* Vol. 3, pp. 151–201.)

them was possible was unthinkable. Nobody had been there. It was a once-in-two-lifetimes shot for a first good look at what until then had been absolutely out of Earth's grasp. Now, after decades of frustration, here was a chance to blow the blur off Uranus and Neptune and see them for what they were. And it was not as if the spacecraft had to be designed and built from scratch for the seventh and eighth planets; the same probes that scouted Jupiter and Saturn could just keep going. Uranus and Neptune were to the solar system exploration community what the Moon had been to everyone in the manned program.

So JPL continued to work on the Grand Tour, but it did so quietly, in the kind of excited conspiracy that went on in private schools of a certain class. There is ample evidence of this. A "Grand Tour Outer Planet Missions Definition Phase," presented by Davies, Hartmann, Owen, Sagan, and others on the project's imaging team on February 1, 1972, specifically mentioned three basic sets of four missions each and favored flights to Jupiter–Saturn–Pluto in 1977; Jupiter–Saturn–Uranus–Neptune in 1978; and Jupiter–Uranus–Neptune in 1979. "The best JSP 77 and JUN 79 trajectories, in terms of favorable encounters with the satellites of Saturn, and Uranus," the report noted, matter-of-factly, "have been identified below in terms of their periapsis [closest approach] times subject to the constraints given in the following table." It was pretty detailed stuff for a mission that was supposedly not being considered.

Murray later noted that Mariner Jupiter Saturn would need a greatly enhanced electronic brain, but designing one was not taken to be a serious problem in 1972 because the spacecraft was "more than five years away from launch." That would bring it into the 1977 Grand Tour window. If there was really no intention of extending the mission beyond Saturn, such a window would not be relevant, and the fact that it continued to be the target period for MJS's launch suggested a mission that would go beyond Saturn. "At the time neither Uranus nor Neptune figured seriously in MJS's plans," Murray would recall. "Seriously" was suitably ambiguous, however. Sending explorers on a route that would take them past Uranus and Neptune without making contingency plans for collecting precious data would have been unimaginable.

At any rate the friend in the White House, a sympathetic guy with an engineering background, sent the message where it needed to go. The now less threatened Space Science Board approved MJS on February 22. By mid-May, so had NASA, the Office of Management and Budget, and Congress, which provided an additional $7 million. The extra money would be used to make the brains of the spacecraft autonomous and reprogrammable and assure very long distance communication even under terrible circumstances. The $7 million, a hiccup even in NASA's budget, would indeed get one of the explorers to Uranus and Neptune.

Pioneering Voyages to Jupiter and Saturn

On March 2, exactly nine days after the Space Science Board issued its approval of Mariner Jupiter Saturn, the first of the two third-generation Pioneers, each also powered by plutonium, roared off the Kennedy Space Center's Launch Complex 36A on an Atlas-Centaur and headed for Jupiter at a record-breaking 32,114 miles an hour. "If Mars were placed on the face of Jupiter," one of the endless number of metaphors would note in a news release, "it would look like a dime on a dinner plate." At closest approach on December 3, 1973, the dinner plate's embrace would increase the bantam explorer's speed to 82,000 miles an hour, another record.

Pioneer F, as the 570-pound pathfinder was called at launch, consisted of a hexagonal body, called a "bus," that contained most of its eleven instruments. The bus was attached to a high-gain antenna that was nine feet in diameter and that would, like the ones on its JPL-developed cousins, both receive instructions and send back data. The antenna would always face home. A message sent to Earth from Jupiter would take forty-six minutes to get to the Deep Space Network at light speed—called "one-way light time"—and an answer would take the same time to reach the spacecraft. This was more important than it seemed because the Pioneers were basically brainless. Onboard computers were left off because they were too heavy, so the robots would have to be "flown from the ground." But it would take more than an hour and a half for the Pioneers to tell their controllers that there was a problem and for the controllers to order them to respond. The delay was dangerous but unavoidable. The instrument package itself weighed only sixty-five pounds, yet another indicator of how far the miniaturization process had come in the United States. Three appendages stuck out of the bus, each a third of the way around it: two booms holding RTGs and a third that carried the ever-present magnetometers.

Unlike JPL's Mariners and the Mariner Jupiter Saturn spacecraft, which were three-axis-stabilized, the Pioneers were spin-stabilized: they spun like giant tops at five rotations a minute. Three-axis stabilization, which was also used on reconnaissance satellites and Landsat, required an onboard brain to turn on the thrusters when the spacecraft tilted. Spinning would reduce the quality of the imagery the Pioneers sent back. But it would enhance the collection of magnetic field, cosmic-ray, charged particle, celestial mechanics, and other data because the spinning would provide 360-degree coverage: the whole picture on every rotation.

The "Peace" Plaque

Knowing that both Pioneers would keep going forever unless they collided with something, and taking it as an article of faith that there is intelligent life out there somewhere, gold-anodized aluminum plaques designed by Sagan, Frank Drake, Sagan's colleague at Cornell, and Linda Sagan, were attached to both Pioneers' buses. Each showed Earth's place in the solar system from which it came, a silhouette of the vehicle itself, when it was launched, a diagram representing fourteen pulsars that were arranged to show a scientifically literate being where the home star of this system was located, and a naked man and woman. The man's right hand was raised as a sign of peace. The woman stood passively beside him.

Whatever other life-forms would make of the two earthlings, they created a storm of controversy on their own planet. The print and television media were confounded by the problem of showing naked people to American families. The *Chicago Sun-Times* reacted by airbrushing out the couple's genitals and her breasts. The *Los Angeles Times* was barraged with letters denouncing the space agency for using taxpayers' money to send smut to space. Outraged feminists complained that the woman was placed slightly behind the man, did not have a hand up, and looked at him with apparent adoration that they insisted amounted to submission. The message that the male-dominated, macho NASA wanted to send to the farthest reaches of the universe, the angry women charged, was that their own sex was subservient. Still others protested that whoever was dominant, the couple represented a very limited group of humans and that, as a British editorial put it, future plaques ought to be designed by a large international ecumenical group of scientists and laypeople. The uproar over the plaque said as much about life on Earth as what it depicted.

The man's peace gesture could conceivably be taken seriously by extraterrestrials. But the spirit of cooperation it tried to convey was stubbornly ignored by Ames and JPL mission planners who were forced to work together. Headquarters had insisted when it assigned Pioneer Jupiter Saturn to Ames not only that JPL handle tracking and communication through the Deep Space Network, but that its experienced navigators help plan the trajectories as well. While the DSN had to be used for all deep-space missions, Hall immediately objected to the part about trajectory planning for what he would later call in absolute candor "competitive" reasons. Ames, after all, was moving into JPL's domain—hostile territory—and felt protective of its important and glamorous mission. At the same time, it rankled the brats of Pasadena that fate—no, headquarters—had awarded the first missions to Jupiter and Saturn not to the institution that had given the world Explorer 1, Ranger, Surveyor, and Mariner but to the upstarts upstate. For its part, headquarters had better things to worry about than a family feud in California. Hall was therefore overruled and ordered to go down to Pasadena and confer with Parks, JPL's senior guidance man, and anyone else who could help get the Pioneers to their destinations.

The sticking point was that neither side wanted to take responsibility for certain phases of the mission, Hall would recall years later. "God, I've never seen four guys who distrusted each other more in my life," he would say, laughing heartily. "They didn't trust us and we didn't trust them. It was like a couple of Arabs trying to sell a dead camel to each other." In the end, though, the relentless pressure of the 1972 window forced a reconciliation of sorts.

By Jupiter

Pioneer F, now renamed Pioneer 10, flew into the asteroid belt in mid-July 1972 and came out unscathed the following February. Seeing that its predecessor had suffered no damage during its seven-month passage, NASA launched Pioneer G, soon renamed Pioneer 11, on April 5, 1973. JPL's Mariner 10 left for Venus and Mercury on the first two-planet mission on November 4 of the same year. Meanwhile, the lab was also gearing up for Mariner Jupiter Saturn. What would soon be called the "golden age of planetary exploration" was under way.

Jupiter has undoubtedly exhilarated every astronomer since Marius and Galileo. It is a glorious sight seen through a telescope against the blackness of night: a spinning colossus of multicolored bands capped by gray hoods that crown both poles. A Great Red Spot larger than Earth was reported in the 1660s by both Robert Hooke, an English scientist, and Giovanni Cassini, the Italian-born director of the Paris Observatory who ensured his place in history by discovering four of Saturn's moons and a great deal about its spectacular rings.

The king of the sky is the dominant planet in this solar system and by itself accounts for two thirds of all planetary matter. It has 1,317 times the volume of Earth, but weighs only 318 times as much, meaning that it is made of something considerably lighter than the home planet's rock and iron. That something, which was revealed by analyzing its radiation through telescopes long before the planetary missions, is compressed hydrogen and helium, the two most abundant gases in the universe, and some methane and ammonia. Both of the first two elements are in roughly the same proportion as they are in the Sun, making Jupiter slightly denser than water. In fact, astronomers have calculated that if Jupiter had been a hundred times the size it is, it could have become a sun. And Marius's and Galileo's telescopes had spotted the first four moons at the dawn of telescopy. Eight more had been added to the original Jovian four by the time Pioneer 10 arrived, and there was every possibility that still others would show up during the encounter. (A NASA news release issued in February 1972 said that Jupiter "has 12 moons." That is a statement no serious scientist would have endorsed, since new objects are always turning up in space. "Has 12 *known* moons" is the way the cautious astronomers would have put it.)

Sagan, as cautious as the next astronomer, was haunted by Jupiter. "It is apparently radiating to space more energy than it is receiving from the Sun, energy perhaps derived from a continuing slow gravitational contraction like that which characterized the Sun before it became capable of thermonuclear reactions," he had said in a lecture in Oregon in 1970 that reflected his frustration. "Visual observation of Jupiter show its clouds to be in seething turbulence. Bubbles float to the cloudtops from the interior and are torn apart by the rapid rotational shear forces. These varying belts and bands and spots in the clouds are marked by an enormous variety of delicate and vivid hues. What are the molecules responsible for this coloration?"

The "Life" Card

Not to leave any possibility of making the morning papers and evening news unexploited, the space agency's public affairs specialists even played the "life" card. The fact that the planet was a mass of deadly radiation did not prevent those who wrote the Pioneer handouts from holding out the possibility that Jupiter, like Mars, just might contain the constituents of life. "Perhaps the most intriguing unknown is the possible presence of life in Jupiter's atmosphere," a hefty press kit issued on February 20, 1972, suggested. "Jupiter's atmosphere contains ammonia, methane, and hydrogen. These constituents, along with water, are the chemical ingredients of the primordial 'soup' believed to have produced the first life on Earth by chemical evolution."

And since Jupiter was known to have an internal heat source, the release continued, "many scientists believe that large regions below the frigid cloud layer are around room temperature." Jupiter, the writer therefore concluded, "could contain the building blocks of life." But even if the building blocks were there, it would in no way have shown that there was life in the massive, swirling poison factory. Europa, one of the Galilean moons, was a far more likely candidate to harbor life in some primitive form because it had a real water ocean and a possible inner heat source. But in 1972 there was no way of knowing that.

Tweaking the Dragon's Tail

Having survived dangers of its own, including a near blinding of the photopolarimeter by the Sun, and then having broken virtually every long-distance space record, Pioneer 10 began its near encounter on November 3, 1973.* During the next twenty-six days, the speeding spacecraft crossed the

* It measured the brightness and polarization of light scattered by interplanetary dust and solid matter and imaged Jupiter itself for the structure and composition of its clouds and to provide details about the amount and nature of the gases in its atmosphere. The heart of the very sensitive photopolarimeter was a photoelectric sensor, which worked on roughly the same principle as a photographic light meter and which would have become totally blinded by a direct look at the Sun.

paths of all seven of the outermost of Jupiter's known moons and was plowing into its fearsome radiation belts. By then it had also penetrated Jupiter's massive bow shock, where the solar wind runs into the planet's magnetic field and separates like water flowing past both sides of a ship's bow. All this and much more poured into Ames via the Deep Space Network. As always happened at encounters, the tightly lidded science and engineering played counterpoint to raw emotion as the explorers at Ames reveled in the history they were making.

"We are really only twelve generations away from Galileo and his first crude look at the planet," an exhilarated, almost disbelieving Charlie Hall told reporters in the main auditorium. "Twelve generations later, we are actually there, measuring many of the characteristics of the planet itself," Pioneer's project manager added as he and his spacecraft sped together toward the gigantic ball of storms. Hall, amazed at the wonder of it all, had just found a place for himself in the annals of time.

At first the images that came back from the photopolarimeter in real time were no better than those taken through telescopes. By December 2, the day before closest approach, they began to surpass anything taken from Earth. The pictures showed not only a growing planet, but Io and Ganymede as well. Pioneer 10's closest approach to Jupiter came within one minute of schedule, with the spacecraft slicing between the planet and Io for an occultation: in effect silhouetting part of Io in a radio beam to Earth that would provide data on the moon's size and the thickness of its atmosphere, if it had one.

By that time, Pioneer 10 was working its mechanical heart out as it slammed into radiation so intense that two of its cosmic-ray detectors became saturated. Others, specially designed to contend with that problem, continued to measure the proton flux around the planet while ultraviolet measurements were recorded and infrared readings were taken of Io. Meanwhile, twenty-three more images were taken of Jupiter, Io, Callisto, and Ganymede.

Less than three hours before closest approach, Pioneer 10 began an hour's imaging of the enigmatic Great Red Spot. The pictures would be the first to suggest that the huge area of turbulence might be a massive storm. They and others were good enough, and were shown to reporters quickly enough, to earn the Pioneer program an Emmy award from the National Academy of Television Arts and Sciences.

The closest approach came at 6:26 P.M. Pacific Standard Time on December 3 as Pioneer 10 skimmed over Jupiter's cloud tops at an altitude of only 81,000 miles. The elated scientists who were running the robot's particles and fields experiment saw that their detector had run into a clearly defined barrage of particles, suggesting that what had long been surmised was probably true:

Jupiter had its own rings. Meanwhile, the planet's atmospheric composition, temperature structure, and thermal balance were being recorded, and so were the temperatures of Io, Ganymede, and Europa.

Then Pioneer 10 suddenly dropped out of sight, disappearing behind Jupiter and causing a sixty-five-minute communication blackout. The tension level at Ames surged even though the break in contact had been expected.

"We watched the PICS [Pioneer Image Converter System, which turned the signals into pictures] image displayed in real-time as the signals came back from the distant planet," Lyn R. Doose, an imaging experimenter from the University of Arizona, would report. Then "a single bright spot appeared, and then another, until a line gradually built up. We knew we were seeing sunrise on Jupiter as the PICS image showed a crescent-like shape. We survived passage through the periapsis," the relieved scientist added, grateful that he and his spacecraft had made it safely through the radiation flux. Pioneer 10 was now the first machine from Earth to make it through both the asteroid belt and Jupiter's blanket of radioactive poison.

"We can say that we sent Pioneer 10 off to tweak a dragon's tail, and it did that and more," said a proud Robert Kraemer, who represented headquarters. "It gave a really good yank and . . . it managed to survive."

"This has been the most exciting day of my life," Pioneer 10's exhausted but exuberant chief scientist, Richard O. Fimmel, added. He spoke for the whole center.

When its encounter with Jupiter was over, Pioneer 10 cruised out of the solar system at about 25,000 miles an hour. Still riding the solar wind, it would cross Pluto's path early in 1990 and then carry its plaque to interstellar space, all the while reporting what it found on increasingly faint signals that took almost twelve and a half hours to reach home. The power of the signals would be so weak, Fimmel explained, that it would have to be collected and saved for 11 billion years to light a 7.5-watt night-light for a millisecond. And by that time, the Sun would be just another dot of light in the vast galaxy. At last report, it was headed for the star Aldebaran in the constellation Taurus.

There would be a party at Ames in June 1988 to mark the fifth anniversary of Pioneer 10's crossing Neptune's orbital path. Someone jokingly asked a representative of the company that had built the tough little record breaker whether it was still under warranty. "TRW's position," he answered, "has been that if you bring it back, we'll fix it."

Pioneer 11 could have duplicated its predecessor's mission, but there were other things to be discovered, and since its predecessor had done so well, it was sent off chasing new possibilities. The second spacecraft to reach an outer planet was therefore ordered to approach Jupiter from its left side and under its southern hemisphere, coming as close as 26,725 miles before hurtling almost straight up and out of the plane of the ecliptic, at least momentarily. It confirmed a great deal of the data sent by Pioneer 10, added information on the Great Red Spot, and provided new details about Jupiter's immense polar regions. Pioneer 11 returned 460 pictures of Jupiter and its four most famous moons, many taken at angles impossible to get from Earth. Then it looped over Jupiter, raced high above its north pole, and leveled out once more as it headed for its rendezvous with Saturn. It was now called Pioneer Saturn.

Lord of the Rings

Having picked up the predicted assist from Jupiter's gravity, Pioneer 11 made its closest approach to Saturn on September 1, 1979, after a two-billion-mile trip that took six and a half years. The United States now put yet another planet on its trophy shelf.

During its ten-day encounter, Pioneer 11 skimmed under the famous ring plane twice, coming as close as 1,240 miles, and streaked past the planet itself at a height of only 13,000 miles. The most fundamental discovery was not about the rings, but that Saturn has radiation belts and a strong magnetic field and magnetosphere, both of which had been suspected but unproved.

The probe also discovered a new moon just beyond the edge of the rings and did photopolarimeter measurements of the known moons—Iapetus, Dione, Tethys, Rhea, and Titan—and ultraviolet observations of Hyperion, Rhea, Dione, Tethys, Enceladus, and Titan. In their book *The Grand Tour,* Ron Miller and William Hartmann took strong exception to what they called the "nine-planet gestalt," arguing that there are many more individual worlds in this solar system than the officially designated planets. It is hard to imagine any scientist disagreeing. Titan, for example, is larger than Mercury and therefore was an irresistible subject for Pioneer 11's atmospheric, temperature, heat balance, magnetic wake, and other experiments.

Carefully following Pioneer 11's trajectory as it sailed past Saturn helped project scientists calculate the planet's shape and gravity more accurately than was possible from Earth. That, together with a temperature profile made from infrared measurements of the heat given off by Saturn's clouds beyond what was absorbed by the Sun, shed light on its interior. Saturn's core, roughly the size of eighteen Earths, was suddenly understood to have two distinct regions. There was an iron-rich, rocky inner core encased by a ball of ammonia, methane, and water. No one suggested the possibility that there was potentially life in there, though the possibility that it existed on Titan was mentioned.

While the resolution of Pioneer 11's imagery was necessarily poor compared to that taken by three-axis-stabilized spacecraft using television cameras, it was good enough to spot two new rings. One, which appeared separated from the A Ring by a gap that was promptly called the Pioneer Division, was named the F Ring. At less than five hundred miles across, it was considered "narrow." The other, named the G Ring, directly adjoined the F Ring and was more than ten times wider. There was much more that would take months to sort out.

When Pioneer 11 left Saturn, it headed across the other side of the solar system from its twin at a speed of about 275 million miles a year. Both of them made three fundamental contributions to the exploration of space. Most obviously, they returned a huge amount of new information about the two largest planets in the solar system and the space between them. They also demonstrated that the outer solar system could be safely navigated. And they did indeed find a path for their successors, both of which were already on their way.

Venus Partially Unveiled

While the Pioneers reconnoitered the two big outer gas balls, JPL's Mariner 10 was looking inward (so to speak). Riding an Atlas-Centaur, it left Florida during the night of November 4, 1973, and had its closest encounter with Venus ninety-three days later as it sailed 3,600 miles over the hot, fog-shrouded sphere.

Whatever else the poisonous mist concealed, it was known to be hiding five Soviet spacecraft (or their scattered remains), one of which, Venera 4, had six years earlier become the first visitor from Earth to transmit data back before it had been crushed by atmospheric pressure at an altitude of seventeen miles. Veneras 5 and 6 had relayed more atmospheric data in 1969 before they, too, had been crushed. Then Venera 7 had made the first soft landing on December 15, 1970, and sent back data for twenty-three minutes. Venera 8, a follow-on soft lander, had repeated the exacting operation in July 1972 with a fifty-minute burst of information showing that Venus was a pressure cooker with a landing-site surface temperature of 860 degrees Fahrenheit and an atmospheric pressure ninety times as great as Earth's. That, as well as an atmosphere that would prove to be 96 percent carbon dioxide, showed beyond doubt that Venus was uninhabitable by anything that lived on Earth. The heat and carbon dioxide atmosphere would also be used to help model what could happen to Earth if the so-called greenhouse effect went out of control. Soviet scientists had not allowed a single Venus launch window to pass without taking at least one shot at it since planetary exploration had begun. And those shots had been very successful. Now it was the Americans' turn.

Mariner 10's camera sent back ultraviolet pictures of features on the top of the atmosphere that could not be seen in the visible spectrum, and which could be used to trace the atmosphere's unique movement. The imagery showed that most of the heat from the Sun is absorbed in the upper atmosphere, making it hotter than the surface, unlike Earth and Mars. No doubt about it: old Venus was now understood to be truly unearthly. And that, for Murray, was why it was important. Certainly building knowledge for its own sake was space science's essential driving force, and one that lent it nobility. Yet many scientists, including Murray, also saw a potentially dire necessity in scouting the solar system. Space scientists in effect lived their professional lives in the implacably hostile desolation and violence of deep space, at least where their psyches were concerned. That much became evident when they anthropomorphized their robots and flew with them in the first person plural. "*We* are ten million kilometers from" Mars (or Jupiter or Ganymede). . . . "*We* have acquired Canopus. . . . *We* are twelve minutes from closest approach" is the way they would invariably describe what their machines were doing. They were in effect out there, too, like the Saltine Warrior. And being out there made them appreciate where they came from. They therefore liked to make love to their planet, which they did by hiking, skiing, mountain climbing, sailing, and camping on it, and swimming deep within its life-giving oceans.

"To avoid accidentally destroying this delicate balance in Earth's atmosphere by mankind's activities," explained Murray (who flew sailplanes and biked for relaxation), "it is mandatory that we understand fully the ecosphere in which we have evolved. This understanding is difficult, if not impossible, without comparisons to other planetary atmospheres." Venus was a warning. Mars was a potential lifeboat. Jupiter and its moons were a miniature solar system that could provide clues to how the larger system worked.

Having returned more than four thousand pictures of Venus, plus a pile of infrared, ultraviolet, charged particle, and other data, Mariner 10 set off for savagely pockmarked Mercury with a boost from Venusian gravity. A few of Aphrodite's veils had finally been pulled off.

Encounter with Mercury

Mysterious little Mercury, named after the son of the mighty Jupiter, was the messenger of the gods, bearer of luck, patron saint of astronauts, and, more to the point, the conductor of souls to Hades' eternal fire. Mysterious because, as the closest planet to the Sun, it is difficult to see through telescopes on Earth. So there had been a great deal of speculation that had been passed off as fact. According to one theory, the little planet did not rotate at all, so the side facing the Sun was therefore as hot as hell: 750 degrees Fahrenheit, according to RAND's authoritative *Space Handbook,* published in 1958. Wrong. Radar observations in 1965 showed that Mercury does indeed rotate, though at a snail's pace compared to Earth. Its dawn-to-dusk day is equal to almost fifty-nine Earth days, while it takes only eighty-eight days to circle the Sun. RAND was right about the temperature, though: it is hot enough to melt lead during the day. But until Mariner 10 got there, Mercury appeared as a dusky blob with no distinguishable surface features. What was known—that it is denser than the Moon and Mars, for example—had been surmised from the blotchy telescope pictures and from the radar data.

On March 24, 1974, with correspondents from around the world gathered in JPL's von Karman Auditorium, Mariner 10 made the first close approach to this system's innermost world. The first picture, taken that day at a distance of more than 3.75 million miles, looked about the way it would have through a telescope on Earth. Even after a computer tried to refine the details, the imagery showed a rough-textured dark globe with the complexion of a splotchy orange. But the fun of it, as it always would be with planetary encounters, was watching the new world grow and became clearer every day as its unique identity came into focus. As Mariner 10's imagery came into JPL, so did data showing how Mercury interacts with space, and particularly with the solar wind. The heavenly breeze does not react the same way on every body it caresses. Earth's protective magnetosphere, which is caused by the planet's rapid rotation, holds the solar wind at bay. The charged ionospheres around Venus and Mars do the same thing. But since the Moon has neither a magnetosphere nor an atmosphere, it bears the breeze's full brunt. Since Mariner 10's scientists thought that Mercury also lacked protection, it was reasonable to believe that it, too, would face right into the great wind. Wrong again. Nineteen minutes before closest approach on March 29, the plasma team was astounded to see the plasma flux jump, indicating that the spacecraft had crossed a bow shock wave. They were left to ponder how a moonlike, atmosphereless planet that barely rotated and that therefore should have no magnetic field could fend off the solar wind.

The answer was not long in coming. Mariner 10's magnetometers showed that Mercury has a magnetic field and that it's powerful enough to deflect the wind. But why? One possibility was that even a slowly rotating planet can have a magnetosphere if it has the right kind of electrically conducive, fluid core.

The answers, as usual, kept raising new questions. What was sometimes difficult for the uninitiated to understand was that the new questions not only required no apology, but were what made the game fun. William K. Hartmann, who was on the Mariner 9 science team that got the first really good look at Mars, explained that the heart of the process is the endless refining of the questions. "It has become a cliché of science reporting that new questions are raised as fast as old ones are answered," he wrote. "How, then, can a voyage of discovery be worthwhile? The answer lies in the existence of a hierarchy of questions. Ideally, the early questions are first-order questions, and as they are answered, we proceed to finer levels of detail."

The magnetosphere was only one case in point. So was the possibility that Mercury had a moon of its own. On March 30, the day after closest approach, the ultraviolet team reported that analysis of spectrometer data collected a few days before seemed to show that ultraviolet sunlight was reflecting off something that was moving away from the planet at regular intervals. A search with the television camera, completed on April 1, turned up nothing. The source of the light was a distant star; a mere twinkle in Mariner 10's ultraviolet eye. April fool!

Meanwhile, the spacecraft's electrical power drain had begun to surge, causing an alarming rise in temperature, which was heating the instrument bay. It was the latest in a series of crises that had the potential to abruptly end the mission. Believing that Mariner 10 was having a mechanical "stroke," its desperate experimenters reacted instinctively by arguing for one last frenzied push to collect as much precious data as possible before their spacecraft expired (along with the papers they wanted to send to journals).

The popular conception of the scientist as an unflappable individual who is always in cool control is misplaced. Good science requires an emotional as well as intellectual investment. And the fleeting nature of the early encounters—quick flybys in which the time to get the data was very limited—made the mission scientists competitive and often irritable. Mariner 10 at Mercury was no exception. James Dunne, a soft-spoken project scientist, thought that the spacecraft could be saved if it got some rest. But his science teams wanted every sensor on Mariner 10 turned on for one last shot at collecting data. It was like wanting to burn out a car's engine in order to win a race. So Dunne became a ship captain trying to stop a mutiny. At one point he rushed into Walter E. "Gene" Giberson's office and, pounding on the project manager's desk, demanded that his scientists be ignored. They were. Instead, Giberson ordered that the television cameras and some other instruments be turned off. The surge subsided, and Mariner 10's sensors began to cool down. So did Dunne.

By carefully nursing its maneuvering gas and working phenomenally difficult trajectory adjustments, Mariner 10's unheralded navigators sent it past Mercury three times, like one horse repeatedly overtaking another as they raced around the Sun. It had its second closest approach on September 12, 1974, and the third and last on March 16, 1975, when it came within 203 miles of Mercury's surface. There were eight trajectory corrections instead of the usual two, amounting to a fourfold increase in navigational accuracy, with Venus's gravity playing a key role. Mariner 10 was finally shut down on March 24, 1975, when it literally ran out of gas.

The first close-up pictures of Mercury and the massive accumulation of other data from both it and Venus were obviously brilliant, with all mission objectives met. Less obviously, Mariner 10 left behind a treasure of in-flight science and engineering experience that was being applied to Pioneer's exploration of the outer planets and to other operations as well. It was a mission, like Pioneer Jupiter Saturn, that the Russians could only envy.

Venus Is Beseiged

By December 1978, the cold war was at its height, and as a result the golden age of exploration was well under way. So intense was the competition between the United States and the Soviet Union to pry secrets out of their planet's neighbors for prestige as well as knowledge that ten separate spacecraft were working at Venus alone that month. Meanwhile, both Viking landers were reporting to Earth from the plains of Mars and Pioneers 10 and 11 were being followed to Jupiter and Saturn by two JPL spacecraft named Voyager 1 and 2. They were the reincarnation of the demised TOPS and one of them was in fact headed for the Grand Tour.

Four of the ten spacecraft that went to Venus were Russian orbiter-landers called Venera 11 and 12. The Venera 11 orbiter arrived on December 21 in time to catch a tremendous storm and managed to spot lightning and record thunder. Its lander came down in one piece and sent data back for ninety-five minutes. Venera 12 followed four days later and ran into the same storm. Its lander transmitted for 110 minutes and reported a surface temperature of 860 degrees.

The most sophisticated of the Venus explorers were developed by Ames and were called Pioneer Venus 1 and 2, or Pioneer 12 and 13. Pioneer 12 was a radar mapper that was supposed to circle Venus and penetrate its soupy envelope. Pioneer 13, its sister ship, was a so-called Multi-Probe consisting of a main spacecraft and four babies (one of them fairly large) that went to Venus together and then parted ways for their individual missions. Those missions, as "probe" implied, were for all five to plunge into the soup and send back data on Venus's wind, temperature, circulation, and the atmosphere's composition and pressure before they slammed into the planet itself and were smashed to bits. (One actually survived the plunge through the thick atmosphere and continued to broadcast for sixty-seven minutes.)

Pioneer 12's radar cut right through the shroud and sent back the first sweeping, detailed views of the surface of Venus. (Venera 9 had sent back the first pictures from the planet's surface in October 1975.) Awestruck mission scientists looking at their television monitors at Ames saw whole continents materialize from out of the mist. One of them, named Ishtar Terra, was the size of Australia; another, Aphrodite, was as big as Africa. Out popped mountain ranges, volcanoes, valleys, impact craters, plateaus, and basins, some of which had been hinted at by Earth-based radar, but were now actually seen for the first time. Pioneer 12 made maps of 90 percent of Venus in detail, including a gravity map that indicated that there had been a lot of "adjustment" on the planet's surface.

Perhaps the most important practical finding would be that Venus's landscape was a lot like Earth's and that its hellish surface temperature almost undoubtedly came not from heat pouring out of its clouds but from a massive greenhouse effect in which carbon dioxide and a tiny amount of water vapor prevent surface heat from being radiated back to space.

Technically, the major part of its observation mission was supposed to be completed by the following August, but the orbiter would actually keep sending data back until Venus's gravity pulled it to a fiery finale in October 1992. Eleven of its twelve science experiments functioned to the end.

The Reincarnation of Voyager

The bitter memories of its first Mars lander project notwithstanding (or maybe because of them), JPL changed Mariner Jupiter Saturn's name to Voyager in 1977, the year two of the spacecraft finally took off for what was still on the books as a two-planet mission. Voyager, unlike Mariner Jupiter Saturn, was an unrestrictive name, a name that would be valid no matter where it went. No one would be able to hold NASA to the two-planet mission on a technicality should a decision be made to extend the trip to Uranus and Neptune.

The two Voyager spacecraft were built in-house and showed their Mariner bloodline. Yet they were designed to be bigger, tougher, and more sophisticated than either their predecessors or Pioneers 10 and 11. These arrows were crafted to reach the outermost planets and return a great deal of data. Their inherent design, in fact, gave the clearest indication that the possibility of making it at least as far as Neptune was under serious consideration at JPL. It does not take an ocean liner to cross a lake, even a Great Lake.

Unlike the Pioneers, for example, Voyagers 1 and 2 were autonomous. Each was fitted with three different computers (each of which in turn had a backup) that steered the spacecraft, processed scientific and engineering data, and coordinated onboard systems. While all three combined did not have the capacity of one of today's laptops, they were the state of the art at the time, and were not used frivolously. With instructions loaded at JPL or transmitted from there and then updated, they could run the science experiments and operate Voyager's own flight systems. And they were even programmed to automatically react to problems or simply changing conditions. The computers could therefore keep the two spacecraft functioning semiautomatically for months if necessary, making them the smartest robots ever to head for deep space.

The heart of each Voyager was a rugged ten-sided bus that was made of ten bays holding science instruments, electrical and communication equipment,

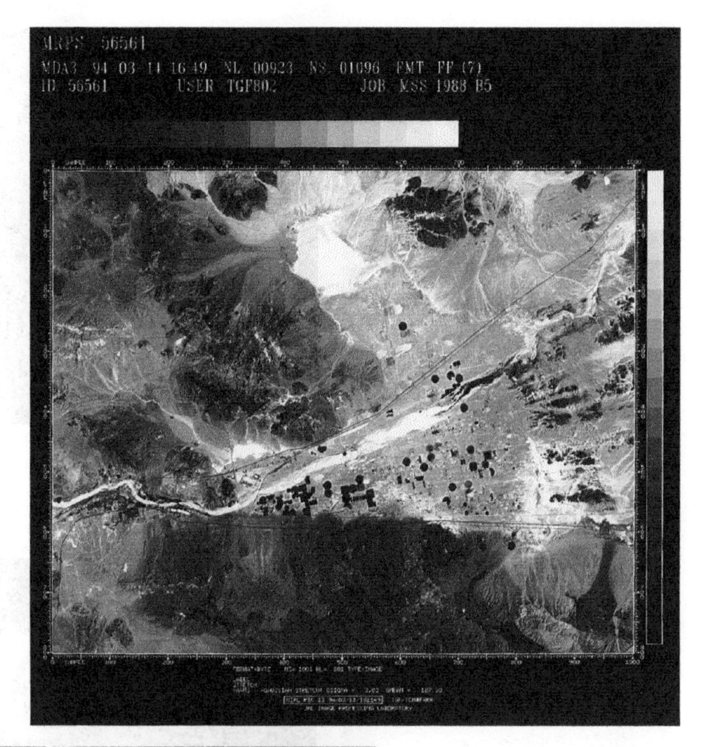

The first Landsat was launched in 1972. Its successors and similar French, Russian, and other civilian remote sensing satellites have revolutionized the way we see and use Earth. This photograph of the Manix Basin area of the Mojave Desert in California was taken in the summer of 1988. The outskirts of Barstow are barely visible in the blowup. The large white area at the top of the picture is Coyote Dry Lake. The dark, ghostly circles immediately to the southeast of the lake are abandoned fields with little vegetation. (NASA)

From the Earth to the Moon. Jules Verne's fiction became a reality on July 16, 1969, when *Apollo 11* was sent to the Moon by this colossal Saturn 5. The photograph of Earth, taken 98,000 miles from home, shows most of Africa, the Middle East, and part of Europe against the icy void of space. It and others like it became the icons of the environmental movement. Four days later, Buzz Aldrin was photographed standing on the Sea of Tranquility by Neil Armstrong. The plaque they left behind said that they represented "all mankind," but they planted the Stars and Stripes. Things did not go so well nine months later, when an oxygen tank exploded on *Apollo 13,* nearly killing Jim Lovell, Jack Swigert, and Fred Haise. Cheering erupted in Houston the instant they splashed down in the Pacific. Gene Cernan took a spin around Taurus-Littrow in a lunar rover during the final, *Apollo 17,* mission in December 1972. (NASA)

HERE MEN FROM THE PLANET EARTH
FIRST SET FOOT UPON THE MOON
JULY 1969, A. D.
WE CAME IN PEACE FOR ALL MANKIND
NEIL A. ARMSTRONG
ASTRONAUT
MICHAEL COLLINS
ASTRONAUT
EDWIN E. ALDRIN, JR.
ASTRONAUT
RICHARD NIXON
PRESIDENT, UNITED STATES OF AMERICA

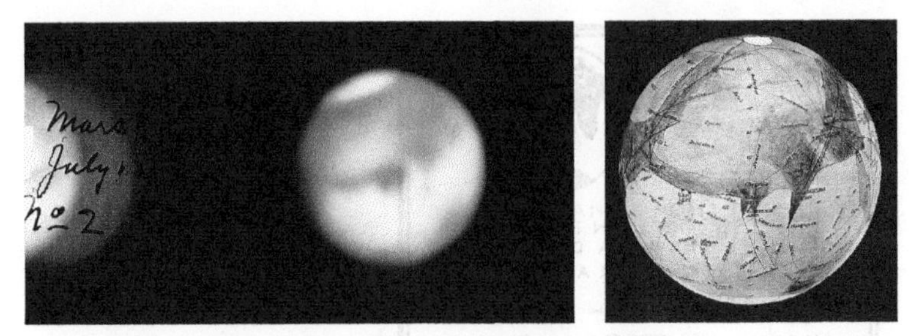

The blurry ball was the best view of Mars obtainable from Earth as late as 1939. It and others were used by astronomers to map the Red Planet. Differences in succeeding images were interpreted as meaning that vegetation was spreading and large bodies of water were changing shape. H. Spencer Jones, Britain's Astronomer Royal, concluded that there was almost certainly vegetation on Mars. (Alton Blakeslee collection)

This "priceless Rembrandt," as JPL's Jurrie van der Woude called it, was the first image returned to Earth from Mars and was ten times better than the best picture taken with a telescope. It was made by Mariner 4 in July 1965. The picture, which hangs at JPL, is in effect a number painting made of thin strips of paper that were glued to a board and colored with pencils according to tiny numbers that corresponded to various tones. (JPL)

Mariner 4, an ungainly cousin of Ranger and ancestor of Viking, looked like a one-eyed bat from this angle. Solar panels extended out from the bus and an antenna sprouted on top of it. (JPL)

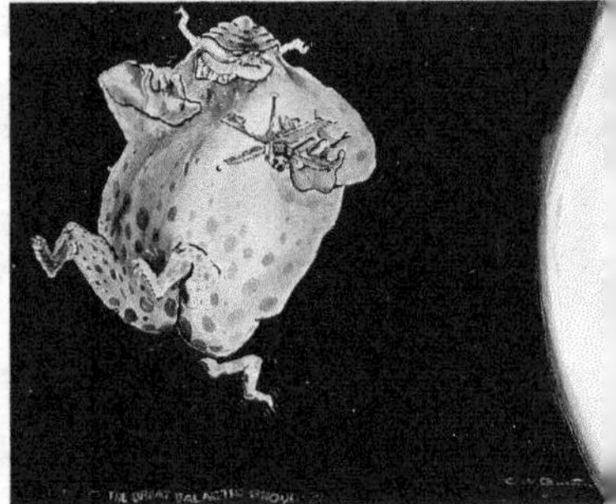

The dreaded Great Galactic Ghoul, seen here about to turn Mariner 7 into an hors d'oeuvre, was conceived by *Time* reporter Donald Neff and painted by contractor artist G. W. Burton. The voracious beast was a metaphor for otherwise unexplainable problems that happened around Mars. It found Soviet spacecraft particularly delicious. (JPL)

Two Vikings, which orbited and landed on the Red Planet in 1976, were the most successful Mars missions and returned an immense amount of data (but no sign of life). The orbiter is seen here on the bottom and is carrying the lander inside a bioshield on its back. Together, they weighed 7,760 pounds and were the last heavy Mars explorers. Viking was also the last billion-dollar unmanned mission. (JPL)

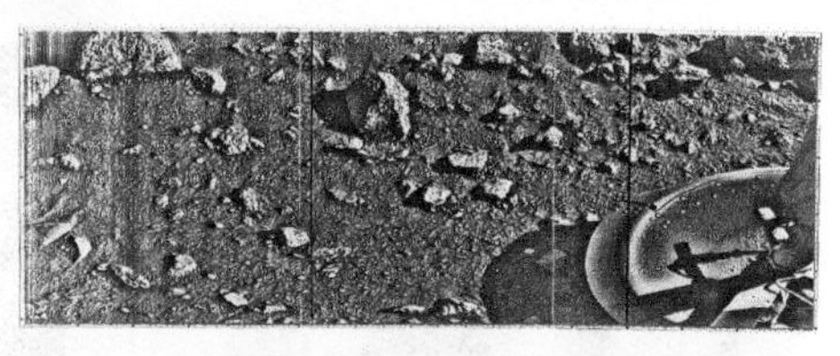

The first complete picture of the Martian surface was taken by the Viking 1 lander minutes after it touched down on Chryse Planitia on July 20, 1976. Before it raised its sights, the camera caught rocks, sand, dust, and one of the lander's own pads. Geologists quickly found evidence of wind. (JPL)

Its sights raised, the camera took in this richly detailed panorama. Meanwhile, the orbiter was photographing Olympus Mons, the highest mountain in the solar system and wider than Missouri; Valles Marineris, a valley the length of the United States; and unmistakable signs that there was once water on the planet. Evidence of a great deal of water moving quickly was found twenty-one years later by Pathfinder's Sojourner, the first rover from Earth to roam the Martian surface. (JPL)

The Viking 1 lander dug this trench and scooped up soil samples for on-board testing for signs of life and then photographed the hole. No life signs were found. (JPL)

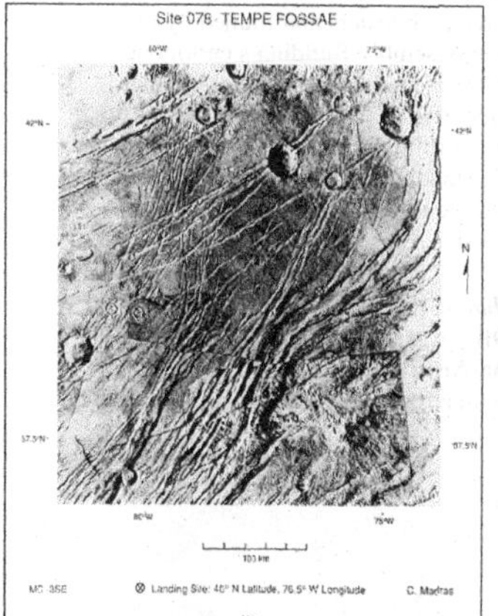

Viking's rich legacy is evident in this image of the Tempe Fossae region of Mars's northern hemisphere sent back by the orbiter and as it appeared in the *Mars Landing Site Catalog*. It not only shows dead volcanoes and fracture lines, indicating a once-active interior, but the possibility of "subsurface water for potential use by manned landing mission." The area was therefore chosen as a good place to send a Mars sample return in the search for evidence of life and a promising landing site for human explorers. (NASA)

NASA Administrator James C. Fletcher and President Nixon posed at San Clemente on January 5, 1972, the day the President formally approved the Space Transportation System. They held a model of the more expensive design, which consisted of two reusable vehicles, and which was therefore not built. (NASA)

Challenger, hanging from a crane in the VAB like a lifeless bird on a string, was lowered for mating with its external tanks and SRBs in early December 1982, in preparation for its maiden flight and the sixth shuttle mission on April 4, 1983. The picture is distorted because the VAB is so big that the photographer had to use a very wide-angle lens. (NASA)

Columbia, attached to its external tank and solid rocket boosters and framed by the Vehicle Assembly Building's enormous doorway, inched toward Pad 39A in late December 1980 for the first manned test flight the following April 12. The size of the people standing on the crawler-transporter around the shuttle gives some idea of the VAB's size. (NASA)

This rare view of two shuttles poised for liftoff—*Columbia* on the Kennedy Space Center's Launch Complex 39A (foreground) and *Discovery* on 39B—was taken in early September 1990. *Discovery* went first, on October 6, carrying the European-American Ulysses solar research spacecraft to orbit. From there, Ulysses went to Jupiter with a rocket booster and then, with gravity assist, on to the Sun. (NASA)

This massive N-1 Moon rocket was launched from Tyuratam at night to hinder Western intelligence. Within two minutes, its first stage failed, sending it back down in a huge ball of fire that could be seen for many miles. The last N-1 test was on November 23, 1972, two weeks before the last Apollo launch, and also ended in disaster. (RSC Energia)

The formidable Energia super-booster, a *Buran* orbiter attached, stood on its pad at Tyuratam in November l988. It was designed to be flexible enough to carry either a shuttle and its crew or unmanned spacecraft. It lifted this unmanned *Buran* to orbit on November 15. The Soviet shuttle was virtually identical with its American counterpart. (RSC Energia)

The unmanned *Buran* launched by the Energia made two orbits and landed safely back at Tyuratam. With the end of the cold war and the collapse of the Russian economy, the shuttle was dropped in both senses of the word. (NASA)

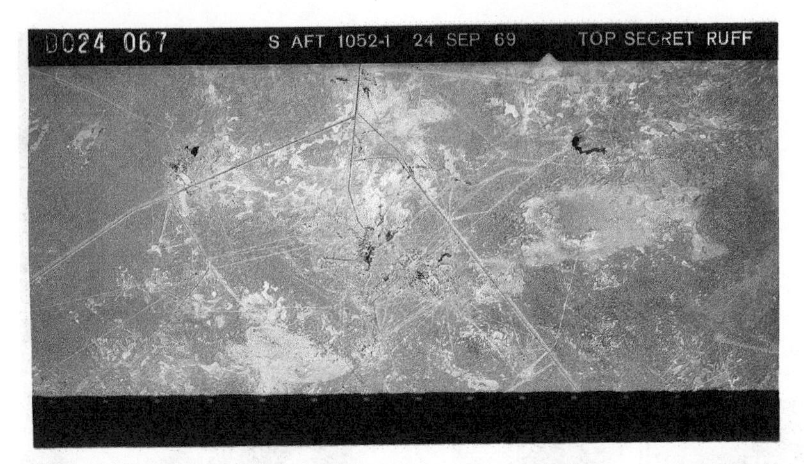

Western intelligence was not hindered, since Tyuratam was routinely photographed from space. This image, taken on September 24, 1969, by the last KH-4A from an altitude of 111 miles, shows the sprawling launch facility exactly as the photointerpreter received it. "1052-1" is the mission number; "Top Secret Ruff" is the classification. (CIA)

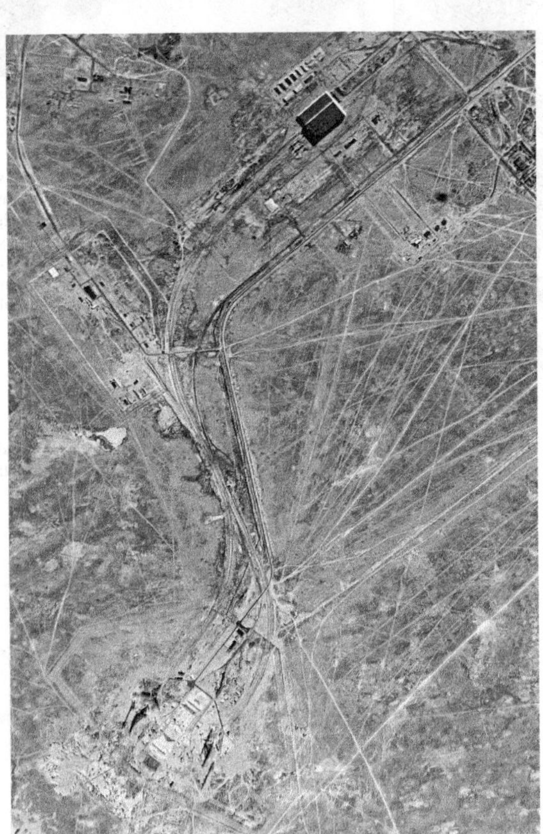

Magnified, the center of the image showed the quarter-mile-long rocket assembly building (dark structure at the top), which was linked by an S-shaped railroad track to a launch complex (bottom left center). Higher magnification revealed N-1s and, in four instances, their charred remains. (CIA)

Public relations has been a NASA staple from the beginning. Here, the crew for *Challenger*'s flight posed for an "F-Troop" portrait while taking a "break in training" in April 1983. The idea, according to a release that accompanied the picture, was that the mission was STS-6 and "F" is the sixth letter of the alphabet. In fact, the elaborate costumes and props suggested that the astronauts were riding across a new and exciting frontier. Donald Peterson, kneeling in front, and Story Musgrave, standing with trumpet, were mission specialists. Paul Weitz, seated, was the commander of the mission. Karol Bobko, standing next to Musgrave, was the pilot. (NASA)

Sharon Christa McAuliffe, a high school teacher from New Hampshire, was one of seven crew members killed when *Challenger* exploded seventy-three seconds after liftoff on January 28, 1986, in the worst single accident of the space age. Her mother later said that Christa told her the day before the explosion that the space agency was under so much pressure to launch that it was going to, no matter how cold it was. (NASA)

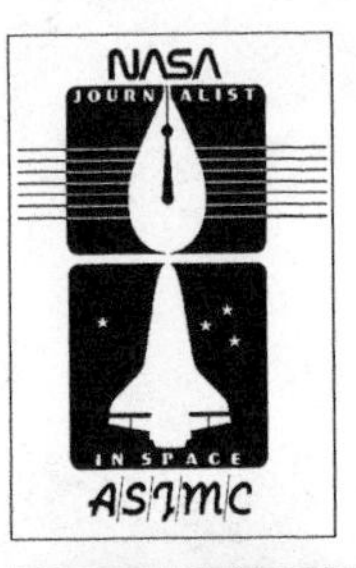

A Journalist in Space was supposed to follow the Teacher in Space. The deadline for applications was thirteen days before the *Challenger* accident. More than seventeen hundred journalists applied for the ride, which was quietly abandoned. (William E. Burrows)

Rockwell tech reps at the Kennedy Space Center who saw this scene from Pad 39B on television monitors likened it to "something out of *Dr. Zhivago*," and specifically to an ice-encrusted house in Siberia, as they urged that the launch be postponed. The temperature at launch was thirty-six degrees Fahrenheit, fifteen degrees colder than the coldest previous launch. (NASA)

The death of *Challenger*. The orbiter and external tank were hidden in the heart of the explosion, while one solid rocket booster raced up and to the left, and debris shot to the right. (NASA)

A skeptical Richard P. Feynman listened to testimony at a session of the Rogers Commission at the Kennedy Space Center. The theoretical physicist and Nobel laureate became an instant celebrity when he dropped a piece of O-ring rubber into his ice water to demonstrate that it "had no resilience whatever when you squeezed it." (NASA)

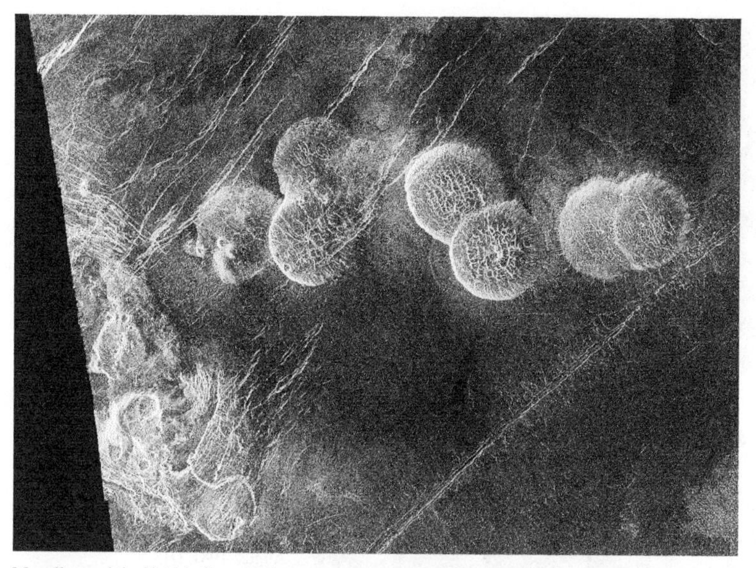

Magellan and the Hubble Space Telescope were two stars of the science program. Magellan's radar collected the first comprehensive imagery under Venus's cloudy shroud. These domical hills, or lava domes, that turned up in the planet's Alpha Regio region in November 1990, have no counterparts on Earth. Though they look as if they belong under a microscope, they average fifteen miles across. (JPL)

A star is born. This spectacular giant star incubator, taken by the Hubble Space Telescope in April 1995, is seven thousand light-years away in the Eagle Nebula. A light-year is the distance light travels in a year: five trillion nine hundred billion miles. Dense hydrogen and dust inside these towering "elephant trunks" condense into lumps and then ignite under their own colossal gravitational pressure, turning into stars. The largest of the pillars is about a light-year high. HST, the "Cyclops in the Sky," can see to the edge of the universe. (Jeff Hester and Paul Scowen/STsI)

The Voyager flights to the outer planets, including Voyager 2's unprecedented "Grand Tour" of Jupiter, Saturn, Uranus, and Neptune from 1979 to 1989, were the most spectacular feats of exploration in history. Saturn and its six largest moons, imaged by Voyager 1 during an encounter in November 1980, were set in this montage at NASA's Jet Propulsion Laboratory, which conducted the tour. (JPL)

Laurence Soderblom, deputy leader of the Voyager imaging team, described the qualities of Jupiter's and Saturn's moons to reporters who covered the Saturn encounter in August 1981. (JPL)

Both Voyagers carried hundreds of images and sounds of the place from which they came, together with its location, for any intelligent beings who find them. One greeting came from the planet's junior citizens. (NASA)

Hello from the
children of the plan
et Earth.

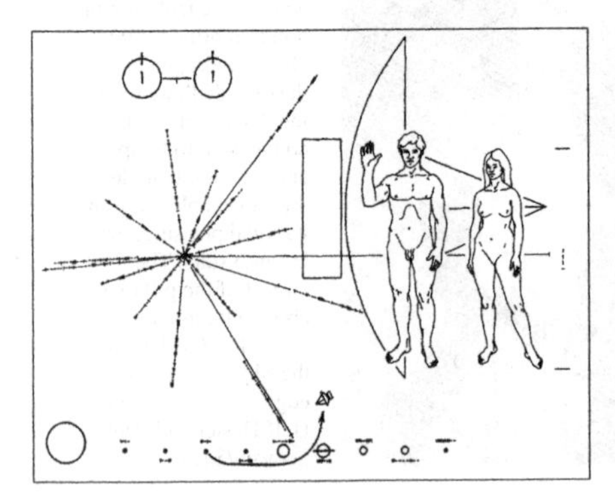

This plaque and one like it, describing where Pioneers 10 and 11 came from, were attached to them for the edification of life-forms somewhere else. But back on Earth, feminists complained that the woman was made to appear subservient and passive, the news media were confounded by how to show genitals and breasts (they were airbrushed out), and a barrage of letters to NASA protested the use of taxpayer money to send "smut to space." (NASA)

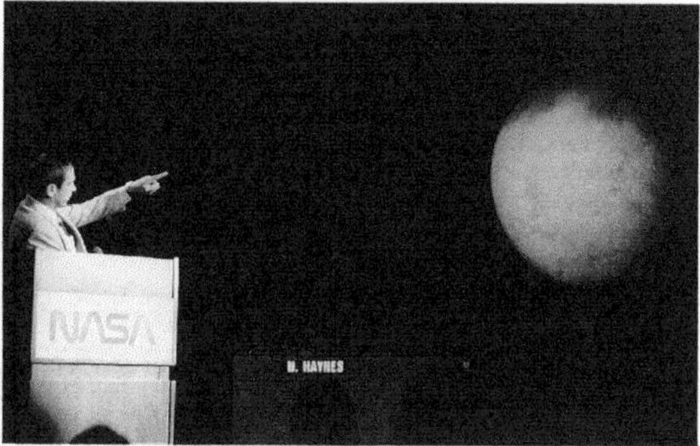

Neptune was the subject of Voyager 2's "last picture show," when it flew past the distant world in August 1989. Typically for a JPL planetary encounter, von Karman Auditorium was packed with reporters and television cameramen in what amounted to a "happening." Onstage, Soderblom explained features of Triton, Neptune's large "blue" moon that had come in during the night. Soderblom and others on the imaging team theorized that the long, straight lines near the center of the moon could have been caused by internal tectonic activity and that the absence of impact craters indicated that the surface had been paved over with lava as recently as a billion years ago. (Reporters and Soderblom, William E. Burrows; Triton, JPL)

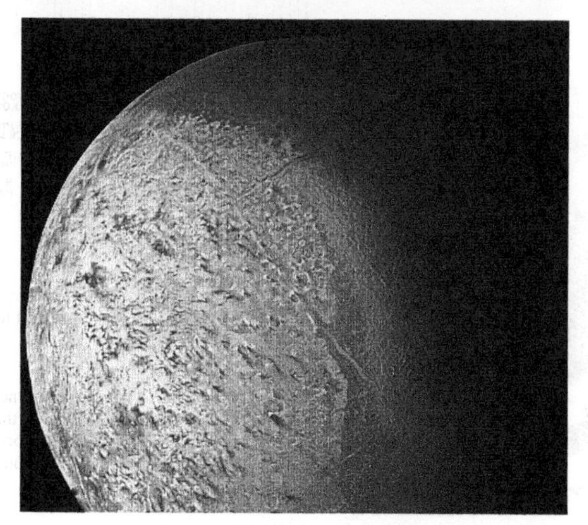

This scene, taken in a corner of the space exhibition area of the industrial pavilion in Moscow in 1992, symbolized the fate of the Soviet space program. Most of the spacecraft and memorabilia had been removed to make room for used American cars, which brought high prices from "new Russian" entrepreneurs. A large plaster bust of Lenin and a discarded model of Sputnik, its antennas bent and lying amid cigarette butts, reflected dramatically changed priorities. (William E. Burrows)

By 1995, the boilerplate *Buran* had been moved to Gorky Park, where it was turned into a tourist attraction, complete with space-type bucket seats. Visitors who paid fifteen thousand rubles ($3) watched a video about Soviet space feats and had their vital signs monitored "using methods of space medicine." (William E. Burrows)

SPACE-EARTH КОСМОС-ЗЕМЛЯ

ОЦЕНКА ФУНКЦИОНАЛЬНОГО СОСТОЯНИЯ ОРГАНИЗМА ПО МЕТОДИКЕ, ИСПОЛЬЗУЕМОЙ В КОСМИЧЕСКОЙ МЕДИЦИНЕ

ИМБП

ESTIMATION OF THE BODY FUNCTIONAL STATE USING METHODS OF SPACE MEDICINE

АО "Космос-земля"
Company "Space-Earth"

Институт медико-биологических проблем
Institute of Biomedical Problems

BILL BURROWZ

Номер кресла (Chair No.) 1

Дата (Date) 31.05.96
Время (Time) 21:02

Частота пульса (Heart rate) 67 уд/мин. (bpm)

Индекс функционального состояния (Functional state index) 1

Функциональдох состояние (Functional state) нормальное (good)

Противопоказания к полету (Closet for flight) Нет

At least one visitor was pronounced fit to fly. (William E. Burrows)

Mir suffered a spate of serious emergencies during the summer of 1997 that also symbolized the state of the program. Spaceflight director Vladimir Solovyov, standing behind a model of the space station at mission control in Moscow on July 17, reflected the situation. (AFP)

Atlantis ferried astronauts and supplies to *Mir* six times between June 1995 and September 1997 in preparation for the construction and habitation of the *International Space Station.* The two spaceships sailed in formation at seventeen thousand miles an hour on missions that signaled the beginning of a new age in space. (RSC Energia)

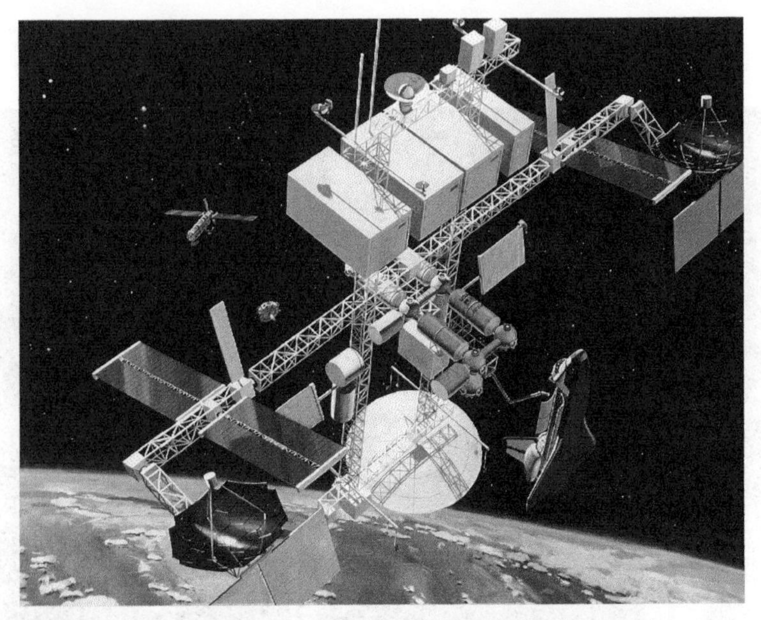

People have wanted to live in space for millennia and have conceived a wide variety of structures in which to do so. The "dual keel" station, above, shown being serviced by a shuttle, was suggested in 1986. It evolved into the *International Space Station,* below, which is planned to bring fifteen nations together in the largest and most daring engineering project in history. (NASA)

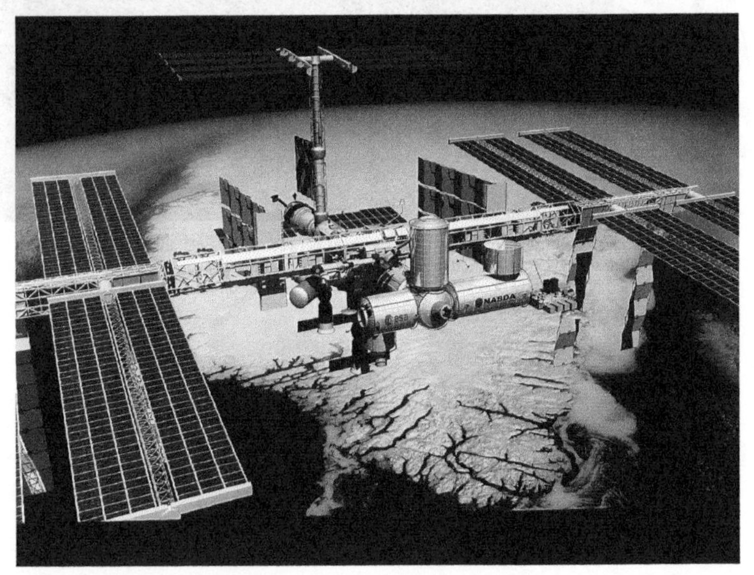

and tape recorders. A sphere in the center of the aluminum doughnut held hydrazine for the steering thrusters. Attached to the bus were a twelve-foot-wide high-gain antenna (three feet wider than Pioneer's), three RTG nuclear power plants stacked on one boom like drums (one more than Pioneer had); a pair of magnetometers attached to another, longer boom; a pair of rabbit-ear radio astronomy and plasma-wave antennas; and a movable scan, or science, platform. If, as Oran Nicks said, exploring machines resemble living creatures, then Voyager's movable scan platform amounted to a head that could be turned and that contained its eyes. The eyes themselves were narrow- and wide-angle television cameras, ultraviolet and infrared spectrometers, a photopolarimeter, and cosmic-ray and plasma detectors. Three-axis stabilization would keep the Voyagers steady for the clearest possible imagery. Stabilizing spacecraft on three axes had by then become so finely engineered that the infinitesimal squirts of hydrazine could adjust the machine's attitude at a rate one tenth to one hundredth the speed of a clock's hour hand.

At a takeoff weight of 1,819 pounds, the Voyagers would be the heftiest planetary explorers except for the Viking Orbiters, which were two and a half times as heavy. But then the Vikings only had to get to Mars.

The spacecraft would also carry messages for extragalactic intellectuals and would be virtual encyclopedias compared to the relatively simple discs fastened to the Pioneers. The new records, twelve-inch gold-plated copper plates, were developed by a Voyager Record Committee whose chairman and executive producer was Carl Sagan and whose members included Drake and Timothy Ferris, an observer who wrote eloquently about cosmology and who believed implicitly that scientists are as obsessed, intuitive, and driven as artists. Ever mindful of the storm of criticism that had been provoked by the Pioneer plaque, the committee spread the widest possible cultural net. The record's greeting, for example, was given in 55 languages including Mandarin, Bengali, Urdu, Farsi, Latin, Welsh, Nguni (Zulu), English, French, Vietnamese, and Japanese. There were 38 sounds, including rain, a barking dog, a Saturn 5 liftoff, Morse code, crickets chirping, heartbeats, laughter, and a kiss, plus 115 images, some of which showed a fertilized ovum, a snowflake, Bushman hunters, rush-hour traffic in India, DNA, and a supermarket. A proposal to send a photograph of a naked man and a pregnant woman looking at each other and holding hands was rejected by NASA, which had been burned by the Pioneer plaque, but Sagan and the others managed to include silhouettes anyway so as not to break the continuity of the human reproductive sequence, as they put it. There were also ninety minutes of eclectic music, including Bach's Prelude and Fugue in C from *The Well-Tempered Clavier,* an initiation song for pygmy girls in the Zaire rain forest, a bamboo flute rendition of the Japanese "Cranes in

Their Nest," Chuck Berry's classic "Johnny B. Goode," and a Navajo night chant.

The record also contained greetings from President Jimmy Carter, who opened by noting that the bearer of the message was made in America but then grew cosmically expansive:

> This is a present from a small distant world, a token of our sounds, our science, our images, our music, our thoughts and our feelings. We are attempting to survive our time so we may live into yours. We hope someday, having solved the problems we face, to join a community of galactic civilizations. This record represents our hope and our determination, and our good will to a vast and awesome universe.

The Voyagers' speed leaving the solar system was predicted to be so slow—a tortoiselike 38,700 miles an hour—that it would take at least forty thousand years before the two caravels from Earth would even get to within a light-year (6 trillion miles) of another star, let alone another planetary system that could harbor intelligent life. Yet everyone wanted to believe that there were intellects out there somewhere. That was a fundamental part of the religion. "The launching of this bottle into the cosmic ocean says something very hopeful about life on this planet," mused Sagan, the irrepressible dreamer. On the other hand, the two "bottles" bore no evidence of the wars, corruption, famine, pestilence, pollution, repression, genocide, and cupidity that are as endemic to Earth as rain, snowflakes, and barking dogs. Certainly they carried no clue about the tortuous political process that the Voyagers themselves would have to go through just to make it out of their own solar system.

The Grand Tour Comes Together

Voyager 2 beat Voyager 1 to space by sixteen days. It left Earth on top of a Titan 3E–Centaur, the most powerful rocket combination in the country, on August 20, 1977, the second anniversary of Viking's departure. Unlike the Mars mission, though, the Voyager twins had a first year that stirred recollections of Odysseus's return from the Trojan Wars. There were two computer failures even before launch and another while one of the spacecraft was still in sight of Launch Complex 41. There were also indications that Voyager 2's scan platform had not fully extended (later proved wrong) and indications that Voyager 1's platform had stuck (later proved right). There was even a maverick thruster firing into the side of Voyager 2, which was like turning a fire hose on a sailboat. A thruster on the other side therefore had to be used to keep the spacecraft on course, using precious hydrazine in the process. Worst of all,

Voyager 2 lost its main radio receiver in April 1978 because of a succession of minor mishaps, making it stone deaf in one ear. Then its backup receiver malfunctioned. Shades of Scylla and Charybdis, Circe and the Cyclops.

"Thrusters fired at inappropriate times, data modes shifted, instrument filter and analyzer wheels became stuck, and the various computer control systems occasionally overrode ground commands," astronomer David Morrison and Jane Samz, a colleague, would later observe in a book about Voyager at Jupiter. "Apparently, the spacecraft hardware was working properly, but the computers on board displayed certain traits that seemed almost humanly perverse—and perhaps a little psychotic." The computers, it turned out, were so sensitive that they overreacted to minor problems. "Ultimately, part of the programming had to be rewritten on Earth . . . to calm them down." "Psychotic" was probably the word Norman Mailer, who believed in the psychology of machines, would have used, too.

A relatively tranquil Voyager 1 reached Jupiter four months ahead of Voyager 2 because it was sent on a shorter, faster trajectory. By January 9, 1979, a stream of pictures better than any taken on Earth were filling the television monitors of the University of Arizona's Bradford A. Smith, the leader of the imaging team, and his dazzled colleagues in Pasadena. One of the most spectacular images ever taken in space came in on February 13, when Voyager 1 was so close to Jupiter—12 million miles—that it could no longer get the whole planet into its cameras. It therefore sent back a detailed picture of Io and Europa in orbit around Jupiter like peas circling a psychedelic basketball, with the clearest view ever of the Great Red Spot in the background behind Io. Europa could be seen in that picture to be ice-covered, while Io was showing brightly colored patterns that would turn out to be part of one of the most amazing of the expedition's trove of discoveries. By the beginning of March, lights burned around the clock in JPL's Space Flight Operations Facility, which controlled the two robots, in the science and engineering areas involved in the mission, and in von Karman, where the usual daily news briefings took place.

Meanwhile, the place crawled with Voyager team members, contractors, technocrats from headquarters, scientists from around the country and elsewhere, some with graduate students in tow, and badged reporters, writers, and television people from almost everywhere, many of them—almost always the younger ones—sporting JPL and Voyager Jupiter T-shirts and pins. The older hands, who worried about appearing biased, would wear no one's uniform, however much they regarded the mission. Special wooden platforms for the batteries of cameras were set up at the rear of the auditorium, facing the stage where the scientists would make their morning presentations during close encounter.

Out beyond the quiet dens of intensity where scientists and engineers pored over data, made sublimely delicate course changes, and monitored the spacecrafts' health, the Fourth Estate was faithfully reporting the string of discoveries and having a party at the same time. Encounters were ethereal adventures painted in false color, world-hopping extravaganzas, the ultimate joyride. There on the screen, at the end of Brad Smith's pointer, was a "pepperoni pizza" named Io that would soon be seen to vomit molten lava, making it only the second place in the solar system with active volcanos. There were eye-popping close-ups of the king of the sky, a surging, stormlike mass of gas that dwarfed the planet they were sitting on. And they got to see it all ahead of everybody else except those on the imaging team as it came in from half a billion miles across the great void. JPL at encounter time was a celestial theme park that no dreary cancer lab, no blizzard of quarks, no endangered owl, toad, or fish could equal. Encounters would come to be working festivals and reunions that were eagerly anticipated by the tribe that covered science. They would be to the space junkies what Grateful Dead concerts were to rock 'n' rollers.

The closest approach came on Monday, March 5, with Voyager 1, now solidly in the lock of Jupiter's gravity, boring in at better than 62,000 miles an hour at a range of 485,000 miles. Governor Jerry Brown spent the previous night on lab (as JPL employees call where they work) and a monitor was set up in the White House so the Carters could watch the show. What they saw up close—a planet alive with pastel bands of swirling gas writhing in every direction and accompanied by a coterie of distinctly different worlds—looked more like modern art, which is probably what at least one reporter had in mind when he said, "Are you sure van Gogh didn't paint that?" Another said it looked like a close-up of a salad, and still another quipped, "They're not showing us Jupiter; that's some medical school anatomy slide."

Unlike van Gogh, the planetary exploration and Earth observation communities used multispectral images, so-called false color, to bring out certain characteristics in some images. The spacecraft's camera system was equipped with colored filters, each of which could highlight a particular range of wavelengths that were registered on separate images. Arranged in the right combination, they would show gases and geological formations that were otherwise invisible. Many color images taken in planetary exploration are in false color, which often accounts for their spectacular beauty.

The first red, orange, yellow, and white close-ups of Io were so startling that even the members of the imaging team called the moon "grotesque," "gross," "bizarre," and "diseased." The ordinarily taciturn Brad Smith, who had watched the incredible imagery come in all night, started the daily 11 A.M. news briefing in a state of rapture. "We're all recovering from what I would call the most exciting, the most fascinating, what may ultimately prove to be the most scientifically rewarding mission in the unmanned space program."

The sequence in which such observations were made could in themselves be a source of bitter contention among the scientists. Unlike orbital missions such as the ones that went around the Moon, Venus, and Mars, in which there was plenty of time to take television pictures and collect ultraviolet, infrared, and other imagery, flybys flew by only once and the scan platform could face in only one direction at a time. Which direction it faced during those precious minutes and hours was often the subject of heated debate between the teams. Voyager 1's passage between Jupiter and Io was a case in point. One team wanted to investigate reports from the Pioneers that there were charged particles around Io, an experiment that would have required the ultraviolet spectrometer's use for half an hour or more. But the imaging team was equally absorbed by the atmospheric dynamics of Jupiter itself and needed the television camera to take a sequence of pictures over the course of several hours.

"When you're far away, it's no big problem," explained Arthur L. "Lonne" Lane, a cheerful member of the Voyager Science Integration Team, whose job it was to arbitrate the sequence in which the data were to be collected. "However, when you get in close—in the case of the Voyager encounter, within, say, about five days—you end up with intense competition." The basic sequence had been set up months before the encounter and after careful deliberation. But as near encounter approached, and intriguing discoveries were made that begged for longer and closer looks, tempers sometimes flared. Lane and others had to weigh the merits of competing experiments under pressure and decide whether or not the sequence would be changed. "This," he once quipped, required "a Solomon and a half." When neither team would budge, the matter went to Edward C. Stone at Caltech, Voyager's chief scientist and JPL's eventual director. "It never came to fisticuffs," Lonne Lane added, laughing, "but there were some pained and pointed discussions."

Voyager 1's encounter with Jupiter was a bonanza, as was Voyager 2's, which followed a different route in July. Between them, they sent back more than thirty-three thousand images of Jupiter and its five major moons. "For the first time, it was possible to see the Jovian satellites as individuals," David Morrison wrote in the introduction to a 972-page tome devoted solely to what was by then the fifteen satellites Jupiter was known to have. As was the case with Viking, project scientists and others would be sifting the mountain of material for years. Using Voyager data and computer simulations, for example, Harvard astronomer and mathematician Philip S. Marcus showed in 1985 that the Great Red Spot was not a storm at all, but a colossal eddy of swirling gas that is driven by Jupiter's turbulent weather.

The biggest single discovery was finding nine active volcanoes on Io, which finally explained the sulfur cloud that hangs in the whole Jovian system and beyond. The Voyagers also found a thin ring of dust around Jupiter, measured winds at up to 300 miles an hour in the planet's turbulent atmosphere, confirmed that Ganymede is the largest moon in the solar system and probably has tectonics similar to Earth's, and spotted three more moons: Metis, Adrastea, and Thebe.

All of this and scores of other "surprises" was presented at the eleven o'clock briefings in von Karman throughout the encounter. What was rarely, if ever, mentioned was the role of the people who were keeping the spacecraft functioning and steering them unerringly toward their destinations. William I. McLaughlin, the manager of JPL's Flight Engineering Office and a potential winner of the Aleksandr Solzhenitsyn look-alike contest, ran that low-key but vital operation for years.

For all of their beautiful scientific and engineering similarities, however, there was a profound difference between Voyager 1's and Voyager 2's encounters with Saturn: they came under different presidents. When Voyager 1 sailed past the ringed wonder on November 12, 1980, Jimmy Carter was in the White House. When the spacecraft that was chasing Voyager 1 arrived the following August 25, Carter was gone and the screen actor turned conservative politician Ronald Reagan was beginning the first of two terms in office.

By the time Voyager 2 reached Saturn, a dark shadow was lengthening over it. The shadow was caused, not by the stately planet's fabled rings, but by David A. Stockman, the new director of the Office of Management and Budget. Like the president he served, Stockman believed in the tenets of the classic free market economy, including the necessity of slashing government spending on extraneous programs by draconian methods, if that was what it took. So during the winter of 1980–81, the first of Reagan's presidency, Stockman worked hard to drastically reduce government spending that he considered unproductive. Although the welfare system was his most conspicuous target, no budget was safe except the military's, which would soar, and NASA's sacred cow, the shuttle (which had a carefully delineated military role). The exploration of the solar system was definitely unproductive, in Stockman's view. So, too, would have been hiring someone to paint pictures on the ceiling of the Sistine Chapel.

While rumors spread that Stockman was boasting that he intended to gut the planetary program altogether, saying that "NASA will be out of the space exploration business by 1984," he slashed Carter's proposed fiscal year 1982 NASA budget of $6.7 billion by $604 million. A Venus Orbiting Imaging Radar spacecraft was killed outright, as was American participation in an International Solar Polar project, which infuriated the project's European partners. A mission to Halley's Comet was in grave danger.

And the situation quickly worsened. NASA Administrator James Beggs learned that summer, even as data from Voyager Saturn were pouring in, that his proposed budget for fiscal year 1983 was $6.5 billion. Pointing out that it would be impossible to complete the shuttle, conduct space science and exploration, and pursue other programs (aeronautical research, for example) with such meager funds, he submitted a request for $7.2 billion. It was rebuffed. Instead, he was given $6 billion, or $0.5 billion less than what he had said he could not live with. Worse still, the projected figure for fiscal year 1984 was even lower, $5.6 billion.

Beggs reacted on September 29 by warning Stockman in a letter—on the record—that the budget reduction would mean the end of the planetary exploration program which, in turn, "would make the Jet Propulsion Laboratory in California surplus to our needs." The real message was that the end of exploration could mean the end of JPL, which was in the president's home state. Stockman stubbornly held his ground.

The Grand Tour Almost Comes Apart

The imagery and other data coming back from Saturn that August 1981 was spectacular. There were exquisitely detailed pictures of its complex ring system showing thousands of dazzling ringlets, some of them appearing to be kinky or braided, that were accompanied by small "shepherd" satellites. Shadowy "spokes" could be seen spread across parts of the ring plain. And the plain itself had incredible dimensions. They were 43,000 miles wide and only 60 feet thick, meaning that if they were reduced in scale to a thickness of about an inch, they would be eight city blocks wide (and Saturn itself would be the height of an eighty-story building). New moons were found, bringing the known total to eighteen, and they were seen to be "icier, whiter, brighter, and junkier" than Jupiter's, in the words of Torrence V. Johnson, a member of the imaging team. One of them, Enceladus, had five different kinds of surfaces and indications of subsurface activity.

The accumulated data, sifted and analyzed, appeared in a richly detailed, 968-page tome from the University of Arizona. Henry S. F. Cooper Jr., *The New Yorker*'s veteran space reporter, wrote a book about the imaging alone. A writer for the magazine *Science* teamed up with an astronomer to produce another just on rings, making the lovely point that in this case, "theory did not lead the way to observation and discovery—rather discovery led theory." *Science* itself devoted its entire issue of January 29, 1982, to Voyager 2's encounter with Saturn, leading off with a short essay by Murray deploring the fact that there was more chaos in space science than there was in space. It would only get worse.

NASA tentatively decided to legitimatize the great conspiracy: it planned to send Voyager 2 on to Uranus and Neptune. In the meantime, the assault on space science came from yet another direction, this one as incongruous as it was painful. George Keyworth, the president's science adviser, weighed in by recommending that all new planetary programs be shelved for a decade. The most important of these was a Jupiter Orbiter-Probe named Galileo, which was to build on the Voyager flyby by circling the great gas ball, dropping a probe into its atmosphere, and studying it and the entire Jovian system. Close behind was the advanced Venus radar mapper, a follow-on to Pioneer Venus, named in honor of Ferdinand Magellan.

Meanwhile, the shuttle's main engines were blowing up on their test stands or otherwise failing with appalling regularity, forcing NASA to raid every budget line it could find for supplemental funding. Keyworth and the White House were by then seriously considering pulling the plug on Voyager 2 and shutting down the whole Deep Space Network, since Voyager now amounted to its sole reason for existence. With the DSN turned off, the "heroic" spacecraft—an adjective that had come into use when its problems had been multiplying and were being solved—might go on to Uranus and Neptune, all right, but no one back home would know what it found. The planetary exploration community's worst nightmare was coming true.

The Revolt of the Boffins

By the time the American Astronomical Society's Division of Planetary Sciences held its annual meeting in Pittsburgh during the week of October 12, 1981, it was clear to the thunderstruck scientists that the new administration intended to kill space exploration outright. The astronomers and planetary scientists were ordinarily a docile lot. The single major exception had been a threatened revolt to save the Large Space Telescope (later named for Edwin Hubble) in 1975 and 1976. But even then there had been a distinct rumble of internecine warfare as supporters of the orbiting eye prepared to challenge their colleagues in the Jupiter Orbiter Probe (JOP) program over congressional appropriations. Space telescope advocates had openly criticized the NASA science budget in 1975, when their own project was near oblivion, and charged that a disproportionate amount of funds were earmarked for solar system exploration at the expense of studying stars and galaxies. "When trouble came, they'd form their wagons in a circle," space historian Joseph N. Tatarewisz would remark about the astronomers. "Then they'd start shooting to the inside."

But the shots between the space telescope faction and JOP's supporters in 1975 were quickly stopped by headquarters. Noel Hinners, who ran NASA's Office of Space Science, had emphatically warned JPL, which was responsible for the big Jupiter spacecraft, not to attack the telescope. Meanwhile, George Field, chairman of the space agency's Physical Sciences Committee, Princeton's John Bahcall, Yale astronomer Lyman Spitzer, and others who had conceived and cherished the powerful orbiting eye had asked their own colleagues not to go after planetary missions. Hinners had openly warned against "fratricide," and it had worked. Both programs—the Hubble Space Telescope and Galileo—would lead almost unbelievably tortured lives, both before and after launch. But they would ultimately distinguish themselves.

Now there was yet another threat. The likelihood that a truly elegant mission—mankind's first close-up look at the outer planets and their environment—would be stopped cold, damaging or wrecking careers in the process, because of the whim of some political toady in Washington clarified the collective mind of everyone in solar system exploration. Men and women whose idea of political combat ordinarily had to do with tenure and promotion cases and picking department heads reacted with rage. The majestic enterprise whose course had been set by the likes of Copernicus, Kepler, Galileo, Newton, Cassini, Hubble, Kuiper, and scores of other giants was being imperiled by something called supply-side economics.

At the Tuscon meeting in 1980, when the threat was still only thunder on the horizon, the astronomers had sniped at Sagan for being a dreaded "popularizer" and getting rich and famous in the process because of the Emmy award–winning *Cosmos* television series and a companion book of the same name. Every segment of the series noted up front that it was "A Personal Voyage." *A Personal Voyage?* Furthermore, the book was copyrighted by Carl Sagan Productions, which no doubt further rankled the university-press crowd. "The Cosmos is all that is or ever was or ever will be," the Cornell astronomer said in the book's opening. "Our feeblest contemplations of the Cosmos stir us—there is a tingling in the spine, a catch in the voice, a faint sensation, as if a distant memory, or falling from a height. We know we are approaching the greatest of mysteries." That might have been schmaltz to the average astronomer, but, however melodramatically, Sagan was trying to get a wide audience to come to terms with humanity's central question: "Where do we fit into the scheme of things?" Solving that mystery, he never tired of saying, depended squarely on exploring the universe: of going out there and looking around.

A year later, in Pittsburgh, no one was barbing at Sagan because most of the scientists suddenly understood that they were about to get shoved aside. David Morrison, chairman of the Division of Planetary Sciences, sent out a letter that got right to the point: "The time has come to politicize the planetary science community." The boffins, as scientists are derisively called in Britain, quickly learned how to lobby congressmen and grumbled about forming their own political action committee. Eugene Levy, chairman of the National Academy of Science's Committee on Lunar and Planetary Exploration, delivered a stinging indictment of a process that allowed space science and exploration to be destroyed by the likes of economists.

The counterattack—over an annual budget of $250 million, or less than the cost of a single B-1 bomber—quickly spread. Edwin Meese, a fellow Californian and presidential adviser, was targeted for a barrage of letters. Morrison and Sagan urged him to get Reagan to answer history's call: A thousand years from now, they carefully explained, "our age will be remembered because this is the moment we first set sail for the planets." Meese told them he would talk to the president. Arnold Beckman, a Caltech trustee, was more pragmatic in his own letter to Meese, warning that the proposed cuts threatened "to create total chaos and a rapid disintegration of a 5,000-person, $400 million Southern California enterprise. . . . There are obvious implications to the support of the President and to his Party should the Administration permit such a catastrophe to take place," Beckman declared. John Rousselot, a conservative California representative, also pelted Meese, as did Thomas Pownall, the president of Martin Marietta, who told him that the firm's work in exploration enhanced its work in national defense. That was brilliant, since the president was always concerned about the state of the nation's armed forces. Mary Scranton, wife of the former governor of Pennsylvania and chairwoman of a new ad hoc Trustees Committee for JPL, yanked congressional strings, as did Marvin Goldberger, the president of Caltech, who wrote to the influential Senate majority leader, Tennessee's Howard Baker. Baker wrote to Reagan, and the letter landed on David Stockman's desk.

The campaign worked. OMB's Budget Review Committee met in December and agreed to add $80 million to $90 million to the planetary exploration budget. The Voyagers and the Deep Space Network were saved. And so was the Jupiter Orbiter with Probe.

Amazing Grace

Voyager 1 and Voyager 2 parted after Saturn. Voyager 1 swung out of the plane of the ecliptic and headed up, past Titan and beyond. On Valentine's Day 1990, it would take the first "family portrait" of the solar system, or at any rate of the Sun and six of its planets, from a distance of almost 4 billion miles. Little Earth finally took its place with most of its sisters.

Meanwhile, Voyager 2, having survived asteroids, radiation, a serious loss of hearing, a stuck scan platform, and David Stockman, headed for Uranus. The spacecraft was four years old and more than a billion miles from home when it sailed through the Saturnian system. When it arrived at Uranus in late January 1986 it would be twice as old and twice as far away. That would create such formidable mechanical problems that the engineers—Earth's Olympic archers—would have to turn it into a virtually new spacecraft. Computer hacks and astronerds, not square-jawed test pilots, would become the heroes of the longest and finest voyage of discovery in history.

14

The Roaring Eighties

The year 1981 marked the seventy-eighth time Earth had circled the Sun since the first heavier-than-air flight, the twenty-fourth since the first machine had taken to space, and the twentieth since a man had first gone there.

Now the space machines were everywhere. Four, bearing greetings from the place from which they came and still reporting back to it, were streaking toward encounters in unimaginable places at an incomprehensible time. Others rested silently on three other worlds, bearing mute testimony to the restless instinct of their creators. Still others, the ones that had tried to reach the Moon, Venus, or Mars but had failed, were lifeless derelicts shrouded in silver- or gold-foil thermoshields that circled the Sun in icy silence. There were hundreds of dead derelicts closer to home, too, tumbling uselessly: yesterday's faithful robots turned into junky navigational hazards. But they were being replaced all the time. Within a mere twenty-five years of the start of the space age, machines that orbited Earth had made the transition from luxury to necessity. They had become an integral part of the life of the planet.

Way out at geosynchronous, U.S. ballistic missile early-warning satellites with frozen eyes peered continuously at Eurasia and the world's oceans through twelve-foot telescopes, watching for a surprise attack that would be answered in kind. Their Soviet counterparts, nine flying at once in three different planes, peered at North America and Western Europe from very elliptical orbits that kept them within constant sight of American missile silos. Other satellites that faced Earth from 22,213 miles out used giant ears to eavesdrop on telephone conversations, radio signals, and missile telemetry. Still others relayed television images of Olympic sprinters and pole-vaulters as they raced and sailed over the crossbars.

Closer to home, there were civilian and military Earth huggers that relayed communication traffic and others that took pictures in a never-ending search for information, most of it secret. Still others conducted a wide range of scientific experiments, tracked shipping, helped navigation, watched the weather, measured the planet, and sniffed at radars. And some kept people high in the sky. The view from Earth orbit was still sensational physically, if not politically.

Stations and Shuttles: The People Imperative

The great enigma that continued to haunt spaceworld had to do with whether people belonged there. Those who practiced the religion, from Lucian of Samosata to Bernard Smith to Arthur C. Clarke, never doubted that they did. Yet the engineering of robots had gotten so good that by 1981, from the standpoint of practical necessity, it essentially preempted a human presence in space. The ostensible reason for killing the Manned Orbiting Laboratory in 1969, for example, had been economic. Yet if it had been the only way to "crack the Communist secrecy curtain," as *Air Force* magazine bluntly put it, the money would have been found. But the robots did it better and cheaper. Starting in December 1976, a top-secret satellite called the KH-11 began sending back high-quality imagery in near real time—within minutes of the pictures being taken—and it did not need the MOL's pricey life-support system (including a Gemini return capsule "lifeboat," to be used like a dinghy on a yacht if the MOL foundered).

Still, there were compelling reasons for sending men and women to space, whether their presence there was physically superfluous or not. First and foremost, there was the tangle of nationalistic motives, including the creation of jobs and keeping the respective space industries healthy. Brenda Forman, a respected and outspoken Washington-based space analyst and a dreamer who was conflicted by an acute sense of reality, maintained that base politics always drove the dream. "Politics, not technology, sets the limits of what technology is allowed to achieve," she insisted, and while that might be unpleasant to the dedicated engineer, it was "perilous" to ignore. By and large, Forman continued, Congress didn't care about a space program's technological or engineering content. "Instead, program funding depends directly on the strength and staying power of its supporters; i.e., its constituency." Put another way: "Jobs *über alles.*" And fabulously expensive, large engineering programs that kept people in space spread jobs and profits to virtually every corner of the country.

More subtly, keeping people in space not only created heroes, but entities with whom ordinary citizens could identify. "We don't give ticker-tape parades for robots," as NASA Associate Administrator Franklin D. Martin aptly put it at a meeting in 1989.

Then, too, sending satellites into orbit in the 1980s was no longer exotic, as France, China, Japan, and India had demonstrated. But sending people to space was infinitely more complicated, and doing so proved in the most dramatic way that both superpowers were holding their lead by a wide margin and, by implication, were robust societies.

And there was no doubt that the pictures of Earth from space, as well as the view shared by the tiny minority that went there, started a gradual shifting of the way civilization as a whole thought about the spaceship upon which it lived and about its place on that ship and in the larger universe. This has been called the overview effect. Astronaut Rusty Schweikart, for example, said that Earth "is so small and fragile and such a precious little spot in that universe that you can block it out with your thumb, and you realize that on that small spot, that little blue and white thing, is everything that means anything to you—all of history and music and poetry and art and death and birth and love, tears, joy, games, all of it on that little spot out there you can cover with your thumb. And you realize from that perspective that you've changed, that there's something new, that the relationship is no longer what it was."

Finally, at the heart of it all, as usual, there remained the core of dreamers—"dreamy-faced loons," as one journalistic detractor called them—who steadfastly believed that it was their race's manifest destiny to leave Earth for both adventure and survival. The dreamers never for a moment believed that the Viking missions to Mars were for the abstract goal of achieving what its own scientists called "pure knowledge." No. The dreamers thought of themselves as utilitarians. It was an article of faith that the exploration of the planet by robots was the necessary prelude to the eventual establishment of a colony. Manned space missions, including living in orbit for long periods, were therefore taken to be logical steps on the road to Mars and beyond. Certainly the dreamers in the two competing space agencies thought so. That is why they justified their stations largely by claiming that they were necessary to see how the human body would react to long flights to Mars. And there was a body of opinion, eventually to be challenged, that the human expedition to Mars would have to leave either from a space station or from the Moon.

Diamonds Are Not Forever

The Russian notion of stations in space—of people living there, as opposed to being tourists—goes back to Tsiolkovsky's floating cities in *Beyond Planet Earth.* But a practical concept of orbiting stations really started in March 1962, when OKB-1 produced a report called "Complex for the Assembly of

Space Vehicles in Artificial Earth Satellite Orbit (the Soyuz)." The document mostly described building a spaceship in orbit that would circle the Moon, but it also dealt with a small station put together with independently launched modules that would hold three cosmonauts who were to observe the planet beneath them from a "science package" module. Three years later, Korolyov's bureau proposed a ninety-ton station that was to be sent up on an N-1 and that would have a docking module with ports for four Soyuz spacecraft.

OKB-1 was by then so busy gearing up for the race to the Moon, though, that the project was given to Korolyov's rival, Vladimir Chelomei. The engineers at OKB-52 then produced plans for a true Battle Star Galactica: a single-launch station equipped with a capsule for as many as three cosmonauts, radar remote-sensing equipment and cameras for imaging Earth, two re-entry capsules for returning data to Earth, and even a cannon for defense against an attack by the Americans. The generic station was designed for launching by Chelomei's own Proton, of course, and was named Almaz (Diamond). It would be the Soviet equivalent of the Manned Orbiting Laboratory; a dedicated, populated, military spacecraft. And like the MOL, it was duly canceled, at least as a cosmonaut carrier. Three unmanned versions would eventually be built with nuclear-powered synthetic aperture radars, one of which was lost when its launcher exploded, and two of which sent down respectable imagery before one of them crashed in Canada.

By February 1970, four Americans had stood on the Moon and several more were training to do so, including two who were scheduled to fly on *Apollo 13* that April. Meanwhile, two N-1s had blown up and an L-1 lunar orbiter test version had turned into rubble when the third stage of the Proton that was carrying it exploded in a shower of flaming fragments. Although there were some in the Soviet space establishment who still wanted to send cosmonauts to the Moon for two-week stays (and who would still be dreaming about it when the Soviet Union disintegrated twenty years later) the race to the Moon was effectively over. The Ministry of General Machine Building therefore looked elsewhere for a competitive edge and quickly decided to push work on stations in the hope of getting at least one in orbit before *Skylab.* Landing people on the Moon, after all, was only a stunt; serious physical and psychological experiments would be done on stations. The fact that General Machine Building wanted to believe that because it had lost the race did not mean it was entirely wrong.

But the effort to get a space station to orbit would also be marked by death and destruction. With work on subsystem components falling dangerously behind during that February of 1970, the ministry transferred Almaz back to Korolyov's bureau while allowing Chelomei's engineers to keep working on it. The idea was to use specially designed Almaz sections for the new station's hull and other major components, plus Soyuz parts for subsystems, and have the thing ready to fly within a year. The first Long-Duration Station, or DOS-1 in the Russian acronym, was the one named *Salyut* in honor of Gagarin's flight and launched on a Proton on April 19, 1971. The main purpose of the forty-eight-foot-long spacecraft was to test the concept of keeping a manned military station—the still-dreamed-of Soviet MOL—in orbit.

An oracle would have named it *Albatross.* Four days after it was launched, three cosmonauts were sent up on *Soyuz 10* to board the station. They docked with it but could not get in because the hatch on their own spacecraft was stuck, so back down they went. Seven weeks later, Georgi Dobrovolsky, Vladislav Volkov, and Viktor Patsayev did manage to climb into *Salyut 1* from their *Soyuz 11,* becoming the first men to fly in a true space station. The trio spent twenty-four days aboard *Salyut 1* testing the ship's systems as well as special "penguin" exercise suits (which made them waddle like the birds) and a treadmill, doing medical tests, making several television broadcasts, communicating with their control center through Molniya 1, a new communication satellite in a very elliptical orbit, observing the heavens with a telescope, and taking pictures of Earth (they had the dubious distinction of photographing the third N-1 blowing up as they passed over Kazakhstan on June 27).

Three days later, the cosmonauts climbed out of *Salyut 1* and started home in the *Soyuz 11.* Down below, the Soviet news media were covering the mission in detail, making the point that it was far more rewarding than Apollo. As was always the case, the returning cosmonauts would be feted and festooned with "Hero of the Soviet Union" medals. But disaster struck on the way down when the pressure valve opened and their blood began to boil. It was at the time the worst flying accident in the history of the space age. With one crew shut out and another killed, *Salyut 1* was officially pronounced a failure. More than a year passed before three cosmonauts were sent to space together.

Salyut 1's follow-on was a failure, too. It dropped into the Pacific on July 29, 1972, minutes after the second stage of its Proton shut down. Having reached an altitude of only forty-seven miles, it was not given a number.

Salyut 2, a dedicated military station, made it to orbit on April 3, 1973, but not without more mishaps. The third stage of the Proton that boosted it to orbit bumped the rear of the space station so hard that it caused an explosion (which was seen and heard by the North American Aerospace Defense Command's radars and the intelligence community's far-flung antennas). Valves snapped, and gas bottles or propellants broke loose and shot away from *Salyut 2,* pushing debris into several different orbits, some of them higher than the station itself. Eleven days later, while another *Soyuz* was being readied for launch, *Salyut 2* began to tumble out of control and break into twenty-five pieces. They burned up re-entering the atmosphere on May 28. Meanwhile, with *Sky-*

lab in its countdown on top of a Saturn 5 at Canaveral, one last desperate attempt was made on April 3 to get a working Salyut into orbit first. But it was not to be. At some point soon after launch, TsUP ("T-soup"), as the flight control center at Kaliningrad was called, lost control of the spacecraft after it suffered a massive power failure. On May 28 it, too, returned to Earth as a disintegrating fireball and was awarded a generic Cosmos number, 557, to disguise the failure. *Skylab* was launched on the fourteenth and was therefore in orbit when *Salyut 2*'s pieces rained down and *Cosmos 557* followed them into the sea.

Success finally came in the early summer of 1974 with the launch of *Salyut 3,* another military station that was tested as a manned reconnaissance craft and orbiting home by Pavel Popovich and Yuri Artyukhin for sixteen days. Now they were getting it right. *Salyut 4,* a modified civilian station similar to *Salyut 1,* went up on December 26, 1974, and was joined by *Soyuz 17* on January 12. When Aleksei Gubarov and Georgi Grechko climbed through the station's hatch, they found a "wipe your feet" sign left by the ground crew at Tyuratam. *Soyuz 17* stayed locked to *Salyut 4* for twenty-nine days before returning to Earth on February 9. The cosmonauts worked at forty experiments during their ride in space, some of which had to do with how their own bodies reacted to weightlessness. *Salyut 4* itself remained in orbit for more than twenty-five months, during which time it was visited by a second crew for sixty days. Its immediate successor, *Salyut 5,* was launched on June 22, 1976, as the Soviet Union's last dedicated military space station. This indicated that the Ministry of Defense had come to the same decision as Robert McNamara had: manned military spacecraft cost more than they were worth. It was also decided to postpone Almaz itself, which was replaced by plans for an ambitious new heavy lifter that would be handled by a new bureau called the Scientific Production Association, or NPO Energia. It would be headed by Valentin Glushko.

Stations in Space

Salyut 6, launched in late September 1977, and *Salyut 7,* which went up in mid-May 1982, were second-generation stations that showed the Soviet Union was serious about populating low Earth orbit with Soviet citizens and their guests. By the time *Salyut 6* went up, the United States was five years into its abandonment of the Moon and was concentrating instead on developing a shuttle.

While *Salyuts 6* and *7* shared roughly the same proportions as their predecessors, they were fundamentally different creatures. They were, in fact, true spaceships because they were capable of truly long duration missions. That meant having the capacity to keep crews resupplied with food, oxygen, and other necessities for long periods, since three people eat and drink up to ten tons of food and beverages a year. They also needed new equipment for experiments and to have the station's maneuvering fuel replenished so its two main engines and thirty-two attitude-control thrusters could resist Earth's pull and keep it in the right position at all times. Both spacecraft would be "gassed up" directly from supply ships named Progress. The advanced Salyuts were therefore given two docking ports that allowed Soyuz ferries carrying cosmonauts and supplies to plug into them simultaneously. In addition, they were six feet wider than their predecessors, which allowed more work and living space; they also had three large, steerable solar arrays that could track the Sun and other more sophisticated hardware.

As the Salyut program developed, so did a whole array of specially designed support craft to service the stations. The Soyuz Ferry, for example, was a crew-transfer vehicle that carried two cosmonauts and life-support equipment. It was replaced in 1976 by the more advanced Soyuz T (for Transport) that would service the space stations until it, too, was superseded by Soyuz TM (Transport, Modified). This latest version was developed exclusively to supply *Mir* (Peace), the third-generation station. The Progress spacecraft, an unmanned version of the ferry, was a remarkable logistics resupply craft that would carry cargo not only to *Salyut*s *6* and *7*—the last of the Salyuts—but to *Mir* as well. Seventy-nine resupply flights would be made by the dependable cosmic pack mules between 1975 and the end of 1997 with no failures. That said something about both the improving state of Soviet computers and a design philosophy wedded to absolute dependability.

The last two Salyuts and the fleet that serviced them were unmitigated successes and, as planned, established a permanent Soviet presence in space that led to *Mir,* which had six docking ports and more advanced technology and was launched in late February 1986. Five of the ports were in a docking hub at one end of the spacecraft, four of them to hold science laboratories and the fifth, the Soyuz cosmonaut vehicle. The sixth port would open for Progress resupply spacecraft. With its full ensemble of visitors plugged in, *Mir* would look like a giant Tinkertoy.

Six long-duration missions in which cosmonauts logged 684 days were flown on *Salyut 6* during its almost five years of operation. Having established the fact in Vostok's earliest days that the first man and woman in space were ethnic Russians, the select circle was expanded to include representatives from other socialist countries. Afghani, Bulgarian, Cuban, Czech, East German, Hungarian, Mongolian, Polish, Romanian, and Vietnamese cosmonauts were therefore treated to flights on *Salyut 6* as a way of rewarding allies with a little gratuitous prestige.

The Most Fragile Part of the Equation

Foreigners and Russians alike participated in often-rigorous physiological experiments to find out how the human body reacted to long periods in space, especially to motion sickness and weightlessness, which can cause serious calcium loss in the skeletal system. Many physiologists, including those in the United States, believed that exercise was vital, so successive crews had to run on a treadmill and peddle a stationary bicyclelike contraption that was in reality far and away the fastest in the world. (Astronauts also jogged in place and peddled to nowhere on *Skylab*.)

What's more, cosmorats flying with the cosmonauts showed that those made to walk in a slowly revolving cage returned to Earth with healthier cardiovascular and endocrine systems and bone and muscle tissue than a control group of what amounted to couch rodents. Hundreds of days in the both *Skylab* and the Salyuts and *Mir* showed that men, women, and other mammals could overcome many physiological problems on long-duration missions through a combination of rigorous exercise and medication.

The mind was a different matter, though. Psychological problems could be worse than physical ones and were far less understood. Sensory deprivation and isolation leading to anxiety, depression, or psychosis had long been known among sailors who spent weeks underwater in nuclear submarines and others who were holed up for long periods in the Antarctic during winter.

It was the same in space, where Russian psychologists eventually identified three phases of adaptation. In the first, lasting about two months, cosmonauts busily adjusted to their new environment. That was followed by a period of increasing fatigue and decreasing motivation in which everything that had once seemed exciting became boring or repetitive. This was seen to worsen during the last phase, when depression and anxiety took hold. Loud noises or unexpected information upset the spacemen, who became testy with each other and their handlers on Earth.

By the hundred sixty-eighth day of his record-breaking 211-day mission on *Salyut 7* in 1982—the Russians continued to chase records even after almost everyone else stopped paying attention—Valentin Lebedev's normally sunny disposition had turned very sour. "My nerves are always on edge. I get jumpy at any minor irritation," he noted in his diary. Like his fellow cosmonaut Sergei Krichevsky, Lebedev wrote poetry, and like Krichevsky's, his was not universally appreciated. "I went on line [to Kaliningrad], hoping to read my poetry expounding on my confidence in our flight," Lebedev continued. "Suddenly, [Dr.] Kobzev poured cold water on me, snapping 'Just keep quiet for today.' Then he turned on the recording of poetry which I read before my flight. He shut me up in front of everyone and devastated my mood." The next day Lebedev "looked in the mirror and was dismayed—my face was covered with red blemishes caused by nerves, and my back is peeling. We're pretty depressed." Asked yet again how he felt during a record-breaking 326 days in space five years later, the otherwise good-natured Colonel Yuri Romanenko snapped, "Leave me alone. I have a lot of work to do." Valeri Ryumin, another veteran cosmonaut, drew an even darker picture: "All the conditions necessary for murder are met if you shut two men in a cabin measuring 18 feet by 20 and leave them together for two months."

Lebedev did not find immediate relief on terra firma, either. He reported feeling sick to his stomach when the rescue team pulled him out of the landing capsule: "When they picked me up, it felt as though they lifted me thirty feet into the air; I knew it was impossible, but my senses overruled my mind. I was terrified they would drop me from such a height. I was still wearing my space suit when I was put into the sleeping bag lined with dog fur. I was lying on the ground, motionless, and began to feel sick again. I asked for a napkin and threw up into it. After I threw up a few more times, I felt better."

A decade later, U.S. astronaut John E. Blaha would become depressed during the first of four months he spent on *Mir.* The retired Air Force colonel blamed the problem on the fact that he lacked both private quarters and a fellow countryman. Both Russians on board had their own compartments. "Isolation is a tough thing," he said, adding, "after I'd been there a month, I was a little bit getting psychologically depressed. So what I finally said to myself is: 'Hey, John, you're here. You may live the rest of your life here.' The more I got into that mode, I became very comfortable." Lebedev, by contrast, felt that spacefarers should share a communal compartment because people should not be separated. "The closeness during long-term missions is crucial," he said, adding that it would help in responding faster to emergencies.

At least three Soviet missions had to be ended early—"aborted"—for reasons that were partly psychological. The *Soyuz 21* mission to *Salyut 5* was abruptly ended, for example, because the cosmonauts complained of smelling an acrid odor in the spacecraft's environmental control system that no one else had ever smelled. The hallucination, if that was what it was, happened to a crew that was not getting along. Other cosmonauts complained of prostate infections and heart trouble that physicians could not find.

Meanwhile, impaired or not, cosmonauts continued to rack up time in space and successively break their own records. On July 25, 1984, Svetlana Savitskaya spent three hours and thirty-five minutes outside of *Salyut 7,* most of it welding and soldering. In the process, she became the first woman to walk in space and upstaged Dr. Kathryn D. Sullivan, who was scheduled to do an EVA while on the shuttle *Challenger* that October. Two months later, three other cosmonauts broke a 211-day record by 26 days. One American trade magazine noted that Russians had spent more than 87,600 hours in orbit. That came to more than ten years. Americans had only a paltry 30,012 hours by comparison. "The previous flights and the present one have shown that two hundred and eleven days is far from being the limit for the human organism," the article quoted Dr. Anatoli Yegorov, a cosmonaut physician, as saying. He was quite right.

Vladimir Titov and Musa K. Manarov lived in *Mir* from December 21, 1987, to December 21, 1988, setting a new record of 366 days, or roughly two and a half times as long as it would take to get to Mars on the shortest possible, 140-day trajectory. Manarov, a dark-eyed Caucasian who sported a mustache, complained in a diary he kept while in orbit that spending seemingly interminable days in space could be trying. It was difficult to get used to sleeping vertically, "on the wall," he wrote, adding that gyros and fire alarms kept going off and also interfered with sleep. He would succinctly sum up his feelings about the mission in 1992—the International Space Year—at a U.S. Space Foundation meeting in Colorado Springs. "I would *never* do that again!" he declared.

The Soviet long-duration missions were technically designed to learn about how people would stand up to the rigors of a mission as long as the one to Mars. But there was another purpose as well: they were also supposed to show the world that landing men on the Moon and then abandoning the place amounted to circus stunts, while a mature and serious space program would keep people in orbit for serious, long-term, scientific purposes. And the successive records amounted to a last, sputtering reach for the prestige that still lingered with Old Glory planted on the Sea of Tranquility.

The Dangers of the Future Conditional

Almost exactly twenty years before Manarov vowed that he would never fly another long-duration mission, during the brief but important period of détente, the two superpowers signed a Strategic Arms Limitation Treaty, SALT 1. The document severely restricted the use of anti-ballistic missiles, or ABMs, and set limits on the number of long-range bombers and ballistic missiles either side could have. Whatever its proponents in the United States thought about the SALT 1 treaty of 1972 as the linchpin of a peace process that included other agreements, many others believed implicitly that it amounted to naively ceding military advantage to the Soviet Union. One of them, a knowledgeable and very alarmed arms and intelligence specialist named Peter N. James, conscientiously wrote a book in what might be called the Paul Revere genre. In it, he tried to warn the American people that they faced a potentially perilous threat from space regardless of peace treaties, including the so-called peaceful uses of outer space agreement that had been signed in 1967 and that prohibited sending weapons of mass destruction into orbit and beyond.*

Selectively using declassified documents and generous amounts of factual information about the Soviet space program and its political and military underpinnings, *Soviet Conquest from Space,* which was published in 1974, tried to show that the Communists were relentlessly moving to grab the high frontier and use it to subjugate Earth. James's book would not find a place among the most penetrating works about the cold war. Yet its author played a small but real part in promoting the very conflict that worried him.

Extrapolating from the fact that the Soviet Union had sent spacecraft to the Moon, Venus, and Mars, for example, James cited "reliable" but unnamed sources as predicting that the Communists intended to go for a five-planet Grand Tour of their own. He conveniently ignored the fact that most of those desperate missions, and particularly to the planets, had been flops. "By comparison," he added, "the U.S. Grand Tour mission was canceled in January 1972 because of budgetary reasons." James also warned that Soviet scientists had successfully tested a "lethal laser weapons system (the death ray)" and planned to place one or more in orbit; that whereas the U.S. space shuttle program was faltering, the enemy had embarked on an extravagant and visionary "total national space program" that made their own shuttles part of a highly integrated system which included orbiting stations and reusable spacecraft. The enemy's shuttles, James warned, would be able to carry up to 100,000 pounds, compared to only 65,000 pounds carried by the puny American lifters, and were scheduled to go into service in the late 1970s.

Furthermore, James continued, a Soviet "orbit-to-orbit" shuttle (a kind of space plane that could maneuver extensively) "*will be* used for reconnaissance and surveillance missions, to deliver and retrieve secret satellites in remote orbits, to inspect and destroy hostile satellites, and provide logistics support for spacecraft and space stations in remote orbits" (italics added). Still writing in the future conditional, James described a 300,000-pound behemoth of a space station that would be propelled to orbit by a "Super Booster" even larger than the Saturn 5. This was an apparent reference to the N-1s, four of which, or at least their charred remains, rested in Tyuratam's grisly boneyard.

James went on to bash the U.S. news media for what he considered to be a lack of interest in the space program that approached treason. By contrast, he went on, a massive amount of literature on the Soviet space program was made available to Russians by a state-controlled press. The result, James con-

* Its official title was Treaty on Principles Governing the Activities of States in the Exploration and Use of Outer Space, Including the Moon and Other Celestial Bodies.

cluded, was that "The Russian space program is wholeheartedly endorsed by the Soviet public and their cosmonauts are literally worshipped."

Eight years later, physicist Robert Jastrow sounded the same warning. He selectively noted that the Soviet Union was developing orbital anti-satellite weapons, lasers, and a permanent manned presence in space through the Salyut stations. The point, he emphasized, was that the Russians were planning to dominate space and fill a void left by supposed American indifference and weakness. "In fact, no matter what the United States does to catch up now, the next few years are almost certain to be a time of Soviet domination in space," he warned in an article in *The New York Times Magazine*. "By undertaking a massive military space program designed to gain control of space, Moscow is attempting to shift the balance of power substantially in its favor. We have ignored these facts at our peril. Twenty-five years ago the Russians surprised us by launching Sputnik. For more than 10 years they have been gearing up to surprise us again. This time, the surprise may have more serious and even deadly consequences," the author of *Red Giants and White Dwarfs* concluded in the mistaken belief that the title of his book applied to international relations as well as to cosmology and evolution.

But that is not the way it looked in the Kremlin. Few Russians would have mistaken their country for being a red giant, much less its opponent for being a white dwarf. Bitter conflicts over the "rockets-versus-butter" question raged quietly for years and would erupt in public only during the first flushes of glasnost—openness, letting all voices be heard—in 1989. There was sharp criticism of spending on space during election debates that year. And once ungagged, the newspapers James credited for their patriotism published stinging attacks that demanded that the once-"sacred [space] cow" be more "oriented toward the needs of [the] national economy." *Izvestia,* the government's own newspaper, complained that the cost of major space projects was still a "state secret." Boris Chertok, who by then was a corresponding member of the Academy of Sciences, was thrown on the defensive. "Space research performed the role of locomotive which has blazed the way for other sectors," the seventy-seven-year-old engineer said, answering critics. "Were it not for astronautics, our lag in such fields as computer engineering would have been much greater than now." The computer-improvement argument would have a familiar ring to anyone who followed the U.S. program's use of "spin-offs," from computers to Teflon to medicine to bedroom-control systems for severely handicapped people, as a rationale for going to space. But computers could have been developed in both countries at far less cost without sending huge rockets to Earth orbit and elsewhere.

Konstantin Gringauz, an IKI scientist, attacked the Ministry of General Machine Building in an article in *Pravda* that would have been unthinkable even two years earlier. "It cannot be ruled out that the main reason for the creation of the Energia-*Buran* system was the desire of the branch to assert itself and not the real needs of the country and science." Energia was the huge and powerful launch vehicle that was designed to carry cargo, unmanned spacecraft, or *Buran* (Snowstorm), the Soviet shuttle. And contrary to what James wrote, *Buran* itself was a near twin of the American shuttle, which preceded it.

Writing in the labor newspaper *Trud* in 1991, as his country was celebrating the thirtieth anniversary of Gagarin's flight and edging toward disintegration, a veteran of the Flight Control Center at Kaliningrad described the space program with unmitigated contempt. While James and others railed at what they took to be their own nation's incompetence, a Russian space flight specialist named Boris Olesyuk, his colleagues, and many of their countrymen had watched with disgust as their own government squandered enormous resources in the search for prestige and to finance what he called futile rivalries among designers and "erratic" priorities and strategies. Satellites and spaceships were being orbited for propaganda, and the Soviet leadership gave practically free rides to foreigners, Olesyuk wrote angrily. The joy rides cost 50,000 rubles an hour (the ruble was then officially roughly equivalent to the dollar), he added, and it cost more than 2.5 million rubles just to train a foreign cosmonaut.

Olesyuk was especially wrathful about *Mir,* charging that there had never been a public accounting of the billions of rubles that were "devoured" by the program, and that virtually everything that was done on it could have been done more cheaply by automated spacecraft. "To conduct astrophysical experiments, the *Mir* complex requires constant, precise orientation. With people on board, it is impossible to ensure such orientation because cosmonauts' movement . . . negatively affects the orbiting complex. Experience shows that extraterrestrial observatories can conduct effective research without human interference," he added.

Nor were there kind words for *Buran.* "For prestige considerations," Olesyuk continued, "the Soviets followed the Americans and plowed over fifteen billion rubles into the *Buran* shuttle without defining a proper use for it." Its use, if that is the word for it, was to counter the American shuttle, which the Soviet aerospace and intelligence communities believed to be a military threat. Five *Buran*s were built. Only one made it to space once, though without humans at its controls, and another was tested suborbitally with jet engines about fifty times.

If there was a lesson to be learned from the continuing anxiety on both sides, it was that the mutual ignorance was not only dangerous, but phenomenally expensive. What became apparent after the cold war was not only how both space programs paralleled each other in the most basic ways but how pervasively they fed off fear, suspicion, and the demonization of the opposition. Americans such as James and Jastrow, and their Russian counterparts, mistakenly allowed their imaginations to override common sense. There was a basic fallacy, for example, in the notion that a country that suffered one embarrassing failure after another on Venus and Mars had the capacity to blithely tour five outer planets. Similarly, the Russians need not have worried about the American shuttle's being designed as a war fighter. The U.S. Air Force hated the thing from the beginning.

Shuttle: The Frequent-Flyer Program

If people were going to stay in space, there were two basic ways to get them there: the so-called cannonball approach that had been pioneered by Vostok and Mercury and which was still used by the Soviet Union, and a kind of spaceplane: a much larger version of the extinct Dyna-Soar that would carry astronauts up and bring them back down. Common sense seemed to indicate, as von Braun had shown in the pages of *Reader's Digest,* that using the same machine over and over again had to be cheaper than discarding it every time. As every accountant knows, all other things being equal, the more an item is used, the cheaper it is to use. That was the core of the argument NASA made to Congress and the White House during Nixon's first term. The problem is that all other things are almost never equal.

Those who had planned the Space Transportation System in the late 1960s, basking in Apollo's Moon glow, envisioned a fully reusable launcher that would carry an equally reusable orbiter, both of them using liquid propellants. The orbiter, about the size of a DC-9 airliner, would carry 65,000 pounds of cargo, be able to land on conventional runways, and fly a hundred missions with only minor refurbishing. The estimated cost of the system topped out at $14 billion. Growing public apathy about space notwithstanding, the heroes of the Moon race believed that sending a Cadillac into orbit instead of a Chevy was their divine right. "Buccaneers stake out and create powerful outposts of stability, sanity, and real future value for mankind in the new unchartered seas of space and global technology," as one NASA memorandum had put it.

Congress, however, was not persuaded that such an extravagant program was necessary. Citing distinguished scientists such as Van Allen and Thomas Gold, one of solar system exploration's true heavyweights, as saying that the United States had no compelling need or use for a shuttle, Senators Walter Mondale, William Proxmire, Jacob Javits, and others responded by telling Paine and then James C. Fletcher, who succeeded him in April 1971, that the Cadillac was out of the question and that even a Chevy would have to be fully justified.* The senators could have pointed out that "buccaneer" is a synonym for "pirate" and that Webster defines one as "an unscrupulous adventurer" who robs the public.

Richard Nixon had compelling reasons to keep Americans in space after Apollo. For one thing, doing so would assure that a very important industry remained robust (profits for Republicans and jobs for Democrats in the most simplistic scenario). For another, leaving all manned space activity to the Russians would have amounted to political suicide. And for still another, ending the manned program would have broken Kennedy's legacy, and Nixon did not want to be vilified for that.

Nixon therefore agreed with one of the Agnew Space Task Group's recommendations and formally approved the Space Transportation System. But there was a caveat: the Office of Management and Budget ruled that STS, as it was called, could cost no more than $5.15 billion to develop.

Out went the reusable lifter and in came two solid rocket "strap-on" boosters and a 154-foot-long external tank to hold the 385,000 gallons of liquid hydrogen and 140,000 gallons of liquid oxygen that would feed the orbiter's three main engines. It was a compromise that would have far-reaching consequences. "That was one of the greatest mistakes that NASA made," a senior agency scientist said years later. Fletcher should have told Nixon that the STS could not be built for $5.15 billion, he added. "If $5 billion is all the nation can afford, then we can't afford the shuttle. . . . I am recommending that we close down the manned space flight centers," he said Fletcher should have warned Nixon. "The program would have been delayed. The Administrator probably would have been fired or forced to resign. But in the long run," the scientist lamented, "Congress and the administration would have found the funds."

Instead, NASA decided to exercise its own form of shuttle diplomacy. NASA resolved to go for the Chevy, and it took two courses of action to make certain the vehicle was delivered. It also made a promise it could in no way fulfill.

First, it had to exaggerate demand to justify the cost. If the cost of the program decreased as the number of missions went up, as the accountants said, the number would have to go way up. The number of launches projected for the period from 1978 to 1990–91 therefore climbed steeply from 440 to 581 to 779 (or five launches a month). At a projected cost of slightly more than $50

* At a Senate hearing in May 1981, Senator Jake Garn, a Utah Republican with military flying experience, asked a NASA official whether he could fly on the shuttle. He was told that he could. At that point Proxmire, who had awarded the space agency a number of his famous "Golden Fleece" awards for wasting taxpayers' money, quipped that if *he* ever went up in the shuttle, he would not return. (*Defense Daily,* May 13, 1981, p. 72.) Garn went on to fly as a "payload specialist" aboard *Discovery* in April 1985 on the program's sixteenth mission. After he became spacesick, the orbiter was temporarily nicknamed "The Vomit Comet."

billion, NASA told Congress, it would be $16 billion cheaper than the $66 billion that would have to be spent on the same number of expendable rockets—the ones that plopped into the ocean and sank, never to be seen again—making the shuttle more "cost-effective." The General Accounting Office called those projections "optimistic" and calculated that the total shuttle cost per pound of payload in orbit would actually be roughly $3,500, or $400 more than the Atlases, Titans, and other expendables in use and $900 more than their successors on the drawing boards cost.

"Telling the Congress that we were building a shuttle that was cost-effective when you could calculate on the back of an envelope that it wasn't. . . . I just don't understand why you would do a thing like that," one of NASA's retired top executives said later. Sure he did. It was to sell the program. That much was obvious. What was not so obvious was the fact that a four-component system—the orbiter, the external tank, and two solid rocket boosters—was more complicated than the two-component system's orbiter and reusable carrier rocket. That, in turn, would lengthen the turnaround time between missions and make logistics more difficult. If anything even approaching the promised launch rate was going to happen, taking otherwise unacceptable risks would become imperative in order to stay on or near schedule, and one of those risks would be to launch in very cold weather. How much it would finally cost to launch a single shuttle depended on what factors went into the computation. Counting not only hardware, launching and landing support, and actual flight operations but civil service salaries, related support systems such as tracking antennas and relay satellites, and ongoing research and development for modifications, it would cost an average of $547 million to launch a shuttle in 1993. That would be $180 million more than the space agency itself estimated.

Second, if the shuttle was going to make the number of flights advertised, it would obviously have to be the only way to get people and payloads, including satellites, planetary probes, and the station, to space. All other civilian and military launch systems, most notably the Atlases and the mighty Titans, would therefore have to be eliminated: phased out by mid-1987.

But the Air Force fought for, and finally won, the right to keep some Titans and other expendables as backups—"options," as they put it—to total dependence on the shuttle. In March 1985 the Air Force announced that it would buy ten huge Titan 34Ds from Martin Marietta at a cost of a little more than $2 billion, and, what was more, it planned to convert a dozen Titan 2s from ballistic missiles to launchers. For the Air Force, the move to the shuttle was only the latest in a series of developments that seemed to complicate its mission as the guardian of the "high frontier" (as the public affairs people never tired of putting it). Having lost reconnaissance to the CIA and MOL to McNamara's budgeteers, the Air Force now faced the prospect of using a civilian system as its primary way to get to space. That was like the Strategic Air Command depending on Pan Am to get its bombs to the target. In the Air Force's view, there was a potential security problem because civilians would be involved. Worse, most shuttle payloads were going to be assigned like ships' manifests: customers would have to queue up and then wait as long as five years to get launched (NRO and some military satellites had priority, however). During a crisis, a reconnaissance satellite could be bolted on top of an Atlas or Titan and tossed into orbit relatively quickly. So the Titans would be there to hedge the bet. If a serious emergency developed anywhere in the world, or there was some problem with the shuttle, the reconnaissance satellites and other national defense spacecraft would not be stuck on Earth.

Understandably, the Air Force also had its own criteria for how the Space Transportation System was to be designed and operated. The price for signing on was the creation of a relatively large spacecraft whose cargo bay alone would have to measure sixty feet by fifteen feet and be able to carry 40,000 pounds into polar orbit and send 5,000 pounds out to geosynchronous on its own rocket, the hydrogen-oxygen-burning minibooster Centaur. The polar orbiters would be KH-11 real-time imaging reconnaissance satellites, then on the drawing board, which were as massive as Greyhound buses (and priced like Rolls-Royces). The ones headed for geosynchronous would have code names such as "Rhyolite," "Magnum," "Vortex," and "Orion." They would eavesdrop on Warsaw Pact communication traffic, monitor telemetry in missile tests, and collect other electronic signals for the National Security Agency's and CIA's analysts and code breakers.

And the Air Force demanded more. Should there be an emergency, the orbiter was going to have to be able to make it back to its launch facility after one go-round. And it would have to have a 1,500-mile cross-range capability as well. That meant an orbiter in trouble would have to be able to fly either east or west anywhere in the world and reach a friendly military air base without passing over enemy territory. The thought of an orbiter coming down in Siberia or Sinkiang with a reconnaissance satellite on board was enough to give a general the vapors. But the law of engineering trade-offs was immutable. In order to build cross-range capability into the orbiters, they would have to have thick delta wings, which would make them heavier. Maneuvering to a safe landing field instead of just coming "straight in" would expose an orbiter to higher temperatures, so the weight of the thermal protection system would have to be doubled, further increasing the orbiter's weight. A heavier orbiter would make it difficult to meet the payload lifting requirement and put an extra strain on the propulsion system.

There were still other concerns. A single man-carrying system would be immensely complicated and vulnerable to accident or attack. A joint study done by the Defense Intelligence Agency and the NSA in 1977 showed that the Space Transportation System faced a daunting array of potential threats: attack by conventional, nuclear, or laser anti-satellite (ASAT) weapons; electronic systems made dysfunctional by nuclear explosions on Earth or in the atmosphere; the jamming of communication links on the shuttle, on Earth, or both by enemy antennas or a "jamming platform" (spacecraft); and attacks on the Kennedy and Johnson Space Centers and Vandenberg Air Force Base. A mothballed MOL launch facility at Vandenberg was turned into the awesomely large, $2.8 billion Shuttle Launch Complex 6 (or "Slick Six") from which shuttles were supposed to carry reconnaissance satellites into near-polar orbits. They never would, though. Slick Six, which took seven years to build, was a monument to bureaucratic incompetence. The blast pit that was supposed to vent steam from the shuttle's engines away from the launchpad, for example, would have trapped any leaked hydrogen and could have caused an explosion when the engines were ignited.

No danger was too remote for the security mavens, who not only claimed that "activities by vandals and pranksters is highly probable" at Kennedy, Johnson, and Vandenberg, but who specifically noted a potentially violent threat from "Puerto Rican Nationalist Groups, Cuban exiles, and Croatian Nationalists," in addition to West German, Japanese, Palestinian, and even domestic environmental groups. "The dissident threat emanates from environmental and other groups which may espouse legitimate concerns but which could conceivably resort to spontaneous or planned violent activity against STS facilities should they perceive an exploitable 'issue.' " Eastern European elements on the West Coast were taken to be a potential danger to Vandenberg. "There is a Romanian establishment in Los Angeles, California, about 90 miles from VAFB," a 140-page Defense Intelligence Agency threat assessment report noted without further elaboration.

NASA and Rockwell International, which built the orbiters and integrated the four main components, were caught in a test of wills between two powerful organizations that did not like the shuttle very much: the Office of Management and Budget, which severely restricted its budget, and the Air Force, which feared sole reliance on a rocket system that was under research and development. Something clearly had to give. And it did. By March 1, 1978—the shuttle's advertised launch date—its main engines were still not performing as they had to and hand-fitting 31,000 heat-resistant ceramic tiles to *Columbia*'s belly, its vertical tail, and the undersides of its wings was taking 670,000 hours, or 335 man-years. *Columbia* was the second orbiter to be built but the first to actually reach orbit. The first orbiter, *Enterprise,* was built without engines and was used only for suborbital glide tests. NASA had wanted to call it *Constitution,* after Old Ironsides, the Navy's first real fighting ship. But a hundred thousand Star Trek fans convinced President Gerald Ford to give it a name that was more relevant to them, so he christened it after the space cruiser that carried Kirk and Spock around the universe. NASA, which was sensitive to jokes about its spacecraft being named after science fiction contraptions on Hollywood sets, was not happy about the name.

But the name of the rocket glider was the least of the problems. The shuttle was turning into an embarrassing fiasco that shamed NASA, angered the Pentagon, and embarrassed Ford's successor, Jimmy Carter. The new president was trying to get a skeptical Congress to ratify the SALT II arms-control treaty in 1977 and 1978 without having a way to get the reconnaissance satellites that were supposed to verify Soviet compliance into space on shuttles. The expendables were launching them when Carter became president, but there were real questions about whether, if ever, shuttles would be able to do it. He therefore made it clear that he wanted the thing to fly no matter what it cost. So it ended up costing $10 billion, almost twice the original estimate. But it flew.

The Gem of the New Ocean

On April 12, 1981—twenty years to the day after Yuri Gagarin's flight—*Columbia,* bolted to its steaming external tank and the two solid-fuel rocket boosters and "twanging" slightly back and forth, roared away from the Kennedy Space Center and into a fifty-four-hour mission. Navy Captain Robert L. Crippen and John W. Young, a civilian who had walked on the Moon in April 1972, were at the controls. Never before, historian Alex Roland would note, had Americans gone to space using solid-propellant rockets, or risked their lives on an engine that was untested in flight.

Two minutes after liftoff, both solid-propellant rockets were jettisoned and floated back down to the Atlantic under parachutes so they could be retrieved and refurbished for another mission. Six minutes later and just short of reaching a velocity that would take it into its 172-mile-high orbit, *Columbia* dropped the external tank, which arced out and disintegrated over the Indian Ocean. A television camera turned on the orbiter's vertical tail early into the flight showed that more than a dozen of the tiles had been ripped off (16 were actually lost and 148 damaged). There was some worry that others would come off during *Columbia*'s searing flight back into the atmosphere, perhaps broiling its crew. But the other tiles stayed glued in place, and Crippen and Young finished their historic flight when they landed at Edwards Air Force base two days and thirty-six orbits after riding *Columbia* to space.

The Majesty of Wings

The first true spaceship, born out of wedlock to NASA and the Air Force after a decade of bruising fights and constant reengineering, had sailed around the world and made it home. Tom Wolfe, an old *New York Herald Tribune* reporter who could be as hard-boiled and cynical as the next guy with a dog-eared press pass buried in his sock drawer and his psyche, rhapsodized about STS-1, as the shuttle's maiden voyage was called. It took him back to his days at Edwards researching *The Right Stuff.* Back to stunt pilot Pancho Barnes and her storied Fly Inn, Happy Bottom Riding Club, dance hall, saloon, motel, and girls. Back to test pilots Chuck Yeager, Jack Ridley, Ivan Kincheloe, Milt Thompson, Joe Walker, Neil Armstrong, Bob White, and Scott Crossfield. Back he went, to the orbiter's oldest true ancestors, the aeronautically improbable streamlined bricks they called lifting bodies. Back to real airplanes like the X-1 that Yeager had ridden to glory, the X-2 that Mel Apt had ridden to his death, and the X-15 that Joe Walker had ridden into the surf of the ultimate ocean. It took him back to the prehistory of manned spaceflight, when pilots had not been strapped onto contoured couches inside computerized cannonballs; when they had not been "Spam in a can."

Wolfe saw something profoundly dignified, even majestic, in the fact that the spacefarers had come home on wings rather than floating down under parachutes as though they had just left the scene of an accident. "To me there was a touch of Rip Van Winkle about it all," he would be moved to write.

> After 54½ hours in earth orbit an airplane—not a capsule or a command module but an airplane, a ship with wings—descends above the high desert in California. It glides toward a landing at Edwards Air Force Base. As in the old days, the mirages of Rogers Dry Lake envelope it like a hallucination. The ship makes a perfect touchdown and rolls to a stop. At last the commander emerges. He is 50 years old. He has grown old and farsighted waiting for this flight. He had to wear glasses to read the instrument panel. He opens his mouth and out comes a drawl that takes me back 25 years, at least, to the cowboy days of Chuck Yeager.

For Wolfe, *Columbia*'s flight continued a story that had been interrupted by Mercury, Gemini, and Apollo, during the frenzied aberration that flung fledgling astronauts at the Moon. Like Mailer, Wolfe knew that good space stories were "people" stories; that for the protagonists, on some level that existed far beneath the intellect, defying death defined life. And that did not start with countdowns on launchpads. It started in cockpits because the stick and pedals were run by the viscera, not by chips made out of sand, and they required instinctive responses every second. The public had always understood that, which is why it had been reading the *Odyssey* for 2,700 years. "It returns the American space program to where it started—which was not Cape Canaveral but the throwback landscape of Edwards Air Force Base, a terrain that evolution left behind, a desert decorated with the arthritic limbs of Joshua trees and memories."

But wherever the space program had started, it was now at Cape Canaveral. In December 1980, perhaps to convince the public that the program was more under control than it really was, the space agency issued the Space Transportation System's first five-year timetable and payload manifest. It called for no less than thirty-seven flights between August 31, 1981, and May 14, 1985, the last being a "secret military payload." All of the missions were listed by their expected launch date. Not only that, but NASA also predicted that there would be five hundred flights by the end of 1991. The time would come when the space agency would gag on that bit of expedient self-indulgence.

The last shuttle to make it under the deadline did not in fact carry a military payload. *Challenger* sailed into orbit on April 29, 1985, carrying a pressurized fifteen-ton modular laboratory called Spacelab 3 that was filled with science experiments put together by the European Space Agency with support from NASA. Scientists worked inside the module in shirtsleeves. The idea was to sell the shuttle to a science community still wedded to cheaper unmanned spacecraft. That was where spin-off came in, not only for computers but even for protein-crystal growth experiments that were touted as possible beginnings of a cure for cancer, radiation poisoning, and AIDS. Of course those things were possible. But they were not very likely. "Possible," "may," and "could" were used like battering rams by space agency and contractor publicists trying to sell the shuttle as a health factory. Health, after all, had mass appeal. And basic research could be so ambiguous that it was unchallengeable. "Astronauts aboard Discovery may help scientists learn to combat one of the most dreadful diseases—AIDS—by conducting experiments in space," *The Orlando Sentinel* reported after interviewing a biochemist who had an experiment aboard the shuttle *Discovery* in October 1988.

So-called Getaway Specials, which offered payload space for science experiments by college students and others for as little as $3,000, were also used to forge links to the community and institutionalize STS. The specials had mixed success, with 40 percent of sixteen experiments failing by October 1983. Some of them produced no data at all.

The first Spacelab, which went up on *Columbia,* flew from November 28 to December 8, 1983, and was pronounced an unmitigated success—a "good omen"—by *Science,* which devoted an entire issue to the results. "The major disappointment," the magazine noted, was a month's delay in launching.

Spacelab 3, which made it under the wire for launches through 1985, was mission 17, not 37. NASA had by then lowered its sights for the shuttle schedule, estimating that there would be only 165 missions through the end of 1991, or 335 short of its prediction. And even that seemed optimistic. "Every now and then I go back and look at those early projections and have to close my eyes and shake my head," sighed L. Michael Weeks, a shuttle official at headquarters.

Weeks made that statement to John Noble Wilford of *The New York Times,* who was not cheerleading for STS (or anything else in space). He quoted Weeks in an article that led his newspaper's science section on May 14, 1985, under the headline "IN HARSH LIGHT OF REALITY, THE SHUTTLE IS BEING RE-EVALUATED." The space junkies were still there, and by nature of the din and danger a launch was still an exciting event. But the honeymoon with NASA had ended on the Moon. Everyone who was paying attention understood that the space agency was under immense pressure to make good on its oversold spaceship and keep even the semblance of a schedule. The shuttle had been concocted by NASA, not its competitors or enemies, and the agency was therefore responsible for being truthful instead of duplicitous about the quagmire it was in. All the public relations lard in the world could not change that.

Reporters covering STS found stacks of fact sheets in the huge permanent tent that served as press headquarters. Standard issue included Reporter's Space Flight Note Pads chock full of statistics and mission summaries and an expensive *National Space Transportation System Reference,* a 1,001-page hardbound loose-leaf that had every conceivable fact about the shuttle those who covered it needed to know—every fact except those having to do with what it really was and how it had gotten there.

So Wilford and a minority of others held NASA to task for the deception. "Discrepancies between promise and performance, combined with growing competition from Europe in the business of launching services," he wrote in the article quoting Weeks, "have led to a major re-evaluation of the manned shuttle program as the centerpiece of American space operations, military, scientific, and especially commercial." (The notoriously self-centered French not only had the temerity to build their own nuclear strike force and leave the NATO military alliance in the 1960s, but they created the Centre National d'Études Spatiales [CNES], a national space agency, in 1962 and developed the fine Ariane launcher, an all-purpose equivalent of Atlas. Now they were aggressively looking for customers who wanted satellites put into orbit. And they were making the satellites, too.)

Alex Roland detailed the shuttle's tortured history and lambasted NASA for its multiple deceptions in a long article titled "The Shuttle: Triumph or Turkey?" that appeared in a special report in the November 1985 *Discover* magazine. In an accompanying article, science writer Dennis Overbye articulated the science-versus-shuttle situation. "Asking scientists what they think about the space shuttle is a little like asking Christians about the crucifixion: it wasn't their idea or even their preference, but they've managed to make something out of it."

The Betrayal of Arthur Rudolph

Arthur Rudolph might have sympathized with the shuttle's various problems. He might have, but he probably did not, since by 1985 he had plenty of problems of his own. Sixteen years after one of his Saturn 5s sent Americans to the Moon for the first time, the master of Mittelwerk languished in self-imposed exile in Hamburg, West Germany, his U.S. citizenship gone.

The year before, in 1984, Nazi hunters had reminded the Department of Justice that Rudolph was a war criminal. A search of his long-dormant file had convinced federal prosecutors that he had indeed been responsible for brutalizing inmates at Dora and specifically for "working thousands of slave laborers to death." They then accused the former Saturn 5 program manager not only of committing the crimes but of concealing them. Rudolph cynically insisted that he was innocent and that he "only deduced there must have been deaths among the prisoners because they were exposed to the same conditions as my Germans and I myself was." But the lawyers who threatened to prosecute him because of the contents of a file none of them would have dared open while he was racing the Russians to the Moon were just as cynical. Rudolph's interrogator's notation that he was a "100% NAZI" and a "dangerous type" who should have been interned had been dropped into his file in 1945 and had remained in it while Armstrong and Aldrin stood on the Moon thanks to his booster.

Rather than face prosecution, the seventy-seven-year-old engineer abandoned his home in San Jose, California, renounced his American citizenship, and fled to Germany. From there he waged a long and futile effort to regain his citizenship. Rudolph died in a coma at the age of eighty-nine in late December 1995 while still in exile from the country that relieved its own conscience only when it was safe to do so.

In the Black

Democracy forced the spymasters of the United States—the "gray" men whose snooping satellites rode to orbit on the shuttles—to develop split professional personalities. On the one hand, and most obviously, they had to be secretive. But on the other, they had to sell their organizations to the White House, to Congress, and indirectly, even to the American people. Like NASA, the institutions that were responsible for collecting the world's secrets were dependent on federal appropriations. And also like the space agency, they were forever lobbying for more. Unlike NASA, however, they could—and did—make the case, with varying degrees of success, that the country's survival hinged on how good they were in finding and reporting threats. This was especially true where spying from space was concerned.

Everything having to do with the satellites, ground facilities, and "product" they delivered in the form of pictures or intercepted signals continued to be scrupulously protected by the sensitive compartmented information (SCI) system that had been used for Corona. Its foundation is the so-called need-to-know principle. That is, the people in the CIA's Directorate of Science and Technology who designed imaging reconnaissance satellites were emphatically not cleared to know operational details, including where they were at any given time. That information stayed in a windowless building called the Satellite Control Facility, or "Big Blue Cube," in Sunnyvale, California, which operated the National Reconnaissance Office's spy satellites. Nor were they allowed to see the digital imagery taken by the satellites, which was analyzed at the National Photographic Interpretation Center in Washington, in the CIA's own Office of Imagery Analysis, and by the military services' and other organizations' analysts.

Like a submarine, the whole intelligence operation was the equivalent of a series of connected watertight compartments, constructed so that one traitor's leaking information, or its being inadvertently spilled some other way, would not sink the whole boat. But fear that secret information would fall into the hands of hostile governments was only one reason to keep the lid on sensitive material. The other was fear that it would fall into the hands of other U.S. intelligence services, in the process perhaps weakening one service's control of the system relative to the others. And since spy satellite technology was very advanced—"cutting edge," as some called it—and therefore horrendously expensive, the budget lines themselves were buried in otherwise innocuous Department of Defense accounts.* While the compartments may have foiled the Soviet Union, Communist China, and other unfriendly nations, they also guaranteed that key parts of the U.S. intelligence community itself would often remain unnecessarily ignorant because the big picture was hidden.

Given the tight security constraint, the intelligence community's salesmen found two ways to advertise their product, and thereby gain favor with the nation's politicians, without going to jail. One was to provide journalists with information whose publication would either point up successes or signal

* It was not until 1997 that the CIA finally disclosed the annual intelligence budget: $26.6 billion. ("For First Time, U.S. Discloses Spying Budget," *The New York Times,* October 16, 1997.)

problems without revealing the so-called sources and methods of the intelligence collection process, for instance, operational details such as the optical resolution of a satellite's telescope or the fact that the "asset" was operated by the National Reconnaissance Office. A classic success story was leaked to the authoritative *Jane's Defence Weekly* in July 1984. It reported that the Soviet Union's Northern Fleet headquarters at Severomorsk on the Barents Sea had been devastated by a series of explosions two months earlier and not only described the normal complement of naval vessels there—190 submarines, for example—but provided an extraordinarily detailed inventory of the missiles that had been destroyed in the blast. The stories, in *Jane's* and then in *The New York Times,* indirectly but effectively made the point that America's fleet of expensive reconnaissance satellites was worth the money because they exposed Soviet secrets.

On the other hand, faced with the possibility of lapses in photoreconnaissance coverage for want of new equipment, the technospooks would often leak that problem to their contacts in the Fourth Estate. This resulted in one article in *Aviation Week & Space Technology* in 1980 warning that U.S. space reconnaissance capability was faltering and another in *The New York Times* seven years later proclaiming the same thing.

The other strategy was and still is less subtle. It required that high-ranking intelligence officers, like other salesmen, show their wares to presidents, to important congressional committees such as Intelligence and Appropriations, and even to friendly and influential foreign leaders. This is done by competing spy organizations to curry favor with the politicians who hold the purse strings.

That is precisely what Allen Dulles and Richard Bissell were doing when they showed Eisenhower the first Corona imagery in August 1960. And it was what E. Henry Knoche—Enno "Hank" Knoche—a career intelligence analyst who happened to be acting director of the CIA, was doing when he paid a courtesy call on Jimmy Carter in the second-floor Map Room of the White House on the afternoon after Carter's inauguration in January 1977. Knoche offered the new president his congratulations and spent the next quarter hour showing him the most remarkable pictures. They were the first of a new type of satellite image, and they had come down early on the previous morning, just before Carter took the oath of office. Although their subject was so mundane Knoche could not remember what it was years later, they were supremely important for the technology they represented. There were three or four photographs made from digital imagery that had come into Fort Belvoir, Virginia, in what engineers called "near real time." That is, they had not been dropped into a bucket the way satellites in the Corona and successor programs had done, but had been sent virtually as the event they recorded happened.

The machine that had done this was a descendent of Samos, the original blurry-eyed near-real-time readout satellite, and was named after its remarkable twenty-foot-long telescope and the sensors at the end of it: Keyhole-11, or simply KH-11. Others knew it by its Byeman code name: "Kennan."* Still others knew it only by its number: 5501. It all depended on the "compartment" they were in. No. 5501, the first in the series and the one that took the pictures carried by Knoche, had been launched the previous December 19 on a huge Titan 3D.

The KH-11 was the carefully nurtured, fiercely protected, and fabulously expensive brainchild of the CIA's Directorate of Science and Technology. General George J. Keegan Jr., who headed Air Force intelligence at the time and who despised the CIA, said that the TRW-built craft had gone $1 billion over budget. Others who did not hate the CIA have generally agreed, but pointed out that by its nature, space reconnaissance was at the very edge of technology and that could be staggeringly expensive. John Pike, a space specialist with the Federation of American Scientists, called the development of reconnaissance satellites in general a "playpen for engineers." Sure it was. But those who objected to the cost of the KH-11 in particular ignored the fact that launching more than a hundred bucket carriers in the Corona program between 1960 and 1972 alone was also horrendously expensive when the costs of sending transport planes and naval vessels to find and snatch each one of them, plus the cost of flying the film to Washington, is taken into account.

Whatever it was, the Directorate of Science and Technology and TRW produced a satellite that could have digital imagery with five-inch resolution on monitors at Fort Belvoir within hours, not days, after an event took place. The Soviet Union would launch its own version, inferior to the KH-11, in 1979. Since the systems were digital, they created a revolution in intelligence analysis because they allowed imagery on computer screens to be manipulated in three dimensions. And their sheer number vastly expanded an already large reference file. Once a particular type of aircraft or missile was in the file, imagery that showed only part of it because the rest was hidden by clouds could be made to fill in the whole weapon. But the revolution that started with the KH-11 had a negative side. Imagery came down in an unending torrent that swamped the CIA's Office of Imagery Analysis, which was responsible for separating what was important from a colossal amount of sheer trivia.

* A "byeman" is someone who works in an underground mine. It is also a classification system that gives names to photographic and signals intelligence satellites in another effort to make identifying them as difficult as possible for the uncleared. Two of the KH-11's predecessors, the KH-7 and KH-8, had been named Gambit; the huge KH-9, sometimes called "Big Bird," had carried the Byeman name Hexagon. When the name Kennan was made public in 1986 in a book called *Deep Black,* the KH-11 was renamed Crystal.

Bud Wheelon, the former deputy director of science and technology for the CIA, maintained after the cold war that in its entirety, the space reconnaissance program was the equivalent of the Apollo program in magnitude and effect. Unlike Apollo, though, it was officially cloaked in absolute secrecy.

Security strictures or not, space reconnaissance was soaked in politics. Wars raged over the costs of the system, with the Air Force almost invariably opposing the dizzingly expensive, high-tech toys that sprang out of the imaginations of the CIA's engineers and physicists. But as soon as the president—any president—claimed to like the machine, or at least its product, the airmen began maneuvering to control it for fear that the civilians would manipulate and misuse it, in the process undermining the nation's defense posture. At the same time, photointerpreters, especially in the armed services, had the understandable tendency to see in the imagery what their superiors wanted to be there: so many MiGs, mobile missiles, and motor patrol boats, for example, that would have to be matched.

It is easy, but simplistic, to criticize the generals and admirals for trying to manipulate the intelligence apparatus to their services' advantage. But it was their responsibility, not the CIA's, to fight their country's wars (at least overtly). Most sincerely believed that there was no such thing as being overprepared for hostilities. But being prepared would accomplish two things: it would dissuade a potential aggressor from attacking and, if attacked, in the view of the general staff, it would ensure the survival of the nation. William E. Colby Jr., a former director of central intelligence who often tangled with General Keegan, was charitable on the issue. "No general ever prepared to win a war by this much," he said, holding his thumb and forefinger about half an inch apart.

Finally, the White House or the Department of Defense can go public with what it knows in order to bring political pressure on the opposition, justify foreign policy, or both. One of several classic cases had to do with a Soviet anti-ballistic missile radar whose existence was revealed to the news media almost five months to the day after President Reagan announced his own plans for research on his own anti-ballistic missile system. The leak was soon supplemented by drawings of the large facility both in the press and in the Pentagon's *Soviet Military Power,* an alleged inventory of Soviet weapons that was started by the Reagan administration, revised annually, and made public. *Soviet Military Power: 1985,* which came out as the battle over the new ABM system crested, showed the phased array radar at Krasnoyarsk, plus another at Pushkino, drawings of Soviet Galosh ABMs, and a map of Moscow pinpointing three ABM radar sites and several missile silos. "The Soviet Union is violating the ABM Treaty through the siting, orientation, and capability of the large phased-array, early warning, and ballistic missile target-tracking radar at

Krasnoyarsk," the report charged with some justification. The ABM material typified how U.S. satellite reconnaissance pictures could be used politically without giving away the secret of how they were taken. The Krasnoyarsk radar started a heated debate that took years to resolve.*

Having set out to violate the ABM treaty (the administration's denials aside), which prohibited the kind of research Reagan wanted, his advisers tried to justify it by arguing that they could prove that the opposition had been first to violate the agreement.

Star Wars (The Weapon)

The most controversial and hotly debated weapon in history after nuclear bombs probably had its roots in the Bible and was spawned, like a whirlwind, by the interaction of three charismatic and devout anti-Communists: a scientist, a politician, and a general.

The scientist was Edward Teller, the godfather of the Lawrence Livermore National Laboratory, the Department of Energy's preeminent nuclear weapons design facility. The bushy-browed Hungarian-born physicist brought to life an implacable belief in both himself and the existential beauty and wonder of the genie he had played a leading role in pulling out of the jug: the thermonuclear explosion. At one time or another, Teller would advocate using nukes to find oil, dig tunnels and harbors, change the course of rivers, and stop both killer rocks from space and Soviet missiles. It was the last that would fate him to become the scientific equivalent of Rasputin to the president of the United States.

The president was Ronald Reagan, an avowed anti-Communist who had been awed by technology at least since his days of selling General Electric products on television and who was obsessed with the biblical prophecy of Ar-

* *Soviet Military Power* first appeared in 1981 and reappeared in an updated form every year through 1990 except 1982. It was published by the Department of Defense to warn of the Soviet military threat on land, sea, air, and in space and made extensive use of drawings derived from satellite imagery, together with photographs taken from aircraft, and colorful bar and pie charts to show that the threat was increasing. The series unwittingly created a conundrum for the administration, which insisted that the United States could not adequately verify arms-control treaties, meaning its national technical means of collection—reconnaissance satellites chief among them—were not able to keep track of Soviet strategic weapons. If the allegation was right, *Soviet Military Power* was wrong. If the annual inventory was accurate, the reconnaissance systems were obviously good enough to verify the treaties. Without getting into the sensitive issue of how good U.S. technical intelligence was, the Soviets maintained that the series was unbalanced and designed to "confuse, intimidate and misinform public opinion in the West." ("Moscow Says the Pentagon Booklet Is Unbalanced," *The New York Times*, September 30, 1981.) The Kremlin responded by issuing its own booklets, the hopelessly unslick and archaically titled *Whence the Threat to Peace,* which depicted U.S. weapons and a map of the world showing them deployed from one end to the other. (See, for example, *Whence the Threat to Peace*, Moscow: Military Publishing House, 1984.)

mageddon: of a final battle between good and evil. "For the first time ever," Reagan told a banquet companion in 1971 when he was governor of California, "everything is in place for the battle of Armageddon and the second coming of Christ." Between then and the presidential campaign that built in 1979 and 1980, Reagan pondered the ultimate firestorm and the policy of mutual assured destruction (MAD), which held that neither superpower would dare strike the other with nuclear weapons for fear of certain, devastating retaliation. Elaborate technologies were then in place, chiefly the DSP ballistic missile early-warning satellites and their Soviet equivalents, to give enough warning so that the nation being attacked could respond, making a first strike suicidal. But Reagan was understandably stunned to learn on a visit to the North American Aerospace Defense Command Headquarters deep inside Cheyenne Mountain that there was absolutely nothing that could be done to stop even one nuclear warhead from coming down on his country. Seizing on that, and undoubtedly knowing the candidate's feeling about the cleansing effect of Armageddon, an aide named Martin Anderson put together a policy memorandum suggesting to Governor Reagan that the American people would probably prefer a system that could shoot down enemy missiles to one that relied on the sanity of those they took to be paranoid despots secreted in the Kremlin.

The concept of ballistic missile defense had been around since the advent of the ballistic missile. As early as 1955, the Air Force and Army had competed to produce a ground-based ABM system, with the latter proposing its sleek, pencil-like Nike-Zeus. Since Nike-Zeus had become old technology by the time Kennedy became president, though, the Army developed a successor with McNamara's approval. It was christened Nike-X. Then it was given a more compelling name: Sentinel. Finally, with Richard M. Nixon in the White House in 1969, a modified version was called Safeguard.

There were two fundamental problems with ballistic missile defense, however, and one of them had to do with physics. The offense had an overwhelming advantage because of the devastation that even one of its warheads could cause if it got through the shield. That was related to the fact that traditional dish radars could not hope to track hundreds of incoming warheads at the same time. And since warheads could not be hit if they could not be tracked, the defense would be saturated if there were very many of them.

The other problem was political. Advocates of missile defense argued that it would be stabilizing because it would be able to repel an attack and therefore discourage one from starting. Opponents argued the opposite. They maintained that even a partially effective missile defense would be dangerously destabilizing because it would give its owner enough of an advantage so that it might be tempted to launch a massive preemptive strike against the enemy in the belief that it could stave off its own annihilation in a counterattack. It would be like two men in an arena, each armed with a sword but only one of whom also held a shield. The one with the shield would be tempted to attack his opponent, knowing that he could ward off an attack himself. Seeing that possibility, the opponent would get his own shield and, in order to break the stalemate, a larger sword—still more nuclear weapons to penetrate the missile shield—until the arms race spiraled ever more dangerously out of control.

That was the way the leadership in both Washington and Moscow saw it. Fearing a runaway nuclear arms race, the two superpowers in effect agreed to hold each other hostage by signing the Anti–Ballistic Missile Treaty as part of the Strategic Arms Limitation Treaty, or SALT 1, in 1972. The agreement stipulated that neither side could have more than two ABM complexes, each with no more than one hundred missiles, to protect its capital and one ICBM site. More to the point, Article 5 was unequivocal in stating that "each party undertakes not to develop, test, or deploy ABM systems or components which are sea-based, air-based, space-based, or mobile land-based." Each side thus held a gun to the other's head. Some called it the doctrine of massive retaliation. McNamara was the one who called it "mutual assured destruction."

But the dream of stopping, or at least decisively blunting, a nuclear attack would not die. Teller believed implicitly in "passive" nuclear defense at the start of the atomic age. "A shelter program would save the great majority of people in the United States even in case of a most ferocious attack," he argued, and the country could recover from an all-out attack "in a small number of years." But in the late 1960s he turned his attention to active defense, specifically to the use of so-called nuclear-pumped lasers. He also turned his attention to courting Ronald Reagan when the actor became governor of California in 1966. Both had a dread of communism and a belief that only massive military strength could deter "Russian nibbling" (as the physicist liked to put it) or all-out attack.

General Daniel Graham believed it, too. The former director of the National Security Agency and deputy director of the CIA was a national security adviser to Reagan during his first presidential campaign. Reagan remained amazed and saddened that no technology existed that could protect the United States from even a single nuclear warhead. So, with Anderson's memorandum in mind, he told Graham to try to come up with one.

The general pondered the problem for months before he came up with a revolutionary idea in 1982. In keeping with the frontier metaphor so beloved by the Air Force (Graham was an Army man), and perhaps also considering his president's penchant for romanticizing the spirit of the Old West, he called his creation "High Frontier." In its sheer scope, what Graham came up with was worthy of Hannibal, Sun Tzu, Clausewitz, or Napoleon. He proposed nothing less than a Global Ballistic Missile Defense System whose heart would be 432 space-based killer satellites, called "trucks," that were to be launched in small batches from shuttles the way scorpions lays eggs. There would also be "high-performance space planes," ground-based lasers, and other weapons off the pages of Flash Gordon that would attack enemy missiles and warheads during every second of their thirty-minute flight to the United States. Graham tried to make his plan more palatable by saying that much of it could be done with "off-the-shelf" components. Maybe off the wizard's shelf.

Freeman Dyson, the physicist who wanted to send explorers all over creation using atomic explosions, said he thought that Graham's idea and other notions of ballistic missile defense crystallized the space soldier's dream. "Enthusiasts for space weaponry," he wrote,

> have always dreamed that their space weapons could dominate ground weapons, that their space forces could impose peace on ground forces. The concrete expression of this dream is a space-based anti-missile system which could reliably destroy any ground-launched or sea-launched missiles in the vulnerable early stages of their flight. In the dream, the space force constantly patrols the planet, ready to kill in a second or two, with a lightning bolt of laser or particle beam energy, every missile which rises from the earth without due notification and authorization. . . . The celestial lightning bolts, without hurting a single hair of a human head, purges the earth of mass-destruction weapons by the controlled and localized force of a superior technology.

Dyson took the position that nothing of the kind could be accomplished in space without its first being accomplished on Earth; that such a system would make the world more dangerous unless it was put in place after serious disarmament.

But the high-frontier concept so appealed to Reagan, who was already embarked on a massive arms buildup to counter the threat from what he called the "Evil Empire," that on March 23, 1983, he announced plans to proceed with research on defense against ballistic missiles. Saying he recognized that defensive systems have limitations and "raise certain problems and ambiguities" including "obligations under the ABM treaty," he publicly called on the scientific community, "those who gave us nuclear weapons to turn their great talent now to the cause of mankind and world peace: to give us the means of rendering these nuclear weapons impotent and obsolete."

The men of the Pentagon, who remembered quite clearly that the old system had not worked and who had been excluded from the tight White House circle that came up with the new one, were taken by surprise and flabbergasted when they heard the speech. "The Pentagon's studies were consistently negative," Graham would later complain. That was about to change.

By October, two ad hoc committees, one headed by Caspar Weinberger, who was now secretary of defense, and the other by James Fletcher, who had left NASA in 1977, concluded that ballistic missile defense showed enough promise to warrant a large research effort. "Powerful new technologies are becoming available that justify a major technology development effort that provides future technical options to implement a defensive strategy," Fletcher assured the House Armed Services Committee's Subcommittee on Research and Development.

Engineers, draftsmen, and artists working for potential contractors and the media wasted no time conceptualizing potential anti-missile weapons. Within a few weeks of Fletcher's statement, *Aviation Week & Space Technology,* other trade journals, and the weekly newsmagazines began running illustrated features that showed a dazzling array of what were described as potential missile killers, including a "mid-infrared chemical laser spacecraft with 25-megawatt, 15-meter-diameter optics," a "neutral particle beam weapon," surveillance and battle management spacecraft, space-based "fighting mirrors," mission mirrors, midcourse advanced tracking and pointing spacecraft, and airborne optical systems.

The star of the group was unquestionably Livermore's own beloved, problem-plagued, nuclear-pumped X-ray laser. Named Excalibur, after King Arthur's renowned broadsword, the weapon was supposed to simultaneously obliterate many warheads with laser beams coming out of a nuclear explosion like multiple lightning bolts coming out of a thundercloud. The idea was to bring the enemy under constant attack from soon after the ICBMs left their silos, through the boost and midcourse phases, until the terminal phase, when missiles were to be shot at them as they came down on their targets. Someone got the idea of launching mirrors and lasers from submarines under the Arctic ice floe. There was even a planned "hypervelocity electromagnetic railgun" that was supposed to fire projectiles at Soviet "hunter-killer" satellites that threatened to attack the defensive spacecraft before they attacked the missiles and warheads that were attacking America. It didn't take the captains of the aerospace industry and grant-hungry chemists and physicists in the universities long to see that a whole new well of government money was starting to gush.

Understandably, skeptics—many of them alarmed—soon named the system "Star Wars" after the futuristic science fiction trilogy that had by then already become a cinematic classic that linked audiences everywhere with what they assumed was their race's future. (Physicist Freeman Dyson has suggested that people may love *Star Wars* because deep down they really love war and the film allows them to indulge their "secret love" with a clear conscience.) The "missile shield's" offended and somewhat defensive supporters responded by calling it the Strategic Defense Initiative, or simply SDI, which was both more pacific-sounding and more dignified.

Star Wars planners envisioned an extraordinarily complex battle management program that was totally dependent on high-speed computers, sensors, and killing machines to spot, track, and destroy a barrage of perhaps ten thousand warheads within half an hour, distinguishing between real ones and decoys, and eliminate nearly all of them in what was usually called a "layered" defense in from three to seven parts: a gauntlet of space-based lasers, railguns, particle accelerators, and other weapons that would hammer at the attacking force all the way to their targets. The idea, put forward almost from the beginning, was that 90 percent of the warheads would be stopped in each successive layer.

Nowhere was there evidence that serious consideration, indeed any consideration, had been given by the Star Warriors to what the Russians were making of SDI and, specifically, to what they could do to counter it. There were in fact more ways to foil the defense than there were defensive measures. The Russians could cold-launch, for example, by using compressed air to force their missiles out of the silos the way submarine-launched ballistic missiles are shot from under water. That would shorten the heat plumes the early warning satellites or their successors were supposed to spot before alerting all the killer things. And the missiles themselves could be spun, making it almost impossible for lasers to bore into them. Warheads could be made both maneuverable and stealthy, which would enormously compound the defenders' job. So would increasing the number of decoys and launching dummy boosters that would confuse the killer satellites that shot at the ICBMs coming out of their silos. The re-entry vehicles themselves could be sent on so-called depressed trajectories, which would make them harder to see and hit, and they could be fitted with heat shields and use chaff, aerosols, and other cheap but effective gimmicks to mask their flight path. And earth- and sea-hugging long-range cruise missiles would be next to impossible to hit from space.

Nor was it reasonable to believe that the Russians would be content to play the role traditionally assigned to Indians in Hollywood's Westerns: single-minded dullards who were somehow reconciled to getting picked off in large numbers while attacking the fort in full view of its defenders and shouting war whoops to help them zero in. To the contrary, the Russians would have been able to take active measures to interfere with or knock out the ABM weapons themselves. The more obvious techniques would have included sending space mines to follow every railgun, mirror, and other weapon and detonating them as a prelude to the attack. The SDI weapons also would have been vulnerable to attack by Soviet anti-satellite weapons, to jamming of their signals, and to colliding with plain junk that was spread in their orbital paths.

The nature of super high technology weapons made them especially vulnerable to low-technology countermeasures. "If you are driving your fancy automobile down the street," IBM's Richard L. Garwin said at a strategic defense conference in New York after Reagan made his famous speech, "you are even more desirous that people don't throw low-technology rocks at it than if you didn't have any such [high-]technology vehicle." Garwin was an outspoken opponent of SDI. Robert Jastrow, who still seemed to believe that the United States was in mortal danger from the Soviets, loved it. They publicly debated SDI on a number of occasions.

Within a week of Reagan's speech, Soviet Premier Yuri V. Andropov counterattacked, saying, "Should this conception be converted to reality, this would actually open the floodgates to a runaway race of all types of strategic arms, both offensive and defensive." In fact, Andropov added, "the strategic offensive forces of the United States will continue to be developed and upgraded at full tilt and along quite a definite line, namely that of acquiring a first nuclear-strike capability." Any thought of winning a nuclear war, the former KGB chief added, "is not just irresponsible, it is insane."

Whatever the validity of his prediction about opening a floodgate, Andropov was on the mark about first-strike capability. By 1985, not only was the United States pouring billions of dollars into finding ways to perfect its missile shield, but it was simultaneously trying to find ways of guaranteeing that its own nukes would make it to their targets no matter what the Russians did. The Advanced Strategic Missile Systems program, as it was called, was designed to develop and test advanced decoys and warheads that would use aerodynamic surfaces such as wings, flaps, and other "penetration aids" to maneuver around any defense the Soviet Union tried to mount. The warhead engineers, who worked on the flip side of SDI, were even designing decoys that reportedly would be able to read the signals coming from enemy radar or infrared sensors and instantly send back a countersignal to convince the defenders that the decoys were real warheads. The idea was to give the Soviet defense the electronic equivalent of a nervous breakdown. The real warheads would then strike with precision.

France was already honing its own offensive technology. Two and a half years after the Star Wars speech, Defense Minister Paul Quiles announced in Paris that his country would respond to space-based missile defense systems—both American and Soviet—not by putting up its own terrifically expensive defense but by improving its warheads to the point where they could penetrate any missile shield. He specifically mentioned the development of miniaturized re-entry vehicles that would be "virtually invisible" to enemy radar. The decision reflected the Quai d'Orsay's need to balance Gallic pride with a restricted budget. But it also reflected two larger, glaringly immutable, facts: nuclear offense was cheaper than defense, and it would always have the advantage.

In that regard, SDI would have made a devastating first strike weapon, a fact that went over neither Andropov's head nor Weinberger's. Not only would Star War's lasers make excellent satellite killers, but they also would adapt well to demolishing "soft" targets on Earth: airplanes, naval vessels and oil tankers, power plants and factories of all kinds, grainfields, and very likely even missile silos. One laser expert and Star Wars advocate bragged that the weapon could "take an industrialized country back to an 18th century level in 30 minutes." Perhaps. But it would have to be the size of the Empire State Building.

Yet that was not the picture that soon appeared on American television. Instead, a high frontier organization called "Peace Shield" that operated out of Washington ran a color cartoon commercial showing stick children and their pets, seemingly drawn by a child, who were shielded from little red bombs that kept hitting a protective shell and disintegrating harmlessly. "I asked my daddy what the Star Wars stuff is all about," an unseen little girl chirped. "He said that right now we can't protect ourselves against nuculer [*sic*] weapons, and that's why the president wants to build a peace shield. It would stop missiles in outer space so they couldn't hit our house. Then, nobody could win a war. And if nobody could win a war, there's no reason to start one. My daddy's smart."

Her daddy probably worked for a weapons lab or a defense contractor. By 1986, a saturnalia, an engineering orgy, was under way across the country as national laboratories such as Livermore, defense contractors, the armed services, the Department of Energy, and physics and chemistry departments in universities and technical institutes went after lucrative appropriations, contracts, or grants in what amounted to a feeding frenzy over an expected $400 billion over the next twenty years. The five-year budget alone was projected at between $25 billion and $30 billion. A study by the Federation of American Scientists showed that Livermore ranked fourth behind Lockheed, General Motors, and TRW in the size of SDI grants with a total of $552 million (Lockheed had a little more than $1 billion). By far the largest number of SDI contracts—more than 45 percent—would go to the president's own state. One Californian, a soft-spoken JPL physicist named Robert W. Carlson, had an infrared experiment on the problem-plagued Galileo mission to Jupiter. He undoubtedly spoke for many in his state when he admitted that he had taken on Star Wars research just so he could keep working. "We thought those [Air Force] guys were off the wall, unrealistic, and more or less crazy," he said. But it was money falling from Heaven. SDI was run by the Department of Defense, not the Air Force, though it was directed by an Air Force general and Air Force research was done.

Meanwhile, an often rancorous debate on SDI raged on. Zbigniew Brzezinski, Jimmy Carter's national security adviser, Max Kampelman, a Reagan

White House arms controller, and Jastrow wrote an article for *The New York Times Magazine* deploring the "theological dimension" of the Star Wars debate and calling for missile defense to be given a fair chance. Hans A. Bethe, a theoretical physicist and Nobel laureate, George Rathjens and Jack Ruina of MIT, Stanford's Sidney D. Drell, and others doggedly attacked the plan in *Scientific American, The New York Times,* public forums, and elsewhere. Garwin said that Reagan was "disconnected" where arms control was concerned. In common with a number of other periodicals, *Foreign Affairs* ran pro and con arguments in the same issue. *The New York Times* came out against SDI, as did the Union of Concerned Scientists, the Federation of American Scientists, and the American Physical Society. The APS issued a statement saying that if even a small percentage of warheads made it through the shield, they would cause unparalleled human suffering and death. The nation's largest organization of physicists added that it would be decades, if ever, before an effective ABM system could be put into place and urged that the "enormous waste of financial and human resources" be stopped. The statement neglected to mention that the system could never be adequately tested, that it would have to be updated and serviced forever at immense cost and, most perniciously, that humans would be left out of the decision-making loop because of the speed of the engagement. It was not clear whether Reagan understood that SDI would therefore subvert his successors' war-making power.

Lowell L. Wood, Teller's messianic understudy at Livermore, counterattacked by insisting that the APS report was "full of errors." The bright young physicist and his colleagues, all of whom were wedded to missile defense and struggling to save it from the withering attacks, would soon come up with a concept they called Brilliant Pebbles. That one would send a swarm of as many as a hundred thousand relatively cheap, three-foot-long, hundred-pound missiles into orbit that could spot ballistic missiles and smash into them on their own with no help from Earth or other spacecraft. Using tiny missile killers got nicely around the argument that big SDI weapons such as fighting mirrors and railguns were vulnerable to being picked off. Brilliant Pebbles would never fly. But the advanced miniaturization that went into the basic design would find a home in lunar and planetary exploration after the cold war.

Proponents of the missile shield claimed that the Russians had their own rocket-propelled ABM program, including nuclear-tipped missiles, and were secretly experimenting with lasers at Sary Shagan. That was true enough. But they also asserted, as did Teller, that the East was well ahead of the West, which was worn-out rubbish.

At about the same time, the autumn of 1986, Reagan reached into the bottom of the political barrel and pulled out his own ultimate defense against attack: jobs. "Just as America's space program created new jobs and industries," he told an audience in Colorado while campaigning for a local congressman, "SDI could open whole new fields of technology and industry, providing jobs for thousands right here in Colorado and improving the quality of life in America and around the world."

Not surprisingly, the nation's civilian space agency weighed in on the side of SDI by proposing to develop spacecraft that would release gas clouds to be used as practice targets for lasers to shoot at, plus other satellites to monitor the tests. It also suffered an ethical lapse when it volunteered to pass data from at least ten planned American and foreign science satellites to Star Wars planners. "NASA space mission data and technical support are becoming an increasingly important part of the SDI space-based missile defense program," one industry journal correctly observed. A year after Reagan's speech, in fact, Air Force Lieutenant General James A. Abrahamson was appointed to manage SDI. His previous assignment had been directing the shuttle program. Two years later, in 1986, Fletcher, who chaired one of the two committees that originally endorsed Star Wars research, became the only man to serve a second term as NASA administrator when Reagan appointed him to that post. As was the case with the shuttle (which by then had a number of SDI tests on its manifest), the civilian and military programs were neatly merging in the always just cause of national security.

As often happens during history's weightier moments, there was no lack of buffoonery and mendacity. Responding to Soviet Premier Mikhail S. Gorbachev's concern about SDI at a recently concluded summit in Iceland, for example, Reagan blithely replied that once the top-secret ABM system had been perfected it would be shared with the Soviet Union. Providing the Kremlin with the defensive shield's most intricate workings would have been like giving a home's alarm code to known burglars. There is no record of either Gorbachev's or Abrahamson's thoughts about the idea, but it very likely caused eyeballs to roll toward the ceiling in both the Kremlin and the Pentagon. Gorbachev, who led a nation of chess players, was left to ponder the remote possibility that the offer represented some new, sublimely intricate stratagem. Or, more likely, that it was simply as stupid and disingenuous as it sounded.

Then there was Abraham Sofaer, the Department of State's legal counsel, who was hired by the Pentagon to show that the apparently ironclad ABM treaty actually allowed the development and testing of anti-missile systems. The main justification for working on the anti-missile system, at least at the outset, was that SDI was merely a research program and that research was not prohibited. Yet the charade was so transparent it was ludicrous. What multi-billion-dollar, decades-long research effort was planned to stop short of development and execution even if it was successful? Sofaer dutifully concluded after going over the classified negotiating record that, in effect, Article 5 covered only systems and components that were current at the time of signing. But the key part of the document, he declared, was Agreed Statement D, which banned deployment of ABM systems but not their development and testing. Sofaer's interpretation of the treaty, then, held that all of the orbital artillery could be developed and then "tested" in orbit for a hundred years.

Sofaer's blessing of SDI deeply angered a number of his colleagues and others who had been closely involved in the SALT 1 and other arms control negotiations. Albert Carnesale, a professor of government at Harvard, and Gerard Smith and John Rhinelander, both lawyers, lost no time in responding. All three had been involved in the treaty work and were sharply critical of Sofaer's position. "Having been through the negotiations myself . . . my understanding of the treaty has always been invariant: Article 5 means what it says, and prohibits development and testing regardless of the nature of the technology," said Carnesale. Smith not only became upset at what he saw as a blatant attempt to transgress the treaty, but he was also apparently somewhat embarrassed for his profession, as well. "When I read the Administration's report, I felt I was reading the work of expert tax lawyers, of people trying to evade the law," he said. "It seems to me [that] we are trying to prepare the ground for a [treaty] breakout and as a lawyer, I would say that constitutes anticipatory breach of contract."

What Smith was saying, almost in so many words, was that one of SDI's intended by-products was the wrecking of arms control, which those on Reagan's far right despised. Nor was Smith the only one who harbored that suspicion. Writing in *Daedalus,* the prestigious journal of the American Academy of Arts and Sciences, three other noted lawyers also vigorously challenged Sofaer. "It is apparent that the SDI enterprise as a whole, its objectives and philosophy, are simply at odds with the purposes and objectives of this [ABM] treaty," they concluded. "Moreover . . . it is inevitable that under the current presidential mandate and Defense Department response, these limits will be breached and the treaty violated outright within a period of time that is relatively short."

There was also the knotty question of what the tests, themselves, actually accomplished. In 1984, as the firestorm over Star Wars was building, and after three unsuccessful attempts to hit a warhead over the Pacific with interceptor missiles, the Army Ballistic Missile Defense System grew desperate enough to fake an intercept. According to informed insiders, Weinberger himself approved the deception, though he denied it. The immediate intention was to deceive the Kremlin into believing that SDI was becoming viable in the hope that it would force the Soviet Union to spend billions, further bleeding it. But the disinformation had a more immediately beneficial effect. It was used to justify continued congressional support for Earth-based defensive systems. In the test, called a "homing overlay" experiment, an interceptor missile was claimed by the Army to have hit a Minuteman re-entry vehicle. In fact, according to a scientist involved in the deception, "We rigged the test. We put a beacon with a certain frequency on the target vehicle. On the interceptor we had a receiver." In effect, the scientist said, the target said to the interceptor, "Here I am. Come get me."

The Strategic Defense Initiative Organization, which ran SDI, also bragged about its embryonic laser weapons program. In early September 1985, General Abrahamson proudly announced the successful completion of the first "laser lethality test" at White Sands. A Navy laser weapon with one of those delightful acronyms—MIRACL, for "mid-infrared advanced chemical laser"—demolished a Titan second stage. The weapon "blasted the thing absolutely apart," the general told American and Canadian reporters, calling it "a very dramatic event." It was true, too, and it made equally dramatic headlines: "LASER DESTROYS MISSILE TARGET IN TEST BY U.S.," *The New York Times* reported, while *Aviation Week & Space Technology* proclaimed, "Missile Destroyed in First SDI Test at High-Energy Laser Facility." What the newspaper and magazine explained under the headlines, however, was that both the killer and its victim were bolted securely to the ground and that it required several seconds for the laser to do its job even then. Relative to conditions on war day, it was like hitting a tin can with a hammer.

A General Accounting Office audit in 1992 would reveal inaccurate claims for the success rate of experiments, the progress of programs, interceptors' ability to distinguish between targets and decoys, the achievement of accuracy and altitude goals, and more. "They have lied about certain functions that their missiles are supposed to perform," said one federal investigator. "They've used things to enhance the target. The fact is that you've got something up there and solving your guidance problem. And you've got an incentive to deceive. That's how you keep your program going."

But Caspar Weinberger took such ruses as a political fact of life. "You always work on deception," he would say almost a decade after the homing overlay hoax. "You're always trying to practice deception. You are obviously trying to mislead your opponents and to make sure they don't know the actual facts." That certainly applied to Congress as well as to the Communists. The words could have come from any of the chief designers or from Dmitri Ustinov himself.

The Military "Arena" Widens

Nor were the SDI people the only ones being deceptive. Since the targets U.S. reconnaissance satellites were supposed to photograph were usually at least a quarter of the way around the world, and because they did not want to store the imagery on board, the National Reconnaissance Office had its KH-11s and their successors beam their digital data to other satellites, which relayed the intelligence home. Initially, this was handled by ambiguously named Satellite Data System spacecraft, the first of which went up in March 1971 to relay other kinds of communication. SDS was replaced in April 1983 by the spaceborne spy network's equivalent of George Smiley, the understated British agent created by novelist John le Carré: a space machine that was as seemingly innocuous as it was audacious. It and its successors were operated in geosynchronous orbit by NASA's Goddard Space Flight Center and one of them hangs in full view of millions of visitors at the National Air and Space Museum in Washington. Officially, their job was to relay communications from NASA's far-flung shuttles and "other orbiting spacecraft" to White Sands for rerouting. It made sense to use a few satellites ringing Earth to keep track of virtually all of NASA's fleet, including the shuttles, and communicate with them directly instead of having to use ground stations around the world that passed on the information individually. And the new satellites had unprecedented capability. Each of them could use its thirty antennas to relay the equivalent of a twenty-volume encyclopedia in a one-second burst.

The spacecraft were called the Tracking and Data Relay Satellites, or TDRS (pronounced TEE-dris by the fraternity brothers); two of the "other" spacecraft they worked for included the KH-11, and a secret radar reconnaissance satellite, named first Indigo and then Lacrosse, which was first launched on the shuttle *Atlantis* on December 2, 1988. Like the KH-11, Lacrosse took well over a billion dollars to develop and survived a long and acrimonious battle between its chief patron, CIA chief William J. Casey, and congressional opponents and the Pentagon. Lacrosse's synthetic aperture radar penetrated clouds and the dead of night to return imagery with resolution of a yard or better. TDRS also relayed telemetry and other intelligence collected during Soviet missile and space launches right to New Mexico in real time. The evident theory behind providing TDRS with a public persona and even hanging one near the entrance to the most popular museum in the world went back to the dime novel mysteries in which the murder weapon was always in plain sight. TDRS, which was also made by TRW, was another classic example of how the civilian and military programs could coalesce into a single, indivisible entity. Goddard, a civilian center, even issued its own Security Classification Guide for the program.

By the spring of 1983, both sides had constellations of reconnaissance satellites that were collecting intelligence, monitoring the arms-control agreements, and finding targets for the ballistic missiles that waited in reinforced concrete silos in the adversaries' heartlands and on nuclear submarines that prowled silently under the high seas. Both had ballistic missile early-warning satellites, together with spacecraft that intercepted communication and telemetry signals and the long-distance relay types. Earth's inhabitants had made it a ringed planet, too.

The Air Force and Navy communication satellite programs—Afsatcom and Fltsatcom—used dedicated Navy satellites or hitchhiked on those in other programs to keep in round-the-clock contact with their far-flung nuclear strike forces. The Air Force system, which started in 1979, rode only on "host" spacecraft, including the Navy's, the Satellite Data System relayers, Defense Support Communication Satellites, and the innovative Global Positioning System spacecraft that would lead space-based navigation into the twenty-first century.

By 1983, both sides also had clusters of ocean surveillance satellites in relatively high orbits, giving them expansive views of the high seas. The U.S. Naval Ocean Surveillance Satellite system sent four White Cloud satellites at a time into orbit so they could accurately fix the location of Soviet naval vessels by collectively picking up communication and other signals from the ships.

The Soviet Union had a more extensive ocean surveillance program because the United States had a more extensive navy. That Navy's heart consisted of the third part of the so-called nuclear triad—submarines loaded with nuclear missiles—and carrier battle groups whose nuclear bomb–laden planes were always poised to strike Warsaw Pact targets from almost all directions. The carriers were therefore tracked wherever they went, at least in theory, so they could be attacked if war broke out. Two kinds of Soviet spacecraft, working together, did the tracking: electronic ocean reconnaissance satellites and radar ocean reconnaissance satellites. The electronic types worked in pairs and just listened to their quarries' electronic and communication signals. The radar satellites bounced their own signals off the ships they tracked, and what came back gave those vessels' positions. The radar reconnaissance satellites, which needed a great deal of power, were also dangerous. They ran on reactors that burned uranium 238 enriched with uranium 235. The combination was potentially lethal if one of the cores came crashing to Earth. And not one but two of them actually did fall back down. The first was Cosmos 954, which skidded across the Canadian tundra in January 1978, leaving a long trail of radioactive debris behind it. Then, almost five years later, another hot satellite called Cosmos 1402 disappeared in the Indian Ocean.

Bombs in Orbit

More dangerous still, at least politically, were nuclear bombs that were actually designed for use in space. Such a system was tested more than a dozen times by the Soviet Union between September 1966 and August 1971, usually with SS-9s that could carry twenty-five-megaton warheads. The Department of Defense named it the Fractional Orbital Bombardment System, or FOBS, because the warheads stayed up for only about 85 percent—a fraction—of one full orbit, which complied with the letter, if not the spirit, of the treaty on peaceful uses of outer space. FOBS was first tested at Kapustin Yar, which allowed the warheads to come down on Soviet territory, not off the coast of the Kamchatka Peninsula, where radars in Alaska could monitor them. The secrecy so alarmed the United States, though, that the Kremlin eventually tested FOBS along the traditional missile route: west to east.

The idea of keeping hydrogen bombs in constant orbit for either very short periods or almost indefinitely, always ready to drop out of the sky and annihilate a city, had captivated science fiction writers and political zealots since the start of the atomic age. But most strategists understood that it would make no sense militarily. Not only would an orbiting bomb take longer to reach its target during an hour-and-a-half orbit around Earth than would a silo-launched warhead, but it would always be vulnerable to attack by satellite killers and would be less accurate. Still, NATO planners feared that with ballistic missile early-warning radars facing north and east—the shortest ICBM routes—the Soviets could use FOBS to sneak in through the back door: from the south. But U.S. ballistic missile early-warning satellites—the Defense Support Program spacecraft parked out at geosynchronous with their powerful telescopes—would still see them coming with eyes that never blinked.

FOBS was first tested in September and November 1966. Both exploded in orbit, sending hundreds of pieces of debris flying in all directions like shrapnel. The first successful test came on January 25, 1967, with Cosmos 139, which was launched at Tyuratam and sent down over the Kapustin Yar test range. The last, Cosmos 433, went up on August 8, 1971. Whatever secrecy surrounded the FOBS program, the purpose of the tests was obvious to the Western intelligence analysts who followed them and who regularly leaked information about them to the civilian space trade, which published it.* That irritating exposure, plus the expense of testing a system that would be illegal if actually orbited, finally convinced the Kremlin to abandon FOBS.

Satellite Killers

ASATs, or anti-satellite weapons, were the other deadly space systems that were tested beginning in the 1960s. If "absolute superiority" in space was essential to the security of the West, as the Air Force's Strangelovian SAC commander, General Thomas S. Power, and others maintained, then the United

* See, for example, *TRW Space Log,* Vol. 19, 1983, p. 49 (Cosmos 139); *RAE Table of Earth Satellites, 1957–80,* p. 268 (Cosmos 433).

States had to be able to "negate" the opposition. That meant it had to be able to clear the sky of every enemy spacecraft at the very beginning of an all-out war, swiftly and relentlessly putting out the opposition's eyes, pulling out its hearing aids, and destroying its ability to command, control, and communicate with its own soldiers and civilians. The plan, essentially, was to attack and destroy its sensory and central nervous system. To do that, the U.S. Space Command kept an enormous computer file, constantly updated, on where everyone's satellites were. It included not only dedicated military spacecraft but civilian ones as well, since the programs overlapped by a wide margin. The Gorizont, Raduga, and Ekrans civilian communication satellites parked out at geosynchronous, as well as the dozen Molniyas that were on highly elliptical orbits, were tracked as methodically as their military counterparts.

Not surprisingly, the owner of those spacecraft—the Soviet Union—saw it the same way. It therefore developed, tested, or tried to develop four kinds of ASATs. The first and crudest were one hundred anti–ballistic missiles that protected Moscow and that could have been used in direct ascent trajectories to explode near targets overhead.

The second was a class of so-called co-orbital satellites that could approach the target and, when in range, either destroy it by firing pellets the way a shotgun would, or simply get close enough to explode, destroying itself and its target in the same instant. The 1983 edition of *Soviet Military Power* carried a drawing purporting to show the shotgun technique being tested. The technique was tested twenty times between October 1968 and June 1982, starting with Cosmos 248 and 249, much to the consternation of U.S. military and intelligence officials who could plainly see Cosmos 249 explode as it passed close to its predecessor as they watched the attack on Cheyenne Mountain's radar. But the co-orbital killers had problems, too. They could be used only against targets that orbited below 600 miles and were ineffective for rapid response. Instead, they had to wait until the enemy flew into view and then ambush it.

In the third type, which the Russians named Radio Electronic Combat, electronic signals would have been used to jam the target spacecraft's uplinks or downlinks or even take control of it. No dedicated installation to direct such an attack was ever found, but the Soviets themselves said that they considered electronic warfare to be the most promising space "defense." Finally, at least in the Pentagon's view, there were the ubiquitous directed energy weapons: lasers. The dreaded death rays haunted the dreams of American generals and admirals, who could easily imagine their precious reconnaissance and command-and-control "assets" being turned into useless, burned-out derelicts in a flash. The Soviets acknowledged in 1988 that they had lasers at the test facility at Sary Shagan, but they told an American delegation touring the place the following year that they were designed to track, not attack, satellites. It was probably true. Since the United States never had any good intelligence on Soviet laser ASATs, it fell into the trap of basing its estimates on its own growing laser weapon capability, which was better than its enemy's.*

The United States had edged into ASATs in the early 1960s, when Soviet leaders implied that orbiting nuclear weapons could be militarily effective. The fact that Khrushchev was brash enough to try to get intermediate-range ballistic missiles into Cuba in 1962 underscored the fact that intimidation of the West was definitely on the Kremlin's agenda. Secretary of Defense McNamara therefore approved testing the Army's Nike-Zeus anti–ballistic missiles as ASAT interceptors in May 1962. Some were deployed in 1964.

Meanwhile the Air Force developed Program 437, which involved putting simulated nuclear warheads on Thors and sending them after dead satellites or debris. Any warhead that got within five miles of the target—and some got a lot closer than that—was ruled a kill. Program 437 was considered operational in 1964, but the passing of the FOBS threat and the knowledge that U.S. satellites would get caught in a deadly radiation flux in any attempt to nuke Soviet spacecraft, ended the program in 1975. But it did not end the work on ASATs. The ability to attack enemy spacecraft remained as central a tenet for the Air Force as the ability to attack enemy aircraft.

On September 1, 1982, the Air Force started its own Space Command, a fully dedicated combat-oriented Air Force organization, along with the Strategic Air Command and the Tactical Air Command, and aggressively pushed the notion that it should be the umbrella command for all U.S. military space operations.† Looking into the future, the Air Force drew its own parallel to the high seas, seeing not only a potential Soviet threat, but threats from other emerging space nations as well (some of them, like China, in the Third World). The service's mission in space, according to a handout released two months before the birth of the new command, was to "keep other nations from exploiting space for their military reasons" and to "ensure free passage" in the new environment. That being the case, the airmen were convinced that they needed to be able to protect

* "We have limited information on the detailed characteristics of Soviet laser systems," one top-secret Defense Intelligence Agency report noted in 1977. "Fluence levels and intensities of Soviet laser threats described in this section are based on U.S. laser technology." ("Space Transportation System [STS] Threat Assessment Report," p. 85.)

† Sticking to the long-standing tradition of resisting the flanking movement, the Navy started the Naval Space Command within ten months and made Captain Richard H. Truly, an astronaut, its first commander. Officially, the Navy's operation was supposed to consolidate the service's various space activities, "exploit opportunities in space, and enhance the Navy's ability to execute worldwide maritime missions." Left out of the rationale was a determination not to allow the Air Force exclusive control of space. Responding directly to the Air Force plan, Vice Admiral Gordon R. Nagler said, "At the present time, I don't see any need for it," and added, lest he be misunderstood, that the Navy depended more on satellite communication than either the Army or the Air Force because its widely scattered carrier battle groups and submarines sent or received forty thousand messages a month. ("Military Divided over Space Policy," *The New York Times*, July 5, 1983.)

their nation's spacecraft with ASATs as surely as the Royal Navy had needed to protect British merchantmen with twenty-four-pounders on frigates.

The Air Force continued to believe in the absolute necessity of ASATs throughout the 1980s and it still does. On September 13, 1985, in a demonstration of technological virtuosity that was unequaled at the time, an F-15 climbing toward space fired a heat-seeking, torpedo-shaped ASAT called a Miniature Homing Vehicle at an Air Force satellite that was passing more than three hundred miles overhead. The missile blew the satellite to smithereens by sheer impact and without exploding. Secretary of Defense Weinberger announced that the experiment had left him "absolutely delighted." Less than delighted were the people whose experiments had been on the scientific satellite, which was called P78-1, because it had still been returning solar and other data when it was attacked. "As a patriotic American, I was glad to see the test succeed," said one dejected scientist. But, he added, "It's really kind of sad. I worked in that research group. It's a definite loss, no doubt about it."

Meanwhile, a Soviet proposal for a moratorium on ASAT testing was scoffed at by supporters of the weapon, including President Reagan, who argued that the Russians wanted to ban testing only because they were ahead. But Congress was not persuaded and voted to stop testing beginning in 1986.

The Bard of Bangor's Ode to Weapons in Space

William S. Cohen, a senator from Maine with long experience in military and intelligence affairs, was highly knowledgeable about the militarization of space. This is what he saw when he looked at the sky:

> Satellites fly,
> Frankenstein birds
> with burning eyes
> that shuttle click
> into the vast night.
>
> Out of the trackless void
> they scan the earth
> searching for prey
> to catch on film then let drop
> down through the long
> darkness, stumbling
> toward the sea.
>
> Now, nightbirds,
> (Gods that we are)
> it is time
> you were given talons

to hold these
mother-of-pearl pistols.

So when the earth
goes red with a
thousand suns, you can
fire your light
into the breast of sky
a thousand times, star-drilled
into all the hydrogen-headed
monsters that rise up
from earth and sea
contemplating great
catastrophe.

Before they unleash
hurricane winds,
Before they breathe
through nostrils red
beyond all Fahrenheit,
Turn them to endless
ash, yes, save us from
their savagery.

Oh, watchful
birds of peace
happy in your
lonely epicycles.

Did you always think
you would be free
from violence, that
the night would be
only for you,
a wildlife refuge,
a cold dark silence?

Cohen was appointed secretary of defense more than a decade after those words were penned, thereby becoming the Pentagon's unofficial poet laureate. By then the old "Evil Empire" would be in shambles and Star Wars would be back in the movies, where it had started in the first place. Or at least it would be back there for a while before reappearing in another form. The truest of all possible code names for ballistic missile defense would have been "Hydra," for the serpent in the Greek myths that kept growing new heads as old ones were chopped off.

15

Downsizing Infinity

The end of the beginning started in 1986. It was marked by two ruptures, widely thought to be technological, but whose causes ran far deeper. One happened in an American spaceplane's solid-rocket booster; the other in the core of a Soviet reactor. Both made the planet wince. Had he been around at the time, John Cabel, the apostle of technology in H. G. Wells's *The Shape of Things to Come,* would have been reviled not just for worshiping a false god but, worse, for worshiping one who betrayed his believers. Yet it was not *Challenger* and the RBMK at Chernobyl-4 that were the culprits, nor even their designers, but the political systems that had produced them. The spacecraft and the reactor were symbols of larger problems that led inexorably to a realignment of the world order, the end of the first space age, and the beginning of the second.

Selling Tickets to Space

There had been twenty-four shuttle flights between *Columbia*'s inaugural mission on April 12, 1981, and its flight on January 12, 1986. During those four years and nine months, *Columbia* and her sisters—*Atlantis, Challenger,* and *Discovery*—had flown roughly 64 million statute miles, orbiting Earth 2,355 times in 152 days. That would have taken astronauts to the Moon and back one hundred and forty times.

Challenger alone flew nine of the missions, taking it around the world 987 times, with an impressive variety of payloads: a Tracking and Data Relay Satellite, communication satellites for Western Union and Indonesia, a Long-Duration Exposure Facility science satellite, a radiation monitoring satellite, three Spacelab packages, assorted Getaway Specials, and even a quarter-million express mail postal envelopes that were carried for a first-day cover. On *Challenger*'s sixth mission, in October 1984, astronauts Kathryn D. Sullivan and Lieutenant Commander David C. Leetsma, decked out in cumbersome Manned Maneuvering Units, spent three hours practicing refueling satellites at the end of the orbiter's cargo bay. Sullivan became the first American woman to leave a spacecraft in orbit, but that was not the purpose of the exercise. One of the shuttle's secret Department of Defense missions was to be an orbital tanker for the maneuverable KH-11 reconnaissance satellites and their successors. Extending maneuverability would add months or years to the phenomenally expensive satellites' lives. While the cost of a shuttle flight—roughly half a billion dollars—approached what it would cost to build and launch replacement KH-11s rather than refuel existing ones, combining a refueling operation with other work could approach cost-effectiveness. That was good.

But it was not good enough. It did not take an Edward Teller to calculate that the Space Transportation System was averaging not fifty-five missions a year, nor even a dozen, but only five. This was because the dream of efficiently using the same vehicles to go up and come down in almost continuous round-trips, like airliners, was running into a hard reality: there were only four reusable man-carrying spacecraft, and they required a great deal of turnaround time.

For all their brutish appearance, the four shuttles were dangerously fragile and temperamental. The airliner analogy might be appropriate at some point in the twenty-first century, with Dr. Heywood Floyd snoozing on a routine flight to the station, in *2001,* but it was inappropriate in the 1980s because there were far fewer shuttles than airplanes and the four that existed were only then evolving from test vehicles to operational ones. More to the point, they were immensely complicated, which made flying them anything but routine. There were hundreds of possible malfunctions, called "criticality 1" failures, which could occur in an orbiter's labyrinthine bowels, in its notoriously finicky engines, in its bomblike external tank, or on its control surfaces and tile skin which would result in disaster. One of the four valves that controlled the flow of hydrogen and oxygen between the external tank and the orbiter's main engines sticking, for example would be a "criticality 1" failure. The term was more engineering jargon. It meant a failure of that particular part would cause the loss of the orbiter and undoubtedly its crew.

By January 1986, in fact, there were what Chief Astronaut John Young would soon call an "awesome" number of potentially catastrophic hazards that were dealt with on a piecemeal basis or temporarily ignored in order to keep even the semblance of a schedule. Logistical problems only made matters worse. Nineteen of the first twenty-four flights landed at Edwards Air

Force Base, which had better weather and more and longer runways than the Kennedy Space Center, where only four landings were made (*Columbia* landed at White Sands' dry lake bed on the third test mission in March 1982). Orbiters that came down at Edwards had to be loaded onto a specially modified 747 and carried piggyback to Cape Canaveral, where they were hoisted off, returned to the Vehicle Assembly Building, once again raised by cables like giant birds with rigor mortis, and carefully refitted and reattached to the external tank and solid-fuel rocket boosters.

Technological and logistical problems were camouflaged by extravagant public relations, while all references to "routine" flights to space quietly evaporated. As the Space Transportation System had originally been given the widest possible industrial and political base, so too were its crews now steadily diversified. Having made certain that women (Sullivan, Sally K. Ride, Anna L. Fisher, Judith A. Resnik, Bonnie J. Dunbar, and Shannon W. Lucid) took the shuttle to space, as well as members of minorities (Ellison S. Onizuka, a Japanese American, and Guion S. Bluford and Ronald E. McNair, African Americans), the pool was widened to include key members of Congress (Jake Garn and then Bill Nelson, whose district contained the space center) and foreigners. Dr. Ulf Merbold, representing the European Space Agency, became the first European to fly on a shuttle when he went up in *Columbia* in November 1983. He was followed not quite a year later by Marc Garneau, a Canadian whose country made the shuttles' Payload Deployment and Retrieval System, the fifty-foot-long mechanical arm that sent some objects on their way from the payload bay and captured others. Four other Europeans would follow Merbold, all of them representing ESA. Sending Garneau and the Europeans to space was by no means only political, since both Canada and ESA would be involved in the international space station, then named *Freedom,* which was starting its own tortured history.

But Prince Sultan Salman Abdel Aziz al-Saud, a rakish scion of Saudi Arabia's royal family, was another matter. The superbly connected son of the desert became the first Arab and the first nobleman to reach space when he got there on *Discovery* in mid-June 1985. Oil, after all, was even thicker than blood. He and others who did not have jobs that were integral to shuttle operations were given the ambiguous title of "payload specialist," which meant attending to whatever cargo was being carried on a given mission or just being there for another reason. Sally Ride, who had a doctorate in physics from Stanford and was a professional NASA "mission specialist" with heavy responsibilities on two shuttle flights, gently but firmly objected to payload specialists being called astronauts.

On August 27, 1984, Reagan announced that the shuttle's first "citizen passenger" would be a schoolteacher. Christa McAuliffe, who taught social studies at Concord High School in New Hampshire and was the mother of two, was selected for the honor from 114 eager applicants. (NASA's publicists, who were as statistics-driven as baseball fans, could point out that Anna Fisher not only became the first mother in space when she flew on *Discovery* in November 1984 but was the first female M.D.)

On October 24, 1985, the space agency added that the second citizen passenger was to be a journalist as part of what was by then being called the Space Flight Participation Program's Journalist-in-Space Project. The Association of Schools of Journalism and Mass Communication, one of journalism education's trade associations, chose a hundred candidates, including Walter Cronkite, Geraldo Rivera, Roger Rosenblatt, Lynn Sherr, Boyce C. Rensberger, Anne "Kathy" Sawyer of *The Washington Post,* and John Noble Wilford of *The New York Times* from more than 1,700 applications. The winner was scheduled to fly sometime in the autumn of 1986. But no shuttle would be flying that autumn, nor in the one that followed.

The most obvious reason for picking teachers and journalists to ride to space as passenger-communicators was for public relations, since they were supposed to be people who could best describe the experience for the average American. On a more subtle level, sending McAuliffes and Cronkites to space signaled a move to break the fighter jocks' hold on the manned space program and open it to Drs. Floyd and Spock and other ordinary folk.

Meanwhile, the Soviet Union was beating its old rival in the capitalistic game of selling tickets to space for both people and products. Having started with honorary Warsaw Pact and Third World cosmonauts, in 1978 the Russians began charging French, German, Chilean, South Korean, and other eager spacefarers hard currency for jaunts to *Salyut* and then *Mir.* Soon, in fact, the Russians would literally take tourism to new heights.

The Death of *Challenger*

As was often the case with launches, the mission that was supposed to carry McAuliffe to orbit in *Challenger* on January 22, 1986, was postponed three times and scrubbed once, partly because *Columbia,* which had made the preceding flight, had launched late and partly because of unseasonably bitter cold weather at the Cape. The overnight temperature on Tuesday, the twenty-eighth, was expected to drop to the low twenties. Throughout the night, as ice developed on the Fixed Service Structure that held *Challenger* in its steel grip on Pad 39B, NASA and contractor officials, including "tech-reps" from Morton Thiokol, which made the two solid-fuel rocket boosters, conferred about whether there would be a "go" or yet another stand-down. The rocket builders, fretting about engineering studies that recommended against using the SRBs if the temperature dropped below fifty-three degrees, first went into their own huddle and came out with a decision not to launch. That was what they told their space agency counterparts, using typed and handwritten charts. But their warning drew rebuttals from the NASA managers who, as one Thiokol engineer put it, wanted him and his colleagues to "prove that we should not launch" rather than accept their informed recommendation. At one point the solid-rocket booster project manager from NASA's Marshall Space Flight Center unwittingly showed that his agency was under tremendous pressure to get orbiters to space. If NASA followed Thiokol's advice, Lawrence B. Mulloy said, chiding the company representatives, "we won't be able to launch until April." That said worlds. Thiokol readily accepted the fact that there would be a launch.

With the Sun rising slowly over the Atlantic, a space agency ice team led by Charles G. Stevenson found icicles hanging from platforms and handrails and sheets of ice covering the gantry, including the route the astronauts would take if they had to scramble away from an emergency. Rockwell representatives who saw the scene on closed-circuit television described it as "something out of *Dr. Zhivago,*" an apparent reference to the scene in the film in which Omar Sharif and Julie Christie were living in exile in an ice-encrusted house on a desolate Russian steppe. The Rockwell tech reps advised against launch. So did a few stubborn representatives of Thiokol. So did Charlie Stevenson, who at one point answered a launch–no-launch question from the firing room by saying, "Well, I'd say the only choice you've got today is not to go. We're just taking a chance of hitting the vehicle," he added, referring to the possibility that heat and vibration from the launch would knock off chunks of ice that could damage the spacecraft.

NASA officials decided to launch *Challenger* anyway. At 11:38 A.M., with a nearby thermometer reading thirty-six degrees—fifteen degrees colder than at any of the previous launches—and with its main engines and solid-fuel rocket boosters blasting, *Challenger* left its pad and headed for space. Seventy-three seconds later, at an altitude of 48,000 feet, the right solid-fuel rocket booster, which was leaking flame from one of its joints, broke loose and slammed into the external tank. The explosion turned the spaceship, its SRBs, and the external tank into a fiery flower that blossomed against the dark blue sky. It was clearly visible to the festive, momentarily disbelieving, crowd in the VIP section.

McAuliffe's parents' and her sister's joy turned to sheer horror as they clutched each other and watched the grim scene. The image of the anguished couple and their daughter clutching each other as they saw Christa's death unfold would flash around the world and then become one of the enduring icons of the space age.

Challenger's crew compartment stayed intact, trailing a mass of umbilical lines like streamers as it continued to climb to 65,000 feet. Ten seconds after the explosion, the crew compartment went into free-fall. It slammed into the Atlantic two minutes and forty-five seconds after the blast at 207 miles an hour.

A report written six months later by Joseph Kerwin, director of life sciences at the Johnson Space Center, said that all of the crew members had remained harnessed to their seats and had been killed on impact. Kerwin noted that the forces acting on the stricken orbiter had been "probably too low to cause death or serious injury to the crew," that the estimated force of the explosion would not have broken *Challenger*'s windows, and that three of four emergency air-supply canisters that were recovered underwater had been activated. While the findings were judged to be "inclusive" and the cause of death of the crew could not be positively determined, the report said, the gravitational forces endured by the seven were "probably not sufficient to cause death or serious injury." And, the carefully worded document concluded, "the crew possibly, but not certainly, lost consciousness in the seconds" after the explosion. In other words, McAuliffe, Commander Francis Scobee, Judy Resnik, Ron McNair, Michael Smith, Ellison Onizuka, and Gregory Jarvis may very well have been conscious all the way down. Robert Crippen, who flew three shuttle missions, including the first on *Columbia,* said that in his opinion, the fact that the three emergency air canisters had been activated indicated that their users had made a "desperate" effort to survive their long plunge to the sea.

The *Challenger* Seven were martyred trying to carry a Tracking and Data Relay Satellite to orbit and do a Halley's Comet experiment. TDRS-B, as the relay satellite was called, was the communication relay satellite that handled civilian and top-secret intelligence traffic. It would have been tasteless to mention it at the moment, but the fact of the matter was that seven people had died trying to get a satellite into orbit that could have been sent there on an expendable.

The explosion traumatized the nation. Like Kennedy's assassination, most people would remember where they were when they first heard about it. They did not see it on commercial television because most of the news media, always hungry for drama and entertainment, had decided that the launches were too routine to cover. But replays of the scene, another wrenching blow from the sky, were aired almost immediately over all of the major networks and they kept airing all day. As the print and television reporters had carefully chronicled the advent of Sputnik and the ultimate triumph on the Moon, they now brought the exquisitely grotesque gnarled flower, its two smoking SRBs turning in on each other and then parting like missiles going berserk, into every home in the country and around the world that took a newspaper or caught the news.

Besides McAuliffe's anguished family watching as their heroine turned into a martyr, there was Jesse W. Moore, the gray-faced and shaken associate administrator for space flight, telling reporters that "all of the people involved in the program, to my knowledge, felt *Challenger* was quite ready to go and I made the decision, along with the recommendation of the team supporting me, that we launched." Moore, who had ordered the launch, had either been appallingly ignorant of all the warnings, which is next to impossible to believe, or he was reflexively spreading the blame for a decision that had monumentally grotesque consequences.

Only two years earlier, with the bungled handling of Grissom's, Chaffee's, and White's deaths still in mind, NASA had asked some reporters on the space beat for their suggestions on how to handle similar disasters. Out of that came a document, issued in March 1984, that emphasized the importance of "a full flow of accurate, timely and factual information to the news media." Yet NASA's response to the *Challenger* disaster was a virtual repetition of its handling of the Apollo fire. More than seven hours passed before the space agency confirmed that everyone on board was dead. Worse, all flight data having to do with the mission were impounded, and even days after the accident, status reports were almost always restricted to a few sentences and contained little new information. Public affairs people would not even tell reporters what the temperature had been before liftoff, forcing them to ask the National Weather Service. And it was five days before the space agency finally released videotapes showing the "abnormal" plume of fire and smoke pouring out of the right SRB. Unconfirmed news reports had long since focused on the strong possibility that the explosion had begun there. One former NASA employee said that there had been an instinctive reaction to "draw the wagons in a circle" and fight off attacks on the shuttle program and the agency in general.

The seemingly endless replays of the explosion, with anchormen like CBS's Dan Rather interviewing experts and trying to pinpoint the first trace of fire on the SRB with stills of the launch and the seventy-three-second ride to oblivion, put more cracks in the altar of technology and laid bare the fragility of trust. A Hollywood version of the disaster would have lasted at least half an hour, allowing plenty of time for desperate heroics, painful good-byes, at least one sacrifice, and soliloquies. But in real life the external tank, still loaded with more than 300,000 gallons of liquid hydrogen and 100,000 gallons of oxygen, turned into a bomb that exploded in a flash. The shuttle had defied gravity, defied the atmosphere, and even defied the motion of Earth itself. Then it defied its masters and the result was terrible.

A blue-ribbon presidential commission headed by former Secretary of State William P. Rogers was convened to find the cause of the worst space accident. Its fourteen members included Neil Armstrong and Sally Ride, an iconoclastic Caltech physicist and Nobel laureate named Richard P. Feynman, Air Force Major General Donald J. Kutyna, and Albert "Bud" Wheelon, whose biographical sketch mentioned that he was executive vice president of Hughes Aircraft Company but carefully omitted the fact that he had also supervised the design of satellites in the Corona and other deep-black intelligence programs, including the superfast SR-71 spyplane, when he had been the CIA's deputy director for science and technology in the early 1960s. Chuck Yeager, by then a retired Air Force brigadier general, was also appointed to the commission. Thanks to *The Right Stuff*—the book and the film—he was widely taken to be the ultimate test pilot, the embodiment of cool courage and selfless dedication. It was therefore ironic that he showed up on the first day the commission met, said, "Go get 'em, Doc," to Rogers, and disappeared until the time came to sign off on the five-volume report.

The commission concluded in June that *Challenger* was lost because an O-ring in the right SRB that was supposed to seal two of its sections had been hardened by the cold and ruptured under the tremendous heat and pressure. At one point Feynman publicly dropped a sample of the O-ring rubber into his glass of ice water and then easily snapped it in two to dramatically demonstrate that it had become brittle. (He was widely credited for thinking up the clever stunt, but he himself would say later that someone else with an agenda had put him up to it.)

Feynman was far more critical of NASA than was Rogers, who ended the commission's report by saying that it applauded the agency's "spectacular achievements of the past and anticipates impressive achievements to come." But the plainspoken, free-spirited, self-proclaimed "explorer" from Brooklyn's Far Rockaway would have none of that. He castigated the space agency for allowing recklessness to override safety and solid engineering. The O-rings that held the SRBs' sections together had a troubled history that was effectively ignored, Feynman charged. In fact, NASA officials at the Marshall Space Flight Center and headquarters had ignored several test reports from Thiokol showing that O-rings had partially burned through, eroded, and became brittle when they were cold.

But the real culprit was not rubber and putty. The pressure that really caused the rocket booster to burst was political, not physical, and it came from the unrelenting need to keep launching in order to justify the program. "It is not clear to the Committee why so many warnings went unheeded by NASA personnel that morning," the House Committee on Science and Technology would note in its own report. Of course it was. It was absolutely clear.

For one thing, the space agency was embarrassed because a program that had been sold as providing what amounted to scheduled airline service was operating like a charter carrier in a banana republic. Like a nuclear industry that had once advertised its product as "too cheap to meter," NASA had sold the Space Transportation System as a low-cost, frequent, and routine way of getting to space. The Fourth Estate, long since grown out of its cheerleading role (and perhaps because of it) had been barbing the space agency, first because of the shuttle's notorious development problems and then because of its inability to stay on schedule, even to the point of making reference to "the little shuttle that couldn't."

A second reason to launch that day had to do with Christa McAuliffe's contribution. She was locked into giving two lessons on the fourth day of the mission, putting it on Friday if *Challenger* lifted off on Tuesday. A twenty-four-hour hold would have effectively killed the lessons and their considerable public relations value, since no one but custodians and janitors would have been in school on Saturday, and not many of them at that.

Finally, Reagan was supposed to deliver his State of the Union Address that night (it was postponed). NASA had already proposed that he mention the mission this way: "Tonight, while I am speaking to you, a young elementary school teacher from Concord, New Hampshire, is taking us all on the ultimate field trip as she orbits the earth as the first citizen passenger on the space shuttle. Christa McAuliffe's journey is a prelude to the journeys of other Americans." In a book published seven years later, McAuliffe's mother, Grace Corrigan, stated flatly that her daughter had said the day before the launch that NASA intended to send *Challenger* up no matter what. "The word was out that today was the day—definitely," she would quote her daughter as having said. The White House, knowing that even a hint that there had been pressure to launch because Reagan wanted to mention McAuliffe in his speech, adamantly refused to comment on whether he planned to do so. As was the case with the orbiter's remains, which were sunk in concrete, not sent to the Smithsonian, the matter was quickly buried.

Isaac Asimov and Feynman both took the occasion to go after the manned space program's deceptive public relations. "It frightens people," the prolific and wide-ranging Asimov said of an accident for which the public had not been adequately prepared. "All of a sudden, space isn't friendly. All of a sudden, it's a place where people can die. . . . Many more people are going to die. But we can't explore space if the requirement is that there be no casualties; we can't do *anything* if the requirement is that there be no casualties."

Feynman would not have quibbled with that. But he did fault the pervasive guile. Annoyed both at what he took to be Rogers's excessively sanguine attitude about NASA, together with the space agency's own relentlessly self-serving and hopelessly unrealistic one-in-a-hundred-thousand chance of a catastrophic accident, he penned his own "personal observations" in an appendix to the commission's report.* "Only realistic flight schedules should be proposed, schedules that have a reasonable chance of being met. If in this way the government would not support them, then so be it. NASA owes it to the citizens from whom it asks support to be frank, honest, and informative," he wrote. ". . . For a successful technology, reality must take precedence over public relations, for nature cannot be fooled."

Challenger was lost because NASA came to believe its own propaganda. The agency's deeply impacted cultural hubris had it that technology—engineering—would always triumph over random disaster if certain rules were followed. The engineers-turned-technocrats could not bring themselves to accept the psychology of machines without abandoning the core principle of their own faith: equations, geometry, and repetition—physical law, precision design, and testing—must defy chaos. No matter that astronauts and cosmonauts had perished in precisely designed and carefully tested machines. Solid engineering would always provide a safety margin because, the engineers believed, there was complete safety in numbers. That made them arrogant. It was arrogance and conceit that persuaded the managers of the manned program that six trips to the Moon and twenty-four shuttle flights around Earth without a deadly accident proved that technology would always triumph over fate; that their numbers were better than God's.

"This," said Feynman, amounted to "a kind of Russian roulette. . . . [The shuttle] flies [with O-ring erosion] and nothing happens. Then it is suggested, therefore, that the risk is no longer so high for the next flights. We can lower our standards a little bit because we got away with it last time. . . . You got away with it, but it shouldn't be done over and over again like that."

What was left of the shuttle fleet was grounded for thirty-two months while design improvements were made by Thiokol and other manufacturers to the three main engines, mainly the solids, the external tank, the wings, and other hardware. And with the loss of the *Challenger* crew still fresh in the space agency's collective mind, a way of getting them out of a stricken spaceship at

* He said that those odds—which, he noted, were management's, not the engineers'—implied that a shuttle could be launched every day for three hundred years with a loss of only one. Feynman liked the engineers' hundred-to-one better and suggested that engineers and others who went into management suddenly became corrupt because they had more at stake in a program's success. (*Challenger* accident report, Vol. 2, p. F-1; Feynman, "Mr. Feynman Goes to Washington," *Engineering & Science* (Caltech), Fall 1987, p. 18.)

relatively low altitude was also developed. They would bail out by sliding down a long pole, much the way firemen do on fire poles, and then parachute the rest of the way. But nothing would save them if another shuttle exploded without warning, like a bomb.

The Public Rallies

A study done by Northern Illinois University's Public Opinion Laboratory and released almost a year to the day after the *Challenger* explosion noted that the subject of space itself was "not a salient topic to most Americans" in spite of the Apollo program and twenty-four successful shuttle flights before *Challenger.* Only a third of the people of the United States had a high level of interest in space activities, the report said. And 12 percent—about 20 million of them—could be called "attentive," meaning that they not only had a high level of interest but considered themselves well informed about space. Thirty million more were interested in the subject but did not consider themselves well informed. They were called "interested." Everyone else—the great majority—was put into a "residual" category. Three quarters of the "attentives" were men and 24 percent of all the "attentives" had college degrees.

Not only had the greatest catastrophe in space not soured the American public, the poll found, but it was actually interpreted positively a week after the accident. More than 70 percent of the space enthusiasts agreed that the shuttle was a complex machine and that accidents would continue. That was not especially surprising. What *was* surprising was that 86 percent of the general public—"residuals" who ordinarily did not care about space one way or another—felt the same way. Similarly, 59 percent of the "attentives" thought the accident had been a minor setback to the space program, while 44 percent of the "residuals" said they believed the same thing. Asked whether the shuttle was still an "outstanding" example of American technology, more than 95 percent in each category answered affirmatively.

Furthermore, while most Americans had said they favored keeping spending constant a year before the accident—a middle position—after the accident there was a clear mood to increase funding. This indicated that the loss of *Challenger* actually stimulated positive attitudes, or possibly because most respondents felt that it would be unpatriotic to do otherwise, possibly because of some kind of a you-don't-quit-when-you're-behind feeling.

A follow-up poll, taken after the Rogers Commission turned in its report in early June, showed that all segments of the public placed primary responsibility for the accident where it belonged: on NASA. It also showed that 72 percent of the "attentive" public and 63 percent of the general public thought that the United States ought to build a space station. And just about half of everybody who was surveyed was for a manned mission to Mars. Indeed, the exploration of the solar system was continuing at that very moment, though the main target was a long way from the Red Planet.

Voyager at Uranus

Voyager 2 made its closest approach to Uranus on January 24, 1986, four days before *Challenger* was lost. The encounter was especially tricky because "the planet that got knocked on its side" and its moons were rotating at roughly a right angle to the rest of the solar system. That meant Voyager 2 was bearing in from the top of the planet and its circling moons, not horizontally through them, so this encounter would be even more challenging than usual. And at nearly 2 billion miles, it would take Voyager 2's faint whispers almost two hours and forty-five minutes to reach the Deep Space Network's giant ears.

Voyager had already plowed into the Uranian magnetic field. Now the whole spacecraft was rolling to keep its lenses trained on cratered moons like a tourist panning from an automobile to photograph a dimly lit monument in the dead of night. And it was imaging mysterious black rings. JPL's flight controllers reported that their probe skimmed over Uranus's hydrogen-, helium-, and methane-soaked bluish-green clouds at 12:59 P.M. Eastern Standard Time on January 24 going more than 50,000 miles an hour. The robot that was giving the people of Earth their first close-up look at the seventh planet from the Sun arrived at its destination slightly more than a minute late on a schedule that had been calculated five years earlier. It took nothing from Voyager's science teams to say that the men and women who navigated and looked after it—the anonymous number crunchers in Bill McLaughlin's Flight Engineering Office—had performed a feat that could only be called extraordinary.

Robert J. Cesarone, one of McLaughlin's "mission designers," was low-key about his trajectory planning. And he took refreshing exception to the worn analogy of sinking a hole in one from New York to Los Angeles. Instead, Cesarone likened keeping Voyager on course to sailing from Hawaii to Los Angeles with a series of course adjustments that stepped the craft along a predetermined route: not a hole in one, but a series of precise putts that had been worked out years in advance.

Burton I. Edelson, a scientifically literate engineer who headed NASA's Office of Space Science and Applications from 1982 to 1987, was less sanguine. He had nothing but praise for the scientists who worked on the Hubble Telescope, Galileo, and Voyager programs. Yet he also believed that Voyager 2, in particular, was an engineering "miracle." And he believed implicitly that the mission planners—the navigators and those who calculated trajectories, power requirements, and communication and imaging techniques, for example—never really got their due.

"It arrived [at Uranus] within seconds of the time it was programmed to arrive years before," Edelson explained. Then it "zips through the system, points a camera at Miranda going forty thousand miles an hour," he added, referring to one of the Uranian moons. ". . . The precision of the tracking of a camera like that in order to get high resolution, and without blurriness of the camera . . . *a billion miles away . . . after eight and a half years . . . !*"*

Edelson was awed. So he allowed himself a moment to savor the sheer elegance of the actual encounter—its majestic precision—before abruptly snapping out of his reverie. "Then the scientists look at it and say, 'Ah, well, I see *this* geological formation; I see *that* geological formation.' The big accomplishment there is an *engineering* accomplishment. It isn't a scientific accomplishment: it's a scientific benefit that they can get out there and look at that. The scientific benefit is a new understanding of how the solar system, and indeed the whole universe, were formed." And, Edelson added, drawing the science and engineering together, "We did it for everybody on the face of the Earth."

Brad Smith, the leader of the imaging team, could not have agreed more. Like Edelson and many others who spent much of their lives prying secrets out of the solar system, the planetary scientist from Arizona thought of himself and his brethren as combination cosmic detectives and impresarios whose mission was to enlighten and dazzle: to turn their spotlight on the rings and moons, craters and ice fields, volcanos, valleys, mountains and storms, great bow shocks and magnetospheres that live out in the vast darkness, beyond human senses. Their calling was essentially to collect nature's secrets, many unknown or unimagined, and bring them home for display the way thirteenth-century Venetians had returned with silk and spices from Cathay.

The Grand Tour was a $750 million mission. That was only a little more than it cost to launch a single shuttle mission. And $750 million amounted to beer money in the defense budget. It was a pittance in the scheme of the history of Earth. Yet it was priceless. No potentate, prince, treasury secretary, or reformed robber baron could go out there and bring back close-up portraits from almost 2 billion miles away. It took the detectives to do that. And they were deeply pleased with their work. Looking at Uranus from Earth, as deputy imaging team leader Larry Soderblom put it, was like looking at a pea from the other side of a football field. Not only were no features on the blue-green ball visible from Earth, but the planet could not even be seen to be rotating.

* He was rounding off the distance: Saturn is roughly a billion miles away, while Uranus is 1.78 billion from Earth.

The thrill of making it to Uranus had to do not only with science and engineering but with the fact, easily lost in the discoveries themselves, that a new world had been reached at all. Edelson, Smith, and their colleagues felt the joy, the exhilaration, that all explorers feel at first contact, when they have lifted another veil.

The pea's mass, radius, temperature, and atmospheric composition had been known by inference and by infrared analysis through telescopes before Voyager 2 arrived. So had the fact that it had five moons and nine narrow rings. But the reprogrammed spacecraft filled in much of the Uranian puzzle.

It reported back constantly as it tore through the Uranian system. And every morning, as was the now-well-established custom, the discoveries were presented by amazed scientists to thoroughly receptive journalists in von Karman Auditorium. Here were the first close-up views of the five known major moons—Oberon, Titania, Umbriel, Ariel, and Miranda. Then the imagery yielded ten other moons, all of them close to the rings. Three of the major moons, Titania, Ariel, and Miranda, were now thought to have had active histories because they showed tectonic activity or volcanic resurfacing. Miranda, only a sixth the size of Earth's Moon and the smallest of the original five, was quickly dubbed "the most bizarre body in the solar system" because of its improbable variety of features. It had specially named "ovoids," for example, which are pockmarked oval trapezoids that are younger than the plains on which they sit and have weird parallel belts of ridges and grooves that touch one another at definitely strange angles. And all five of the originals could now be seen to have water ice. That news dazzled the exobiologists and others because where there's water there could be life.

And there was much more. Two new rings were added to what was now clearly seen as a family of dark rings that was much different from the rings around Jupiter or Saturn. Unlike Saturn's immense, flat ring plane, the Uranian rings were very narrow, made of chunks of charcoal-dark methane ice and minerals ranging in size from pebbles to boulders. And they were not all shaped the same way, either. One of them, called the Epsilon ring, was being shepherded by two of the tiny new moons. The ten new moons themselves were small and dark like the rings and merged with them at their outer edge. So it was possible that the rings and moonlets were intimately connected.

The true surprise came when scientists at JPL looked at pictures of the rings taken as Voyager 2 photographed them on the far side of the planet, when it was backlit by the Sun. Carolyn C. Porco, a ring specialist from Arizona, and others saw that the sunlight highlighted a wide, flat gleaming plane of dustlike particles, much the way it would show dirt on a windshield. The plane seemed to be made of disintegrating boulders that eroded and then fell into the planet itself in a continuing cycle of birth and death.

The magnetic field and atmosphere people were gorged on data as well. Uranus's magnetic field was clearly inclined fifty-five degrees from true north, by far the largest divergence in this solar system (at least at that point). This hinted at a mass of electrically conducting fluids that acted like a dynamo. And since the Sun shines on one of Uranus's poles, it was anticipated that the atmosphere there would be warmer than at the equator. Wrong. The temperatures turned out to be about the same, indicating that the heat was distributed from the pole to the equator. The wind pattern and cloud bands were similar to Jupiter's and Saturn's, flowing parallel to the equator and showing that atmospheric circulation was affected more by the planet's rotation than by sunlight.

All that and more was supposed to go into a summary of Voyager 2's historic rendezvous with Uranus as the close encounter ended for the von Karman regulars—the scientists and the scribes—on January 28, 1986. It was supposed to go into a summary that day, but it did not. At 8:38:17 that morning, as Voyager's various teams were going over their newest data and preparing to end their near encounter with a final flourish, *Challenger* exploded three time zones away. Voyager 2 was precisely 1,840,858,437 miles from home and 3,175,901 miles beyond Uranus the instant the orbiter blew apart over the edge of the Atlantic. The robot explorer was looking back and getting ready to start a seven-and-a-half-hour real-time observation when the deadly white flower bloomed. *Challenger*'s end was seen and reseen on the NASA select channel by just about every thunderstruck member of the Voyager team who was at JPL at that hour.

There were tears in Pasadena, as elsewhere. There were feelings of sorrow for the shuttle's crew and sadness for the greater space community. But there was also bewilderment and vexation, disappointment and frustration over how the accident was going to affect the lives of those who explored the solar system. Everyone knew that the shuttle was the only way for science to get to space. Now that route was closed, at least temporarily, for missions returning to Venus, Jupiter, and elsewhere. The shuttle had been the nemesis of science and exploration from the start, and now, here it was, even stealing the explorers' thunder from its grave in the Atlantic.

"We knew that there were a bunch of dead people now, and that the space program was in deep trouble. And we knew we had been cheated, intentionally or not, out of our encounter," Bobby Brooks, who worked on Voyager's complicated sequencing, recalled, his eyes burning with anger. "Here we were, flying high, and they took it away from us." He spoke for many of his colleagues who delivered their finale the next day to a considerably thinned group of reporters. The story had moved to Florida and Texas. Soon it would move to the Soviet Union.

Thinking Great Thoughts

The heavy thinkers in the space science and academic establishments reacted to the shuttle catastrophe, to an angry, resentful, and recriminating Congress, to the Reagan administration's relentless push for Star Wars at the expense of science, and to what they took to be increasing public apathy by doing what intellectuals always do when they are frustrated: they organized committees and commissions to study the situation and write reports. By 1989, bookshelves would fairly creak under their weight.

As early as 1983, the Solar System Exploration Committee, which was part of the NASA Advisory Council, formally answered the threat to solar system exploration by bemoaning the fact that there was virtually nothing to follow Surveyor, Pioneer, Mariner, Viking, and Voyager. After claiming (without elaboration) that Soviet planetary missions had achieved "remarkable successes," the blue-ribbon committee urged that a fresh philosophy be brought to space exploration and that a coherent and integrated plan was needed. Like most of the other reports that would soon start flowing out of Washington, this one spiked the central argument—that determining the "origin, evolution and present state of the solar system" was a noble goal—with the need for the United States to "preserve its leading role" in space exploration. This was the de rigueur, though oblique, warning about the ubiquitous Russians.

"An emphasis on overall program cost should replace the arbitrary rationing of the rate of new mission activity: the control of mission costs depends upon increasing mission frequency well above the present depressed level so that the economies of heritage in hardware and software can be realized," the committee asserted in language that was reminiscent of the pitch used to sell the shuttle. The group, which was loaded with the cream of U.S. space science, went on to recommend what it called a "core" program that would keep basic exploration going for about $300 million a year. The first four missions would send a radar mapper to Venus, a geoscience and weather orbiter to Mars, another radar mapper to Titan, and still another spacecraft out to rendezvous with a comet. Three years later, the committee turned out an "augmented program" report that called for more elaborate exploration, including sample-return missions to Mars and to a comet, and sending robotic rovers to the Red Planet.

In May 1986, an ad hoc National Commission on Space, which was appointed by Reagan and which counted Nobel Laureate Luis W. Alvarez, Neil Armstrong, former NASA Administrator Tom Paine, former U.N. Ambassador Jeane J. Kirkpatrick, and Yeager among its fifteen voting members, issued its own report, "Pioneering the Space Frontier." It, too, sounded a warning about competition, though not only Russian. America's major commercial competitors in space in the future, the report warned, would be Western Europe and Japan, plus the Soviet Union and China. So it suggested three mutually supportive objectives. The first was the familiar one of advancing an understanding of Earth and the rest of the universe. The other two called for exploring, prospecting, and settling the solar system and creating and operating systems and institutions to provide low-cost access to space. Both Earth's Moon and Phobos, which belongs to Mars, were named as likely candidates for prospecting.

"Pioneering the Space Frontier" was a carefully thought out, balanced blueprint for the move to space over the next half century. Unlike many other studies, it was not hardware oriented, but instead laid out a philosophy for "humankind to move outward from Earth as a species destined to expand to other worlds." The idea was to explore, prospect, and settle the solar system in cooperation with other nations and in a way that was steady, logical, and progressive. To that end, the commission called for spending "a small but steady fraction of our national budget" on space, rather than the wasteful roller-coaster approach that followed Apollo. There were repeated references to settling Earth's frontier—chiefly the American West—and to the crucial role of both government and private enterprise in supporting a vast venture that would ultimately generate many times the value of the investment put into it. The beauty of "Pioneering the Space Frontier" was its healthy mix of scientific and economic goals, making the point that understanding the evolution of the universe and the solar system, for example, were no less important than mining Earth's Moon and others and establishing a colony on Mars.

Six months later, another NASA Advisory Council group, the Space and Earth Science Advisory Committee, weighed in with "The Crisis in Space and Earth Science." Warning that the *Challenger* accident had been only one aspect of a larger crisis that was sapping the morale of American scientists, the SESAC lamented the space program's lack of focus, systematic planning, and long-term objectives. The answer, this report concluded, was to tighten the system and make it more responsive to the needs of basic science and expand flight opportunities for all scientists.

There was a singular victory for the United States in 1987, however, that went unnoticed by spacewatchers everywhere. That year, the nation broke the all-time world record for printing reports about space. In June, the Solar System Exploration Committee churned out another report complaining that there had not been a planetary launch since Pioneer Venus in 1978 and that the budget for exploration was due to be slashed from an already anemic $358 million to $307 million. "The U.S. planetary exploration program is in worse shape now than when the SSEC was formed seven years ago. NASA no longer leads the world in planetary exploration," the committee's chairman, Dave Morrison, told Fletcher, the recipient of the report.

That summer, a workshop commissioned by NASA Administrator Fletcher and led by Sally Ride came out with "Leadership and America's Future in Space," better known as the Ride Report. It, too, called for tightened management and a long-term commitment to study Earth in a so-called Mission to Planet Earth program, explore the solar system, build an outpost on the Moon, and colonize Mars.

Yet once again there sounded a warning, this one high-pitched, that the Russians were coming. While no American spacecraft had gone to Mars since Viking in 1976, the report said, the Soviets had announced an extensive Martian exploration program beginning in 1988 and continuing through the 1990s. In addition, the document warned, *Skylab* had been abandoned in 1974, whereas the Soviet Union had a succession of space stations in orbit since the mid-1970s, the latest being *Mir.* "The United States has clearly lost leadership in these two areas," the report concluded, "and is in danger of being surpassed in many others during the next several years."

While the part about Viking was true enough, "Leadership and America's Future in Space" used some highly selective reporting to make its case, a fact that would not have displeased Fletcher. The Soviets could have announced that they were planning to send the Bolshoi Ballet and trained bears on a tour of the Jovian moons, but that did not necessarily mean they would have been able to do so. Their record on Mars was wretched. And if they could not even make it there with robots, sending cosmonauts was hopeless, particularly since they could not even get their countrymen onto the Moon. And if they could not get people to Mars, *Salyut* and *Mir* were a waste of time and resources. Meanwhile, the Pioneer missions to Jupiter and Saturn and Voyager 2's sensational flyby of Uranus the previous year proved that outer-planet exploration, which was far more difficult than going to Mars, belonged to the United States. Yet there was no mention of those missions in the report.

The NASA Advisory Council Task Force on International Relations in Space issued its own report on October 12 (Columbus Day) called "International Space Policy for the 1990s and Beyond." This one warned that the "Soviet Union is organizing a full-court commercial and marketing press, with new organizations, sales force, and slick brochures, offering demonstrable launch capability, schedule reliability, pricing flexibility, and security of users' proprietary technology." It also made the oversimplistic and misleading point that the Soviet civilian and military programs were inseparable, while America's were "separate." Commercial threats loomed in every direction, including from Western Europe, China, Japan, and even India, which, the report said, had a $170 million annual space budget and had paid the Russians to send a satellite named Aryabhata into orbit in April 1975.

The Space Science Board of the National Academy of Science's National Research Council, which would not deign to mention the threat from the East, issued its own seven-volume report in 1988: "Space Science in the Twenty-First Century." It did, however, mention the threat from the manned program right up front, in a letter from Frank Press, the NRC's chairman, to Fletcher. "Any portrayal of the future of space science presupposes successful solutions to the severe problems that our nation's space science program faces today. The *Challenger* accident, coupled with our over-reliance on manned launch capabilities, has, to all intents and purposes, crippled our space science programs by depriving us of access to space." That was a real stinger directed, as it was, to the man who signed off on a shuttle that used solid rocket boosters and who also ordered an end to alternate ways of getting civilian payloads into space.

As would be expected, "Space Science in the Twenty-First Century" was an elaborate and thoughtful program for doing the Mission to Planet Earth, which would study Earth as a global system, as well as solar system exploration, space physics, astronomy and astrophysics, fundamental physics and chemistry (including microgravity studies), and life sciences. The report accepted the fact that humans belonged in space. At the same time, the document made it absolutely clear that they should not get into the way out there.

Discovery returned astronauts to space with what NASA described as a "flawless" four-day mission that started on September 29, 1988. "Whew! America Returns to Space," a *Time* cover declared breathlessly a week later. It was a telling comment because it still equated a presence in space with men and women being there. But it missed the bigger picture. The United States was very much in space while its astronauts were stuck on Earth. The whole fleet of Earth huggers was continuing to relay communications (and intercept them), watch the weather, monitor resources, photograph the Soviet Union and other places, guide ships and planes, and do science. Pioneers 10 and 11, still whispering about what they were coming across, were streaking out of the solar system. And Voyager 2 was bearing down on Neptune.

Twenty-three months earlier, in October 1986, *National Geographic* magazine had run a long cover story about the Soviet space program titled "Are the Soviets Ahead in Space?" With U.S. astronauts and big deep-space missions such as the Galileo Jupiter probe grounded because of *Challenger,* it was a fair question. The lavishly illustrated article showed the vast Tyuratam complex photographed from space, drawings of *Buran* and the Energia superbooster being raised on its erector, a progression of ten rockets from the original R-7 to the mighty Proton, and cosmonauts floating in orbit (three of them in *Mir,* which had been launched in February, right after the *Challenger* explosion). "Experts" were quoted as saying they believed that *Buran* was an improvement over the U.S. shuttle. The author of the article came to a sober, reflective conclusion that was based on his own extensive research in the U.S.S.R. and elsewhere: "Their strengths, analysts agree, lie in their methodical, building-block approach and the breadth of their commitment: strong military and manned programs, imaginative space science goals, and a busy launch schedule—all the while developing a shuttle and medium- and heavy-lift rockets. It's a race without a finish line, but they're running hard." It was true as far as it went. The problem was that the runners on the Red team were training on the political and economic equivalent of candy bars. The glucose gave them enough energy for sprinting and hurdles but not enough for the marathon. And as much could be said about the whole system.

The Decline and Fall of the Soviet Empire

Four months after the destruction of *Challenger,* one of four graphite-moderated power reactors close to the town of Pripyat in northwest Ukraine had a catastrophic accident. At 1:24 on the morning of April 26, 1986, an ill-advised test to see whether a turbogenerator could provide power in the Chernobyl-4 reactor after the steam produced by the reactor was turned off resulted in what nuclear engineers call a "power excursion." That is, the reactor literally blew its top because the steam that built up in its core decreased its water's ability to slow neutrons. Within seconds, the neutron bombardment became so intense—so hot—that the reactor turned into a bomb. Two thousand cubes that constituted the "biological shield" over the reactor's core, each of them weighing 770 pounds, started to "jump up and down," in the words of Russian nuclear engineer Grigori Medvedev, like people "tossing their hats in the air."

As panic-stricken technicians ran for their lives, Chernobyl-4 exploded with a thunderclap that sounded like a jet plane going through the sound barrier, sending red-hot fragments of steel, concrete, and graphite into the air. According to a watchman standing nearby, a great, swirling fireball rose into the sky. It turned into a plume of deadly radioactivity that quickly blanketed the area and spread as far as Germany and Sweden. Typically, Chernobyl-4's operators would be blamed for what turned out to be a calamity. But it was the reactor design itself that was the problem; its American counterparts were not allowed to go on such "excursions."

As it happened, a Landsat earth resources monitoring satellite passed over the area shortly after the explosion and sent back false color imagery showing radiation, in red, pouring out of the demolished facility. That digital image was broadcast around the world within seventy-two hours of the blast, absolutely preventing Moscow from denying that there had been a monumental accident. While ham radio operators in the area were broadcasting in English that as many as a hundred people had died, hospitals were crowded, and food was contaminated, Soviet authorities maintained that only two had died and that the situation was under control. Meanwhile, U.S. intelligence put the number of deaths at two thousand and warned that a second reactor looked as though it, too, was about to blow up. This was deduced—erroneously—from imagery sent back by the single KH-11 that was in orbit at the time. (Another of the multimillion-dollar reconnaissance satellites had been destroyed eight days before Chernobyl erupted when a Titan 34D exploded at Vandenberg.)

Chernobyl was the worst nuclear accident in history. It almost immediately surpassed even *Challenger* as a dreadful example of what could happen when high technology turned on its creators. It was soon woven into the philosophical banner under which the anti-nuclear movement marched.

But there was more to the episode than that. For one thing, the American eye in the sky was a witness to the immediate aftermath of the accident and sent pictures of it around the world. It therefore effectively prevented any attempt to deny that the event had happened or minimize its seriousness. The Kremlin might have refuted the Swedes, Germans, and others who insisted that the radiation carried on the wind from Chernobyl was rising to alarming levels in their countries. But there was no refuting the photographs. Landsat (with its French SPOT 1 counterpart to follow within a month) had moved space reconnaissance out of the intelligence community's protected realm and pushed it into broad daylight. From that day on, every government on Earth knew that an increasing number of its secrets could be exposed to public scrutiny. In October 1987, for example, the top-secret Soviet strategic-defense laser weapons facility at Nurek, pictures of which were locked in equally secret Air Force and CIA archives, made it to the pages of *Aviation Week & Space Technology.* (No one in the Strategic Defense Initiative Office was heard to object to pictures showing that the Soviets were trying to do what *they* were trying to do.)

More important, Chernobyl was a monument—a tombstone—to the system that created it. Both the stricken reactor and the political and engineering cultures that had conceived it were poorly designed and were operated with dangerous expedience and a pervasive, ultimately destructive political correctness. The nuclear energy and weapons programs, and certainly the space program, reflected that larger culture. Obsessive secrecy, ostensibly kept to protect technology from a hostile world, had led to the total control of science by a government that did not fundamentally understand it. And the secrecy protected the managers themselves from all outside criticism and the risk of blame.

Far from being the culturally egalitarian society Lenin and the others had promised, and which sympathetic Westerners believed to be true, the Soviet Union was in fact obsessively racist, nationalistic, and conservative. And that directly impacted on science. Ethnically pure scientists and engineers were officially encouraged and promoted, particularly if they played politics, and so were conservative thinkers. Those of "dubious" ethnic origin, or who were truly inventive, were routinely screened out or dismissed because the party hacks, who had no way of understanding what they did, assumed they were potentially troublesome. "By favoring the obedient and subservient over the challenging and inventive, the industry and the country suffered," Sergei Kapitsa maintained. "Of those who remained, many were highly trained and specialized experts, but ingenuity and initiative cannot flourish in a compartmentalized and centrally controlled atmosphere. In such a system," the Russian physicist added, "it is better to fail according to the rules than succeed by breaking them."

Mikhail S. Gorbachev became general secretary of the Communist Party in 1985 and president four years later. He could travel to any place in the empire during those years and see the fruits of six decades of virulent socialism in theory and practice. There were gaunt, soul-deadening patriotic art and architectural monstrosities such as Stalin's notorious seven Gothic skyscrapers and the Hotel Rossiya in Moscow. There was spiritual and physical decay, from chronic alcoholism, absenteeism, and indifference to shoddy merchandise, shortages of food, disintegrating buildings, broken sidewalks, rampant pollution, and an unparalleled assault on nature, including the rape of the nation's rivers, forests, and coastal waters (where old nuclear submarines, complete with "hot" reactors, were routinely deep-sixed). Almost no one under the age of seventy really believed in Marxism, but they were playing out the charade because the system was controlled by what the disillusioned Yugoslavian politician Milovan Djilas called the "new class" of privileged Communists; a class that provided just enough bread, potatoes, and vodka to its workers and peasants to prevent a counterrevolution. The workers and peasants ate their food, drank their vodka, and told endless "Armenian Radio" jokes:

> Ivan tells Aleksei that he will give him five thousand rubles if he shouts that Brezhnev is an idiot in Red Square at high noon.
>
> Aleksei reasons that if he does it, he will be arrested for hooliganism, get three months in prison, and will then be able to enjoy all the money. So he and Ivan go to Red Square. "Brezhnev is an idiot!" he shouts at noon.
>
> Sure enough, three KGB men emerge from the crowd, handcuff poor Aleksei, and stuff him into the back of their unmarked car.
>
> "So," Aleksei says, "you're going to put me in prison for three months, huh?"
>
> "No," answers one of the KGB men, "we're going to put you in prison for twenty years."
>
> "Twenty years?" a frightened and dumbfounded Aleksei cries. "But the penalty for hooliganism is three months!"
>
> "We're not throwing you in prison for hooliganism, comrade," the KGB man says. "We're putting you there for revealing state secrets."

Gorbachev did not have to compare his country to the United States to see that it was drifting dangerously close to history's reef. There were nations on both sides of his that were far outstripping the Soviet Union in the quality of both life and the economy. So he decided to begin changing his nation's direction. This would be called glasnost and it would have repercussions that no one, including Gorbachev himself, could foresee. A process soon began that set the stage for weakening the party's hold on power, invigorating local government, running the first free elections since 1917, encouraging the first flushes of free enterprise and a free press, and starting the process of bringing down the Iron Curtain and breaking up the union. It started slowly enough. Then, inexorably, it began to lurch out of control.

Phobos et Deimos (Fear and Terror)

What was happening in space as the last decade of the millennium approached still reflected what was happening on Earth. The cold war had spawned and nourished both the Soviet and American programs. When it ended, and the Soviet Union collapsed, the Russian program began showing the effects of political indifference or outright antagonism and a chronically, desperately weak economy. It was Chernobyl in slow motion. With the economy in shambles, using the treasury to subsidize the training of hundreds of spacemen and women and the launching of colossal rockets and shuttles—shuttles for what?—struck increasing numbers of Russians as being outrageous. Funds therefore started to dry up even as the empire disintegrated. And whether of a living creature or an institution, malnourishment inevitably leads to disease and death.

The first real signs of trouble came in September 1988, when both the manned and planetary programs suffered problems simultaneously that reflected systemic illness. On September 6, a Soviet cosmonaut and his Afghani crewmate were stranded in orbit for twenty-five tense hours when the retro-rocket that was supposed to slow *Soyuz TM-5* failed to start. A malfunctioning infrared sensor told the spacecraft's computer that it was incorrectly oriented. Re-entry was therefore abruptly called off, and, like airline pilots who have missed the approach to a runway, the two cosmonauts went around again. And again. And again. In all, there were two sensor failures, one computer failure, and another missed landing because the retro-rocket burned for just six seconds instead of 230 seconds. They finally made it down with little air and no food left.

While some controllers were trying to get the unlucky cosmonauts down safely, others were feverishly working on another problem. On July 7, 1988, an ambitious and highly original Mars explorer named Phobos 1 had set out to study its namesake, the larger of Mars's two moons. Phobos and Deimos, the other moon, were thought to be captured asteroids. Scientists at Roald Sagdeev's Space Science Institute designed the spacecraft to orbit Phobos, which is a fourteen-mile-wide dark gray "potato," and do radar and chemical analyses from high altitude. The spacecraft was also supposed to record Martian weather.

Then, according to the meticulously conceived plan, the mission would really get interesting. Phobos the spacecraft was supposed to swoop down to within fifty yards of the moon's surface. There, according to IKI's imaginative and daring plan, it was to fly in formation with the moon and not only fire a harpoon into it for soil study but drop a battery-operated "hopper" that would use its spring-loaded legs to jump twenty yards at a time like a hundred-pound mechanical cricket for close inspection of Phobos's surface. On July 12, five days after Phobos 1 left Tyuratam, a second seven-ton explorer, Phobos 2, followed it to Mars. In keeping with the practice at the time, Sagdeev and his colleagues wanted a redundant spacecraft. Ideally, that would not only increase the data sent home but would salvage the mission if harm came to one of the explorers.

And whatever its scientific potential, the Phobos mission was politically important for at least three reasons. It was supposed to show the Russian public that the space program was still an important national asset and therefore deserved continued support. And since its thirty-seven experiments came from eleven European nations, the European Space Agency, and the United States, as well as the Russian Federation, it was intended to demonstrate that international cooperation in space science was important and that Russia was a reliable and still innovative partner. Finally, it was to be one of several reconnoitering missions to the Red Planet's neighborhood that were to end with cosmonauts landing there in the twenty-first century. Training for the expedition to Mars by collecting physiological and psychological data on *Mir* was almost the space station's sole justification. Bruce Murray, Brad Smith, and James W. Head, a leading planetary scientist at Brown University, were among a dozen or more Americans who were involved with Phobos. Their collaboration with Sagdeev and his colleagues was a scientific and political milestone; the beginning, it was hoped, of close cooperation on a path that would lead earthlings to Mars together once and for all.

But disaster struck on August 29. The software that sent commands to Phobos 1 was supposed to be checked by a computer before transmission. But the computer failed. Instead of waiting for it to be fixed, however, a technician overrode the check-out procedure and, violating the rules, sent the operating command sequence to the spacecraft anyway. The sequence had the kind of small but grotesque error that had bedeviled planetary exploration since Mariner 1 had been lost at sea in 1962 because a minus sign had been left out of its launcher's guidance sequence. The tiny mistake—the omission of a single number—switched off Phobos 1's attitude-control system. This meant the tiny steering thrusters that kept it balanced and flying in the right position were abruptly told to shut down. Without their little corrective spurts of gas, Phobos 1 became dizzy and slowly turned until its solar panels no longer faced the Sun. With its solar energy cut off, the spacecraft's battery soon drained, leading to further disorientation and cutting off all communication with the controllers back home who now were trying to snap Phobos 1 out of its tumble to oblivion. Sagdeev, in evident exasperation, could say only that leaving out the last digit of the command amounted to ordering the spacecraft to "commit suicide." It did, sailing mutely past Mars, the nemesis of Russian planetary exploration, and heading for open space like "a dazed acrobat," as Sagdeev put it.

Then, incredibly, it happened again. By the time Phobos 2 swung into orbit around Mars at the end of January 1989 in preparation for a landing on Phobos, one of its three redundant computers was dead and one was gravely ill. The logic built into the Phobos computing system required all three to "vote" on the spacecraft's operation, with two thirds carrying the decision. But with one computer dead and another too sick to vote, the healthy survivor, again in Sagdeev's words, was "unable to overwhelm the 'no' vote of two 'dead souls.' " By the end of March, the fragile communication link was broken and contact lost. Phobos 2 did manage to send back some data, including several good pictures of the Martian surface and thirty-seven color shots of the moon that was its main target before it, too, tumbled out of control and disappeared.

As happened after Chernobyl, the operators were blamed for the embarrassing failure. But it was really a mission design problem. "The management system on one of our spacecraft would have said: 'Tell me that again; I didn't believe it the first time,' " said Sagdeev's friend Bruce Murray. Murray and the others had nothing but compassion for the Russians, their now-hemorrhaging program, and their computers' forlorn souls. It was worthy of Tolstoy (or Gogol).

"No one can describe the state of shock in the Space Research Institute—especially among the young men whose only activity throughout their short careers had been to prepare the experimental devices flown on the Phobos mission," Sagdeev said later. He also claimed that the same sort of hubris that infected NASA led to the loss of the Phobos mission. Successful shots at Venus and a rendezvous with Halley's Comet in 1986, he said, had brought a "dangerous kind of comfort" to the Soviet program.

But a high comfort level at IKI—what Stalin might have called being "dizzy with success"—would be hard to fathom. Mars had been a graveyard for Soviet spacecraft from the beginning; the Ghoul definitely had it in for the

Soviet Union. With the single exception of Mars 5, which successfully orbited the Red Planet in 1974, the Russians had a string of ten out-and-out failures and two partial failures going back to 1960. Then came the Phobos twins. And it would only get worse.

Quayle Airs His Thoughts on Mars

For its part, NASA came out of the cold war with two major long-range programs: the so-called Mission to Planet Earth, which would study the planet with unparalleled thoroughness (at least by civilian standards), and the venerable, constantly redesigned space station.

Mission to Planet Earth was one of two space projects that came out of President George Bush's White House in 1989. The other, which was called the Space Exploration Initiative during its briefly flickering life, was announced by Bush on the steps of the National Air and Space Museum on the morning of July 20, 1989, to mark the twentieth anniversary of the first Apollo Moon landing. Borrowing a page from Kennedy, Bush called for an audacious long-term commitment that would send astronauts back to the Moon and on to Mars for a permanent settlement in space. "What was once improbable," he said, "is now inevitable."

Inevitable, perhaps, but not imminent. The exploration initiative quickly faltered because it came with no timetable, plan, or anything else that suggested the White House itself was taking it seriously. What it did come with was an estimated $400 billion price tag, more than twice as high as most industry experts would have calculated. Perhaps even worse, it fell to the National Space Council, nominally headed by Vice President Daniel Quayle, to make those estimates.

Twenty-two days after Bush's speech, as Voyager 2 began its encounter at Neptune, Cable News Network asked the vice president for his thoughts on sending astronauts to Mars. His answer was stunning. "Mars is essentially in the same orbit [as Earth]," the chairman of the space council explained as a prelude to one of his most memorable gaffes. "Mars is somewhat the same distance from the Sun, which is very important. We have seen pictures where there are canals, we believe, and water. If there is water," he continued, straight-faced, "that means there is oxygen. If oxygen, that means we can breathe." And where earthlings could breathe, Quayle continued in his earnest, boyish way, they could function quite nicely. The Space Exploration Initiative soon disintegrated, and so, in a manner of speaking, did Dan Quayle. Congress, in no mood to put $400 billion into space exploration without Russian competition, killed Bush's plan outright. Meanwhile, Quayle headed to JPL to take advantage of what was sure to be the momentous (and heavily reported) arrival of Voyager 2 at Neptune, its last stop on the Grand Tour.

The Eagle Is Grounded

The loss of *Challenger* three years earlier had forced the U.S. space program into limbo for thirty-two months. Like a line of railroad cars stalled on a track by a broken engine, a number of ambitious missions, including the forays to Venus and Jupiter and the launching of the extravagantly planned Hubble Space Telescope, were stopped cold. All three projects in fact suffered double setbacks because of the Space Transportation System.

First, they were threatened and delayed by the shuttle in a variety of ways, many of them budgetary, during their own creation. Writing in *Scientific American* early in 1986, Van Allen—the space shuttle's most implacable and outspoken foe—called the siphoning of funds from the Galileo mission to Jupiter and other science programs to feed the reusable man carrier a "slaughter of the innocent." Galileo's political and economic tribulations forced one expensive design change after another while its manager at JPL, John Casani, practically commuted to Washington like a lawyer fighting for a client's stay of execution. Van Allen muttered that the program reminded him of *The Perils of Pauline,* the silent film serial whose heroine repeatedly escaped death by the narrowest of margins. Then there was the accident itself, which grounded the explorers that had survived the budgetary shoals, further thwarting the science community. Both Galileo and the Venus Radar Mapper, eventually renamed Magellan, were first designed to be launched from the shuttle (Galileo, in fact, was supposed to be the shuttle's first interplanetary passenger).

But the civilians were not the only ones who were frustrated by the long grounding of the Space Transportation System, during which time some four hundred changes were made to the SRBs and the orbiters at a cost of $2 billion. Having been pressured to support the shuttle as the primary launch vehicle for their spy, relay, early warning, navigation, communication, and other high-priority satellites, the Air Force and the various intelligence organizations now saw that what was supposed to have been their main route to space was far more tenuous than those in NASA who concocted the hundred-thousand-to-one accident scenario had them believe. That was why the Air Force had demanded it be allowed to keep using Titan 34Ds, the largest and most powerful of the expendables, particularly for the critically important and very heavy imaging reconnaissance satellites.

Yet the Titans were blowing up too. One of the $250 million to $350 million launchers exploded on August 28, 1985, taking a half-billion-dollar KH-11 with it. Then, on April 18, 1986—not quite three months after the *Challenger* disaster—the next attempt to orbit one of the large reconnaissance satellites ended in a fiery explosion seconds after liftoff at Vandenberg. The horrendous blast not only demolished the spacecraft, the launchpad, and several automobiles and trailers in the vicinity but released a cloud of toxic gas that forced teachers to keep schoolchildren on the base indoors.

There was a successful Titan 34D launch in October 1987, lofting another reconnaissance satellite into polar orbit and prompting Secretary of the Air Force Edward C. Aldrich Jr. to declare that the event "allows us to resume launching critical national security payloads on a regular basis." Not quite. Another failure occurred on September 2, 1988, this time when the upper-stage rocket booster, or trans stage, which was supposed to send a top-secret communication intercept satellite into geosynchronous orbit, malfunctioned after it separated from a Titan 34D out of Canaveral.

By then the space shuttle *Discovery* was being readied for a launch that would come twenty-seven days later. Late on the morning of September 29, the twenty-sixth shuttle mission got under way from Canaveral as more than a million people, most of them stuffed onto and beside roads the way they had been when *Apollo 11* left for the Moon, watched five astronauts carry another Tracking and Data Relay Satellite to orbit. There were shouts, tears, and applause as *Discovery* disappeared into high clouds, trailing a thunderous wake that everyone it washed over wanted to last longer than seventy-three seconds.

Following doctrine, the TDRS was launched six hours into the mission—as soon as possible—so that it would be safely on its way should the orbiter have an emergency. Charles Addams, the cartoonist whose ghoulish humor had appeared in *The New Yorker* for five decades, died of a heart attack that day in New York. He, for one, would have appreciated the prudence of anticipating catastrophe. So did the generals of the Air Force, who had long since written off the shuttle and decided to stick with their expendables.

A Famous Victory

But with the demise of the Soviet Union and the shriveling of its vaunted rocketry, the term "expendable" took on a new and disquieting meaning for the opposition. Suddenly, in the absence of a competitor, the U.S. space program seemed to be edging toward a kind of expendability of its own as the last vestiges of the great race abruptly ended. The Eagles now sat on their bench (to use a metaphor that the football-loving John Kennedy would have appreciated), looked across the gridiron, and were dumbfounded to see that not only was the Bears' bench emptying, but the players were selling their helmets to buy lunch.

Suddenly, the game was over. But with the exception of those whose dreams were stalked by mushroom clouds—the arms controllers—there was a strange sense of emptiness. There was, in fact, a perceptible and disquieting void. Those who define themselves through competition necessarily get disoriented and confused when it ends. In "The Battle of Blenheim," Robert Southey, an English poet and friend of Coleridge, touched on the vague but discernable sense of ennui and lack of fulfillment (and perhaps nostalgia) that came with the end of the cold war:

> "And everybody praised the Duke,
> Who this great fight did win."
> "But what good came of it at last?"
> Quoth little Peterkin.
> "Why, that I cannot tell," said he;
> "But 'twas a famous victory."

Triumph at Neptune

Thirty-two days later, on Monday, August 21, it was overcast and unseasonably chilly in the foothills of the San Gabriel Mountains as the Sun came up. A thick haze, dappled with puffs of smoky gray and white, alternately obscured the rising Sun or turned it into a disk the color of a blood orange. Just outside JPL's main entrance and near von Karman, a line of trailers and mobile television studios sprouting dish antennas and beaded with the night's dew was coming to life.

Calling it the prelude to "an extremely exciting week . . . a week of great drama," Lew Allen welcomed the throng in von Karman at a quarter after ten. Allen, the director of JPL, was a tall, bespectacled, soft-spoken former Air Force chief of staff who had a doctorate in physics. "This is the last first encounter of a planet for a very long time," he said, knowing that nobody was going to mount a mission to study Pluto anytime soon. The icy rock was so scorned by some astronomers that one of them, Brian Marsden, even suggested that it be "demoted" to the status of a minor planet or asteroid.

"I suspect many of you are here for your sixth and final Voyager encounter," Edward Stone, the project scientist, next told the reporters and cameramen. He referred to Voyager 1's flybys of Jupiter and Saturn in addition to its sister's visits to the two huge gas balls, then Uranus, and now Neptune. "In a certain sense, this is the final movement of the Voyager symphony of the outer planets. . . . And I think in true symphonic form, the tempo appears to be accelerated as we approach encounter, as we all hoped it would be. So," the Caltech physicist added, yet again putting everybody onto the spacecraft, "I think we're in for a good ride."

A good ride and a very long one. Voyager 2 had traveled 4.5 billion miles since its launch twelve Augusts earlier. It was now so far from home that it took four hours and six minutes for instructions to reach it at the speed of light and that much time for its discoveries to make it back to Earth. Now, having survived massive doses of radiation, frigid temperatures, micrometeorite hits, partial deafness, a damaged clock, an arthritic scan platform, and David Stockman, it was a tired old bird. Voyager 2's afflictions and limitations were already in the folklore. Its 1972-vintage computers had so little memory that they had to be spoon-fed instructions. The twenty watts of power that sent its discoveries home in the barest of whispers arrived at an energy level one twenty-billionth that of the battery that powers a digital watch. And it was so dark at Neptune that taking pictures was like photographing the inside of Yankee Stadium by the light of a single match. No wonder the battered robot lent itself to anthropomorphisms: hardy, intrepid, resilient, tough, indomitable, a survivor, and, in the words of Norman R. Haynes, Voyager's project manager, the "doughty little spacecraft that could."

Even before Voyager 2 got to Neptune, it added four more moons to the two that were known to circle the planet and spotted what at the time looked like partial rings. It also sent back pictures of a huge storm, soon named the Great Dark Spot, that was the size of Earth and similar in proportion to Jupiter's Great Red Spot. This and other swirling spots, plus a second, smaller storm and a large, moving cloud mass that was immediately named the Scooter, showed beyond doubt that Neptune had a dynamic atmosphere. That was surprising since the atmosphere gets only a twentieth as much internal energy and sunlight as Jupiter and one three-hundred-fiftieth that of Earth. And unlike Uranus, which looked bland, Neptune was very beautiful.

At 8:03 P.M. Pacific Daylight Time on August 24, Voyager 2 streaked past Neptune's fragile ring plane and, drawn by the planet's gravity to a top speed of 61,000 miles an hour, skimmed only 3,000 miles above its methane-shrouded north pole fifty-seven minutes later. This was the closest of Voyager 2's four closest planetary approaches. It was like flying a plane under a low bridge; one tiny steering error could have led to disaster. Five hours and fourteen minutes after it looped over the pole, it shot past Triton, the largest of Neptune's moons, at a distance of almost 24,000 miles. Voyager 2 penetrated the mist of time and brought an eighth planet and a wondrous new moon into the fold. Then it flew into eternity.

Back home, in the Voyager Project Office, the first of twenty-three pictures of Triton started coming in at 3:40 A.M. the next day. With empty champagne bottles, Styrofoam coffee cups, and pizza crusts scattered everywhere, Stone and the imaging team peered at their monitors. The resolution of the pictures was marvelous. Smith, Soderblom, Sagan, Davies, Torrence Johnson, Toby Owen, Carolyn Porco, Andy Ingersoll (the "weatherman"), Sasha Basilevsky, and others stared in utter amazement. What they saw on the other side of the glass was the stuff that turns scientific methodologists into giggling children.

"Here we go!"

"Oh!"

"Look at that!"

"Holy cow!"

"Wow," Soderblom declared, "what a way to leave the solar system."

What a way indeed. Part of Triton had the complexion of a cantaloupe. But other parts were covered with liquid nitrogen and solid islands of frozen methane. There were vast ice basins and ridgelines, fault lines that looked like highways, a blue streak, and what seemed to be smoke marks from geysers, all pointing in the same direction. Four actual geyserlike plumes would soon be discovered near Triton's south pole. And the most thrilling part was that Triton also had an atmosphere, its own magnetic field, and photochemical smog. Then a geologist rushed into the room clutching a black-and-white picture and announced, "This is real evidence of volcanic features with multiple flooding incidents." Volcanos and floods. Internal energy and water. As Jim Head would later put it, "Triton has one of just about everything we've ever seen." In fact, with a single obvious exception, Triton had all the makings of a full-fledged planet. The exception was that it did not circle a star. But maybe it had once orbited the Sun and been pulled away and captured by Neptune's powerful embrace. It is, after all, the only large moon in the solar system that is in a retrograde orbit: it flies backward. "Triton," Larry Soderblom declared, "is the most curious thing we've ever seen."

Neptune was a stunner. Voyager 2 sent home nine thousand images, with observations made in the visible, infrared, and ultraviolet wavelengths. Radio emissions from the planet itself showed that it rotates once in a little more than sixteen hours. Voyager 2 turned up three more major rings composed of dark matter, with the outermost in turn made of three arclike segments that were not complete circles. It also found six new satellites, all of them inside Triton's orbit and four of them nestled within the rings themselves. And, like Uranus's, Neptune's magnetic field was found to be very tilted and offset from the planet's poles. The stream of data provided a richly detailed look at the planet's violent weather system at all altitudes, temperature and pressure readings, complex radiation belts, and much more.

"As exciting as these and other discoveries we've made at Neptune have been," Ed Stone said at the last news conference of the Grand Tour, "this encounter has really been just the fourth and final planet that Voyager has revealed in an unprecedented decade of discovery. . . . The Voyager discoveries from this survey will serve as an encyclopedia for the giant outer planets for decades to come. And without question, for those of us fortunate enough to have been involved, this has been the journey of a lifetime."

The "encyclopedia" began appearing by the end of the year, when the journal *Science* published a richly detailed eighty-five-page special report, complete with photographs and charts, by Voyager's science teams. Almost a year later, after more data had been digested, *Science* took its readers back to the Neptunian system in a special series of reports on Triton itself. Four of the scientists who analyzed the geyserlike plumes concluded that they were probably the result of trapped solar radiation—a greenhouse effect similar to the one that seems to be starting on Earth—with a separate underground heat source possibly also playing a role. But two others, including Ingersoll, decided that they had nothing to do with subsurface activity and instead were caused by atmospheric conditions that made them the equivalent of Earth's dust devils. It was just one of the delightful disagreements that would go on for years.

In an editorial titled "Voyager the Intrepid," *Aviation Week & Space Technology* heaped praise on the mission and those who undertook it, in some cases devoting the better part of seventeen years to what could honestly be called more a way of life than a job. "It begs the obvious to say that this was a breathtaking achievement. But it is especially true of a program conceived as a compromise after NASA management and congressional bean counters killed an early 1970s plan for a 'grand tour' mission to Jupiter, Saturn, Uranus and Neptune. . . . It was a shortsighted, unimaginative decision. Fortunately, a team of dedicated scientists and engineers at the Jet Propulsion Laboratory in Pasadena, Calif., refused to take no for an answer." For less than the flyaway cost of three B-2 bombers, the editorial continued, the men and women who took the world on the Grand Tour "have made historic contributions to science and civilization. . . . Voyager has rewritten the planetary astronomy books and added new chapters of human understanding to the solar system." At a cost of less than a billion dollars, it concluded, "that is the biggest bargain of the space age to date."

Jurrie van der Woude, the godfather of the space writers and JPL's master of public affairs, sat in the photo trailer parked outside von Karman as the television crews packed up on the afternoon of August 29, the last day of the last encounter. With his fists pushed into the pockets of his gray-green fighter pilot's jacket, he barely managed to choke back a sob in midsentence as he talked about the end of the tour. Outside, perched on a bench in the sunshine in front of von Karman, Bobby Brooks watched the TV trucks being loaded and the reporters heading for the parking lot across the street. Then, with a quiet intensity that played sharp counterpoint to the festival that had just ended, he said something that echoed Burt Edelson: "We've given the people of this planet the solar system." Brad Smith could not have agreed more. "It's humanity's first look at the outer solar system," he said, "and we're so privileged to be the people who did that."

The same point had been made at the Omni Shoreham Hotel in Washington three months earlier, when NASA had staged a two-day "Pathway to the Planets" conference that dealt with human exploration of the solar system in considerable detail. At one point, Lennard Fisk, who headed the space agency's Office of Science and Applications—the "practical" operation—good-naturedly chided Franklin Martin, whose Office of Exploration ran Voyager, the Galileo mission that would leave for a return to Jupiter that October, and the other far-flung science collectors. Martin, who was exuberant because Voyager 2 was already sending back data on Neptune and Galileo was ready to go to Jupiter, was not in the mood to hear that there was no compelling reason for solar system exploration. "Your job is to save Earth, Lenny," Frank Martin said at last. "Our job is to find a place to go if you screw up."

Voyager's historic achievement and the imperative to explore were celebrated in widely varying kinds of music. Michael Lee Thomas wrote a New Wave tribute to the mission in 1990 and called it *Voyager Grand Tour Suite.* Its ten parts included "Jupiter Encounter: the Majesty of Size," "Saturn Encounter: Worlds Among the Rings," "Uranus Encounter: Father of Miranda," and "Neptune Encounter: The Conquest of Beginning." Interludes had titles such as "Pictures from Forever" and "To the Stellar Eternal."

Two years later, in 1992, a new Philip Glass opera called *The Voyage* made its debut at the Metropolitan Opera in New York. It was commissioned to celebrate the five hundredth anniversary of Columbus's discovery of the New World. But it said less about Columbus himself than about the human drive to explore, including in space, and implied that the urge to do so had been planted 15,000 years ago, when four aliens had crash-landed here. In Columbus's deathbed scene, he and Isabella discuss the purity of the explorer's motive. The great navigator insists that exploration is done to further the Kingdom of God and is encouraged by the Almighty himself. The queen counters that Columbus had treasure—gold—on his mind when he set sail and that he was really guided not by the wind and stars, but by Lucifer. Like Fisk and Martin, the composer and his librettist did not resolve the issue, which is moot.

For Sagan, who was on Voyager's imaging team, the best was yet to come. When the irrepressible astronomer from Cornell looked in the direction Voyager was headed, he was delighted to imagine a fantastical place that was a combination Big Rock Candy Mountain and Oz:

> There is a place with four suns in the sky—red, white, blue, and yellow; two of them are so close together that they touch, and star-stuff flows between them. I know of a world with a million moons. I know of a sun the size of the Earth—and made of diamond. There are atomic nuclei a few miles across which rotate thirty times a second. There are tiny grains between the stars with the size and atomic composition of bacteria. There are stars leaving the Milky Way, and immense gas clouds falling into it. There are turbulent plasmas writhing with X- and gamma-rays and mighty stellar explosions. There are, perhaps, places which are outside our universe. The universe is vast and awesome, and for the first time we are becoming part of it.

Tragedy in Russia

Boris Yeltsin was elected president of Russia in July 1991, survived an attempted coup d'état by hard-line Communists the following month and, impatient with the slowness of change, quickly consolidated his power. While the United States was preparing to honor the explorer whose arrival had heralded the European migration to a New World, the empire that stretched from East Germany to the Bering Strait came apart. The Union of Soviet Socialist Republics that Stalin and his successors had held together at gunpoint ceased to be a union, ceased to be soviet, and certainly ceased to be socialist. It was replaced by a so-called Commonwealth of Independent States that was loosely tied by a free-market economy but that was plagued by mutual distrust that grew out of Russian imperialism and ethnic hatred.

The freeing of prices of most goods early in 1992 turned the economy into chaos, with yet another new class, this one of gangsters and every shade of freestyle entrepreneur, starting or grabbing businesses. Fortunes were soon being made in Russia and quietly invested elsewhere. Meanwhile, civil authority weakened, taxes went largely uncollected, and pensioners and government workers went unpaid. The police and militia, all of whose members carried sidearms, tended to make up for the shortfall by shaking down their fellow citizens for such patently false infractions as going through stop signs that were nowhere in sight. Hapless drivers had no choice but to pay the unfair fines the way they would pay tolls.

Eastern Europe broke away and, drawn by the allure of the West's freedom, stability, and treasures, headed in that direction at top speed both politically and economically. Russia, by now the Russian Federation, was in turmoil. Its vaunted Army—the force that had brought Napoleon and Hitler to their knees—was chased out of Afghanistan by the feared and despised Muslims and then bloodied by Chechnyan separatists: irregulars who mined roads, attacked Russian tanks savagely, and shot helicopter gunships out of the sky. Russia was hemorrhaging.

In September 1991, far above it all, two cosmonauts aboard *Mir* passed over a homeland that was undergoing a coup, a countercoup, and a worsening constitutional crisis. It was a place that would be radically different from the one they had left four months earlier. There was a real possibility—soon to be a fact—that Tyuratam, where they had lifted off, would belong to another country when they landed. That raised an obvious question, at least to the spacefarers: To whom would *Mir* belong if the Republic of Kazakhstan became a separate country? "Is it true that the Russian Federation plans to sell the space station *Mir*?" one of them asked Kaliningrad. "And we're to be sold with the ship?" the other joked.

It was not so funny a year later, though, when Sergei Krikalev's tour of duty on *Mir* was extended from 180 days to 313 because of money problems and political changes on Earth. When he was hoisted out of the Soyuz-TM capsule that returned him to Earth in late March 1992 and given smelling salts and boullion, the familiar "C.C.C.P." (U.S.S.R.) patch on the woozy flight engineer's space suit was a meaningless relic. The nation he had left the previous May was gone, replaced by eleven sovereign states that were in an uncomfortable alliance, so the Kazakh steppe upon which he stood was foreign territory. Not only that, but the cosmonaut's home city, Leningrad, had vanished, at least on maps, and had reverted to the old St. Petersburg. Perhaps as amazing, underpaid flight controllers at Kaliningrad were threatening the previously unthinkable—a job action—and word was out that the nation's *Buran* shuttle and twenty-story-high Energia, the monster rocket that could carry more than a hundred tons to low earth orbit, were to be abandoned. Energia flew twice, in 1987 and 1988, followed by an unsuccessful effort to sell its services to the West.

Mir would not be sold. Nor, of course, would its occupants. Yet the humor was closer to the mark than the man who uttered it could possibly know. Gorbachev decided in 1988 and 1989 that his country's industry should produce more consumer goods, and the two most obvious areas to squeeze so that goal could be achieved were the military and space programs. While the old guard resisted what it saw as the horrific dismantling of the nation's most prestigious technology—former Minister of General Machine Building Oleg Baklanov was one of the leaders of the coup against Gorbachev—most of the leaders of the nation's republics were vehemently against the space program in 1991 and called it a blatant waste of money. And that was exactly how the program's scientists felt about the missions on *Salyut* and then *Mir.* On that score, they had much more in common with their Western colleagues than with their own leadership.

"In many ways everything we space scientists were able to get came independently of the government," Roald Sagdeev, the head of the Space Research Institute, complained. "Sometimes we got what we needed despite the big bosses and we took the risk of pushing for cooperation, while official circles were involved in the rhetoric of the Cold War." (Joseph Veverka, a Cornell planetary scientist, had vented his own anger eight years earlier when he had told a reporter that "the institutions that control funds clearly regard science as not useful, and believe that applications, especially military applications, are sacred.") Yeltsin himself, dueling with Gorbachev during those turbulent months, at least agreed with his rival on that point. He demanded that all space programs be suspended for six to eight years.

At the same time, knowledgeable political and military leaders pointed out that the old empire's space facilities were so spread out among the republics—

notably in Kazakhstan and Ukraine—it would be virtually impossible to maintain a coherent program with the union disintegrated. A joint program patterned on the European Space Agency was suggested, but bickering over its control, as well as over larger military and economic questions, doomed it to failure. The Kremlin reacted in February 1992 by creating the Russian Space Agency for civilian programs, including scientific ones. The politicians created their country's own Ministry of Defense the following month and in August started the Space Forces of the Russian Federation, a military space arm that was put under the control of the Ministry of Defense. Furthermore, and most significantly, the legislative branch was included in the space budget process, with the president submitting a budget to a parliament that either accepted it or returned it with corrections. Ukraine reacted to all this by turning off communications between its tracking stations and the rest of the network.

The Biggest Yard Sale in History

Like a formerly well off uncle now on the skids, Russia was reduced to selling its possessions—some die-hard Communists and nationalists would say its soul—by 1993. Its only full-sized nuclear-powered aircraft carrier was put on the auction block as either a battle-ready naval vessel or scrap metal. Almost anybody who was not obese and who had five thousand dollars could get a ride in a MiG-25 or in a top-of-the-line MiG-29 ("Fly a MiG in Moscow," one ad beckoned). And not only were formerly top-secret high-resolution space reconnaissance imagery taken by the excellent KVR-1000 cameras, technical marvels of military intelligence, put on the market, but time on the spy satellites themselves was offered by Sovinformsputnik, an arm of the Russian Space Agency, in partnership with small American firms.

The test model of *Buran,* one of the five knockoffs of the American space shuttle, was turned into an attraction in Gorky Park, an amusement park and woods on the bank of the Moscow River. For 15,000 rubles (by then three dollars), customers sitting in padded chairs got to watch a film celebrating Soviet space triumphs and had their "heart rhythm" tested by a representative of an "Institute of Biomedical Problems" who performed an "estimation of the body functional state using methods of space medicine" that were allegedly used to screen cosmonauts.

Toyohiro Akiyama, the foreign news editor of the Tokyo Broadcasting System, got to ride in *Mir* for $12 million. The intrepid newsman spent much of his flight feeling sick and describing the challenge of relieving himself in zero g to viewers back home, who tended to switch channels whenever the subject came up. The *Soyuz TM* that carried Akiyama to the space station also carried paintings on its hull that depicted the Japanese rising sun, as well as the logos of a pharmaceutical company and a manufacturer of sanitary napkins.

On Saturday morning, December 11, 1993, with more television cameras watching than were at professional football games, 238 remnants of Soviet and Russian space programs, both hardware and memorabilia, went on the block at Sotheby's, one of New York's most prominent auction houses. The fork and can opener used by Titov on *Vostok 2* turned up, as did Khrushchev's congratulatory telegram to Gagarin, Aleksei Leonov's bulky space suit, three tiny Moon rocks brought back by Luna 16, a specially designed nonmagnetic chess set that had been used on *Soyuz*es *3* and *4,* the charred carcass of the *Soyuz TM-10* that had returned Akiyama to Earth (*TM-11* stayed attached to *Mir*), and assorted gloves, wristwatches, notebooks, logbooks, training manuals, autographed photographs, flight jackets, patches, training shoes, and kitschy items that included a celestial globe, a space hammer, signal rockets, matryoshka dolls, survival kits (including fishing tackle), and a charred, "highly secret" military recovery capsule that had probably carried spy camera film from *Salyut 5* to Earth.

There were two motives for putting the relics up for sale. One, obviously, was their owners' need for hard currency during lean times. But the other was an understanding that their country had moved into a turbulent and uncertain period and that its space program remained the crowning achievement of its technology (hydrogen bombs were perhaps necessary, but they were nothing to brag about).

The Kremlin had used space for political ends, but there would have been nothing to use had the rocketeers not been able to deliver. And deliver they had. The Sputniks and Gagarin, Lunas 3 and 16, *Vostok, Soyuz, Salyut,* and *Mir,* Leonov and Tereshkova, Venera, Proton, Energia, and all the rest might have been used by the politicians, but they had not been created by them. It was not the party cadres who had gotten their nation into space first and most often, but brave men and women and world-class scientists and engineers, some of them geniuses and all of them overcoming appalling obstacles at every turn.

The builders of the Soviet space program believed implicitly that they, together with a military that had beaten back the Nazis and emerged from the war as a large, disciplined, and nuclear-armed fighting force, had finally overcome the centuries-old inferiority complex and shown the world what Slavic mettle really was. There was a single, indisputable point that could be made to those who snickered because the Soviet Union disintegrated while Krikalev was in *Mir:* the Soviet Union might be gone, but the space station was still there, defying political, as well as physical, gravity. The space program had outlived the propagandists who used it. So the Russians who sent pieces of their history to New York had more than money on their minds. They were proud enough of their country's role in space to want as many important artifacts as possible from the program to be displayed in the West, and especially in the United States, their old and worthy rival. They wanted posterity to remember what they and Tsiolkovsky, Korolyov, Glushko, Tsander, Chelomei, Tikhonravov, and all the others had done. Mankind had gotten off its home planet for the first time exactly once, and it had done so on Russian rockets. No matter what happened to their homeland now, whatever new turmoil lay ahead, they wanted that singular feat to be remembered forever.

The *TM-10* went to H. Ross Perot, a third-party presidential candidate, for $1.7 million. Vasily Mishin's thirty-one-volume diary, much of it written obliquely in what he called "code" because keeping private records had been forbidden, was also bought by Perot, as were several other items. The Texas millionaire would donate them to the National Air and Space Museum in Washington for a permanent exhibition on the space race, which could not have pleased their former owners more. "I think this is a very reasonable action. It permits an exchange of cultural values for humanity," Aleksei Leonov remarked stoically as he watched the historic event in a business suit from behind dark glass in Sotheby's boardroom.

By the spring of 1995, Zvezdny-Gorodok—the Yuri A. Gagarin Cosmonaut Training Center, or Star Town—looked like a campus whose students and faculty had mostly departed for an endless summer someplace else. The long building that contained the training versions of *Salyut, Soyuz TM,* and *Mir* still functioned, and so did the immense blue pool in which cosmonauts, streams of bubbles rising from their helmets, still practiced on another *Mir* under the watchful eyes of lifeguards in wet suits who were underwater with them. There were employees in the cafeteria and an occasional technocrat or military officer on some chore or other.

The cosmonauts, their families, and other permanent residents still lived in the isolated cluster of ten-story apartment buildings that were monuments to the glory days of privileged austerity and, compared with their decaying counterparts in Moscow, models of luxury. But the tidy apartments looked out on grass that had gone far too long without cutting, weeds that sprouted defiantly between the paving stones in the main plaza, and buildings with silent corridors and empty rooms, stripped of decoration, that were essentially abandoned. Sergei Krichevski, the cosmonaut-poet, who lived in one of the apartments with his wife and young daughter, observed with grim resignation, but no trace of sadness, that Zvezdny-Gorodok had become a Potemkin village.

Valentina Gagarina, her daughters long since departed to make their own lives, also kept an apartment in Star Town. She was now the aging and respected doyenne of the cosmonaut corps and the lone guardian of her husband's ghost. Gagarina, still attractive, lived with photographs and other mementos of the old days. But the best mementos were locked in her memory. She knew the unforgettable sensations of that exhilarating time when the cosmonaut corps had ridden a wave of acclaim around the world and carried the nation's prestige on its shoulders. There had been the smell of scorched kerosene and bouquets of roses; the sounds of thunderous applause and motorcades and endless toasts at state dinners and private parties. And there had been the medals and flags and confetti and laughter. Gagarina remembered very clearly when the seedy place she now occupied had been an Olympic Village of space and one record after another had been set by its cosmic vaulters.

The other old-timers were gone. Leonov was now a vice president of Alpha Bank in Moscow and occupied a second-floor office that, foot for foot, was better guarded than Tyuratam. German Titov, like John Glenn, was a politician. The venerated sixty-year-old retired colonel general outpolled ten rivals, including a well-known model named Yelena Mavrodi, to win a parliamentary seat in the Kolomna district, fifty miles outside Moscow. The fact that an attractive woman was running for public office said something about the new Russia. So did the fact that the seat was vacant because the last man to hold it, Sergei Skorochkin, had been kidnapped and shot three months before the election.

And on April 2, 1996, PepsiCo unveiled video shots of two orbiting *Mir* cosmonauts holding a sign that read: "Even in space, Pepsi is changing the script." Maybe. But the original script had featured two cans of Coca-Cola, which had been brought to *Mir* aboard the Progress M-8 ferry in August 1991. Still another commercial hawked Israeli milk. It showed cosmonaut Vasily Tsibliyev swallowing a floating glob of the stuff. "Hmmm," the cosmonaut proclaimed, "why not Israeli milk?

Even Mikhail Gorbachev, the last Communist leader, abandoned his longstanding rule never to lend his name to product endorsements when he starred in a Pizza Hut commercial in November 1997, reportedly for close to a million dollars. The film showed "Gorby" arriving at the fast-food establishment in a limousine with his granddaughter. "Because of him, we have economic confusion," grumbled one patron. "Because of him, we have economic opportunity," said another. "Because of him," chirped yet another, "we have things like Pizza Hut." Gorbachev, whose wife, Raisa, had gained notoriety for carrying an American Express card during his presidency, did not actually eat a slice but sipped coffee as he watched his granddaughter happily digging in. The commercial was made for the United States, not Russia, where Gorbachev was widely reviled as the man who had propelled the Soviet Union toward disintegration and chaos.

By then, Muscovites and their countrymen were pulling apart like the underside of some heaving iceberg that was breaking up. A small minority of entrepreneurs, called "new Russians," quickly adapted to the free-market economy and made instant fortunes in several lucrative enterprises, some honest, many not. They were the power wielders who built palatial dachas, bought roses for five hundred dollars each to give to favored showgirls, and made Moscow's BMW dealership the busiest in the world.

Meanwhile, the vast majority of their countrymen and women—the part of the iceberg that was over its head economically—struggled to stay alive. Jobs disappeared and pensions dried up because taxes did, forcing dignified-looking middle-aged ladies to spend day after humiliating day trying to sell puppies and kittens at Metro entrances. Apartment buildings deteriorated into darkened caves, and violent crime soared.

Mira Ivanova, an undignified, weary old woman who shared a small apartment with a daughter she despised, was typical of a class that had no future at all. She spent her days prowling the streets, picking up anything that could be turned into cash, and then returned to her decrepit building, which was always dark and cold and smelled of cat urine. Speaking to an American journalist early in 1997, Ivanova mentioned that the coming Saturday was Cosmonauts Day, marking the anniversary of Gagarin's flight. "How I remember that day," she said dreamily. "It was thirty-six years ago and my baby was three weeks old. Everything seemed possible." Then she started to cry.

16

The Rings of Earth

The idea of a permanent outpost circling Earth, a kind of electron faithfully orbiting its nucleus to help keep it together, goes back at least to Conrad Haas's flying house and then to Edward Everett Hale's "Brick Moon" in 1869. It successively captivated Tsiolkovsky, Oberth, and, in the 1950s, von Braun and the artist Chesley Bonestell. "When man first takes up residence in space," Willy Ley wrote about von Braun's design in *Collier's* in 1952, "it will be within the spinning hull of a wheel-shaped space station rotating around the earth much as the moon does." The plucky German expatriate went on to describe von Braun's 250-foot-diameter station in considerable detail, right down to the fact that its eighty-man crew would need 240 pounds of bottled liquid oxygen a day that would have to be delivered by shuttles.

With a single exception, Ley's grasp of English matched his keen grasp of astronautics, but that exception was a crucial one. It had to do with the recurrent danger of applying the future tense of the verb to be—"will be"—to any space project, and certainly to one as big as the station. That machine's prospects, let alone its final design, would be anything but perfect. The space station would be the subject of the longest and most bitterly fought embroglio in the history of the U.S. space program. Politicians would spend nearly two decades hacking at one another over whether it deserved to exist, and if so, in what form. It extended the war between the manned and unmanned factions to an even higher plane, drew proponents of so-called big science into conflict with those who favored "little" science, set off a nasty competition between the NASA centers themselves, and even soured relations between the American space community and its increasingly anemic Russian counterpart.

The huge spinning roulette wheel that von Braun had designed when the cold war had been young and money no real object had become impossibly expensive by the time Ronald Reagan became president in 1981. On the Fourth of July 1982, he went to Edwards Air Force Base to personally welcome the crew of *Columbia* back from the fourth shuttle mission. There, with a vicious behind-the-scenes battle raging between NASA Administrator James M. Beggs, Presidential Science Adviser George Keyworth, and others in the White House and on Capitol Hill, Reagan publicly urged that the nation "look aggressively to the future by demonstrating the potential of the shuttle and establishing a more permanent presence in space." While this was virtually off the cuff, it showed that the space station had the presidential imprimatur, much to NASA's delight. Keyworth, trying to carry out the policy of frugality that he believed his president wanted, had fought hard to keep "permanent presence" out of the speech. But he was finally defeated by his own leader who, captivated by the spirit of the moment and perhaps thinking of Kennedy's Moon speech, issued the dramatic call for the creation of the station.

It was named *Freedom* as an in-your-face gesture to the Evil Empire, but *Anarchy* would have been closer to the mark. As was the case with Apollo and the shuttle, the space station developed an inexorable inertia of its own within a space agency whose central canon called for mega-projects that, once under way, would be difficult to cancel. So in May 1982—two months before Reagan gave the station his blessing at Edwards—an "informal" Space Station Task Force was formed to consider various possibilities. At the very outset, there was internecine competition over its basic design, with the Johnson Space Center favoring a large, permanently manned—and therefore expensive—"space operation center," while the Marshall Space Flight Center took a more evolutionary approach in which unmanned "space platforms" would be man-tended; that is, they would at first be visited by astronauts and only later permanently occupied. Understanding that the movement of people in the vacuum of space would upset scientific experiments—astronomy, for example—Daniel H. Herman, who directed advanced planning for the planetary program, came out for several "units" flying in formation, one of which would be manned, while the others would carry science packages.

Meanwhile, David Stockman, who had nearly wrecked planetary exploration, staunchly maintained that the budget could not stand such an extravagant machine. He said at a meeting with the president and others in December 1983 that while the station was a fine idea, it needed to be deferred until the nation's financial situation improved. "You know, David, I'll bet that the Comptroller of Ferdinand and Isabella made exactly the same speech when Columbus proposed to them that he sail west to reach the Far East," Attorney General William French Smith said to Stockman, chiding him. "Yes, Bill, you're right," Reagan interrupted, "but you remember that the Comptroller won the argument: Isabella had to pawn her jewels in order to pay for Columbus' trip." The rejoinder drew laughter, but the president was as misinformed as Eisenhower, who had made the same reference about Apollo. The queen had pledged her jewels to bankroll the expedition, but funding had been found in the country's treasury.

The Pentagon, still ruffled over the loss of the Manned Orbiting Laboratory two decades earlier, came out flatly against the station because it allegedly would have drained funds from vital military programs. (Though in 1986 the Pentagonians briefly flirted with the notion of using the station to conduct Star Wars experiments, which incensed Canada, Japan, Germany, and other partners in the "international" enterprise.) Predictably, Edward Teller thought that the very idea of building a space station reflected a pitifully dangerous lack of will. What the United States really needed, he believed, was to build a lunar base before the Communists did.

Reagan nonetheless remained convinced that the station was a good idea. He told a packed meeting on the subject in the White House in August 1983 that forty years earlier, Buster Crabbe had played the part of Flash Gordon on the Hollywood set adjoining his and that he had never thought he would see the day when such science fiction became reality. Reagan officially endorsed the station in his State of the Union Address on January 25, 1984, and invited other nations to participate. This was not only good politics, but it spread the expense of the project, then estimated to cost $8 billion. *The New York Times* maintained sourly that the station lacked technical challenge and that microgravity experiments, manufacturing in zero g, and other operations supposedly destined to ride on it could better be performed without people. It called *Freedom* "An Expensive Yawn in Space."

Some Assembly Required

NASA's Space Station Program Office was established three months later, and by March 1986 eight contractors submitted proposals from which a "dual-keel" layout was chosen. The heart of this basic design, which would be continuously reworked for more than a decade, consisted of a rectangle made of metal trusses that enclosed pressurized modules holding up to eight astronauts and from which sprouted immense solar panels, antennas, and related equipment. The trusses soon turned into a pair of long tubes held closely together. By 1988, NASA estimated that the station would cost $16 billion, including the eleven shuttle launches that would be necessary to lift its parts into orbit, since they would have to be connected in orbit because of their weight. The more credible National Academy of Science's National Research Council put the real figure at $27.5 billion in 1984 dollars.

The program was nearly killed that summer by congressmen who were persuaded by scientists and generals that it amounted to a waste of precious funds at a time of enormous budget deficits caused, in large part, by the administration's unprecedented peacetime military buildup. The station's opponents were particularly irked by the space community's penchant for hiding ballooning costs and overplaying the spaceship's benefits, just as it had done to sell the Space Transportation System. George Keyworth, by then out of the White House, charged publicly that NASA routinely lied about the costs of its programs. The station was saved only by intensive lobbying by Alabama Senator Howell Heflin and others who pushed the familiar jobs issue.

By the end of 1989, the entire program was muddled, some believed permanently. The space agency was caught in an apparently hopeless conundrum. It responded to threatened budget cuts on Capitol Hill by shrinking and simplifying the station—in effect dismantling it—twice. For this it was sternly rebuked by exasperated members of the House Science, Space, and Technology Committee, who wanted the design frozen once and for all. Robert Roe, a New Jersey Democrat and the chairman of the committee, warned NASA in hearings held on Halloween that the design had to be fixed once and for all, and if it wasn't in 1990, his committee would have the "guts" to ask Congress to put the program out of its misery. Roe seemed to ignore the fact that several redesigns had been ordered by Congress itself in an effort to save money. But the micromanagement had undoubtedly added billions of dollars to the station's cost. As the journal *Science* noted, during the station's five-year history, it had run through four directors, four agency chiefs, and eleven planning reviews, costing more than $2 billion in the process.

But it got worse. Early in 1990, NASA's own engineers turned up a nightmare when they discovered that even before the orbiting outpost was completed, astronauts would have to turn from assembling it to an elaborate program of inspections, repairs, and preventive maintenance requiring a staggering 2,200 hours of extravehicular activity a year. The total time logged by American spacefarers working outside their spacecraft during the previous thirty years had come to about four hundred hours. The problem, the engineers discovered, was that there were 5,578 parts outside the inhabited modules, including solar panels, antennas and other electronic devices, and structural elements that would fail in time and require inspections and replacement. Admiral and former astronaut Richard H. Truly, by then the NASA administrator, played down the problem and, typically, blamed the news media for what he called their failure to understand that "problems are uncovered in the design process, which is moving along in an orderly fashion." But it was, indeed, back to the drawing board.

The station narrowly escaped being killed by Congress again in June 1991, when the House voted 240–173 to appropriate $1.9 billion for the next fiscal year after a heated six-hour debate and President George Bush's threat to veto any vote that killed the station. The money would be taken from space science and shuttle operations, which would raise other hackles. As far as *The New York Times* was concerned, killing the station would have been just fine, since it believed that *Freedom* was in reality a purposeless boondoggle, an expensive dog, that was unconnected to reality. "What a mess!" the *Times* barked on the eve of the vote. "The grandiose project, designed to be the next centerpiece of the ailing space program, has become ever more pinched in scope and vision, yet it remains dreadfully costly." The editorial went on to charge that such "breathtaking goals" as using it as an orbiting medical laboratory, a staging outpost for manned missions to the Moon and Mars, an astronomical observatory, and a gravity-free factory for making exotic drugs and materials had mostly become irrelevant because the cost of the station had escalated and schedules were stretched. The article all but said explicitly that NASA had grossly oversold *Freedom* and that the program amounted to a quagmire.

Indeed, the station was taken in many quarters, including both the news media and Congress, to be a symbol of a bigger, more pernicious problem. The journalists and the congressmen turned on the space agency like a lover who suddenly concludes that his or her sweet-talking counterpart has been perfidious all along. There had been the selling of the shuttle under patently false pretenses, the *Challenger* disaster, and the stonewalling that had followed it. The same basic line had been used to sell the space station, followed by a series of its own foul-ups and by persistent management problems, cost overruns, and what could only be taken as institutionalized deception. It was coasting on Sputnik and Apollo long after they had become irrelevant. "NASA is not connecting with the American people, and the agency is losing its relevance," Representative Alan Mollohan, a West Virginia Democrat, warned the American Astronautical Society at a meeting in March 1993. Unless it could "prove it contributes to U.S. economic competitiveness, politicians will view it as a cold war anachronism."

The Congressional Budget Office also sounded an alarm. In a report called "Reinventing NASA" that was issued exactly one year later, the CBO concluded that the space agency was trying to do too much with too little. Instead, the report suggested, NASA should more narrowly focus its activities: placing greater emphasis on manned spaceflight and exploration; emphasizing space science and using manned flights for scientific purposes; or emphasizing technology and missions with commercial potential and science with applications value while eliminating the manned program altogether. NASA would basically ignore that advice, but it would try to shift more responsibility to the private sector, which was another of the CBO's suggestions.

That same year, 1994, a slew of companies including Boeing, Martin Marietta, McDonnell Douglas, Rockwell, General Electric, Rocketdyne, TRW, Grumman, IBM, and Thiokol were working with various NASA centers to finalize the station design. Some were also placing advertisements in newspapers and magazines that said in so many words that the station would create jobs, or *could* help cure terrible illnesses, and would be wonderful for science. Boeing, a prime contractor with 730 employees working on the station and a contract worth more than $1.5 billion, waged a prolonged and aggressive public relations campaign for the *International Space Station,* as it was being called in 1994 (the Russians had by then contracted to supply the vehicle's service module and other hardware, plus help to service it, so the name *Freedom,* now a provocative holdover from the cold war, was quietly dropped.)

A "Progress Report" issued by Boeing early in 1994 raved about the science that would be performed on the station. It said that more than six hundred experiments had been proposed for the orbiting laboratory. A number of them, including the isolation of proteins in perfect crystals for use in a technique called crystallography, which uses X rays to examine the structure of atoms and molecules, promised a "significant reward" in fighting disease. Knowing that the station was taken to be a cannibal by most of the science community, the public relations people in the Boeing Defense & Space Group crafted a twenty-six page "Media Advisory" packed with the purported scientific benefits that they claimed would come from the station. The report was passed out at the important American Association for the Advancement of Science annual meeting in San Francisco that year. So were single-page flyers printed for individual states (Texas, Florida, Arkansas, Georgia, and California) asserting that "75,000 to 100,000 jobs depend directly or indirectly on Space Station nation-wide" and that it would ensure the country's economic competitiveness, inspire young people, bring breakthroughs in health care, monitor the environment and, as usual, provide a slew of "commercial spin-offs."

There were even comic books. *Adventures on Space Station Freedom,* which appeared in 1989 as the Evil Empire was starting to come apart, not only described the scientific and exploratory panacea the station represented but claimed that it would bring the peoples of Earth together (with the notable exception of Communists from the U.S.S.R., China, Cuba, and elsewhere). In one story, four young international science fair winners flew to the station on the shuttle *Discovery* and met their adult astronaut counterparts: Michelle, from the European Space Agency; John, a black from the District of Columbia; Juanita, a payload scientist from Houston; Divinity, from Canada; Elton S. Ono, a scientist from Japan; and others. (The similarity between Elton S. Ono and Ellison S. Onizuka, the Japanese American who had been killed on *Challenger,* was unmistakable.)

But no amount of public relations could obscure the fact that by early 1997, mounting delays caused mainly by cost overruns and by the Russians' inability to deliver the station's service module for lack of $100 million in appropriations by the Duma, was all but killing the possibility of doing serious experiments until well after 2000. The overruns forced NASA to dip into its own science fund and pull out almost half a billion dollars for use on other hardware, while the Kremlin and the legislature struggled not only to come up with the Russian Federation's share of the funding but to find ways of explaining to their devastated countrymen why they were willing to spend precious rubles on yet another space venture but not take immediate steps to reduce the misery and squalor on the ground.

The explanation was that partnership in joint space operations not only would continue to bring prestige to their country but would make more than it cost by leading to the sale of rocket engines, Proton launches, reconnaissance imagery, and other goods and services. Like Slavic Willy Lomans, the new Russian space merchants set off on capitalism's dangerous road trying desperately to get their collective foot into the world market's front door.

But the cold war had been hideously expensive, and its bill was now due from winner and loser alike. Most of the winners, many thousands of whom were being painfully "downsized" out of their jobs as the economy adjusted, tended to think that sending people to space in fabulously expensive machines for no compelling reason was as wasteful as it was irrelevant.

"It is disconcerting to look up from the bottom line of Form 1040 to read in your news article on April 5 [1997] that the Government I am financing has just launched another shuttle shot," a Columbia University law professor complained in a letter to *The New York Times.* "The primary scientific mission of this one, I gather, is to light a match in a weightless environment. The appetite of science is voracious, and a good thing, too, I'm sure. But how much (in this impoverished world) is it worth to play with fire while floating around in a spacecraft? You have written that the cost to us 1040 writers is between $300 million and $500 million per flight. . . . I wish you would keep telling us just how much each of these space games costs—especially those like the latest that seem to have been lofted mainly for the fun of it."

It was not for fun. The professor referred to *Columbia*'s sixteen-day science mission that had begun on April 4; he had possibly been put off by reading that a launch controller had told the crew, "Enjoy your on-orbit spring break." But it was no break. Of thirty-three experiments conducted during the flight, the most important involved setting scores of small fires in insulated chambers to see how they behaved in weightlessness.

This was anything but frivolous. Rapid decompression and fire are spacefarers' two worst nightmares. Flame acts far differently, and much more dangerously, in a weightless environment than it does in gravity. A candle that burns for five to ten minutes on Earth can turn into a ball of flame that radiates heat and light for forty-five minutes in orbit because of simple diffusion. Fire and its effects were tested in space as early as 1975 on *Skylab* and on shuttles in 1990, 1992, and 1996. The tests showed that fire spreads especially fast on paper and plastic in orbit and that polyurethane foam easily smolders with little exposure to flame and gives off much higher levels of carbon monoxide than on Earth. In February 1997—only six weeks before the fire tests on *Columbia*—a flash fire caused by a ruptured oxygen generator on *Mir* sent one-to-four-foot-long flames into the cabin, chunks of molten metal splattering against a bulkhead, and noxious smoke into the air. Since astronauts and a cosmonaut had died in fires even before they got to space, both the Soviet and American programs studied the matter intensively. NASA still does.

At any rate, the losers—the Russians—whose tattered ranks included frail old women begging for money on street corners, soldiers and scientists who went months without being paid, and a soaring adult mortality rate brought on by alcoholism, poor nutrition, and a health care system in near collapse, thought that sending people to space was contemptible if they thought about it at all. The *International Space Station* was taken to be an obscene indulgence by a cynical elite that was out of touch with ordinary people.

Cyclops in the Sky

The Hubble Space Telescope had a tortured history to match the station's. It was the brainchild of Princeton astrophysicist Lyman Spitzer Jr., who had first proposed it in a RAND report in 1946, and was named after the American astronomer who did pioneering work on the expansion of the universe.*

During the forty-four years before it was carried into orbit by *Discovery* on April 24, 1990, the largest, most complex, and most powerful observatory ever sent to space survived enough near-death experiences to make it the hero of one of the old dime novels. (Astronomer R. S. Richardson's idea to put a three-hundred-inch telescope on the Moon for the same reason Spitzer wanted to put one into orbit was in fact published in *Astounding Science Fiction* in 1940.)

By the early 1960s, shrewd NASA technocrats were supporting what was then called the Large Orbital Telescope and then the Large Space Telescope as an integrated component of their cherished shuttles and space station, but soon ran into the usual budget problems on Capitol Hill. Seeing the intensity of the politics that erupted over the telescope, largely because it was a vastly expen-

* Oberth suggested lofting a powerful telescope above the distorting atmosphere in 1923, but it was Spitzer who developed a concrete concept and assigned it four goals: measuring the extent of the universe; studying the structure of galaxies and globular clusters; and providing details on the nature of the other planets. Edwin P. Hubble proved that the universe is expanding in all directions as though its pieces had been shot out of a cosmic bomb like shrapnel. This became known as the Big Bang theory, which has been crucial in dating the approximate age of the universe.

sive "big science" enterprise, Fred Whipple quipped that it should be called the "Great Optical Device," which would have given it the politically appealing acronym GOD. Instead, it was renamed for Hubble in 1983, after nearly two decades of intermittent fighting in nearly every corner. There were battles with Congress over funding, within the space agency itself over its design and operation, and with the Department of Defense, which vehemently insisted that related Keyhole satellite technology such as optics and solar panels could not be shared with uncleared civilians. The engineers working on Keyhole knew that their counterparts working on Hubble would have problems—with its solar arrays, for example—but never told them, even though doing so would have had no bearing on national security. It was a classic example of how compartmenting information could drive up the cost of a program.

NASA's Office of Advanced Research and Technology fought with its Office of Space Science for control of Hubble's development. And the Space Telescope Science Institute itself, which would operate the telescope, was to be headed by Riccardo Giacconi, a brilliant, fiercely independent X-ray astronomer who engendered what one historian called a "love-hate relationship" between the institute and headquarters. Giacconi, whose boss at Harvard called him "aggressive even to the point of being difficult to deal with," would easily match Pickering as a source of irritation and frustration in Washington.

The white hat–black hat division was particularly ludicrous because the telescope was manufactured by the Lockheed Missiles and Space Company in Sunnyvale, California, which also built KH-9 "Big Bird" reconnaissance satellites at the same site (but in different clean rooms). In fact, the forty-three-foot-long, 25,500-pound orbiting observatory is roughly the size of the KH-9 and uses the same basic primary mirror–secondary mirror Cassegrain design as the KH-11s and their successors.

But NASA itself managed to turn what would ultimately become one of science's most extraordinary devices into a self-inflicted wound. By insisting that the telescope be both launched and serviced by shuttles, it not only made the device dependent on the fragile lifter to reach orbit (Hubble's launch was set back four years because of the *Challenger* disaster) but kept it leashed to a relatively low 330-nautical-mile-orbit because that was an altitude the shuttle could reach for house calls. A higher orbit would have doubled the observation time.

In 1976, when the space agency bent to congressional pressure to leave Hubble out of its budget altogether, the noted Princeton astrophysicist John N. Bahcall wrote to Fletcher to complain that it looked as if NASA was more interested in supporting its own institutional needs (the shuttle) than "wider goals." As it turned out, the cost of designing and building Hubble between 1977 and 1986—roughly $1.54 billion—was triple the original estimate. A big chunk of that money went to the Perkin-Elmer Corporation of Danbury,

Connecticut, which submitted a ludicrously low bid of $70 million to build what would turn out to be a badly flawed primary mirror. And in an acrimonious book appropriately titled *The Hubble Wars,* Eric J. Chaisson, an astrophysicist who worked on Hubble, charged that NASA both neglected sharing space telescope data with the nation's schoolchildren and outlandishly hyped its capability. He labeled as "patently false" the space agency's claim that Hubble would "see seven times farther than other telescopes." Hubble's great advantage, Chaisson knew, would be to operate so far above the atmosphere that it would see more clearly than ground-based telescopes, not farther.

If the Hubble Space Telescope's gestation period was convulsive, its debut was an out-and-out calamity. A month after it was launched, when it effectively first opened its eye in what astronomers call "first light" and peered at a star cluster known as NGC 3532 in the constellation Carina, it sent back not a clear image but a sharp spike of light surrounded by a huge halo and strange tendrils of light coming out of the central core. The picture, it soon became apparent to the thunderstruck scientists at the Space Telescope Science Institute in Baltimore and other insiders, was hopelessly out of focus. Six months later, a NASA board headed by JPL Director Lew Allen found that the "spherical aberration" had happened because Perkin-Elmer not only had polished the 94.5-inch primary mirror incorrectly but had not tested it enough. So much for what one of the firm's losing competitors called its "sinful" lowball bidding. It got $437 million for the flawed mirror.

Chaisson called what followed a "public relations cruise through hell" as the "techno-turkey"—the orbital Mr. Magoo—became the target of cartoonists, politicians, headline writers, and late-night comedians from coast to coast. "Pix nixed as Hubble sees double," one newspaper proclaimed, while another called the telescope "one sick puppy!" "*Hubble* is working perfectly," said the host of NBC's *Tonight Show,* "but the universe is all blurry." One cartoonist drew six dopey-looking people holding a sign upside down that read: "Hubble Space Telescope Optical Design Group." A *New York Times* editorial writer called the HST a "myopic chunk of orbiting glass." The House of Representatives was now as convinced as much of the public that big science was irredeemably quirky, wasteful, and mostly irrelevant because of ghastly blunders like Hubble. It was equally convinced that NASA was incompetent, so it put the kibosh on George Bush's year-old, $100 billion plan to send Americans back to the Moon and then on to Mars by draining all start-up money out of the space agency's budget tank.

The Ghoul Strikes Again

But, as usual, it was infinitely worse in Russia. On November 16, 1996—eight months to the day after a second Russian space auction was held at Sotheby's—

the Russians were hit yet again trying to get to Mars. Mars 96, a $300 million mission that involved twenty nations and a whole decade of planning (mostly because of budget problems), abruptly ended when its Proton booster's fourth stage failed to restart after its first burn, dumping the complex spacecraft into the Pacific.

The French and Germans were consternated. The members of the Russian Space Agency, who were by then desperately trying to reestablish at least some of the old credibility, and who thought of partnerships with the United States, European nations, and some others the way drowning people think of life preservers, went into collective shock. A number of them would have known—but would definitely not have told their foreign partners—that parts of Mars 96 had been integrated at Tyuratam in the glow of kerosene lamps because the Kazakhs had cut off the electricity in exasperation over a pile of unpaid bills. Nursultan Nazarbayev, Kazakhstan's president, said in 1997 that Moscow owed the equivalent of $460 million for four years' rent. That was only one symptom of a deep and widespread poverty and demoralization that was killing the program.

Russia ranked second to last in spending (after India) out of twenty nations that were active in space research in 1996, the Russian Space Agency's general director, Yuri Koptev, announced bitterly. The United States spent a little more than $12 billion on civilian space programs that year, while France and Japan each spent roughly $2 billion and Russia roughly $700 million. Of twenty-seven civil missions the space agency had hoped to conduct in 1996, he told Parliament while pleading for a $1.1 billion budget for 1997, only eleven had actually been launched. And between 1990 and 1996, Koptev added, still pelting the legislators with statistics, almost half of the engineers and technicians in the space program had deserted it because they could not live on an average salary of one hundred dollars a month. The sorry state of the Russian space program could have been the sole subject of a course in a business school. It showed in the clearest way where space goes on the priority list of a country that is severely squeezed financially. Like supporting the arts or caring for the environment, space was taken to be a luxury affordable only in good times.

The picture Koptev painted did not bode well for his country's participation in the *International Space Station* by then under advanced development, since it was supposed to contribute 38 percent of the hardware and the equivalent of $3.3 billion (versus $17 billion by the United States, $3.2 billion by ESA, and $3 billion by Japan). If funding did not increase, Koptev warned, there was a distinct possibility that Russia would bail out of the international station and abandon *Mir* and human spaceflight altogether by 2000.

It was no idle threat. As he spoke, *Atlantis* was being loaded with nearly 3,600 pounds of supplies for *Mir,* 43 percent of it food and clothing, plus 1,400 pounds of water. The Russian space station was now dependent on the American shuttle for resupply. That is how bad it was.

Then it got worse. Early in 1997, time and the severely shrunken budget began to catch up with the world's only space station, then in its eleventh year in orbit (it had been designed to fly for five). Life on *Mir* had never been especially easy. It had logged about 1,500 malfunctions since its launch in 1986, most of them routine, while orbiting Earth roughly 65,000 times. Michael Foale, an American astronaut who was in orbit with Vasily Tsibliyev and Aleksandr Lazutkin at the time, compared living on *Mir* to camping out in a dirty old car. The United States paid Russia $470 million to let astronauts build up flying time on *Mir* in preparation for crewing the *International Space Station.* Dr. Norman Thagard became the first American to fly on *Mir* when he went up on a *Soyuz TM* in March 1995. He was collected by *Atlantis* 115 days later. Shannon W. Lucid came back to Earth on *Atlantis* in September 1996 after setting a six-month record, and the same ship delivered her replacement, John E. Blaha, who started his own four-month mission that same day.

But *Mir* ran into the Ghoul in February and was severely pummeled. On the twenty-third, a backup oxygen generator, called a "candle" because its lithium perchlorate burns like one as it releases oxygen, caused a serious fire that sent molten metal flying and filled the station with smoke. The Russians downplayed the accident. But astronaut Jerry Linenger, who was on board at the time, said later that the long flames had threatened to burn through the station's wall. In March, the main oxygen generator failed and leaking coolant loops sent the temperature of the main living module up to a toasty eighty-six degrees Fahrenheit. In April, the main air-purification system began leaking antifreeze, overheated, and had to be shut down for fear of another fire.

And Again

Then, on June 25, the tempo of the orbital soap opera picked up dramatically when a Progress supply ship smashed into *Mir*'s Spektr science module as Tsibliyev guided it through a docking maneuver with a joystick on *Mir.* The unmanned spacecraft had brought up food and other supplies and was scheduled to be sent out of orbit and incinerated in the atmosphere after it and *Mir* parted. But Tsibliyev was unable to brake the approaching spacecraft's speed. It quickly caught up with the station and veered to the left, smashed into one of the Spektr's solar panels, and then hit the module itself, puncturing its skin. Such an accident would have been a fender-bender on Earth. But as Tsibliyev, Lazutkin, and Foale knew when they felt the whole station shudder and their ears pop during the rapid decompression, it was potentially deadly. More than two dozen cables that ran through the hatchway connecting the module and the rest of the station had to be disconnected in order to close and seal the airtight hatch between them.

Besides losing solar power from the Spektr's panels, the collision—the worst in the history of the space age—knocked *Mir* out of alignment with the Sun, so its other panels were not generating much electricity either. Without power, the computers that fired the thrusters that kept *Mir*'s attitude stabilized shut down, so it started to drift very slowly, like a tipsy derelict. The loss of electricity also quickly shut down the climate-control systems, again sending the temperature past eighty degrees and increasing the humidity. Then the carbon dioxide scrubbers and the oxygen generators stopped, also for lack of electricity.

Working with flashlights held in their mouths and breathing air from the finicky oxygen candles, the two Russians and the American soon managed to restore more than half of the ship's power and most of its life-support systems. And they realigned it by using the maneuvering jets on the attached *Soyuz TM* escape ship the way ocean liners are eased into docks by tugboats. But all three still lived precariously on a crippled station that was Russia's last great achievement in space and that would therefore not be abandoned unless there was no alternative.

Now one of the strangest episodes in the history of the space age—some would call it pathetic, some a farce—began to unfold while much of the world looked on, courtesy of the communication satellites, with growing interest. A few days after the collision, Tsibliyev and Lazutkin were ordered to repair the damage based on a quickly devised technique that cosmonauts working in the *Mir* mockup underwater at Star Town had devised. Special equipment, including a laptop computer, would quickly be sent up to them on another Progress. The plan called for the two cosmonauts to crawl through a hatch to the damaged and unpressurized Spektr (while wearing bulky space suits) and reconnect the cables to a modified hatch. Meanwhile, Foale would climb into the *Soyuz-TM* and prepare it for a fast getaway in case something terrible happened.

Knowing perfectly well that the assignment could cost him his life, Tsibliyev, a forty-three-year-old former Air Force pilot (and Israeli milk salesman), reacted with uncharacteristic bluntness. "I have never done this kind of work," he warned the controllers. "Without training, it will not be possible to do this job." On July 14, after the Progress emergency supply ship had safely docked, he complained about his heavy workload, a loss of sleep, and, more ominously, about an irregular heartbeat. It would turn out to be stress-induced. The next day, *Mir*'s exasperated controllers asked NASA's permission to allow Foale to do the tricky repair, and two days later it was granted.

Meanwhile, *Life* played the family angle, just as in the good old days. "On a clear night in June, a woman and two children wave at a light in the sky: 'I love you, Daddy,' the kids call out. Daddy is Michael Foale, 41, the astrophysicist and astronaut who, 237 miles above this Texas backyard, is weathering crisis after crisis on the 12-year-old Russian space station *Mir*. The woman is his wife, Rhonda, and the children are Jenna, six, and Ian, three." It ran a picture of Rhonda Foale watching the news of her husband's precarious mission on the floor of their living room while Jenna and Ian slept beside her. "That night, we never did pull in NASA's feed of Mike's spacewalk," she told a *Life* reporter. "The walk was the one thing that worried me. He's in a beat-up old spacecraft, and then he's getting out of it. But I just have to pray about things I can't control."

On the following day, the sixteenth, there was yet another bizarre accident. A cosmonaut accidentally unplugged one of the cables that had originally been reconnected by flashlight, again knocking out the computer that kept the solar panels pointed in the right direction. "This," muttered Vladimir Solovyov, the exhausted and thoroughly disgusted mission director, "is a kindergarten." The remark would find a place in the folklore of the space age.

That did it. A two-man relief crew was dispatched to *Mir* on August 5 and docked two days later (manually, since there was yet another computer failure). Leaving Foale behind, Tsibliyev and Lazutkin—the "hapless" crew, as one newspaper account put it—climbed into the *Soyuz* and pulled away from the stricken station. "Thank God," the forlorn Tsibliyev said when he was sure that his *Soyuz TM* and *Mir* had separated. He and his relieved flight engineer landed safely on the Kazakh steppe three hours later (but only after the landing, or touchdown, rockets on the *Soyuz* misfired).

And it did not end there, either. Like the Chernobyl explosion, the Phobos fiasco, and innumerable other accidents, blame was immediately placed on those who were operating the equipment, not the equipment itself. "We have found not a single fault in the systems of either *Mir* or Progress during the collision," deputy mission director Viktor Blagov, who became the operation's unofficial spokesman, said at a hastily called news conference within hours of the cosmonauts' arrival back on Earth. "A logical conclusion follows: technology was not at fault there," he added, all but saying that responsibility for the embarrassing fiasco had been Tsibliyev's. Yeltsin, also covering for his country's tarnished technology, went a step further. The problems on *Mir,* he said publicly and without having the slightest idea of what he was talking about, had been caused by "human error."

Tsibliyev knew perfectly well that he had commanded an antique—"a valuable space jalopy," as one American editorial put it—and he also knew that shifting the blame to someone else had been a staple of his country's culture since at least 1917. But he was not about to let himself be made to look like an incompetent fool, to have his career tarnished for all time, by some bunker-bound technocrat whose idea of personal danger was riding the Metro during rush hour. So he reacted with wrath at his own news conference two days later.

"It has always been a tradition here in Russia to look for a scapegoat," said the defiant cosmonaut, who wore a red, white, and blue tracksuit with "Rossiya" printed on the back. "Of course it's always easier to put the blame on the crew. But in this case, that's not fair," he added in a patient but firm voice. Tsibliyev reminded the reporters that *Mir* was a jury-rigged flying tool kit that had gone six years beyond its design life. Then he himself assigned blame for his ship's problems. "It all comes from Earth. From our economy, our affairs, our poor lives. Even the equipment needed to live aboard the station that we requested to be sent—and we're not talking about coffee, tea, and milk for us—it just doesn't exist on Earth. Simply, the factories don't work, or have insufficient supplies, or the prices they want are too high for us to afford. It can be crazy. The truth is," Tsibliyev continued, "that many things we need on the station just aren't there. Day in and day out we are doing everything we can. It is hard to understand the attempts that were made to accuse us. People said I complained about my health. It wasn't a complaint; it was a real problem. You can't give orders to your heart. We should have abandoned the station three separate times," he added, "but we never thought about jumping ship. Not once."

A few days later, after still another computer failure again sent *Mir* wobbling and forced its new crew to steer by the light of the stars "like medieval sailors," Blagov got sick and tired of his work being treated like an ongoing cosmic joke. He knew that he and his colleagues, including the cosmonauts, had really all been sabotaged by the system in which, as a matter of pride, they tried so hard to function. So he closed ranks with Tsibliyev. Publicly. "We used to change *Mir*'s computer parts after their technical life expectancy ran out," he complained bitterly. "Now, because of money, we must use each part until it dies," he said, adding that some of the parts had not been changed or even repaired during *Mir*'s entire life. "We are saving a lot of money the way we work. But soon we really have to decide if we need safety or whether we need to save money."

The new crew soon got *Mir* under control again, and by August 22 the major repairs were completed. The station's daunting physical problems finally ended, at least temporarily. Tsibliyev's and Lazutkin's, however, continued. They could not even get paid for their mission until December, though at a reported $100,000 and $80,000 respectively, they did finally find some consolation in having suffered through their ordeal.

"Peace" at Any Price

While Tsibliyev and his crewmates struggled to keep their crippled station livable, the political soup caused by the accident—or rather the series of them—bubbled on Earth. Descriptions of the collision, usually accompanied by vivid diagrams, made page one of most serious Western newspapers and were played on radio and television. Under a boldface headline that screamed "Moscow, We Have a Problem," *Newsweek* solemnly reported that "in orbit, three men—two Russians, one American—struggle to survive a terrifying mishap. Inside the debacle."

Lost in those immediate reports of the "debacle," and in editorial cartoons that portrayed the spaceship named for peace as the scene of one cliff-hanging ordeal after another, was the fact that many other, similar problems had taken place over the years but had gone unreported. More important, in the rush to report the new mishaps, *Mir*'s considerable accomplishments were usually overlooked. They made for boring reading in the morning paper, but, taken together, they filled in a vital part of the blueprint that was supposed to send humans to Mars and beyond. An enormous store of information on humans' physical and psychological reactions to long flights in space had built up during its 65,000 orbits: data that showed, for example, how nutrition and exercise could substantially reduce or eliminate changes in body fluids, bone marrow, heartbeat, and the other effects of prolonged weightlessness. There were even experiments showing that wheat and other nutritionally crucial plants could be grown in space for generations. That knowledge would be invaluable, not only for extended flights, but for use on giant man-made space platforms and in terraforming Mars and any of a number of moons: that is, building large, shielded colonies on other worlds. But the library of accomplishment was lost in the noise.

Representative James Sensenbrenner, the Wisconsin Republican who headed the House Science Committee, quickly called on NASA to "re-examine the balance of value versus risk" in sending more Americans to fly on *Mir.* Two more were in fact scheduled to go. There was both more political value and more risk than met the eye. Certainly the death of an American and his or her Russian crewmates were obvious risks: three more road kills on the highway to the stars. But like a matryoshka doll, there were political reasons within reasons to keep *Mir* flying and to keep Americans on it.

The smallest reason, but one continuously pitched by NASA, was that every problem that developed and was solved on *Mir* was a learning experience that would be invaluable in case similar problems developed on the *International Space Station.* By working together under normal conditions and in moments of severe strain, the argument went, astronauts and cosmonauts were inventing a modus vivendi that would become the linchpin of civilian space operations in the new millennium. That was unquestionably true. But it also managed to put a healthy spin on all of the dire dilemmas that were arising in orbit. It prepared the public for the possibility of catastrophe, an idea that would have softened the blow from *Challenger.*

The next-biggest reason had to do with keeping the Russian Federation involved in the *International Space Station* itself, and therefore cooperating with the international community in space. It was no time to isolate a broken and humiliated giant (particularly one that still clutched a nuclear club). Russia's part of the station was the crucial service module. While no one seemed able to set a final, realistic price on the thing, it would probably cost on the order of a half billion dollars. But their devastated economy had so badly eroded work on the module that the desperate—and always embarrassed—Russians alternately proposed a cheaper module based on an advanced *Mir* design or simply stonewalled when they were pushed on an original design by their partner nations. None of the other participants much liked the modified *Mir* idea, partly because it would use old technology as an integral part of a new system, partly because there would be still more time slippage that could end up killing the project altogether. As it was, launching the first hardware had slipped from 1997 to mid-1998 and would slip some more. The Khrunichev State Research and Production Center, the service module's prime contractor, started work on the heavy main tube thinking it was going to get a $55 million payment to complete its part of the $17 billion project.* But work ground to a halt when the Russian Space Agency paid it only $10 million.

That sent a ripple through Washington, where annoyed congressmen, including Sensenbrenner, a longtime supporter of the space station, worried that NASA would be left holding the funding bag. The irritated stir on Capitol Hill in turn worried NASA because of the fear that if the Russians actually pulled out of the station, Congress might kill it outright rather than see the United States absorb the additional cost of the service module itself. As it was, Boeing, the station's prime contractor, was itself sliding precariously down the financial slope. A General Accounting Office report issued in September 1997 said that the company's cost overrun had quadrupled to $355 million between April 1996 and July 1997. "The station program's financial reserves have also significantly deteriorated, principally because of program uncertainties and cost overruns," the report concluded.

But American taxpayers would absorb the additional cost of the service module and other hardware anyway through an elaborate series of subsidies that totaled $600 million by early 1996, $200 million of which Boeing paid Khrunichev to build the craft's power hub. The most important reason for paying the Russian Space Agency to fly seven U.S. astronauts on *Mir* and carry supplies to it in the process was not, in fact, so they could train for the *International Space Station,* though that was obviously important. It was to launder

* That figure is from the General Accounting Office report, "Space Station: Cost Control Difficulties Continue," p. 2. Other estimates have reached $27 billion ("Station Problems Test U.S. Russian Resolve," *Aviation Week & Space Technology,* p. 20), $35 billion ("Russia, Vexing Partners, Asks for Changes in Space Station," *The New York Times,* December 10, 1995), and even as high as $73 billion ("Moscow, We Have a Problem," *Newsweek,* July 7, 1997). The last is very improbable. As is the case with the shuttle, the total cost of the station depends on what is included.

money that would be used to complete the service module so there would *be* an *International Space Station.* Yet even pumping in all that funding was not helping. By March 1997, with *Mir*'s problems unfolding remorselessly, the Russians were still breaking agreements as they desperately tried to juggle an expensive presence in space with a sputtering economy.

The third, still bigger reason, was to shore up Russia's self-esteem. Involvement in the space station played out against the larger theme of political isolation, the loss of the empire, the disintegration of the military (highlighted by the embarrassing quagmire in Chechnya), NATO's senseless and provocative expansion to the Russian frontier, and a standard of living that was sinking. At the same time, the country was being overrun by evangelists of every persuasion, Buddhists and other religious zealots who knew a spiritual vacuum when they saw one (or thought they did), art collectors who scoured the markets for bargain icons and other precious cultural treasures, salesmen from Coke, Pepsi, and scores of breweries and distilleries from Milwaukee to Manila, and the ubiquitous junk-food emporiums. Moscow had the ignominious distinction of having the biggest McDonald's (Mac-DON-Aldz) on the planet. Having to import French art, German scientists, and Italian architects had been demeaning enough. Now they were importing Beeg MHAK . . . FeeLET O Feesh . . . and de CERT.

It was therefore considered important to get the country through capitalism's first, ugly flushes with as much dignity as possible lest it succumb to the ever-present sirens on the far right. And the space program, which carried the last vestige of nobility, was the logical place to start. So cooperation began to move in a number of directions, most of them having little or nothing to do with the space station itself. On February 3, 1994, a cosmonaut named Sergei K. Krikalev rocketed off Cape Canaveral in *Discovery,* making him the first Russian ever to be launched on an American spacecraft.

Coincidentally, but significantly, Japan launched its own H-2 heavy lifter for the first time the next day. The 164-foot-high rocket, with "Nippon" painted in large letters on its side, was the first big rocket built solely with Japanese technology. It took off from that country's sleek launch facility at Tanegashima, an island in the south, and sent the clear message that Japan had its own long-term plan for a future in space. Generally missed by the public at large was the significant fact that three nations, not two, had reached the Moon. A pair of small orbiters from Japan reached the Moon early in 1990. The spacecraft, Muses-A and Hagoromo (Veil of an Angel) sent no data home because of transmission failure. But they did prove that Japan had the wherewithal to make long strides in Earth orbit and beyond. Indeed, by mid-decade, plans were under way for Japan to be the third nation to reach Mars, with a "Planet B" probe that was scheduled to be launched in July 1999. The famously ambitious Japanese understood that the prestige of reaching the Red Planet would benefit their competitive position on Earth. Their fleet of busy communication and Earth observation satellites was enlisted for more practical uses such as television relay and weather prediction. And by 1994, the national defense institute was recommending that Japan launch three reconnaissance satellites in conjunction with other Asian nations.

Meanwhile, joint business deals were struck between Lockheed and Khrunichev-Energia to build Proton launchers that would compete with France's Arianespace. Other American firms, including McDonnell Douglas, which went to Russia in 1993 to get experience building lunar and planetary rovers, did the same. Pratt & Whitney, the venerable jet engine producer, teamed with NPO Energomash to market two highly advanced and reusable Russian-designed rocket engines, one of which ran on liquid oxygen and kerosene and developed more than 1.5 million pounds of thrust. Klaus Riedel, Johannes Winkler, and the rest of the VfR would have been stupefied. Shuttle-*Mir* missions were started, as was the practice of flying each other's supplies to orbit. In 1995 the United States took the nearly unthinkable step of sharing some reconnaissance satellite imagery of Bosnia with the Russians since Russian soldiers were there with Americans. It was enough to send any number of old cold warriors after their smelling salts.

Elsewhere, the United States was listening attentively while Russians pitched a joint missile early-warning system, and both space agencies were studying the possibility of launching a "Mars Together" Russian rover and American orbiter in 2001. The idea behind it all, including participation in the station, was not only to tap Russia's considerable expertise in rocketry but to keep it actively engaged in the world community rather than risk its turning into an isolated renegade.

And that led to the most important long-range reason for keeping Russia on board the space station: political leverage in arms control. Even as the imbroglio with Russia on the station and the string of fiascos on *Mir* continued, Washington was squeezing Moscow to keep long-range ballistic missile technology out of Iran. Help with the missiles, which would be able to reach Israel, Saudi Arabia, and U.S. forces in the Persian Gulf, was coming from the scientific institutes and companies that had been an integral part of the state-owned Soviet military complex and that had begun to starve after the cold war ended and contracts dried up. (Pressure to get them to stop selling weapons, the Russians noted ruefully, came from a superpower whose economy was booming in part because it was the biggest arms exporter in the world.) But involvement in the station and other programs, it was hoped, would keep bread and butter on the scientists' and engineers' tables without their having to sell weapons, including nuclear ones.

17

The Second Space Age

The cold war was over. The great space race was over. And the first space age was over, too. It was an age born out of the unlikely coincidence of a political and military confrontation between two Goliaths and the ripening of the rocket. It was marked by often brilliant technical innovations like the reconnaissance satellite, by colossally expensive and wasteful posturing—going to the Moon for the wrong reason and then abandoning it, for example—and by daring and soul-stirring feats of exploration to both the Moon and the planets. While going to space brought a scientific windfall and science was used as a justification for the trip, most scientists knew that their participation was basically a respectable cover for dark objectives. It is unlikely that the exploration of Mars and the outer planets would have happened so soon without competition between the gladiators. And what James Van Allen and his colleagues learned from the V-2s and Vikings at White Sands went straight to warhead designers as well as to meteorologists. Certainly billions upon billions of taxpayer dollars would not have been spent to send twenty-four government employees to another world except in a near-desperate race for prestige. The competition was as good as it was wasteful. Somewhere along the way, though, the road from Tsiolkovsky to Tomorrowland took a detour.

Yet there were three lasting developments that grew out of the first space age and set the basis for the second. For one thing, people from the Australian outback to Greenland accepted the fact that the planetary civilization had not only taken to space, but was going to stay there. A presence in space, whether human or robotic, had become a universal norm approaching the acceptance of travel on land and sea. Second, where the military was concerned, space was seen as a new environment for war fighting, and it therefore needed to be controlled. The surviving gladiator, its opponent vanquished, was determined to be master of what it appropriately called the "new arena." Finally, the use of space, and particularly low Earth orbit, had become fundamentally institutionalized. In other words, the use of space to serve society's needs had undergone a transition from luxury to necessity.

Securing the High Frontier

From the military's point of view, access to space, like the ability to make nuclear weapons, was a rapidly spreading capability, and no one was going to change that. There were tremendous economic incentives to have both many kinds of satellites in orbit and to be able to get them there. That is why competitive consortiums sprang up among American, European, and Russian companies and why the Chinese and Japanese went into the launch business on their own with the powerful Long March series and H-2 boosters, respectively. With China, France, India, Israel, Japan, and, of course, Russia already able to reach space on their own, and with many more countries eager to follow, Space Command and its constituent organizations took it as doctrine that even if future wars were not actually fought in space, space would certainly figure decisively in their outcome.

When the military looked into the future, it saw replays of the short but vicious war against Iraq in 1990–91 following Saddam Hussein's invasion of Kuwait. The Air Force quickly took to calling that confrontation the "first space war" to get it into congressional heads that the combat arena now and forever extended to space and that the operative goal was total control. "Unless we have a sound space-control capability, we may find ourselves in a conflict with a nation with space forces while we have no means to prevent space-supported attacks on ourselves and our allies," Air Force General Donald Kutyna, the head of Space Command, warned the Senate Armed Services Committee three months after the Persian Gulf War ended.

"In Your Face from Outer Space"

At the same time, understandably, the Air Staff believed that its service was uniquely qualified to control space. By 1994, the Air Force had crawled under the covers with its old nemesis, the CIA, and was trying to create a Joint Space Management Board to be cochaired by both organizations, leaving the Army, Navy, and Marine Corps to participate only on an ad hoc basis. That year it also put the finishing touches on a Space Warfare Center in Colorado Springs whose job was to "support combat operations through control and exploitation of space": in other words, to fully use space to help the nation's armed services win wars. One of the hundreds of ways to do that, for example, would be to relay real-time reconnaissance satellite imagery right into the cockpits of fighter-bombers so their pilots could see targets beyond their line of sight and therefore fire their missiles before missiles were fired at them. The center's motto reflected its leaders' attitude: "In Your Face from Outer Space."

The Space Warfare Center was invented because of Desert Storm, when Iraqi forces were in effect blinded, decapitated, and obliterated by orbiting spacecraft as much as by bombs ("smart" and otherwise) and Tomahawk cruise missiles. Reconnaissance satellites located targets, including the overrated Republican Guard, exposing them to punishing aerial bombardment and then to General Norman Schwarzkopf's renowned flanking action, or left hook. Other satellites eavesdropped on signals transmitted by the Iraqis. Still others provided continuously updated data on the weather; instantaneously relayed communication between field commanders and their headquarters and between headquarters and the national command authority in Washington; guided both cruise missiles and manned aircraft to their targets; and, from way out at geosynchronous, tracked the Scud ballistic missiles launched at Israel and Saudi Arabia and sent warnings to those countries almost instantly. In military jargon, the satellites were "force multipliers."

But if the United States could use space to inflict devastating damage on an enemy, the reasoning went, the proliferation of launchers and satellites worldwide meant it was only a matter of time before some other nation would be able to do the same to the United States. Had Iraq been able to use space reconnaissance to spot the VII Corps's left hook and a simultaneous feint by sea, for example, the outcome of Desert Storm would have been profoundly different. "My biggest nightmare is that one day I may end up watching [Cable News Network] as an entire Marine battalion landing team is wiped out," Air Force General Charles A. Horner, another head of Space Command, said in 1993. "Why? Because we were unable to deny the enemy space-based intelligence and imagery." Horner's replacement, General Joseph W. Ashy, was even more emphatic (if a bit Strangelovian): "The United States will—and I'm not trying to promote war here, I'm just saying that's what humankind is like—eventually fight from space and into space" to protect its vital military and commercial systems. The general and many of his comrades in arms increasingly drew parallels between space and the high seas in the eighteenth and nineteenth centuries, when merchant ships had to be protected from marauding pirates and enemy nations; when the sea-lanes had to be kept open for commerce and defended against predators. In 1997, with two hundred active American satellites worth more than $100 billion in orbit (out of a total of some five hundred), the unified U.S. Space Command proposed to the Department of Defense that space formally be turned into a separate area of military responsibility, after land, sea, and air: a "fourth medium," as another Air Force general put it. As the Pentagonians saw it, freedom to conduct commerce on land, on sea, and in the air had historically led to the creation of the Army, Navy, and Air Force to protect those realms. Space was an extension of them. Referring to both his country's civilian and military satellites, Air Force General Howell M. Estes III predicted that "somebody is going to threaten them. And when they [do], we [should] have armed forces to protect them. So, it's a natural evolution."

The way to prevent the targeting of U.S. civilians and military forces was clear: develop the capacity to disable or destroy any spacecraft that threatened them. Work on ASATs—anti-satellite weapons, especially lasers—that could clear the sky of hostile satellites as soon as a threat materialized gathered momentum during the late 1990s. Ironically, part of that threat was America's own doing. The Clinton administration decided in the winter of 1994 to make the nation's depressed aerospace industry more competitive with those of France, Russia, and other spacefaring nations by allowing it to build reconnaissance satellites for countries that were judged to be friendly to the United States. The contracts were potentially worth hundreds of millions of dollars. This raised the bizarre possibility that in time of war the United States would not only destroy satellites that it itself had built and sold but in the process create a market for their eventual replacements. Capitalism triumphant.

A year later, on February 24, 1995, an executive order signed by President Clinton authorized the release of some 800,000 Corona pictures taken from the beginning, in August 1960, to 1972. Making the imagery public not only helped to legitimatize building spy satellites for other countries—all of those detailed pictures were now widely available anyway, the argument went—but it was useful for studying the environment. One shot, which went out in a press kit, showed a dramatically shrunken and damaged Aral Sea.

An invitation-only conference, "Piercing the Curtain: Corona and the Revolution in Intelligence," was sponsored by the CIA and George Washington University on May 23–24, 1995. It was attended by Mert Davies of RAND; Bud Wheelon, who had been the CIA's first deputy director for Science and Technology; Dino Brugioni, the crack photointerpreter who had managed the National Photographic Interpretation Center during the Corona years; Richard Garwin, who had been on the president's Science Advisory Committee during that time; and other fathers of the world's first space espionage system. Also present were the uncleared intelligence junkies who had spent years diligently trying to pry out the secrets. The first great coming-out party for space reconnaissance was a true happening; the reconnaissance world's equivalent of Viking's landing on Mars or Voyager's flyby of Jupiter.

The CIA was less upset about shedding light on its early spy satellite operations than might be imagined. The new reality was that, with the Soviet threat gone, it faced a challenge that was unique since its founding in 1947: justifying its existence. "Langley," as the agency is obliquely called by most of those who work there (it is actually located in McLean), had gotten a series of highly publicized black eyes, the worst of which was inflicted by Aldrich Ames, its director of operations, who was arrested in 1994 after passing important secrets to the KGB and its successor, the Foreign Intelligence Service, for nine years. The spy agency's vexing public relations problem was that its blunders invariably landed on page one, while its successes had to be kept secret. With the cold war over, and with tales of incompetence and betrayal floating to the surface like dirty oil, there were calls for the CIA to be shut down. The agency therefore welcomed the release of the Corona pictures because they helped to justify its existence and bore testimony to its technical prowess.

Similarly, the National Reconnaissance Office went public on September 18, 1992. Within four years, it had done something that would have been unimaginable to the intelligence junkies, not to mention its own employees, only a few years before. It issued a slick brochure calling itself "The Nation's Eyes and Ears in Space" that not only briefly described its history, mission, and organizational structure but listed its first sixteen directors, asserted that the digital techniques it had developed to analyze real-time satellite imagery had a spin-off in detecting breast cancer, and urged anyone who wanted more information—including new "customers" in government—to write or use the NRO's Web site (www.nro.odci.gov).

The first civilian "spy" satellite, Earlybird 1, was built by Earthwatch Inc. of Longmont, Colorado, and launched at Svobodny on a Russian commercial rocket on December 24, 1997. Earthwatch, which was in partnership with Japan's Hitachi Ltd., kept the spacecraft for itself rather than sell it to another country and planned to market its ten-foot-resolution imagery for $300 to $725 a picture. This marked the beginning of an era in which anyone with a credit card could look at virtually anything, from a military installation to a rival's factory. Experts predicted that Earlybird 1 would be followed by at least half a dozen others within three years. Its taking to space, said Earthwatch's president, was "a vivid reminder that the cold war is over." And that a new age was under way. Next, Lockheed-Martin would land a contract to sell a spy satellite to Japan.

Meanwhile, work also accelerated on ASAT's first cousin, anti-ballistic missiles, which took on their third life in large measure because of the Iraqi Scud attacks, one of which killed twenty-nine U.S. servicemen and -women in a barracks in Saudi Arabia. By 1997, Reagan's Strategic Defense Initiative Organization had been scrapped by the Clinton White House, which wanted to distance itself from John Wayne–style jingoism and the Republican right. Yet the specter of an accidental Russian launch or a deliberate attack by a crazed and fanatical Islamic madman—invariably portrayed as a kind of cross between Saddam Hussein, Libya's Muammar Qaddafi, and Iran's Ayatollah Ruhollah Khomeini—so haunted the military that a Ballistic Missile Defense Organization was established to study so-called theater defense: weapons that could be used in the battlefield to protect U.S. and friendly forces against ballistic and cruise missile attack from hundreds of miles away. A number of research programs were under way by then, some of them concentrating on high-altitude "exoatmospheric" defense, while others were focused on knocking out lower-flying missiles. An infrared-guided "exoatmospheric kill vehicle" was under development. Naturally, the leap from theater defense to national defense was only a matter of degree, so a National Missile Defense—NMD—system to protect entire countries slowly began to coalesce. The idea this time was to produce between twenty and one hundred interceptor missiles to knock out from five to twenty warheads. Star Wars was playing again in two theaters: tactical and strategic.

And Dyna-Soar was dug up, too. If the United States was going to be able to control space (or achieve "superiority" there, as the more diplomatic euphemism had it), it allegedly would also need a manned spaceplane: a fighter-bomber that could fly to any place on Earth or into low orbit in less than an hour for reconnaissance or to attack with the deadly speed of a cobra. A team of Air Force and NASA engineers was working on the concept even as the ASAT and missile defense work accelerated. The space agency was involved because it was already planning a transatmospheric replacement for the shuttle. Called the X-33, or reusable launch vehicle, the new spacecraft would be smaller than the shuttle. But unlike either Dyna-Soar or the shuttle, it would be a seed-shaped lifting body that could take off and land like an airplane. The Air Force's spaceplane, which was supposed to finally turn into reality the old dream of extending the domain of warships from the water ocean to the air ocean to the space ocean, would also be a lifting body. And there would be no Russians around to challenge it, either.

The United States prepared to enter the new millennium with "milspace" being not merely an adjunct to ground, sea, and air forces—supplying communication links, weather forecasts, and intelligence on an ad hoc basis, for instance—but as a fully integrated and always available part of the whole war-fighting system. "There was a time when, if you said 'space weapons,' you were politically incorrect," said General Thomas S. Moorman Jr., a former Air Force vice chief of staff. "But now, people talk about it . . . and it's accepted that we should be pursuing enabling technologies so we're better prepared [if] and when the threat materializes." His colleague, General Estes, the head of Space Command, added that sheer military ambition was not the reason the United States was arming the heavens. Rather, it was the explosive growth of commerce up there. "There's a lot of money going into space, and a lot of it is U.S. dollars. There's such an economic investment in space that it will soon be a vital national interest—and certainly an economic center of gravity—for the U.S.," he said. Controlling space, the general added, was therefore vital to the national security of the United States. A constant military presence in space was becoming, as he put it, "part of the normalization process."

Virtual High

NASA, meanwhile, was finally caught in the conundrum that had stalked it for years. Its bread-and-butter programs—the shuttle and station—were drawing yawns from citizens who were so used to seeing astronauts cavorting in weightlessness that they hardly paid attention. "Though there are actually some NASA astronauts up in the sky this very minute," *New York Times* columnist Frank Rich observed during an *Atlantis-Mir* rendezvous in the summer of 1995, "they are hardly the ones captivating the country. The astronauts of choice are Tom Hanks and Kevin Bacon—not whatstheirnames. The space vehicle of the moment is the ill-fated Apollo 13 of 1970—not the trouble-free shuttle *Atlantis,*" he added, referring to the film that depicted Lovell's, Haise's, and Swigart's brush with death on their way to the Moon. "The sense of national purpose and shared sacrifice that first ignited the space program [has] vanished, replaced by an ethos in which the immediate gratification of one's own self is a far higher priority than any joint venture, to the moon or anyplace else. In an America that values inner space over outer, the solitary technology of virtual reality—a simulated space trip via movie, Disneyland ride or computer—is what makes us high, not the inconvenient and costly national pursuit of the real thing."

True. Yet virtual reality was better than no reality at all; even seeming to be there was preferable to being utterly grounded. With the patriotic card now as useless as a spent booster, NASA had to find other ways to bring space and the citizenry together, and do so with fewer people and a plunging budget. Even as Rich reflected in the glow of space programs past, a newly elected Republican majority in Congress was demanding a balanced budget.

Daniel S. Goldin, who became NASA's ninth administrator in April 1992, and who understood that he would have to let blood or have it done for him, went on record as saying that the agency would indeed restructure: it would cut more than 3,500 civil service jobs by the end of the century, reducing NASA employees from 21,000 to 17,500. He also promised the irate young conservatives and their congressional colleagues that another 25,000 contractor employees would be downsized, all ten centers would be made leaner and more cost-efficient, and a process would be started that would eventually turn the shuttle program over to private industry. Allowing for inflation, which was very low, the budget would drop from $14.3 billion to $11.7 billion in five years. After the cuts, Goldin said, his agency would be the size it had been in 1961, before Apollo. Then, reacting to a call for even deeper cuts, the embattled administrator foreshadowed Koptev's dire warning two years later when he told the House of Representatives that still more trimming would force NASA to consider "shutting down a combination of enterprises, programs and centers," putting his agency on a "road to disaster." It was the same political tactic that had been used successfully to head off the dismantling of solar system exploration in 1981. What California congressman was going to vote to close JPL, Ames, or the huge antenna at Goldstone? Who was going to bleed TRW and Lockheed-Martin?

It should be said of Goldin, who came to NASA from a quarter of a century at TRW, that he inherited a sluggish, free-spending agency in which self-satisfaction was deeply impacted because of dazzling victories on the Moon, Venus, Mars, and the outer planets. He understood implicitly that cost overruns were endemic at NASA and, worse, that they were continuing because of a combination of hubris and inertia that had to be reversed. To that task he brought a tough, explosive management style that was almost reminiscent of the adage about not making Stalin's eyebrows move. Those who worked closely with him knew that his brusque but composed manner could suddenly turn into rage. "When crossed," one intimate confided, "he's like a jealous housewife." But it worked. When it became clear that heads would really roll unless the budget process was brought under control, it was brought under control. The "jealous housewife" got his managers' attention.

At the same time, the beleaguered space agency forged ahead on the station with the debilitated Russians in tow while beefing up science and exploration in the hope of satisfying the science community and capturing the popular imagination at the same time. As the end of the millennium neared, science and exploration were certainly doing their part, even with depleted budgets.

Space the Institution

By the end of the last century of the millennium, the satellites in orbit around the Earth could be said to be its central nervous system and sensory organs.

In 1997, with shuttle service in its sixteenth year and the space station's parts being assembled in Europe, Canada, Japan, and the United States, a formidable number of satellites formed a vast electronic shell around Earth that extended from low orbit to geosynchronous. And many more were shortly to come. Very many more. No one in spaceworld any longer referred to the cumulative number as a group, or even a fleet. The new buzzword was "constellation," as in sky full of stars.

There were watchers, civilian and military (or both) from the United States, the Russian Federation, China, India, Israel, France, Japan, and the European Space Agency, several of which routinely photographed virtually every square mile of the planet. They oversaw wildlife management, weather forecasting, civil engineering projects, urban planning, pollution control, natural disaster assessment, man-made disaster assessment (the fires set by withdrawing Iraqis in Kuwaiti oil fields and related damage done to Manifah Bay early in 1991, for example), archaeological investigation, political and military espionage, targeting, and arms-control compliance.

NASA had pioneered remote sensing with the launch of the Earth Resources Technology Satellite, later renamed Landsat, in 1972. Three years later, stunningly detailed Landsat pictures were published in an extraordinary *Photo-Atlas of the United States* that for the first time showed broad expanses of the country not just imagined from space, but actually photographed from there in a detail that seemed miraculous. One of the never-seen-before images, for instance, showed what were then incredible views of Seattle, in which the University of Washington, Mercer Island, Sea-Tac International Airport, and other sites were clearly visible.

In that spirit, and knowing that many in Congress wanted a more immediate return on investment than the country was getting from astronauts in orbit, the space agency had conceived an audacious program to monitor Earth's vital signs in unprecedented detail in 1989 after George Bush made a multibillion-dollar Earth Observing System the heart of a new U.S. Global Change Research Program. The idea, a very ambitious one, was to orbit a fleet of large spacecraft loaded with instruments that would thoroughly monitor Earth and send down data on thousands of changes in its biosphere the way a battery of medical instruments described a patient's condition. The key was to do the monitoring over decades so that global warming, deforestation, atmospheric and ocean pollution, and other changes and dangers could be measured and analyzed.

But Mission to Planet Earth, as the Global Change Research Program was soon called, like much of the rest of the space program—including the military—ran a budgetary gauntlet for the next eight years as it was attacked, mainly by Republicans, for being poorly defined, lacking immediate benefit, and not returning what Senator Larry Pressler of South Dakota called "a dollar back for every dollar we put in." The mission was restructured three times by 1995 as appropriations were progressively cut from $18 billion to $7.25 billion. That year and the next, a new Congress, dominated by conservatives who were determined to balance the federal budget and return more decision making to the states, attacked Mission to Planet Earth. By the end of 1997, it was still in the planning stage, while Landsat, a series of less ambitious NOAA monitoring satellites, and counterparts from Europe, Canada, Japan, and Russia were still flying.

Flash Gordon, CEO

The dedicated eavesdropping spy satellites were still there, too, and so were the ballistic missile early-warning types and the clunky military weather, navigation, and communication spacecraft. But a true revolution was under way. It was led, not by hawkish generals who wanted to turn enemy spacecraft and facilities into smoking rubble, but by greedy capitalists who wanted to turn a profit. And a very fat one at that. With communism vanquished and capitalism triumphant and spreading, private investment in space shot up dramatically except for military operations, the big-ticket *International Space Station,* and science and exploration. And even the Pentagon, under pressure to keep the cost of occupying space down, was relying increasingly on the private sector.

Investment opportunities seemed to be almost everywhere in the sky. The most lucrative area was the new global information superhighway, where information and communication would be treated as the commodities they were and would spread around the globe. By 1995, the availability of Western capital for private enterprise instead of defense spending, the remarkable miniaturization of satellite components, and a worldwide drive to get into the new planet-girdling communication network or become a permanent backwash, converged in space.

Early that May, Ronald H. Brown, the secretary of commerce who was soon to be killed in an air crash, told space industry representatives at a meeting in Virginia that "information access in all its forms—computer data on the Internet, mobile voice communication over [satellite] constellations, remote sensing images from space, and location data from Global Positioning System satellites—will be a critical element in the success of the U.S. aerospace industry." An industry executive who spoke at the same meeting brushed aside fiber optics and telephone poles as so many antiques. Global handheld telephone systems using satellites would bring the most dramatic change to mass communication in history, he said, adding that mobile wireless telephones would soon reach almost 4 billion people around the world who did not have telephones. "That's a staggering number," he continued. "There are fifty million people in the world who are on waiting lists for up to two years to get fixed wire-line telephones." Satellite telephone systems will be able to provide instant communication infrastructures anywhere, he added. Everyone at the meeting was looking at the future. And they called what they saw a global information infrastructure, or "infostructure," that would deliver high-quality video, voice, data, and multimedia services anywhere in the United States and, eventually, in the world. It depended upon advanced antennas, high-speed data processing on board satellites, laser communication among satellites, digital compression of data into short, intense bursts, and other technology that had been born in the cold war for military and intelligence use.

By then, the race to plug in the global village very early in the twenty-first century had started in earnest. By April 1996, well over a hundred civilian satellites providing voice, data, and video services were in orbit, including dozens built and launched for nonspacefaring nations: Arabsat, Asiasat, Brasilsat, Cansat, Hispasat, Koreasat, Thaicom, Turksat, Optus (Australia), and Palapa (Indonesia). China, India, Japan, and Russia had their own, sometimes sizable, fleets.

William H. Gates and Craig O. McCaw, two tycoons who made billions in computers and telecommunications, had by then come up with a plan to launch 840 satellites, plus up to 84 spares, in multiple orbits, or "planes," which would relay data back, forth, and sideways and spread it all over the world. They called their company Teledesic. In its application to the Federal Communications Commission, Teledesic portrayed itself as a kind of Internet for Montagnard tribesmen. By 1997, the number of satellites had been scaled back to a still-sizable 288, plus spares, flying along a dozen orbital planes.

Gates and McCaw faced stiff competition. Globalstar, owned by Loral Space and Communications and Qualcomm, planned to send six satellites and a spare on each of eight circular orbital planes. A company called Iridium, which had started the race to low-orbit megacommunication, wanted to fly a dozen 1,500-pound Motorola-built heavyweights on six orbital planes that would almost be polar. By 1997, it had invested $5 billion in the project. Other satellites, large and small, were called Ellipso, ECCO, Odyssey, Sativod, Inmarsat, Carcom, ORBCOMM (at eighty-eight pounds, the lightest of the competitors), and ICO (at 5,400 pounds, the heaviest). Industry analysts predicted that the satellite communication market was potentially worth $15 billion a year.

The United States, for one, had become so dependent on the vast array of spaceborne robots that when a communication relay satellite named Galaxy 4 malfunctioned on May 19, 1998, it caused instant pandemonium. Pagers belonging to an estimated forty million people suddenly went stone dead, as did National Public Radio's *All Things Considered* and a number of news and financial data services whose subscribers depend on up-to-the-second information. Those who relied on the services suddenly found themselves completely in the dark because a $250 million satellite more than 22,000 miles away went on the blink. A man in Chicago was stuck in an elevator for two hours because Galaxy 4 went haywire, since the repair company was unaware of the problem. Meanwhile, the Newark, New Jersey, police department's bomb squad could not answer a bomb threat. The director of engineering at WVPN, a public radio station in Charleston, West Virginia, called the failure devastating. "We not only obtain our programming from NPR via satellite," he said, "we also distribute our signal throughout West Virginia via the same satellite. So we're dead in the water." The president of a firm that sold financial research on-line summed it up this way: "We go up to the roof of our building and see four satellite dishes and think, 'Hey, we're redundant.' But we never thought that they were all pointed at the same satellite and that it might fail. We all have power backups, server backups, computer room backups, even vendor backups, and this is just one thing that no one ever thought about."

Beyond the tremendous educational and financial impact of transnational satellite communication at the end of the century, satellites were already well into starting a political revolution of profound importance, as coverage of Chernobyl showed. As far as exposure to foreign news, culture, and ideas were concerned, the signals that rained down everywhere and in all directions were sending customs officers and censors the way of the saber-toothed tiger. To take one example, independent television coverage of the brutal war in Chechnya, brought to homes throughout Russia by satellite, undoubtedly played a role in Yeltsin's willingness to negotiate a truce. In Saudi Arabia, where censors still blacked out pictures in magazines that showed too much cleavage, small dish antennas sprouting on rooftops in Jidda and elsewhere plucked sexy American movies out of thin air for private viewing behind private walls. "Satellite television is here," one Saudi official was quoted as saying, "and the Government is looking the other way." Well, not entirely looking the other way. Arabsat satellites that were launched by the United States and operated for years by a consortium of Middle Eastern companies were put into service by the Middle Eastern Broadcasting Center as early as 1991. The idea was to reach the millions of viewers around the world who speak Arabic. MEBC's chief investor was Walid al-Ibrahim, a brother-in-law of King Fahd.

Even in Iran, the repressive Islamic leadership had to supplement programming that featured a mullah intoning on the right way to pray, Ayatollah Khomeini denouncing "the Great Satan," and a deadly dull carpenter instructing on the correct way to make bookshelf corners, along with coverage of the World Cup soccer matches. That was because the roofs of Teheran and other cities had turned into antenna farms where $700 Iranian-made dishes were pulling in everything from late-night soft porn from Turkey to news from the British Broadcasting Corporation's World Service. "We got the dish eight months ago," one exhilarated Persian said, "and we haven't watched Iranian television since."

And in Afghanistan, the even more repressive Taliban religious fanatics reacted to what they saw as a dire technological threat by ordering that all television sets, videocassette players, and stereo systems be publicly "hanged."

Position Is Everything

Even the Taliban would have had no problem with an array of satellites that told them exactly where they were all the time. A Taliban with $200 could have walked into an electronics store in any number of other places and left with a gadget the size of a mobile telephone that told him exactly where in the world he was within a few feet.

He would have been using the Global Positioning System, or GPS, as it was almost universally known. Not long after Sputnik went up, some scientists figured that radio signals coming from several satellites in precisely defined orbits to a receiver could tell that receiver where it was by measuring its position relative to the satellites. It would be a very sophisticated version of the three-antenna system used in World War II movies to "triangulate" the location of enemy spies by pinpointing their clandestine broadcasts. Since accurately fixing position had great advantages for navigating on the high seas, the Navy designed a primitive satellite system called Transit in the 1960s to help its far-flung sailors know roughly where they were at all times. This was followed in the 1970s by the Department of Defense's Navstar Global Positioning System, a far more refined, $10 billion operation that eventually used two dozen satellites carrying atomic clocks broadcasting the time a thousand times a second to tell anyone on Earth with a receiver where he or she was. Basically, any four of the satellites could pinpoint a location on Earth because they were at different distances from the receiver and their clocks therefore broadcast infinitesimal differences in the exact time. Those differences—say, 3:05.54.56745 versus 3:05.54.56748—converged at the receiver. The discrepancies among the four told the receiver where it was in three dimensions.

The Global Positioning System had obvious uses in guiding ships and positioning planes so they could drop bombs down Iraqi smokestacks. But by August 1997, while *Mir*'s crew was trying to get their bearings like medieval sailors using sextants (they were unable to use their country's own first-rate Glonass satellite navigation system because of incessant computer and power problems), the GPS was growing as fast as its imaging and communication counterparts. The government's $10 billion was being leveraged into a $20-billion-a-year industry that fed a growing legion of people who wanted to know exactly where they were. Campers who used to debate the best way to collect rainwater now chatted about how to get better GPS signals under dense tree cover. Scientists were using it to track desert tortoise migrations, check the grazing patterns of sheep, and chart the slow but steady shifting of the Earth's crust and the movement of glaciers. GPS receivers were guiding motorists, finding people with medical emergencies who dialed 911, helping to protect endangered species, and even being considered to guide the blind and track the wanderings of ex-jailbirds who were supposed to stay in a particular area. Like the computer, the radio, and pasta, the Global Positioning System was developing its own distinctive culture and an almost infinite number of uses.

And venture capitalists were also looking well beyond the region around Earth. Marco Polo, Columbus, and other explorers, they never tired of pointing out, had been chasing riches, not advances in cartography, when they had set off on their long journeys. Now, with the new space age under way, it was time to do it again. By 1997, plans to mine asteroids and the Moon, start tourism in space, and launch other enterprises had made the transition from wishful thinking to serious planning. That year, James W. Benson, a retired software entrepreneur, founded Space Development Corporation to launch a private probe to a near-Earth asteroid. As he did so, a NASA mission called Near Earth Asteroid Rendezvous was returning data from the asteroid Mathilde and racing toward another asteroid, Eros. NEAR was part of the space agency's Discovery Program, which had been created to start low-cost solar system exploration missions, none of them priced at more than $150 million or taking more than three years to develop. As potentially important as NEAR's scientific return was, the fact that it was run not by NASA per se but by academic and industrial partners was significant. It represented an historic shift away from government support of space ventures to support by the private sector. Benson and others saw the change in emphasis as foreshadowing a true revolution.

"We're going to be the first private exploration company," he was quoted as saying. To do that, he brought together volunteer aerospace engineers to design a Near Earth Asteroid Prospector, the private equivalent of the Lunar and Mars Prospectors that NASA planned to send out on its Discovery missions. Allen B. Binder, Lunar Prospector's principal investigator, jumped in by announcing he intended to send his own missions to the Moon and sell the data they collected to NASA. By the time he made that announcement, a company in Virginia called LunaCorp had gone into partnership with Carnegie Mellon University's Robotics Institute to send two small rovers to the Moon, where they would serve as the central attraction for a theme park. Both organizations had managed to interest the Walt Disney Company in the venture. Disney backed off when one of the launch vehicles of choice, Russia's Proton, disappeared into the Pacific carrying the Mars 96 spacecraft. But David P. Gump, LunaCorp's president, still hoped to interest Japanese television in the project.

Fodor's Guide to Low Earth Orbit

And the notion of transporting tourists was in the air again. The concept had taken hold, however whimsically, when Pan Am had been bombarded with requests for reservations to the Moon at Apollo time. By April 1985, it had become serious enough to be put on the agenda of the Fourth Annual Space Development Conference that was held in Washington. Charles D. Walker, a space station specialist from McDonnell Douglas Astronautics in St. Louis, pitched the idea that space was not merely a place but a unique physical and psychological experience that would attract large enough numbers of people to be profitable. The obvious way to start taking tourists to orbit, it came out at the meeting, was on shuttles.

The always handy Russians were invoked as competitors, even in tourism. *Buran* "has a payload capacity of almost a hundred tons, more than three times the shuttle's capacity, and it might be an interesting vehicle for space tourism," one panelist warned. (It had about the same capacity.) "It would be unfortunate if the Soviets beat the Americans to the space tourism market, but that definitely is an option." He had probably never flown on Aeroflot, the Soviet national airline. At any rate, *Challenger* blew up nine months later, effectively ending talk about tourists, at least for a while.

The subject had been resurrected by 1995, when the Space Transportation Association of Arlington, Virginia, proposed to NASA using a two-stage spaceplane to carry tourists to orbiting hotels for between $10,000 and $20,000 each. The proposal cited market surveys in Europe, Japan, and the United States that found tickets to space costing $72,000 for a five-day stay in an orbiting hotel would attract 150,000 passengers a year and generate nearly $11 billion annually. At the other end of the scale, $2,000 tickets would attract 5 million tourists a year and bring in $10 billion annually. The space agency, which by then was under pressure to reduce its budget drastically because the competition had effectively disappeared—to make its operations "faster, better, cheaper," as Goldin put it—agreed to study the matter. "Space tourism will begin ten years after people stop laughing [at the concept]," David M. Ashford, director of England's Bristol Spaceplanes Ltd. remarked. Now, he added, "people have stopped laughing." Konstantin Tsiolkovsky certainly would not have laughed. It had been his idea.

Had the great rocket theorist been a member of the Stanford Alumni Association, in fact, the winter travel brochure that came in the mail early in 1998 would have offered him a trip to space on or about December 1, 2001, for $98,000 (pending Federal Aviation Administration approval). "If the journey to space has been one of your life's great dreams," the brochure predicted, "this is the trip for you." Neither the university's alumni travel director nor any of several potential spacecraft builders doubted that the time for commercial space travel was at hand. To encourage such flights by latter-day Lindberghs, the X Prize Foundation of St. Louis was offering a $10 million prize to the first private team that successfully sent three adults to at least sixty-two miles twice in two weeks. "When Lindbergh crossed the Atlantic, he changed the mind-set of almost every human on the planet," said Peter H. Diamandis, the chairman of the foundation. Meanwhile, NASA was planning to open an office of space travel and tourism later in the year to encourage entrepreneurs. "Make no mistake, it is the private sector that will finally build the machines and provide the access to space that makes the dream a reality for all Americans," Administrator Goldin said in May 1996, when the X Prize was announced. Zegrahm Space Voyages of Seattle, the producer of the Stanford tour, planned to use a mother ship called a Sky Lifter to carry a corporate-jet-sized, six-passenger spaceplane to 50,000 feet, where the smaller craft's own jets and rocket would fling it out to sixty-two miles. The whole flight would last three hours, during which the passengers would briefly be weightless.

What went largely unmentioned in all this talk about shooting tourists into space, however, was the physiological reaction of the uninitiated. Even veteran astronauts occasionally became disoriented and were hit by nausea. The experience could be ghastly for newcomers, as several college students found out during their spring break in April 1997. In order to turn them on to space, NASA gave them a ride in a KC-135A transport that it used to simulate zero g for astronauts. It accomplished that by flying roller-coaster parabolas in which it climbed to 34,000 feet at a forty-five-degree angle, then plunged two miles at 20,000 feet a minute, abruptly pulled out, and repeated the maneuver thirty-nine more times. The students and a hitchhiking reporter were weightless for twenty-five seconds each time the plane went over the top of its climb. "The scene at 34,000 feet," the reporter wrote, "could be dubbed the Science Fair from Hell. About half of the 16 college students aboard NASA's experimental zero-gravity plane have spent at least part of the past hour throwing up. The worst off sit slumped in their olive green jumpsuits, ashen like shell-shocked soldiers." It was an experience, he observed, "that would give even the most iron-gutted business traveler white knuckles." NASA had experienced that reaction so often during the plane's thirty years of operation that it nicknamed it the "Vomit Comet." Drugs that eased or prevented motion sickness would be one way to solve the problem, as would a launch that approached airliner smoothness. It was another problem for the entrepreneurs to solve.

By the time the students took to the air for their wrenching experience, Space Transportation System operations had been turned over to the private sector, which was another sign that the first space age's big-government involvement in programs was giving way to the second age's entrepreneurial interests. Effective October 1, 1996, operation of the shuttles was given to a company called United Space Alliance in a $7 billion, six-year contract. The company itself was a joint venture between Lockheed-Martin and Rockwell Aerospace. The idea, which hung heavy in the air by 1996, was to reduce NASA's role in overseeing day-to-day operations while making the system more efficient and cost-effective. At the time the contract was signed, the space agency employed 2,300 of its own workers in shuttle operations, plus another 23,000 who worked for contractors, and was spending about $3 billion a year to operate its four spacecraft.

The contract called for United Space Alliance to keep 35 percent of whatever it saved and turn 65 percent back to the government. Not surprisingly, United Space Alliance decided within a year of taking over shuttle operations that it wanted to use the vehicles to retrieve dead or dormant satellites in low orbit. Since most of the valuable satellites were only useless because they had run out of maneuvering fuel, explained James C. Anderson, United Space's chief operating officer, it made financial sense to bring them back to Earth for refurbishing, refueling, and relaunching. Ultimately, the plan called for the shuttles to be completely privatized early in the twenty-first century, with the government spending nothing on them except for specific missions. The transition from public to private operations, which was evolving elsewhere in the space program, truly heralded the beginning of a new era.

Mining the Moon for Fun and Profit

Other entrepreneurs, like the prospectors who swarmed westward in the nineteenth century, were planning to strip-mine the Moon. They were convinced that one element alone, helium 3, would make the venture eminently worthwhile. Knowing that the demand for energy on an increasingly populated and technologized planet would almost undoubtedly be the most important issue in the next century, researchers were concentrating on fusion even before the 1990s began. The fusion of atoms—forcing them together, which is the opposite of nuclear fission, which splits them—is what powers the Sun and other stars. There are two basic types of fusion: hydrogen and helium. Hydrogen fusion, which combines deuterium and tritium, gets by far the most research funding. But it has two serious drawbacks. While deuterium is found in water and is safe, tritium produces deadly radioactivity. And fusing the two requires temperatures of 100 million degrees Centigrade and enormous pressure before any energy is released. And the amount of energy that is released from the squeezing, at least so far, is less than what is required to create the high temperature.

Helium 3, on the other hand, is clean. Fusing helium 3 and deuterium would be much safer and cleaner than using tritium. And fusing it with more helium 3 would be the safest and cleanest of all since it would produce charged particles that would be converted directly to electricity and no deadly neutrons. And helium 3 fusion is not only an attractive energy source on Earth but could be used to propel rockets on long-distance missions. Yet it has not been taken as seriously as hydrogen because it, too, has a serious drawback: there is not much around. Unlike hydrogen, which is found in great quantities in the ocean, the only naturally occurring helium 3 on Earth is primordial amounts left over from when the planet was formed that are buried deep in its mantle, plus trace amounts in cosmic dust particles that hit Earth.

But the lunar surface is loaded with helium 3. When W. M. Braselton Jr. thinks of the Moon, he sees a base whose main function would be to mine the vast amount of helium 3 that is spread over its surface and bring it home. In an address to the United States Space Foundation in Colorado Springs in April 1993, the Harris Corporation vice president made a straightforward case for mining the Moon. "Only deuterium and helium 3 has the potential of direct electrical conversion, very low radiation and heat waste, and is clearly economically viable. One space shuttle load of twenty-five metric tons will electrically power the U.S. for one year. It would have a market value of seventy-five billion dollars today and there is in excess of one million tons on the Moon," he said, adding that there is a thousand-year supply of helium 3 on the Moon "with a total market value of three quadrillion dollars at today's oil prices."

A wide-ranging report issued two years earlier by a so-called Synthesis Group led by former astronaut Thomas Stafford, which worked with both the American Institute of Aeronautics and Astronautics and RAND, came to the same conclusion. *America at the Threshold,* as this last of the major studies was called, matter-of-factly described working conditions on the Moon that would be needed for the mining operation, including facilities to condense the helium 3, liquify and store it, and move it around in transporters that would themselves be fueled with the operation's "volatile by-products." There seemed to have been no doubt by whomever wrote that section of the report that helium fusion and solar power would replace fossil fuel and fission soon after 2020. No doubt at all. The message was clear: men (and women) were going back to the Moon. This time, however, the voyage would be an economic and environmental necessity, not a political stunt. "Space is a unique store of resources: solar energy in unlimited amounts, materials in vast quantities from the surfaces of the Moon and Mars, gases from the Martian atmosphere, and the vacuum and zero gravity of space itself," the Synthesis Group concluded. "With suitable processing, these raw resources are transformed into useful products."

"The space age fizzled because the grand dreams turned out to be too expensive," physicist Freeman Dyson wrote in 1995. "From now on, space technology will thrive when it is applied to practical purposes, not when it is pursued as an end in itself." Yet the distinction, as far as the exploration of the Moon, Mars, and some planetary satellites is concerned, can be hard to make. Mining the Moon and settling Mars might one day be very practical indeed. For the immediate future, however, they are still fundamentally out of reach. The dream is way ahead of reality.

Science in Orbit

Science, with communication and navigation, was among the oldest of the space cultures, and by the end of the millennium it, too, had proliferated to an extent that was virtually unknown to the general public. In 1995, NASA had more than fifty separate science missions under study, in development, or in space, many of them in cooperation with other nations, including Russia. Most of the working spacecraft were in Earth orbit doing esoteric studies that added knowledge in small increments.

The Interplanetary Monitoring Platform, for example, was typical of the world's unheralded, workaday scientific data collectors that quietly returned their investment many times over. IMP-8 was a nine-hundred-pound, sixteen-sided drum that flew an orbit taking it as far as 150,000 miles from Earth, where its field, particle, and plasma instruments continuously collected data on what NASA called the interplanetary environment: cosmic rays, the solar wind, and the magnetosphere. And in this case, "continuously" was pretty impressive. It was launched in October 1973 and was still faithfully sending data to its science team, which was scattered at the Goddard Space Flight Center, Caltech, MIT, Johns Hopkins, the University of Chicago, and elsewhere twenty-two years later.

The Cosmic Background Explorer, on the other hand, turned into an overnight superstar that made the pages of *Science, Discover, Air & Space, Scientific American, The Economist,* and newspapers around the world, including *The New York Times.* COBE, as it was called, accomplished that by using an instrument called a far infrared absolute spectrophotometer, which looked back—or far out—almost to the beginning of time and virtually proved that the Big Bang really happened. For years, cosmologists had been plagued by an apparent inconsistency between theory—that the explosion must have been uniform—and observation: that the universe is bumpy. After analyzing hundreds of millions of precise measurements made over the course of ten months, a team led by George F. Smoot of the Lawrence Berkeley Laboratory and John C. Mather of Goddard found "hot" and "cold" regions—ripples—stretching more than 100 million light-years across and going back to within 300,000 years after the Big Bang. Maps of the far reaches of the universe made with COBE data showed minuscule temperature variations, which were the cosmic ripples. This was electrifying because it provided the strongest indication to date that the universe was indeed started by a primeval explosion of unimaginable force and that it was not perfectly uniform. The ripples, Smoot and the others calculated, could have created the gravity that attracted more and more matter into what evolved into stars, galaxies, clusters of galaxies, and the great void in between over the next fifteen-or-so billion years.

Smoot's announcement of his team's findings was made at a meeting of the American Physical Society in Washington on April 23, 1992, and created an instant sensation. And the theoretical physicist set off a smaller, secondary uproar among some of his colleagues when he added, "If you're religious, it's like looking at God." "It is a mystical experience, like a religious experience," he was quoted as saying several days later as he reflected on the immensity of the explosion and what it created. "It really is like finding the driving mechanism for the universe, and isn't that what God is?" There was, after all, an almost unbroken two-thousand-year-old mutual antagonism between religion and science. That was what the persecution of Galileo and others was about. Nor, for the most part, were the scientists less parochial. In 1860, Thomas Huxley, the first popularizer of Darwinism, growled that "extinguished theologians lie about the cradle of every science, as strangled snakes beside that of Hercules." Not much had changed more than a century later.

The Space Telescope Delivers the Universe

COBE's orbiting cousin, the disgraced Hubble telescope, also finally distinguished itself by collecting the seeds of cosmic creation for all of Earth to see. Ironically, the road to rehabilitation and triumph started on December 7, 1993, when spacewalking astronauts from *Endeavour* pulled out the telescope's Wide Field Planetary Camera the way some car radios are removed from dashboards and replaced it with one that had been specially modified to correct the primary mirror's distortion. They then attached a second corrective optical system the size of a phone booth. Together, the two new instruments amounted to a $151 million pair of prescription eyeglasses. The total cost of the operation was $664 million.

What happened next was literally astronomical. During the next three years, and less than four centuries since Galileo's flimsy tube had first pierced the mysteries of the night sky, its orbiting descendent lavished Earth with knowledge so profound that even its users were stupefied. It looked back 11 billion years—to a time and distance that are ungraspable, almost to the dawn of creation—and saw whole galaxies forming. Pointed to a region of sky no bigger than a grain of sand held at arm's length, Hubble unveiled new layers of galaxies as far as the eye could see, providing rich new data on the creation of the universe.

In all, the flying Cyclops revealed a mind-numbing 40 billion new galaxies—five times the previously estimated number—most of which had gone unseen by Earth-based telescopes because of their dimness. "As the images have come up on our screens," pondered Robert Williams, director of the Space Telescope Science Institute, "we have not been able to keep from wondering if we might somehow be seeing our own origins in all of this."

The images Hubble has collected lend themselves to superlatives. Peering at the core of galaxy M87 in the Virgo cluster less than a year after it was repaired, Hubble found a black hole, turning Einstein's prediction that the all-consuming cosmic colossuses existed into accomplished fact. Then, staring into the heart of galaxy NGC 4261, it spotted an eight-hundred-light-year-wide disc of dust that was whirling so fast it could only be in the clutches of still another black hole, this one 1.2 billion times the size of the Sun. Looking seven thousand light-years in another direction, the telescope sent back eerie, breathtaking images of about fifty stars being born in immense columns of hydrogen and dust—a churning, phantasmagoric eruption whose sheer size defied meaningful description—that rose with infinite majesty out of a dark and turbulent molecular cloud towering 6 trillion miles high. Elsewhere, it turned up two entire galaxies colliding in a cosmic fireworks display that produced millions of new stars. And still elsewhere, the HST spied colossal discs that were probably planets in the making. It also spotted ozone on Ganymede and provided the first good look at Pluto's surface (understandably, there is a lot of ice). It played a role in tracking and imaging the sensational Shoemaker-Levy string of twenty comet fragments that slammed into Jupiter in July 1994, leaving a huge dark smudge in the gas ball's otherwise pristine atmosphere. The comets were named after their discoverers: Eugene M. Shoemaker, Carolyn S. Shoemaker, and David H. Levy.

No one any longer doubted that Hubble had been worth the pain, or that "big science" could bring rewards to match its name. The Hubble Space Telescope undoubtedly went furthest in profoundly changing humanity's view of itself in the universe and how that universe itself came to be. By the end of 1997, for example, the astronomers who used Hubble to measure the expansion of the universe and the great engine that drives it were collaborating with subatomic physicists at the Brookhaven National Laboratory on Long Island on a joint project of unprecedented dimension. The physicists planned to use a heavy-ion collider—a super atom smasher—to smash "heavy" atomic nuclei together at such high velocity that their protons and neutrons disintegrate into quarks, the most elementary building blocks of matter, while the so-called gluons, which glue the quarks together, come unglued. The temperature of the resulting explosion would theoretically be hotter than the core of the hottest sun and would finally re-create the first few microseconds of the Big Bang.

And Hubble was by no means the only observatory in space. Twenty-three telescopes, including Hubble, were there in 1996. One of the others, named Ulysses, had gone into orbit around the Sun after swinging past Jupiter and was also returning unprecedented data. Another, a European Space Agency satellite named Hipparcos, sent back data that narrowed the age of the universe to 10 to 13 billion years. All of them were in varying degrees helping to answer Carl Sagan's ultimate question about humanity's place in the scheme of things. And before he died on December 20, 1996, at the age of sixty-two, he had the pleasure of knowing that.

Sagan and his colleagues knew that the rocket had carried their science over a phenomenally important threshold. Cosmology was finally making an historic transition whose consequences were profoundly exciting. It was moving from theory to observation and therefore from conjecture to fact.

And it would keep getting better. By the time Sagan died, work was well along on an Advanced X-ray Astrophysics Facility, scheduled for launch on a shuttle before the end of 1998. AXAF, as it was called, was the third of four Great Observatories (after Hubble and the Compton Gamma-Ray Observatory, which was launched in 1991, and before the Space Infrared Telescope, which was due to go to orbit in 2001). AXAF's job is to observe the violent world of exploding and collapsing stars, expanding superhot gas clouds, and the immense, voracious black holes that eat everything in their neighborhood. Since those phenomena are visible only on the X rays they send out, they cannot be seen by Hubble and other electro-optical telescopes, and are invisible to the human eye. AXAF is called a "facility" rather than a "telescope" because in 1988, when it was proposed to Congress, the HST was running into serious cost overruns and NASA did not want knotty comparisons to be made. But AXAF by any other name is the Hubble of X-ray astronomy and is supposed to continue the profound revolution started by its predecessor.

Return to the Planets

The Venus Orbiting Imaging Radar mission, renamed Magellan and launched from *Atlantis* on May 4, 1989, slipped into a polar orbit around the sulfuric acid–shrouded planet on August 10, 1990. Five weeks later it began systematically radar mapping Venus's surface in exquisite, unprecedented detail. (In 1984, the Soviet Venera 15 and 16 radar mappers had returned impressive imagery of 25 percent of Venus with a resolution of a mile; Magellan imaged 98 percent of the planet with a resolution of a little better than eighty yards, meaning that it could define features the size of the Rose Bowl.) That imagery unveiled terrain on the ultimate greenhouse that astounded its science team and the rest of the world. What was almost as remarkable as the science return was the fact that Magellan, which had been killed during Stockman's war on solar system exploration, had been resurrected under so much financial pressure that many of its parts had been scavenged from other missions (proving, perhaps, that there might have been some merit in the budget cutters' case and anticipating Goldin's belt tightening). Its main antenna, which was used both to do the mapping and communicate with Earth, for example, was a Voyager spare. Using a single antenna for both operations demanded special inventiveness by its masters at JPL.

While almost everything that had been found on the other terrestrial planets and moons could be set in the context of features on Earth—volcanos, faults, craters, and dry riverbeds, for instance—Venus forced the geologists to expand their vocabulary. At a news conference in von Karman two months after Magellan reached Venus, the project's chief scientist showed awed reporters pictures of seven pancake-shaped lava domes that were half a mile high, ten to fifteen miles across, and strung out in a remarkably straight line. They were "features never seen before" on any planet, R. Stephen Saunders noted before guessing that they were probably formed by especially thick molten lava.

"Like kids watching a baseball game through a crack in the fence, planetary scientists have eagerly and impatiently watched the action unfold as crisp radar images from the Magellan spacecraft progressively unveil the surface of Venus," *Scientific American* cooed. "At the conclusion of its first year of mapping, Magellan has revealed a vast array of stunning and often enigmatic landscapes." The images, which were turned into the same hues of yellow, orange, and ocher that had been used for Io, were spectacularly beautiful, especially when they were digitally manipulated on video.

During its four years of operation, Magellan sent back a large library of information that profiled Venus in extraordinary detail. There was sand blown into the shape of giant horseshoes, a thin, plastic crust, rugged mountains and plateaus, clear evidence of a complete volcanic resurfacing, and other hints of recent volcanic activity on Maat Mons, which, at five miles, was the planet's second highest mountain. Venus seemed alive with volcanoes and other subsurface activity, including a mysterious lava channel that meandered 4,200 miles across the planet. There were erosion patterns that suggested that sand had been moved by high winds and there were impact craters distorted by the atmosphere. (Like all other features on Venus, the craters were named for famous women or female mythological characters. One set of three craters was named for Gertrude Stein and another for Cleopatra.)

"We probably have a better global map of Venus now than we have of Earth, because most of the ocean basins on Earth are so poorly mapped," Saunders said in October 1991, a year after Magellan went to work. He was right. Mission to Planet Earth was designed to correct that problem, but the same shrunken budget that ended Magellan's map making, gravity experiments, and other work by the end of 1994 sapped the Earth studies program as well.

On October 9, 1992, Magellan's predecessor, Pioneer 12, plunged to a fiery end in the deadly Venusian atmosphere after fourteen years of faithful operation. "This has been an enormously productive mission," Richard O. Fimmel, Pioneer Mission manager, said at what amounted to a wake for his own explorer at Ames. "In spite of the fact that we are sad at the loss of an old friend, we have the satisfaction of a job well done." Scientists were quoted in one newspaper account as saying that the 810-pound spacecraft had continued to send back valuable information about the Venusian atmosphere until radio contact was lost. "Near the end," the obituary said, "Pioneer 12 was 79.5 miles above the planet's surface and was making history as the first orbiting spacecraft to sample the high atmosphere of any planet."

Two years later almost to the day, Magellan followed Pioneer 12 down to oblivion. But even in the throes of its own demise, it did an aerobraking maneuver as it plowed into the hot carbon-dioxide soup and disappeared that would help other spacecraft when they returned to Mars. Typically, it was the first time the maneuver, which uses a planet's atmosphere to slow the spacecraft, was done. "It went out in a blaze of glory," its mission manager, Douglas Griffith, said by way of an epitaph. Glory indeed.

The King of the Sky Is Revisited

Unlucky Galileo, which was launched five months after Magellan, reached Jupiter in December 1995 by a circuitous, gravity-assisted route that sent it looping around Venus and Earth before making its final, long-distance dart to the great gas ball. Even before it left, radical environmentalists raucously warned that its nuclear power plant would contaminate a large part of Florida if it crashed, or spray the whole planet with deadly radioactivity if it came in too low on its flight around Earth, hit the atmosphere, and disintegrated. NASA's argument was that the explorer's RTGs were built to contain their plutonium-238 even if *Atlantis,* the shuttle that carried it to space on October 18, 1989, blew up the way *Challenger* had three and a half years earlier. The picketers who carried placards outside Canaveral's main gate were no more convinced than their fictitious counterparts who had tried to stop the launch in *Things to Come.*

Then, with the mission barely under way, Galileo's handlers at JPL were horrified to discover that its furled high-gain antenna was stuck partly closed, severely limiting the rate of data that could be sent home. But the robot explorer ultimately maneuvered around Venus and Earth without mishap (collecting a great deal of data in the process) and made it to Jupiter in early December 1995. There, it dropped its instrument-filled capsule—the first Jovian penetrator—into the atmosphere and then settled into the first close orbital inspection of the giant planet. Besides a steady flow of richly detailed data about the Great Red Spot and other surface features, Galileo returned information about how Jupiter interacts with its moons—the miniature solar system—and much about the Galilean satellites themselves.

The most intriguing news from Ganymede came in 1996, when it was discovered to have many planetary qualities, including a magnetic field of its own and the same kind of dynamic forces that move continents on Earth. But the big news came from Europa, where the most detailed pictures ever taken of that world convinced mission scientists that liquid water or slush lay beneath its six-mile-thick crust of cracked ice. An ocean whose surface was frozen would account for the fracture lines that were seen to crisscross its surface and the huge ice blocks that appeared to be shoved around like moving icebergs. And all that, in turn, amounted to compelling evidence that there is an internal volcanic heat source. Water plus heat could equal life on Europa, at least in the view of some enthusiastic scientists, who quickly raised the possibility that it could exist in the depths of the Europan ocean as it exists under Earth's. "We found building blocks that are very conducive to life on Earth," said Torrence V. Johnson, the Galileo project scientist. "But we don't know that that always leads to life."

The "Life" Card (Again)

It certainly led to life in the space program, one of whose most reliable funding staples is the possibility that there is life out there. The news from Europa came on top of other discoveries, all of them controversial, which raised the ancient prospect yet again that Earth's creatures are not alone in a vast, dark closet. In 1992, the first sighting of two planets orbiting a far-off star called PSR B1257+12 was announced, and it was confirmed three years later. In 1995, two teams of astronomers working independently found another, this time orbiting a sun similar to ours and much closer. The discovery of two more planets early in 1996, one of them 200 trillion miles away in the Big Dipper and the other in the constellation Virgo, convinced the editors of *Time* to run a cover story that asked this ill-focused but suitably dramatic question: "Is Anybody Out There?"

While the astronomers were turning up galaxies by the billion and sighting their first planets, researchers led by a NASA geologist at the Johnson Space Center announced another stunner: they had found three different minerals, an organic residue, and bacterialike structures in a potato-sized meteorite that had been blasted off the surface of Mars in ancient times and found its way to the Antarctic.* David S. McKay, who had taught basic geology to the Apollo astronauts and studied the soil samples they returned, led the team that scrutinized AHL84001, as the chunk of debris was called. "We are not claiming that we have found life on Mars," McKay explained with appropriate caution.

* It was one of only eight fragments found in the Antarctic in 1984.

"And we're not claiming that we have found the smoking gun, the absolute proof, of past life on Mars. We're just saying that we have found a lot of pointers in that direction." McKay and his colleagues were convinced that the rock had come from Mars because it closely matched soil chemical analyses sent back by the two Vikings.

Goldin reacted by calling it a "startling discovery." President Clinton reacted as Goldin hoped he would. He asked Vice President Al Gore to launch a "bipartisan space summit" to pursue the matter and added that his administration was committed to the aggressive robotic exploration of Mars. It was already under way. The first of those missions, part of the new and streamlined Discovery Program, was called Mars Pathfinder. It was scheduled to leave early in December and arrive at Mars on the following Fourth of July.

Happy Birthday from Mars

The Fourth of July 1997. It is the United States of America's two hundred and twenty-first birthday, and NASA has a present: Mars. For the first time in twenty-one years, a visitor from Earth is returning to the Red Planet. Pathfinder is closing at 16,600 miles an hour, and the dreamers in Pasadena are winding up for another celebration.

"This Independence Day Earth Invades Mars!" small posters taped around the Caltech campus shout in a clever play on the names of two popular films in which Martians invade Earth. The Planetary Society is throwing a three-day "Planetfest '97" at the Pasadena Convention Center. More than two thousand space buffs from around the world have gathered to meet and get autographs from astronauts Sally Ride, Buzz Aldrin, Story Musgrave, and Franklin Chang-Diaz, science fiction writers Greg Bear, Jerry Pournelle, Kim Stanley Robinson, and others, and see a pilot plant for making rocket fuel on Mars, a full-scale mockup of the Mars ascent vehicle, lively lectures, exhibits, "the best computer graphics of the solar system ever seen on Earth," and more. All of it will be wrapped around the heart of the happening: a direct, live, continuous hookup through JPL's monitors to Pathfinder's eyes, 119 million miles away.

Four miles away, on the other side of Pasadena, the spacecraft's masters are intently following its long, graceful arc into the thin Martian atmosphere. They are mostly young and new to exploration. Gone are Homer Joe Stewart, Jack James, Bud Schurmeier, Bill Pickering, Brad Smith, Toby Owen, and Bill McLaughlin. They have been replaced by a new generation of mission specialists, most of them in their thirties: Tony Spear, the project manager; Brian Muirhead, a senior citizen of forty-five who is the flight system manager and Spear's deputy; Matthew Golombek, the chief project scientist; Rob Manning, the chief engineer; and others. Many of the three hundred or so people who designed, tested, and who are now guiding Pathfinder to its distant target were just entering kindergarten or were in grade school when the Viking twins set off to Mars. They are so young in July 1997 that one newspaper cartoonist draws three angry kids saying to a guard barring their way into the mission control room: "*What* school tour group?! We're the mission directors!!"

Pathfinder is new, too, and represents the next generation of Earth huggers and explorers. It consists of a three-foot-high spacecraft that weighs a little under a ton and has three solar panels folded over its miniaturized bus like tulip petals that are supposed to open once it is on Mars. Attached to one of the panels is a twenty-three-pound rover named Sojourner whose flat top is itself a solar panel. Sojourner, a six-wheeled surface crawler the size of a microwave oven, carries a pair of its own tiny cameras, solar cells to collect energy from the Sun to run itself and to measure dust accumulation, and an alpha proton X-ray spectrometer. When Sojourner bumps into rocks and mounds of soil, the theory goes, its spectrometer will determine their chemical composition. But while this mission has science objectives, those who run it often take pains to point out, they are very limited. They insist that it is primarily a technology demonstrator, not a science collector.

Yet of all the places Pathfinder could have been told to land—seventy-eight of them are listed in detail in the NASA manual with the marvelously optimistic title *Mars Landing Site Catalog*—the spot chosen was Ares Vallis, a rocky plain over which a flood of water flowed eons ago. Having been burned by the well-publicized search for life on Viking's otherwise sumptuous mission, neither headquarters nor JPL wants to play that card again, at least so blatantly. But the reporters in von Karman (most of whom are the new counterparts of the young mission team) know about the water-life connection, so the life angle is just under the surface (politically, if not physically). What many of the newcomers in the press section do not notice, nor could put in context even if they did, is the fact that Pathfinder's white skin carries an American flag and JPL's orange logo, not the red, white, and blue NASA "meatball." And Sojourner also wears a small "license plate" that goes unseen in the diagrams of it that appear in newspapers and magazines. It, too, says "JPL," not "NASA." If Keith Glennan's ghost is here, and if it notices this small but unmistakable sign of lingering egotism, it is almost certainly shaking its head and sighing in exasperation. The current administrator notices and he is fuming.

Pathfinder's landing is going to be more difficult than Viking's. Viking dropped out of Martian orbit. Pathfinder, on the other hand, has taken a direct route—a straight shot—from Florida to Ares Vallis, slamming into the thin Martian atmosphere at a 14.2-degree angle. Seventy seconds after it enters the atmosphere, and while it is pulling 20 gs at an altitude of about six miles and its heat shield is braking its speed, its parachute is supposed to pop open. Pathfinder should then be floating almost straight down. When the radar altimeter tells the spacecraft's computer than it is a mile above Mars, air bags inflate until they look like giant globs of molecules, retro-rockets fire, and the bridle that is connected to the parachute and the back shell that holds the rockets is jettisoned at two hundred feet. If everything goes right up to this point, Pathfinder, snuggling inside its protective shield of air bags, is calculated to hit Mars at twenty-two miles an hour, bounce as high as a ten-story building, then bounce again and keep bouncing until it has rolled to a dead stop. Then the air bags should slowly deflate so Pathfinder can open its petals to collect solar energy and check out the terrain with its camera. If the coast is clear, Sojourner will then roll down a short ramp and into history as the first vehicle from Earth to roam over another planet.

"Every time the Russians have tried this, they've failed," a chronicler of the event reminds Skip McNevin of JPL's Public Affairs Office two days before the scheduled landing.

"We're not Russians," McNevin answers resolutely.

And he is right. That is, the mission goes exactly as planned.

A little before 10 A.M. on Independence Day, within three minutes of rolling to a stop, Pathfinder tells JPL that it is safely down. Less than an hour later, Muirhead, sitting in front of mock-ups of Pathfinder and Sojourner and beside Dan Goldin and Ed Stone, tells the reporters in von Karman what Pathfinder has told him. The news draws applause from the young reporters. While they are clapping and cheering, the older hands take notes. The older hands never applaud anything while they are working. Muirhead thanks the news media for their support. There is more applause. The older hands are still taking notes. "This is a revolution just of the type of revolution that caused our country on the date of our birthday," Goldin tells the press, adding that "this is going to change the history of the space program." He congratulates Muirhead and Spear. There is a third round of applause.

A telephone sits on a small table beside each of the three men, who are now basking in the glare of the television lights. A woman from public affairs interrupts a question-and-answer session to tell everyone to stand by.

"The phone is ringing here," the administrator mumbles.

"We might have a phone call," says the woman from public affairs. Dan Goldin, squinting in the lights, picks up his phone.

"Hello?" says the administrator, the phone to his ear. A loud metallic shriek and then a short hiss comes out of von Karman's speakers.

"You guys just stand by?" says the woman from public affairs.

"That mean me, too?" asks the administrator. The young reporters laugh and applaud some more.

"You all stand by for a minute?" says the woman from public affairs. She is getting a little tense. The administrator is getting a little tired of the lights and is still squinting. He is also playing the game, as he has many times before.

"Hang up. He'll call back," shouts a reporter. Von Karman breaks into laughter.

Then someone says that the vice president is on the line.

"Hello, Mr. Vice President," says the administrator.

"Hey, Dan," says Al Gore. "Congratulations to all of you out there on behalf of President Clinton and all the people of our country." Wesley T. Huntress, who heads NASA's space science operations, is in von Karman, too. He tells everyone that Pathfinder's safe arrival marks the beginning of the second era in the exploration of Mars.

The first pictures, black-and-whites showing Pathfinder and its still-attached rover, with the Mars-scape behind them, start coming in at about 4:30 in the afternoon. They soon turn to color panoramas of a salmon sky and a plain strewn with rocks, most of them rusty and many of them obviously having been tossed around by once-raging water. Late the next night Sojourner, reminding some of its controllers of Neil Armstrong's slow, immortal descent down *Eagle*'s ladder almost exactly twenty-eight years earlier, slowly makes its way down the short ramp to start its own historic odyssey.

Three Earth days after Pathfinder has landed and one day after its little prospector has rolled around, studying a rock its controllers name Barnacle Bill, the geologists in Pasadena start to put a picture together. The shapes, colors, and tilt of rocks, long chains of pebbles, surface textures, and wavelike ridges indicate that there was once a tremendous flood of water. It seems to have poured in from the southwest—from the area called Lunae Planum and, even beyond it, from the Grand Canyon–like Valles Marinaris—and formed a sea hundreds of miles across and hundreds of feet deep on what is now bone-dry terrain. "This was huge," says Michael Malin, one of the geologists, in delighted wonder. "But we don't know where the water went." Figuring it out, as usual, is the fun of it. Mariner 4's descendants are relentlessly rolling back the ignorance. They are shining their light into dark corners, turning a once mysterious and fearsome place into the old neighbor.

Exploration was making the neighbor so familiar, in fact, that some inhabitants of Earth were already staking claims there. Two Yemeni men who insisted they owned Mars through an inheritance from ancient ancestors became so angry because Pathfinder had landed there that they filed a lawsuit against NASA for trespassing with Yemen's prosecutor general. Mohammad al-Bady, the prosecutor general, threatened to have them arrested if they did not withdraw the case. They did. "The two men are abnormal," al-Bady was quoted as saying.

Meanwhile, another of Mariner's descendants, Mars Global Surveyor, was heading for Mars even as little Sojourner scrutinized more rocks with its spectrometer: an instrument that had evolved from the Russian Phobos mission and was identical to one lying at the bottom of the Pacific in Mars 96. Making it past the Ghoul, Mars Global Surveyor arrived in September, swung into orbit, and got ready to begin high-resolution photomapping of the planet's surface in March 1998. The purpose of the mission was to get fine details of physical features and mineral composition, more information on where the water was, and data on geological features.

Mars Surveyor 98, a small orbiter-lander, was supposed to follow in the next twenty-six-month window, late in 1998 or early 1999. The lander would be aimed at the south polar region, where the planetary scientists think water ice forms. Meanwhile, the orbiter would be studying Martian weather. The Discovery plan called for Mars 01, an orbiter-lander-rover, to leave during the next window so it could collect samples and send back a detailed picture of the entire basic composition of the Martian surface. Mars 03, another orbiter-lander-rover, was supposed to collect information specifically related to human exploration, study more samples, and set up communication and navigation facilities for later missions. The series's finale was planned for 2005 and called for two rovers to collect samples and get them on board a return ship that would climb into an orbit around the Red Planet, make a rendezvous with another spacecraft, and transfer the samples to it for the trip back to Earth. Given the size of Mars, the big question was whether they would take the search for life beyond what was found in ALH84001, the Antarctic potato. In 1997, NASA put the probable cost of all the missions at $1.5 billion, or the price of fourteen and a half of the three hundred and thirty-nine $100 million F-22 "Raptor" fighters the Pentagon wanted.

The space agency had become so convinced by 1998 that the search for life out there would ultimately be fruitful that it started an Astrobiology Institute at Ames. It was responsible for encouraging interdisciplinary research on all manner of possible life beyond Earth, as well as studying the effects of living in space on creatures from Earth and how to create stable ecosystems on Mars for human survival there. The plan was to provide access to manned and unmanned space missions and link scientists around the country on the world wide web in a "virtual institute," as NASA astronomer David Morrison put it. The idea, he said at a symposium at the American Association for the Advancement of Science annual meeting in Philadelphia in February 1998, was to "break down the barriers between life and physical scientists" so they could unify to search the universe—certainly the planetary systems that were starting to show up—for any manner of life-forms, including robust and ubiquitous bacteria. And an integral part of the process would be studying human physiology in orbit, including what were euphemistically called "senior citizens." The first of them, in fact, was already poised to fly.

The Oldest Astronaut

While the space agency was calculating the cost of sending robots back to Mars, Senator John Glenn was petitioning it to send him back into Earth orbit. For two years the bespectacled, balding former astronaut had been asking NASA to get him a seat on a shuttle mission, and in mid-January 1998, he got a "go" for a shuttle that was scheduled to leave that October, when he would be seventy-seven and out of the Senate. The ostensible reason for the flight was to do experiments relating space to the aging process. But in approving it, NASA scored a brilliant public relations victory, particularly at a time when millions of baby boomers had reached middle age and there were more coherent "senior citizens" than ever. "I'll give it my very best try," Mr. Right Stuff assured a packed news conference in Washington. No one doubted that.

Ad Astra

Early in 1986, as Voyager 2 was streaking past Uranus on the last leg of the Grand Tour and Magellan, Galileo, and Ulysses were being readied for their own historic encounters with Venus, Jupiter, and the Sun, Freeman Dyson peered into the future and saw a flock of tiny successors to those ponderous explorers. It was a future, the zany and immensely imaginative theoretical physicist decided, that had no place for robotic wanderers that were too large, too few, and too slow to provide all the knowledge that mankind wants about the worlds beyond Earth.

Dyson's idea was to combine twenty-first-century genetic engineering, advanced artificial intelligence, and solar electric propulsion—all of which, he predicted, would be in place by the end of the second decade of the twenty-first century—and "grow" hundreds of spacecraft, each weighing just two pounds. These Astrochickens, as he whimsically called them, would be as maneuverable as hummingbirds, would sustain themselves by munching on the ice and hydrocarbons found in planetary rings, and would reproduce with DNA blueprints they themselves carried. Their mission would be to use their own advanced intelligence to dart around the solar system and well beyond it, landing where they pleased, all the while describing to the humans back home how they felt and what they discovered.

"Birds and dinosaurs were cousins, but birds were small and agile while dinosaurs were big and clumsy," Dyson explained. Mainframe computers and the shuttle are dinosaurs, he continued, while microprocessors and Astrochickens are birds. "The future," Dyson concluded, "belongs to the birds."

The last of the dinosaurs, a 12,800-pounder named after the astronomer Giovanni Cassini, left in early October 1997 for a long encounter with Saturn and its moons in 2004. Cassini was the only member of a class of heavyweights originally called Mariner Mark 2. Another, named CRAF (Comet Rendezvous Asteroid Flyby), was dropped in its tracks in January 1992—killed in the development stage—because NASA's shrinking budget could not handle it. Cassini's job was to do at Saturn what Galileo was doing at Jupiter: return for a long, close inspection. Like Galileo, it carried a probe, this one named Huygens, after the seventeenth-century Dutch astronomer Christiaan Huygens, which is supposed to be sent to Titan.

And like its unlucky but successful predecessor, it was attacked even before launch by anti-nuclear protestors who claimed that the seventy-two pounds of plutonium in its three radioisotope thermoelectric generators would produce hundreds of thousands of casualties if the spacecraft blew up during launch or crashed into Earth when it skimmed five hundred miles over the planet in August 1999 for a gravity boost before heading to its ringed destination. What was on Goldin's mind at the time of the demonstrations, however, was not the threat from radioactive contamination. It was the threat from Congress that absolutely ended more upscale projects like Cassini, which took eight years to develop and cost almost $3.5 billion.

The first of the new birds, to use Dyson's analogy, was named Clementine. It was basically a Department of Defense project conceived in 1992 and constructed by the Naval Research Laboratory. Its sensors were designed and built by the Lawrence Livermore National Laboratory in cooperation with industry and were based on the miniature, self-contained hardware that was supposed to have gone into the Star Wars Brilliant Pebbles armada, which had also been a Livermore project. Clementine was a triumph of miniaturization. Its ten scientific instruments together weighed only seventeen and a half pounds. The ultraviolet-visible camera, for example, weighed a little more than a pound and ran on five watts of power (a refrigerator bulb uses eight times as much). The whole spacecraft weighed only five hundred pounds. Loaded with another five hundred pounds of maneuvering fuel, it still had only one-thirteenth the mass of Cassini.

Clementine was launched in late January 1994—barely two years after it was conceived—and was controlled not from the concrete castle at JPL but from a room in an abandoned armory in Alexandria, Virginia. The place was named "the Bat Cave" after one of its original inhabitants, which had to be evicted with a broom. Clementine proved its stuff a year later when it sent back detailed pictures that were carefully turned into a high-resolution mosaic of the Moon. What's more, a close study of radar imagery turned up what looked like water ice in the permanently cold shadows in a basin near the Moon's south pole. No one seriously believed that there could be traces of life there. But if water really did exist, it might not only help support a colony but be convertible to hydrogen and oxygen and therefore to rocket fuel. (Building an outpost near the south pole would not be a good idea, though, because it is more difficult to reach than the equatorial region and far colder. As usual, there was a downside.)

And a second bird—or rather, a class of them—was hatching at JPL in a program called New Millennium. The name stood not for particular spacecraft but for a concept of designing and "validating" very advanced technology that would lead to small, exceptionally smart machines sharing fundamental similarities while being able to do many kinds of Earth orbit and exploration missions. Three of the first six missions would involve circling Earth and would therefore be designed and run by Goddard. Three others were to be designed for deep space exploration, with one flying by an asteroid and another sinking a pair of six-and-a-half-inch-long harpoons called "penetrators" into the Martian crust. The two "smart dart" penetrators were to report back on the chemicals they found in the Martian soil. Sarah Gavit, the Mars mission's flight team leader, said she could see the day when Mars would be salted with networks of penetrators and rovers that would search for signs of life, listen for tremors, and take simultaneous weather readings from several locations.

Meanwhile, rovers—the successors to Sojourner—were being designed and tested at JPL's Microdevices Laboratory, MIT's "Insect Lab," NASA's Telerobotics Research Program, Carnegie-Mellon, and elsewhere. They had names such as Dante, Genghis, Rocky 4, and Squirt, and many were being programmed to think for themselves, move over rough terrain, and report home. Dante weighed almost a thousand pounds, but it was an exception. Most were so much smaller than Sojourner that they made it look elephantine by comparison. Squirt, for example, got its name from its size: one cubic inch. Rocky 4, a JPL creation, was less than two and a half inches long and was designed specifically for a Mars survey. It will not go any time soon, however, since it and the other microrovers are still more theory than fact. The rovers that are ticketed to ride to the Red Planet in 2001, 2003, and 2005 will in fact be bigger and heavier than Sojourner.

But the new birds will not be birdbrained. They will be provided with enough artificial intelligence so they can guide themselves to their destinations, collect what they were sent to get, and handle trouble without help from home, just as Dyson predicted. "A lot of us have gotten really excited by the idea of these autonomous spacecraft: spacecraft that can basically do everything on their own," Ellen R. Stofan, New Millennium's program scientist, explained. "It can make its own trajectory corrections and maneuvers and doesn't have to send data back and forth to the ground. It does that all on board." In other words, it is supposed to run its own mission without calling home. "From a scientific standpoint," said Stofan, "it blows your mind."

Twenty-two deep-space missions were being planned or were under way as the end of the century neared, four of them to the Moon, nine to Mars, five to comets and asteroids, a "Pluto Express," and the Galileo, Cassini, and Ulysses flights. Pioneer 10, which was launched on March 2, 1972, and sent the first close-ups back from Jupiter, was turned off at the end of March 1997 after traveling more than 6 billion miles in twenty-five years of continuous service. It would then cruise silently through interstellar space on a course that would take it past ten stars in the next million years. And both Voyagers were still sending whispers home from what NASA called extended missions.

Mushing to Mars and Beyond

Setting up an outpost on Mars, and then colonizing it, has fixated the space community as a matter of manifest destiny for decades. In 1951 Arthur C. Clarke, then chairman of the British Interplanetary Society, called for terraforming Mars. An illustration in his book *The Exploration of Space* showed a city, complete with green parks, protected by a huge bubble dome. (It is a measure of where we have come since then that an accompanying map of the planet was so blurred it looked Lowellian and a photo of astronauts in space suits was taken from the set of *Destination Moon,* which came out that year.)

Since then, a vast literature has grown on the subject of human travel to Mars, much of it highly detailed and scientifically grounded work by NASA, its constituents, and related groups. The space agency held a Mars Conference at the National Academy of Science in July 1986, for example, at which scrupulously researched papers were presented on necessary technologies for an expedition to Mars, life support for a Mars mission, and living and working there. That year JPL published its own study, *The Case for Mars: Concept Development for a Mars Research Station,* which considered everything from spaceship design to extracting breathable air by reducing carbon dioxide in the atmosphere to get oxygen, and dehumidifying the air to get water. This, the report noted, would require huge amounts of energy. "While atmospheric supplies [of water] may be adequate for the initial base, its scarcity and energy expense may eventually limit agricultural expansion."

Meanwhile, a dreamer who never doubted that it was mankind's destiny to colonize Mars was taking on von Braun's ghost. But this dreamer, like von Braun himself, also happened to be an aerospace engineer and a space evangelist. Robert Zubrin turned von Braun's *Mars Project* on its head with his own book, *The Case for Mars.* Far from making Mars accessible, Zubrin contended, von Braun's ponderous interplanetary wagon train—a "Club Med interplanetary cruise ship"—would be so expensive that it would never be launched. The former science teacher and senior engineer at Martin Marietta Astronautics noted that during the nineteenth century large flotillas of steam-powered British warships loaded with coal and supplies had battled ice packs for years in futile, and sometimes perilous, attempts to explore the Canadian Arctic. But explorers working for fur companies had done it on dogsleds by traveling light and living off the land. While working on advanced concepts for interplanetary missions in 1990, Zubrin came up with "Mars Direct": in effect, sending dogsleds to Mars and living off the land. *"It is the richness of Mars that makes the Red Planet not only desirable, but attainable,"* Zubrin believed implicitly.

The plan was to send a single unmanned and unfueled Earth Return Vehicle to Mars in 2005, perhaps landing on Utopia Planitia, where there is thought to be ground ice. There, it would use a pump to suck in carbon dioxide from the atmosphere, combine it with hydrogen brought from Earth, and produce enough methane-oxygen rocket fuel to make it back to Earth. In October 2007—the next window—a crew would be launched. It would land in April 2008. Since their return ride would be waiting for them, the explorers would not have to carry an immense amount of fuel. They would bring their own food. But those who followed would use extra oxygen produced by the Earth Return Vehicle to turn the ice and permafrost into water and grow their food in inflatable greenhouses. They would also build their own habitat by using geothermal power and make their own bricks, ceramics, glasses, plastics, metals, wires, and tubes. Eighteen months after they landed, the earthlings would come home on the Earth Return Vehicle, leaving behind a habitable outpost for the next crew to occupy and expand.

The problem faced by Mars Direct, in Zubrin's opinion, is political—or rather, economic—not technical. The hardware is available now, and the total initial mission cost would be about $20 billion, plus $2 billion more for each trip. That, said Zubrin, comes to about 7 percent of the civilian and military space budgets. The real barrier to such missions comes from "powerful forces within NASA linked to the Space Station program" who oppose Mars Direct because the plan does not need the station: it "dejustifies" it, in NASAese. It also does not require any assembly in orbit, which is most of the justification for all the space walks. And it even bypasses the Moon. It would, however, require making perhaps hundreds of thousands of gallons of propellant on Mars and loading them onto the return vehicle. More to the point, Saunders Kramer pointed out, it would require the American public to agree to "stuff" four to eight spacefarers into an Apollo-like module and send them off on a trip that could last eight months. He did not think the dogsled idea was politically feasible, and, he went on, it would be quite a bit more expensive that Zubrin thought.

Zubrin sees terraforming Mars as the real long-range goal, and it is an intensely practical one. Yet it would also be the most awesome technical challenge ever confronted by mankind. It would require raising Mars's temperature by 100 degrees Fahrenheit, turning its carbon dioxide atmosphere into breathable air, constructing farms, factories, and cities under Clarke's biospheric bubbles, and transporting thousands of people on space arks that would be the precursors of starships. Clarke, among other hard-core dreamers, never doubted that earthlings and Martians would inhabit neighboring planets before the end of the twenty-first century. Neither do the editors of the *Journal of the British Interplanetary Society,* which has devoted whole issues and even series of them to articles on terraforming—"Terraforming Mars: Conceptual Solutions to the Problem of Planet Growth in Low Concentrations of Oxygen," to take one example among many—exobiology, and practical robotics for interstellar missions.

Nor did they doubt that living spacecraft—starships, or colony ships—would one day strike out to carry the human seed to other star systems and even galaxies. Certainly Tsiolkovsky never doubted it. It is just what he had in mind when he conceived of whole civilizations floating in space. The British Interplanetary Society did a pioneering study of what it would take to send people to Barnard's Star, six light-years away, and published it in 1978 as Project Daedalus. The American Astronautical Society, which is as scientifically clearheaded as its British counterpart, took star travel seriously enough to publish a detailed study in 1992 of what would be required. The book, *Prospects for Interstellar Travel,* was written by John H. Mauldin, a Cornell University graduate with degrees in physics and a veteran of early work on Voyager. It went into considerable detail about every aspect of such odysseys. He discussed the engineering and operation of starships as entire life support systems, down to "astrogation," waste recycling, communication over immense distances, keeping social and political order, and other challenges that would come from spending entire generations and whole lifetimes on pilgrimages to destinations that are inconceivably distant. "Large colony 'worldlet' ships are starships for thousands of people with extensive growing areas maintained not just for food but for enjoyment of nature during long journeys requiring tens or hundreds of generations," he wrote.

Mauldin explained that travel to a nearby star would cover a distance a million times farther than puddle jumping to Mars and would have a time period thousands of times greater. "Costs," he calculated, "could be thousands to millions of times greater than all that has been spent on space thus far. The effort seems so large," Mauldin added, "that no one can make a realistic estimate of when a world society might be ready to plan an interstellar program." It would surely require such a society, with every nation able to do so committing to the common enterprise: the ultimate, fantastic human odyssey. That, in fact, would be one of its greatest attractions. Konstantin Tsiolkovsky, for one, never doubted that it would happen. Neither did Hermann Oberth, Robert Goddard, or any of the other dreamers.

"When the capability of travel to the stars is near to hand," Mauldin wrote at the end of his book, "then humans may ask: If we can go to the stars, will we and should we? Enthusiasts will by definition think so, but the billions of unidentified people who support the effort must be prepared to help make the decision. The long haul starts here at home now."

Saunders Kramer has no doubt that intergalactic missions will come. Indeed, they will have to come because this planetary system, like all the others that are almost certainly out there, is doomed. The Hubble Space Telescope has taken detailed pictures of stars in their death throes and shown that those the size of the Sun make flamboyant exits. With their hydrogen fuel exhausted, they expand like balloons into red giants, sending out waves of blistering heat. "After the Sun has become a red giant and burned the Earth to a cinder," one astronomer predicted, "it will eject its own beautiful nebula and then fade away as a white dwarf star. We have another six billion years to get out of town before that happens."

Kramer, the classic embodiment of science and the religion—the practitioner of technology in the service of the ancient dream—thinks that his race will eventually have to migrate over truly immense distances or vanish. It will have to ride the heavenly breezes or become extinct without leaving so much as a trace (except for the plaques and the derelicts carrying them) that it ever was or ever could be. Not striking out for other worlds, not spreading the seeds, would mean that those tortured souls at the end of the line would have to accept that their civilization willingly abandoned nature and ultimately counted for absolutely nothing: that Earth, soon to be the wayward corpse of another burned-out, unremarkable star, was utterly irrelevant, its death as unnoticed as its birth. They would agonize over the fact that, where the universe was concerned, they never happened.

There is no doubt in Kramer's mind, nor in the minds of the others, that their race has a different fate. They have no doubt that the journey has already begun; that the ultimate voyage is under way. "Would I volunteer to go on a starship?" Kramer asks. "Give me a microsecond to think about it." Saunders Kramer and the Saltine Warrior.

Notes

Note: All sources referred to in shortened form can be found referenced in full in the sources section.

Foreword

ix Sagan: 1970 London Lecture at Eugene, Oregon.

1. Bird Envy

3 *Saltine Warrior* was sculpted by Luise Meyers Kaish, a Syracuse alumna, and presented to the university by the class of 1951.

3 Gravity hater: Kowitt and Kaplan, "The Wings of Daedalus: The Convergence of Myth and Technology in 20th Century Culture," p. 440.

4 Awareness: White, *The Overview Effect,* p. 190.

4 "arts unknown": Mayerson, *Classical Mythology in Literature, Art, and Music,* p. 319.

5 "not the regions of the air": Holme, *Bulfinch's Mythology,* p. 187.

5 Homer and Socrates: *Classical Mythology in Literature, Art, and Music,* p. 319.

6 Ley on Aristotle: Ley, *Rockets, Missiles, and Space Travel,* p. 8.

6 Icaromenippus: Crouch, "To Fly to the World in the Moon: Cosmic Voyaging in Fact and Fiction from Lucian to Sputnik," in Emme, *Science Fiction and Space Futures,* p. 10.

6 Icaromenippus's description: Cain, *Luna Myth & Mystery,* p. 154. He is sometimes called Menippus.

7 "not science fiction": "To Fly to the World in the Moon," p. 10.

7 Elixir and Cheng Ssu-yüan: Fang-toh Sun, "Rockets and Rocket Propulsion Devices in Ancient China," in Skoog, *History of Rocketry and Astronautics,* Vol. 10, p. 29.

8 First firecrackers: Baker, *The Rocket,* p. 9.

8 "fire drug": "Rockets and Rocket Propulsion Devices in Ancient China," p. 29.

8 Siege of Tzu T'ung: *The Rocket,* p. 8.

8 Flying fire lance: "Rockets and Rocket Propulsion Devices in Ancient China," p. 32.

8 "running fire": Chae, "A Study of Early Korean Rockets," pp. 4–5.

8 Magical machine arrows and launcher: Ibid., pp. 6–14.

9 The popular formula: Skoog and Winter, "The Swedish Fire Arrow: The Oldest Rocket Specimen Extant," p. 46.

9 Battle in Silesia: Subotowicz, "Analysis of Rocket Construction, Described in Manuscripts and Printed Books During the 16th and 17th Centuries," p. 3.

9 *Rochetta,* the Italians, and the eclipse: "The Swedish Fire Arrow," p. 46.

10 Birds and swimmers: MacCurdy, *The Notebooks of Leonardo Da Vinci,* Vol. 1, p. 410.

10 Flying the ornithopter: Ibid., p. 498.

11 Copernicus: Shapley, Wright, and Wright, *A Treasury of Science,* pp. 56–57.

12 Haas: Baker, *The History of Manned Space Flight,* pp. 8, 13; "The Swedish Fire Arrow," pp. 51–53; "Analysis of Rocket Construction," pp. 3, 5.

13 People thrown off Earth: Galileo, *Dialogue on the Great World Systems,* p. xv.

13 Kepler the astrologer: Goodstein and Goodstein, *Feynman's Last Lecture,* p. 28.

13 Brahe and Kepler: Jay M. Pasachoff, *Contemporary Astronomy* (New York: Saunders College Publishing, 1981), pp. 242–43.

14 Bruno: Reston, *Galileo,* pp. 59–60; Zubrin, *The Case for Mars,* pp. 20–21.

14 Galileo's "spyglass": Chaisson, *The Hubble Wars,* p. 27.

14 Galileo's telescopes: *A Treasury of Science,* p. 58.

15 Conclusion about the Jovian moons: Ibid., pp. 60–61.
16 Marius and the moons: Burgess, *By Jupiter,* p. 33.
16 Galileo's description of the Moon: Morrison and Morrison, *The Ring of Truth,* p. 28. The book is dedicated "To Galileo, the Lynx with the glass eye."
16 Morrison and Morrison: Ibid., p. 27.
17 Cigoli: Ibid., p. 36.
17 "not very biblical": Ibid., p. 35.
17 "Lutheran" and thought police: Reston, *Galileo,* p. 59.
17 Magini: Galileo, *Dialogue on the Great World Systems,* p. xii.
18 Donne: Galileo, Ibid., p. xxi.
18 The historian of science was Giorgio de Santillana, writing in the introduction of ibid., p. xxi.
18 Kepler and Mercury's wand: Ibid., p. xiii.
18 Establishing the astronomy: Lear, *Kepler's Dream,* pp. 106–07.
19 *Somnium:* Robert Lambourne, Michael Shallis, and Michael Shortland, *Close Encounters?* (Bristol, England: Adam Hilger, 1990), p. 3.
19 The Jesuits: *Dialogue on the Great World Systems,* p. xxxix.
19 *Dialogue on . . . :* Stephen F. Mason, *A History of the Sciences* (New York: Collier Books, 1962), p. 161.
20 Pope John Paul II: "Vatican Science Panel Told by Pope: Galileo Was Right," *The New York Times,* November 1, 1992.
20 Telescope continues: *The Ring of Truth,* pp. 36, 40.
21 Moulton: *A Treasury of Science,* p. 66.
22 Wan Hu: Winter, "Who First Flew in a Rocket?" pp. 275–76.
22 Celebi: Ibid., p. 276.
22 "On Fire Balloons": Ibid., pp. 276–77.
23 Law: Ibid., pp. 278–80.
23 Bacon: *Rockets, Missiles, and Space Travel,* pp. 53–55. His contemporary, Albertus Magnus, is also mentioned.
23 Bayly: von Braun and Ordway, *History of Rocketry and Space Travel,* p. 30.
24 "sanitizing" statement: Becklake and Millard, "Congreve and His Works," p. 282.
24 Breakthroughs: Ibid., p. 282.
24 "above 2,000 rockets": *History of Rocketry and Space Travel,* p. 31.
25 Locke's series: French, *The Moon Book,* p. 41.
26 Life on the planets, rings, and moons: Dick, *Celestial Scenery,* p. 305.
26 The population of Mercury: Ibid., pp. 71–72.
26 Life in the Sun: Ibid., pp. 239–40.
26 Life on Mars: Ibid., pp. 137–39.
27 Schiaparelli and Lowell: Sheehan, *Planets and Perception: Telescopic Views and Interpretations, 1609–1909,* pp. 177–79; Hoyt, *Lowell and Mars,* pp. 80–83.
27 Fascinated: Ley and von Braun, *The Exploration of Mars,* p. 63.
27 Goddard and Lowell: Sagan, *Cosmos,* p. 111.
28 Verne, Nadar, and marriage: Costello, *Jules Verne,* pp. 69, 79–80.
28 Society for Aerial Locomotion and *Five Weeks in a Balloon:* Ibid., pp. 69, 75–76.
28 The five other space books: *Rockets, Missiles, and Space Travel,* pp. 39–40.
29 "Instead of yielding": Ibid., p. 40.
29 "the first mechanics": Verne, *From the Earth to the Moon,* p. 547.
29 "From the time of": Ibid., p. 568.
30 "The travelers' sleep": Ibid., p. 700.
30 "It is clear": Miller, *The Annotated Jules Verne: From the Earth to the Moon,* p. 21.
30 Poor translations: Ibid., p. x.
31 Not for children: Ibid.
31 Niagara and Miller: Ibid., p. 87.
32 Tsiolkovsky on Verne: Kosmodemyansky, *Konstantin Tsiolkovsky,* p. 37.
32 Oberth on Verne: *The Annotated Jules Verne,* p. 115.
32 Goddard on Verne: Lehman, *Robert H. Goddard,* p. 28.
33 Hale through Graffigny: Ordway and Liebermann, *Blueprint for Space,* pp. 53–56.
34 Dime novels and trips to the Moon and Mars: Winter and Liebermann, "A Trip to the Moon," p. 63.
34 *Luna*'s journey: Ibid., pp. 62–66.
35 Chronicle of early aviation: Kaempffert, *The New Art of Flying,* p. v.

2. Rocket Science

36 "no such thing as gravity.": Kosmodemyansky, *Konstantin Tsiolkovsky,* p. 9.
36 "It estranged me": Riabchikov, *Russians in Space,* p. 92.
36 Brown bread: *Konstantin Tsiolkovsky,* p. 11.
37 Yellow stains: Ibid., p. 13.
37 Fyodorov: Lytkin, Finney, and Alepko, "The Planets Are Occupied by Living Beings: Tsiolkovsky, Russian Cosmism and Extraterrestrial Civilizations," pp. 2–3. Also see White, *The Overview Effect,* p. 96.
38 Tsiolkovsky's monographs and atomic philosophy: "The Planets Are Occupied by Living Beings," pp. 5, 13, 14.
38 "[T]he observed body": *Russians in Space,* pp. 95–96.
38 Flying cask: *Konstantin Tsiolkovsky,* pp. 15–16.
39 Metal dirigibles: Ibid., pp. 18–19.
39 "Bird-like (Aviation) Flying Machine": *Konstantin Tsiolkovsky,* p. 26; *Russians in Space,* p. 97.
39 The two novels: *Russians in Space,* pp. 97–98.
41 The query letter: Ibid., p. 99.
42 Communist utopia: Stites, "World Outlook and Inner Fears in Soviet Science Fiction"; Graham, *Science and the Soviet Social Order,* pp. 299–301.
42 Tsiolkovsky's imagination: *Russians in Space,* pp. 94–95.
42 The 470-ruble grant: *Konstantin Tsiolkovsky,* pp. 31–32.
43 Supplements and rewards: Baker, *The Rocket,* p. 21.
43 All-Union Program: Medvedev, *Soviet Science,* pp. 61–62.

43 Angels in orbit: Tsiolkovsky, *Beyond the Planet Earth,* p. 159.
44 Goddard's maladies and self-education: Lehman, *Robert H. Goddard,* pp. 19, 28.
44 "No one would have believed": Wells, *The War of the Worlds,* p. 11.
45 The letter to Wells: *Robert H. Goddard,* p. 23.
45 The rocket and its mission: Goddard, *A Method of Reaching Extreme Altitudes,* pp. 1, 3, 6.
45 "never return": Ibid., p. 54.
46 Rocket to the Moon: Ibid., pp. 55–57.
46 Press release and headlines: *Robert H. Goddard,* pp. 108 and 104, respectively.
46 The *Times*' story: "Believes Rocket Can Reach Moon," *The New York Times,* January 12, 1920.
46 Bronx Exposition, Inc.: Goddard and Pendray, *The Papers of Robert H. Goddard,* Vol. I, pp. 408–09.
47 The damning editorial: Topics of the Times, *The New York Times,* January 13, 1920.
47 "Too much attention": "Goddard Rockets to Take Pictures," *The New York Times,* January 19, 1920.
47 "Well, Robert": *Robert H. Goddard,* p. 108.
47 American public's fancy: Ibid., pp. 110–11.
47 The three Russian students: *The Papers of Robert H. Goddard,* Vol. 2, p. 666.
48 Oberth's childhood: von Braun and Ordway, *Space Travel, A History,* p. 56.
48 Liquid-propelled bombardment missile: Ibid., p. 57.
49 Oberth's letter: *Robert H. Goddard,* p. 132.
49 Goddard and Oberth: Ibid., pp. 132–33.
49 The rejected dissertation: Oberth, "My Contributions to Astronautics," in Durant and James, *First Steps Toward Space,* p. 136.
50 Oberth's four main points: *Rockets, Missiles, and Space Travel,* p. 109.
50 "explained": "My Contributions to Astronautics," p. 136.
50 Details of Oberth's book: *Space Travel: A History,* p. 57; *Rockets, Missiles, and Space Travel,* pp. 109–10.
50 "completely independently": *Space Travel: A History,* p. 57. Goddard's work on liquid rockets in 1920 is from his foreword to a reissue of *A Method of Reaching Extreme Altitudes* written in May 1945.
50 Goddard's dismay: *Robert H. Goddard,* p. 133.
51 Letter to the secretary: *The Papers of Robert H. Goddard,* Vol. 1, pp. 497–98.
51 "possible to think.": *Rockets, Missiles, and Space Travel,* p. 108.
51 Tsiolkovsky's distribution: Winter, *Prelude to the Space Age/The Rocket Societies: 1924–1940,* p. 22.
52 "a theorist, not an engineer": *Rockets, Missiles, and Space Travel,* p. 125.
52 Nebel's failed rocket: Ibid., pp. 128–30; *Space Travel: A History,* p. 58.
52 Valier's call to space: *Robert H. Goddard,* pp. 134–35.
53 Valier's death: *Rockets, Missiles, and Space Travel,* pp. 135–36.
53 Opel and the rocket stunts: Neufeld, *The Rocket and the Reich,* pp. 7–8.
53 Goddard's flight statistics: *The Papers of Robert H. Goddard,* Vol. 2, p. 580.
54 The second entry: Ibid., Vol. 2, p. 581.
54 "get the hell out of here!": *Robert H. Goddard,* p. 143n.
55 Hohmann Transfer: Schulz, "Walter Hohmann's Contributions Toward Space Flight: An Appreciation on the Occasion of the Centenary of His Birthday," pp. 290–95; Damon, *Introduction to Space,* p. 40; Glasstone, *Sourcebook on the Space Sciences,* pp. 69–71.
55 Warning to Esnault-Pelterie: Blosset, "Robert Esnault-Pelterie: Space Pioneer," in Durant and James, *First Steps Toward Space,* p. 10.
55 *L'Astronautique:* Ibid., pp. 11–12.
55 The word "astronautics": *Prelude to the Space Age/The Rocket Societies: 1924–1940,* p. 25.
55 Oberth the winner: "Robert Esnault-Pelterie: Space Pioneer," p. 11.
56 Sänger: Baker, *The Rocket,* p. 90; Ordway and Liebermann, *Blueprint for Space,* p. 65.
56 Mail rockets: Schmiedl, "Early Postal Rockets in Austria: A Memoir," in Hall, *History of Rocketry and Astronautics,* pp. 109–12.
57 Potocnik: *Blueprint for Space,* p. 110; *Rockets, Missiles, and Space Travel,* pp. 369–71; Miller, "Herman Potocnik—*Alias* Hermann Noordung," pp. 295–96.
57 The 1926 edition: *Robert H. Goddard,* p. 135.
58 Tikhomirov and his laboratory: *Prelude to the Space Age,* p. 55.
58 Tikhomirov and the crater: Baker, *The Rocket,* p. 77.
58 Exhibitions: Ibid.
59 GIRD branches: Merkulov, "Organization and Results of the Work of the First Scientific Centers for Rocket Technology in the USSR," in Ordway, *History of Rocketry and Astronautics,* p. 74.
59 Rynin to Goddard: Goddard and Pendray, *The Papers of Robert H. Goddard,* Vol. 2, p. 585.
59 Perelman: *Space Travel: A History,* p. 61.
60 "On to Mars!" (and Mercury): *Russians in Space,* p. 106.
60 Tsander, Glushko, and Tikhonravov: *Space Travel: A History,* p. 62.
60 "She really flew": *Russians in Space,* p. 113.
60 "rockets must conquer space!": *The Rocket,* p. 80.
60 Model 212: *Russians in Space,* photo caption facing p. 64; Golovanov, *Korolev,* pp. 194–95.
61 RNII: Pobedonostsev, Shchetinkov, and Galkovsky, "A History of the Organization and Activity of the Jet Propulsion Research Institute (RNII), 1933–1944," in Lattu, *History of Rocketry and Astronautics,* pp. 68–69. This paper, delivered at a history symposium at Baku in 1973, put the number of workers at 260. Frank H. Winter, an American space scholar, put it at about 1,000 and reported that Glushko himself claimed that LenGIRD alone had more than 400 employees by 1932 (*Prelude to the Space Age,* p. 63).
61 Tikhonravov at RNII: Yu. V. Biryukov, "The Role of Mikhail K. Tikonravov in the Development of Soviet Rocket and Space Technology,"

in Lattu, *History of Rocketry and Astronautics,* p. 347.
61 The purge: *Prelude to the Space Age,* p. 63.
61 The purge of Keldysh: Medvedev, *Let History Judge,* p. 441.
62 Kapitsa on Beria: Author interview.
62 Beria and science: Interviews with Khozin, Kapitsa, and Khrushchev; Rhea, *Roads to Space,* p. 341; Bethe, Gottfried, and Sagdeev, "Did Bohr Share Nuclear Secrets?" pp. 88–89.
62 Testimony by Glushko and the charge: Author interview with Arthur M. Dula, July 13, 1996; Harford, *Korolev,* p. 49.
62 Tupolev and Korolyov: Harford, *Korolev,* p. 57.
63 "golden cage": Author interview.
63 Korolyov in the gulag: Oberg, *Red Star in Orbit,* pp. 20–21.
63 Korolyov's afflictions: Harford, *Korolev,* p. 51.
63 Korolyov's cynicism: Vladimirov, *The Russian Space Bluff,* p. 146.
63 Meteorological rockets: Polyarny, "On Some Work Done in Rocket Techniques, 1931–38," in *First Steps Toward Space,* p. 187.
63 Korolyov to Perelman: *Prelude to the Space Age,* p. 66.
64 Ley on the origin of the VfR: *Rockets, Missiles, and Space Travel,* p. 139.
65 Scrounging: *Prelude to the Space Age,* p. 41.
65 Welfare checks: Ibid.
65 Dark suits and the tuxedo: *Rockets, Missiles, and Men in Space,* fifth photo page after p. 174.
65 Marianoff: *Prelude to the Space Age,* p. 42.
65 von Braun joins the VfR: Ibid., p. 39.
65 First Mirak: Ibid., p. 40.
65 May 10, 1931, launch: Ibid., p. 43.
65 Riedel to Ley: Ordway and Sharpe, *The Rocket Team,* p. 15.
66 Flight of the Repulsor: *Rockets, Missiles, and Space Travel* (1957), p. 148.
66 Who's Who of rocketry: Ibid., p. 117.
66 Records, tests, and flights: *Prelude to the Space Age,* p. 43.
67 Powder blue uniforms: *Rockets, Missiles, and Space Travel,* p. 160.
67 British Interplanetary Society: Thompson and Shepherd, "The British Interplanetary Society: The First Fifty Years (1933–1983)," in Sloop, *History of Rocketry and Astronautics,* Vol. 12, pp. 37–55.
67 American Interplanetary Society: *Prelude to the Space Age,* p. 73.
67 Pendray's VfR report: Pendray, "The German Interplanetary Society and the *Raketenflugplatz,*" pp. 5–12.
68 Lasser: "David Lasser, 94, a Space and a Social Visionary," *The New York Times,* May 7, 1996.
68 Lasser on rocket war: "The Rocket and the Next War," AIS *Bulletin,* No. 13, November 1931, p. 7.
68 First AIS *Bulletin: Bulletin,* No. 1, June 1930.
69 Pickering: "Presence of Lunar Life Debated," AIS *Bulletin,* No. 12, September 1931, pp. 9–10.
69 Martian vegetation: Schachner, "Can Man Exist on Other Planets?" p. 9.
69 Esnault-Pelterie: "Two Thousand at Museum Meeting," AIS *Bulletin,* No. 7, February 1931, pp. 1–5.
69 "lousy planet": *Prelude to the Space Age,* p. 77.
69 Smith's launch: Ordway, "Some Vignettes from an Early Rocketeer's Diary: A Memoir," in Sloop, *History of Rocketry and Astronautics,* pp. 70–71.
70 China Clipper: Gwynn-Jones, "Farther: The Quest for Distance," in Greenwood, *Milestones of Aviation,* p. 69.
71 Gernsback: Ordway and Liebermann, *Blueprint for Space,* pp. 72, 76.
73 *Things to Come:* The film. Also see Ordway, "Space Fiction in Film," in Emme, *Science Fiction and Space Futures,* pp. 32–33; Verne's *Paris in the Twentieth Century* was published by Random House in 1996.

3. Gravity's Archers

74 Oberth's predictions: "War with Rockets Pictured by Oberth," *The New York Times,* January 31, 1931.
75 Shirer's description: Shirer, *The Rise and Fall of the Third Reich,* pp. 137–38.
75 Election results: Ibid., p. 138.
76 Seeckt's doctrine: Wheeler-Bennett, *The Nemesis of Power: The German Army in Politics, 1918–1945,* pp. 96–97.
76 Becker's interest in rockets: Neufeld, *The Rocket and the Reich,* p. 16.
77 Attitude toward Nebel and secrecy: Ibid., pp. 18, 21.
77 Dornberger's mandate: Ordway and Sharpe, *The Rocket Team,* p. 18.
77 " 'circus-type approach' ": *Spaceflight,* a documentary that was produced by KCET in Los Angeles and which aired on public television in May 1985 (Part 1).
77 Flight of the Repulsor: *The Rocket and the Reich,* pp. 19–20; *The Rocket Team,* pp. 18–19.
78 Nebel's complaint: Ley, *Rockets, Missiles, and Space Travel,* p. 199.
78 Riedel's decision: Ibid., p. 21.
78 "vulgar": Ibid., p. 22.
79 Milking the cow: Ibid.
79 Goebbels's decree: Ibid., p. 28.
79 A-1: *The Rocket,* pp. 35–36.
80 Flight of the A-2s: *The Rocket and the Reich,* p. 38.
80 Von Braun's dissertation and diploma: Ibid., p. 37.
82 Baroness von Braun's suggestion: Ibid., p. 49.
82 The division of Peenemünde and pledges: Ibid., p. 50.
82 Becker's warning: Ibid.
82 "How much do you want?": *The Rocket Team,* p. 24; *The Rocket and the Reich,* p. 50.
82 "big time": *The Rocket and the Reich,* p. 50.
82 More than $70 million: *The Rocket Team,* p. 31.
82 "arrow stability": *The Rocket and the Reich,* p. 67.
83 Lindbergh and Guggenheim: Lehman, *Robert H. Goddard,* pp. 173–74.

84 The Carnegie grant: Durant, "Robert H. Goddard and the Smithsonian Institution," *First Steps Toward Space,* p. 63.

84 Goddard meeting Lindbergh: *The Papers of Robert H. Goddard,* Vol. 2, pp. 723–24.

84 Equipment to New Mexico: *Robert H. Goddard,* p. 178.

84 Mescalero and "High Lonesome": Ibid., pp. 179, 180–81.

84 Goddard's patents: *The Papers of Robert H. Goddard,* Vol. 3, p. 1651.

84 Goddard to the admiral: Ibid., Vol. 1, pp. 442–43.

85 From the second monograph: Goddard, *Liquid-Propellant Rocket Development,* p. 2.

85 Goddard to Harry Guggenheim: *The Papers of Robert H. Goddard,* Vol. 2, p. 923.

86 Malina's proposal and Millikan's response: Malina, "On the GALCIT Rocket Research Project, 1936–38," in *First Steps Toward Space,* pp. 114–15; von Karman with Lee Edson, *The Wind and Beyond,* p. 235.

87 The invasion of the Martians: Wilford, *Mars Beckons,* pp. 36–39; "Excerpts from the 'War' Broadcast," *The New York Times,* November 1, 1938.

88 Newark to Pittsburgh: *Mars Beckons,* p. 41; "Radio Listeners in Panic, Taking War Drama as Fact," *The New York Times,* October 31, 1938.

88 The press conference: *The Battle over Citizen Kane,* WGBH (Boston PBS), February 1996.

88 Parsons and Forman: *The Wind and Beyond,* p. 235; Koppes, *JPL and the American Space Program,* p. 3.

89 "Suicide Club": *The Wind and Beyond,* p. 240.

89 Millikan to Goddard: *The Papers of Robert H. Goddard,* Vol. 2, p. 1012.

89 Parsons and Goddard: Ibid., Vol. 2, pp. 931–32, 938.

89 Malina on Goddard: "On the GALCIT Rocket Research Project, 1936–38," p. 117.

90 Goddard to Millikan: *The Papers of Robert H. Goddard,* Vol. 2, pp. 1012–13.

90 Von Karman on Goddard: *The Wind and Beyond,* pp. 241–43.

90 How Goddard did not help the Germans: Neufeld, *Rockets and the Reich,* p. 53.

91 Goddard in the *Bulletin:* See, for example: "Goddard Describes New Stratosphere Plane," *Bulletin,* No. 12, September 1931, and his letter to the editor in the June–July 1931 issue.

91 Lobbied for contracts: *Robert H. Goddard,* pp. 294–306.

91 "trench mortars": *The Papers of Robert H. Goddard,* Vol. 3, p. 1311.

92 Goddard to Abbot: Ibid., Vol. 3, pp. 1313–14.

92 Goddard at du Pont: Ibid., Vol. 2, p. 715.

93 Goddard's talk with Boushey: *Robert H. Goddard,* pp. 314–17.

93 Boushey's Moon base: Allen, "Early Lunar Base Concepts of the U.S. Air Force," pp. 4–5.

94 Flying Fortress analogy: Huntington, "V-2: The Long Shadow," p. 85.

95 Aerodynamics institute: *The Rocket and the Reich,* p. 88.

95 Goddard's demonstration: *The Papers of Robert H. Goddard,* Vol. 1, pp. 291–316.

95 Ideologically dedicated Nazis: *The Rocket and the Reich,* pp. 87 (Hermann), 101 (Steinhoff), 180 (Rudolph). Beier: Hunt, *Secret Agenda,* p. 44.

95 Rudolph's lament: Ordway and Sharpe, *The Rocket Team,* p. 22.

95 Rudolph and Dora: *Secret Agenda,* p. 218.

96 Porsche: "Volkswagen's History: The Darker Side Is Revisited," *The New York Times,* November 7, 1996.

96 Dornberger on Nazism: *The Rocket and the Reich,* p. 183.

96 Zanssen: Ibid., pp. 180–84.

96 Von Braun and the SS: Ibid., p. 178.

96 Himmler and Peenemünde: *The Rocket,* p. 52.

96 The rank of major: *Secret Agenda,* p. 44.

96 Porsche and the SS: "Ferdinand Porsche, Creator of the Sports Car That Bore His Name, Is Dead at 88," *The New York Times,* March 28, 1998.

97 Scientists and engineers: *The Rocket and the Reich; The Rocket Team.*

98 Propellant tanks and combustion chamber: *The Rocket,* pp. 44–45.

98 The nozzle: *The Rocket and the Reich,* p. 78.

98 Rudders, gyroscopes, and the brains: *The Rocket,* pp. 45–46.

99 A-11: Gatland, *Project Satellite,* p. 48.

100 Ehricke on the launch: *Spaceflight,* Part 1, *NOVA,* May 1985.

100 Dornberger on the first successful launch: *The Rocket and the Reich,* pp. 164–65.

100 Hitler to Speer: Speer, *Inside the Third Reich,* p. 368.

101 The *A-4 Primer:* McGovern, *Crossbow and Overcast,* pp. 80–81. An English translation was published in 1957 Reports Publications Section of the then–Army Ballistic Missile Agency, Redstone Arsenal, Huntsville, Alabama.

102 Total launches and deaths: *The Rocket Team,* pp. 251, 252. V-2 deaths: Huntington, "V-2: the Long Shadow," p. 87.

102 Murrow on V-2s: Bliss, *In Search of Light,* p. 89.

102 Reconnaissance of Peenemünde: Babington-Smith, *Air Spy,* pp. 204, 209, 210.

103 The raid on Peenemünde: McGovern, *Crossbow and Overcast,* pp. 29–30; *The Rocket,* p. 54.

103 Irreplaceable targets: Johnson, *The Secret War,* p. 133.

103 Himmler in the pie: Speer, *Inside the Third Reich,* p. 369.

103 V-3: Irving, *The Mare's Nest,* pp. 121, 178.

103 "antechambers of hell": Michel, *Dora,* p. 202.

104 Conditions at Dora: *The Rocket and the Reich,* pp. 210–11.

104 Total number of deaths: *Dora,* p. 95.

104 V-1 and V-2 numbers: *The Rocket and the Reich,* p. 263.

104 Mass hangings: *Dora,* p. 96.

104 Sabotage and Rudolph: *Secret Agenda,* pp. 74–75.

104 Von Braun and the "detainees": *The Rocket and the Reich,* p. 228.

105 Baldridge on Dora: Baldridge, *Victory Road,* p. 179.
105 On von Braun's role: Ibid., p. 180.
105 Michel on scientists and Dornberger: *Dora,* p. 99.
105 Michel on the V-2's uselessness: Ibid., p. 206.
106 Speer on the V-2's uselessness: *Inside the Third Reich,* p. 366.
106 V-2's estimated cost: *The Rocket Team,* p. 253.
106 HS-293: *Rockets, Missiles, and Space Travel,* p. 241.
106 Missiles from submarines: Baldwin, *The Great Arms Race,* p. 27.

4. Missiles for America

108 "that leaves the Americans.": Ordway and Sharpe, *The Rocket Team,* p. 274.
109 Huzel's rationale: Ibid., p. 261.
109 Pots and pans: Ibid., p. 267.
109 Crates of documents: McGovern, *Crossbow and Overcast,* pp. 113–16.
109 Vengeance Express: Ibid., p. 112.
110 Kammler's hostages: von Braun and Ordway, *Space Travel: A History,* p. 115.
110 Liberation of Nordhausen and something "fantastic": *Crossbow and Overcast,* p. 120.
111 "death from the stratosphere": DeVorkin, "War Heads into Peace Heads: Holger N. Toftoy and the Public Image of the V-2 in the United States," p. 439.
111 12-K: Dushkin, "Experimental Research and Design Planning in the Field of Liquid-Propellant Rocket Engines Conducted Between 1934–1944 by the Followers of F. A. Tsander," p. 81.
111 The Malina–Hsue-shen report and JPL: Malina, "America's First Long-Range-Missile and Space Exploration Program: the ORDCIT Project of the Jet Propulsion Laboratory, 1943–1946," p. 343.
112 Creation of ORDCIT: Ibid., pp. 344–46.
112 Blacklist: *Crossbow and Overcast,* p. 96.
112 "good riddance.": *The Rocket Team,* p. 278.
112 Staver in London: *Crossbow and Overcast,* p. 99.
113 Evacuation of the V-2s: *The Rocket Team,* pp. 278–82.
114 Korolyov at Mittelwerk: Neufeld, *The Rocket and the Reich,* p. 267; Harford, *Korolev,* pp. 64–70.
114 Chertok's mission in Germany: Rhea, *Roads to Space,* p. 473; author interview with Dula.
114 All-Union Design Bureau: *Roads to Space,* p. 36.
114 Soviets in Germany: Ibid., biographical sketches, and interview with Mishin.
115 Isayev in the nozzle: Harford, *Korolev,* p. 65.
115 Korolyov at Cuxhaven: *The Rocket Team,* p. 306.
115 Von Karman and Pickering at Cuxhaven: Ibid., pp. 307–08.
116 Hochmuth's conversation: Ibid., pp. 306–07.
116 Magnus surrenders: *Crossbow and Overcast,* p. 144; *The Rocket Team,* pp. 1–2.
116 "talking Chinese": *Crossbow and Overcast,* p. 146.
117 "never make it to America": *The Rocket Team,* p. 274.
117 Broadcasting jobs: *The Rocket and the Reich,* p. 268. Bribing the cooks: *Inside the Third Reich,* p. 505. Kidnapping von Braun: *Korolev,* p. 68.
117 Grottrup: *The Rocket and the Reich,* pp. 268–69.
117 "we got four or five": *Rockets, Missiles, and Space Travel,* p. 244. The six thousand figure is from Medvedev, *Soviet Science,* p. 53.
117 Axster: *Secret Agenda,* p. 44.
118 Von Braun's goal: *The Rocket Team,* p. 274.
118 Hsue-shen Tsien and von Braun: Hall, "Earth Satellites, a First Look by the United States Navy," in *History of Rocketry and Astronautics,* Vol. 7, p. 253.
118 Von Braun's report: *Crossbow and Overcast,* pp. 147–48.
119 Solar space station: "The Station in Space: Sun Power Stations Planned By Germans," *Journal of the American Rocket Society,* pp. 8–9.
119 Regener and the Tonne: DeVorkin, *Science with a Vengeance,* pp. 26–36.
119 Patterson's concern: *Crossbow and Overcast,* p. 193.
120 Hoover's warning: *Secret Agenda,* p. 263.
121 Whitman on Nazis: Whitman, *The Obituary Book,* p. 8.
121 Unit 731: Williams and Wallace, *Unit 731,* pp. 69–70, 210; "Japan Confronting Gruesome War Atrocity," *The New York Times,* March 17, 1995.
122 "wards of the Army": *Crossbow and Overcast,* p. 205.
122 The Fort Bliss seven: Ibid., p. 207.
122 Experts to Wright Field and the Navy: *Secret Agenda,* p. 31.
122 Prisoners of peace: *The Rocket Team,* p. 347.
122 At least 1,600 and altered files: *Secret Agenda,* pp. 1, 3.
122 Dornberger's interrogation: *Rockets and the Reich,* p. 269.
122 Dornberger and Ehricke at Bell: *The Rocket Team,* p. 373.
122 "100% NAZI": *Secret Agenda,* p. 30.
123 Wallops Island: Shortal, "Rocket Research and Tests at the NACA/NASA Wallops Island Flight Test Range 1945–1959: A Memoir," in Skoog, *History of Rockets and Astronautics,* p. 206.
123 V-2s to White Sands: Powell and Scala, "Historic White Sands Missile Range," p. 87. Tonnage: DeVorkin, "War Heads into Peace Heads: Holger N. Toftoy and the Public Image of the V-2 in the United States," p. 440.
124 ORDCIT/JPL: Koppes, *JPL and the American Space Program;* Pickering and Wilson, "Countdown to Space Exploration: A Memoir of the Jet Propulsion Laboratory, 1944–1958," in *History of Rocketry and Astronautics,* Vol. 7, Part 2, pp. 385–91.
124 Inertial guidance at MIT: Draper, "The Evolution of Aerospace Guidance Technology at the Massachusetts Institute of Technology, 1935–1951: A Memoir," in *History of Rocketry and Astronautics,* pp. 232–50.
124 X-1: Guenther and Miller, *Bell X-1 Variants,* pp. 6–8.
124 Committee for Evaluating the Feasibility of Space Rocketry: "Earth Satellites, a First Look by the United States Navy," *History of Rocketry and Astronautics,* pp. 254–57.
125 $5 million to $8 million: Ibid., p. 258.

125 LeMay: Ibid.
125 Bombing the *Utah:* Copp, *A Few Great Captains,* pp. 392–98.
126 Spaatz and Doolittle: McDougall, . . . *the Heavens and the Earth,* p. 88.
127 Space as an extension of airpower: Stares, *The Militarization of Space,* p. 26.
127 Background of the RAND study: Davies and Harris, *RAND's Role in the Evolution of Balloon and Satellite Observation Systems and Related U.S. Space Technology,* pp. 3, 7.
128 RAND's satellite predictions: *Preliminary Design of an Experimental World-circling Spaceship,* pp. 1–2.
128 RAND engineering analysis: Ibid., pp. 30–37.
128 A meteoroid hit: Ibid., p. 156.
128 Ridenour: Ibid., pp. 9–16.
129 The February 1947 report: *RAND's Role in the Evolution of Balloon and Satellite Observation Systems and Related U.S. Space Technology,* p. 10.
129 The birth of the RAND Corporation: Ibid., p. 18.
129 The RDB ruling: Bulkeley, *The Sputniks Crisis and Early United States Space Policy,* p. 81.
129 Ridenour on space travel: *Preliminary Design for an Experimental World-circling Spaceship,* p. 16.
129 Rocket Sonde Research Section's purpose: Bergstralh and Krause, "Early Upper Atmospheric Research with Rockets," p. 135.
130 Krause on rockets: DeVorkin, "Scientists and the Space Sciences," in Collins and Kraemer, *Space: Discovery and Exploration,* pp. 85, 87.
130 Van Allen and the use of rockets: Author interview.
130 The V-2 panel: "Early Upper Atmospheric Research With Rockets," p. 135.
131 Van Allen on V-2 experiments: Author interview.
131 Toftoy on civilian scientists: Devorkin, *Science with a Vengeance,* p. 61.
132 A-9/A-10 objective: Neufeld, *The Rocket and the Reich,* p. 283.
132 A-9 and A-10 details: Lipp, "Interrogation of Professor von Braun on Missile Design," p. 3.
132 Lipp's report: *RAND's Role in the Evolution of Balloon and Satellite Observation Systems and Related U.S. Space Technology,* p. 9.
132 DeVorkin's observations: *Science with a Vengeance,* pp. 125 and 128.
132 V-2 failures: "Historic White Sands Missile Range," pp. 89–90.
133 Attack on Ciudad Juárez: *The Rocket Team,* p. 355; *Spaceflight* (KCET), Part 1.
133 Blockhouse: "Historic White Sands Missile Range," p. 83; *Science with a Vengeance,* p. 110.
133 Ultraviolet data: *Science with a Vengeance,* pp. 143–44.
133 Havens: Newell, *Beyond the Atmosphere,* pp. 36–37.
133 The mouse and the monkeys: "Historic White Sands Missile Range," p. 89.
133 "peace head": *Science with a Vengeance,* p. 154.
134 Aerobee: "Historic White Sands Missile Range," p. 95; Baker, *The Rocket,* p. 231.
134 Origin of Viking: Murphy, "Liquid Rocket Propulsion System Advancements 1946–1970," p. 259.
134 Viking the missile: *Science with a Vengeance,* p. 177.
134 Cost of Viking and Aerobee: "Historic White Sands Missile Range," p. 92.
135 Viking 11 imagery: Baumann and Winkler, "Rocket Research Report No. XVIII: Photography from the Viking 11 Rocket at Altitudes Ranging up to 158 Miles," picture 127.
135 March 1947 V-2 imagery: T. A. Bergstralh, "Photography from the V-2 at Altitudes Ranging up to 160 Kilometers," in Newell and Siry, *Upper Atmosphere Research Report No. IV,* pp. 119–20.
135 Soviet atomic bomb: Truman, *Years of Trial and Hope,* p. 306.
136 Clandestine Soviet attack: "Soviet Capabilities for Clandestine Attack Against the US with Weapons of Mass Destruction and the Vulnerability of the US to Such Attack (Mid-1951 to Mid-1952)," pp. 1–2.
136 Mishin and clandestine attack: Author interview.
137 Nike and Bomarc: *Jane's All the World's Aircraft, 1962–63,* pp. 409–10, 389–90.
137 MX-774: *The Rocket,* p. 231.
138 The ARS, Porter, and Stehling: *The Sputniks Crisis and Early United States Space Policy,* pp. 83–84.
139 "hooey": Ibid., p. 83.
139 MOUSE: Ley, *Rockets, Missiles, and Space Travel,* p. 329.
140 Haviland's paper: "TV Rocket Kept Secret by Builders," *New Haven Register,* October 14, 1951.
140 UFOs: "Fans Hang On As UFOs Spin into 3d Decade," *The Washington Post,* July 2, 1967.
141 The UFO phenomenon: Peebles, *Watch the Skies!,* pp. 244–50; McAndrew, *The Roswell Report: Case Closed;* "For U.F.O. Buffs, 50 Years of Hazy History," *The New York Times,* June 14, 1997.
142 Von Braun versus Rosen: "2 Rocket Experts Argue 'Moon' Plan," *The New York Times,* October 14, 1952; "Space Rockets with Floating Base Predicted," *New York Herald Tribune,* October 14, 1952.
143 Bonestell: Miller, "Chesley Bonestell's Astronomical Visions," *Scientific American,* pp. 76–81.
143 Space rhetoric: "What Are We Waiting For?" *Collier's,* March 22, 1952, p. 23.
144 Lunar base: Ley, "Inside the Lunar Base," *Collier's,* October 25, 1952, p. 46.
144 Whipple on Mars: Whipple, "Is There Life on Mars?" *Collier's,* April 30, 1954, p. 21.
144 First "Tomorrowland": "Man in Space" was directed by Ward Kimball, who had read the *Collier's* series and brought it to Walt Disney's attention.
144 "practical passenger rocket": Smith, "They're Following Our Script," *Future,* p. 58.
145 Von Braun's lecture: Armed Forces Staff College, "Challenge of Outer Space," 1955.

5. The Other World Series

149 *Freedom and You: Wings over Water* documentary, WNET, October 22, 1986.
149 "Brother-Son Network": Klehr, Haynes, and Firsov, *The Secret World of American Communism,* pp. 205–34.
150 27 million Soviet losses: "Allied Victory in Europe Commemorated in Moscow," *The New York Times,* May 10, 1995.
151 A vast Peenemünde: Bulkeley, *The Sputniks Crisis and Early United States Space Policy,* p. 65.
151 Chicago to Tokyo and LeMay's observation: Rhodes, *Dark Sun,* pp. 22–23.
152 The "Strategic Chart:" Ibid., pp. 23–24.
152 Kramer and the Soviet targets: Memorandum to the author, December 8, 1997.
152 The bomber force and long-range operations: Polmar and Laur, *Strategic Air Command,* pp. 36, 38.
153 LeMay: Brugioni, *Eyeball to Eyeball,* pp. 263, 265.
153 LeMay's nuclear policy: Rhodes, "The General and World War III," pp. 53, 55, 59; *Dark Sun,* pp. 574–75.
154 LeMay's oath: Conversino, "Back to the Stone Age: The Attack on Curtis E. LeMay," p. 66. The other historian, cited by Conversino, was R. Cargill Hall.
154 Khrushchev's fear of attack: McDonald, *Corona,* p. 41.
155 Kistiakowsky: Author interview.
156 Tokayev, TU-4, Mya-4, TU-16, and TU-95: Prados, *The Soviet Estimate,* pp. 39–40.
156 Bissell: Author interview.
157 Aerial penetrations: Burrows, "Beyond the Iron Curtain."
157 Ike and a surprise attack: Killian, *Sputnik, Scientists, and Eisenhower,* p. 68.
157 "from the sky": Author interview, February 11, 1985.
158 TCP predictions: *Sputnik, Scientists, and Eisenhower,* pp. 73–74.
158 Offense and defense recommendations: Ibid., pp. 76–78.
159 What intelligence can uncover: Ibid., p. 80.
159 TCP letter to the Department of State: "Report to the President on the Threat of Surprise Attack," March 14, 1955, p. 3.
160 Eisenhower's attitude toward the Air Force and the U-2 program: Author interview with Killian.
160 Briefing Stalin: *The Rocket,* p. 115.
161 Tokayev: *The Sputniks Crisis and Early United States Space Policy,* p. 65; Kapitsa: Author interview.
161 Stalin's eyebrows: Golovanov, *Korolev,* p. 352.
161 BON: Ibid., pp. 360–61.
161 Grottrup's assignment: "Scientific Research Institute and Experimental Factory 88 for Guided Missile Development, Moskva/Kaliningrad," pp. 2–3 (subsequently referred to as OSI report of March 1960).
162 Korolyov and NII 88: Rhea, *Roads to Space,* p. 129.
162 R-10/R-15: OSI report of March 1960, pp. 7–8.
162 R-14 report: "The R-14 Project, a Design of a Long Range Missile at Gorodomyla [*sic*] Island." "Korolov" is mentioned on p. 3.
162 Kapustin Yar: Harvey, *Race into Space,* p. 22; *Roads to Space,* p. 128.
163 The special train: Golovanov, *Korolev,* pp. 362–63.
163 Glushko, Isayev, and rocket techniques: Prishchepa, "History of Development of First Space Rocket Engines in the USSR," pp. 95–100.
163 Khrushchev's introduction to rockets: Oberg, *Red Star in Orbit,* p. 25.
164 Thrust improvement and clustering: "History of Development of First Space Rocket Engines in the USSR," pp. 97–100.
164 Tikhonravov and rocket clusters: Author interview with Sergei Khrushchev; *Roads to Space,* p. 486.
164 Tikhonravov's multistage rocket: Tikhonravov, "How to Make Rockets Fly Farther," pp. 3–25.
164 RD-107/RD-108: Glushko, *Rocket Engines GDL-OKB,* pp. 25–26; Bilstein, *Stages to Saturn,* pp. 386–87.
165 Creation of Tyuratam: Harvey, *Race into Space,* p. 27.
165 The four R-7 launches: Varfolomeyev, "Soviet Rocketry That Conquered Space," p. 261.
165 Russian threats: Prados, *The Soviet Estimate,* p. 55.
166 Malinovsky and Khrushchev: McDougall, . . . *the Heavens and the Earth,* p. 240.
166 Seventeen and a half hours: Kramer memorandum of December 12, 1997.
166 Expense of the R-7 as an ICBM: Author interview with Sergei Khrushchev.
166 NSC 5520 and a large surveillance satellite: Day, "A Strategy for Space," p. 309.
167 Quarles and Waterman: Ibid., p. 308; NSC 5520, p. 309.
167 "The national interest of the United States": *The Sputniks Crisis and Early United States Space Policy,* p. 102.
168 NSC 5522: National Security Council NSC 5522, June 8, 1955, "Comments on the Report to the President by the Technological Capabilities Panel," pp. S5, S23.
168 "waning rather rapidly": *The Sputniks Crisis and Early United States Space Policy,* p. 90.
169 MOUSE: Clarke, *The Making of a Moon,* pp. 33–34; Gatland, *Project Satellite,* 56–57.
169 The IGY meeting and the lobster: Sullivan, *Assault on the Unknown,* pp. 20–22.
170 The Air Force's megaton warhead: NSC 5522, p. S7.
170 Minimum Satellite Vehicle (Orbiter): Hall, "Origins and Development of the Vanguard and Explorer Satellite Programs," pp. 104–05.
171 Selecting Vanguard: "A Strategy for Space," p. 310.
172 The Army-Navy meetings and Vanguard's technical problems: "Origins and Development of the Vanguard and Explorer Satellite Programs," pp. 107–110; *Assault on the Unknown,* p. 86. Budget busting: *Assault on the Unknown,* pp. 85–86.

173 Puffery and "first space vehicles": Bergaust and Beller, *Satellite!*, p. 46.

173 Dulles bankrolling the satellite: "A Strategy for Space," p. 312.

174 NSC Planning Board in November 1956: Ibid., p. 8.

175 Two-volume 1951 reconnaissance study: J. E. Lipp et al., "Utility of a Satellite Vehicle for Reconnaissance," and Greenfield and Kellogg, "Inquiry into the Feasibility of Weather Reconnaissance from a Satellite Vehicle," Project RAND.

175 "Feed Back": Davies and Harris, *RAND's Role in the Evolution of Balloon and Satellite Observation Systems and Related U.S. Space Technology*, pp. 53–56.

175 Feed Back and Raymond's design: Richelson, *America's Secret Eyes in Space*, pp. 8, 10, 15; Davies and Harris, *RAND's Role in the Evolution of Balloon and Satellite Observation Systems and Related U.S. Space Technology*, p. 83.

175 WS-117L: Day, "A Strategy for Reconnaissance: Dwight D. Eisenhower and Freedom of Space," pp. 98–100; Hall, "Post-War Strategic Reconnaissance and the Genesis of Project CORONA," pp. 12–13; Richelson, *America's Secret Eyes in Space*, pp. 8–9.

176 The Russian donation: *The Jules Verne Omnibus* (*From the Earth to the Moon*), p. 599.

176 The literature of space: Krieger, *A Casebook on Soviet Astronautics*, Parts 1 and 2.

177 Nesmeyanov's statement: Krieger, *Behind the Sputniks*, p. 3.

177 Interdepartmental Commission on Interplanetary Communication: *Assault on the Unknown*, p. 56.

178 Tikhonravov and work on Sputnik: Tikhonravov, "The Creation of the First Artificial Earth Satellite: Some Historical Details," pp. 207–09.

178 The Sputnik designs: Hart, *The Encyclopedia of Soviet Spacecraft*, pp. 121–27; Tikhonravov's role in the creation of the lightweight Sputnik was related by Yevgeni V. Shabarov, a former test manager at NPO Energiya, in *Roads to Space*, p. 278.

179 Three dogs in a rocket: *Behind the Sputniks*, p. 9.

179 Bardin's letter: *Assault on the Unknown*, p. 57.

180 June 1957 *Radio:* V. Vakhnin, "Information for Radio Amateur Observers," pp. 290–300, and A. N. Kazantsev, "Scientific Value of Radio Signals," pp. 301–07, in *Behind the Sputniks.*

180 July and August *Radio: Behind the Sputniks*, p. 11.

180 TASS report: Ibid.

180 Two ICBM tests: Kramer memorandum of December 12, 1997.

181 Exhibit at the National Academy: *Assault on the Unknown*, pp. 62–63.

181 The substitution of sputniks: Author interviews with Birylikov and Kantemirov.

182 No cackling: Ibid., p. 1.

183 The TASS/*Pravda* story: Harford, *Korolev*, p. 121; Krieger, *Behind the Sputniks*, pp. 311–12.

184 The call to Sullivan: *Assault on the Unknown*, p. 1.

184 Sullivan and Berkner's announcement: Ibid.; "Soviet Embassy Guests Hear of Satellite from an American as Russians Beam," *The New York Times*, October 5, 1957.

184 Blagonravov's briefing and the forlorn Vanguard: *Assault on the Unknown*, p. 67.

185 Typographical error: Author interview with Kapitsa.

6. To Race Across the Sky

186 Kaplan: "Device Is 8 Times Heavier Than One Planned by U.S.," *The New York Times*, October 5, 1957.

186 Fulbright: Oberg, *Red Star in Orbit*, p. 35.

186 Fulton: Sullivan, *Assault on the Unknown*, p. 77.

187 Mills: Divine, *The Sputnik Challenge*, p. xvii.

187 Holaday: "Device Is 8 Times Heavier Than One Planned by U.S."

187 Quarles on legality: Day, "A Strategy for Space," p. 311.

187 Celestia: "Owner of 'Celestia' Sees Soviet Trespass," *The New York Times*, October 6, 1957.

188 Bennett: "Satellite Belittled," *The New York Times*, October 5, 1957.

188 Randall: Baldwin, *The Great Arms Race*, pp. 11–12.

188 Von Braun and McElroy: Koppes, *JPL and the American Space Program*, p. 83.

188 The Redstone-Sergeant combination: *Assault on the Unknown*, pp. 83–84.

189 Jackson and Symington: Bulkeley, *The Sputniks Crisis and Early United States Space Policy*, p. 5.

189 Democratic Advisory Council: *The Sputnik Challenge*, p. xv.

189 LBJ on Sputnik: Killian, *Sputnik, Scientists, and Eisenhower*, p. 9.

190 Ike's Oklahoma City speech: *The Challenge of the Sputniks*, pp. 36–40.

190 Baldwin on Eisenhower: *The Great Arms Race*, pp. 82–83.

190 The launcher as long-range missile launcher: Ibid., p. 15.

191 Killian on politics: *Sputnik, Scientists, and Eisenhower*, p. 9.

191 Bridges: *The Sputniks Crisis and Early United States Space Policy*, p. 5.

191 Eisenhower and sacrifices: *The Sputnik Challenge*, p. 17.

191 Baruch's observations: *The Challenge of the Sputniks*, pp. 19, 20.

192 Murrow on science and Sputnik: Bliss, *In Search of Light*, pp. 310–11.

192 Williams's poem: *Sputnik, Scientists, and Eisenhower*, p. 8.

192 Rinehart's prediction: Ibid., pp. 7–8.

192 Teller and Pearl Harbor: *The Sputnik Challenge*, p. xvi.

192 Teller and the Moon: *Sputnik, Scientists, and Eisenhower*, p. 8.

192 Spies and loose talk: *The Sputnik Challenge*, p. xvii.

194 Welch on Sputnik: Welch, *The Blue Book of the John Birch Society*, pp. 33–34. Welch claimed to have named the society after a young fundamen-

talist Baptist preacher who had been murdered by Chinese Communists "because of the powerful resistance he would have been able to inspire against them" (p. 158).

194 Smuggled A-bombs: Ibid., p. 23.

194 "just another . . . launch" and *Pravda*'s headline: Harford, *Korolev,* pp. 121–22.

195 Khrushchev to Reston: McDougall, . . . *the Heavens and the Earth,* p. 237.

195 "sausages" and R-7 deployment: Rhea, *Roads to Space,* p. 375.

195 The Gallup Poll: . . . *the Heavens and the Earth,* p. 144.

196 Etzioni: Needell, *The First Twenty-five Years in Space,* pp. 33–34.

197 Tikhonravov on Sputnik: Tikhonravov, "The Creation of the First Artificial Earth Satellite: Some Historical Details," p. 191.

198 TASS dispatch: *Russians in Space,* p. 149.

198 Sputnik 2's science package: Shternfeld, *Soviet Space Science,* p. 216.

198 Laika's experiments and state: *Assault on the Unknown,* pp. 74, 76.

198 Poisoning of Laika: Hart, *The Encyclopedia of Soviet Spacecraft,* p. 124. A week later: Kramer memorandum of December 12, 1997.

199 Killian on secrecy: *Sputnik, Scientists, and Eisenhower,* pp. xvii, xviii.

200 Ike on Americans: *The Sputnik Challenge,* p. 17.

201 Jackson, Johnson, Rickover, HEW, and *U.S. News & World Report: The Sputnik Challenge,* pp. 52–53.

202 The chemists' report: "Some General Conclusions Regarding Russia," p. 69.

202 Russian engineer: *The Sputnik Challenge,* p. 53.

203 Ike's October 9 news conference: Ibid., pp. 7–9.

203 Full-fledged orbital attempt: . . . *the Heavens and the Earth,* p. 154.

203 Gates and McElroy: *The Sputnik Challenge,* p. 71.

204 Vanguard launch sequence: Sullivan, *Assault on the Unknown,* pp. 90–91.

205 Wolfe on Vanguard: *The Right Stuff,* pp. 73–74.

205 Kilgallen: Gray, *Angle of Attack,* p. 43.

205 "Disaster": *Assault on the Unknown,* p. 90.

205 "COLD WAR PEARL HARBOR": Diamond, *The Rise and Fall of the Space Age,* p. 14.

205 "Rearguard": *Assault on the Unknown,* p. 91.

205 Aid to the United States: . . . *the Heavens and the Earth,* p. 154.

205 "the tomb": *The Sputnik Challenge,* p. 47.

206 The Super Jupiter plan: "A National Integrated Missile and Space Vehicle Development Program."

206 Medaris's decision: Koppe, *JPL and the American Space Program,* p. 86.

207 Pickering's decision: Author interview.

207 Explorer and its three experiments: Burrows, *Exploring Space,* pp. 78–79.

208 Explorer secrecy: *JPL and the American Space Program,* p. 88.

209 Juno's launch: *Assault on the Unknown,* pp. 94–95.

209 The prelaunch news blackout: "Newsmen Agreed to Delay Reports," *The New York Times,* February 1, 1958.

209 Ike's reaction to Explorer: Divine, *The Sputnik Challenge,* pp. 94–95.

209 Huntsville celebration: *Assault on the Unknown,* p. 95.

209 Van Allen's description of the War Room: Van Allen, *Origins of Magnetospheric Physics,* pp. 59–60.

210 The scene in the Great Hall: Burrows, *Exploring Space,* p. 78.

210 Pickering's remark about Explorer's apogee: "Satellite Height over 1,000 Miles," *The New York Times,* February 2, 1958. Medaris's remark: "Eisenhower Announced Good News," *New York Herald Tribune,* February 1, 1958.

211 Media reaction to Explorer: *The Sputnik Challenge,* p. 95.

212 Van Allen's discovery: *Exploring Space,* pp. 79–80.

212 Up to a million years: "Million Year Life Seen," *The New York Times,* October 6, 1957. The astronomer was I. M. Levitt.

213 Vanguard's history: "Origins and Development of the Vanguard and Explorer Satellite Programs," p. 111.

213 Sputnik 3's radiation package: *Soviet Space Science,* p. 348.

213 "satellites the size of oranges": . . . *the Heavens and the Earth,* p. 175.

214 "National Space Establishment": Newell, *Beyond the Atmosphere,* pp. 46–47, 430–32 (Appendix D).

214 Truax's letter and the report: *The Sputnik Challenge,* p. 97.

214 PSAC endorsement of a space agency: Ibid., p. 102.

215 Killian on Medaris: *Sputnik, Scientists, and Eisenhower,* pp. 127–28.

215 Killian on the Air Force: Ibid., p. 128.

215 Boushey's lunar base: Allen, "Early Lunar Base Concepts of the U.S. Air Force," pp. 4–5.

215 RAND and lunar bases: "An Annotated Bibliography of RAND Space Flight Publications," pp. 13, 14, 15.

216 "Introduction to Outer Space": *Sputnik, Scientists, and Eisenhower,* pp. 123–24.

216 "deluge" of proposals: Newell, *Beyond the Atmosphere,* p. 90.

217 Civilian-military battles: Ibid., pp. 116–17.

217 McDougall: . . . *the Heavens and the Earth,* p. 174.

218 Creation of NACA: Bilstein, *Orders of Magnitude,* p. 3.

218 Newell on NACA: *Beyond the Atmosphere,* pp. 90–92.

7. War and Peace in the Third Dimension

220 Nuking the Moon: Koppe, *JPL and the American Space Program,* p. 85; author interview.

220 Kramer's calculation: Kramer memorandum of December 16, 1997.

220 Kellogg's explosion: Kellogg, "Observations of the Moon from the Moon's Surface."

220 Russians nuking the Moon: "Soviet Moon Shots Signal New Ventures," *Air Intelligence Digest,* p. 29.

221 The football-to-intelligence analogy: Wheelon and Graybeal, "Intelligence for the Space Race," p. 1.
222 Kapustin Yar tests: Special National Intelligence Estimate 11-10-57, "The Soviet ICBM Program," December 17, 1957, pp. 5–6.
222 NII 88 report and Korolyov: "Scientific Research Institute and Experimental Factory 88 for Guided Missile Development, Moskva/Kaliningrad," pp. 11, 16, 29.
224 The Lunik operation: Finer, "The Kidnaping [*sic*] of the Lunik."
225 Norway, Germany, and Iran: Sources who spoke on condition of anonymity.
226 Origins of WS-117L: Peebles, *High Frontier,* pp. 5–15.
226 Douglas's memo: Douglas, "Thor and WS-117L Program," p. 1.
227 Midas/DSP: "Military Space Projects, March–April–May 1960," pp. 27–37.
227 Billings: Kistiakowsky, *A Scientist at the White House,* p. 45.
227 Television and line scanning: Richelson, *America's Secret Eyes in Space,* p. 19.
227 The RAND study: Huntzicker and Lieske, "Physical Recovery of Satellite Payloads—A Preliminary Investigation."
228 The interim reconnaissance satellite: Ruffner, *Corona: America's First Satellite Program,* pp. 4–5; "An Earlier Reconnaissance Satellite System," (Project RAND).
228 Origin of "Corona": Kramer memorandum of December 16, 1997. Also see Madden, "The Corona Camera System," p. 14; McDonald, *Corona,* p. 58.
228 Corona and KH designations: McDonald, *Corona,* p. 694.
228 CIA–Air Force cooperation on Corona: Hall, "Post-War Strategic Reconnaissance and the Genesis of Project Corona," in Day, Logsdon, and Latell, *Eye in the Sky,* p. 16.
229 Bissell and Ritland's relationship: Haines, *The National Reconnaissance Office,* p. 15.
229 "power and control": Author interview with Keegan, September 16, 1981.
230 Mice and monkeys: *Corona: America's First Satellite Program,* p. 17.
230 The Discoverer pamphlet: "Summary: U.S. Air Force Discoverer Satellite Recovery Vehicle Program," p. 1.
231 Fake contracts: Madden, "The Corona Camera System," p. 17.
231 Southern Pacific: Greer, "CORONA," in *Corona: America's First Satellite Program,* pp. 11–12.
232 Corona recovery technique: Ravnitsky, "Catch a Falling Star: Parachute System Lessons Learned During the USAF Space Capsule Mid-Air Recovery Program, 1959–1985," pp. 5–6; *Corona: America's First Satellite Program,* p. 12.
233 The Corona mishaps: *Corona: America's First Satellite Program,* p. 18.
233 The five lost capsules: "Summary [of] U.S. Air Force Discoverer satellite recovery vehicle program I thru XIV," pp. 1–2; U.S. Air Force, "History of the 6593rd Test Squadron, Aug–Dec 1958," Maxwell Air Force Base, Alabama, microfilm, undated, p. 3; Discoverer 1 at the South Pole: *Corona: America's First Satellite Program,* p. 16.
233 Bissell's heartbreak: Author interview.
234 First space photo: CIA, "Historical Imagery Declassification Fact Sheet" and accompanying photographs distributed at the "Piercing the Curtain: CORONA and the Revolution in Intelligence" conference, George Washington University, May 23–24, 1995.
234 What a stunned Eisenhower saw: *Corona: America's First Satellite Program,* p. 2; Hall, "Eisenhower and the Cold War," p. 68.
234 "darkened warehouse": Wheelon, "Lifting the Veil on CORONA," p. 254.
235 *Aviation Week* and Pied Piper: "Test Firings for Pied Piper Due Soon," p. 145.
235 "advanced reconnaissance systems": "Agena B to put Samos, Midas in Orbit," p. 73. Renaming WS-117L: Peebles, *High Frontier,* p. 11.
235 "close examination": "First Capsule Recovered from Satellite," p. 33.
235 The Soviet warning: "Soviets Warn U.S. on Spy Satellites," *The New York Times,* November 14, 1960.
236 Highest-priority targets: *Corona: America's First Satellite Program,* pp. 49–58.
236 Zenit's first flight: Gorin, "Zenit—The First Soviet Photo-Reconnaissance Satellite," p. 7; Oberg, "The First Soviet Spy Satellite."
236 The Moscow embassy parking lot: Author interview with Davies, January 4, 1989.
236 The car in front of the embassy: Author conversation with Dino Brugioni.
236 Soviet attitude toward U.S. spy satellites: "Probable Reactions to U.S. Reconnaissance Satellite Programs," Special National Intelligence Estimate Number 100-6-60, pp. 1, 2.
237 McNamara and McCone: Wheelon, "Lifting the Veil on CORONA," p. 252.
238 Wheelon and McCone: Wheelon, Ibid., p. 253.
238 Kistiakowsky et al. and Samos: Hall, "The Eisenhower Administration and the Cold War," pp. 67–68.
238 The Kistiakowsky group's plan and Ike's response: Ibid.
238 Sharpe's memo: "Memorandum for the Chief of Staff, USAF" from Dudley C. Sharpe.
239 Artificially grown human cells: "Capsule Carried Cells of Humans," *The New York Times,* November 17, 1960.
239 Discoverer 17's real mission: Greer, "CORONA," p. 26.
239 Behind the Ford dealership: Burrows, "Stakeout," p. 83.
239 Kistiakowsky's observations: *A Scientist at the White House,* p. 182.
240 NRO, secrecy, and the Special Projects Office: Burrows, *Deep Black,* pp. 199–200.
240 The NRO's programs: Haines, *The National Reconnaissance Office,* p. 22. Haines reported that Program D included Air Force aircraft.
240 Wheelon and the analysis center: Ibid., p. 33 (note 39).

240 Schriever's complaint: Letter to General Curtis E. LeMay of December 26, 1963, marked "Secret."
240 Wheelon and McMillan: Haines, *The National Reconnaissance Office,* p. 22.
241 In-house engineering: Ibid., p. 33 (note 42).
241 Keegan on submarines: Author interview.
241 "Cloud Reconnaissance": Katz, "An Air Force Weather Satellite—Why and How," p. 5.
241 Improved resolution: Peebles, *High Frontier,* p. 15.
241 Satellite names and numbers: *Corona: America's First Satellite Program,* pp. xiv–xv; Richelson, *America's Secret Eyes in Space;* Burrows, *Deep Black,* p. 227.
242 Number of Corona satellites and KH5/KH6: *Corona: America's First Satellite Program,* pp. xiv–xv.
242 Talent-Keyhole: Ibid., p. 59; Richelson, *America's Secret Eyes in Space,* p. 66.
242 Eisenhower on security: *Corona: America's First Satellite Program,* p. 75.
242 Nondisclosure agreement: Sensitive Compartmented Information Nondisclosure Agreement, FORM DD 1847-1, 83 JAN.
242 Wheelon: Author interview.
243 Johnson: Author interview of October 18, 1996.
243 Brugioni and dirty words: Author interview. Toilet shot: Brugioni, *Eyeball to Eyeball,* last picture in the picture section.
243 Analysis details and heartbeats: Sources who spoke on condition of anonymity.
244 Ten thousand observations a day: Johnson, "U.S. Space Surveillance," p. 5.
244 Soviet photoreconnaissance capability: National Intelligence Estimate 11-1-67, March 2, 1967, p. 7.
245 The Soviet article on U.S. satellites: Yarevand and Shevchuk, "The Pentagon's Global Espionage Network," pp. 111–12.
245 Tricks of the interpreter's trade: Brugioni, "The Art and Science of Photoreconnaissance," pp. 82, 84.
245 "cratology": Goodell, "Cratology Pays Off," pp. 2, 7.
245 The test at Lop Nor: "The Art and Science of Photoreconnaissance," p. 83.
246 NSA, Huntsville, and NPIC: Author interview with Johnson, October 18, 1996; *Deep Black,* pp. 20–21.
246 Battle over the SS-8: A highly knowledgeable source who spoke on condition of anonymity.
247 "Bloody" battles: Author interviews with Keegan on June 2, 1984, and February 11, 1985, and with Scoville on April 12, 1984.
248 Arnold on space: *The War Reports of General George C. Marshall, General H. H. Arnold, Admiral Ernest J. King,* pp. 452–56.
248 Vandenberg: Perry, "Origins of the USAF Space Program, 1945–1956," p. 23.
248 Navy's withdrawal: Ibid., August 1962 revision, p. 2.
248 White and the December memo: Bowen, "The Threshold of Space: The Air Force in the National Space Program, 1945–1959," pp. 17–18.
248 Conception of its domain: "Origins of the USAF Space Program, 1945–1956," p. 2.
249 Some of the fifteen missions: Bowen, "The Threshold of Space," p. 18.
249 A million feet: Guenther, Miller, and Panopalis, *North American X-15/X-15A-2,* p. 5.
250 Dornberger: Ibid., pp. 1–2.
250 Heat and the X-15: Stillwell, *X-15 Research Results,* p. 20.
250 Thompson on the X-15: Thompson, *At the Edge of Space,* p. 35.
251 X-15 flights: *North American X-15/X-15A-2,* pp. 17–23.
251 Bomi: Baker, *The History of Manned Spaceflight,* p. 24.
251 Robo, Brass Bell, and Hywards: Bowen, "The Threshold of Space," p. 35.
252 Dyna-Soar planning and the letter: Ibid., p. 35.
252 "inspector" and interceptor: "Review and Summary of X-20 Military Application Studies," p. 4.
252 "guided glide weapons": "Phase O Study of X-20 for Program 706," p. 12, in "Review and Summary of X-20 Military Application Studies."
253 Storms and LeMay: Gray, *Angle of Attack,* p. 41.
253 MISS and the Moon: Swenson, Grimwood, and Alexander, *This New Ocean,* pp. 91, 97–98.
253 Titan 1 explosion: Berger, "The Air Force in Space: Fiscal Year 1961," p. 52.
254 Appropriations House Committee: "The Air Force in Space: Fiscal Year 1961," p. 56.
254 McNamara's reasoning: Stares, *The Militarization of Space,* p. 130.
254 McNamara's news conference: "Dyna-Soar Space Glider Scrapped; New Project Set," *The Wall Street Journal,* December 11, 1963.
255 LeMay rejecting MOL: Peebles, *High Frontier,* p. 21.
255 MOL's goal: Butz, "MOL: The Technical Promise and Prospects," p. 43.
256 Ferguson: U.S. Congress, House, *Department of Defense Appropriations for 1963: Hearings Before a Subcommittee of the Committee on Appropriations,* 87th Cong., 2nd sess., 1962, Part 2, pp. 477–88.
256 No MOL overflights: Peebles, *High Frontier,* p. 22.
256 The Navy MOL: Ezell, *NASA Historical Data Book,* p. 94.
256 The nine practical military uses of MOL: "MOL: The Technical Promise and Prospects," p. 46.
256 MOL budget cuts: "MOL Delayed by Funding Cut," *Aviation Week & Space Technology,* April 21, 1969, p. 17.
256 The cost and termination: Compton and Benson, *Living and Working in Space: A History of Skylab,* p. 109.
258 "vital to the prestige": "Chronology of Early Air Force Man-in-Space Activity, 1955–1960," 65-21-1, AFSC Historical Publications Series, Office of Information, Space Systems Division, Air Force Systems Command, January 1965, p. 14.
259 NACA assets to NASA: Bilstein, *Orders of Magnitude,* p. 49; *. . . the Heavens and the Earth,* p. 196.

259 Vanguard and the lunar probes: *Orders of Magnitude,* p. 49.
260 Glennan on Medaris: Hunley, *The Birth of NASA: The Diary of T. Keith Glennan,* p. 9.
260 Brucker's tirade: *The Birth of NASA,* p. 10.
261 Early development of Saturn: Bilstein, *Stages to Saturn: A Technological History of the Apollo/Saturn Launch Vehicles,* pp. 25–28.
261 Tail between his legs: *The Birth of NASA,* p. 10.
261 JPL to NASA: Koppes, *JPL and the American Space Program,* pp. 98–99.
262 James's recollections: "The Early History of JPL and the Evolution of Its Matrix Organization."
263 "half a loaf": *The Birth of NASA,* p. 12.
264 Johnson and Rich: Author interview with Johnson on January 26, 1983, and with Rich on June 19, 1984; Burrows, "How the Skunk Works."
264 "go to hell": Author interview with Pickering, June 15, 1987.
264 Glennan on "outsiders": *The Birth of NASA,* p. 183.
264 Glennan to Newell: Glennan's pencil notations are on pp. 799 and 827 of Newell's comment draft of June 1978. His covering letter was written on January 1, 1979. Both are in the NASA History Office Archive.
266 NASA facilities: Van Nimmen, Bruno, and Rosholt, *NASA Historical Data Book,* Vol. 1, pp. 241–539.
266 Employment at the end of 1961: Ibid., Vol. 1, p. 70.
266 Contractor jobs: Hirsch and Trento, *The National Aeronautics and Space Administration,* p. 57.
266 The budget: *NASA Historical Data Book,* Vol. 1, pp. 128, 125.
266 "like Isabella": Emme, "Presidents and Space," in Durant, *Between Sputnik and the Shuttle,* p. 27.
267 Space Task Group briefings: Brooks, Grimwood, and Swenson, *Chariots for Apollo,* p. 15.
267 Silverstein and "Apollo": Ibid.
267 Clarke on radio satellites: Clarke, "V-2 for Ionosphere Research and Extraterrestrial Relays."
267 Pierce on communication satellites: Edelson, "Communication Satellites: The Experimental Years," p. 96.
268 Bell Labs to Paris: Ibid., p. 99.
269 Comsat and Intelsat: Laskin, "Communicating by Satellite," pp. 29, 39.
269 Telstar: "Communication Satellites: The Experimental Years," pp. 96–97, 102.
270 ATS-6: Canby, "Satellites That Serve Us," pp. 291, 294.
271 NRO weather satellites: Peebles, *High Frontier,* pp. 28–29.
271 "UNITED STATES": "Presidents and Space," p. 27.
272 SAC coveting reconnaissance: "The History of the Strategic Air Command," p. 79.
272 Dulles on Corona: Author interview with Bissell.
272 ARPA, Corona, and the Air Force: *Corona,* pp. 6, 8.

8. The Greatest Show on Earth

275 "favorite child": Gagarin, *Road to the Stars,* p. 122.
275 Oberth at home: Oberg, *Red Star in Orbit,* p. 37.
275 Konstantinov and Wells: Vladimirov, *The Russian Space Bluff,* p. 22. "Professor K. Sergeev": Harford, *Korolev,* p. 242.
276 Protection from hostile agents: Harford, *Korolev,* p. 91.
277 Mishin on the military and "wooden rubles": Author interview.
278 Basic structure of the space program: Tarasenko, "Transformation of the Soviet Space Program After the Cold War," pp. 342–44.
278 MOM's vital interests: Bogdanov, "The Organization and the Structure of the Management of Space Affairs in Russia," pp. 4, 6.
278 OKB-52, the Moon, and Yangel: Johnson, *The Soviet Reach for the Moon,* p. 10.
278 OKB-586: Rhea, *Roads to Space,* p. 106.
279 "Black Raven": Ibid., p. 84.
279 Sagdeev on secrecy: Sagdeev, *The Making of a Soviet Scientist,* p. 234.
280 Hiring Sergei Khrushchev: Vladimirov, *The Russian Space Bluff,* pp. 53–54.
280 Khrushchev on Chelomei: Harford, *Korolev,* pp. 258–59.
280 Mishin on Glushko: Rhea, *Roads to Space,* p. 96; Mishin, "Why We Didn't Get to the Moon," p. 18.
280 Glushko and Chelomei: *The Soviet Reach for the Moon,* p. 11.
281 The designer of Vostok: Kirimov, *How All This Started,* p. 28; *Roads to Space,* p. 383.
281 March 14, 1960: *Roads to Space,* p. 495.
281 Cosmonauts: Harvey, *Race into Space,* p. 41.
282 Krichevski: Author interview; Krichevski, *Earth Soul with Wings,* p. 48.
283 Lebedev's complaint: Lebedev, *Diary of a Cosmonaut,* p. x.
283 The imperative to be Russian: Vladimirov, *The Russian Space Bluff,* pp. 92–93.
283 "plain Russian people": *Road to the Stars,* p. 5. The book, an "as-told-to," was written by two *Pravda* correspondents and edited by an Air Force lieutenant general.
283 Lebedev on Gagarin: *Diary of a Cosmonaut,* pp. 10–11.
283 Titov: *The Russian Space Bluff,* p. 93.
283 Slavic Frank Merriwells: Graham, *Science and the Soviet Social Order,* pp. 180–81.
285 Faget, the Mercury paper, and the lifting body: Baker, *The History of Manned Space Flight,* pp. 34–35. The Ames engineers were Thomas J. Wong, Charles A. Hermach, John O. Reller, and Bruce E. Tinling.
285 Lifting bodies: *To Space and Back* (video).
286 "Objectives" report: *This New Ocean,* pp. 110–11.
287 Koelle's presentation: *Chariots for Apollo,* pp. 4–5.
287 Meeting at Dolley Madison House: Ibid., pp. 4–6.
287 "impossibly complex": *Stages to Saturn,* p. 28.

287 The four rockets: *The History of Manned Space Flight,* p. 37.
287 Redstone and Atlas: *This New Ocean,* pp. 122–23.
287 Naming Mercury: Ibid., p. 132.
288 The contract: Ibid., pp. 121, 137; Baker, *The History of Manned Space Flight,* p. 38.
289 Qualified pilots: Flaherty, *Psychophysiological Aspects of Space Flight,* p. 97.
289 Astronaut selection criteria: *This New Ocean,* pp. 130–31.
290 Kuettner and Pierce: Ibid., pp. 171–72, 174.
290 "*doing nothing* under stress": Wolfe, *The Right Stuff,* pp. 180–81.
291 Slayton: *This New Ocean,* p. 177.
291 Reducing the number of astronauts: Flaherty, *Psychophysiological Aspects of Space Flight,* pp. 97–98.
291 Outstanding background and knowledge: Ibid. The tester was Robert B. Voas, a Navy psychologist.
292 Training: Ibid., p. 100 (Voas).
292 "lowest bidder": *Spaceflight,* NOVA, Part 2.
292 "out of the damned thing": *Spaceflight* (video), Part 2.
294 Glenn and Heaven: Aldrin, *Men from Earth,* p. 42.
294 Galling chimp: Shepard and Slayton, *Moon Shot,* p. 86.
294 Press debut and the monkeys: Hunley, *The Birth of NASA,* p. 20.
295 Kennedy and the chimp: *Spaceflight* (video), Part 2.
295 "link between monkey and man": *Moon Shot* Part 1, TBS, July 11, 1994.
295 Contemptuous and angry: Mayer, *Making News,* p. 278.
295 "great headlines": *The Birth of NASA,* p. 276.
295 Help from the press: Ibid., p. 286. The editors were, respectively, Ben McKelway and J. Russell Wiggins.
296 Getting into Glennan's hair: Ibid., pp. 20–21.
296 Barr's article: Barr, "Is Mercury Program Headed for Disaster?" pp. 12–14.
296 Careless work on capsules: *This New Ocean,* p. 256.
297 Low and UPI: Ibid., p. 283.
297 "reliability engineering": Ibid., pp. 178–79.
298 Sidey and Yeager: *Spaceflight,* (video), Part 1.
298 Yeager: Rolex ad, for example, in *Air & Space,* November 1995, inside cover and p. 1. The game was "Chuck Yeager's Advanced Flight Trainer," Electronic Arts, San Mateo, California.
298 The astronauts' children: The Astronauts, *We Seven,* pp. 66–67.
298 "charm school": *Spaceflight* (video), Part 2.
299 "premedicated murder": Author interview with King.
300 "Wickie": Author conversation with John N. Wilford and interview with King.
300 "hotshot pilots": Author interview with King.
300 "ladies in Cocoa Beach": *Moon Shot* (video), Part 1. The voice-over was Slayton's.
300 Cronkite and Schweickart: *Spaceflight* (video), Part 1.
300 Young girls: Wolfe, *The Right Stuff,* pp. 167–74.
301 Glenn versus Shepard: Ibid., pp. 170–73.
302 Cosmonauts' escapades: Hendrickx, "The Kamanin Diaries 1960–1963," p. 39; interviews with Grigori Khozin and Serge Khrushchev.
302 LeMay and Power: Rhodes, "The General and World War III."
303 "disastrous situation": *The Russian Space Bluff,* p. 71.
304 Kennedy's remarks: Berger, "The Air Force in Space: Fiscal Year 1961," p. 2.
304 "long, slow afternoon": Logsdon, *The Decision to Go to the Moon,* p. 65.
304 Kennedy on control of space: "The Air Force in Space: Fiscal Year 1961," p. 2.
304 Nixon: Ibid., p. 3.
304 "cooperate": *The Birth of NASA,* p. 131.
305 "Here we are again": Ibid., p. 146.
306 X-15 accomplishments: Stillwell, *X-15 Research Results,* pp. 2–8.
306 X-15 and Mercury-Redstone: Thompson, *At the Edge of Space,* p. 6.
307 Glennan and Echo: Hunley, *The Birth of NASA,* pp. 204–05.
307 Korabl Sputnik 2: Newkirk, *Almanac of Soviet Manned Space Flight,* p. 22; Harvey, *Race into Space,* p. 45.
308 Khrushchev on Grechko: Author interview.
308 Turbopump problems: Kramer memorandum of December 16, 1997.
309 Fuel that exploded: Rhea, *Roads to Space,* p. 374.
309 Clothes on fire: A color film of the disaster, routinely shot of the preparation for launch, was later screened by representatives of the U.S. technical intelligence and space communities.
309 Death of Nedelin: *Roads to Space,* pp. 374–75; *Almanac of Soviet Manned Space Flight,* pp. 9–10.
309 Worsening morale: Medvedev, *Soviet Science,* p. 100.
309 Dolgov and Vostok improvements: *Almanac of Soviet Manned Space Flight,* pp. 8–9; *The Russian Space Bluff,* pp. 88–89.
309 Nedelin's "plane crash": *Red Star in Orbit,* p. 40.

9. A Bridge for Galileo

311 Korolyov's health: Harvey, *Race into Space,* pp. 53, 55.
311 Interrupted telemetry: Matson, *Cosmonautics: A Colorful History,* p. 34. E-rocket: *From First Satellite to Energia-Buran and Mir,* p. 51.
312 Monitoring *Vostok 1:* Kramer memorandum of December 21, 1997; Plaster, "Snooping on Space Pictures," p. 34; Wheelon and Graybeal, "Intelligence for the Space Race," p. 5.
312 Gagarin's observations: Gagarin, *Road to the Stars,* pp. 152, 154.
313 Twelve cosmonauts before Gagarin: Kramer memorandum of December 16, 1997.
313 Glennan and the May 15 shot: Hunley, *The Birth of NASA,* p. 239.
313 "little eagles": *The Russian Right Stuff* (video), Part 1, "The Invisible Spaceman"; Vladimirov, *The Russian Space Bluff,* p. 94.

313 No truth to the rumor: Newkirk, *Almanac of Soviet Manned Space Flight,* p. 19; author interview with Sergei Khrushchev.
314 Nullifying the feat: Rhea, *Roads to Space,* p. 418.
314 Gagarin's description of the "ballet": Rebrov, *Space Catastrophes,* p. 23.
314 "Malfunction!!!": Scrawled notes by Yevgeni A. Karpov, *Russian Space History* catalogue, Sotheby's, March 16, 1996, pp. 38–40; "Russian Space Mementos Show Gagarin's Ride Was a Rough One," *The New York Times,* March 5, 1996.
315 Moscow Radio: *Race into Space,* p. 56.
315 "immortal": "Russian Orbited the Earth Once . . . ," *The New York Times,* April 13, 1961.
315 Gagarin's welcome: *Race into Space,* p. 58.
316 Russian inferiority complex: Kliuchevsky, *A Course in Russian History: The Seventeenth Century,* pp. 275–78.
317 Ramo on the Russians: "The Practical Dimensions of Space," Needell, *The First Twenty-five Years in Space,* p. 51.
317 Bark sandals and cabbage soup: John Noble Wilford, "First into Space, Then the Race," p. 343.
318 Ham's flight: Swenson, Grimwood, and Alexander, *This New Ocean,* pp. 312–18; Shepard and Slayton, *Moon Shot,* p. 90.
318 The problem with the Redstone: *This New Ocean,* p. 328.
319 Shepard's anger: *Moon Shot,* p. 91.
319 The Joint Chiefs and Laos: Schlesinger, *A Thousand Days,* p. 332.
320 "Occupying" space: Author interview with Sorensen.
320 "unchallenged world leadership": Sorensen, *Kennedy,* p. 524.
320 White and NASA: Hunley, *The Birth of NASA,* p. 283.
320 Delayed appointment of NASA's head: Author interview with Sorensen; Logsdon, *The Decision to Go to the Moon,* pp. 82–83.
320 Limited confidence: A source requesting anonymity.
321 Baldwin's angst: "Flaw in Space Policy," *The New York Times,* April 17, 1961.
321 Toy store: Author interview with Sorensen.
321 "How long is this going to continue?": Ibid.
322 Meeting in the Cabinet Room: *The Decision to Go to the Moon,* p. 106.
322 "We're going to the Moon": Murray and Cox, *Apollo: The Race to the Moon,* p. 79.
323 Kennedy to LBJ: *The Decision to Go to the Moon,* pp. 109–11.
323 Wiesner rarely asked: Young, Silcock, and Dunn, "Why We Went to the Moon," p. 37.
323 Science's role in the lunar program: Jastrow and Newell, "Why Land on the Moon?," pp. 42–43.
324 Wiesner on Kennedy: *The Decision to Go to the Moon,* pp. 110–11.
325 The suit's features: Flaherty, *Psychophysiological Aspects of Space Flight,* p. 27.
325 Colley: "Russell Colley, Designer of Spacesuits, Is Dead at 97," *The New York Times,* February 8, 1996; *This New Ocean,* pp. 230–31.
325 Wetting the diaper: *Moon Shot* (video), Part 1.
326 "Ahhh . . . Roger": *This New Ocean,* p. 341.
327 The flight and the physical: Ibid., pp. 352–58.
327 Eyeballs and sockets: Related by John Glenn in *Spaceflight* (video), Part 2.
327 *Millyet: This New Ocean,* p. 361.
327 Khrushchev infuriated: Baker, *The History of Manned Spaceflight,* p. 79.
328 "A-OK": Ibid.
328 Rejection of additional funds: "A 3-Man Trip to Moon by 1967 Projected by White House Aides," *The New York Times,* May 26, 1961.
329 McNamara-Webb memo: "Why We Went to the Moon," pp. 37–38.
329 Bundy and the Moon and JFK's response: Author interview.
331 Kennedy's speech: "Kennedy Asks $1.8 Billion This Year to Accelerate Space Exploration, Add Foreign Aid, Bolster Defense," *The New York Times,* May 26, 1961.
331 Ad-libs and the ride with Sorensen: *Kennedy,* p. 526; author interview with Sorensen.
331 Kennedy and national resolve: Author interview with Sorensen.
332 "dumb accord": "Why We Went to the Moon," p. 38.
333 The bomber force in 1961: Polmar and Laur, *Strategic Air Command,* p. 66.
334 Weight of the guidelines: "Why We Went to the Moon," p. 41.
334 Qualifications for Apollo and the SEC decision: *Chariots for Apollo,* pp. 41–44.
335 Bobby Baker and associates: Mailer, *Of a Fire on the Moon,* p. 173.
335 North American's contract: "Why We Went to the Moon," pp. 39–40.
335 Apollo contract to pork barrel: Ibid., p. 40.
335 "superficial entertainment": Author interview with Van Allen.
336 Helfand: Author interview.
336 Gallup Poll: "Why We Went to the Moon," p. 36.
337 The hatch: *This New Ocean,* p. 367.
338 Grissom's flight: Ibid., pp. 370–75.
338 Korolyov on Titov's mission: Daniloff, *The Kremlin and the Cosmos,* p. 108.
339 Titov's routine: Riabchikov, *Russians in Space,* pp. 166–69.
339 "observing thunderstorms": Hendrickx, "Korolev: Facts and Myths," in *Spaceflight,* February, 1996, p. 46.
339 Tikhonravov, Korolyov, and reconnaissance: Gorin, "Zenit: Corona's Soviet Counterpart," p. 86.
339 Zenit details: Ibid., pp. 88–90.
339 Captain to major: Newkirk, *Almanac of Soviet Manned Space Flight,* p. 27.
340 Dress rehearsal: *This New Ocean,* pp. 386–89.
340 Little people displaced: Beschloss, *Taking Charge,* p. 281.
340 STG to MSC: *This New Ocean,* p. 392.
340 One hundred orbits: Ibid., p. 426.
340 The flight of *Friendship 7:* Ibid., pp. 422–32.
342 The hero and Kennedy's remarks: Ibid., p. 434.
342 Carpenter, Schirra, and Cooper: Ibid., pp. 640–41.
343 Slayton: Ibid., pp. 440–42.

343 The psychology of machines: *Of a Fire on the Moon,* p. 168.
344 Ponomareva's remark: Author interview.

10. To Hit a Moving Target

346 Fifty percent reliable and cutting corners: Author interview with Ponomareva.
346 Ponomareva: Author interview.
347 Wondering if she was still alive: Harvey, *Race into Space,* p. 79.
347 Tereshkova's complaint: Rhea, *Roads to Space,* pp. 441–42 (Ivan Spitsa).
347 Luce on Tereshkova: Baker, *The History of Manned Space Flight,* p. 167.
347 Leonov on Tereshkova: Author interview.
347 Ponomareva on Korolyov: Author interview.
348 Golden Eagle to Seagull: Rhea, *Roads to Space,* p. 442.
348 Smith: Smith, *The Russians,* p. 179.
348 Cobb and Hart: Levine, *Appointment in the Sky,* pp. 146–47.
349 Shatalov, Leonov, and Tereshkova: Oberg, *Red Star in Orbit,* pp. 70–71.
349 Neyfach: Baker, *The History of Manned Space Flight,* p. 167.
350 Brill: *Appointment in the Sky,* p. 47.
351 Vladimirov's assertion: Vladimirov, *The Russian Space Bluff,* p. 124.
351 Korolyov's warning: Author interview with Kantemirov, June 5, 1996.
351 Less than 50 percent: Ibid.
351 Rocket under the shroud: Newkirk, *Almanac of Soviet Manned Space Flight,* p. 33.
351 Nose rocket problem: Author interview with Kantemirov.
352 The end of Khrushchev: Vladimirov, *The Russian Space Bluff,* p. 137.
352 Brezhnev and the military: Author interview with Mishin.
352 "a vastly improved . . . spaceship": Shelton, *Soviet Space Exploration: The First Decade,* p. 161.
352 "Swim" in space: Matson, *Cosmonautics: A Colorful History,* p. 39.
353 Leonov's EVA and its aftermath: Newkirk, *Almanac of Soviet Manned Space Flight,* p. 37.
355 Lysenko: Graham, *Science and the Soviet Social Order* (introduction by the editor), p. 2. Palchinsky: Graham, *The Ghost of the Executed Engineer.*
355 Beria and Kapitsa: Author interview with Sergei Kapitsa.
355 Khozin: Author interview.
355 Ustinov foiling Korolyov: Author interview with Kantemirov, June 5, 1996.
355 Ustinov and the *kremlevka:* Sagdeev, *The Making of a Soviet Scientist,* p. 247.
355 Ustinov versus Grechko: *Roads to Space,* pp. 111–12.
356 Yeltsin's observer: Michael Specter, "Yeltsin's Kremlin: A Mystery Wrapped in a Riddle," *The New York Times,* July 18, 1996.
356 Yeltsin's style: The speaker was former Deputy Prime Minister Gennadi Burbulis, as quoted in ibid.
356 "festive inauguration ceremony": "Russia Improvises an Inauguration," *The New York Times,* August 3, 1996.
357 On being clever: "Russian Rockets Get Lift in U.S. from Cautious and Clever Design," *The New York Times,* October 29, 1996.
357 "storybook launch": *Gemini Press Reference Book,* pp. 1–2.
358 "random successes": McCurdy, *Inside NASA,* p. 63.
358 *Titanic* and *Molly Brown: The History of Manned Space Flight,* p. 198.
358 *Faith 7*'s name: Brooks, Grimwood, and Swenson, *Chariots for Apollo,* p. 292.
358 OAMS operation: *The History of Manned Space Flight,* p. 200.
359 White's flight: Ibid., p. 206.
359 Food on Gemini: Ibid., p. 204.
360 Waste disposal: Ibid.
360 *Gemini 8: Inside NASA,* p. 151.
361 Gemini mission data: Ezell, *NASA Historical Data Book,* Vol. 2, pp. 161–70.
361 Using Jodrell Bank: Lovell, *The Story of Jodrell Bank,* pp. 209–10.
362 First three Russian failures: Johnson, *The Soviet Reach for the Moon,* p. 6.
362 Pioneer 3: Koppes, *JPL and the American Space Program,* p. 92.
362 Lunik details and Mechta: *Cosmonautics: A Colorful History,* p. 26.
362 Pioneer 4: *JPL and the American Space Program,* p. 93.
363 Lunas 2 and 3: *The Soviet Reach for the Moon,* p. 8.
363 Luna 3 operation: *Cosmonautics: A Colorful History,* pp. 26–27.
363 Johnson's statement: Author interview, October 11, 1988.
364 Robot explorers as living creatures: Nicks, *Far Travelers,* p. 19.
365 Rangers 1–4: Hall, *Lunar Impact,* pp. 103, 108, 143–44, and 153–54, respectively.
365 Hammers and sickles: Johnson, *Handbook of Soviet Lunar and Planetary Exploration,* p. 14.
365 Webb, Khrushchev, and Pickering: *Lunar Impact,* pp. 154–55.
366 The end of Mariner 1: Burrows, *Exploring Space,* pp. 128–30.
366 Schurmeier's order: *Lunar Impact,* p. 179.
367 Television narration: "News Conference on Ranger VI Impact on Moon," NASA Release No. 2-2509, pp. 3, 5 (transcript made in Washington from voices received over telephone lines).
368 Cunningham's and Pickering's reactions: *Lunar Impact,* p. 237.
368 Ranger's end and the Avon message: "News Conference on Ranger VI Impact on Moon," pp. 8–9.
368 "Significant achievement": Ibid., p. 11.
368 JPL postmortem: "RA-6 Investigation Committee Final Report," p. v.
369 NASA's board of inquiry: "Final Report of the Ranger 6 Review Board," pp. 11–12.
369 The committee's recommendations: *Lunar Impact,* p. 254.

369 The *Times*' report: "Ranger's Failure Spurs Drive to Delay Lunar Trips," *The New York Times,* February 4, 1964.
369 Rebuffed in Houston: *Lunar Impact,* p. 232.
370 One-way trip: *Chariots for Apollo,* p. 65.
370 EOR: Ibid., p. 62.
371 LOR: Ibid., pp. 72–79.
371 Cost saving of LOR: Levine, *Managing NASA in the Apollo Era,* p. 20; *Chariots for Apollo,* p. 83.
371 Von Braun's reasoning: *Chariots for Apollo,* p. 82.
371 "DX" category: *NASA Historical Data Book,* Vol. 2, p. 182.
371 Apollo budget: Ibid., Vol. 2, p. 128.
372 NASA's coordination: *Managing NASA in the Apollo Era,* p. xix.
372 Contractors: *NASA Historical Data Book,* Vol. 2, pp. 192–93.
373 Von Braun's ruminations: Bilstein, *Stages to Saturn,* p. 261.
373 Juno 5 and Dyna-Soar: Ibid., p. 35.
373 Saturn 5 capability: Ibid., pp. 58–59.
374 Crawlers and the VAB: NASA/KSC Release No. 114-85, "Crawler/Transporter"; "Launch Complex 39 Facilities," NASA/KSC Release No. 6-81.
374 Renaming the Launch Operations Center: Nimmen, Bruno, and Rosholt, *NASA Historical Data Book,* Vol. 1, p. 332.
375 "Spider": *Chariots for Apollo,* p. 292.
375 Mission profile: *NASA Historical Data Book,* Vol. 2, p. 174.
376 Problem of Three Bodies: Whittaker, *A Treatise on the Analytical Dynamics of Particles and Rigid Bodies,* p. 339; *Encyclopaedia Britannica,* 1947, Vol. 22, p. 161.
376 Draper: Gray, *Angle of Attack,* p. 177.
376 Swamped computer: Ibid., p. 191.
377 LBJ's ranch and ocean landing: Ibid., p. 185.
377 McCarthy's ire: Ibid., p. 191.
378 S-2 welding problem: *Stages to Saturn,* p. 217.
378 Shifting thickness: Ibid., p. 218.
378 Failed S-2 test: *Angle of Attack,* pp. 196–97; *Stages to Saturn,* pp. 224–25.
378 O'Connor's memo: *Stages to Saturn,* p. 224.
379 Rees's memo: Ibid., p. 227.
379 Collapsing: *Angle of Attack,* p. 194.
379 Matthew's question: Ibid., p. 158.
379 Pump problem: *Chariots for Apollo,* p. 209.
379 The lemon: Ibid.
379 Cost through 1966: *Managing NASA in the Apollo Era,* p. 155.
380 Picasso and Koestler: "Prophecy of Exploration of Space Left Unfulfilled After 25 Years," *The New York Times,* July 20, 1994.
380 Polls: Gallup, *The Gallup Poll,* Vol. 3, p. 1952.
381 Potlatch: Diamond, *The Rise and Fall of the Space Age,* pp. 1, 7. Diamond was a colleague of the author's at New York University until his death in July 1997.
381 Diamond's complaint: Ibid., pp. 4–5.
382 Moon on the Potomac: Etzioni, *The Moon-Doggle,* pp. 164–65.
382 Mumford: "Prophecy of Exploration of Space Left Unfulfilled After 25 Years," *The New York Times,* July 20, 1994.
382 "Barbarians" and the publicity stunt: *The Moon-Doggle,* pp. 151, 153.
384 Mark and people in space: Mark, *The Space Station,* pp. 44–45.

11. From the Earth to the Moon

388 KH-4 "take": NIE 11-8-64, "Soviet Capabilities for Strategic Attack," in Steury, *Intentions and Capabilities: Estimates on Soviet Strategic Forces, 1950–1983,* pp. 197, 200; SNIE 13-4-64, "The Chances of an Imminent Communist Chinese Nuclear Explosion," in Ruffner, *Corona: America's First Satellite Program,* p. 239. U.S. bomber and missile numbers: Polmar and Laur, *Strategic Air Command,* p. 91.
388 "Vela Hotel": Ball, "The U.S. Vela Nuclear Detection Satellite (NDS) System: The Australian Connection," pp. 3–4; U.S. Congress, Joint Committee on Atomic Energy, *Developments in the Field of Detection and Identification of Nuclear Explosions (Project Vela) and Relationship to Test Ban Negotiations,* pp. 49–50.
388 Weather and communication satellites: Yenne, *The Encyclopedia of U.S. Spacecraft,* p. 153 (TIROS), p. 110 (Nimbus), pp. 151–52 (Telstar), p. 73 (Intelsat), p. 38 (DSCS).
389 Screws and a washer: Hall, *Lunar Impact,* p. 257.
389 Lunar and planetary tally: Burrows, *Exploring Space,* pp. 419–21.
390 The picture from 1,500 feet: Heacock, "Ranger: Its Mission and Its Results," p. 10.
390 Final minutes of Ranger 7 and the press: *Lunar Impact,* pp. 267–70, 272.
390 Kuiper: Ibid., p. 273.
391 Surveyor on a back burner: Newell, *Beyond the Atmosphere,* pp. 270–71.
392 Glennan on Centaur: Nicks, *Far Travelers,* p. 90.
392 Military-civilian Centaur work: Sloop, "Liquid Hydrogen as a Propulsion Fuel, 1945–1959," pp. 200–1.
392 Open-ended trouble: Kloman, *Unmanned Space Project Management,* p. 10.
392 Doubts about Centaur: Ibid.
392 Faget on Surveyor: *Far Travelers,* p. 91.
393 Lunar Survey Module: Ibid., p. 92.
393 Description of Surveyor: Ezell, *NASA Historical Data Book,* Vol. 2, p. 326.
393 Faget's call to Nicks: *Far Travelers,* p. 95.
394 Surveyor details: *NASA Historical Data Book,* Vol. 2, pp. 326–31.
394 Lunar Orbiters: Ibid., Vol. 2, p. 319.
395 Restricting imagery: *Far Travelers,* p. 94.
396 Russians using Lunar Orbiter imagery: Author interview with Basilevsky.
396 Korolyov's approaches and Chelomei's rocket: Johnson, *The Soviet Reach for the Moon,* pp. 10–11.
396 The R-56: Ibid., p. 13.
397 Lebed and dog analogy: "The Quotable Lebed," *The New York Times,* January 23, 1997.
398 Brezhnev's remarks: Weiss, "The Farewell Dossier," p. 121.
398 "no control levers": Mishin, "Why Didn't We Get to the Moon?," pp. 35–36.

398 The L-1 and competition: Author interview with Mishin.
399 Korolyov and the L-1: *The Soviet Reach for the Moon,* p. 12.
399 Soyuz and the end of EOR: Ibid.
399 N-1 specifications: Covault, "Soviet Union Reveals Moon Rocket Design That Failed to Beat U.S. to Lunar Landing," pp. 58–59.
400 Approval of the lunar landing and selection of Chelomei: *The Soviet Reach for the Moon,* pp. 13–14; Portree, *Mir Hardware Heritage,* p. 5.
400 Highly classified: Author interview with Basilevsky.
401 Soyuz described: *Mir Hardware Heritage,* p. 3; Newkirk, *Almanac of Soviet Manned Space Flight,* p. 65.
401 Victory over Chelomei: *The Soviet Reach for the Moon,* p. 15.
402 L-3 mission profile: Covault, "Russians Reveal Secrets of Mir, Buran, Lunar Landing Craft," pp. 38–39; *Mir Hardware Heritage,* p. 20.
402 Space suit hoop: *The Soviet Reach for the Moon,* p. 28.
403 Korolyov's intimate recollections: Author interview with Leonov.
403 "Ten will be enough.": Author interview with Filina.
404 Medvedev on Petrovsky: Medvedev, *Soviet Science,* p. 88.
404 Diagnosis and surgery: Author interview with Filina; Harford interview with Korolyov's daughter in *Korolev,* pp. 278–79; Vishnevsky in Oberg, *Red Star in Orbit,* pp. 88–89.
404 Lamentation of the technical specialist: Smith, *The Russians,* p. 331.
405 Chertok's strengths and liabilities: Author interview with Dula.
405 Keldysh's approval: *The Soviet Reach for the Moon,* p. 16.
406 "all-up" testing: Murray and Cox, *Apollo: The Race to the Moon,* pp. 158–62; Levine, *Managing NASA in the Apollo Era,* p. 6.
406 Faget on nitrogen: Brooks, Grimwood, and Swenson, *Chariots for Apollo,* pp. 239–40.
407 Kramer on oxygen: Memorandum of December 21, 1997.
407 Gemini's oxygen: *Gemini-Titan II Press Handbook,* Part 2, p. 29.
407 Explosive hatch bolts and Grissom: Gray, *Angle of Attack,* pp. 139–41.
408 Fire scene: *Chariots for Apollo,* pp. 214–15.
408 Minute and a half: Bergaust, *Murder on Pad 34,* caption on p. 7 of the picture section.
409 Fighting the fire and the autopsies: *Chariots for Apollo,* p. 217.
409 *Life*'s eulogy: "Put Them High on the List of Men Who Count," p. 18.
409 ASNE and other reaction: *Murder on Pad 34,* pp. 150–51.
409 *The Boston Globe:* Ibid., p. 153.
410 Thompson report: *Chariots for Apollo,* pp. 221–22.
410 Slayton, Kraft, and blame: *Moon Shot* (video), Part 2, TBS, July 13, 1994.
410 Roaring drunk: Ibid.
410 Borman and Stafford: Kramer memorandum of December 21, 1997.
411 Bergaust: Bergaust, *Murder on Pad 34,* pp. 175–76.
411 Mueller: *Chariots for Apollo,* p. 222.
411 Webb and the bomb: *Angle of Attack,* p. 252.
411 Death of Bondarenko: Author interview with Leonov; Harford, *Korolev,* p. 242.
412 The widows' suit: *Chariots for Apollo,* p. 224n.
412 Necessary improvements: *Angle of Attack,* pp. 245, 253–55.
412 HAL's betrayal: Frederick I. Ordway III, "*2001: A Space Odyssey* in Retrospect," in Emme, *Science Fiction and Space Futures,* pp. 100–1.
413 "baptism": McCurdy, *Inside NASA,* pp. 69–70.
413 "we'll help you out": Oberg, *Red Star in Orbit,* p. 90.
413 Soyuz launch and East Berlin radio: Newkirk, *Almanac of Soviet Manned Space Flight,* pp. 58–59.
413 AP story: *Red Star in Orbit,* p. 91.
413 The end of *Soyuz 1: Almanac of Soviet Manned Space Flight,* pp. 58–63; Rhea, *Roads to Space,* pp. 89–90 (Mishin oral history).
414 Design imperfections: *Roads to Space,* p. 89.
414 *Apollo 4* test: *Chariots for Apollo,* pp. 232–33.
414 Von Braun: Bilstein, *Stages to Saturn,* p. 357.
415 Press metaphors: Ibid., pp. 354–55. U.N. Secretariat: Aldrin, *Men From Earth,* p. 141.
415 *Apollo 5: Chariots for Apollo,* pp. 241–44.
416 "Gagarin is not with us": Rebrov, *Space Catastrophes,* p. 94.
417 *Apollo 6: Chariots for Apollo,* pp. 248–52.
418 "dancing bears": Michener, *Space,* p. 415.
418 *Apollo 7*'s problems: Shepard and Slayton, *Moon Shot,* pp. 221–22; *Moon Shot* (video), Part 2.
418 *Apollo 7: NASA Historical Data Book,* Vol. 2, p. 185; *Chariots for Apollo,* pp. 267–71.
419 Intelligence reports: *Moon Shot,* pp. 226–27.
419 Zond 5: *Almanac of Soviet Manned Space Flight,* pp. 51, 69; Matson, *Cosmonautics,* pp. 47–47.
419 Slayton: *Moon Shot,* p. 228.
419 Paine is persuaded: *Chariots for Apollo,* pp. 258–59.
419 "single greatest gamble": *Moon Shot,* p. 229.
420 Lovell to Houston: Ibid., p. 232.
420 Lovell's description: "Apollo 8 Heads Back to Earth After Circling Moon 10 Times," *The Washington Post,* December 25, 1968.
420 '68 special issue: *Life,* January 10, 1969.
420 Greetings from *Apollo 8: Moon Shot,* p. 233.
421 "Santa Claus": Ibid., p. 234.
421 Long-distance record: *NASA Historical Data Book,* 1958–1968, Vol. 2, p. 189.
421 Kaminin's diary: *Moon Shot,* p. 230.
421 The three Soviet accidents: *The Soviet Reach for the Moon,* p. 32.
422 *Apollo*s *9* and *10:* Ezell, *NASA Historical Data Book,* Vol. 3, pp. 81–84.
422 Mailer: Mailer, *Of a Fire on the Moon,* p. 471.
423 Vonnegut: Kurt Vonnegut, Jr., "Excelsior! We're Going to the Moon. Excelsior," *The New York Times Magazine,* July 13, 1969.
423 Toynbee: Author interview.

423 Niebuhr: Author interview.
423 Van Doren: Author interview.
423 Asimov: Author interview.
424 Mead: Author interview.
424 Morrison: Author interview.
425 McLuhan: Author interview.
425 Cost of Apollo: *NASA Historical Data Book,* Vol. 3, p. 61.
425 Esther Goddard: "Widow of a Pioneer in Rocketry Recalls 41-Foot Flight in '26," *The New York Times,* July 17, 1969.
425 Journalists: "Apollo's Great Leap for the Moon," *Life,* July 25, 1969, p. 23.
425 VIPs: "The V.I.P. Guests Can Hardly Find Words," *The New York Times,* July 17, 1969.
425 Agnew's crimes: "Spiro T. Agnew, Point Man for Nixon Who Resigned Vice Presidency, Dies at 77," *The New York Times,* September 19, 1996.
426 Space Task Group: "The Post-Apollo Space Program: Directions for the Future," pp. 5, 21.
426 Agnew: *Angle of Attack,* p. 268; "Agnew Proposes a Mars Landing," *The New York Times,* July 17, 1969.
426 Nixon's proclamation: "Nixon Calls for a Holiday So All Can Share in 'Glory,' " *The New York Times,* July 17, 1969.
427 Mailer and the VIPs: *Of a Fire on the Moon,* pp. 90–91.
427 Crowds: "Some Applaud As Rocket Lifts, but Rest Just Stare," *The New York Times,* July 17, 1969.
427 Boats: "Armada of Small Craft Clogs River to See Apollo Launching," *The New York Times,* July 17, 1969.
427 Abernathy: "Some Applaud As Rocket Lifts, but Rest Just Stare."
427 TV coverage: Diamond, *The Rise and Fall of the Space Age,* p. 96; "Apollo 11 TV Coverage to Engage Many Earthlings," *The New York Times,* July 14, 1969; "Apollo Coverage—The Networks," *The New York Times,* July 20, 1969.
427 Soviet metalworkers: "Earthly Worries Supplant Euphoria of Moon Shots," *The New York Times,* July 20, 1994.
428 Commission of inquiry: Author interview with Basilevsky; "Luna Mission Ends," *The New York Times,* July 22, 1969.
428 Nuclear explosion: *Stages to Saturn,* p. 357.
428 The launch: Aldrin, *Men from Earth,* pp. 227–28.
428 Abort: Ibid., p. 237.
429 The landing: Ibid., pp. 238–41.
429 Lapps and the judge: "All the World's in the Moon's Grip," *The New York Times,* July 22, 1969.
429 Television coverage: "Apollo Coverage—The Networks."
429 Feoktistov: "Soviet Television Shows the Moon Walk 3 Times," *The New York Times,* July 22, 1969.
430 The view from the Moon: *Men from Earth,* p. 243.
430 Nixon's greeting: "Nixon Sees Crew," *The New York Times,* July 25, 1969.
430 Departure from and arrival on Earth: *NASA Historical Data Book,* Vol. 3, p. 85; *Men from Earth,* p. 246.
431 Nixon and Moon rocks: "World Leaders to Get a Piece of the Moon," *The New York Times,* July 28, 1969.
431 Origin of the Moon: Taylor, "The Scientific Legacy of Apollo," pp. 42–43. A rich and readable scientific account of the Moon was published in 1977: French, *The Moon Book.*
431 Reagan and the Moon: "Prophecy of Exploration of Space Left Unfulfilled After 25 Years," *The New York Times,* July 20, 1994.
431 Passengers to the Moon: "200 on Pan Am Waiting List Are Aiming for Moon," *The New York Times,* January 9, 1969.
431 Von Braun's prediction: "Prophecy of Exploration of Space Left Unfulfilled After 25 Years."
432 Barbara Day: "Moonstruck, Even in Remote Arkansas," *The New York Times,* July 25, 1969.
432 Luna 16: Johnson, *Handbook of Soviet Lunar and Planetary Exploration,* pp. 54–62.
432 No lives at risk: Vladimirov, *The Russian Space Bluff,* pp. 157–58.
432 Wolfe: "Prophecy of Exploration of Space Left Unfulfilled After 25 Years."
433 Logsdon's thesis: Logsdon, "Apollo Exemplifies What Not to Do," *Space News,* July 18–24, 1994, p. 10.

12. Destination Mars

434 Morrison on the Moon: Author interview.
435 Life on Venus: *Encyclopaedia Britannica,* Vol. 23, 1947, p. 72.
435 De Vaucouleurs: Kieffer et al., *Mars,* p. 72.
435 RAND and vegetation: Buchheim, *Space Handbook,* p. 20.
437 NASA organization: Ezell, *NASA Historical Data Book,* Vol. 2, pp. 200–02; Vol. 3, pp. 130–31.
437 Budget through 1966: Ibid., Vol. 2, p. 204.
437 Budget for the 1970s and percentage of the entire budget: Ibid., Vol. 3, p. 132.
437 Helfand's remarks: Author interview.
438 Lunar Orbiter imagery: Author interview with Basilevsky.
439 Murray gloating: *Destination Mars,* Horizon, BBC.
440 Agnew and sustaining interest: *The Post-Apollo Space Program: Directions for the Future,* p. 5.
440 "commonality and reusability": Ibid., p. ii.
441 Von Braun on the shuttle: Manuscript dated October 6, 1969, in the NASA History Office archive.
441 Mueller numbers: Roland, "The Shuttle: Triumph or Turkey?"
441 Mathematica: Biddle, "The Endless Countdown."
441 NASA report: *Goals and Objectives for America's Next Decades in Space* (Washington, D.C.: National Aeronautics and Space Administration, September 1969), pp. 8–9, 30.
442 Budget drops: Ezell, *NASA Historical Data Book,* Vol. 3, p. 8; Baker, *The History of Manned Space Flight,* p. 451.
443 "downright dangerous": Lovell and Kluger, *Lost Moon,* p. 13.

443 Haise's and Lovell's exhausting vigil: "The Joyous Triumph of Apollo 13," pp. 30–31.
443 Deadly effects on the body: *Lost Moon,* p. 1.
444 Apollo Applications Program and MOL: *NASA Historical Data Book,* Vol. 3, p. 98.
445 Deaths of the *Soyuz 11* crew: Harford, *Korolev,* p. 298; Oberg, *Red Star in Orbit,* pp. 102–03; Newkirk, *Almanac of Soviet Manned Space Flight,* pp. 100–04.
445 *Skylab* details: *NASA Historical Data Book,* Vol. 3, pp. 103–08.
445 "oneness": *The Post-Apollo Space Program: Directions for the Future,* pp. 6–7.
446 Kennedy, Khrushchev, and the United Nations: Ezell and Ezell, *The Partnership,* pp. 38–52.
447 *Marooned:* Ibid., pp. 9–10.
447 Kissinger, Low, Keldysh, and competition: Ibid., pp. 126–27.
448 Public relations and *Aviation Week:* Ibid., pp. 306–07.
448 Tyuratam at night: Johnson, *Handbook of Soviet Lunar and Planetary Exploration,* p. 227.
449 Bergman: *The Partnership,* p. 280.
450 Dobrynin: Dobrynin, *In Confidence,* p. 342.
450 Greetings and Metz: *The Partnership,* p. 329.
450 Inside *Apollo-Soyuz:* Shepard and Slayton, *Moon Shot,* pp. 355–56.
452 Sagdeev: Sagdeev, *The Making of a Soviet Scientist,* p. 160.
453 Murray on exploration: Author interview, July 2, 1987.
454 Von Braun on Mars: Von Braun, *The Mars Project,* p. 1.
454 Discouraging students: Ezell and Ezell, *On Mars,* p. 10.
454 Struve: Struve, "The Astronomical Universe."
455 Gingerich: Author interview.
455 Sea Dragon: Dowling et al., "The Origin of Gravity-propelled Interplanetary Space Travel," p. 3.
455 Atomic energy to the Moon: Gatland and Kunesch, *Space Travel,* p. 94.
456 Orion: Dyson, *Weapons and Hope,* pp. 65–66.
456 Ion engine: Author interview with Brophy.
456 Increasing onboard velocity. Sutton, "Rocket Propulsion Systems for Interplanetary Flight"; Stearns, "Electro-propulsion System Applications"; Newgard and Levoy, "Nuclear Rockets"; Yaffee, "Electric Rocket Program Being Reshaped"; Langmuir, "Electric Spacecraft—Progress 1962"; Spencer et al., "Nuclear Electric Spacecraft for Unmanned Planetary and Interplanetary Missions" (technical report); Beale, Speiser, and Womack, "The Electric Space Cruiser for High-Energy Missions" (technical report); von Karman, "Introductory Remarks on Space Propulsion Problems," in Casci, *Advances in Astronautical Propulsion;* Seifort, "Propulsion for Space Vehicles," in Brown, *Space Logistics Engineering.*
456 Hohmann Trajectory: Glasstone, *Sourcebook on the Space Sciences,* pp. 69–70.
456 "Least fuel path": Munick, McGill, and Taylor, "Analytic Solutions to Several Optimum Orbit Transfer Problems," p. 77.
456 Hohmann Transfer: Lawden, *Optimal Trajectories for Space Navigation,* p. 107; Lawden, "Interplanetary Orbits," in Bates, *Space Research and Exploration* (1958), p. 174; Lawden, "Transfer Between Circular Orbits," *Jet Propulsion,* July 1956, p. 555; Bryson and Ouellette, "Optimum Paths to the Moon and Planets," *Astronautics,* Sept. 1958, pp. 18–19; Dowling et al., "The Origin of Gravity-propelled Interplanetary Space Travel," IAF paper, October 1990, p. 2.
457 "It's a killer": Author interview.
458 Solving the Problem of Three Bodies: Dowling et al., "The Origin of Gravity-propelled Interplanetary Space Travel," p. 7.
458 Discovery of gravity propulsion: Ibid., pp. 7–9.
458 Other astrodynamicists: Tsander, *Flights to Other Planets,* 1925, pp. 237–302; Lawden, *Optimal Trajectories for Space Navigation;* Lawden, "Interplanetary Orbits," in Bates, *Space Research and Exploration,* pp. 164–84; Herrick, "Earth Satellites and Related Orbit and Perturbation Theory," in Seifert, *Space Technology,* pp. 5-02–5-30; Ehricke, *Space Flight,* pp. 963–85.
459 Lawden and the 1954 article: Murray and Burgess, *Flight to Mercury,* p. 11.
459 The Soviet paper: Duboshin and Okhotsimsky, "Some Problems of Astrodynamics and Celestial Mechanics," p. 9.
460 Minovitch and the outer planets: Minovitch, "Utilizing Large Planetary Perturbations for the Design of Deep-Space, Solar-Probe, and Out-of-Ecliptic Trajectories"; Dowling et al., "Gravity Propulsion Research at UCLA and JPL, 1962–1964," pp. 29–31.
460 Flandro: Flandro, "Utilization of the Energy Derived from the Gravitational Field of Jupiter for Reducing Flight Time to the Outer Solar System."
460 Sagdeev on competition: *Venus Unveiled,* NOVA.
460 Automatic Interplanetary Station: Kieffer et al., *Mars,* p. 73.
461 Soviet planetary failures: *Handbook of Soviet Lunar and Planetary Exploration,* pp. 136, 180.
461 First to Venus and Mars: Matson, *Cosmonautics: A Colorful History,* pp. 28–29.
462 Choice of Venus: *On Mars,* p. 40.
462 Mariner 2: Smith, *Planetary Exploration,* pp. 69–70; *On Mars,* pp. 40–41; Koppes, *JPL and the American Space Program,* p. 128.
463 Morrison on life: Author interview.
463 JPL, NASA, and the press on Mars: *On Mars,* pp. 76–77.
464 Mariner 4 data from Mars: Hartmann and Raper, *The New Mars: The Discoveries of Mariner 9,* pp. 22–23; *On Mars,* pp. 76–79; *Mariner-Mars 1964: Final Project Report,* pp. 6–7, 265–311.
464 "priceless Rembrandt": Author interview.
464 Murray on Mars: Wilford, *Mars Beckons,* p. 59.
466 Leighton on Mars: *On Mars,* pp. 175–76.
466 Cosmos 419: *Mars,* pp. 85–86.
467 Johnson's observation: Author interview, October 11, 1988.
467 Davies on responding: Author interview, January 4, 1989.
468 The Ghoul: Author conversation with John Casani, July 23, 1997; Helms, "The Great

Galactic Ghoul," p. 52; Oberg, "The Great Galactic Ghoul," pp. 10–11.

468 Mariner 9's pictures and data bits: *The New Mars: The Discoveries of Mariner 9,* p. 162.

468 Mariner 9 data: *Mars,* pp. 89–94. Also see *The New Mars: The Discoveries of Mariner 9.*

469 Murray and Burgess on Mars: *Flight to Mercury,* p. 2.

469 Bradbury: *The New Mars: The Discoveries of Mariner 9,* p. 162.

470 Viking's cost: *NASA Historical Data Book,* Vol. 3, p. 142. The overrun: *On Mars,* p. 219. TRW overrun: Ibid., p. 229; Kramer memorandum of January 7, 1998.

471 Mutch: Viking Lander Imaging Team, *The Martian Landscape,* p. 3.

471 Viking's first picture: Ibid., pp. 34–35.

471 Landing sites: Greeley, *Mars Landing Site Catalog.*

472 Water from volcanoes: Greeley, "Release of Juvenile Water on Mars: Estimated Amounts and Timing Associated with Volcanism," pp. 1653–54.

472 Horowitz's observations: Horowitz, *To Utopia and Back,* pp. 129, 141.

473 Murray on Viking: Murray, *Journey into Space,* pp. 69–73.

13. A Grand Tour: The Majestic Adventure

474 Minovitch and Flandro: Minovitch, "Utilizing Large Planetary Perturbations for the Design of Deep Space, Solar Probe, and Out-of-Ecliptic Trajectories"; Flandro, "Utilization of Energy Derived from the Gravitational Field of Jupiter for Reducing Flight Time to the Outer Solar System."

474 Stewart: Stewart, "New Possibilities for Solar-System Exploration."

475 The 1962 discovery: Murray and Burgess, *Flight to Mercury,* p. 12.

475 Minovitch's multiple flybys: Minovitch, "The Determination and Characteristics of Ballistic and Interplanetary Trajectories Under the Influence of Multiple Planetary Attractions."

475 Flandro and Bourke: Interoffice Memorandum to T. A. Barber and P. W. Haurlan, 312.5-201, October 10, 1966, p. 3.

475 Numbers of scientists and engineers: Levine, *Managing NASA in the Apollo Era,* p. 116.

476 General Electric and Corona: "First Capsule Recovered from Satellite," p. 33 (to take only one example among scores).

476 Fostering competition: Author interview with Hall.

476 Solar flare data: NASA/Ames Educational Data Sheet No. 503, "Pioneers 6 to 9," July 1971, p. 1.

476 Galactic Jupiter Probes: Fimmel, Van Allen, and Burgess, *Pioneer: First to Jupiter, Saturn, and Beyond,* pp. 26–27.

477 NASA release: "Grand Tour" of Planets, Release No. 69-84, June 2, 1969.

478 Hidden surcharge: Murray, *Journey into Space,* p. 138.

478 Naugle's invitation: "Invitation for Participation in Mission Definition for Grand Tour Missions to the Outer Solar System," NASA headquarters letter, October 15, 1970.

479 The All-Stars: "Grand Tour Scientists," NASA Release No. 71-56, pp. 5–11.

479 NASA's funding for Fiscal 1972 and shuttle funding: Ezell, *NASA Historical Data Book,* Vol. 3, pp. 9, 69.

479 Fletcher to Weinberger: Letter of December 22, 1971.

480 Murray and his friend: *Journey into Space,* p. 141.

480 Eneyev's plan: "Russian Criticizes U.S. Plan for a Tour of the Outer Planets by Unmanned Craft," *The New York Times,* May 9, 1971.

481 The imaging team's report: Belton et al., "Grand Tour Outer Planet Missions Definition Phase," pp. 28–30.

481 Not serious about Uranus and Neptune: *Journey Into Space,* p. 141.

481 Improved brains and communication: Ibid.

482 "dime on a dinner plate": "Pioneer Will Reach Jupiter December 3," NASA News Release No. 73-243, November 7, 1973, p. 8.

482 "flown from the ground": Burgess, *By Jupiter,* p. 20.

483 The plaque controversy: Sagan et al., *Murmurs of Earth,* pp. 57–58.

484 Selling a dead camel: Author interview.

484 Twelve moons: "Pioneer F Mission to Jupiter," NASA Release No. 72-25, for release on February 20, 1972.

485 Sagan and Jupiter: Sagan, "Planetary Exploration," p. 58.

485 Possibility of life: "Pioneer F Mission to Jupiter," Release No. 72-25, p. 24.

486 Hall, Galileo, and the reporters: *Pioneer: First to Jupiter, Saturn, and Beyond,* p. 80.

487 Pioneer 10 at closest approach: "Pioneer Will Reach Jupiter December 3," pp. 38–44.

487 Doose's, Kraemer's, and Fimmel's remarks: *Pioneer: First to Jupiter, Saturn, and Beyond,* pp. 84, 90.

487 Fimmel's night-light: "NASA Expects Pioneer 10's Useful Lifetime Will Be Extended by Decade," p. 43.

487 TRW's position: Roundup, *Aviation Week & Space Technology,* June 20, 1988, p. 15.

488 Nine-planet gestalt: Miller and Hartmann, *The Grand Tour,* p. 8.

488 Data on the core: *Pioneer: First to Jupiter, Saturn, and Beyond,* p. 155.

488 Life on Titan: "Pioneer Saturn Results Are Summarized," p. 2.

489 F Ring: *Pioneer: First to Jupiter, Saturn, and Beyond,* pp. 152–53.

489 G Ring: "New Saturn Discoveries by Pioneer."

489 Veneras 4–8: Johnson, *Handbook of Soviet Lunar and Planetary Exploration,* pp. 147–58. His observation on the frequency of shots is on p. 155.

490 Understanding the ecosphere: Murray and Burgess, *Flight to Mercury,* p. 69.

490 Same side toward the Sun: Buchheim, *Space Handbook,* p. 18.

491 Mercury and the solar wind: *Flight to Mercury,* pp. 85–87.

491 Magnetometers: Ibid., p. 98.
492 The order of questions: Hartmann and Raper, *The New Mars,* p. 163.
492 Mercury's "moon": *Flight to Mercury,* pp. 109–10.
492 The power surge: *Flight to Mercury,* pp. 108–09.
494 Pioneer and the surface of Venus: Fimmel, Colin, and Burgess, *Pioneer Venus,* pp. 116–27.
494 Formation of the planets and the greenhouse: "Early Findings from Pioneer Venus," NASA Release No. 79-06, pp. 4–5, 7.
494 Pioneer-Venus 1's end: "Pioneer 12 Is Sinking into the Venusian Clouds and Burning Up," *The New York Times,* October 10, 1992.
495 Voyager's structure and innards: JPL's "Voyager Uranus Travel Guide," pp. 57–73; Morrison and Samz, *Voyage to Jupiter,* pp. 26–32.
495 Photograph of the nude couple: Sagan et al., *Murmurs of Earth,* p. 74.
496 Voyager's plaque: Ibid., passim.
496 Carter's message: "Voyager Will Carry 'Earth Sounds' Record," unnumbered and undated NASA release, p. 9 (copy of White House letter dated June 16, 1977).
497 Computer and other problems and solutions: *Voyage to Jupiter,* pp. 48–50.
497 Io, Europa, and the Great Red Spot: Ibid., p. 60.
498 Brown, the Carters, and Van Gogh: Ibid., pp. 74–75.
499 Smith enraptured: Ibid., p. 76.
499 Sequencing: Author interview with Lane.
499 Great Red Spot as an eddy: "Jupiter's Baffling Red Spot Loses Some of Its Mystery," *The New York Times,* November 12, 1985.
500 Rumors about Stockman: Washburn, *Distant Encounters,* pp. 221–22; Davis, *Flyby,* p. 76.
501 Funding for 1983 and 1984: John M. Logsdon, *The Survival Crisis of the U.S. Solar System Exploration Program* (Washington, D.C.: NASA History Office, Sept. 1989), pp. 13–14.
501 Beggs's warning: Ibid., p. 15.
501 The rings' dimensions: Morrison, *Exploring Planetary Worlds,* p. 164.
501 Johnson on the moons: Waldrop, "The Puzzle That Is Saturn," *Science,* p. 1347.
501 Discovery led theory: Elliot and Kerr, *Rings,* p. ix.
501 Murray's lament: "The End of the Beginning," *Science,* January 29, 1982, p. 459.
502 Keyworth's statement: *The Survival Crisis of the U.S. Solar System Exploration Program,* pp. 25–26.
502 Shooting to the inside: Author interview with Tatarewicz.
503 Telescope versus planetary science: Smith, *The Space Telescope,* p. 184.
503 "the greatest of mysteries": Sagan, *Cosmos,* p. 4.
503 "Where do we fit?": *The News Hour with Jim Lehrer,* PBS, December 20, 1996.
503 Morrison's letter: Washburn, *Distant Encounters,* p. 222.
504 Beckman, Rousselot, Pownall, Scranton, Goldberger, and the Budget Committee: *The Survival Crisis of the U.S. Solar System Exploration,* pp. 30–37.

14. The Roaring Eighties

505 Soviet early-warning satellites: Johnson, *The Soviet Year in Space: 1978,* pp. 66–68.
506 Cracking the curtain: Leavitt, "MOL: Evolution of a Decision," p. 35. Lifeboat: Memorandum, "Comments on Need for Space Rescue," from Cyrus Vance to Bill Moyers, May 29, 1965.
506 Forman on politics: Forman, "The Political Process in Systems Architecture Design," pp. 24–25.
507 No parades for robots: Martin made the remark at the "Pathways to the Planets" conference in Washington on May 31 and June 1, 1989. It was sponsored by NASA.
507 Schweickart on Earth: White, *The Overview Effect,* p. 38.
507 "Loons": Collins, *Mission to Mars,* p. 185.
508 Almaz: Portree, *Mir Hardware Heritage,* pp. 63–65.
508 Two weeks on the Moon: "Russians Reveal Secrets of Mir, Buran, Lunar Landing Craft," pp. 38–39.
508 Stations are more serious: Oberg and Oberg, *Pioneering Space,* p. 201.
509 *Salyut 1* and the *Soyuz* problems: Newkirk, *Almanac of Soviet Manned Space Flight,* pp. 96–104; Rhea, *Roads to Space,* p. 91.
510 Watching and listening to *Salyut 2* and the fate of *Cosmos 557:* Kramer memorandum of January 7, 1998.
510 *Salyuts 3-5: Almanac of Soviet Manned Space Flight,* pp. 125–46.
510 The end of Almaz": *Mir Hardware Heritage,* p. 64.
511 Ten tons of food: Matson, *Cosmonautics,* p. 60.
511 *Salyut* improvements: *Mir Hardware Heritage,* p. 75.
511 *Soyuz Ferry:* Ibid., p. 24.
511 *Soyuz T* and *TM:* Ibid., pp. 47–59.
511 *Progress:* Ibid., p. 36.
511 Foreign cosmonauts: Ibid., pp. 79–87.
512 Running and peddling: Ibid., p. 85.
512 Peddling on *Skylab:* Compton and Benson, *Living and Working in Space,* NASA SP-4208, p. 341.
512 Cosmorats: "Soviet Space Prescription: Artificial Gravity for Long Flights," *The New York Times,* May 17, 1988.
513 Lebedev's problems: Lebedev, *Diary of a Cosmonaut,* p. 264.
513 Romanenko's irritation: "Long Stay in Space Proves 'Very Difficult' for Soviet Commander," *The New York Times,* December 29, 1987; "Record Soviet Manned Space Flight Raises Human Endurance Questions," p. 25.
513 Ryumin on murder: Reviews and Previews, *Air & Space,* June–July 1997, p. 84.
513 Lebedev's physical illness: *Diary of a Cosmonaut,* p. 270.
513 Blaha's depression: "Astronaut Tells of Down Side to Space Life," *The New York Times,* January 22, 1997.
513 Lebedev on closeness: *Diary of a Cosmonaut,* p. 257.

513 Psychological problems: Cooper, "The Loneliness of the Long-Duration Astronaut," pp. 37–38.
513 Savitskaya's EVA: "Russian Astronaut Becomes First Woman to Walk in Space," *The New York Times,* July 26, 1984.
514 Ten years in orbit: "Salyut 7 Crew Prepared to End Record Flight," p. 29.
514 Manarov's diary entries: Sotheby's, *Russian Space History,* Lot 347.
514 "*Never . . .* again!": Said to the author.
516 James's treatise: James, *Soviet Conquest From Space,* Grand Tour, p. 111; death ray, p. 126; shuttles, pp. 136, 141; orbit to orbit, p. 150; station and Super Booster, p. 156; Russian public interest, p. 159.
516 Jastrow's warning: Jastrow, "The New Soviet Arms Buildup in Space," p. 100.
517 *Izvestia,* Chertok, and Gringauz: "Space Rockets vs. Butter Is the Talk of Russia," *The New York Times,* April 17, 1989.
517 Energiya-*Buran:* Johnson, *The Soviet Year in Space: 1988,* pp. 18, 108–14.
517 Olesyuk's complaint: Kidger, "Mir's Effectiveness Questioned," p. 189.
517 Russian fear of the shuttle: Author interview with Dula.
518 Buccaneers memo: Roland, "The Shuttle: Triumph or Turkey?," pp. 34–35.
519 $5.15 billion: McCurdy, *Inside NASA,* p. 86.
519 The scientist commenting on Fletcher: Ibid., p. 87.
520 Projected launch rate and cost: General Accounting Office, Report to Congress, "Analysis of Cost Estimates for the Space Shuttle and Two Alternate Programs," B-173677, June 1, 1973, p. 1.
520 Cost per pound of payload in orbit: Ibid., p. 30.
520 Using the back of an envelope: *Inside NASA,* p. 86.
520 Cost of a shuttle launch: "Launch Cost of a Shuttle: Take Your Pick," *Space News,* November 30–December 6, 1992, pp. 4, 29.
521 Air Force hedging its bet: "Defense Dept. to Retain Expendable Launchers as Backup to Shuttle," p. 115.
521 KH-11 and Sigint satellites: Burrows, *Deep Black,* pp. 225–51; Lindsay, *The Falcon and the Snowman,* pp. 57–64; Richelson, *The U.S. Intelligence Community,* pp. 173–76.
521 Air Force shuttle requirements: Logsdon, "The Space Shuttle Program: A Policy Failure?," pp. 1100–01.
522 Slick Six: "USAF Prepares West Coast Site for Space Shuttle Processing," p. 42; Kramer memorandum of January 7, 1998.
522 The joint DIA-NSA report: "Threat Assessment Report—Space Transportation System (STS)"; vandals and pranksters, p. 106; Romanian establishment, p. 107.
523 Tile problems and verification: Roland, "The Shuttle: Triumph or Turkey?," pp. 40–41. $10 billion: "The Shuttle: America Poised for a Return to Space," *The New York Times,* April 7, 1981.
523 Never before: "The Shuttle: Triumph or Turkey?," p. 29.
523 Launch sequence: *National Space Transportation System Reference,* Vol. 1, p. 11.
525 Wolfe on the flight of *Columbia:* Wolfe, "Columbia's Landing Closes a Circle," p. 474.
525 Spacelab details: "STS-9/Spacelab 1 Press Kit," Release No. 83-173, ESA and NASA, November 1983. A better one, "Spacelab News Reference," 14M983, was published by the Marshall Space Flight Center at about the same time, though it was undated. Also see Froehlich, *Spacelab: An International Short-Stay Orbiting Laboratory.*
525 Cancer, radiation, and AIDS research: "Shuttle Crystal Growth Tests Could Advance Cancer Research," p. 18; "Shuttle May Carry Bone Marrow into Space for Study on Radiation," *The New York Times,* February 11, 1985. AIDS: "STS-26 Press Information," Rockwell International, Office of Public Relations, September 1988, p. 52.
525 AIDS experiments in space: "Experiments Tied to AIDS Battle," *The Orlando Sentinel,* October 2, 1988.
525 Getaway Specials: "NASA's Get Away Specials," NASA Facts, Greenbelt: Goddard Space Flight Center, August 1989.
525 Mixed success: " 'Getaway Specials' Seen As Less Than a Great Success," *The New York Times,* October 11, 1983.
525 The *Science* issue on Spacelab 1: July 13, 1984. Editorial comment was on p. 127.
526 Flights projected and flown: "In Harsh Light of Reality, the Shuttle Is Being Re-evaluated," *The New York Times,* May 14, 1985.
526 Wilford's discrepancies: Ibid.
527 "the crucifixion": Dennis Overbye, "Success amid the Snafus," p. 54.
527 Rudolph's claim of innocence: "Rocket Expert Regrets Giving Up Citizenship," *The New York Times,* October 1, 1985.
527 The death of Rudolph: "Arthur Rudolph, 89, Developer of Rocket in First Apollo Flight," *The New York Times,* January 3, 1996; "Scientist Accused as Ex-Nazi Is Denied Citizenship," *The New York Times,* February 20, 1993.
529 Severomorsk explosion: "U.S. Says Blast Hit Soviet Arms Base," *The New York Times,* June 23, 1984; "Soviet Naval Blast Called Crippling," *The New York Times,* July 11, 1984.
529 Articles citing the need for new equipment: "Space Reconnaissance Dwindles," pp. 18–20; "Pentagon Nursing an Aging Network of Key Satellites," *The New York Times,* July 20, 1987.
529 Knoche's visit to the White House: Author telephone interview.
530 $1 billion over budget: Author interview.
530 "playpen for engineers": Author interview.
530 More than a hundred bucket carriers: Ruffner, *Corona,* p. xv.
531 Soviet real-time reconnaissance: Author interview with Anatoli I. Savin.
531 Winning by "this much": Author interview.
532 Soviet ABM details: *Soviet Military Power: 1985* (Washington: Department of Defense, 1985).

532 The first ABM radar leak: "U.S. Scrutinizing New Soviet Radar," pp. 19–20. For a drawing of the site made from satellite imagery, see the illustration in the issue of May 7, 1984, p. 115, and in *Soviet Military Power: 1985*, p. 46.

532 Teller's uses for nuclear bombs: Broad, *Teller's War*, pp. 46–47.

533 Armageddon, MAD, and Anderson's memo: Ibid., pp. 98–101.

533 Nike-X, Sentinel, and Safeguard: Schwartz, "Past and Present: The Historical Legacy," in Carter and Schwartz, *Ballistic Missile Defense*, pp. 335–41.

534 Article 5: *Arms Control and Disarmament Agreements*, p. 140.

534 Teller on passive defense: Teller, "The Feasibility of Arms Control and the Principle of Openness," in Brennan, *Arms Control, Disarmament, and National Security*, p. 123.

535 Global Ballistic Missile Defense: Graham, *High Frontier*, pp. 135–90; 432 trucks, p. 152; off-the-shelf hardware, p. 25.

535 Dyson and the dream: Dyson, *Weapons and Hope*, pp. 69–70.

535 Reagan's speech: "Reagan Proposes U.S. Seek New Way to Block Missiles," *The New York Times*, March 24, 1983.

535 Graham on the Pentagon: *Visions of Star Wars* (video), NOVA, April 22, 1986.

536 Fletcher to the House: Fletcher, "The Strategic Defense Initiative," Statement Before the Subcommittee on Research and Development, Committee on Armed Services, House of Representatives, March 1, 1984.

536 Star Wars hardware: "Panel Urges Defense Technology," *Aviation Week & Space Technology*, October 17, 1983, pp. 16–18; "Shuttle May Aid in Space Weapons Test," *Aviation Week & Space Technology*, October 31, 1983, pp. 74–77; "Panel Urges Boost-Phase Intercepts," *Aviation Week & Space Technology*, December 5, 1983, pp. 50–52; "Defense Dept. Developing Orbital Guns," *Aviation Week & Space Technology*, July 23, 1984, pp. 61–67.

536 Dyson on *Star Wars: Weapons and Hope*, p. 65.

538 Garwin's high-tech automobile: Made at a conference on "Strategic Defense: Technical Problems and Prospects," New York Academy of Sciences, October 12, 1983.

538 Andropov's statement: "Andropov Says U.S. Is Spurring a Race in Strategic Arms," *The New York Times*, March 27, 1983.

538 Aerodynamic warheads: Bunn, "The Next Nuclear Offensive," pp. 28–36.

538 Penetration aids: "U.S. Seeks Missiles to Evade Defenses," *The New York Times*, February 11, 1985.

538 French super-warheads: "Paris to Try to Counter a Russian Space Shield," *The New York Times*, November 13, 1985.

539 BMD as an offensive weapon: "Dark Side of 'Star Wars': System Could Also Attack," *The New York Times*, March 7, 1985.

539 Peace Shield's commercial: *Visions of Star Wars* (video).

539 Budget estimates: "$400 Billion Price Is Seen for Antimissile Plan," *The New York Times*, October 27, 1984; "Cost of Missile Defense Put at $70 Billion by 1993," *The New York Times*, February 12, 1985.

539 Livermore and Lockheed: Federation of American Scientists, *Public Interest Report*, April 1987, p. 7.

539 Forty-five percent: "Star Wars Risky Business for Aerospace Contractors Factor in Push for Early Deployment of System, Concludes New Study by Federation of American Scientists," press release, April 20, 1987.

539 Carlson: Author interview.

540 The deploring of theology: Brzezinski, Jastrow, and Kampelman, "Defense in Space Is Not 'Star Wars,' " pp. 28–29 ff.

540 Reagan's being "disconnected": "Group of Top Scientists Close to Government Fighting Space Weapons Plan," *The New York Times*, November 16, 1983.

540 SDI in the press: Bethe et al., "Space-based Ballistic Missile Defense," pp. 39–49; Rathjens and Ruina, "Group of Top Scientists Close to Government Fighting Space Weapons Plan," *The New York Times*, November 16, 1983; Payne and Gray, "Nuclear Policy and the Defensive Transition"; Burrows, "Ballistic Missile Defense: The Illusion of Security," pp. 820–56.

540 APS statement: "APS Statement Urges No Early Commitment to SDI Deployment," press release, April 24, 1987.

540 The *Times*, APS, and Lowell Wood: "The Strategic Black Hole" (editorial), *The New York Times*, February 14, 1986; "Weapons Designers Challenge SDI Report," p. 127; "SDI Zapped," p. 18.

540 Brilliant Pebbles: " 'Brilliant Pebbles' Missile Defense Concept Advocated by Livermore Scientist," *Aviation Week & Space Technology*, June 13, 1988, pp. 151–52, 155.

540 The Soviet laser ABM: Stevens, "The Soviet BMD Program," in Carter and Schwartz, *Ballistic Missile Defense*, pp. 189–201. Teller: *Visions of Star Wars* (video).

541 Jobs in Colorado: "Reagan Asserts 'Star Wars' Plan Will Create Jobs and Better Life," *The New York Times*, October 31, 1986.

541 NASA and SDI: "NASA's Link to Strategic Defense Effort Grows with New Spacecraft," p. 18.

541 Sharing SDI: "Reagan Asserts 'Star Wars' Plan Will Create Jobs and Better Life."

542 Sofaer's opinion: "Reagan Reinterprets the ABM Treaty," p. 644.

542 Carnesale's rebuttal: Ibid.

542 Smith's opinion: " 'Star Wars' Tests and the ABM Treaty," *Science*, July 5, 1985, pp. 29–30.

542 *Daedalus* assessment: Chayes, Chayes, and Spitzer, "Space Weapons: The Legal Context," p. 215. The magazine devoted a double issue to "Weapons in Space," airing both sides.

542 Weinberger's role in the deception: "Lies and Rigged 'Star Wars' Test Fooled the Kremlin, and Congress," *The New York Times*, August 18, 1993; Weinberger's denial: "General Details Al-

tered 'Star Wars' Test," *The New York Times,* August 27, 1993.

543 The rigged test: "Lies and Rigged 'Star Wars' Test Fooled the Kremlin, and Congress," *The New York Times,* August 18, 1993.

543 The MIRACL test: "Laser Destroys Missile Target in Test by U.S.," *The New York Times,* September 13, 1985; "Missile Destroyed in First SDI Test at High-Energy Facility," p. 17.

543 Weinberger on deception: "Missile Destroyed in First SDI Test at High-Energy Facility."

544 Satellite Data System: Industry Observer, *Aviation Week & Space Technology,* March 2, 1987, p. 13.

544 TDRS's peaceful mission: "TDRS: NASA's Tracking Data and Relay Satellite," KSC Release No. 63-88 (rev. December 1992), NASA, John F. Kennedy Space Center.

544 Twenty-volume encyclopedia: Richelson, *America's Secret Eyes in Space,* p. 222.

544 Lacrosse: Ibid., pp. 220–28; "Atlantis' Radar Satellite Payload Opens New Reconnaissance Era," pp. 26–28.

544 TDRS's intelligence mission: Charles, "Spy Satellites: Entering a New Era"; a source who requested anonymity.

544 Goddard security: "Security Classification Guide: Tracking and Data Relay Satellite System," Goddard Space Flight Center, April 15, 1982.

545 Afsatcom and Fltsatcom: Desmond Ball, "The US Air Force Satellite Communications (Afsatcom) System: The Australian Connection," and "The U.S. Fleet Satellite Communications (Fltsatcom) System: The Australian Connection."

545 White Cloud: Ball, "Ocean Surveillance."

545 EORSATs and RORSATS: Johnson, *The Soviet Year in Space, 1983,* pp. 31–32.

546 FOBS: Baker, *The Shape of Wars to Come,* pp. 103–05.

547 General Power and "absolute superiority": Ibid., p. 113.

547 Soviet communication satellites: *The Soviet Year in Space, 1983,* pp. 16–19.

547 The shotgun test: *Soviet Military Power, 1983,* pp. 64–65.

547 Soviet co-orbital tests: Johnson, *The Soviet Year in Space, 1990,* p. 95; *The RAE Table of Earth Satellites, 1957–1980,* p. 178n.

547 Radio Electronic Combat: *The Soviet Year in Space, 1990,* p. 96.

547 Laser ASATs: Ibid., p. 96.

548 STARFISH, Nike-Zeus, and Squanto Terror: Karas, *The New High Ground,* pp. 150–51.

548 Mission in space: "Air Force Mission in Space," Fact Sheet 82-23, pp. 1, 3.

549 The F-15's ASAT: "USAF Vehicle Designed for Satellite Attack," p. 21; "Air Force Missile Strikes Satellite in First Test," *The New York Times,* September 14, 1985; "Working Solar Monitor Shot Down by ASAT," p. 44.

549 Reagan and ASAT: "Reagan Announces a New ASAT Test," p. 946.

550 Cohen's poem: Cohen, *A Baker's Nickel,* pp. 57–58.

15. Downsizing Infinity

552 Extravehicular refueling: "Shuttle Plan Emphasizes Earth Survey," pp. 38–47; "2 Aboard Shuttle Get Go-ahead to Try a Satellite Refueling Test," *The New York Times,* October 11, 1984.

552 Young's statement: "Astronauts' Chief Says NASA Risked Life for Schedule," *The New York Times,* March 9, 1986.

553 Ride on payload specialists: Author interview.

553 Reagan and the schoolteacher: "First Shuttle Ride by Private Citizen to Go to Teacher," *The New York Times,* August 28, 1984.

554 Journalist-in-Space: Undated application package from the ASJMC, c/o College of Journalism, University of South Carolina; the one hundred: "Journalist-in-Space Project" release, College of Journalism, University of South Carolina, April 16, 1986, p. 5.

554 Foreigners on *Mir:* "Negotiations to Bring More Foreign Crews to Mir," *Space News,* April 27–May 3, 1992, p. 3.

554 Temperature in the low twenties: *Report of the Presidential Commission on the Space Shuttle Challenger Accident,* Vol. 1, p. 17 (hereafter referred to as the *Challenger* accident report).

555 Thiokol-NASA interchange: Ibid., Vol. 1, pp. 90–94.

555 Rockwell and *Dr. Zhivago: Investigation of the Challenger Accident,* Committee on Science and Technology, House of Representatives, pp. 71–72.

555 Stevenson's recommendation: "NASA Official Advised Against Liftoff," *The New York Times,* November 5, 1986.

556 The *Challenger* explosion: *Challenger* accident report, Vol. 1 pp. 19–21.

556 Death of the astronauts: Letter and report from Joseph P. Kerwin to Admiral Richard M. Truly and accompanying NASA News Release No. 86-100, "NASA Releases Challenger Transcript and Report on Cause of Death," July 28, 1986.

556 Crippen's comment: "The Fate of Challenger's Crew," Challenger Remembered, MSNBC (World Wide Web), February 8, 1997, p. 3.

557 Moore's statement: "Transcript of NASA News Conference on the Shuttle Disaster," *The New York Times,* January 29, 1986; tape of the conference.

557 NASA's handling of the disaster: "Challenger, Disclosure and an 8th Casualty for NASA," *The New York Times,* February 14, 1986; "Journalists Say NASA's Reticence Forced Them to Gather Data Elsewhere," *The New York Times,* February 9, 1986.

558 Yeager's absence: Sally Ride, Bud Wheelon, and Saunders Kramer (Memorandum of January 7, 1998).

558 Someone with an agenda: *The Last Journey of a Genius,* WGBH (Boston) for NOVA, 1990.

558 Rogers applauding the agency: *Challenger* accident report, Vol. 1 p. 201.

558 Thiokol tests: Ibid., pp. 138–43.

559 So many unheeded warnings: *Investigation of the Challenger Accident* (House report), p. 72.

559 Press barbs: Mayer, *Making News,* p. 279.
559 The State of the Union and Mrs. Corrigan's recollection: "Mother Recounts Challenger Blast," *The New York Times,* September 8, 1993.
559 The White House stonewalling: The author's request for information was ignored.
559 Asimov on casualties: "Space Shuttle," *48 Hours,* CBS, April 21, 1988.
560 Feynman on public relations: *Challenger* accident report, Vol. 2, p. F-5.
560 "Russian roulette": *Challenger* accident report, Vol. 1, p. 148.
561 Numbers of people interested in space: Miller, "The Impact of the Challenger Accident on Public Attitudes Toward the Space Program," pp. 4–5.
561 Continuing accidents and outstanding technology: *Impact of the Challenger Accident Survey,* p. 18.
562 Responsibility for the accident: *Impact of the Challenger Accident Survey,* p. 28; space station, p. 44; Mars, p. 45.
562 Cesarone on navigating: Author interview.
563 Edelson on navigating: Author interview.
563 Soderblom and the pea: Waldrop, "Voyage to a Blue Planet," p. 916.
563 No features visible: Ingersoll, "Uranus," p. 38.
564 Three active Uranian satellites and ovoids: Johnson, Brown, and Soderblom, "The Moons of Uranus," pp. 50–55.
564 Uranian rings: Cuzzi and Esposito, "The Rings of Uranus," pp. 52–66; Kerr, "Voyager Finds Uranian Shepherds and a Well-behaved Flock of Rings," pp. 793–96; *The Planet That Got Knocked on Its Side* (video), NOVA.
565 Magnetic field and atmosphere: Smith, "Voyager 2 in the Uranian System: Imaging Science Results," pp. 44–64; Lane et al., "Photometry from Voyager 2: Initial Results from the Uranian Atmosphere, Satellites, and Rings," pp. 65–70.
565 Distance from Voyager 2 to Earth and Uranus and observation: Robert Cesarone at JPL, June 23, 1987, and Robert Brooks on July 1, 1987.
565 Brooks's observations: Author interview.
566 "remarkable successes": "Planetary Exploration Through Year 2000," Part 1, Executive Summary, p. 5.
566 The SSEC's cost-control strategy: Ibid., Part 1, p. 14.
566 Core missions: Ibid., pp. 16–20.
566 Augmented Program: Ibid., An Augmented Program, pp. 58–125.
567 Commercial competition in space: "Pioneering the Space Frontier," p. 158.
567 Goals and mining the Moon and Phobos: Ibid., pp. 5, 85–87.
567 The commission's goals: Ibid., pp. 3–4.
567 Morrison to Fletcher: "Scientists Urge Immediate Change in Planetary Exploration Policies," p. 68.
568 Alleged U.S. leadership problems: "Leadership and America's Future in Space," p. 11.
568 Commercial threats from the Soviet Union to India: "International Space Policy for the 1990s and Beyond," pp. 1, 10.
569 Press to Fletcher: "Space Science in the Twenty-first Century," Overview, p. iii.
569 *Buran* an improvement: Canby, "Are the Soviets Ahead in Space?," p. 425.
570 Soviets running hard: Ibid., p. 458.
570 Chernobyl explosion: "Report of the U.S. Department of Energy's Team Analyses of the Chernobyl-4 Atomic Energy Station Accident Sequence," p. i.
570 People "tossing their hats": Medvedev, *The Truth About Chernobyl,* p. 74.
570 The watchman's description: Ibid., p. 83.
571 Ham radio operators and official casualty claims: *NBC Nightly News,* April 30, 1986.
571 The KH-11 and Chernobyl: "Threat to Soviets Grows, U.S. Spy Photos Indicate," *The Miami Herald,* April 30, 1986; "Single U.S. Spy Satellite Sending Pictures," *The Miami Herald,* May 2, 1986.
571 SPOT Nurek imagery: "Soviet Strategic Laser Sites Imaged by French Spot Satellite," p. 26; "Civilians Use Satellite Photos For Spying on Soviet Military," *The New York Times,* April 7, 1986 (SPOT 1 had only been launched six weeks earlier).
572 Kapitsa's observation: Kapitsa, "Lessons of Chernobyl," p. 9.
573 Stranded cosmonauts: "Soviet Crew Lands Safely After Crisis in Space," *The New York Times,* September 7, 1988; "How a Soviet Capsule Became Stuck in Space," *The New York Times,* September 11, 1988.
574 Phobos operation and experiments: "Soviets Planning 1988 Mission to Study Martian Moon Phobos," pp. 66–67; "Phobos Dynamics Experiment," JPL fact sheet, undated; "Loss of a Soviet Mars Spacecraft Shakes Project to Explore Planet," *The New York Times,* March 30, 1989.
575 Phobos 1's problem: Johnson, *The Soviet Year in Space: 1989,* pp. 118–20; Sagdeev, *The Making of a Soviet Scientist,* pp. 315–16; "Soviets See Little Hope of Controlling Spacecraft," *The New York Times,* September 10, 1988.
575 The plight of Phobos 2: *The Making of a Soviet Scientist,* pp. 318–19; *The Soviet Year in Space: 1989,* pp. 119–20; "Soviets Lose Contact with Phobos 2 Spacecraft," p. 22; "Loss of a Soviet Mars Spacecraft Shakes Project to Explore Planet," *The New York Times,* March 30, 1989.
575 Murray on Phobos: Second interview with the author.
575 Dangerous comfort: *The Making of a Soviet Scientist,* p. 315.
576 Bush's SEI speech: "Remarks by the President at 20th Anniversary of Apollo Moon Landing" (Washington, D.C.: The White House, July 20, 1989), p. 2.
576 Quayle and Mars: The CNN interview aired on August 18. The quotation appeared in "Scientists Ponder Quayle's Remark About Water on Mars," *San Jose Mercury News,* September 1, 1989 (from *The Washington Post*).
577 Van Allen's slaughter of the innocent: Van Allen, "Space Science, Space Technology, and the Space Station," p. 36.
577 *The Perils of Pauline:* Author interview.
577 First planetary passenger: Yeates et al., *Galileo: Exploration of Jupiter's System,* p. 2.

578 Two Titan 34D explosions: "Titan Rocket Explodes over California Air Base," *The New York Times,* April 19, 1986; "Titan Accident Disrupts Military Space Program," *Science,* May 9, 1986, pp. 702–04.

578 Aldrich: "2 Years of Failure End As U.S. Lofts Big Titan Rocket," *The New York Times,* October 27, 1987.

578 The third Titan failure: "Titan 34D Upper Stage Failure Sets Back Pentagon Intelligence Strategy," *Aviation Week & Space Technology,* September 12, 1988, p. 26.

579 Demoting Pluto: Beatty, O'Leary, and Chaiken, *The New Solar System,* p. 172.

580 Neptune's atmosphere's energy: Smith et al., "Voyager 2 at Neptune: Imaging Science Results," p. 1422.

581 Stone's parting remarks: Author's tape recording and notes.

582 *Science* special report: December 15, 1989, pp. 1361–1582.

582 Geyserlike plumes: Brown et al., "Energy Sources for Triton's Geyser-Like Plumes," pp. 431–35.

582 Dust devils: Ingersoll and Tryka, "Triton's Plumes: The Dust Devil Hypothesis," pp. 435–37.

582 *Aviation Week* editorial: "Voyager the Intrepid," *Aviation Week & Space Technology,* August 28, 1989, p. 7.

582 Brooks's gift to the world: Author interview.

582 Smith at Neptune: *Neptune's Cold Fury* (video), NOVA, WNET (New York), April 9, 1991.

583 Martin to Fisk: Author's notes.

583 Sagan's soliloquy: Sagan, "Planetary Exploration," p. 15.

584 "sold with the ship": "From Soviet Space Station: Queries on State of the Union," *The New York Times,* September 7, 1991.

585 Krikalev's return: "After 313 Days in Space, It's a Trip to a New World," *The New York Times,* March 26, 1992.

585 Job action and the end of *Buran* and Energiya: "For Two (Ex-Soviet) Astronauts, an Alien World Lies Ahead," *The New York Times,* March 25, 1992.

585 Selling Energiya to the West: "Russian Seeks U.S. Buyer for World's Biggest Rocket," *The New York Times,* July 9, 1991.

585 Sagdeev's complaint about science and the Soviet government: *The Making of a Soviet Scientist,* p. 174.

585 Veverka on the shuttle: "Worthy Projects Suffer Because of the Shuttle, Critics of NASA Charge," *The New York Times,* April 7, 1981.

585 Opposition to space and Yeltsin's demand: Tarasenko, "Transformation of the Soviet Space Program After the Cold War," p. 345.

586 Transformation of the space program: Ibid., pp. 345–46.

586 Sovinformsputnik: "Ventures Offer Time on Russian-Operated Satellite," *Space News,* October 9–15, 1995.

586 Akiyama's ride: "Soviets Send First Japanese, a Journalist, into Space," *The New York Times,* December 3, 1990; "A Japanese Innovation: The Space Antihero," *The New York Times,* December 8, 1990; "Japanese Journalist Reaches Soviets' Space Station Aboard Commercial Flight," p. 23.

587 Items at auction: *Russian Space History* (Sotheby's catalogue).

588 *TM-10*'s price: "Space Artifacts of Soviets Soar at a $7 Million Auction," *The New York Times,* December 12, 1993.

588 Mishin's "coded" diary: Author interview.

588 Leonov's observation: "Space Artifacts of Soviets Soar at a $7 Million Auction."

588 Potemkin village: Author interview in the apartment, June 3, 1996.

589 Titov and politics: "New Star German Titov," *Pravda,* May 16, 1995 (in Russian); "Titov Defeats Mavrodi, Vedenkin," *The Moscow Tribune,* May 16, 1995.

589 Pepsi in space: "Pepsi Ad Taps Mir," *Space News,* April 8–14, 1996, p. 25.

589 Coca-Cola on *Mir:* Johnson and Rodvold, *Europe and Asia in Space: 1991–1992,* p. 69.

589 Israeli milk: "Got Milk?" picture caption, *The New York Times,* August 22, 1997; "Mir Bobbles Dim the Evil Empire's Aura," *The New York Times,* August 24, 1997.

589 Gorbachev and Pizza Hut: "From Perestroika to Pizza: Gorbachev Stars in TV Ad," *The New York Times,* December 3, 1997.

590 Mira Ivanova: Specter, "Moscow on the Make," p. 75.

16. The Rings of Earth

591 Ley on the space station: Joseph Kaplan et al., *Across the Space Frontier,* p. 98.

592 Reagan's speech and Keyworth: Mark, *The Space Station,* pp. 149–51.

592 Johnson versus Marshall: Ibid., pp. 136–37.

592 Herman's plan: Ibid., p. 140.

592 Reagan and Isabella's jewels: Ibid., p. 186.

592 Keyworth and the Pentagon: Ibid., pp. 144–45.

593 The station and Star Wars: "Pentagon Seeks 'Star Wars' Role for Space Station," *The New York Times,* December 20, 1986.

593 Reagan and Flash Gordon: *The Space Station,* p. 177; State of the Union: "President Backs U.S. Space Station As Next Key Goal," *The New York Times,* January 26, 1984 (and text of address).

593 "Expensive Yawn": "An Expensive Yawn in Space," *The New York Times,* January 29, 1984.

593 NASA's and NRC's budget estimates: "Space Station Realities," *Aviation Week & Space Technology,* August 8, 1988, p. 7; "Space Station Price Climbs Higher," *Science,* July 17, 1987, p. 237.

594 Keyworth and lying: "Why a Space Station That Costs $25 Billion May Never Leave Earth," *The Wall Street Journal,* September 1, 1988.

594 Heflin and jobs: Ibid.

594 Roe and the station's travails: "Space Station Science: Up in the Air," *Science,* December 1, 1989, p. 1110.

594 Extra maintenance: "Major Flaw Found in Space Station Planned by NASA," *The New York Times,* March 19, 1990.

594 Truly's rebuttal: "No 'Horrible Design Flaw' in the Space Station" (Letters), *The New York Times,* April 23, 1990.

595 The vote and space science: "House Approves Rescue of Space Station," *The New York Times,* June 8, 1991; "House Vote Sets Stage for Conflict Between Two Allies in Space Program," *The New York Times,* June 8, 1991.
595 The *Times'* editorial: "Space Yes; Space Station No," *The New York Times,* June 6, 1991.
595 Mollohan's remarks: "Image Change Needed for NASA To Gain Support," *Space News,* March 15–21, 1993, p. 5.
595 The CBO's suggestions: *Reinventing NASA* (Washington, D.C.: Congressional Budget Office, 1994), pp. 35, 25–26.
596 Boeing's employees and contract: *Space Station Freedom Media Handbook* (Washington, D.C.: National Aeronautics and Space Administration, 1989), p. 100; Boeing "Progress Report," April 1994, p. 1.
596 "Media Advisory": "Space Station Freedom: Science, Technology, and our National Laboratory in Space," AAAS News Package, American Association for the Advancement of Science Annual Meeting, Boston, February 11–16, 1993 (Huntsville, Ala.: Missiles and Space Division).
597 The Russians and the space station: "NASA Prepares to Drop Russian Station Module," *Space News,* December 9–15, 1996; "Station Problems Test U.S.-Russian Resolve," *Aviation Week & Space Technology,* March 31, 1997, p. 20.
597 Letter to the *Times:* "Space Games Cost Taxpayers Too Much," April 10, 1997.
597 Fires in *Columbia:* "Shuttle Lifts Off on Mission to Explore Lack of Gravity," *The New York Times,* April 5, 1997.
598 Fire in space: "Mir Prompts New Effort to Quench Threat of Fire in Space," *The New York Times,* October 14, 1997.
598 Richardson and pulp fiction: Smith, *The Space Telescope,* p. 28.
599 GOD: Ibid., pp. 56–57.
599 Unshared technology: Chaisson, *The Hubble Wars,* pp. 207–08.
599 Fighting over Hubble and Giacconi: *The Space Telescope,* p. 342.
599 Too low an orbit: "Great Telescope, Bad Service Plan," *Science,* December 22, 1989, p. 1551.
599 Bahcall's complaint: Letter from Bahcall to Fletcher dated January 22, 1976 (NASA History Archive).
599 Cost overrun: Chaisson, *The Hubble Wars,* p. 114.
600 Perkin-Elmer's bid: Ibid., p. 150.
600 Neglecting children and hyping: Ibid., pp. 107, 111.
600 First light: Petersen and Brandt, *Hubble Vision,* p. 5.
600 The Hubble board's finding: *The Hubble Space Telescope Optical Systems Failure Report* (Washington, D.C.: National Aeronautics and Space Administration, November 1990), pp. 6-1–7-9.
600 Lowballing: *The Hubble Wars,* p. 149.
600 "cruise through hell": Ibid., p. 192.
600 "techno-turkey" through "sick puppy": *Hubble Vision,* p. 10.
600 *Tonight Show* and dopey-looking people cartoon: *The Hubble Wars,* p. 202.
600 The *Times'* editorial: "Calamities in Space," July 1, 1990.
600 The end of Bush's plan: *The Hubble Wars,* pp. 192–93.
601 The end of Mars 96: "Confusion Marks Mars 96 Failure," *Aviation Week & Space Technology,* November 25, 1996, p. 71; "Scientists to Discuss Life After Mars 96," *Space News,* December 2–8, 1996.
601 Kerosene lamps at Tyuratam: "Station: Wheat and Chaff" (letter from Robert G. Oler to the Editor), *Space News,* January 13–19, 1997, p. 13.
601 Tyuratam's rent: "Russian Isn't Celebrating Sputnik," AP, October 4, 1997.
601 U.S., French, Japanese, and Russian budgets for 1996: "Mir Shadow Falls on Russian Role in Space," *The New York Times,* August 2, 1997.
601 Koptev's statistics and warning: "Russian Space Chief Voices Dire Warnings," *Aviation Week & Space Technology,* January 6, 1997, p. 26.
601 *Atlantis* and *Mir:* "Russia Dependent upon U.S. Resupply of Mir," *Aviation Week & Space Technology,* January 6, 1997, p. 22.
601 Foale on *Mir:* "Mir Astronauts Now Facing Uncharted Territory," *The New York Times,* June 30, 1997.
603 Repair plans: "Russia and U.S. Chart Daring Mir Salvage," pp. 22–24.
603 Tsibliyev's lament: "Mir Astronauts Now Facing Uncharted Territory."
603 Irregular heartbeat: "Mir Commander's Heart Ills Cast Doubt on Repair Effort," *The New York Times,* July 15, 1997.
604 Rhonda Foale: "Space Man of the Year," *Life* (Collector's Edition), January 1998, pp. 92, 94.
604 "kindergarten": "Astronaut Error Deepens Anxieties on Space Station," *The New York Times,* July 18, 1997.
604 Relief crew: "After One More Mishap, Relief Craft Links to Mir," *The New York Times,* August 8, 1997.
604 "Thank God": "Hapless Crew of Mir Station Back on Earth," *The New York Times,* August 15, 1997.
604 Blagov's accusation: Ibid.
604 Yeltsin and "human error": "Russian Astronauts Insist Errors, While Human, Were on Earth," *The New York Times,* August 17, 1997.
604 "space jalopy": "A Valuable Space Jalopy" (lead editorial), *The New York Times,* August 27, 1997.
605 Tsibliyev's response: "Russian Astronauts Insist Errors, While Human, Were on Earth."
605 Blagov's comments: "Jeers Sting Mir Mission Control, Which Bemoans a Money Pinch," *The New York Times,* August 20, 1997.
605 Tsibliyev's and Luzutkin's back pay: "2 Ex-Mir Astronauts Are Paid for Mission," *The New York Times,* December 26, 1997.
606 *Newsweek* coverage: July 7, 1997, pp. 24–27.
606 Sensenbrenner: "Moscow, We Have a Problem," *Newsweek,* July 7, 1997, p. 27.
607 Russian changes using *Mir:* "Russia, Vexing Partners, Asks for Changes in Space Station," *The New York Times,* December 10, 1995.
607 GAO report: "Space Station: Cost Control Problems Are Worsening," p. 3.

607 $335 million: "U.S., Russia Negotiating More Money," *Space News,* February 4, 1996.

608 Krikalev in *Discovery:* "Shuttle Lifts Off on First Mission with a Russian," *The New York Times,* February 4, 1994.

608 The H-2 launch: "First Big Space Rocket Is Launched by Japanese," *The New York Times,* February 4, 1994.

608 Japanese Moon craft: "Joining Space Race, Japan Launches Rocket to Moon," *The New York Times,* January 25, 1990.

609 Lockheed-Khrunichev: "Lockheed-Khrunichev-Energiya Wins European Launch Order," p. 28. Rovers: "McDonnell Douglas, Russia Team on Rover," *Space News,* January 18–24, p. 6.

609 Sharing satellite imagery: "Out of the Cold: U.S. and Russian Spies Share Cloaks in Bosnia," *The New York Times,* January 19, 1996.

609 Joint early-warning system: "Russian Pitches Common Early Warning Network," pp. 46–47.

609 "Mars Together": "U.S., Russia Define Joint Mars Mission," p. 32.

609 Russian missile aid to Iran: "U.S. Telling Russia to Bar Aid to Iran by Arms Experts," *The New York Times,* August 22, 1997.

17. The Second Space Age

611 Kutyna's warning: "Military Calls Space Superiority Essential," *Space News,* May 6–12, 1991, p. 6.

611 The Air Force and the CIA: James T. Hackett, "Military Battle over Space Control" (column), *Space News,* August 29–September 4, 1994, p. 15.

612 Space Warfare Center: "Space Warfare Center Supports 'Warfighters,' " *Aviation Week & Space Technology,* March 28, 1994, pp. 64–65.

612 Horner's nightmare: "Air Force ASAT Research Goes On Unabated," *Space News,* April 26–May 2, 1993, p. 10.

612 Ashy's prediction: "A.F. Space Chief Calls War in Space Inevitable," *Space News,* August 12–18, 1996, p. 4.

613 Estes and the fourth medium: "Pentagon Considers Space as New Area of Responsibility," *Aviation Week & Space Technology,* March 24, 1997, p. 54.

614 Earlybird 1: "First Civilian Spy Satellite Soars into Space, Launched in Russia by a U.S. Company," *The New York Times,* December 25, 1997.

615 EKV: "EKV Contractor Selection Targeted for Fiscal 1999," p. 52. NB: The *Aviation Week & Space Technology* issues of February 24 and March 3, 1997, had extensive special reports on ballistic missile defense.

615 National Missile Defense: "Missile Defense Soon, But Will It Work?," p. 39; "Mix of Simulation, Flight Testing Troubles BMDO Leaders," pp. 64–67.

615 Moorman and Estes: " 'Milspace' Maturing into Warfighter Roles," pp. 46–47.

616 Rich on space: "The Lost Frontier" (Op-Ed), *The New York Times,* July 5, 1995.

616 The shrunken budget: "Space Agency Plans Layoffs, Shrinking to Pre-Apollo Size," *The New York Times,* May 20, 1995.

617 Goldin's reaction to cuts: "Goldin: House Budget 'Road to Disaster," *Space News,* May 22–28, 1995.

618 Landsat photo book: *Photo-Atlas of the United States,* p. 18 (Seattle).

618 Earth Observation System: "NASA Mission Gets Down to Earth," pp. 1208–10.

619 Brown's and the executive's remarks: "Satellites to Expand Information Highway," *Space News,* May 8–14, 1995, p. 6.

620 International communication satellites: "Fixed Satellite Service Reference Chart," *Space News,* April 22–28, 1996, p. 10.

620 Teledesic's 840 satellites: "Cyberspace Cadets," *Scientific American,* June 1994, pp. 98, 100.

620 Stiff competition: Dichmann, "The Constellations in LEO," pp. 40–41.

620 $15 billion a year: "Communications Will Drive Space Sector's Growth," pp. 83–84.

621 Looking the other way: "Saudis Feast Freely from TV Dishes," *The New York Times,* January 23, 1996.

621 MBC: "Tapping the Power of Satellite TV," *The New York Times,* April 15, 1996.

621 Iranian television: "Satellite Dishes Adding Spice to Iran's TV Menu," *The New York Times,* August 16, 1994.

621 Publicly "hanged": "From Cold War, Afghans Inherit Brutal New Age," *The New York Times,* February 14, 1996.

622 GPS: Herring, "The Global Positioning System," pp. 44–50; "GPS Technology Ripens For Consumer Market," p. 50; "Glonass Nears Full Operation," pp. 52, 54; Gleick, "A Sense of Where You Are," p. 19; "GPS Against Crime," *Aviation Week & Space Technology,* May 6, 1996, p. 13; "Sonic Device for Blind May Aid Navigation," *The New York Times,* September 6, 1994.

623 Benson, prospecting, and LunaCorp: "Buck Rogers, CEO," pp. 34, 36.

624 Space tourism meeting and Walker: David et al., "The Space Tourist," pp. 179–92.

624 Soviet competition in tourism: Ibid., p. 203.

624 Space tourism: "NASA Begins Space Tourism Enterprise Assessment," *Space News,* September 18–24, 1995; "Studies Claim Space Tourism Feasible," pp. 58–59.

625 Stanford tours to space: "Ambitious Entrepreneurs Planning to Send Tourists Into 'Astronaut Altitude,' " *The New York Times,* February 17, 1998.

625 The "Vomit Comet": "Did Armstrong Turn This Shade of Green on His First Flight?," *The Wall Street Journal,* April 16, 1997.

626 United Space Alliance: "Private Contractor to Manage NASA Space Shuttle Program," *The New York Times,* October 1, 1996.

626 Anderson and space salvage: "New Work Proposed for Shuttles: Salvage in Space," *The New York Times,* September 16, 1997.

627 Helium 3 details: Kulcinski, "Helium-3 Fusion Reactors—A Clean and Safe Source of Energy in the 21st Century" (paper); Sullivan and McKay, "Using Space Resources," pp. 9–10

(paper); Anderson, "Helium-3 from the Mantle: Primordial Signal or Cosmic Dust?" (article); Wittenberg, Santarius, and Kulcinski, "Lunar Source of 3He for Commercial Fusion Power"; "Who Will Mine the Moon?" *The New York Times,* January 19, 1995.

627 Braselton on helium 3: Braselton, "Space Power for an Expanded Vision."

627 Synthesis Group: *America at the Threshold: America's Space Exploration Initiative,* p. A-33.

627 "unique store of resources": Ibid., p. 52.

627 Dyson on practicality: Dyson, "21st-Century Spacecraft," p. 88.

628 Interplanetary Monitoring Platform: *Flight Project Data Book, 1995,* pp. 87–88.

628 COBE and the Big Bang: Smoot et al., "Structure in the COBE DMR First Year Maps."

629 Smoot and God: "In the Glow of a Cosmic Discovery, a Physicist Ponders God and Fame," *The New York Times,* May 5, 1992.

629 The repair mission: "Space Team Fixes Flaw in Telescope," *The New York Times,* December 8, 1993; "Flight to Fix Hubble Pays Off," *Aviation Week & Space Technology,* December 13–20, 1993, pp. 24–26.

629 Total cost: "Telescope Repair Costs," *Aviation Week & Space Technology,* December 13–20, 1990, p. 26.

629 Forming galaxies: "Worlds Without End," *National Geographic,* April 1997, pp. 10–12.

629 Forty billion galaxies: "Suddenly, Universe Gains 40 Billion More Galaxies," *The New York Times,* January 16, 1996.

629 Williams: "Hubble Discovers Ancient Galaxies," *Aviation Week & Space Technology,* January 22, 1996, p. 26.

630 Black holes: "Space Telescope Confirms Theory of Black Holes," *The New York Times,* May 26, 1994; "Hubble Finds Proof That Black Hole Exists," *Aviation Week & Space Technology,* May 30, 1994, p. 28.

630 NGC 4261: "Just Another Billion-Sun Black Hole," *Science,* December 15, 1995, p. 1759.

630 Stellar incubators: "A Stunning View Inside an Incubator for Stars," *The New York Times,* November 3, 1995.

630 Colliding galaxies: "Fireworks in Deep Space as Two Galaxies Collide," *The New York Times,* October 22, 1997.

630 Ganymede and Pluto: "Hubble Discovers Ozone on Ganymede," *Aviation Week & Space Technology,* October 16, 1995, p. 26; "Hubble Provides First Glimpses of the Surface of Distant Pluto," *The New York Times,* March 8, 1996.

630 HST and Shoemaker-Levy: Weaver et al., "Hubble Space Telescope Observations of Comet P/Shoemaker-Levy 9 (1993e)," pp. 787–91; Levy, Shoemaker, and Shoemaker, "Comet Shoemaker-Levy 9 Meets Jupiter."

631 AXAF: "Telescope Will Offer X-ray View of Cosmos," *The New York Times,* March 31, 1998.

631 Scavenged Magellan parts: Saunders, "Magellan: The Geologic Exploration of Venus," p. 16.

632 Lava domes and Saunders: "On Venus, Pancakes for Volcano Domes," *The New York Times,* November 17, 1990.

632 Kids at a ball game: "Venus Revealed," p. 16.

632 Saunders on the better Venus map: "Spacecraft's Maps Show Evidence of Active Volcanoes on Venus," *The New York Times,* October 30, 1991.

633 Pioneer 12: "Pioneer 12 Is Sinking into the Venusian Clouds and Burning Up," *The New York Times,* October 10, 1992.

633 Griffith: "Magellan Craft Dies in Plunge to Venus," *The New York Times,* October 13, 1994.

634 Europa's surface and the possibility of life: "Europa Water Evidence Hints Life Possible," pp. 32–33.

634 Two planets at PSR B1257+12: Wolszczan, "Confirmation of Earth-Mass Planets Orbiting the Millisecond Pulsar PSR B1257+12," pp. 538–42.

634 The closer planet: "2 Sightings of Planet Orbiting a Sunlike Star Challenge Notions That Earth Is Unique," *The New York Times,* October 20, 1995.

634 The *Time* cover story: "Is There Life in Outer Space?," pp. 51–57.

634 The meteorite: David S. McKay et al., "Search for Past Life on Mars: Possible Relic Biogenic Activity in Martian Meteorite ALH84001," pp. 924–30.

635 McKay's statement: "Ancient Life on Mars?," *Science,* August 16, 1996, p. 864.

635 Goldin on the "potato": "Clues in Meteorite Seem to Show Signs of Life on Mars Long Ago," *The New York Times,* August 7, 1996.

635 Clinton's reaction to the "potato": "President Clinton Statement Regarding Mars Meteorite Discovery," Office of the Press Secretary, the White House, August 7, 1986.

636 The cartoon: *Editorial Humor,* July 23–August 5, 1997, p. 5.

636 Pathfinder and Sojourner: "Mars Pathfinder Landing" press kit, pp. 28–32.

636 Possible landing sites: Greeley, *Mars Landing Site Catalog.*

638 Malin: "Picture Emerges of Mars, Swept by Surging Waters," *The New York Times,* July 8, 1997.

638 Yemeni lawsuit: "Get Off Mars, It's Ours, Yemeni Men Tell U.S.," Reuters, July 24, 1997.

639 Cost of the Discovery missions: "Mars Pathfinder Landing" press kit, p. 20.

640 Glenn returns to space: "Glenn to Slip Bonds of Age in Space," *The New York Times,* January 17, 1998.

640 Dyson and the Astrochickens: Dyson, *Infinite in All Directions,* pp. 196–200.

641 Cassini's plutonium and its cost: "Saturn Mission's Use of Plutonium Fuel Provokes Warnings of Danger," *The New York Times,* September 8, 1997.

642 Clementine: Nozette et al., "The Clementine Mission to the Moon: Scientific Overview," pp. 1835–39.

642 Water on the Moon: "The Moon May Have Water, and Many New Possibilities," *The New York Times,* December 4, 1996; "Seeking Water on Moon in New NASA Mission," *The New York Times,* January 4, 1998.

642 Microrobots: "Fast, Cheap, and Out of Control," pp. 959–61.
642 Rocky 4: "Swarms of Mini-Robots Set to Take on Mars Terrain," p. 1621.
642 Stofan's remarks: Author interview.
643 The end of Pioneer 10: "Well Done, Pioneer 10," *The New York Times,* March 4, 1997.
643 Clarke, *The Exploration of Space,* pp. 136–37; pp. 120–21.
643 Reiber, *The NASA Mars Conference.*
643 Welch and Stoker, *The Case for Mars,* p. 100.
644 The Arctic and the richness of Mars: Zubrin, *The Case for Mars,* pp. xvi–xvii.
644 Mars mission cost: Ibid., p. 3.
644 Political opposition: Ibid., pp. 65–66.
645 Kramer on Zubrin: Memorandum of January 8, 1998.
645 *JBIS* and terraforming Mars: Fogg, "Terraforming Mars," pp. 427–34.
645 Interstellar missions: Mauldin, *Prospects for Interstellar Travel.*
645 Distance, time, and cost of star travel: Ibid., pp. 5–6.
646 "will we and should we?": Ibid., p. 310.
646 The end of Earth: "Hubble Takes Gaudy Photos of Dying Stars," *The New York Times,* December 18, 1997.

Sources

Books

Aldrin, Buzz, and Malcolm McConnell. *Men from Earth.* New York: Bantam Books, 1989.
Arms Control and Disarmament Agreements. Washington, D.C.: U.S. Arms Control and Disarmament Agency, 1980.
Babington-Smith, Constance. *Air Spy.* New York: Harper & Brothers, 1957.
Baker, David. *The History of Manned Space Flight.* New York: Crown Publishers, 1981.
———. *The Rocket.* New York: Crown Publishers, 1978.
———. *The Shape of Wars to Come.* Cambridge, England: Patrick Stephens Ltd., 1981.
Baldridge, Robert C. *Victory Road.* Bennington, Vt.: World War II Historical Society, 1995.
Baldwin, Hanson W. *The Great Arms Race.* New York: Frederick A. Praeger, 1958.
Beatty, J. Kelly, Brian O'Leary, and Andrew Chaikin. *The New Solar System.* Cambridge, Mass.: Sky Publishing Corporation, 1982.
Bergaust, Erik, and William Beller. *Satellite!* Garden City, N.Y.: Hanover House, 1956.
Berkner, L. V., and Hugh Odishaw. *Science in Space.* New York: McGraw-Hill, 1961.
Berman, Arthur I. *Astronautics: Fundamentals of Dynamical Astronomy and Space Flight.* New York: John Wiley and Sons, 1961.
Beschloss, Michael R. *Taking Charge: The Johnson White House Tapes, 1963–1964.* New York: Simon & Schuster, 1997.
Bliss, Edward, Jr., ed. *In Search of Light: The Broadcasts of Edward R. Murrow, 1938–1961.* New York: Alfred A. Knopf, 1967.
Brennan, Donald G., ed. *Arms Control, Disarmament, and National Security.* New York: George Braziller, 1961.
Broad, William J. *Teller's War.* New York: Simon and Schuster, 1992.
Brugioni, Dino A. *Eyeball to Eyeball.* New York: Random House, 1990.
Brzezinsky, Zbigniew K. *The Permanent Purge.* Cambridge, Mass.: Harvard University Press, 1956.
Buchheim, Robert W. *Space Handbook: Astronautics and Its Applications.* New York: Random House, 1958.
Bulkeley, Rip. *The Sputniks Crisis and Early United States Space Policy.* Bloomington: Indiana University Press, 1991.
Burgess, Eric. *By Jupiter.* New York: Columbia University Press, 1982.
Burrows, William E. *Deep Black: Space Espionage and National Security.* New York: Random House, 1986.
———. *Exploring Space.* New York: Random House, 1991.
Cain, Kathleen. *Luna Myth & Mystery.* Boulder, Colo.: Johnson Publishing Co., 1991.

Carpenter, M. Scott, et al. *We Seven.* New York: Simon and Schuster, 1962.
Carter, Ashton B., and David N. Schwartz, eds. *Ballistic Missile Defense.* Washington, D.C.: Brookings Institution, 1984.
Chaisson, Eric J. *The Hubble Wars.* New York: HarperCollins, 1994.
Chapman, J. L. *Atlas: The Story of a Missile.* New York: Harper & Brothers, 1960.
Clarke, Arthur C. *The Exploration of Space.* New York: Harper & Brothers, 1951.
———. *The Making of a Moon.* New York: Harper & Brothers, 1957.
Cohen, William S. *A Baker's Nickel.* New York: William Morrow & Co., 1986.
Collins, Martin J., and Sylvia K. Kraemer, eds. *Space: Discovery and Exploration.* Southport, Conn.: Hugh Lauter Levin Associates, 1993.
Collins, Michael. *Mission to Mars.* New York: Grove Weidenfeld, 1990.
Cooper, Henry S. F., Jr. *Imaging Saturn.* New York: Holt, Rinehart and Winston, 1981.
Copp, DeWitt S. *A Few Great Captains.* Garden City, N.Y.: Doubleday & Company, 1980.
Costello, Peter. *Jules Verne.* London: Hodder and Stoughton, 1978.
Damon, Thomas D. *Introduction to Space.* Malabar, Fla.: Orbit Book Company, 1989.
Daniloff, Nicholas. *The Kremlin and the Cosmos.* New York: Alfred A. Knopf, 1972.
Davis, Joel. *Flyby: The Interplanetary Odyssey of Voyager 2.* New York: Atheneum, 1987.
Day, Dwayne A., John M. Logsdon, and Brian Latell, eds. *Eye in the Sky: The Story of the CORONA Reconnaissance Satellite.* Washington, D.C.: Smithsonian Institution Press, 1998.
DeVorkin, David H. *Science with a Vengeance.* New York: Springer-Verlag, 1992.
Diamond, Edwin. *The Rise and Fall of the Space Age.* Garden City, N.Y.: Doubleday & Company, 1964.
Dick, Thomas. *Celestial Scenery, or, the Wonders of the Planetary System Displayed.* New York: Harper & Brothers, 1838.
Divine, Robert A. *The Sputnik Challenge.* New York: Oxford University Press, 1993.
Dobrynin, Anatoly. *In Confidence.* New York: Times Books, 1995.
Durant, Frederick C., III, ed. *Between Sputnik and the Shuttle,* AAS History Series, Vol. 3. San Diego: American Astronautical Society, 1981.
Durant, Frederick C., III, and George S. James, eds. *First Steps Toward Space,* AAS History Series, Vol. 6. San Diego: American Astronautical Society, 1985.
Dyson, Freeman. *Infinite in All Directions.* New York: Harper & Row, 1988.
———. *Weapons and Hope.* New York: Harper & Row, 1984.
Ehricke, Krafft A. *Space Flight.* New York: D. Van Nostrand, 1962.
Elliot, James, and Richard Kerr. *Rings.* Cambridge: MIT Press, 1987.
Emme, Eugene M., ed. *Science Fiction and Space Futures,* AAS History Series, Vol. 5. San Diego: American Astronautical Society, 1982.
Etzioni, Amitai. *The Moon-Doggle.* Garden City, N.Y.: Doubleday & Company, 1964.
Feinberg, Gerald, and Robert Shapiro. *Life Beyond Earth.* New York: William Morrow, 1980.
Firsoff, V. A. *Strange World of the Moon.* New York: Basic Books, 1959.
Flaherty, Bernard E. *Psychophysiological Aspects of Space Flight.* New York: Columbia University Press, 1961.
Flight Project Data Book, 1995. Washington, D.C.: NASA Office of Space Science, September 1995.
French, Bevan M. *The Moon Book.* New York: Penguin, 1977.
From First Satellite to Energia-Buran and Mir. Kaliningrad, Russian Republic: S. P. Korolev Space Corporation Energia, 1994.
Gagarin, Yuri A. *Road to the Stars.* Moscow: Foreign Languages Publishing House, undated.
Galilei, Galileo. *Dialogue on the Great World Systems.* Chicago: University of Chicago Press, 1953.
Gallup, George H. *The Gallup Poll,* Vols. 2 and 3. New York: Random House, undated.
Gatland, Kenneth W., ed. *Project Satellite.* New York: British Book Centre, 1958.
Gatland, Kenneth W., and Anthony M. Kunesch. *Space Travel.* New York: Philosophical Library, 1953.
Gehrels, Tom, and Mildred Shapley Matthews, eds. *Saturn.* Tucson: University of Arizona Press, 1984.

Glasstone, Samuel. *Sourcebook on the Space Sciences.* New York: Van Nostrand, 1965.
Glushko, V. P. *Rocket Engines GDL-OKB.* Moscow: Novosti Press Agency Publishing House, 1975.
Goddard, Esther C., and G. Edward Pendray, eds. *The Papers of Robert H. Goddard* (3 vols.). New York: McGraw-Hill Book Company, 1970.
Golovanov, Yaroslav. *Korolev: Facts and Myths.* Moscow: Nauka, 1994 (in Russian).
Goodstein, David L., and Judith R. Goodstein. *Feynman's Lost Lecture.* New York: W. W. Norton, 1996.
Graham, Daniel. *High Frontier.* New York: Tom Doherty Associates, 1983.
Graham, Loren R. *The Ghost of the Executed Engineer.* Cambridge, Mass.: Harvard University Press, 1993.
———, ed. *Science and the Soviet Social Order.* Cambridge, Mass.: Harvard University Press, 1990.
Gray, Mike. *Angle of Attack.* New York: W. W. Norton, 1992.
Greenwood, John T., ed. *Milestones of Flight.* Southport, Conn.: Hugh Lauter Levin Associates, 1989.
Guenther, Ben, and Jay Miller. *Bell X-1 Variants.* Arlington, Texas: Aerofax, 1988.
Guenther, Ben, Jay Miller, and Terry Panopalis. *North American X-15/X-15A-2,* Arlington, Texas: Aerofax, 1985.
Hall, R. Cargill, ed. *History of Rocketry and Astronautics,* AAS History Series, Vol. 7, Part 2. San Diego: American Astronautical Society, 1986.
Hall, Stephen S. *Mapping the Next Millennium.* New York: Random House, 1992.
Harford, James. *Korolev.* New York: John Wiley & Sons, 1997.
Hart, Douglas. *The Encyclopedia of Soviet Spacecraft.* New York: Exeter Books, 1987.
Harvey, Brian. *Race into Space.* Chichester, England: Ellis Horwood Limited, 1988.
Hecker, Frank, ed. *Proceedings of the Fourth Annual L5 Space Development Conference,* Science and Technology Series, Vol. 68. San Diego: American Astronomical Society, 1987.
Hirsch, Richard, and Joseph John Trento. *The National Aeronautics and Space Administration.* New York: Praeger Publishers, 1973.
Holme, Bryan. *Bulfinch's Mythology.* New York: Viking Press, 1979.
Horowitz, Norman H. *To Utopia and Back.* New York: W. H. Freeman, 1986.
Hoyle, Fred, and Chandra Wickramasinghe. *Diseases from Space.* New York: Harper & Row, 1979.
Hoyt, William Graves. *Lowell and Mars.* Tucson: University of Arizona Press, 1976.
Hunt, Linda. *Secret Agenda.* New York: St. Martin's Press, 1991.
Irving, David. *Mare's Nest.* London: William Kimber, 1964.
James, Peter N. *Soviet Conquest from Space.* New Rochelle, N.Y.: Arlington House, 1974.
Jane's All the World's Aircraft, 1962–63. Great Missenden, England: Jane's All the World's Aircraft Publishing Company, 1962–63.
Johnson, Brian. *The Secret War.* New York: Methuen, 1978.
Johnson, Nicholas L. *Handbook of Soviet Lunar and Planetary Exploration.* San Diego: American Astronomical Society, 1979.
———. *The Soviet Reach for the Moon.* Washington, D.C.: Cosmos Books, 1995.
Kaempffert, Waldemar. *The New Art of Flying.* New York: Dodd, Mead and Company, 1911.
Karas, Thomas. *The New High Ground.* New York: Simon and Schuster, 1983.
Killian, James R., Jr. *Sputnik, Scientists, and Eisenhower.* Cambridge, Mass.: MIT Press, 1982.
Kirimov, Lieutenant General Kirim. *How All This Started* (tentative title), unpublished manuscript.
Kistiakowsky, George B. *A Scientist at the White House.* Cambridge, Mass.: Harvard University Press, 1976.
Klehr, Harvey, John Earl Haynes, and Fridrikh Igorevich Firsov. *The Secret World of American Communism.* New Haven, Conn.: Yale University Press, 1995.
Kliuchevsky, V. O. *A Course in Russian History: The Seventeenth Century.* London: M. E. Sharpe, 1994.
Kohlhase, Charles, ed. *The Voyager Neptune Travel Guide,* JPL 89-24. Pasadena, Calif.: Jet Propulsion Laboratory, June 1, 1989.

Koppes, Clayton R. *JPL and the American Space Program.* New Haven, Conn.: Yale University Press, 1982.
Kosmodemyansky, Arkadii. *Konstantin Tsiolkovsky.* Moscow: Foreign Languages Publishing House, 1956.
Krieger, F. J. *Behind the Sputniks.* Washington, D.C.: Public Affairs Press, 1958.
———. *A Case Book on Soviet Astronautics.* Santa Monica, Calif.: RAND Corp. (Part I, RM-1760, June 21, 1956; Part II, RM 1922, June 21, 1957).
Lattu, Kristan R., ed. *History of Rocketry and Astronautics,* AAS History Series, Vol. 8. San Diego: American Astronautical Society, 1989.
Lawden, D. F. *Optimal Trajectories for Space Navigation.* London: Butterworths, 1963.
Lear, John. *Kepler's Dream.* Berkeley: University of California Press, 1965.
Lebedev, Valentin. *Diary of a Cosmonaut: 211 Days in Space.* New York: Bantam Books, 1990.
Lehman, Milton. *Robert H. Goddard.* New York: Da Capo Press, 1963.
Levine, Sol. *Appointment in the Sky: The Story of Project Gemini.* New York: Walker and Company, 1963.
Lewis, Richard S. *The Voyages of Apollo.* New York: Quadrangle, 1974.
Ley, Willy. *Rockets, Missiles, and Men in Space.* New York: Viking Press, 1968. (The fifth edition of the book below, which was itself an updated version of *Rockets,* published in 1944.)
———. *Rockets, Missiles, and Space Travel.* New York: Viking Press, 1957.
Ley, Willy, and Wernher von Braun. *The Exploration of Mars.* New York: Viking Press, 1956.
Lindsey, Robert. *The Falcon and the Snowman.* New York: Pocket Books, 1979.
Littmann, Mark. *Planets Beyond: Discovering the Outer Solar System.* New York: John Wiley & Sons, 1988.
Lovell, Bernard. *The Story of Jodrell Bank.* London: Oxford University Press, 1968.
Lovell, Jim, and Jeffrey Kluger. *Lost Moon.* New York: Houghton Mifflin Company, 1994.
MacCurdy, Edward. *The Notebooks of Leonardo Da Vinci,* Vol. 1. New York: George Braziller, 1958.
Mailer, Norman. *Of a Fire on the Moon.* Boston: Little, Brown and Company, 1969.
Mark, Hans. *The Space Station: A Personal Journey.* Durham, N.C.: Duke University Press, 1987.
Matson, Wayne R. *Cosmonautics: A Colorful History.* Washington, D.C.: Cosmos Books, 1994.
Mauldin, John H. *Prospects for Interstellar Travel.* San Diego: American Astronautical Society, 1992.
Mayer, Martin. *Making News.* Boston: Harvard Business School Press, 1993.
Mayerson, Philip. *Classical Mythology in Literature, Art, and Music.* New York: Xerox Corporation, 1971.
McAndrew, Captain James. *The Roswell Report: Case Closed.* Washington, D.C.: Headquarters, United States Air Force, 1997.
McCurdy, Howard E. *Inside NASA.* Baltimore: Johns Hopkins University Press, 1993.
McDonald, Robert A., ed. *Corona: Between the Sun & the Earth.* Bethesda, Md.: The American Society for Photogrammetry and Remote Sensing, 1997.
McDougall, Walter A. *. . . the Heavens and the Earth.* New York: Basic Books, 1985.
McGovern, James. *Crossbow and Overcast.* New York: William Morrow & Company, 1964.
Medvedev, Roy. *Let History Judge.* New York: Columbia University Press, 1989.
Medvedev, Zhores A. *Soviet Science.* New York: W. W. Norton, 1978.
Michel, Jean. *Dora.* New York: Holt, Rinehart and Winston, 1979.
Michener, James A. *Space.* New York: Random House, 1982.
Miller, Ron, and William K. Hartmann. *The Grand Tour: A Traveler's Guide to the Solar System.* New York: Workman Publishing, 1981.
Miller, Walter James. *The Annotated Jules Verne: From the Earth to the Moon.* New York: Thomas Y. Crowell, 1978.
Morrison, David. *Exploring New Worlds.* New York: Scientific American Library, 1993.
Morrison, Philip and Phylis. *The Ring of Truth.* New York: Random House, 1987.
Morritt, J. B. S. *A Grand Tour: Letters and Journeys, 1794–96.* London: Century Publishing, 1985.
Murray, Bruce. *Journey into Space.* New York: W. W. Norton, 1989.
Murray, Bruce, and Eric Burgess. *Flight to Mercury.* New York: Columbia University Press, 1977.

Murray, Charles, and Catherine Bly Cox. *Apollo: The Race to the Moon.* New York: Simon and Schuster, 1989.

Needell, Allan A., ed. *The First Twenty-Five Years in Space.* Washington, D.C.: Smithsonian Institution Press, 1983.

Neufeld, Michael. *The Rocket and the Reich.* New York: Free Press, 1995.

Newkirk, Dennis. *Almanac of Soviet Manned Space Flight.* Houston, Texas: Gulf Publishing Company, 1990.

Oberg, James E. *Red Star in Orbit.* New York: Random House, 1981.

Oberg, James E., and Alcestis R. Oberg. *Pioneering Space.* New York: McGraw-Hill, 1986.

Ordway, Frederick I., III, ed. *History of Rocketry and Astronautics,* AAS History Series, Vol. 9. San Diego: American Astronautical Society, 1989.

Ordway, Frederick I., III, and Randy Liebermann, eds. *Blueprint for Space: Science Fiction to Science Fact.* Washington, D.C.: Smithsonian Institution Press, 1992.

Ordway, Frederick I., III, and Mitchell R. Sharpe. *The Rocket Team.* New York: Thomas Y. Crowell, 1979.

Peebles, Curtis. *Watch the Skies!* Washington, D.C.: Smithsonian Institution Press, 1994.

Petersen, Carolyn Collins, and John C. Brandt. *Hubble Vision.* New York: Cambridge University Press, 1995.

Photo-Atlas of the United States. Pasadena, Calif.: Ward Ritchie Press, 1975.

Polmar, Norman, and Timothy M. Laur. *Strategic Air Command,* 2nd edition. Baltimore: Nautical and Aviation Publishing Company of America, 1990.

Rebrov, Mikhail. *Space Catastrophes.* Moscow: Izdat, 1993.

Reston, James, Jr. *Galileo.* New York: HarperCollins, 1994.

Rhea, John, ed. *Roads to Space: An Oral History of the Soviet Space Program.* New York: Aviation Week Group (McGraw-Hill), 1995.

Rhodes, Richard. *Dark Sun.* New York: Simon & Schuster, 1995.

Riabchikov, Evgeny. *Russians in Space.* New York: Doubleday & Company, 1971.

Richelson, Jeffrey T. *America's Secret Eyes in Space.* New York: Ballinger, 1990.

———. *The U.S. Intelligence Community.* New York: Ballinger, 1989.

Ryan, Cornelius, ed. *Across the Space Frontier.* New York: Viking Press, 1952.

Sagan, Carl. *Cosmos.* New York: Random House, 1980.

———. *Murmurs of Earth.* New York: Random House, 1978.

Sagdeev, Roald Z. *The Making of a Soviet Scientist.* New York: John Wiley & Sons, 1994.

Salisbury, Harrison E., ed. *The Soviet Union: The Fifty Years.* New York: Harcourt, Brace & World, 1967.

Shapley, Harlow, Samuel Wright, and Helen Wright, eds. *A Treasury of Science,* New York: Harper & Row, 1963.

Sheehan, William. *Planets & Perception: Telescopic Views and Interpretations, 1609–1909.* Tucson: University of Arizona Press, 1988.

Shepard, Alan, and Deke Slayton. *Moon Shot.* Atlanta: Turner Publishing, 1994.

Shirer, William L. *The Rise and Fall of the Third Reich.* New York: Simon and Schuster, 1960.

Shternfeld, Ari. *Soviet Space Science.* New York: Basic Books, 1959.

Skoog, A. Ingemar, ed. *History of Rocketry and Astronautics,* AAS History Series, Vol. 10. San Diego: American Astronautical Society, 1990.

Sloop, John L., ed. *History of Rocketry and Astronautics,* AAS History Series, Vol. 12. San Diego: American Astronautical Society, 1991.

Smith, Arthur. *Planetary Exploration.* Wellingborough, England: Patrick Stephens, 1988.

Smith, Hedrick. *The Russians.* New York: Ballantine, 1984.

Smith, Robert W. *The Space Telescope.* New York: Cambridge University Press, 1989.

Sorensen, Theodore C. *Kennedy.* New York: Harper & Row, 1965.

Sparks, Major James C. *Winged Rocketry.* New York: Dodd, Mead & Company, 1968.

Speer, Albert. *Inside the Third Reich.* New York: Macmillan Company, 1970.

Stares, Paul B. *The Militarization of Space.* Ithaca, N.Y.: Cornell University Press, 1985.

Sullivan, Walter. *Assault on the Unknown.* New York: McGraw-Hill, 1961.

Thompson, Milton O. *At the Edge of Space.* Washington, D.C.: Smithsonian Institution Press, 1992.

Toland, John. *Adolf Hitler,* Vol. 2. Garden City, N.Y.: Doubleday & Company, 1976.

Truman, Harry S. *Years of Trial and Hope.* Garden City, N.Y.: Doubleday & Company, 1956.

Tsiolkovsky, Konstantin. *Beyond the Planet Earth.* New York: Pergamon Press, 1960.
Van Allen, James A. *Origins of Magnetospheric Physics.* Washington, D.C.: Smithsonian Institution Press, 1983.
Verne, Jules. *From the Earth to the Moon, The Jules Verne Omnibus.* New York: J. B. Lippincott, undated.
Vladimirov, Leonid. *The Russian Space Bluff.* London: Tom Stacey, 1971.
von Braun, Wernher. *The Mars Project.* Urbana: University of Illinois Press, 1991.
von Braun, Wernher, and Frederick I. Ordway III. *Space Travel: A History.* New York: Harper & Row, 1985.
Von Däniken, Erich. *Gods from Outer Space.* New York: G. P. Putnam's Sons, 1971.
von Karman, Theodore, with Lee Edson. *The Wind and Beyond.* Boston: Little, Brown and Company, 1967.
Welch, Robert. *The Blue Book of the John Birch Society.* N.p.: 1961.
Wells, H. G. *The War of the Worlds.* New York: Pocket Books, 1988.
Wheeler-Bennett, J. W. *The Nemesis of Power: The German Army in Politics, 1918–1945.* New York: Viking Press, 1964.
White, Frank. *The Overview Effect.* Boston: Houghton Mifflin Company, 1987.
Whitman, Alden. *The Obituary Book.* New York: Stein and Day, 1971.
Whittaker, E. T. *A Treatise on the Analytical Dynamics of Particles and Rigid Bodies.* New York: Dover Publications, 1944.
Wilford, John Noble. *Mars Beckons.* New York: Alfred A. Knopf, 1990.
Williams, Peter, and David Wallace. *Unit 731.* London: Hodder & Stoughton, 1989.
Winter, Frank H. *Prelude to the Space Age/The Rocket Societies: 1924–1940.* Washington, D.C.: Smithsonian Institution Press, 1983.
Witkin, Richard, ed. *The Challenge of the Sputniks.* Garden City, N.Y.: Doubleday & Company, undated.
Wolfe, Tom. *The Right Stuff.* New York: Farrar, Straus & Giroux, 1979.
Zubrin, Robert, and Richard Wagner. *The Case for Mars.* New York: Free Press, 1996.

Reports, Catalogues, Special Histories, Hearings and Press References

"Adventures on Space Station Freedom" (comic book). Austin, Texas: Custom Comic Services, 1989.
"Analysis of Cost Estimates for the Space Shuttle and Two Alternate Programs," B-173677. Washington, D.C.: U.S. General Accounting Office, June 1, 1973.
"An Annotated Bibliography of RAND Space Flight Publications," RM-2113-1. Santa Monica, Calif.: RAND Corporation, March 1, 1959.
"An Earlier Reconnaissance Satellite System," Recommendation to the Air Staff (Project RAND), Santa Monica, Calif.: RAND Corporation, November 12, 1957.
"Anti-Satellite Weapons: Arms Control or Arms Race?" Cambridge, Mass.: Union of Concerned Scientists, June 30, 1983.
Baumann, R. C., and T. A. Bergstralh. "Photography from the V-2 at Altitudes Ranging Up to 160 Kilometers." In H. E. Newell and J. W. Siry, eds., *Upper Atmosphere Research Report No. IV.* Washington, D.C.: Naval Research Laboratory, October 1, 1947.
Baumann, R. C., and L. Winkler. "Rocket Research Report No. XVIII: Photography from the Viking 11 Rocket at Altitudes Ranging Up to 158 Miles," NRL Report 4489. Washington, D.C.: Naval Research Laboratory, February 1, 1955.
Beale, Robert J., Evelyn W. Speiser, and James R. Womack. "The Electric Space Cruiser for High-Energy Missions," Technical Report No. 32-404. Pasadena, Calif.: Jet Propulsion Laboratory, June 8, 1963.
Belton, Michael J. S., et al. "Grand Tour Outer Planet Missions Definition Phase," Report of the Imaging Science Team, Part 1, February 1, 1972.
Berger, Carl. "The Air Force in Space: Fiscal Year 1961." Washington, D.C.: USAF Historical Division Liaison Office, 1966.
Bourke, R. D., and G. A. Flandro. "Comments on Proposed Grand Tour Mission Study," Interoffice Memorandum 312.5-201, Jet Propulsion Laboratory, October 10, 1966.

Bowen, Lee. "The Threshold of Space: The Air Force in the National Space Program, 1945–1959." Washington, D.C.: USAF Historical Division Liaison Office, September 1960.

"Chronology of Early Air Force Man-in-Space Activity: 1955–1960," 65-21-1. Los Angeles, Calif.: Office of Information, Space Systems Division, Air Force Systems Command, January 1965.

"Comments on the Report to the President by the Technological Capabilities Panel," National Security Council NSC 5522, White House Office, Office of the Special Assistant for National Security Affairs, June 8, 1955.

"Crawler/Transporter," NASA/KSC release No. 114-85, August 1988.

"The Crisis in Space and Earth Science," Space and Earth Science Advisory Committee, NASA Advisory Council, November 1986.

Davies, Merton E., and William R. Harris. *RAND's Role in the Evolution of Balloon and Satellite Observation Systems and Related U.S. Space Technology.* Santa Monica, Calif.: RAND Corporation, 1988.

Developments in the Field of Detection and Identification of Nuclear Explosions (Project Vela) and Relationship to Test Ban Negotiations. Washington, D.C.: U.S. Congress, Joint Committee on Atomic Energy, 87th Cong., 1st Sess., July 25–27, 1961.

Fletcher, James C. "The Strategic Defense Initiative," statement before the Subcommittee on Research and Development, Committee on Armed Services, House of Representatives, 98th Cong., 2nd Sess., March 1, 1984.

Froehlich, Walter. *Spacelab: An International Short-Stay Orbiting Laboratory,* EP-165. Washington, D.C.: National Aeronautics and Space Administration, October 1983.

Gemini-Titan II Press Handbook, 2nd ed. Baltimore: Martin Company; St. Louis: McDonnell Aircraft, July 1, 1965.

"Goals and Objectives for America's Next Decades in Space." Washington, D.C.: National Aeronautics and Space Administration, September 1969.

Greeley, Ronald, ed. *Mars Landing Site Catalog* (NASA Reference Publication 1238). Washington, D.C.: National Aeronautics and Space Administration, August 1990.

Greenfield, S. M., and W. W. Kellogg. "Inquiry into the Feasibility of Weather Reconnaissance from a Satellite Vehicle" (Project RAND), R-218, Santa Monica, Calif.: RAND Corporation, April 1951.

Greer, Kenneth E. "Corona." In Ruffner, *Corona: America's First Satellite Program.*

Haines, Gerald K. *The National Reconnaissance Office: Its Origins, Creation, & Early Years.* Washington, D.C.: National Reconnaissance Office, September 1997.

Hartmann, W. K., and O. Raper. *The New Mars: The Discoveries of Mariner 9* (SP-337). Washington, D.C.: National Aeronautics and Space Administration, 1974.

"The History of the Strategic Air Command." Historical Study No. 72, January 1, 1958–June 30, 1958. Omaha, Neb.: Headquarters, Strategic Air Command, Historical Division.

"The Hubble Space Telescope Optical Systems Failure Report." Washington, D.C.: National Aeronautics and Space Administration, November 1990.

Huntzicker, J. H., and H. A. Lieske. "Physical Recovery of Satellite Payloads—a Preliminary Investigation," RM-1811 (Project RAND), Santa Monica, Calif.: RAND Corporation, June 26, 1956.

"International Space Policy for the 1990s and Beyond." Washington, D.C.: Task Force on International Relations in Space, NASA Advisory Council, October 12, 1987.

Investigation of the Challenger Accident, Committee on Science and Technology, House of Representatives, 99th Cong., 2nd Sess., October 29, 1986.

Johnson, Nicholas L., and David M. Rodvold. *Europe and Asia in Space: 1991–1992,* DC-TR-2191.103-1. Colorado Springs, Colo.: Kaman Sciences Corporation, undated.

Johnson, Nicholas L. *The Soviet Year in Space: 1982.* Colorado Springs, Colo.: Teledyne Brown Engineering, 1983.

———. *The Soviet Year in Space: 1983.* Colorado Springs, Colo.: Teledyne Brown Engineering, 1984.

———. *The Soviet Year in Space: 1987.* Colorado Springs, Colo.: Teledyne Brown Engineering, 1988.

———. *The Soviet Year in Space: 1988.* Colorado Springs, Colo.: Teledyne Brown Engineering, 1989.

———. *The Soviet Year in Space: 1989.* Colorado Springs, Colo.: Teledyne Brown Engineering, 1990.

———. *The Soviet Year in Space: 1990.* Colorado Springs, Colo.: Teledyne Brown Engineering, 1991.

Katz, Amrom H. "An Air Force Weather Satellite—Why and How," RAND Document 93-1146. Santa Monica, Calif.: RAND Corporation, March 31, 1959.

Kellogg, W. W. "Observations of the Moon from the Moon's Surface," RM-1764. Santa Monica, Calif.: RAND Corporation, July 27, 1956.

Laskin, Paul R. "Communicating by Satellite." New York: Twentieth Century Fund, 1969.

"Launch Complex 39 Facilities," NASA/KSC Release No. 6-81, Kennedy Space Center, Fla., August 1982.

"Leadership and America's Future in Space" (Ride Report). Washington, D.C.: National Aeronautics and Space Administration, August 1987.

Lipp, J. E. "Interrogation of Professor von Braun on Missile Design" (memorandum to F. R. Collbohn with a copy to RAND), July 26, 1946.

Lipp, J. E., et al. "Utility of a Satellite Vehicle for Reconnaissance" (Project RAND), R-217, Santa Monica, Calif.: RAND Corporation, April 1951.

Logsdon, John M. *The Survival Crisis of the U.S. Solar System Exploration Program.* Washington, D.C.: NASA History Office, September 1989.

Madden, Frank J. "The Corona Camera System: Itek's Contribution to World Stability." Lexington, Mass.: Hughes Danbury Optical Systems, October 1996.

Magellan at Venus. Washington, D.C.: American Geophysical Union, 1992.

Mariner-Mars 1964 Final Project Report, NASA SP-139. Washington, D.C.: National Aeronautics and Space Administration, 1967.

"Mars Pathfinder Landing" press kit, Release No. 96-205. Washington, D.C.: National Aeronautics and Space Administration, July 1997.

"Military Space Projects, March–April–May 1960," Report No. 10. Washington, D.C.: Department of Defense, Office of the Director of Defense Research and Engineering, August 16, 1960 (Eisenhower Library).

Miller, Jon D. "The Impact of the Challenger Accident on Public Attitudes Toward the Space Program," De Kalb: Public Opinion Laboratory, Northern Illinois University (for the National Science Foundation), January 25, 1987.

Morrison, David. *Voyages to Saturn,* NASA SP-451. Washington, D.C.: National Aeronautics and Space Administration, 1982.

Morrison, David, and Jane Samz. *Voyage to Jupiter,* NASA SP-439. Washington, D.C.: National Aeronautics and Space Administration, 1980.

"NASA's Get Away Specials," NASA Facts. Greenbelt, Md.: Goddard Space Flight Center, August 1989.

National Aeronautics and Space Act of 1958, As Amended, and Related Legislation, U.S. Senate, Committee on Commerce, Science, and Transportation, 95th Cong., 2nd Sess., December 1978.

National Commission on Space. "Pioneering the Space Frontier." New York: Bantam Books, May 1986.

National Space Transportation System Reference, Vol. 1, *Systems and Facilities.* Washington, D.C.: NASA, June 1988.

"New Saturn Discovery by Pioneer," NASA Release No. 79-45. Washington, D.C.: NASA, November 21, 1979.

"News Conference on Ranger VI Impact on Moon," NASA Release No. 2-2509. Washington, D.C.: National Aeronautics and Space Administration, February 2, 1964.

Peebles, Curtis. *High Frontier: The U.S. Air Force and the Military Space Program.* Washington, D.C.: Air Force History and Museums Program, 1997.

Perry, Robert L. "Origins of the USAF Space Program, 1945–1956," U.S. Air Force Systems Command, AFSC Historical Publication Series, 62-24-10, 1961.

"Pioneer F Mission to Jupiter," NASA News Release No. 72-25. Washington, D.C.: National Aeronautics and Space Administration, February 20, 1972.

"Pioneer Saturn Results Are Summarized," NASA Release No. 79-42. Washington, D.C.: National Aeronautics and Space Administration, October 24, 1979.

"Pioneer Will Reach Jupiter December 3," NASA News Release No. 73-243. Washington, D.C.: National Aeronautics and Space Administration, November 7, 1973.

"Planetary Exploration Through Year 2000: A Core Program." Solar System Exploration Committee, NASA Advisory Council, Washington, D.C.: National Aeronautics and Space Administration, 1983.

"Planetary Exploration Through Year 2000: An Augmented Program." Solar System Exploration Committee, NASA Advisory Council. Washington, D.C.: National Aeronautics and Space Administration, 1986.

Portree, David S. F. *Mir Hardware Heritage,* NASA Reference Publication 1357. Houston: Johnson Space Center, March 1995.

"The Post-Apollo Space Program: Directions for the Future" (Agnew Report). Washington, D.C.: Space Task Group Report to the President, September 1969.

Preliminary Design of an Experimental World-Circling Spaceship, Report No. SM-11827. Santa Monica, Calif.: Douglas Aircraft Company, May 2, 1946.

"Ranger 6 Review Board, Final Report," 2-2472. Washington, D.C.: National Aeronautics and Space Administration, March 17, 1964.

"RA-6 Investigation Committee Final Report," EPD 205. Pasadena, Calif.: Jet Propulsion Laboratory, February 14, 1964.

Reiber, Duke B., ed. *The NASA Mars Conference,* Science and Technology Series, Vol. 71. San Diego: American Astronautical Society, 1988.

"Reinventing NASA." Washington, D.C.: Congressional Budget Office, March 1994.

Report of the Presidential Commission on the Space Shuttle Challenger Accident (5 vols.). Washington, D.C., June 6, 1986.

"Report of the U.S. Department of Energy's Team Analyses of the Chernobyl-4 Atomic Energy Station Accident Sequence," DOE/NE-0076. Washington, D.C.: U.S. Department of Energy, November 1986.

"Review and Summary of X-20 Military Application Studies," 63ASZR-2099, X-20 System Program, Aeronautical Systems Division, Air Force Systems command, Wright-Patterson Air Force Base, December 14, 1963.

Rosenberg, Max. "The Air Force in Space: 1959–1960." Washington, D.C.: USAF Historical Division Liaison Office, June 1962.

Ruffner, Kevin C., ed. *Corona: America's First Satellite Program.* Washington, D.C.: CIA Center for the Study of Intelligence, 1995.

Russian Space History, catalog for Sale 6516. New York: Sotheby's, December 11, 1993.

Russian Space History, catalog for Sale 6753. New York: Sotheby's, March 16, 1996.

"Scientific Research Institute and Experimental Factory 88, Moskva/Kaliningrad," OSI-C-RA/60-2. Washington, D.C.: Central Intelligence Agency (Office of Scientific Intelligence), March 4, 1960.

Sloop, J. L. "Liquid Hydrogen as a Propulsion Fuel, 1945–1959," N79-16994. Washington, D.C.: National Aeronautics and Space Administration, 1978.

"Soviet Space Shuttle Analysis: Engineering Directorate Report." Washington, D.C.: National Aeronautics and Space Administration, May 25, 1989.

"Spacelab News Reference," European Space Agency and National Aeronautics and Space Administration, 14M983. Huntsville, Ala.: Marshall Space Flight Center, 1983.

Space Science Board, National Research Council. "Space Science in the Twenty-First Century" (7 vols.). Washington, D.C.: National Academy Press, 1988.

"Space Station: Cost Control Difficulties Continue," GAO/NSLAD-96-135. Washington, D.C.: U.S. General Accounting Office, July 1996.

"Space Station: Cost Control Problems Are Worsening," GAO/NSLAD-97-213. Washington, D.C.: U.S. General Accounting Office, September 1997.

Space Station Freedom Media Handbook. Washington, D.C.: National Aeronautics and Space Administration, April 1996.

Spencer, Dwain F., et al. "Nuclear Electric Spacecraft for Unmanned Planetary and Interplanetary Missions," Technical Report No. 32-281. Pasadena, Calif.: Jet Propulsion Laboratory, April 25, 1962.

Steury, Donald P., ed. *Intentions and Capabilities: Estimates on Soviet Strategic Forces, 1950–1983.* Washington, D.C.: CIA Center for the Study of Intelligence, 1996.

"STS-9/Spacelab 1" press kit, ESA-NASA, November 1983.

"Summary [of] U.S. Air Force Discoverer Satellite Vehicle Recovery Program, I thus XIV," Washington, D.C.: U.S. Air Force, undated.

Synthesis Group. *America at the Threshold: America's Space Exploration Initiative.* Washington, D.C.: U.S. Government Printing Office, May 3, 1991.

Viking Lander Imaging Team. *The Martian Landscape* (SP-425). Washington, D.C.: National Aeronautics and Space Administration, 1978.

Watts, A. Frank, D. J. Dreyfuss, and H. G. Campbell. "The Economic Impact of Reusable Orbital Transports on the Cost of Planned Manned Space Programs, 1970–1999," P-3465. Santa Monica, Calif.: RAND Corporation, November 15, 1966.

"Weapons in Space," *Daedalus* (2 vols.) Spring 1985.

Welch, S. M., and C. R. Stoker. *The Case for Mars,* JPL Publication 86-28. Pasadena, Calif.: Jet Propulsion Laboratory, April 15, 1986.

Yeates, C. M., et al. *Galileo: Exploration of Jupiter's System,* NASA SP-479. Washington, D.C.: National Aeronautics and Space Administration, 1985.

NASA History and Technology

Bilstein, Roger E. *Orders of Magnitude: A History of the NACA and NASA, 1915–1990,* NASA SP-4406. Washington, D.C.: National Aeronautics and Space Administration, 1989.

———. *Stages to Saturn: A Technological History of the Apollo/Saturn Launch Vehicles,* NASA SP-4206. Washington, D.C.: National Aeronautics and Space Administration, 1980.

Brooks, Courtney G., James M. Grimwood, and Loyd S. Swenson, Jr. *Chariots for Apollo,* NASA SP-4205. Washington, D.C.: National Aeronautics and Space Administration, 1979.

Compton, David W., and Charles D. Benson. *Living and Working in Space: A History of Skylab,* NASA SP-4208. Washington, D.C.: National Aeronautics and Space Administration, 1983.

Ezell, Edward Clinton, and Linda Neuman Ezell. *On Mars: Exploration of the Red Planet, 1958–1978,* NASA SP-4212. Washington, D.C.: National Aeronautics and Space Administration, 1984.

———. *The Partnership: A History of the Apollo-Soyuz Test Project,* NASA SP-4209. Washington, D.C.: National Aeronautics and Space Administration, 1978.

Ezell, Linda Neuman. *NASA Historical Data Book,* Vol. 2, *Programs and Projects 1958–1968,* NASA SP-4012. Washington, D.C.: National Aeronautics and Space Administration, 1988.

———. *NASA Historical Data Book,* Vol. 3, *Programs and Projects 1969–1978,* NASA SP-4012. Washington, D.C.: National Aeronautics and Space Administration, 1988.

Fimmel, Richard O., James Van Allen, and Eric Burgess. *Pioneer: First to Jupiter, Saturn, and Beyond,* NASA SP-446. Washington, D.C.: National Aeronautics and Space Administration, 1980.

Fimmel, Richard O., Lawrence Colin, and Eric Burgess. *Pioneer Venus,* NASA SP-461. Washington, D.C.: National Aeronautics and Space Administration, 1983.

Hall, R. Cargill. *Lunar Impact: A History of Project Ranger,* NASA SP-4210. Washington, D.C.: National Aeronautics and Space Administration, 1977.

Hartmann, William K., and Odell Raper. *The New Mars: The Discoveries of Mariner 9,* NASA SP-337. Washington, D.C.: National Aeronautics and Space Administration, 1974.

Hunley, J. D., ed. *The Birth of NASA: The Diary of T. Keith Glennan,* NASA SP-4105. Washington, D.C.: National Aeronautics and Space Administration, 1993.

Levine, Arnold S. *Managing NASA in the Apollo Era,* NASA SP-4102. Washington, D.C.: National Aeronautics and Space Administration, 1982.

Logsdon, John M. et al., eds. *Exploring the Unknown: Selected Documents in the History of the U.S. Civil Space Program* (NASA SP-4407, 3 vols.). Washington, D.C.: National Aeronautics and Space Administration, 1995, 1996, 1998.

Newell, Homer E. *Beyond the Atmosphere,* NASA SP-4211. Washington, D.C.: National Aeronautics and Space Administration, 1980.

Nicks, Oran W. *Far Travelers,* NASA SP-480. Washington, D.C.: National Aeronautics and Space Administration, 1985.
Stillwell, Wendell H. *X-15 Research Results,* NASA SP-60. Washington, D.C.: National Aeronautics and Space Administration, 1965.
Swenson, Loyd S., Jr., James M. Grimwood, and Charles C. Alexander. *This New Ocean* (SP-4201). Washington, D.C.: National Aeronautics and Space Administration, 1966 (TL 789 .8. .U6. M.484 c.1).
Van Nimmen, Jane, Leonard C. Bruno, and Robert L. Rosholt. *NASA Historical Data Book,* Vol. 1, NASA SP-4012. Washington, D.C.: National Aeronautics and Space Administration, 1988.

Articles and Excerpts from Anthologies

Science and Technology

Angerson, Don L. "Helium-3 from the Mantle: Primordial Signal or Cosmic Dust?" *Science,* July 9, 1993.
Becklace, E. J., and D. Millard. "Congreve and His Works." *Journal of the British Interplanetary Society,* July 1992.
Belton, M. J. S., et al. "Galileo's First Images of Jupiter and the Galilean Satellites." *Science,* October 18, 1996.
Brown, R. H., et al. "Energy Sources for Triton's Geyser-Like Plumes." *Science,* October 19, 1990.
Brugioni, Dino A. "The Art and Science of Photoreconnaissance." *Scientific American,* March 1996.
Burrows, William E. "Beyond the Iron Curtain." *Air & Space,* September 1994.
———. "The Military in Space: Securing the High Ground." In Collins and Kraemer, *Space: Discovery and Exploration* (anthology).
———. "The New Millennium." *Air & Space,* September 1996.
———. "Stakeout." *Air & Space,* August–September 1997.
Canby, Thomas Y. "Are the Soviets Ahead in Space?" *National Geographic,* October 1986.
———. "Satellites That Serve Us." *National Geographic,* September 1983.
Clarke, Arthur C. "V-2 for Ionosphere Research and Extraterrestrial Relays." *Wireless World,* February 1945.
Cooper, Henry S. F., Jr. "The Loneliness of the Long-Duration Astronaut." *Air & Space,* June–July 1996.
Cuzzi, Jeffrey N., and Larry W. Esposito. "The Rings of Uranus." *Scientific American,* July 1987.
DeVorkin, David H. "Scientists and the Space Sciences." In Collins and Kraemer, *Space: Discovery and Exploration.*
Dichmann, Donald J. "The Constellations in LEO." *Launchspace,* August–September 1997.
Dyson, Freeman J. "21st-Century Spacecraft." *Scientific American,* September 1995.
Edelson, Burton I. "Communication Satellites: The Experimental Years." In Sloop, *History of Rocketry and Astronautics,* Vol. 12.
"Enhanced GPS Spawns Innovative Applications." *Aviation Week & Space Technology,* November 21, 1994.
"Europa Water Evidence Hints Life Possible." *Aviation Week & Space Technology,* April 14, 1997.
"Fast, Cheap, and Out of Control." *Science,* May 25, 1990.
Fogg, Martyn J. "Terraforming Mars: Conceptual Solutions to the Problem of Planet Growth in Low Concentrations of Oxygen." *Journal of the British Interplanetary Society,* October 1995.
Forman, Brenda. "The Political Process in Systems Architecture Design." *Program Manager,* March–April 1993.
Forward, Robert L. "Ad Astra!" *Journal of the British Interplanetary Society,* January 1996.
Gleick, James. "A Sense of Where You Are." *The New York Times Magazine,* February 4, 1996.
"Goddard Describes New Stratosphere Plane." AIS *Bulletin,* No. 12, September 1931.
Gohlke (no first name). "Thermal-Air Jet Propulsion." *Astronautics,* May 1942.

Goodell, Thaxter L. "Cratology Pays Off." *Studies in Intelligence,* Fall 1964.
Goodman, Billy. "Ancient Whisper." *Air & Space,* April–May 1992.
"GPS Technology Ripens for Consumer Market." *Aviation Week & Space Technology,* October 9, 1995.
Greeley, Ronald. "Release of Juvenile Water on Mars: Estimated Amounts and Timing Associated with Volcanism." *Science,* June 26, 1987.
Grinspoon, David Harry. "Venus Unveiled." *The Sciences,* July–August 1993.
Gwynn-Jones, Terry. "Farther: The Quest for Distance." In Greenwood, *Milestones of Flight.*
Hall, R. Cargill. "The Eisenhower Administration and the Cold War: Framing American Astronautics to Serve National Security." *Prologue,* Spring 1995.
Helms, Harry L. "The Great Galactic Ghoul." *Saga,* July 1976.
Herrick, Samuel. "Earth Satellites and Related Orbit and Perturbation Theory." In Howard S. Seifert, ed., *Space Technology,* New York: Wiley, 1959.
Herring, Thomas A. "The Global Positioning System." *Scientific American,* February 1996.
Huntington, Tom. "V-2: The Long Shadow." *Air & Space,* February–March, 1993.
Ingersoll, Andrew P. "Uranus." *Scientific American,* January 1987.
Ingersoll, Andrew P., and Kimberly A. Tryka. "Triton's Plumes: The Dust Devil Hypothesis." *Science,* October 19, 1990.
"Is There Life in Outer Space?" *Time,* February 5, 1996.
Jastrow, Robert, and Homer E. Newell. "Why Land on the Moon?" *The Atlantic,* August 1963.
Johnson, Torrence V., Robert Hamilton Brown, and Laurence A. Soderblom. "The Moons of Uranus." *Scientific American,* April 1987.
Kargel, Jeffrey S., and Robert G. Strom. "Global Climatic Change on Mars." *Scientific American,* November 1996.
Kerr, Richard A. "Ancient Life on Mars?" *Science,* August 16, 1996.
———. "Voyager Finds Uranian Shepherds and a Well-behaved Flock of Rings." *Science,* February 21, 1986.
Kowitt, Mark E., and Michael S. Kaplan. "The Wings of Daedalus: The Convergence of Myth and Technology in 20th Century Culture." *Journal of the British Interplanetary Society,* November 1993.
Laeser, Richard P., William I. McLaughlin, and Donna M. Wolff. "Engineering Voyager 2's Encounter with Uranus." *Scientific American,* November 1986.
Lane, Arthur L., et al. "Photometry from Voyager 2: Initial Results from the Uranian Atmosphere, Satellites, and Rings." *Science,* July 4, 1986.
Langmuir, David B. "Electric Spacecraft—Progress 1962." *Astronautics,* June 1962.
Lasser, David. "The Rocket and the Next War." AIS *Bulletin,* No. 13, November 1931.
Lawden, D. F. "Interplanetary Orbits." In D. R. Bates, ed., *Space Research and Exploration.* New York: William Sloane Associates, 1958.
———. "Transfer Between Circular Orbits." *Jet Propulsion,* July 1956.
Levy, David H., Eugene M. Shoemaker, and Carolyn S. Shoemaker. "Comet Shoemaker-Levy 9 Meets Jupiter." *Scientific American,* August 1995.
Ley, Willy. "Inside the Lunar Base." *Collier's,* October 25, 1952.
"Lockheed-Khrunichev-Energia Wins European Launch Order." *Aviation Week & Space Technology,* March 21, 1994.
McKay, David S., et al. "Search for Past Life on Mars: Possible Relic Biogenic Activity in Martian Meteorite ALH84001." *Science,* August 16, 1996.
Miller, Ron. "Chesley Bonestell's Astronomical Visions." *Scientific American,* May 1994.
———. "Herman Potocnik—*Alias* Hermann Noordung." *Journal of the British Interplanetary Society,* July 1992.
Murray, Bruce. "The End of the Beginning." *Science,* January 29, 1982.
Newgard, John J., and Myron Levoy. "Nuclear Rockets." *Scientific American,* May 1959.
Nozette, Stewart, et al. "The Clementine Mission to the Moon: Scientific Overview." *Science,* December 16, 1994.
Oberg, James E. "The Great Galactic Ghoul." *Final Frontier,* October 1987.
Oberth, Hermann. "My Contributions to Astronautics." In Durant and James, *First Steps Toward Space.*
Peebles, Curtis. "Satellite Photograph Interpretation." *Spaceflight,* April 1982.

Pendray, G. Edward. "The German Interplanetary Society and the *Raketenflugplatz*." AIS *Bulletin,* No. 9, May 1931 (in American Interplanetary Society, Nos. 1–18, June–April 1930–32, Aeronautics Library, California Institute of Technology).
Powell, Corey S. "The Golden Age of Cosmology." *Scientific American,* July 1992.
Reichhardt, Tony. "Gravity's Overdrive." *Air & Space,* February–March 1994.
Saunders, R. Stephen, and Gordon H. Pettengill. "Magellan: Mission Summary." *Science* (Magellan issue), April 12, 1991.
Saunders, Steve. "Magellan: The Geologic Exploration of Venus." *Engineering & Science* (Caltech), Spring 1991.
Schachner, Nathan. "Can Man Exist on Other Planets?" AIS *Bulletin,* No. 15, January 1932.
Schwartz, David N. "Past and Present: The Historical Legacy." In Carter and Schwartz, *Ballistic Missile Defense.*
Smith, B. A., et al. "Voyager 2 at Neptune: Imaging Science Results." *Science,* December 15, 1989.
———. "Voyager 2 in the Uranian System: Imaging Science Results." *Science,* July 4, 1986.
Smith, David E., and Maria T. Zuber. "The Shape of Mars and the Topographic Signature of the Hemispheric Dichotomy." *Science,* January 12, 1996.
Smith, David R. "They're Following Our Script." *Future,* May 1978.
"The Station in Space: Sun Power Stations Planned by Germans." *Journal of the American Rocket Society,* September 1945 (unsigned).
Stearns, John W. "Electro-propulsion System Applications." *Astronautics,* March 1962.
Stewart, H. J. "New Possibilities for Solar-System Exploration." *Astronautics & Aeronautics,* December 1966.
Stone, E. C., and E. D. Miner. "The Voyager 2 Encounter with the Neptunian System." *Science,* December 15, 1989.
———. "The Voyager 2 Encounter with the Uranian System." *Science,* July 4, 1986.
Sutton, George Paul. "Rocket Propulsion Systems for Interplanetary Flight." *Journal of the Aero/Space Sciences,* October 1959.
"Swarms of Mini-Robots Set to Take on Mars Terrain." *Science,* September 18, 1992.
Taylor, G. Jeffrey. "The Scientific Legacy of Apollo." *Scientific American,* July 1994.
Varfolomeyev, Timothy. "Soviet Rocketry That Conquered Space." *Spaceflight,* August 1995.
"Venus Revealed." *Scientific American,* January 1992.
Waldrop, M. Mitchell. "The Puzzle That Is Saturn." *Science,* September 18, 1981.
———. "Voyage to a Blue Planet." *Science,* February 28, 1980.
Weaver, H. A., et al. "Hubble Space Telescope Observations of Comet P/Shoemaker-Levy 9 (1993e)." *Science,* February 11, 1994.
Whipple, Fred L. "Is There Life on Mars?" *Collier's,* April 30, 1954.
Wilford, John Noble. "First into Space, Then the Race." In Salisbury, *The Soviet Union: The Fifty Years.*
Winter, Frank H. "Who First Flew in a Rocket?" *Journal of the British Interplanetary Society,* July 1992.
Winter, Frank H., and Randy Liebermann. "A Trip to the Moon." *Air & Space,* October–November 1994.
Wittenberg, L. J., J. F. Santarius, and G. L. Kulcinski. "Lunar Source of 3He for Commercial Fusion Power." *Fusion Technology,* September 1986.
Wolszczan, Alexander. "Confirmation of Earth-Mass Planets Orbiting the Millisecond Pulsar PSR B1257+12." *Science,* April 22, 1994.
Yaffee, Michael L. "Electric Rocket Program Being Reshaped." *Space Technology International,* April 1964.

Russia and the Soviet Union

Bethe, Hans A., Kurt Gottfried, and Roald Z. Sagdeev. "Did Bohr Share Nuclear Secrets?" *Scientific American,* May 1995.
Bogdanov, Anatoly V. "The Organization and the Structure of the Management of Space Affairs in Russia." *Space Bulletin,* Vol. 1, No. 1, 1993.
Brugioni, Dino A. "The Tyuratam Enigma." *Air Force Magazine,* March 1984.

Covault, Craig. "Russians Reveal Secrets of Mir, Buran, Lunar Landing Craft." *Aviation Week & Space Technology,* February 10, 1992.
———. "Soviet Union Reveals Moon Rocket Design That Failed to Beat U.S. to Lunar Landing." *Aviation Week & Space Technology,* February 18, 1991.
Finer, Sydney Wesley. "The Kidnaping [sic] of the Lunik." *Studies in Intelligence,* Winter 1967.
"Glonass Nears Full Operation." *Aviation Week & Space Technology,* October 9, 1995.
Gorin, Peter A. "Zenit: Corona's Soviet Counterpart." In McDonald, *Corona: Between the Sun & the Earth.*
Hendrickx, B., "The Kamanin Diaries 1960–1963," *Journal of the British Interplanetary Society,* January 1997.
———. "Korolev: Facts and Myths." *Spaceflight,* February 1996.
"Japanese Journalist Reaches Soviets' Space Station Aboard Commercial Flight." *Aviation Week & Space Technology,* December 10, 1990.
Jastrow, Robert. "The New Soviet Arms Buildup in Space." *The New York Times Magazine,* October 3, 1982.
Kapitsa, Sergei P. "Lessons of Chernobyl." *Foreign Affairs,* Summer 1993.
Kidger, Neville. "Mir's Effectiveness Questioned." *Spaceflight,* June 1991.
Milov, Yuri G. "The Basic Elements of Russia's Space Program." *Space Bulletin,* Vol. 2, 1993.
"Mir Accident Imperils U.S.-Russian Cooperation." *Aviation Week & Space Technology,* June 30, 1997.
Mishin, V. P. "Why Didn't We Get to the Moon?" *Cosmonautics and Astronomy,* Moscow: Znanije, 1990 (in Russian).
Oberg, James. "The First Soviet Spy Satellite." *Air Force,* July 1995.
"Phobos Loss—Spacecraft Designers Blamed." *Spaceflight,* July 1989.
"Record Soviet Manned Space Flight Raises Human Endurance Questions." *Aviation Week & Space Technology,* January 4, 1988.
"Russia and U.S. Chart Daring Mir Salvage." *Aviation Week & Space Technology,* July 7, 1997.
"Russians Pledge to Put Station Back on Track." *Aviation Week & Space Technology,* April 22, 1996.
"Salyut 7 Crew Prepared to End Record Flight." *Aviation Week & Space Technology,* October 1, 1994.
Sanders, Berry. "An Analysis of the Trajectory and Performance of the N-1 Lunar Launch Vehicle." *JBIS* (Journal of the British Interplanetary Society), July 1996.
"Some General Conclusions Regarding Russia." *Chemical & Engineering News,* April 14, 1958.
"Soviets Lose Contact with Phobos 2 Spacecraft." *Aviation Week & Space Technology,* April 3, 1989.
"Soviet Moon Shots Signal New Ventures." *Air Intelligence Digest,* November 1959.
"Soviet Strategic Laser Sites Imaged by French Spot Satellite." *Aviation Week & Space Technology,* October 26, 1987.
"Soviets Planning 1988 Mission to Study Martian Moon Phobos." *Aviation Week & Space Technology,* October 28, 1985.
Specter, Michael. "Moscow on the Make," *The New York Times Magazine,* June 1, 1997.
"Station Problems Test U.S.-Russian Resolve." *Aviation Week & Space Technology,* March 31, 1997.
Tarasenko, Maxim V. "Transformation of the Soviet Space Program After the Cold War." *Science & Global Security,* Vol. 4, 1994.
"U.S. Scrutinizes New Soviet Radar." *Aviation Week & Space Technology,* August 22, 1983.
Varfolomeyev, Timothy. "Soviet Rocketry That Conquered Space." *Spaceflight,* August 1995.
Weiss, Gus W. "The Farewell Dossier." *Studies in Intelligence,* Annual Unclassified Edition, 1996.

United States

"Agena B to Put Samos, Midas in Orbit." *Aviation Week,* February 8, 1960.
"ASAT Test Stalled by Funding Dispute." *Aviation Week & Space Technology,* July 1, 1996.
"Atlantis Radar Satellite Payload Opens New Reconnaissance Era." *Aviation Week & Space Technology,* December 12, 1988.

Barr, James. "Is the Mercury Program Headed for Disaster?" *Missiles and Rockets,* August 15, 1960.
Bethe, Hans A., et al. "Space-based Ballistic Missile Defense." *Scientific American,* October 1984.
Biddle, Wayne. "The Endless Countdown." *The New York Times Magazine,* June 22, 1980.
Brandwein, David S. "Telemetry Analysis." *Studies in Intelligence,* Fall 1964.
Brzezinski, Zbigniew, Robert Jastrow, and Max M. Kampelman. "Defense in Space Is Not 'Star Wars.' " *The New York Times Magazine,* January 27, 1985.
Bunn, Matthew. "The Next Nuclear Offensive." *Technology Review,* January 1988.
Burrows, William E. "Ballistic Missile Defense: The Illusion of Security." *Foreign Affairs,* Spring 1984.
———. "How the Skunk Works Works." *Air & Space,* April–May 1994.
———. "Securing the High Ground." *Air & Space,* December 1993–January 1994.
Butz, J. S., Jr. "MOL: The Technical Promise and Prospects." *Air Force,* October 1965.
Charles, Daniel. "Spy Satellites: Entering a New Era." *Science,* March 24, 1989.
Chayes, Abram, Antonia Handler Chayes, and Eliot Spitzer. "Space Weapons: The Legal Context." *Daedalus,* Weapons in Space special issue, Vol. 2, Summer 1985.
"Communications Will Drive Space Sector's Growth." *Aviation Week & Space Technology,* March 14, 1994.
Conversino, Mark J. "Back to the Stone Age: The Attack on Curtis E. LeMay." *Strategic Review,* Spring 1997.
Day, Dwayne A. "A Strategy for Reconnaissance: Dwight D. Eisenhower and Freedom of Space." In Day, Logsdon, and Latell, *Eye in the Sky.*
———. "A Strategy for Space." *Spaceflight,* September 1996.
"Defense Dept. to Retain Expendable Launchers as Backup to Shuttle." *Aviation Week & Space Technology,* March 18, 1987.
DeVorkin, David H. "War Heads into Peace Heads: Holger N. Toftoy and the Public Image of the V-2 in the United States." *Journal of the British Interplanetary Society,* November 1992.
Duff, Brian. "The Great Lunar Quarantine." *Air & Space,* February–March 1994.
"EKV Contractor Selection Targeted for Fiscal 1999." *Aviation Week & Space Technology,* March 3, 1997.
Feynman, Richard P. "Mr. Feynman Goes to Washington." *Engineering & Science* (Caltech), Fall 1987.
"First [Corona] Capsule Recovered from Satellite." *Aviation Week,* August 22, 1960.
"From Surveyor: The Stark and Airless Beauty of the Moon." *Life,* July 1, 1966.
Hall, R. Cargill. "Origins and Development of the Vanguard and Explorer Satellite Programs." *Airpower Historian,* October 1964.
———. "Strategic Reconnaissance in the Cold War." *Prologue,* Summer 1996.
Heacock, Raymond L. "Ranger: Its Mission and Its Results." *TRW Space Log,* Summer 1965.
Johnson, Nicholas L. "U.S. Space Surveillance." *Advances in Space Research,* Vol. 13, No. 8, COSPAR, 1993.
"The Joyous Triumph of Apollo 13." *Life,* April 24, 1970.
Leavitt, William. "MOL: Evolution of a Decision." *Air Force,* October 1965.
Logsdon, John M. "The Space Shuttle Program: A Policy Failure?" *Science,* May 30, 1986.
McDonald, Robert A. "Corona: Success for Space Reconnaissance, a Look into the Cold War, and a Revolution for Intelligence." *PE & RS* (Photogrammetric Engineering & Remote Sensing), June 1995.
" 'Milspace' Maturing into Warfighter Roles." *Aviation Week & Space Technology,* September 1, 1997.
"Missile Defense Soon, but Will It Work?" *Aviation Week & Space Technology,* February 24, 1997.
"Missile Destroyed in First SDI Test at High-Energy Facility." *Aviation Week & Space Technology,* September 23, 1985.
"NASA Expects Pioneer 10's Useful Lifetimes Will Be Extended by Decade." *Aviation Week & Space Technology,* June 27, 1988.
"NASA Mission Gets Down to Earth." *Science,* September 1, 1995.

"NASA's Link to Strategic Defense Effort Grows with New Spacecraft." *Aviation Week & Space Technology,* August 26, 1985.
Overbye, Dennis. "All Aboard for Outer Space," *Discover,* March 1981.
———. "Success Amid the Snafus," *Discover,* November 1985.
Payne, Keith B., and Colin S. Gray. "Nuclear Policy and the Defensive Transition." *Foreign Affairs,* Spring 1984.
Plaster, Henry G. "Snooping on Space Pictures." *Studies in Intelligence,* Fall 1964.
Powell, J. W., and K. J. Scala. "Historic White Sands Missile Range." *Journal of the British Interplanetary Society,* March 1994.
"Put Them High on the List That Counts." *Life,* February 3, 1967.
"Reagan Reinterprets the ABM Treaty." *Science,* November 8, 1985.
Rhodes, Richard. "The General and World War III." *The New Yorker,* June 19, 1995.
Roland, Alex. "The Shuttle: Triumph or Turkey?" *Discover,* November 1985.
"Scientists Urge Immediate Change in Planetary Exploration Policy." *Aviation Week & Space Technology,* June 15, 1987.
"SDI Zapped." *Scientific American,* June 1987.
"Shuttle Crystal Growth Tests Could Advance Cancer Research." *Aviation Week & Space Technology,* February 25, 1985.
"Shuttle Plan Emphasizes Earth Survey." *Aviation Week & Space Technology,* September 24, 1984.
"Space Man of the Year." *Life* (Collector's Edition), January 1998.
"Space Reconnaissance Dwindles." *Aviation Week & Space Technology,* October 6, 1980.
" 'Star Wars' Tests and the ABM Treaty." *Science,* July 5, 1985.
"Test Firings for Pied Piper Due Soon." *Aviation Week,* June 16, 1958.
"USAF Prepares West Coast Site for Space Shuttle Processing." *Aviation Week & Space Technology,* May 5, 1986.
"USAF Vehicle Designed for Satellite Attack." *Aviation Week & Space Technology,* January 14, 1985.
Van Allen, James. "Space Science, Space Technology, and the Space Station." *Scientific American,* January 1986.
"Weapons Designers Challenge SDI Report." *Science,* July 10, 1987.
Wheelon, Albert D. "Lifting the Veil on CORONA." *Space Policy,* November 1995.
Wheelon, Albert D., and Sidney N. Graybeal. "Intelligence for the Space Race." *Studies in Intelligence,* Fall 1961.
Wolfe, Tom. "Columbia's Landing Closes a Circle." *National Geographic,* October 1981.
"Working Solar Monitor Shot Down by ASAT." *Science,* October 4, 1985.
Yarevand, Colonel-Engineer Y., and Major-Engineer L. Shevchuk. "The Pentagon's Global Espionage Network." *Aviatsiya I Kosmonavtika (Aviation and Cosmonautics),* May 1972.
Young, Hugo, Bryan Silcock, and Peter Dunn. "Why We Went to the Moon." *The Washington Monthly,* April 1970.

Papers

Allen, R. D. "Early Lunar Base Concepts of the U.S. Air Force," 43rd Congress of the International Astronautical Federation, Washington, D.C., August 28–September 5, 1992.
Ball, Desmond. "Ocean Surveillance," Reference Paper No. 87. Canberra: Research School of Pacific Studies, Australian National University, November 1982.
———. "The U.S. Air Force Satellite Communications (Afsatcom) System: The Australian Connection," Reference Paper No. 76. Canberra: Research School of Pacific Studies, Australian National University, March 1982.
———. "The U.S. Fleet Satellite Communications (Fltsatcom) System: The Australian Connection," Reference Paper No. 69. Canberra: Research School of Pacific Studies, Australian National University, July 1981.
———. "The U.S. Vela Nuclear Detection Satellite (NDS) System: The Australian Connection," Reference Paper No. 70. Canberra: Research School of Pacific Studies, Australian National University, October 1981.

Bergstralh, Thor, and Ernst Krause. "Early Upper Atmospheric Research with Rockets." In Ordway, *History of Rocketry and Astronautics,* Vol. 9.

Biryukov, Yu. V. "The Role of Mikhail K. Tikhonravov in the Development of Soviet Rocket and Space Technology." In Lattu, *History of Rocketry and Astronautics,* Vol. 8.

Blosset, Lise. "Robert Esnault-Pelterie: Space Pioneer." In Durant and James, *First Steps Toward Space.*

Braselton, W. M., Jr. "Space Power for an Expanded Vision." Melbourne, Fla.: Harris Corporation, April 1993.

Burrows, William E. "Imaging Space Reconnaissance Operations During the Cold War: Causes, Effects and Legacy," Conference on U-2 Flights and the Cold War in the High North, Bodø, Norway, October 7, 1995.

———. "Spyglass to Eyeglass: Space Reconnaissance Comes in from the Cold," Science, Technology and Government Colloquium (Carnegie Corporation), New York University, December 4, 1995.

Chae, Yeon Seok. "A Study of Early Korean Rockets (1377–1600). In Sloop, *History of Rocketry and Astronautics,* Vol. 12.

Crocco, Luigi. "Early Italian Rocket and Propellant Research." In Durant and James, *First Steps Toward Space.*

Crouch, Thomas D. " 'To Fly to the World in the Moon': Cosmic Voyaging in Fact and Fiction from Lucian to Sputnik." In Emme, *Science Fiction and Space Futures.*

Danne, Harold A. "Across the Atlantic in a Rocket Plane." *Bulletin* of the American Interplanetary Society, No. 10, June–July 1931.

David, Leonard, et al. "The Space Tourist." In Hecker, *Proceedings of the Fourth Annual L5 Space Development Conference.*

Dowling, Richard L., et al. "Gravity Propulsion Research at UCLA and JPL, 1962–1964," IAA-91-677, 42nd Congress of the International Astronautical Federation, Montreal, October 1991.

———. "The Origin of Gravity-Propelled Interplanetary Space Travel," IAA-90-630, 41st Congress of the International Astronautical Federation, Dresden, October 1990.

Draper, C. Stark. "The Evolution of Aerospace Guidance Technology at the Massachusetts Institute of Technology, 1935–1951: A Memoir." In Ordway, *History of Rocketry and Astronautics,* Vol. 9, pp. 89–104.

Duboshin, G. N., and Dimitry Y. Okhotsimsky. "Some Problems of Astrodynamics and Celestial Mechanics." In Edmond Brun and Irwin Hersey, eds., *XIVth International Astronautical Congress,* Paris: Gauthier-Villars, 1965.

Durant, Frederick C., III. "Robert H. Goddard and the Smithsonian Institution." In Durant and James, *First Steps Toward Space.*

Dushkin, Leonid S. "Experimental Research and Design Planning in the Field of Liquid-Propellant Rocket Engines Conducted Between 1934–1944 by the Followers of F. A. Tsander." In Hall, *History of Rocketry and Astronautics,* Vol. 7, Part 2.

Emme, Eugene M. "Presidents and Space." In Durant, *Between Sputnik and the Shuttle.*

Fang-toh Sun. "Rockets and Rocket Propulsion Devices in Ancient China." In Skoog, *History of Rocketry and Astronautics,* Vol. 10.

Flandro, Gary A. "Utilization of Energy Derived from the Gravitational Field of Jupiter for Reducing the Flight Time to the Outer Planets," Space Programs Summary No. 37-35. Pasadena, Calif.: Jet Propulsion Laboratory, October 31, 1965.

Goddard, Robert H. *Liquid Propellant Rocket Development,* Smithsonian Miscellaneous Collections, Vol. 95, No. 3. Washington, D.C.: Smithsonian Institution, March 16, 1936.

———. *A Method of Reaching Extreme Altitudes,* Smithsonian Miscellaneous Collections, Vol. 71, No. 2. Washington, D.C.: Smithsonian Institution, 1919.

Gorin, Peter A. "Zenit—The First Soviet Photo-Reconnaissance Satellite," *Journal of the British Interplanetary Society,* Nov. 1997.

Hall, R. Cargill. "Earth Satellites, a First Look by the United States Navy." In Hall, *History of Rocketry and Astronautics,* Vol. 7, Part 2.

Kulcinski, Gerald L. "Helium-3 Fusion Reactors—A Clean and Safe Source of Energy in the 21st Century." Madison: University of Wisconsin Fusion Technology Institute, April 1993.

Lytkin, Vladimir, Ben Finny, and Liudmila Alepko. "The Planets Are Occupied by Living Beings: Tsiolkovsky, Russian Cosmism and Extraterrestrial Civilizations," presented at the 1995 International Conference on SETI and Society, Chamonix, France, June 1995.

Malina, Frank J. "America's First Long-Range-Missile and Space Exploration Program: The ORDCIT Project of the Jet Propulsion Laboratory, 1943–1946." In Hall, *History of Rocketry and Astronautics,* Vol. 7, Part 2.

———. "On the GALCIT Rocket Research Project, 1936–38." In Durant and James, *First Steps Toward Space.*

Mason, C. P. "The Principles of Interplanetary Navigation." *Bulletin* of the American Interplanetary Society, No. 16, February 1932.

Merkulov, I. A. "Organization and Results of the Work of the First Scientific Centers for Rocket Technology in the USSR." In Ordway, *History of Rocketry and Astronautics,* Vol. 9.

Minovitch, M. A. "The Determination and Characteristics of Ballistic Interplanetary Trajectories Under the Influence of Multiple Planetary Attractions," TR 32-464. Pasadena, Calif.: Jet Propulsion Laboratory (Caltech), 1963.

———. "A Method for Determining Interplanetary Free-Fall Reconnaissance Trajectories," Technical Memo #312-130. Pasadena, Calif.: Jet Propulsion Laboratory, August 23, 1961.

———. "Utilizing Large Planetary Perturbations for the Design of Deep Space, Solar Probe, and Out-of-Ecliptic Trajectories," Technical Report No. TM 312-514. Pasadena, Calif.: Jet Propulsion Laboratory, February 15, 1965.

Munick, H., R. McGill, and G. R. Taylor. "Analytic Solutions to Several Optimum Orbit Transfer Problems," 11th International Astronautical Congress, Stockholm, August 1960.

Murphy, J. M. "Liquid Rocket Propulsion System Advancements, 1946–1970." In Skoog, *History of Rocketry and Astronautics,* Vol. 10.

Murray, Bruce C., and Merton E. Davies. "A Comparison of U.S. and Soviet Efforts to Explore Mars," P-3285, Washington, D.C.: National Aeronautics and Space Administration, January 1966.

Ordway, Frederick I., III. "Some Vignettes from an Early Rocketeer's Diary: A Memoir." In Sloop, *History of Rocketry and Astronautics,* Vol. 12.

———. "Space Fiction in Film." In Emme, *Science Fiction and Space Futures.*

Pendray, G. Edward. "The Conquest of Space by Rocket." *Bulletin* of the American Interplanetary Society, No. 17, March 1932.

———. "Definition and History of the Rocket." *Bulletin* of the American Interplanetary Society, No. 5, November–December 1930.

———. "Recent Worldwide Advances in Rocketry." *Bulletin* of the American Interplanetary Society, No. 14, December 1931.

Pickering, William H., and James H. Wilson. "Countdown to Space Exploration: A Memoir of the Jet Propulsion Laboratory, 1944–1958." In Hall, *History of Rocketry and Astronautics,* Vol. 7, pp. 385–421.

Pobedonostsev, Yuri A., Ye. S. Shchetinkov, and V. N. Galkovsky. "A History of the Jet Propulsion Research Institute (RNII), 1933–1944." In Lattu, *History of Rocketry and Astronautics,* Vol. 8.

Polyarny, A. I. "On Some Work Done in Rocket Technology, 1931–38." In Durant and James, *First Steps Toward Space,* Vol. 6.

Prishchepa, V. I. "History of Development of First Space Rocket in the USSR." In Ordway, *History of Rocketry and Astronautics,* Vol. 9.

Ravnitzky, Michael J. "Catch a Falling Star: Parachute System Lessons Learned During the USAF Space Capsule Mid-Air Recovery Program, 1959–1985," AIAA-93-1243, American Institute of Aeronautics and Astronautics conference and seminar, London, May 10–13, 1993.

Sagan, Carl. "Planetary Exploration," Condon Lectures, Oregon State System of Hugher Education, Eugene, Oregon, 1970.

Schmiedl, Friedrich. "Early Postal Rockets in Austria." In Hall, *History of Rocketry and Astronautics.*

Schulz, Werner. "Walter Hohmann's Contributions Toward Space Flight: An Appreciation on the Occasion of the Centenary of His Birthday." In Skoog, *History of Rocketry and Astronautics,* Vol. 10.

Shortal, Joseph Adams. "Rocket Research and Tests at the NACA/NASA Wallops Island Flight Test Range, 1945–1959: A Memoir." In Skoog, *History of Rocketry and Astronautics,* Vol. 10.

Skoog, A. Ingemar, and Frank H. Winter. "The Swedish Fire Arrow: The Oldest Rocket Specimen Extant." In Skoog, *History of Rocketry and Astronautics,* Vol. 10.

Smoot, G. F., et al. "Structure in the COBE DMR First Year Maps." Berkeley, Calif.: Lawrence Berkeley Laboratory, April 17, 1992 (presented at the American Physical Society meeting on April 23).

Stewart, H. J. "New Possibilities for Solar System Exploration." *Astronautics & Aeronautics,* December 1966.

Stites, Richard. "World Outlook and Inner Fears in Soviet Science Fiction." In Graham, *Science and the Soviet Social Order.*

Struve, Otto. "The Astronomical Universe," Condon Lectures, Oregon State System of Higher Education, Eugene, Oregon, 1958.

Subotowicz, Mieczyslaw. "Analysis of Rocket Construction, Described in Manuscripts and Printed Books During the 16th and 17th Centuries." In Ordway, *History of Rocketry and Astronautics,* Vol. 9.

Sullivan, Thomas A., and David S. McKay. "Using Space Resources." Houston: Johnson Space Center, 1991.

Teller, Edward. "The Feasibility of Arms Control and the Principle of Openness." In Brennan, *Arms Control, Disarmament, and National Security.*

Thompson, G. V. E., and L. R. Shepherd. "The British Interplanetary Society: The First Fifty Years (1933–1983). In Sloop, *History of Rocketry and Astronautics,* Vol. 12.

Tikhonravov, Mikhail K. "The Creation of the First Artificial Earth Satellite: Some Historical Details." In Lattu, *History of Rocketry and Astronautics,* Vol. 8. (Also in the *Journal of the British Interplanetary Society,* May 1994, pp. 191–94.)

———. "How to Make Rockets Fly Farther," Report to the Artillery Academy, July 14, 1948, as reprinted in *The History of Aviation and Cosmonautics,* 1967, and the Russian Academy of Sciences, Moscow, 1995 (in Russian).

National Intelligence Estimates and Related Documents

"The Clandestine Introduction of Nuclear Weapons into the US," NIE 4-70, July 7, 1970.

Douglas, James H. "Thor and WS-117L Program" (memorandum for the secretary of defense), February 14, 1958.

"A National Integrated Missile and Space Vehicle Development Program," Report No. D-R-37, Development Operations Division, Army Ballistic Missile Agency, December 10, 1957. (Eisenhower Library.)

"Probable Reactions to U.S. Reconnaissance Satellite Programs," Special National Intelligence Estimate Number 100-6-60, August 9, 1960.

"The R-14 Project, a Design of a Long Range Missile at Gorodomyla Island," Information Report CS-G-14851, August 26, 1953. (National Archive.)

"Report to the President on the Threat of Surprise Attack," March 14, 1955. (Three-page Technological Capabilities Panel summary letter to the Department of State.)

"Scientific Research Institute and Experimental Factory 88 for Guided Missile Development, Moskva/Kaliningrad," OSI-C-RA/60-2, Central Intelligence Agency, Office of Scientific Intelligence, March 4, 1960.

Sharp, Dudley C. "Memorandum for the Chief of Staff, USAF" (establishing the Office of Missiles and Satellite Systems,) August 31, 1960. (Manuscript Division, Library of Congress.)

"Soviet Capabilities and Intentions to Orbit Nuclear Weapons," NIE 11-9-63, July 15, 1963.

"Soviet Capabilities for Clandestine Attack Against the US with Weapons of Mass Destruction and the Vulnerability of the US to Such Attack (Mid-1951 to Mid-1952)," NIE-31, September 4, 1951.

"Soviet Capabilities in Guided Missiles and Space Vehicles," NIE 11-5-58, August 19, 1958.

"Soviet Capabilities in Guided Missiles and Space Vehicles," NIE 11-5-59, November 3, 1959.

"The Soviet ICBM Program," Special NIE 11-10-57, December 17, 1957.

Soviet Military Power: 1983. Washington, D.C.: Department of Defense, 1983.

Soviet Military Power: 1985. Washington, D.C.: Department of Defense, 1985.

"The Soviet Space Program," NIE 11-1-67, March 2, 1967.
"Threat Assessment Report—Space Transportation System (STS)," SAMS-CR-12-02-77. Washington, D.C.: Defense Intelligence Agency, November 1977.

Press Releases

"Air Force Mission in Space," Fact Sheet 82-23. Washington, D.C.: Secretary of the Air Force, June 1982.
"APS Statement Urges No Early Commitment to SDI Deployment." New York: American Physical Society, April 24, 1987.
"Early Findings from Pioneer Venus," NASA Release No. 79-06, February 8, 1979.
Gemini-Titan II Press Handbook, 2nd ed. Baltimore: Martin Company; St. Louis: McDonnell Aircraft, July 1, 1965.
"Grand Tour of Planets," NASA Release No. 69-84, May 29, 1969.
"Grand Tour Scientists," NASA Release No. 71-56, April 4, 1971.
"Mariner-Venus '73 Flight Genesis," NASA Release No. 70-112, July 5, 1970.
"NASA Releases Challenger Transcript and Report on Cause of Death," NASA Release No. 86-100, July 28, 1986.
"News Conference on Ranger VI Impact on Moon," NASA Release No. 2-2509, February 2, 1964.
"Phobos Dynamics Experiment," JPL Fact Sheet, undated.
"Pioneer F Mission to Jupiter," NASA Release No. 72-25, for release February 20, 1972.
"Pioneer Will Reach Jupiter December 3," NASA Release No. 73-243, November 7, 1973.
"Pioneer Saturn Results Are Summarized," NASA Release No. 79-42, October 24, 1979.

Videos

Apollo 13: To the Edge and Back, WNET (New York), July 20, 1994.
The Battle over Citizen Kane, NOVA, WGBH (Boston), February 1996.
Challenger News Conference, NBC, January 28, 1986.
Challenge of Outer Space, produced by the Armed Forces Staff College, 1955 (Film Archives, National Air and Space Museum, No. VB 00716).
Destination Moon, Horizon, BBC (London), July 5, 1997.
The Early History of JPL and the Evolution of Its Matrix Organization, a talk by Jack N. James at JPL on February 12, 1987.
The Last Journey of a Genius, NOVA, WGBH (Boston), 1990.
Lifting Bodies and the Shuttle, Discovery Channel, October 4, 1993.
Mission to the Moon, NOVA, WNET (New York), July 20, 1994.
Moon Shot, two-part series, TBS, July 11 and 13, 1994.
Neptune's Cold Fury, NOVA, WNET (New York), April 9, 1991.
The Planet That Got Knocked on Its Side, NOVA, WNET (New York) October 29, 1986.
Rescue Mission in Space: The Hubble Telescope, WNET (New York), December 13, 1994.
Rocket Power: Goddard and von Braun, Frontiers of Flight, Discovery Channel, April 25, 1994.
The Rocky Road to Jupiter (Galileo), NOVA, WNET (New York), April 7, 1987.
The Russian Right Stuff, NOVA, WGBH (Boston), three-part series, February 26–28, 1991.
Sail On, Voyager!, NOVA, WQED (Pittsburgh), November 28, 1990.
Spaceflight, four-part series, NOVA, KCET (Los Angeles), May 1985.
Space Shuttle, 48 Hours, CBS, April 21, 1988.
Space Shuttle Challenger Accident Investigation, Jet Propulsion Laboratory, July 2, 1987.
Space Warriors, A&E, March 11, 1994.
To Boldly Go, NOVA, WGBH (Boston), April 16, 1991.
To Space and Back (lifting bodies), Discovery Channel, October 4, 1993.
Venus Unveiled, NOVA, WGBH (Boston), October 18, 1995.
Visions of Star Wars, NOVA, WNET (New York), April 22, 1986.
Where Are All the UFOs?, A&E, September 22, 1996.

Interviews

Mentioned in the Book

Asimov, Isaac	Cambridge, Mass.	May 30, 1969
Basilevsky, Aleksandr	Vernadsky Institute	May 15, 1995
Biryukov, Yuri V.	Russian Academy of Science	May 17, 1995
Bissell, Richard M., Jr.	CIA (ret.)	May 23, 1984
Brooks, Bob	JPL	Feb. 18, 1988
Brophy, John	JPL	June 26, 1987
Brugioni, Dino A.	Ret.	Jan. 3, 1997
Bundy, McGeorge	Carnegie Corporation of New York	Feb. 1, 1995
Carlson, Robert W.	JPL	Feb. 23, 1988
Cesarone, Robert J.	JPL	June 22, 1987
Colby, William E., Jr.	CIA (ret.)	Apr. 17, 1984
Davies, Merton E.	RAND	Jan. 4, 1989
Dula, Arthur M.	Houston	July 13, 1996
Edelson, Burton I.	Johns Hopkins University	Jan. 29, 1988
Filina, Larisa A.	Korolyov Memorial Museum	May 17, 1995
Gingerich, Owen	Harvard University	May 31, 1969
Hall, Charles F.	Ames Research Center	Mar. 1, 1988
Helfand, David J.	Columbia University	May 10, 1989
Johnson, Nicholas L.	Teledyne Brown Eng.	Oct. 11, 1988
	Johnson Space Center	Oct. 18, 1996
Kantemirov, Boris N.	Memorial Museum of Cosmonautics	May 17, 1995
		June 6, 1996
Kapitsa, Sergei	Moscow	May 16, 1995
Keegan, General George J.	U.S. Air Force (ret.)	Sept. 16, 1981
		June 2, 1984
Khozin, Grigori S.	Russian Diplomatic Academy	June 5, 1995
Khrushchev, Sergei	Brown University	Nov. 27, 1995
Killian, James R., Jr.	MIT	Feb. 11, 1985
King, John W.	NASA (ret.)	Mar. 21, 1995
Kistiakowsky, George B.	Harvard	Apr. 12, 1981
Knoche, E. Henry	CIA (ret.)	July 1, 1985
Krichevski, Sergei	Gargarin Training Center	May 14, 1995
Lane, Arthur L.	JPL	Feb. 19, 1988
Leonov, Alexei	Moscow	June 8, 1995
McLuhan, Marshall	Toronto	July 11, 1969
Mead, Margaret	American Museum of Natural History	May 28, 1969
Minovitch, Michael A.	Phaser Telepropulsion	Mar. 16, 1995
Mishin, Vasily P.	Moscow Aviation Institute	June 4, 1996
Morrison, Philip	MIT	May 31, 1969
Murray, Bruce	Caltech	July 2, 1987
		Jan. 10, 1989
Niebuhr, Reinhold	Stockbridge, Mass.	May 30, 1969
Pickering, William H.	JPL (ret.)	June 15, 1987
Pike, John	Federation of American Scientists	Aug. 25, 1981
Ponomareva, Valentina	Moscow	June 5, 1996
Ride, Sally K.	Stanford University	Mar. 1, 1988
Savin, Anatoli I.	Moscow	May 12, 1995
Scoville, Herbert, Jr.	CIA (ret.)	Apr. 12, 1984
Stofan, Ellen R.	JPL	Mar. 17, 1995
Tatarewicz, Joseph N.	National Air and Space Museum	Jan. 28, 1988
Toynbee, Arnold	London	June 9, 1969
Van Allen, James A.	University of Iowa	Feb. 14, 1987
van der Woude, Jurrie	JPL	Feb. 22, 1988

Van Doren, Mark	Columbia University (ret.)	June 5, 1969
Wheelon, Albert D.	CIA (ret.)	Jan. 6, 1997

For Background

Alexander, Joseph K.	NASA	Jan. 29, 1988
Allen, Lew	JPL	June 18, 1987
		Jan. 6, 1989
Beichman, Charles	Caltech	June 25, 1987
Bowen, Fred W.	NASA	June 15, 1987
Bourke, Roger D.	JPL	June 24, 1987
Bundy, McGeorge	NYU	June 28, 1984
Casani, E. Kane	JPL	Mar. 13, 1995
		Dec. 15, 1995
Casani, John R.	JPL	June 24, 1987
Chahine, Moustafa T.	JPL	June 18, 1984
		June 19, 1987
		Jan. 6, 1989
Colby, William E.	CIA (ret.)	Apr. 17, 1984
Collins, Richard F.	JPL	June 29, 1987
Collins, Stewart A.	JPL	June 25, 1987
Davies, Merton E.	RAND	Mar. 14, 1995
		Dec. 19, 1995
Diaz, Alphonso	JPL	Jan. 27, 1988
Draper, Ronald F.	JPL	Aug. 28, 1989
Dunne, James A.	JPL	June 18, 1987
Fimmel, Richard O.	Ames	Mar. 4, 1988
Ford, John P.	JPL	June 23, 1987
French, Bevan M.	NASA	Jan. 28, 1988
Friedman, Louis D.	Planetary Society	Feb. 13, 1988
Gavit, Sarah	JPL	Dec. 14, 1995
Giberson, W. E.	JPL	June 30, 1987
Goldfine, Milton	JPL	June 29, 1987
Haynes, Norman	JPL	June 23, 1987
Jacobson, Allan S.	JPL	June 26, 1987
James, J. N.	JPL	Jan. 22, 1988
Janesick, James	JPL	June 14, 1984
Johnson, Nicholas L.	Teledyne Brown	June 22, 1984
Johnson, Torrence V.	JPL	June 17, 1987
Katz, Amrom	RAND (ret.)	June 16, 1984
		June 2, 1984
Khrushchev, Sergei	Brown University	Nov. 27, 1995
Kukkonen, Carl A.	JPL	June 16, 1987
Land, Edwin	Polaroid (ret.)	Oct. 27, 1984
Lane, Arthur L.	JPL	June 22, 1987
		June 29, 1987
Ledebuhr, Arno G.	Livermore National Lab	Dec. 11, 1995
Lehman, David H.	JPL	Dec. 21, 1995
Lovell, Bernard	Jodrell Bank	June 13, 1969
Lyman, Peter T.	JPL	June 29, 1987
Mark, Hans	University of Texas	May 6, 1985
McLaughlin, William I.	JPL	June 25, 1987
		Feb. 23, 1988
		Jan. 4, 1989
Meeks, Willis G.	JPL	June 17, 1987
Meinel, Aden	JPL	July 2, 1987

Meinel, Marjorie	JPL	July 2, 1987
Morrison, David	Ames	Jan. 18, 1989
Murray, Bruce	Caltech	Dec. 21, 1995
Nelson, Jerry	UC Berkeley	June 6, 1984
Pickering, William H.	JPL (ret.)	Feb. 18, 1988
		Jan. 5, 1989
Pike, John	Federation of American Scientists	June 27, 1983
Pleasance, Lyn	Livermore National Lab	Dec. 11, 1995
Rabi, I. I.	Columbia University (ret.)	May 15, 1987
Rea, Donald G.	JPL	June 19, 1987
Scoville, Herbert, Jr.	CIA (ret.)	Apr. 12, 1984
Shapiro, Robert	NYU	Sept. 16, 1987
Shirley, Donna	JPL	Dec. 19, 1996
Smrekar, Suzanne	JPL	Dec. 18, 1995
Standish, E. Myles	JPL	July 1, 1987
Stewart, Homer J.	Caltech (ret.)	July 2, 1987
		Jan. 5, 1989
Stofan, Ellen R.	JPL	Mar. 17, 1995
Tatarewicz, Joseph N.	National Air and Space Museum	Mar. 7, 1988
Vane, Gregg	JPL	Dec. 14, 1995
Wagoner, Paul D.	USAF	June 21, 1984
Wasserburg, Gerald	Caltech	Feb. 24, 1988
Wilson, Barbara	JPL	Dec. 18, 1995

Taped Meetings

AIAA/JPL Conference on Solar System Exploration, May 19–21, 1987.

Fisk, Lennard A. "The Role of Robotic Precusors in Human Exploration," "Pathways to the Planets" conference, Washington, D.C., June 1, 1989.

"Future of Space Sciences in the United States" (organized by James A. Van Allen), American Association for the Advancement of Science annual meeting, Chicago, 1991.

Helfand, David J. "Astropolitics: Science in the Backseat at NASA," New York Academy of Sciences, April 27, 1989.

Index

ABOUT THE AUTHOR

WILLIAM E. BURROWS has reported on aviation and space for *The New York Times, The Washington Post,* and *The Wall Street Journal,* and has had articles in *The New York Times Magazine, The Sciences,* and other publications. He is a contributing editor at *Air & Space/Smithsonian* and the author of seven other books, including *Deep Black,* the classic work on spying from space, and *Exploring Space,* an award-winning history of solar system exploration.

Mr. Burrows is a professor of journalism at New York University and the founder and director of its graduate Science and Environmental Reporting Program.

ABOUT THE TYPE

This book was set in Times Roman, designed by Stanley Morison specifically for *The Times* of London. The typeface was introduced in the newspaper in 1932. Times Roman has had its greatest success in the United States as a book and commercial typeface, rather than one used in newspapers.